TIMING AND TIME PERCEPTION

ANNALS OF THE NEW YORK ACADEMY OF SCIENCES
Volume 423

TIMING AND TIME PERCEPTION

Edited by John Gibbon and Lorraine Allan

The New York Academy of Sciences
New York, New York
1984

Library of Congress Cataloging in Publication Data

Main entry under title:

Timing and time perception.

(Annals of the New York Academy of Sciences ; v. 423)
Bibliography: p.
Includes index.
1. Time perception—Congresses. I. Gibbon, John, 1934– . II. Allan, Lorraine, 1940– . III. Series.
Q11.N5 vol. 423 [QP445] 500 s [153.7'53] 84-11443

SP
Printed in the United States of America

ISBN 0-89766-240-7 (cloth)
ISBN 0-89766-241-5 (paper)
ISSN 0077-8923

ANNALS OF THE NEW YORK ACADEMY OF SCIENCES

Volume 423

May 11, 1984

TIMING AND TIME PERCEPTION*

Editors and Conference Co-Chairmen

JOHN GIBBON AND LORRAINE ALLAN

CONTENTS

Part I. Time Perception

Part II. Timing of Motor Programs and Temporal Patterns

*This volume is the result of a conference on Timing and Time Perception held on May 10–13, 1983 by The New York Academy of Sciences.

Part III. Timing in Animal Learning and Behavior

Part IV. Time in Cognitive Processing and Memory

Part V. Rhythmic Patterns and Music

Part VI. The Internal Clock

Poster Papers

Financial assistance was received from:

- INSTITUTE FOR COGNITIVE RESEARCH through COLUMBIA UNIVERSITY
- McMASTER UNIVERSITY
- NATIONAL INSTITUTE OF MENTAL HEALTH—NATIONAL INSTITUTES OF HEALTH
- NATIONAL SCIENCE FOUNDATION
- OFFICE OF NAVAL RESEARCH

PART I. TIME PERCEPTION

Introduction

JOHN GIBBON

New York State Psychiatric Institute
New York, New York 10032; and
Department of Psychology
Columbia University
New York, New York 10027

LORRAINE ALLAN

Department of Psychology
McMaster University
Hamilton, Ontario, Canada L8S 4K1

The study of timing and time perception has a venerable dual history in experimental psychology. Animal psychologists studying learning and conditioning have investigated extensively the timing and temporal patterning of behavior under instrumental control. In the psychophysical tradition, the psychophysics of time perception in humans also had early attention within the classical study of the limits of human sensory capacity. From these beginnings, the techniques and results in animal and human work have evolved and broadened so that currently there is a burgeoning literature in animal psychophysics dealing with time perception, and another in human timing dealing with the production and organization of temporal patterns.

The first part of this *Annal* exemplifies perhaps the greatest overlap of these two traditions. The papers in this section by Stubbs *et al.*, Eisler, Gibbon *et al.*, and Wasserman *et al.* present studies of time perception in rats and pigeons that bear quite directly on similar procedures or similar theories studied previously in the human. It is comforting to find that at least occasionally similar phenomena emerge. For example, Stubbs *et al.* report that pigeons, like humans, judge auditory durations as longer than visual durations. Further extensions of this trend, perhaps also including the study of human time perception with animal techniques, may be expected in the future.

Two papers are from the classical human psychophysical tradition and deal with perceived or subjective duration: Jamieson concentrates on providing a theoretical framework for the time-order error, and Allan describes two contingent duration aftereffects. She shows that these differ in important ways from the McCollough effect.

Another paper stems from the study of human information processing. This is Schweickert's analysis of critical path networks. His representations allow a determination of the temporal position of decision stages during processing.

Three of the papers develop models of time perception. They are good examples of the ways in which the two traditions are beginning to meld. Kristofferson's real-time criterion model for human duration discrimination postulates a minimal time "quantum" that determines all of the variance in the discriminal process. The model presented by Hopkins is similar in postulating central, deterministic (quantum) delays for response–stimulus synchronization performance in humans. The information-processing model developed by Gibbon *et al.* is applied to three different timing tasks performed by animal subjects. This account regards the procedures as differing in decision and memory processes, but utilizing the same central clock process. The pacemaker they propose for the internal clock bears structural similarities to the quantum process.

Quantal and Deterministic Timing in Human Duration Discrimination

ALFRED B. KRISTOFFERSON

Department of Psychology
McMaster University
Hamilton, Ontario, Canada L8S 4K1

When we began to study duration discrimination, we expected it to give us rather direct information about time perception. That expectation has not been fulfilled. Instead, our "thresholds" for duration seem to be determined by our ability to produce a time interval, to time it out internally. The size of a threshold is wholly determined by the extent to which repeated attempts to time out a fixed time interval are variable.[1]

A set of duration stimuli is shown at the top of FIGURE 1. Each stimulus consists of two 10-msec auditory pulses separated by the stimulus duration, D. The stimuli differ from each other only in D, and the values of D are symmetrically arranged around a midpoint value. The midpoint of the set is the base duration. A single stimulus is presented on a trial and the subject is asked to categorize it as "long" (R_L) or "short" (R_S). Values of D greater than the midpoint are called long, and the decision on each trial is whether P_2 occurred before or after the midpoint value of D. In the experiments to be reported here, the subject is instructed to respond as quickly as possible, and the data consist of response probabilities and response latencies for each stimulus duration.

The general hypothesis is pictured in FIGURE 1B. On each trial, P_1 triggers an internally timed interval, I, which terminates as the criterion event, C. P_2 triggers a sensory event B_2. If, as shown here, C occurs before B_2, then a long response is determined. Short responses are triggered by B_2 whenever it occurs first. The discrimination mechanism is a race between the two response triggers, C and B_2.

The two kinds of responses, therefore, have different causal histories, as shown in FIGURE 1C, R_L being linked to P_1 and R_S being linked to P_2. Long responses should be time-locked to P_1 and should occur at the same time for all stimuli, that is, regardless of the time of occurrence of P_2. R_S, on the other hand, should be time-locked to P_2. These time-locking predictions are a major test of the hypothesis and they have been confirmed experimentally.[1]

Since responding is speeded, short responses are direct reactions to P_2 and might resemble simple reaction times. Long responses are similar, except that the responses must be delayed, and R_L latencies might resemble time estimation latencies. These expectations are also confirmed, and speeded duration discrimination appears to be a combination of simple reaction time and time estimation, in which one kind of response occurs on a trial, the other being countermanded.

Practice with the stimulus set adjusts I so that C falls near the midpoint of the stimulus set. The lower panel in FIGURE 1 displays a specific model in which the times of occurrence of C are assumed to form an isosceles triangle. The variability in C is due solely to variance in I, the afferent latencies having zero variance. Therefore, for each D, B_2 is a fixed point which divides the triangle into two parts. The proportion of the area under the triangle to the left of B_2 represents the probability of R_L for that stimulus. Knowing the probability of R_L for two different stimuli, both of which have a B_2 within the triangle, enables one to calculate the quantum size, q, in msec, and also

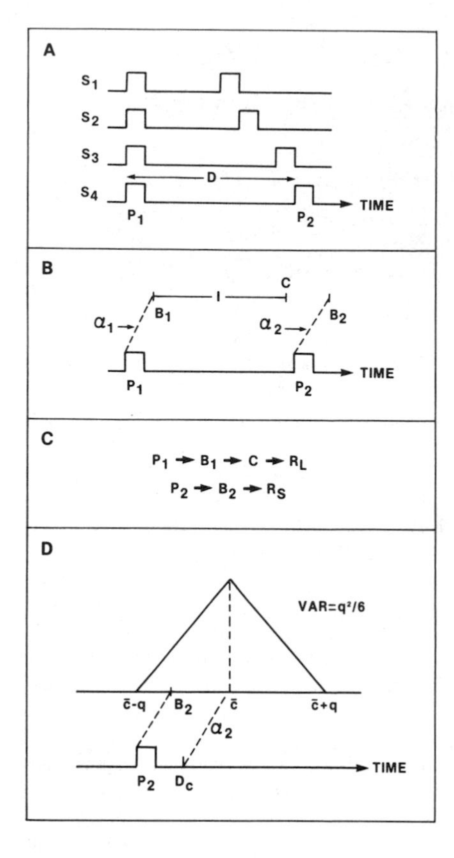

FIGURE 1. Schematic aids for the real-time criterion theory of duration discrimination. (**A**) a set of duration stimuli. Each pair of brief auditory pulses defines a stimulus duration, *D*. The midpoint of the set (MP) is halfway between D_2 and D_3 and is the base duration. (**B**) the general hypothesis. A race between *C* and B_2 determines which response will be made. (**C**) causal chains for the responses "long" and "short." (**D**) a specific quantal model. The times of occurrence of C are assumed to be distributed as an isosceles triangle having a base equal to twice the size of the time quantum.

D_C, the stimulus duration for which B_2 falls at the mean of the triangle. With a little practice, D_C is placed near the midpoint of the stimulus set.

We have shown in several experiments that the cumulative form of the triangle, which is a fully bounded sigmoid, is a satisfactory description of the psychometric function, at least as satisfactory as a normal ogive.[2–4]

The quantum size can also be calculated from the response latencies, giving a completely independent estimate of that parameter. This involves assuming that the two causal chains show in FIGURE 1 include the same variance sources except for the component C, an assumption already partially stated above. Consequently, the response latency variance for R_L is greater than that for R_S by an amount $q^2/6$.

The experimental tests that have been reported so far confirm all of the expectations described above.[1] Those experiments employed a single stimulus set, with a midpoint of 1150 msec, and one purpose of the experiments to be described now was to perform the same tests using other base durations. If the value of q is different at another base duration, then the magnitude of the change in q should be the same when it is calculated from response latencies as it is when calculated from response probabilities.

The value of q is a function of base duration,[5] and the form of the function is embodied in a doubling rule: Doubling or halving base duration a given number of times, doubles or halves q the same number of times. This is illustrated in FIGURE 2.

In these experiments responding was not speeded, latencies were not measured, and the values of q were calculated from response probabilities only. Twenty consecutive sessions were conducted at one base duration before changing to another base duration. The dashed line shows performance during the first five sessions at each base duration, with the data points omitted here. Early in practice, q is directly proportional to base duration. Since the standard deviation of the discriminal process is $0.41\,q$, the dashed line indicates that the ratio of S.D. to mean is constant at 0.053 during the first few sessions. This agrees closely with the ratio of 0.05–0.06 obtained by Getty.[6]

The data points and the line segments fitted to them in FIGURE 2 show performance during sessions 18–20. There is a practice effect which is specific to each base duration and which is large at some base durations and small at others. As a consequence of practice, steps unfold from the dashed line in such a way that the doubling rule is preserved. There appear to be steps at 200, 400, and 800 for this subject, with the quantum size doubling at each step.

However, if in the limit this is a step function, then that limit has not been reached in 20 sessions because the steps clearly slope upward slightly, and all by about the same amount. A second major purpose of the present experiments was to determine whether the steps become flat with even larger amounts of practice.

It is necessary to insert a methodologic note at this point. In measuring latencies, one finds two additional powerful sources of variance which complicate matters. One is competition between responses when two overt responses are used. We use a "go–no go" procedure in which the response may signal "long" and no response "short" in a particular experiment, or the reverse. This doubles the running time, but solves the problem. The second source of extraneous variance arises from properties of the response triggers C and B_2. Each of them triggers its own response, and each also inhibits the opposite response. When C and B_2 occur in close succession, the first one determines the response outcome, but the second one perturbs the latency of that outcome. Hence, a minimum time separation must be assured between the two triggers in order to solve this problem, as will be shown.

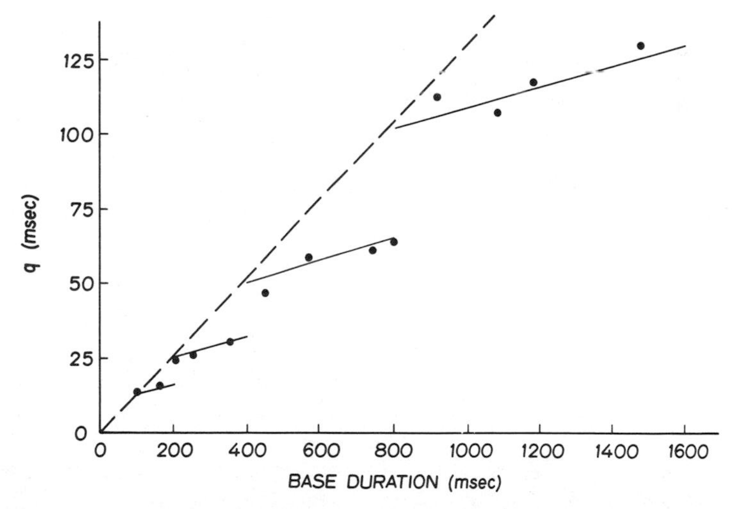

FIGURE 2. Quantum size as a function of base duration. The *dashed line* represents the results obtained with no prior practice. The *points* show the results that follow 17 sessions of practice at each base duration. (From Kristofferson.[5] Reprinted by permission.)

Extensive data have been published that were obtained with me as the experimental subject.[1,5] I am also the subject in the experiments to be described now, and it is possible to make quantitative predictions against which to compare these new data. In the 1977 experiment, the base duration was 1150 msec and speeded responding was used. The obtained value of q calculated from the response probabilities was 95.3. The measured latency variance for R_L was 1799 and that for R_S was 279 and the difference between these yields the second estimate of q, which is 95.5. Therefore, with an estimated $q = 95.4$, the two previously measured latency variances, the doubling rule, and the supposition that the steps are flat in the quantal step function, a basis is provided for calculating predicted values at other base durations.

When the base duration is 280 msec, as it is in the first three experiments, q should be on the 25-msec step and its predicted value is 23.9. FIGURE 3 shows the predicted distribution of C centered at 280 and with a base of 48 msec. The durations of S_2 and S_3 were fixed throughout, as indicated by the small arrows. In Experiment 1, the response was R_L and the duration of S_4 was varied over the indicated range, with S_1 fixed at 245. In Experiment 2, the response was R_S and S_1 was varied, with S_4 fixed at 305.

After sixty sessions of practice under the conditions of Experiment 1, 38 sessions were done with D_4 set at a different value from session to session so that each of the 19 values of D_4 were measured twice. For Experiment 2, the response was changed to R_S and 80 practice sessions were run with D_1 set at 255, five sessions at 250, and so on, down to 220. The parameters of the triangle calculated from the response probabilities actually obtained in Experiments 1 and 2 are also plotted in FIGURE 3. The span of the triangle is unaffected either by the variation in S_1 and S_4 or by the change in the meaning of the response. The centering of the triangle is slightly affected, but the total effect, amounting to only 1%, can be ignored for present purposes.

The predicted value of q is indicated by the dashed line in FIGURE 3. It agrees almost exactly with the mean of the obtained values, both being 23.9 msec.

The latency variances are shown in FIGURE 4. When S_1 or S_4 are close to the triangle, their latency variances are greatly inflated due to the temporal proximity of C

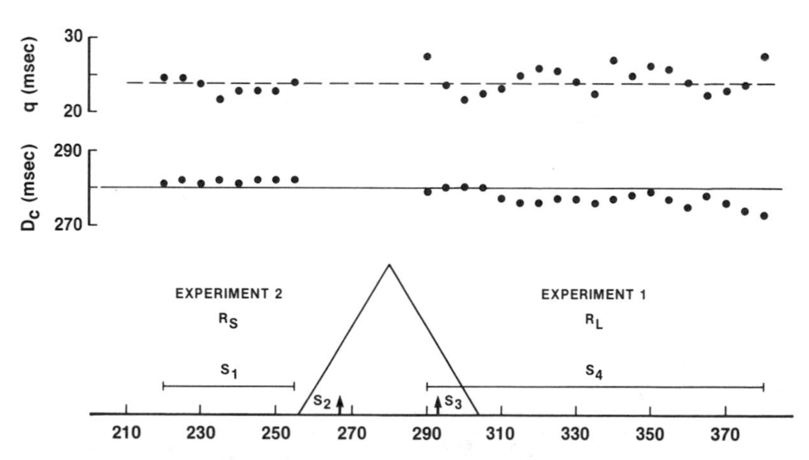

FIGURE 3. Experiments 1 and 2. The base duration was 280 msec and the triangle is the predicted distribution of C, centered at 280 and having a base of 48 msec. $D_2 = 267$, $D_3 = 293$. Experiment 1 varied only the duration of S_4 over the range shown. In Experiment 2, only D_1 was varied. The obtained values of D_C and q, calculated from the response proportions for S_2 and S_3, are shown above. The *dashed line* is the predicted value of q.

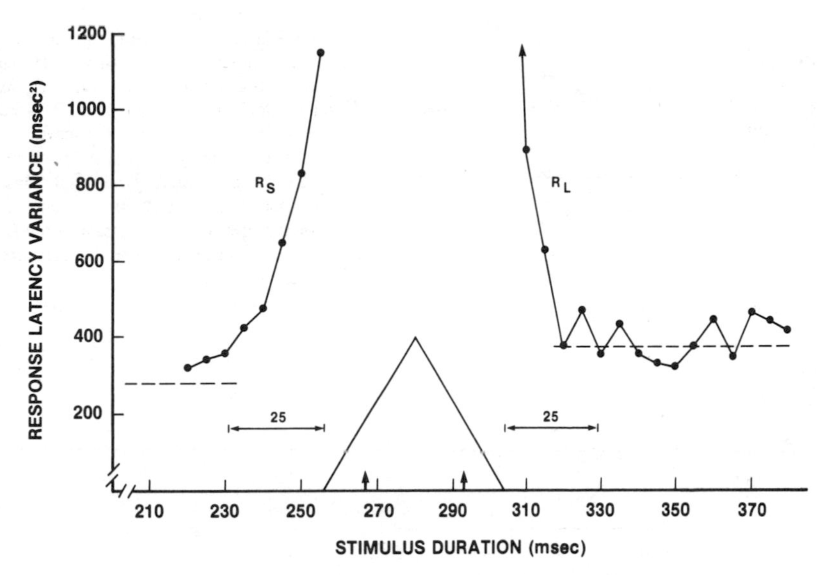

FIGURE 4. The same as FIGURE 3, except that the data plotted here are the response latency variances for R_S/S_1 on the **left** and R_L/S_4 on the **right.** The four values for $S_4 < 310$ are not shown; they are off the scale. As B_2 approaches the triangle from either direction, latency variance is greatly increased, but when B_2 is more than 25 msec from the nearest C, the latency variance is close to the predicted level, shown by the *dashed* lines.

and B_2. The size of the effect is roughly the same above and below the triangle and it extends about 25 msec from the triangle in both directions. This suggests that the two response triggers, C and B_2, must be separated by 25 msec or more to avoid latency interaction between them. It is important to add that within the two 25-msec windows, the latencies are affected, but response errors are not; the correct responses are given 100% of the time to stimuli within these windows.

The predicted latency variances are indicated by the dashed lines. The R_L variances were 1799 when the base duration was 1150; here, when the base is 280, they are much smaller, averaging 388 for the points more than 25 msec from the triangle. This agrees well with the predicted variance, which is 374.

The short response variance is affected very little by the change in base duration from 1150 to 280. It was very low at 1150,[1] and it appears to be slightly greater at 280. To determine its level more precisely, an additional 75 sessions were added to Experiment 2 with the duration of S_1 fixed throughout at a value that places it exactly 25 msec below the left corner of the triangle. The main results are presented in FIGURE 5. The value of q given by the response probabilities is stable and averages 23.8. There is a small, slow additional practice effect on the latency variance. After 35 sessions, the variance becomes extremely stable around a mean value of 280, the same as that reached when the base duration was 1150.

These experiments confirm the real-time criterion theory for short durations. Changing base duration from 1150 to 280 msec reduces the variance of the discriminal dispersion and the variance of the long-response latencies by the same large amount, an amount equal to that expected due to the change in quantum levels. The variance of the short-response latencies is unaffected. However, these results do not inform us about the flatness of the quantal steps. The next experiment was done for that purpose.

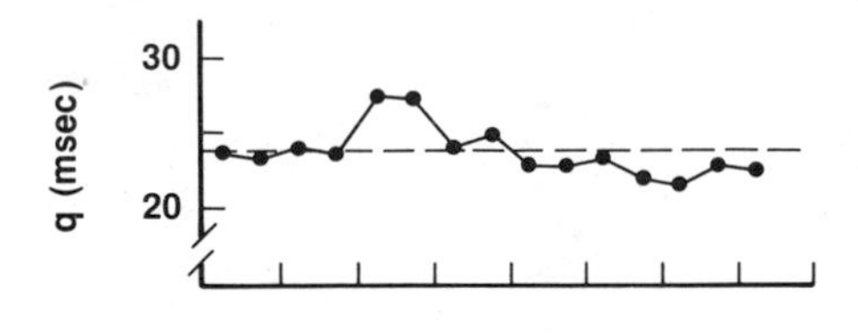

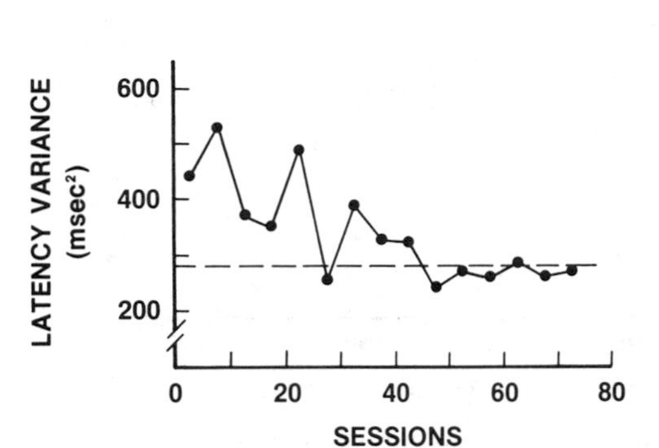

FIGURE 5. Experiment 3 (a continuation of Experiment 2 with all stimulus durations fixed for 75 sessions). D_4 was set at 305. D_1 was set at 235, which placed B_2 25 msec from the lower corner of the triangle when S_1 was presented. **Upper graph:** q calculated from response proportions for S_2 and S_3. **Lower graph:** Latency variances for R_S/S_1. The *dashed lines* are the predicted values.

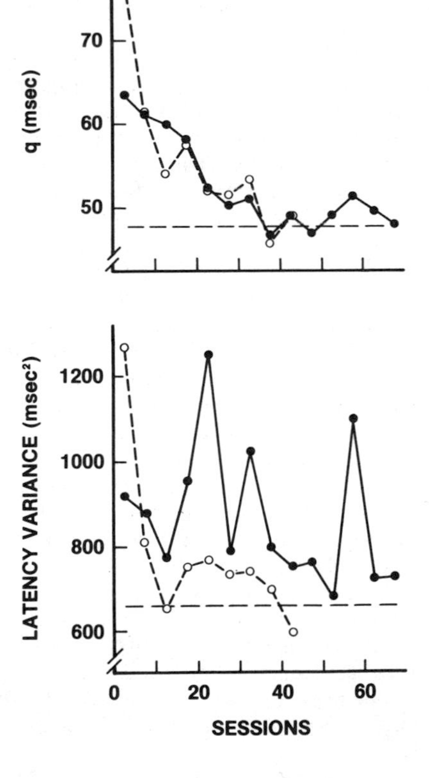

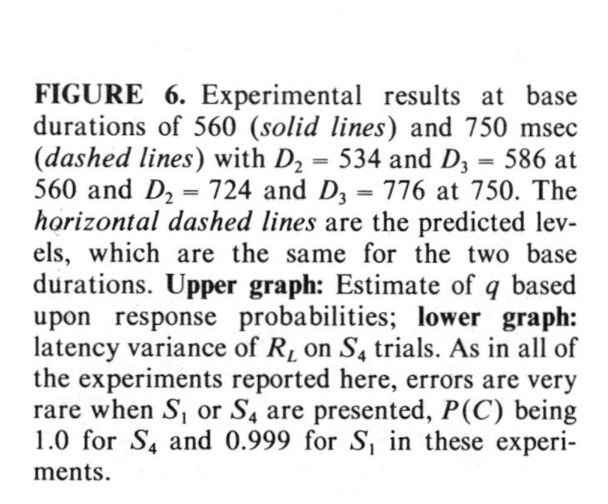

FIGURE 6. Experimental results at base durations of 560 (*solid lines*) and 750 msec (*dashed lines*) with $D_2 = 534$ and $D_3 = 586$ at 560 and $D_2 = 724$ and $D_3 = 776$ at 750. The *horizontal dashed lines* are the predicted levels, which are the same for the two base durations. **Upper graph:** Estimate of q based upon response probabilities; **lower graph:** latency variance of R_L on S_4 trials. As in all of the experiments reported here, errors are very rare when S_1 or S_4 are presented, $P(C)$ being 1.0 for S_4 and 0.999 for S_1 in these experiments.

Two new base durations, 560 and 750 msec, both selected from the q_{50} step, but widely separated on that step, were examined next. Seventy sessions were conducted at 560, followed by 45 sessions at 750 msec. The response was R_L throughout.

Once again, prolonged practice produced large changes in performance The upper panel of FIGURE 6 shows that q, as obtained from the response probabilities, is the same for the two base durations after a few sessions of practice. It diminishes in value for about 35 sessions, thereafter averaging 48.6 at 560 and 47.3 at 750 msec, as compared to the predicted level of 47.7 indicated by the horizontal dashed line. The response latency variances (in the lower panel) are more erratic, but after 35 sessions they average 791 for the shorter base duration and 646 for the longer, compared to the predicted level of 658.

For the next experiment the base duration was 1500 msec, a duration that is both

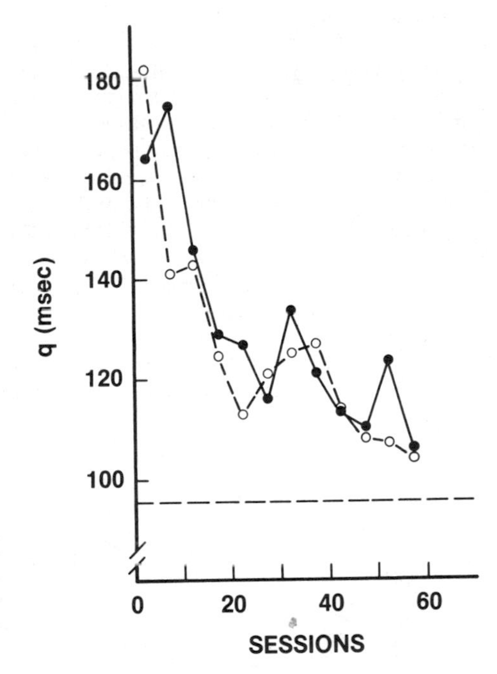

FIGURE 7. Results for 60 consecutive sessions at a base duration of 1500 msec. *Solid line:* Estimate of q calculated from response probabilities for S_2 and S_3; *dashed line:* estimates calculated from latency variance of R_L on S_4 trials, using 279 msec^2 as the latency variance of R_S (see text). The *horizontal dashed line* is the predicted level. $D_2 = 1448$; $D_3 = 1552$.

long and is also situated near the high end of its quantal step, in this case the q_{100} step. A large practice gain is needed to unfold this location on the step from the Weber's law line.

The results of a 60-session series at 1500 are presented in FIGURE 7. The solid line shows the response probability, q. The latency variance for S_4 has been transformed into its associated value of q and that is indicated by the dashed curve. The practice effect is large and the two independent measurements of q are very similar throughout practice. This suggests that the practice effect is solely a reduction in the variability of I.

While the practice gain is large and relatively rapid during the first 25 sessions, it is obviously not complete after the usual 35 sessions, or for that matter even by session 60. The predicted level is approached, but not attained. During the final five sessions, the error of prediction is 11%.

The final experiment in this series had two purposes. There is the obvious question about the existence of a q_{200} level and a step up to that level near a base duration of 1600. To find out, a base duration of 1800 msec was studied next. This base duration would be close to the lower boundary of the step, if one exists, and the practice effect should be minimal for that reason, even though the base duration is even longer than the 1500 of FIGURE 7.

Twenty sessions were sufficient to answer both questions in the affirmative, as FIGURE 8 demonstrates. Unlike previous figures, FIGURE 8 plots single-session values. The solid circles show q as calculated from response probabilities, and the open dots indicate q as calculated from the latency variance.

A stable limit is achieved after only two sessions of practice. The line in the figure is fitted to all of the data points after session 2, and its slope is 0.077. The two measurements of q agree closely, both averaging 204.

This result suggests that the doubles set of q should be expanded to include q_{200} and that the quantal principle need not be confined only to very brief time periods.

The doubling hypothesis predictions discussed above have fared quite well for the various base durations that have been investigated. Taken together, the results show that in the limit, the quantal steps of the base duration function do become flat. This is summarized in FIGURE 9, where the results of the various base durations, including 1150,[1] are brought together on a single step, the q_{50} step, by applying the doubling rule. For example, the points plotted here at the 450-base duration are from the base duration of 1800, with both 1800 and the obtained values of q halved twice.

Only those experiments in which the response was R_L, and in which D_4 placed B_2 more than 25 msec above the triangle, are included. In all cases, the amount of prior practice at the base duration is large, at least 35 sessions, except for the 1800-base duration, as pointed out above.

The solid circles show the value of q_{50} calculated from the response probabilities for S_2 and S_3. The open circles show the values calculated from the R_L latency variance for

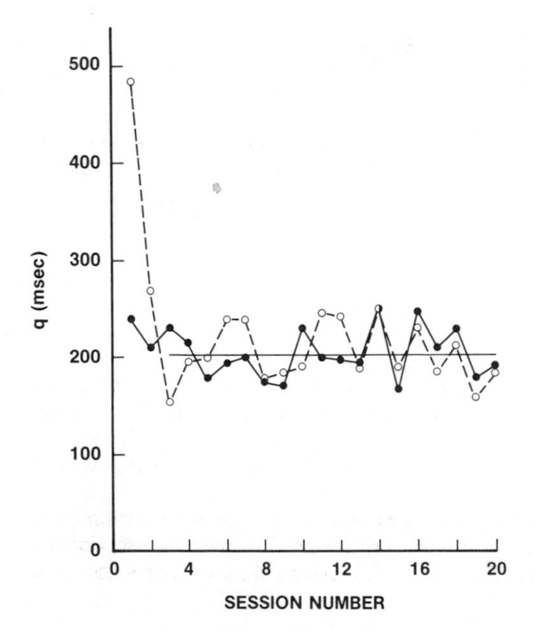

FIGURE 8. Twenty sessions at a base duration of 1800 msec. *Solid circles:* q calculated from response probabilities for S_2 and S_3; *open circles:* q calculated from response latency variance on S_4 trials. The *solid line* is the least-squares line fitted to all of the data after session 2; it has a slope of +0.077. D_2 = 1700; D_3 = 1900.

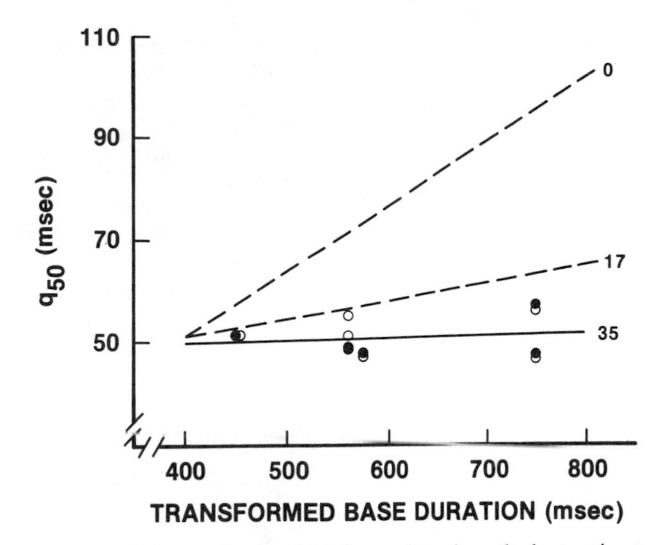

FIGURE 9. Summary of all experiments. The data points show the best estimate of q after at least 35 sessions of practice. *Solid circles:* q from response probabilities; *open circles:* q from latency variances. All base durations are included by doubling or halving both q and base duration so as to bring all within the Q_{50} range. The *solid line* is fitted to all of the data and has a slope of almost zero. The *upper dashed line* shows performance with no practice and the *lower dashed line* shows performance with 17 practice sessions (both from FIGURE 2).

S_4, using 279 as the estimate of R_S latency variance. These two measurements of q agree closely, confirming the real-time criterion theory over the full range of base durations.

The solid line is fitted to all of the data points and its slope is close to zero, indicating that the steps do become flat with sufficient practice. The upper dashed line shows the result obtained with little or no practice and the middle line the result obtained after 17 sessions of practice.[5] With increasing practice, the steps unfold progressively and finally become flat. The hinge is very close to 400 msec. These experiments were done at various times during a 10-year period and the mechanisms that are involved appear to be quite stable.

DISCUSSION

When the time quantum concept was proposed in 1967, the supporting experiments included several different ways to measure the quantum size, each of which gave a value near 50 msec.[7] Later experimental work gradually revealed that if the concept were to be retained with some degree of generality, values other than 50 would have to be accepted. For example, under certain conditions, successiveness discrimination functions appeared to require a mixture of quantum sizes of 50 and 100.[8,12] Also, our initial work on duration discrimination gave us values of 25 as well as 50 (Ref. 2). There were other such instances.

The step function in duration discrimination described above is the first single function to reveal a full range of quantum sizes. The permissible values form a doubles set extending from 12 to 200 msec. The two end values of this set are not yet firmly established by multiple operations and their status is tentative.

The step function has a second important implication. Since the steps have flat treads, deterministic timing appears to be involved. For example, for all base durations between 200 and 400 msec, $q = 25$. Changing the base duration changes the mean value of the internally timed interval, I (see FIGURE 1). Since q expresses the variability of I, the conclusion is that the mean of I can be changed without affecting its variance. Within limits, the timing of I is deterministic. When $q = 25$, deterministic timing up to at least 400 msec is possible.

This is our second report of deterministic timing. The first appeared several years

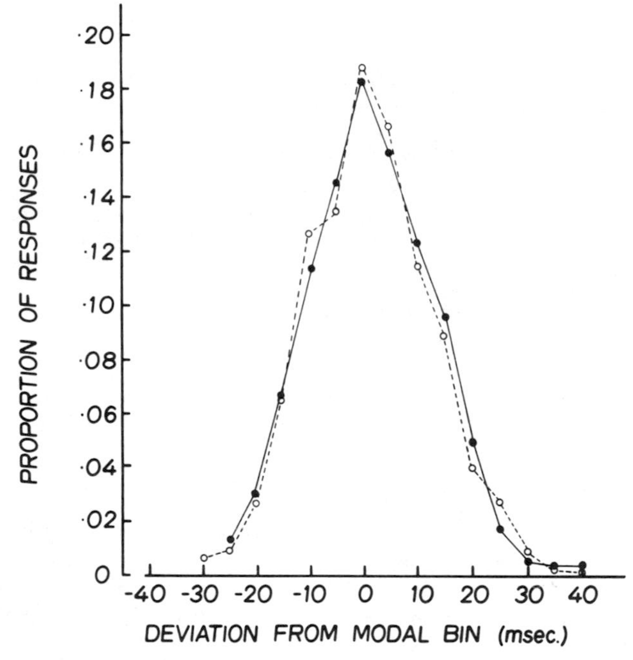

FIGURE 10. Two distributions of response latencies in response–stimulus synchronization. *Solid line:* mean = 547, *S.D.* = 12, *N* = 600; *dashed line:* mean = 307, *S.D.* = 12, *N* = 900. (From Kristofferson.[9] Reprinted by permission.)

ago in experiments on response–stimulus synchronization.[9] There too, the upper limit was found to be 400 when q is 25. Sample data are shown in FIGURE 10.

Two distributions of latencies for a response initiated by a brief auditory pulse are plotted in FIGURE 10. The two distributions appear to be identical, but actually they differ in one respect: one has a mean of 300 msec, while the mean of the other is 550 msec. The mean latency can be as short as the reaction time limit (here, about 160 msec), or the response can be delayed some additional time, up to 400 msec, without changing the variance. This internally timed interval can be set at any value between zero and 400 and give this result.

The quantum size here is 25, and the distributions are the convolution of an isosceles triangle having a base of 50, with a low variance, efferent delay distribution having a standard deviation of about 6 msec. A separate means of assessing the efferent delay variance, which agrees with this estimate, is provided by the theory and data developed by Alan Wing concerning interresponse timing.[10]

These distributions describe the latencies of an element S–R chain. They have variance less than that of simple reaction times by a factor of 4 or more.[11] They have been obtained for quantum sizes of 50 (Ref. 9), 25 (shown above), and 12 (Ref. 11). The last value provides some additional evidence for a 12-msec quantum size. Gordon Hopkins will pursue this in greater depth, including the convolution interpretation mentioned above, in this volume.

The duration discrimination step function extends our description of deterministic timing by showing that the upper limit of deterministic timing is not merely the single value of 400 msec. Instead, the upper limit changes, depending upon the value of q that is in effect. For example, when q is 12, the upper limit is 100, and when q is 100, deterministic timing occurs out to 1600 msec. On the basis of these data, the rule seems to be that the upper limit of deterministic timing is 16 times the current quantum size. This encourages the view that quantal timing and deterministic timing are not the result of completely separate mechanisms. However, I resist the temptation to propose a mechanism for two reasons. The first reason is obvious: extensive data for the step function are available so far only for the single subject discussed above. The second reason is that, while the 16-q rule did also hold in the initial synchronization experiments, it has also failed. Hopkins and Kristofferson[11] found deterministic timing to an upper limit of at least 400 with q in the 12-msec range.

There is another reason to be cautious when generalizing from the data presented above. The difficulty level of the discrimination is determined by the size of ΔD, the difference in duration between S_2 and S_3. While I am confident that the triangle is a good description of the discriminal dispersion, and therefore that q is independent of ΔD for a given distribution of C, I cannot say that the value of q attained asymptotically with practice is independent of ΔD. When ΔD is fixed throughout practice, as it was here, the asymptotic q might possibly be affected by the value of ΔD that is used. This possibility must be investigated.

The quantum size, therefore, can no longer be considered a single value. Instead, it may take on any value from the doubles set. There is no compelling evidence that it is variable beyond that, however. I point this out because some of the graphs presented herein, such as those that show the apparent gradual decrease in q with practice, may be taken to imply that q varies continuously. The data were presented this way for ease of exposition, with q as calculated for preasymptotic sessions meant to be only a rough estimate of variability. Actually, we do not yet have anything more than guesses concerning the reasons for the variance reductions observed to occur as a result of practice.

A few empirical principles of timing are beginning to take shape. My own list has only three entries, followed by many blank spaces. Two are the quantal and the deterministic principles that I have been discussing, and there is also a statistical principle, embodied in Weber's law, which I have only mentioned, but for which one can find a substantial amount of support.

On the level of theory we are still quite unconstrained so far as general directions are concerned. My own preference is to postulate a periodic process as a generator of quantal variability and to use the deterministic nature of the periodic process to explain deterministic timing, thereby linking the first two principles. Since the standard deviation of some quantal distributions is directly proportional to quantum size, the third principle, Weber's law, could also be brought within such a theory by postulating continuously variable quanta in some parts of the system.

Other theoretical directions are possible, of course. To take an extreme example, an essentially probabilistic model could be constructed to explain deterministic timing. One could postulate that the variability of the added delays does increase as the mean increases but that there is also a gradient of increasingly negative correlation which just balances the variance increase. That does not seem plausible to me, but it is possible. It would be difficult to extend such a model to explain quantal doubling and the doubling of the upper bound on deterministic timing.

It is obvious that the mere fact that a system displays temporal regularities does not mean that the system contains dedicated timing mechanisms. As far as we know, the system with which we deal may contain any number of clocks, including none at all. The empirical principles that we adduce might all describe nothing more than temporal properties of the system. Whether some subset of the principles reflects the operation of a clock or clocks, we do not know.

The proposition that time quanta are generated by a periodic process does not imply the existence of a clock. The hypothesis that there is exactly one clock that controls timing at all loci in the system seems to me to be the most likely candidate for rejection. One reason is that we repeatedly find that quantal delays at different loci fluctuate independently of each other. (For example, the triangular distribution is the distribution of the sums of two identically distributed quantal delays, each uniformly distributed, only if the two quantal delays are statistically independent). How could that be so if a single clock were controlling both loci?

Another reason is to be found in the analysis of the response "long" in the real-time criterion theory. There appear to be two pairs of quantal loci in the chain between the stimulus (P_1) and R_L (see FIGURE 1). One pair precedes C and the other follows C. Changing base duration can change the quantum level of the first pair, but does not change that of the second pair. A single clock would have to be able to generate two frequencies "almost" simultaneously. To strengthen this argument, we need to find out whether two real-time criteria can be timed during overlapping periods of time and at different quantum levels.

I believe we are being pushed in the direction of having to postulate multiple clocks. The greater the number, the less point there seems to be to talk of dedicated mechanisms, and we will be left where perhaps we should agree we now are: talking about temporal properties of the system.

REFERENCES

1. KRISTOFFERSON, A. B. 1977. A real-time criterion theory of duration discrimination. Percept. Psychophys. **21**(2): 105–117.
2. ALLAN, L. G., A. B. KRISTOFFERSON & E. W. WIENS. 1971. Duration discrimination of brief light flashes. Percept. Psychophys. **9:** 327–334.
3. ALLAN, L. G. & A. B. KRISTOFFERSON. 1974. Judgments about the duration of brief stimuli. Percept. Psychophys. **15:** 434–440.
4. KRISTOFFERSON, A. B. & L. G. ALLAN. 1973. Successiveness and duration discrimination. *In* Attention and Performance: IV. S. Kornblum, Ed. Academic Press, New York, NY.
5. KRISTOFFERSON, A. B. 1980. A quantal step function in duration discrimination. Percept. Psychophys. **27**(4): 300–306.
6. GETTY, D. J. 1975. Discrimination of short temporal intervals: A comparison of two models. Percept. Psychophys. **18:** 1–8.
7. KRISTOFFERSON, A. B. 1967. Attention and psychophysical time. Acta Psychol. **27:** 93–100.
8. KRISTOFFERSON, A. B. 1967. Successiveness discrimination as a two-state quantal process. Science **158:** 1337–1339.

9. KRISTOFFERSON, A. B. 1976. Low-variance stimulus-response latencies: Deterministic internal delays? Percept. Psychophys. **20**(2): 89–100.
10. WING, A. M. & A. B. KRISTOFFERSON. 1973. Response delays and the timing of discrete motor responses. Percept. Psychophys. **14**(1): 5–12.
11. HOPKINS, G. W. & A. B. KRISTOFFERSON. 1980. Ultrastable stimulus-response latencies: Acquisition and stimulus control. Percept. Psychophys. **27**(3): 241–250.
12. ALLAN, L. G. & A. B. KRISTOFFERSON. 1974. Successiveness discrimination: Two models. Percept. Psychophys. **15**: 37–46.

Ultrastable Stimulus–Response Latencies: Towards a Model of Response–Stimulus Synchronization

GORDON W. HOPKINS[a]

Department of Psychology
University of Alberta
Edmonton, Alberta, Canada T6G 2E9

Over the past few years I have been investigating the remarkable ability of human beings to accurately anticipate the time of occurrence of a predictable sensory event and to synchronize an overt response to that event. Presumably, this type of response–stimulus synchronization behavior is mediated by central temporal mechanisms which time the delay required to trigger the response such that it occurs temporally coincident with stimulus onset. In our examination of the nature and functioning of these human temporal mechanisms, a major aim was to develop special procedures for minimizing response latency variances to facilitate mathematical simulation and modeling of the information-processing stages involved in this type of stimulus–response chain.

The basic task, modeled after Kristofferson,[1] is diagrammed in FIGURE 1. On each trial, contact of the subject's index finger with a small touch-sensitive switch (R_i) generated a short, variable foreperiod followed by presentation of two brief auditory stimuli, P_1 and P_2, separated by a short time interval, referred to as the P_1P_2 interval. This interstimulus interval was fixed for each subject, but varied across subjects from 310 to 550 msec. The subjects were instructed to anticipate the second stimulus, timing from the first (P_1), in order to trigger a finger withdrawal or synchronization response (R_s) which would be manifested in synchrony with onset of the second stimulus (P_2). Thus, response latency was measured from P_1 onset to the moment when the subject's finger broke electrical contact with the response button. Immediate perceptual feedback regarding the accuracy of response was available to the subject by attending to the temporal order relationship between R_s and the onset of P_2. In addition, a delayed feedback signal was provided, in the form of a third auditory pulse, which had a duration equal to the error of synchrony, the time difference beween R_s and P_2 onset. The direction of error was signaled by a light accompanying the delayed feedback pulse when the error was positive, that is, when R_s occurred after P_2 onset.

This procedure that I have outlined institutes several modifications of the basic response–stimulus synchronization procedure used by Kristofferson[1]; these include the use of subject-paced trials and provision of highly salient feedback. Also, a special technique was developed that involved manipulation of the foreperiod between R_i and P_1 onset in conjunction with independent deletion of each of the stimulus components in a trial sequence. This kind of control allowed unambiguous determination of the role played by each of the stimuli in maintaining synchronization performance.

These modifications resulted in a significant reduction in the lowest previous estimate of response latency variance. Minimum variances under 35 msec^2 were

[a]Present address: Communications Research Centre, P.O. Box 11490, Station "H," Ottawa, Ontario, Canada K2H 852.

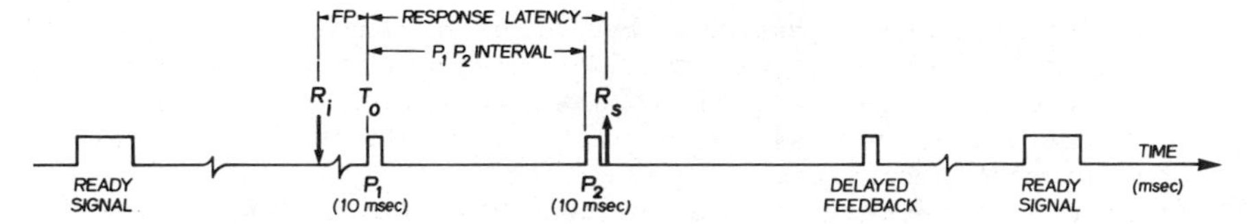

FIGURE 1. Diagram of a typical synchronization trial. Spacing of trials is paced by the subject's initiation response, R_i, which can be made at any time following ready signal. FP refers to the foreperiod duration.

obtained and the data indicated that response latency variance was independent of mean latency over the range of synchronization intervals from 310 to 550 msec. Within this range, latency distributions were the same, symmetrical, and sharp-peaked, with all responses contained within a 50-msec time window. A typical response latency distribution exhibiting these characteristics is shown in FIGURE 2. This relative frequency distribution combines 1500 response latencies obtained at a P_1P_2 interval of 460 msec and is plotted using a bin size of 3 msec. The mean is 461 msec with an overall variance of 35 msec.[2] The role of feedback in accurate synchronization performance was also examined, using the manipulations outlined previously. This provided data that indicated feedback to be one of the most important factors responsible for producing and maintaining the ultrastable, low-variance stimulus–response latencies observed. The other important factor appears to be prolonged practice at a particular synchronization interval.

Results of these studies all provide support for Kristofferson's[1–3] notion of nonvariable, centrally timed delays which can be inserted into a stimulus–response

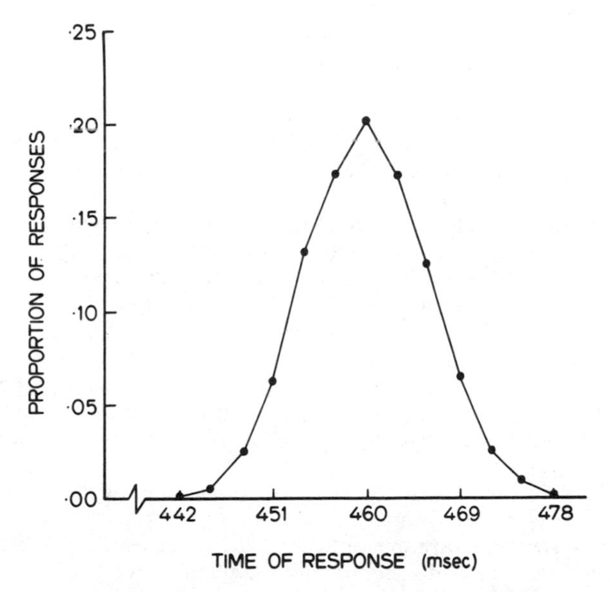

FIGURE 2. Low-variance, relative frequency distribution containing 1500 response latencies for subject G.H. on sessions 266–270 with a P_1P_2 interval of 460 msec (mean = 461, total variance = 35, mean within-block variance = 34, variable foreperiod). Bin size is 3 msec.

chain. These delays are easily adjustable, but once set, are deterministic. The aim of this presentation is to examine this notion in the response–stimulus synchronization context, to formulate a mathematical model on the basis of the data obtained from carefully controlled experiments, and to try to develop a theoretical framework for conceptualizing this type of behavior.

Several models have been proposed to account for the various timing capacities of the human central nervous system. Many of the mechanisms incorporated into these models hypothesize a "time base" of some sort which generates a succession of temporal cues that can be used by the central nervous system for response timing[4] and for controlling the gating of information flow from one central stage to another.[5]

In some models the internal clock varies somewhat in rate, causing successive temporal judgments to be variable. The mechanism for timing in these models involves accumulating clock pulses, during the duration to be judged, from a source with identically distributed interpulse delays. In Creelman's[6] model this source is assumed to be Poisson-distributed, whereas for Triesman,[7] the nature of the distribution is not specified. In both cases, however, the models predict increasing variance in temporal judgments as a function of the mean interval to be represented. In some experiments the results suggest an increasing linear function between variance and mean,[6] whereas other results suggest a similar relationship, but between standard deviation and mean.[8,9] In any event, neither model is appropriate for describing the current response–stimulus synchronization data in which variance and mean response latency are independent over a substantial range of temporal intervals, suggesting that some kind of deterministic timing mechanism might be involved.

Kristofferson[1] proposed a deterministic type of model to account for the response–stimulus synchronization performance he observed in his experiments. Latency distributions revealed a simple, homogeneous stimulus–response unit that was the same whether the mean was 160 msec or 550 msec. Kristofferson proposed that the elementary response latency distributions observed resulted from the convolution of three independent sources of variance inherent in the stimulus–response chain.

One of these component distributions is normally distributed and represents variability in the efferent delay between the time when the response is triggered internally and when the overt response is produced. Afferent latencies, on the other hand, are assumed deterministic or nonvariable. Support for this assumption comes from temporal order discrimination[10] and duration discrimination data.[2]

The other two sources of variance are assumed to be identical and independent uniform distributions spanning a range of one time quantum. When convoluted, these produce a triangular distribution spanning two time quantum units. These delays, according to the model, represent variable delays in the processing of the stimulus–response chain, but the exact nature or locus of these delays is left unspecified. Also, there is no mechanism proposed to account for the assumption of independence between the quantal units. If these delays result from the operation of a single central mechanism, which gates information through the central information processor, then it is difficult to explain how the two delays, assumed quantal in nature, and their associated variances can be considered independent because they presumably are both dependent upon the same nonrandom underlying process.

Despite these criticisms, Kristofferson's model does provide quite accurate predictions of the asymptotic response latency variance obtained using the modified response–stimulus synchronization procedure described earlier. These predictions were based on an estimate of a minimum time quantum of 12 msec, suggested by some[2,3] duration discrimination work, and a minimum efferent delay variance of 10 msec,[2] based on interresponse interval timing experiments.[11] The model, however, was never tested for its "goodness-of-fit" to the data. Thus, the aim of the rest of this discussion is to further examine new empirical data relevant to specifying characteristics of a revised model of anticipatory timing, to outline the resulting model in detail, and to provide mathematical support for the model's ability to accurately represent the data.

In response–stimulus synchronization, variability on the temporal axis can arise from several possible sources. These include inconsistency in the afferent delay between P_1 onset and its central registration, variability associated with the timekeeping process itself, and variance in the efferent delay between response trigger and overt response. However, as mentioned earlier, data from several experiments suggest that afferent latencies in this task can be considered deterministic, which means that although the afferent delay has some non-zero value associated with it, the variance is negligible. Thus, only variability in central timing and output of the response need be considered.

In the model being developed, the central and motor components are assumed, on the basis of several pieces of evidence, to be independent. For example, an analysis of carefully collected interresponse timing data conducted by Wing[11] indicated that response latency variance was basically a constant and was independent of mean interresponse interval. Total variance increased with interresponse interval, but this was attributed to increases in variability of the central processing component responsible for triggering the overt responses.

Further support comes from several simple and delayed reaction time (RT) experiments, in which electromyographs (EMGs) were taken while response times were measured, after which calculations were made of the correlation between pre-motor time (time from onset of the action stimulus to EMG onset) and response time versus the correlation between motor time (time from EMG onset to the overt response) and response time. The results revealed correlation coefficients close to zero between motor time and response time, indicating independence between central and motor components.[12] Also, in the delayed RT situation, the EMG activation preceded the overt response by a relatively constant interval regardless of the actual response latency produced.[13]

A similar result was obtained by Michaels[14] using a response–stimulus synchronization paradigm and a countermanding procedure. The subject's task was to withhold the synchronization response if a third signal occurred during the P_1P_2 interval. The data were proportions of correctly countermanded responses as a function of the time between the countermand signal and P_2 onset. From this data an estimate of response trigger timing was derived and the results indicate that the trigger always precedes P_2 onset by a fixed time period, independent of the P_1P_2 interval.

These findings lead to the conclusion that manipulations involving anticipatory response affect only the central, pre-motor component of a stimulus–response chain and argue against efferent stages' having any major participation in timekeeping, at least in this rather simplified paradigm. Obviously, for extended chains of motor behavior, timing may become more complex. Tyldesley and Whiting[15] among others[16–19] have suggested that in these situations some "timing" is simply a byproduct of efferent delays generated by each motor component. However, this discussion will not address that type of timing.

The last piece of evidence in support of the notion of independence between central and efferent stages comes directly from our synchronization experiments. Findings of independence between response latency variance and mean over a wide range of P_1P_2 intervals strongly suggest the existence of an adjustable, nonvariable, central delay mechanism combined with a constant mean efferent latency. This is so because if response latency was lengthened by increasing the number of motor components in the S–R chain, each with its own inherent variability, then there should be a commensurate increase in variance. But this is not the case. Consequently, we assume that efferent delays are distributed with constant mean and standard deviation despite relatively large changes in the overall mean response latencies produced during synchronization at different P_1P_2 intervals.

The last point to be discussed, before formulation of the model, concerns the accuracy and variance associated with the motor component of response latencies. The overt response is simply a finger withdrawal, but actually several motor elements are involved in producing this movement. The interesting point is that although each of the underlying elements, when measured separately, exhibits a rather large temporal variation, the outcome of their joint action produces a response well defined in time. Meijers and Eijkman[20] have examined this apparent paradox and offer an explanation based on the macro-activity required for the motor system to initiate elementary movements. They show that execution of an overt response requires the joint effort of many elements and it is this requirement of joint activity that allows the remarkably small stochastic variation observed to be obtained. This is accomplished by summation of element activities, thereby providing a better time definition than that produced by any of the individual activities. In other words, averaging the behavior of several elements can cancel the effect of individual temporal inaccuracies.[20]

From the foregoing discussion, it is obvious that several factors must be considered in formulating a mathematical model that will not only provide a good representation of the data, but will also have parameters that are psychologically relevant. The latter stipulation is important because without it the model will have little utility in generating the testable predictions needed for furthering our understanding of the internal mechanisms that underlie response–stimulus synchronization behavior, thereby extending existing theory.

THE MODEL

Consequently, a rather traditional approach was taken in formulating the model. The approach is based on the premise that if the data are stable, then any stochastic processes associated with delays accrued in each of a series of independent processing stages would be reflected in an overall distribution of response latencies given by the convolution of all of the component distributions.

The response–stimulus synchronization model that is being developed is similar to Kristofferson's,[1] but includes some modifications and extensions in an attempt to provide a locus for the central delays incurred in processing the stimulus–response chain as well as to account for the assumption of independence of the central, stochastic components. In describing the model, I will employ some computer metaphors to facilitate an understanding of the mechanisms involved. However, the use of these metaphors is purely for eclectic reasons and should not be construed as implying any direct analogy. I would also like to note that the first part of this discussion assumes that a steady-state condition exists in the central information processor. Violation of this assumption will be dealt with subsequently.

FIGURE 3 diagrams the modified response–stimulus synchronization model. The onset of P_1 is a sharply defined external event, although its sensory effect is extended over time, as shown by the interval labeled *afferent latency*. This latency refers to the time from onset of the peripheral stimulus until an internal state has developed, as a result of P_1 stimulation, that is sufficient to exceed some criterion and trigger the next stage in the information-processing chain. This process is similar to filling an input buffer and setting a flag that indicates that information is available for further processing. In this context, the afferent latency is a combination of the transduction time at the peripheral receptor, the conduction time from periphery to the central system, plus the time required to represent this information in one of the registers of the central processing unit.

Although the afferent latency certainly has some non-zero value, its variability is assumed to be negligible on the basis of several pieces of evidence presented earlier. Therefore, afferent latency can be considered a constant, contributing nothing to the variance and shape of the observed distribution of response latencies.

Once the stimulation produced by P_1 onset is registered centrally, in a buffer, the information must wait for a period represented by W_1 before gaining access to subsequent processing stages. This waiting time for information transfer from input buffer to the deterministic timekeeping mechanism results because the contents of the

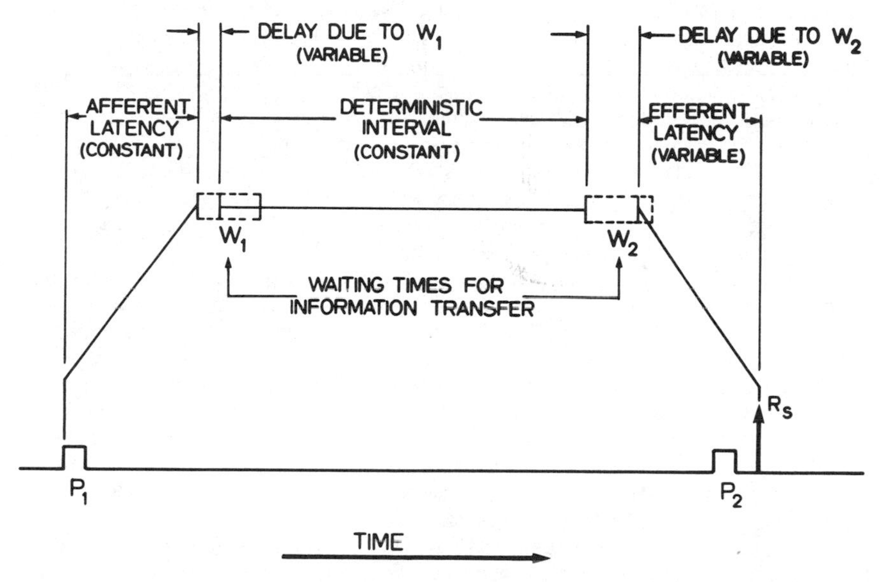

FIGURE 3. Diagram of response–stimulus synchronization model. P_1, P_2, and R_s are external, observable events, while W_1 and W_2 are hypothesized internal events. The diagram represents the time course of various components of a stimulus–response chain on a typical trial.

input buffer are only accessed periodically. Every n units of time the central processor reads the contents of the input buffer and performs operations based on this information. Because this cycle time, or scheduling of access time-points, is independent of peripheral stimulation, the delay due to W_1 is variable and uniformly distributed over a range from zero to W_{1max} milliseconds. For example, sometimes stimulus information will reach the input buffer just prior to the start of a new cycle and thus will gain access to the central processor with very little delay. On the other hand, stimulus information loaded into the input buffer just after the start of a cycle will have to wait almost an entire cycle, W_{1max}, before gaining access to the central processor.

After access of stimulus information to the central processor, or in this case, the deterministic timekeeper mechanism, a delay appropriate to the synchronization interval and state of the organism is assumed to be generated. Since this delay has negligible variance associated with it, the term deterministic interval is used. Physiological mechanisms capable of producing such delays are not forthcoming from the literature, but the rationale for assuming the existence of such mechanisms is clear from the preceding discussion and experiments.

The termination of this deterministic delay produces information that is then loaded into an output buffer and generates a flag in a fashion similar to that described for the input buffer. This time, however, the information is waiting to gain access to the response processor. This processor also has a fixed cycle time, W_{2max}, which is similar to that of the central processor, W_{1max}, thus generating uniformly distributed delays over a range from zero to W_{2max}. Although W_{1max} and W_{2max} are similar, they are not exactly the same. As a result the two processors cycle in and out of phase relatively frequently during the synchronization interval. Consequently, the two waiting times can be considered independent since the initiation of a synchronization interval (P_1 onset) is totally independent of any phase relationship existing between central and response processor cycle times.

An example may help to clarify this mechanism that I am proposing to allow the assumption of independence between W_1 and W_2. Suppose W_{1max} is 12 msec and W_{2max} is 13 msec. These values can be considered the periods, or cycle times, of the periodic processes responsible for the waiting times W_1 and W_2, respectively. Thus, these two processes will pass in and out of phase every 156 (12 × 13) msec. Now, consider how W_1, W_2, and their phase relationship are related to the stimulus sequence used in response–stimulus synchronization. The occurrence of P_1 can be considered independent of W_1 for several reasons. First, trials are subject-paced such that the intertrial interval varies greatly with respect to the cycle time responsible for W_1. Second, there is no evidence to suggest that the conscious decision to elicit an initiation response, R_i, is dependent in any way on the central process responsible for W_1. And third, even if one postulated a relationship between W_1 and R_i triggering, its characteristics would be lost because of two sources of temporal variability interposed between the triggering of R_i and the occurrence of P_1. One source is due to the variable efferent delay between the central response trigger and the overt response, R_i, and the other is due to the experimentally introduced, variable foreperiod between R_i and P_1 onset. Both sources are random and relatively large compared with the cycle time of W_1. Therefore, the occurrence of P_1 onset can be considered independent and random with respect to the W_1 cycle, resulting in a uniform distribution of W_1 waiting times.

How is W_2 independent of W_1? Well, since P_1 onset is registered centrally at some random point in time with respect to the W_1 cycle and since W_1 and W_2 have slightly different cycle times, knowing at what point in the W_1 cycle P_1 is registered provides no information about what part of the W_2 cycle will be intersected after the fixed deterministic delay. Consequently, W_1 and W_2 can both be considered independent and uniformly distributed. As a result, the convolution of these two distributions will generate a unit of central temporal variability that is basically triangular, that is, as long as the values of W_{1max} and W_{2max} are not too dissimilar. A slight difference in their periodicities only produces a slight bluntness in the peak of the triangular distribution. Finally, since P_1 onset, W_1, and W_2 can all be considered independent, no autocorrelation should exist between trials, which is consistent with the data.

After information transfer to the output buffer, the output stage takes control, triggering the appropriate action (finger withdrawal), which, after an efferent delay, is manifest in the overt response, R_s. As discussed earlier, these efferent latencies are assumed to be approximately normally distributed with a relatively small variance.[20] In the model, the logistic distribution is substituted for the normal because of its marginal superiority in mathematically representing this kind of variability.

The ultimate goal of this sequence of processing stages is to produce a response that is perfectly synchronous with P_2 onset. However, because of variability incurred at various stages, the best performance that can be realized involves centering the response latency distribution about the time-point corresponding to P_2 onset and minimizing the variance of the various stochastic components.

The overall response latency distribution is given by the convolution of the component distributions. In the model this involves convoluting the distributions associated with W_1, W_2, and the efferent latency. Although W_{1max} and W_{2max} are not exactly the same, for purposes of the initial modeling they were considered identical. Thus, the convolution of the distributions of waiting times produces a triangular distribution spanning a range of $2 \times W_{max}$. When this is further convoluted with a relatively low-variance, logistically distributed component, it produces a distribution that is still basically triangular, but with short tails and a slightly blunted peak.

The general shape of distribution generated by the model seems to characterize the data quite well, but a more rigorous test of the model's ability to represent the data was obtained by mathematically testing the goodness-of-fit between the distribution function generated by the model and the cumulative probability of response distribution derived from the data.

Parameters of the model, representing the variable components, were estimated by allowing them to vary over a calculated range while repeatedly testing for goodness of fit via the minimizing χ^2 technique. The two parameters estimated consisted of W, which represented the average of W_{1max} and W_{2max} (the maximum times required for information transfer), and b, which represents the standard deviation of the efferent response latency distribution. Values for W and b were covaried because they were constrained by the overall variance of the observed distribution. Variance of the triangular distribution is given by $W^2/6$ and variance of the logistic distribution is represented by b. Thus, the equation for the overall response latency variance V_{total} is:

$$V_{total} = W^2/6 + b^2$$

Therefore, the constraints are clear. If $b = 0$, then:

$$W_{max} = \sqrt{6 \times V_{total}}$$

and conversely, if $W = 0$, then:

$$b_{max} = \sqrt{V_{total}}$$

Moreover, for any W chosen in the range from 0 to W_{max} the value of b is fixed by the following equation:

$$b = \sqrt{V_{total} - W^2/6}$$

FIGURE 4 shows the excellent representation of the data provided by the model. The points on the figure are the actual data obtained from a well-practiced subject and the line is the best fit to this data provided by the model. It is evident that the model provides an excellent description of the data with a χ^2 value of 4.81 on 9 degrees of freedom. In this case, the estimates of W and b are 11.4 and 3.6 msec, respectively. This produces an estimate of overall variance of 34.7 relative to 34.6 msec^2 obtained from the data.

FIGURE 5 presents the results of a similar analysis based on the lowest variance data obtained in the experiments. I have added quartile lines to emphasize the amazing accuracy with which humans can perform response–stimulus synchronization. Mean response latency is only 0.3 msec longer than the P_1P_2 interval (460 msec) and the

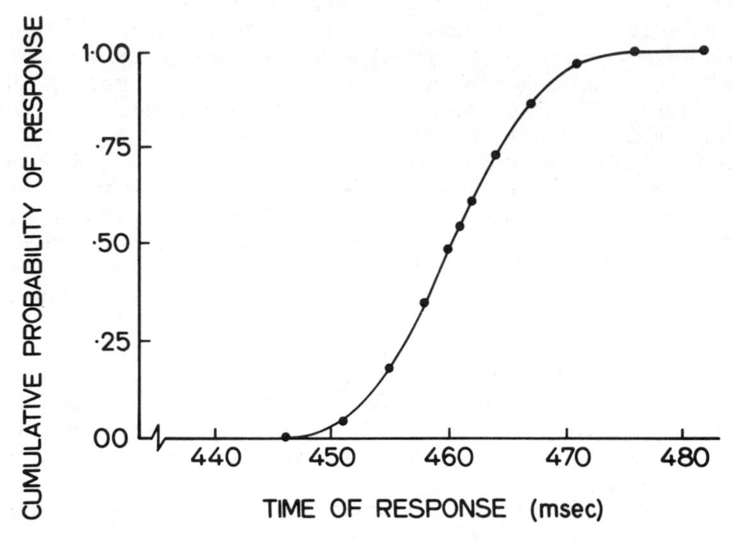

FIGURE 4. Graph showing goodness-of-fit of model to the data from sessions 266–270 for subject G.H. The *solid circles* are the data points and the *line* is the psychophysical function predicted by the model. Parameter values are given in the text. P_1P_2 interval is 460 msec.

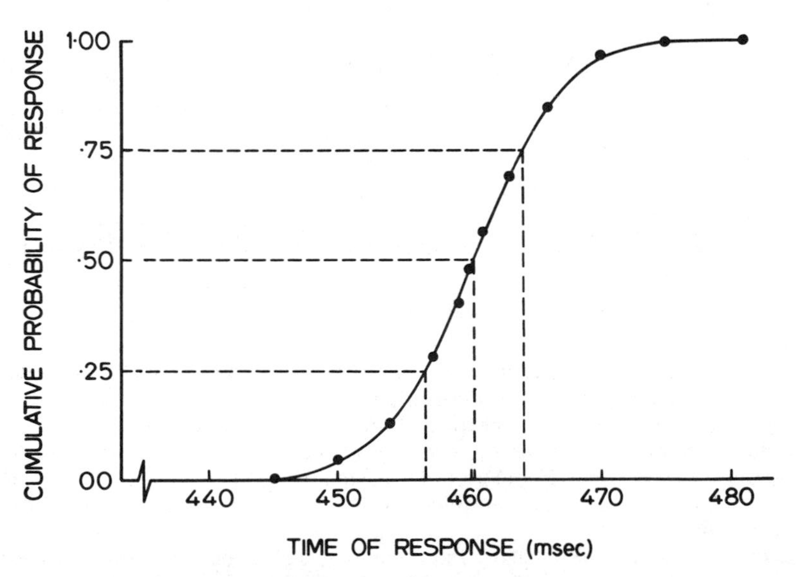

FIGURE 5. Goodness-of-fit of model to the data from five lowest-variance sessions for subject G.H. The *solid circles* show data points and the *line* the psychophysical function predicted by the model. P_1P_2 interval is 460 msec. *Dotted lines* indicate the quartile response times.

spread of the distribution of response latencies is small, with 50% of all responses falling within a 7-msec time window. Expanding the time window to 19 msec accounts for more than 90% of the responses and, in fact, all responses in this analysis fall within ±16 msec of the mean. This distribution represents five sessions worth of data or 1500 response latencies. The estimates for W and b for this set of data are 9.1 and 4.18, respectively. These values provide an estimate of overall variance of 31.3 msec^2, which is exactly what was obtained from the data.

Parameters were extracted not only from stable data obtained after much practice, but also from data obtained during acquisition when the response latency variances were decreasing. Results of these analyses suggest that the value of W changes little while b exhibits substantial decreases to account for the reduced variances.

In spite of the fact that the model fits the data extremely well, the method of extracting parameters seemed to lack power because there was usually a range of W and b values that produced near optimal fits to the data. For example, in one analysis b values ranging from 3.5 to 5.0, in combination with their corresponding values for W, all provided reasonably acceptable representations of the data. As a result, I was concerned about the ability of the minimizing χ^2 procedure to extract meaningful estimates of the underlying parameters involved.

To further exploit the model, I decided to simulate subjects' response latency distributions, using parameters extracted from the original data, by randomly generating response latencies which summed the delays associated with each of the model's variance components. This simulated data was then analyzed to try and recover the parameters used in generating it. This would provide some indication of the power of the parameter-extraction procedure employed.

Results of these simulations were quite rewarding. Parameter estimates based on the simulated data never exceeded 10% error from those parameters used to generate the data, even with the apparent lack of power associated with the parameter-extraction technique, and typically the correspondence was almost perfect. Two examples are shown in FIGURE 6. The left panel shows a simulated response latency distribution generated with $W = 11.4$ and $b = 3.6$. The corresponding parameters extracted from this data were 11.6 and 3.5, respectively. The χ^2 statistic for the fit is 4.7 on 11 degrees of freedom. The second example, shown in the right panel, was based on a set of data generated with $W = 11.6$ and $b = 3.5$. In this case, the recovered value for W was identical at 11.6 and the b value was very close at 3.35, with a χ^2 value of 5.27 on 10 degrees of freedom. The overall variances associated with these two distributions are 35.1 msec^2 for the left-hand distribution and 33.7 for the other.

All the analyses to this point have been using an average value for W_1 and W_2. However, as pointed out earlier, it is imperative to the assumption of independence between these two components that they be slightly different. To test whether this restriction would affect theorizing, several sets of data were simulated using different values for W_1 and W_2. As also pointed out earlier, it is only necessary for W_1 and W_2 to differ by a millisecond or less for the argument of independence to hold. Consequently, data sets were generated in which W_1 and W_2 differed by 1 msec. The results were almost identical to those obtained when the average values of W_1 and W_2 was used. In fact, FIGURE 6 could also be used to present these data because the differences in the distributions and parameters extracted are less than 1 percent.

Thus, the method used for parameter extraction does not appear to impose any serious restrictions on our ability to estimate W and b. Moreover, it is interesting to note that the estimates of b that were calculated, where b^2 represents the efferent delay variance, agree well with estimates provided by Wing and Kristofferson,[9] and the estimate of W is in the same range as the minimum unit of quantal time that Kristofferson has proposed.[3] However, this latter correspondence does not imply a single "clock." In fact, the mechanism proposed to produce independence assumes two different periodicities.

One consistent finding was that efferent latency variance, estimated by b, tends to contribute proportionately less to the total variance as a function of practice. In terms of the response latency distribution, it means that the shape should become more triangular as practice continues and, to some extent, this can be seen in the data. The coefficients of kurtosis are generally around 3.2 at the beginning of practice and decrease over time to values under 3.0. The coefficient of kurtosis for the logistic distribution is 4.2 and for the triangle is 2.4. Therefore, the trend towards a reduction in the coefficient of kurtosis, especially in the range observed, is consistent with the notion of a transition in shape from logistic to triangular. However, this information should only be taken as corroborating evidence because the change in coefficient values was rather small and inconsistent in a few instances.

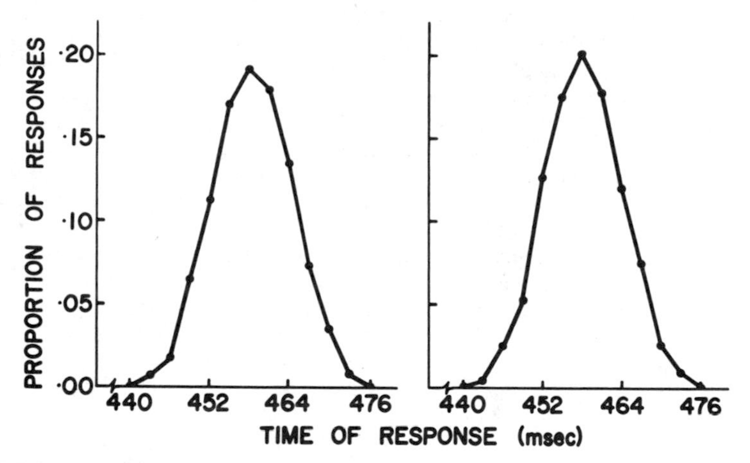

FIGURE 6. Relative frequency distributions for two sets of simulated data. Left panel shows distribution generated from the model (see text), with $W = 11.4$ and $b = 3.6$. Extracted parameter values based on these data were $W = 11.6$ and $b = 3.5$. Right panel shows similar distribution generated with $W = 11.6$ and $b = 3.5$. Recovered parameter values were $W = 11.6$ and $b = 3.4$. χ^2 for goodness-of-fit of model to the left and right distributions was 4.7 on 11 d.f. and 5.3 on 10 d.f., respectively.

Up to this point I have been referring to the parameter b as representative of efferent variability. But the analyses indicate that most of the improvement with practice is reflected in reductions in b. Because several thousand trials are required to attain asymptotic performance, it is difficult to understand how a simple finger withdrawal response could continue to be refined after such a number of trials. Consequently, I would like to introduce a further extension to the model that addresses this problem as well as that of violating the assumption of steady-state behavior, as mentioned earlier.

In reality, the state of the central mechanism governing response trigger timing probably changes slowly over time, being affected by concurrent cognitive activity and changes in the physiological state of the organism. Thus, internal conditions could be viewed as if they were in a state of continual flux. This would necessitate a dynamic process for fine-tuning, or updating, the central timing stage based on feedback about recent successes and/or failures in synchronization.

The result of this continual updating of the system introduces a new source of variance that was not previously considered. As the internal state of the organism changes, timing will be affected such that the response triggers will begin to occur too early or too late on the average. This information, provided by the available feedback, allows the central stage to alter the timing process accordingly in order to maintain accurate synchronization.

Obviously the accuracy of such a feedback loop is determined, in part, by the amount of information considered in determining the extent of alterations to be made. To be most accurate the subject should try to integrate, or average, as much information from preceding trials as possible. Such a notion is supported by the sequential dependency analyses that were applied to the experimental data. Those analyses indicate that feedback information becomes more reliable with practice, suggesting the experienced subject integrates increasingly more information into the updating decision process. Further increases in updating accuracy probably also stem from the subject's learning to interpret the feedback information better.

If this new variance component is considered to be logistically distributed, then it would be combined in our estimate of b. Thus, b may actually encompass two sources of variation that cannot really be separated because they are similarly distributed. Probably the most reasonable explanation of decreases in the value of b with practice would share the responsibility between decreases in efferent latency variance and increases in accuracy of updating the central components to maintain synchronization with the "real world." For example, the large decreases in b that are observed early in practice could be largely a function of response learning, while after some initial training, continued decreases in b could reflect the improved updating abilities of subjects. However, when asymptotic performance is attained, indications are that b provides a good estimate of minimum efferent latency variance because the updating process has been so refined as to approximate a steady-state system.

What happens without the feedback? Well, the data show that performance deteriorates rapidly and strong sequential dependencies begin to emerge. In terms of the model, lack of feedback prevents the updating process from operating, which results in mean response latency varying over time as a function of changes in the internal state of the organism. This produces a slow wandering of the mean, thereby inflating overall response latency variance and introducing autocorrelations between successive responses. Decrements in performance are probably further enhanced by degradation of the memorial representation of the P_1P_2 interval. Without feedback, there is no way to refresh the memory for the synchronization interval.

If feedback helps to synchronize the central timekeeping process with the passage of physical time, then one would expect individual differences to be minimized when feedback is available. This is exactly what was observed in these experiments. When feedback was present, all subjects, independent of sex and P_1P_2 interval, exhibited very similar performance. However, in the absence of feedback, large individual differences developed in terms of mean errors of synchrony, response latency variances, and autocorrelations.

SUMMARY

To summarize this discussion, I have presented several pieces of evidence in support of the proposed model. Substantial support comes from the fact that when parameter estimates obtained from independent areas of research are substituted into the model, it predicts asymptotic variance levels almost identical to those observed. The model also provides a remarkable mathematical representation of the data and can account for independence of mean and variance over the range of P_1P_2 intervals tested. Extensions made to the model provide a locus for the central variance components in the S–R chain and also provide a mechanism to produce independence between the central components by assuming two different periodicities. Finally, not only did the model provide a good mathematical representation of the data, but also the parameters were theoretically appealing for further postulating about central mechanisms and processes underlying response–stimulus synchronization. This last point is important because we also tested several other models, which I have not discussed, that provided acceptable mathematical representations of the data, but none had parameters that were theoretically meaningful.

As a final point, I would like to reexamine the traditional distinction made between subjective and objective time. Subjective time has often been thought of as a dimension of experience only, in which the nature of the activities occurring during a period is the major determinant of the phenomenal duration, rather than movements of hands on a clock.[21] In fact many investigators, back to the time of William James,[22] have been intrigued by these alterations in perceived duration produced by varying the physical events generating our subjective experiences. Obviously these are important aspects of cognitive functioning to understand, but it is also intriguing to find out that under some circumstances there is no transformation made between physical and psychological time. The two are the same. The characteristics of these circumstances, however, have just recently begun to emerge from synchronization and duration discrimination studies. Consequently, it is important that this mode of information processing receive further investigation because possibly it provides the crucial link between our minds and our environment in "real" time.

REFERENCES

1. KRISTOFFERSON, A. B. 1976. Low variance stimulus-response latencies: Deterministic internal delays? Percept. Psychophys. **20:** 89–100.
2. KRISTOFFERSON, A. B. 1977. A real-time criterion theory of duration discrimination. Percept. Psychophys. **21:** 105–117.
3. KRISTOFFERSON, A. B. 1980. A quantal step function in duration discrimination. Percept. Psychophys. **27:** 300–306.
4. MICHON, J. A. 1967. Timing in Temporal Tracking. Institute for Perception (RVO-TNO). Soesterburg, The Netherlands.
5. KRISTOFFERSON, A. B. 1967. Attention and psychophysical time. Acta Psychol. **27:** 93–100.
6. CREELMAN, C. D. 1962. Human discrimination of auditory duration. J. Acoust. Soc. Am. **34:** 582–593.
7. TRIESMAN, M. 1963. Temporal discrimination and the indifference interval: Implications for a model of the "internal clock." Psychol. Monogr. Vol. **77** (13, whole no. 576).
8. GETTY, D. J. 1975. Discrimination of short temporal intervals: A comparison of two models. Percept. Psychophys. **18:** 1–8.
9. WING, A. M. & A. B. KRISTOFFERSON. 1973. Response delays and the timing of discrete motor responses. Percept. Psychophys. **14:** 5–12.
10. ALLAN, L. G. 1975. The relationship between the perception of successiveness and the perception of order. Percept. Psychophys. **18:** 29–36.
11. WING, A. M. 1974. The timing of interresponse intervals by human subjects. Diss. Abstr. Int. **35:** 4237B.
12. BOTWINICK, J. & L. W. THOMPSON. 1966. Premotor and motor components of reaction time. J. Exp. Psychol. **71:** 9–15.

13. SASLOW, C. A. 1972. Behavioral definition of minimum reaction time in monkeys. J. Exp. Anal. Behav. **18:** 87–106.
14. MICHAELS, A. A. 1977. Timing in Sensorimotor Synchronization. Unpublished Master's thesis, McMaster University, Ontario, Canada.
15. TYLDESLEY, D. A. & H. T. A. WHITING. 1975. Operational timing. J. Hum. Movement Stud. **1:** 172–177.
16. ADAMS, J. A. & L. R. CREAMER. 1962. Proprioception variables as determiners of anticipatory timing behavior. Hum. Factors **4:** 217–222.
17. JONES, B. 1972. Outflow and inflow in movement duplication. Percept. Psychophys. **12:** 95–96.
18. KEELE, S. W. 1973. Attention and Human Performance. Goodyear Publishing Company. Pacific Palisades, CA.
19. SCHMIDT, R. A. & R. W. CHRISTINA. 1969. Proprioception as a mediator in the timing of motor responses. J. Exp. Psychol. **81:** 303–307.
20. MEIJERS, L. M. M. & E. G. J. EIJKMAN. 1974. The motor system in simple reaction times experiments. Acta Psychol. **38:** 367–377.
21. ORNSTEIN, R. E. 1969. On the Experience of Time. Penguin Books. Harmondsworth, England.
22. JAMES, W. 1908. Principles of Psychology, Vol. 1. Holt. New York, NY.

The Perception of Temporal Events[a]

D. ALAN STUBBS, L. R. DREYFUS,
AND J. G. FETTERMAN

Department of Psychology
University of Maine
Orono, Maine 04469

Much of the research on animals' temporal discrimination has an implied framework that treats the phenomena from a sensory process point of view.

Recent years have seen the development of a variety of psychophysical techniques.[1,2] For example, animals may be presented with different durations of a stimulus, with one choice response reinforced if the duration had been short and a second choice response reinforced if the duration was long. In addition, there has been an increase in research using reinforcement schedules that is psychophysical in orientation; orderly relations between schedule value and behavior have been observed under a wide variety of temporally defined schedules.[3,4] This line of research has provided valuable information about Weber's law, logarithmic and power relations, and subjective time scales. While this sort of research has produced basic and necessary information, most of it has treated stimulus duration as a simple dimension that is similar to sensory dimensions such as frequency or intensity. In most of the research, the durations to be discriminated are composed of a single unchanging light or tone. The stimulus component has been de-emphasized so that orderly relations could be obtained and so that the operation of a sensory-based timing process could be assessed.

Not only the procedures, but also current theories, imply a sensory framework. Recent theories are concerned with the timing process and, in general, all attempt to separate the timing process and factors affecting it from other processes and aspects of the situation.[3,5,6] The different theories have different emphases. Gibbon's scalar expectancy theory, for example, emphasizes the orderly psychophysical relations that transcend individual procedures in an attempt to delineate the temporal scaling process that is common to the different situations. The internal clock theory of Church and his students has emphasized the "clock" and has focused on factors that influence the way in which the clock functions. Although no one argues that the timing process is simple, the treatment of timing does indicate a sensory process framework. The theories are typically described in cognitive terms and the theories take an information-processing approach, but the roots seem to come from a sensory-based orientation. In particular, the internal clock theory seems to treat the clock as a receptor that registers the passage of time. With sensory dimensions such as brightness there are correlated physiological processes; the clock seems to serve a similar though hypothetical function as a time sense.

An alternative orientation places temporal discrimination within the framework of perception. This framework places animals' time perception within the context of other complex perceptual phenomena such as space perception, motion perception, and form perception. A perceptual framework suggests analogies between animal time perception and other perceptual processes and indicates that the study of animal timing might

[a]Authorship of this paper is equal. The research was funded in part by a Faculty Research Fund Award from the University of Maine to D. Alan Stubbs.

well profit from the lessons learned in these other areas. Stimulus durations may be arranged easily (more easily than most stimulus continua), orderly psychophysical findings might occur when durations are used, and inferences about a timing process can be made. But the ease of arranging the stimulus dimension and the orderly relations may be misleading and deceive us into viewing the problem in a particular way. With space, movement, and form perception, continua have also been established, elegant psychophysical experiments have been conducted, and valuable results have been obtained, but more is involved in all of these areas. With space perception there are the many "cues," both stimulus and organismic; prior experience tells us about the distance of things; different information is provided to stationary and moving observers; and there are higher-order invariants and relations when observers move about in a world of textured surfaces. In light of all of these factors, the space perception literature does not have an analogue to timing; that is, researchers do not talk about a "spacing" process. Research such as that with the Howard–Dolman apparatus (in which people judge the relative distance of two rods against a homogeneous background) demonstrates the role of accommodation, convergence, and disparity. These organismic or internal cues provide the closest analogy to timing. However, the space perception literature has shown that these cues tell only a part of the story; much more is involved when humans and nonhumans move about in less restricted visual environments.

The adopting of a perceptual framework leads to the conclusion that the implied sensory framework that guides most current thinking may be limiting the questions that are asked and the research that is done. A perceptual approach argues for a change in emphasis and for increased attention in several areas. First, greater emphasis should be given to the stimuli constituting the durations to be discriminated. Most researchers recognize that temporal discrimination tasks do not measure discrimination of time, but instead the discrimination of the duration of stimuli. It would seem proper that a greater emphasis be given to the stimulus side of stimulus duration. While human research has employed a wide variety of stimulus situations, most animal research has used durations marked by lights and tones that do not change. The function of these stimuli seems to be simply that of a time marker; the stimulus does not appear to be given much importance, but rather serves as a way of bounding abstract time or setting a clock in operation. Second, most of the research has attempted to eliminate environmental events in duration tasks. With people, time judgments are influenced by watches, the sun, the amount of work done, the variety and structure of events, and the like. But these conditions are often eliminated in research, particularly that with animals; they are eliminated in part because they are viewed as contaminating features that might provide extraneous cues to interfere with "pure" timing. The perception of time is more properly considered as the perception of temporal events and as such resembles in some ways the perception of movement and perception of change.[7,8] Unfortunately, different research strategies separate the areas: with movement and change, researchers are searching for the relevant stimuli; with time, researchers (particularly animal researchers) typically hide or limit the relevant temporal events. Use of unchanging stimuli in duration tasks reduces temporal structure in the environment so that the only events remaining are internal, which inevitably leads to a view of these events as crucial. The situation may be like that with space perception in the past, when accommodation, convergence, and disparity assumed a role of importance when experiments eliminated most of the features of a textured world. Third, the perceptual framework suggests that a wider range of tasks, including tasks that are more complex, be used. Although a variety of procedures are currently available, these procedures are primarily variations on the basic human psychophysical techniques. These sensory-like techniques contribute to the sensory process framework for considering research findings. What is needed is a more varied set of tasks to deal with questions like those raised in human time perception and like those raised in other areas of perception.

The two experiments to be reported have their origins in our consideration of temporal discrimination within the perceptual framework.

EXPERIMENT 1: THE AUDITORY–VISUAL DIFFERENCE IN PIGEONS

The purpose of the first experiment was to see whether different stimuli would influence performance in a duration task. A robust finding in the area of human time judgment is that people judge sounds as longer than lights of equal duration.[9] This finding holds across different procedures and variations in method and thus seemed to be a good starting point in the investigation of stimulus factors as they affect temporal judgments of animals.

The procedure used for the task was based on the free-operant scaling procedure of Stubbs.[10] The scaling procedure was selected since it permits a greater flexibility of response than do discrimination procedures and thus seemed to offer a greater likelihood of differences in response when lights and sounds were presented.

Three pigeons were trained in a standard operant chamber, with daily sessions lasting until a pigeon had received 50 reinforcers. FIGURE 1 gives the outline of the procedure. Intertrial intervals separated the duration periods. The intertrial intervals were variable in length, with an average of 15 sec. In addition, a reset contingency was used such that a peck to either key would prolong the intertrial interval by 5 sec. During the intertrial interval, the houselight was on and the right key was lit by white light. After the intertrial interval, there were two types of trials, with each occurring equally. On light trials, the onset of the trial was signalled by the onset of a red light behind the left key (the houselight and white light remained on). When the red light was on, red-key responses were sometimes reinforced between 5 and 6 sec after light onset. If food was produced, an intertrial interval followed and then a new trial began. The pigeon could also peck the white key, with the result that the stimulus conditions would change with green replacing red on the left key. Once the green light was on, pecks to white had no further consequences (that is, the pigeon could not change back

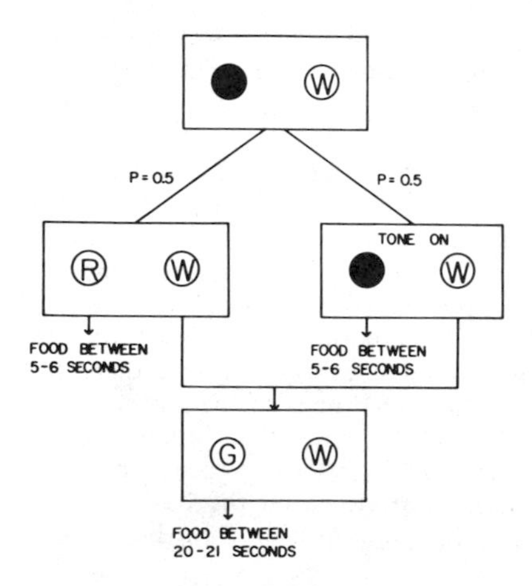

FIGURE 1. Outline of the procedure for Experiment 1. The *top* portion represents the intertrial interval, which was variable in duration. A trial began with either the onset of a red light or a tone. Food was produced intermittently for left-key responses 5–6 sec after stimulus onset. A peck to the white key changed the stimuli and, in the presence of green, left-key responses produced food intermittently 20–21 sec after trial onset. Trials ended with food or if 24 sec elapsed without food, and then a new intertrial interval began. A houselight was on at all times except during food delivery.

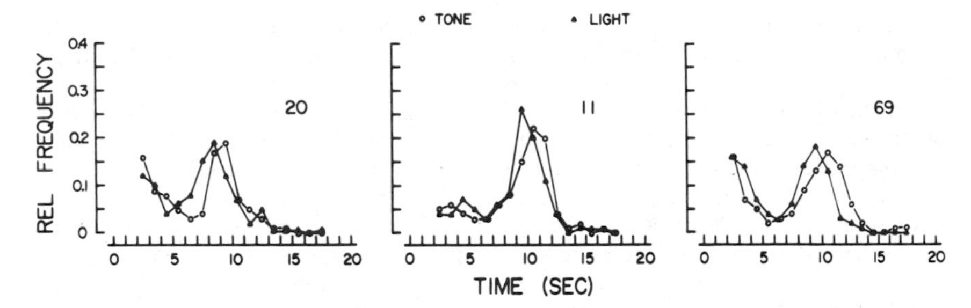

FIGURE 2. Relative frequency of a changeover as a function of time since a stimulus period began. Data were computed separately and are shown for both light and tone trials. Food was delivered between 5 and 6 sec in the presence of red light or the tone and between 20 and 21 sec in the presence of green. Data are based on the total number of changeovers in the last five sessions.

to red). Pecks to the green key were sometimes reinforced between 20 and 21 sec of trial onset. If food was not produced, the trial ended after 24 sec. The contingencies for tone trials were the same except that a tone (produced by a Sonalert) was presented in place of the red light. Trials began with the onset of the tone, which was the only change in stimuli after the intertrial interval. The houselight and white key light were kept on during intertrial intervals so that changes in visual stimuli would not accompany the onset of the relevant auditory stimulus. On tone trials, a peck to the dark, "tone" key sometimes produced food between 5 and 6 sec after tone onset. A peck to the white key turned on the green key light and turned off the tone. The contingencies in the presence of green were the same as on light trials.

Food was made available on some trials only. Once available, food was assigned equally for short and long responses to ensure an equal distribution of reinforcers. For reinforcers assigned for short responses, half were assigned during tone trials and half during light trials. If a reinforcer was assigned but missed, the reinforcer was assigned again on the next appropriate trial. This reinforcement procedure fixed relative reinforcement frequency under the different conditions to make sure that any performance differences would not be due to differences in obtained number of reinforcers.

Basically the procedure provided reinforcement at a short and a long time after stimulus onset. The major difference in trials was whether a light or a tone was associated with the early portion of the time period, while a light was always associated with the later portion. The question was whether the light-tone difference early in the time period would affect the time at which the animals would change to green.

FIGURE 2 shows the relative frequency of a changeover from red to green and from tone to green in different time classes. The general finding, in agreement with previous research, is that the frequency of changing over was highest at times that approximated the geometric mean (10–11 sec) of the two intervals at which reinforcement was provided (5–6 sec [short] and 20–21 sec [long]). The forms of the distributions were similar on light and tone trials, except that the distributions are shifted to the right under tone conditions. The difference in the distributions suggests that the time to food in the presence of the tone was judged as longer than that in the presence of the red light. Although red-key and "tone"-key responses both produced food at the same interval after stimulus onset, perception of time to food in the presence of the tone as longer would result in a longer time spent before changing to green.

The results agree with the findings on human time judgments and suggest that the auditory–visual difference holds for pigeons as well as people. While the results suggest an auditory–visual difference, we must be cautious in the interpretation. No attempt was made to equate the intensity of the light and sound. Although the auditory-visual difference holds for people even with changes in stimulus intensity,[11] our results may have depended on differences in intensity. Also, light and tone trials differed with regard to whether the pigeons pecked on a red or dark key (although response rates were similar in both cases). Probably these factors are not responsible for the difference in performance, but the matter needs future corroborative research.

Whatever the reason for the difference, the results demonstrate that the stimuli that demarcate temporal events affect performance. The results join a small body of literature showing that stimuli can have varied effects on temporal judgments. Mantanus,[12] for example, trained pigeons on a duration-discrimination task and found that accuracy was higher when durations were "filled" than when they were "unfilled." Similarly, Spetch and Wilkie[13] observed higher accuracy when durations of food access were used as opposed to durations of a light. Stubbs *et al.*[14] found that discrimination performance was affected not just by the stimuli that make up durations but also by the stimuli that bound durations. They trained pigeons on a schedule in which fixed-interval performance resulted sometimes in food and sometimes in a briefly presented stimulus. Occasionally the schedule was interrupted by a choice situation: one response was reinforced if only a short portion of the interval had elapsed, while a second response was reinforced if a longer portion had elapsed. The main result was that accuracy of choice was lower if the interval was preceded by the brief stimulus rather than food. The few available results suggest that the stimuli that make up temporal events have varied effects that include changes in accuracy and changes that suggest differences in perception of different temporal events. Only detailed attention to a whole range of stimulus conditions will reveal the range and extent of stimulus factors.

EXPERIMENT 2: TEMPORAL INTEGRATION

There are numerous procedures for the study of animals' temporal discrimination, but in most the consequences depend on the passage of a single interval. Fetterman and Dreyfus developed an alternative procedure in which performance depended on more than the passage of a single interval of time.[15] The procedure involved the comparison of two durations with different choice responses reinforced depending on which duration was longer. Pigeons were presented with a red light of one duration followed by a green light of a second duration; then two choice-key lights were turned on and a response to one key was reinforced if the duration of red had been longer than green, while a response to the second was reinforced if the duration of green had been longer than red. The pigeons were accurate in discriminating many different durations, with accuracy a function of the relative difference in the durations.

The present experiment is derived from the method of Fetterman and Dreyfus and represents an increase in the complexity of the task. Pigeons were given four rather than two durations, in the order red–green–red–green, with the duration of each color changing from trial to trial. As before, the pigeons' task was to give one response if the duration of red was longer than green and another if the duration of green was longer than red; the added feature was that the pigeons were required to integrate or add together the two red durations and the two green ones and to respond on the basis of whether the total red duration was shorter or longer than that of the total green.

FIGURE 3 outlines the procedure. For each trial, the center key was white initially and the side keys dark. One response on the center key started the series of stimulus durations on the center key with each changing independently of behavior. Each series consisted of four durations: red, green, red, green. Each duration was arranged by a probability gate set at a value of 0.10, which was pulsed once every 0.5 sec. The average time of each duration was thus 5 sec, with a range between 0.5 and 16 sec. This system provided for more than a million different combinations of durations and allowed for the total of the four durations to be as short as 2 sec or as long as 64 sec. After completion of the fourth duration of the series, the center-key stimulus went off and the side-key stimuli came on. If the total of the two red durations was longer than that of green, a left-key response was correct; if the total of green was longer than red, a right key response was correct. Correct responses either resulted in a 10-sec intertrial interval (during which all lights were off) or intermittently produced food. Food was assigned randomly from reinforcer to reinforcer either for a correct left-key or for a

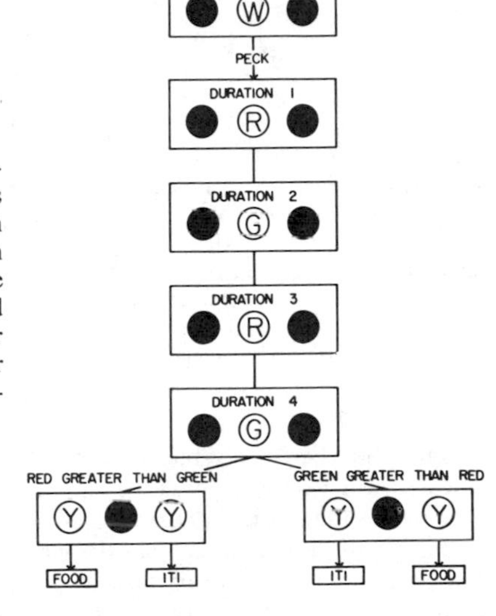

FIGURE 3. Outline of the procedure for Experiment 2. A response to the white key started a series of four durations, with the series changing from trial to trial. At the completion of the duration series, two side-key stimuli were lit. One response was reinforced if the total of two red durations had been longer than the total of green; the other response was reinforced if green had been longer than red. Incorrect responses produced an intertrial interval during which all lights were off.

correct right-key response. If a reinforcer was assigned to the left (right) only a correct response to that key could produce food; when a reinforcer was assigned to the left (right), correct responses to the right (left) key simply produced the intertrial interval. This procedure was instituted to hold the number of reinforcers equal for left-key and right-key responses. The net result of this procedure is that roughly half of the correct responses resulted in food. All incorrect responses produced the 10-sec blackout. In addition, in those cases in which the total red and green durations were equal, choices would be neither correct nor incorrect, so choices always produced a blackout under these circumstances. Each session consisted of a series of trials that lasted until a pigeon received 75 reinforcers.

FIGURE 4 shows the probability of a right-key response (reporting green as longer than red) as a function of the relative difference in the durations of red and green and shows orderly functions relating probability of a right-key response to changes in the relative durations of red and green. The functions are not symmetrical around the one

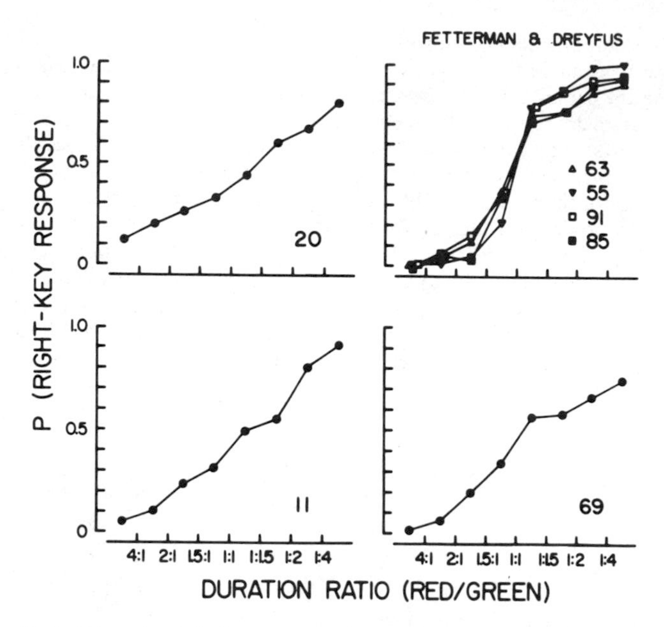

FIGURE 4. Probability of a right-key response ("green longer than red") as a function of the duration ratio of red to green. The first point shows performance when the duration of red was greater than four times that of green; the second shows performance when red was two to four times longer than green, and so forth. The data comprise many different duration series in which data from the last five sessions were totaled; data included approximately 1000 problems per pigeon. The *upper right panel* shows equivalent data from an experiment by Fetterman and Dreyfus in which performance depended on two rather than four durations.

to one ratio; there was a bias to respond left or, stated differently, accuracy was higher in cases when red was longer than green as compared with similar cases when green was longer than red. Calculations on the few instances (5–6 per person) in which red and green durations were equal show that the pigeons responded more often on the left key in these cases. The pattern of performance persisted in spite of the procedural arrangement of holding the relative number of reinforcers equal for the two choice responses.

The upper right hand panel of FIGURE 4 gives the comparable data from the experiment by Fetterman and Dreyfus, who used a similar procedure except with only two durations for comparison rather than four. Comparison of performance in the two experiments reveals steeper ogival functions in their experiment, indicating that performance was more accurate. This difference is not surprising considering the greater difficulty of the present experiment.

FIGURE 4 included performance across a wide range of combinations of durations. A reasonable question is whether performance was similar when the durations were short and when they were long. Notions of memory might suggest that accuracy would be lower when the durations were long since a greater time would elapse between the early durations in the series and a choice. FIGURE 5 categorizes the data into two classes of duration that depend on the total time of the duration series, with the classes being divided in terms of the average combined time of the four stimuli.

FIGURE 5 indicates that the two sets of data were comparable and that accuracy was similar whether the durations were short or long. The two functions are similar for pigeons 20 and 69, with a great deal of overlap. For pigeon 11, the open circles lie above the solid circles in most comparisons. The difference in performance appears to be one of bias rather than sensitivity. Pigeon 11 was relatively more likely to respond left when the total duration was long and right when it was short, but accuracy was comparable in both cases.

Another question concerned the relative contribution of the first and second red–green duration pairs. The reinforcers were arranged with respect to the total red and total green durations, but the question can be raised whether choice responses were controlled to a greater degree by one red–green pair than the other. Consider the second red–green pair. Sometimes the red–green relation was the same in both red–green pairs (for example, red longer than green in both pairs) and sometimes the relation was opposite in the two pairs. In addition, the duration of the second pair was sometimes longer than that of the first pair and sometimes shorter. If the second pair was longer, the contribution of this pair to the total red–green duration would be greater than the first; as a result, the second pair, although opposite to the first, would provide the greater contribution to the total red–green ratio and hence the second red–green relation would probably correspond to that of the total. The point is that a pigeon, responding on the basis of only the second red–green pair of a duration series, could be correct more often than not due to the interrelations between the different durations.

FIGURE 6 provides the relevant data by showing probability data when the red–green relation of the first red–green pair (top portions) or of the second pair (bottom) was in the same or the opposite direction as the relation of the total red and green. In the top portions, for example, the solid circles show performance when red was longer than green both in the first pair and total relation, and when green was longer in both cases; the open circles show performance when the first red was longer than green while the total green was longer and when the first green was longer than red while the total red was longer.

The most general finding is that performance changed as a function of the total red–green duration ratio regardless of whether a similar or opposite relation occurred in the first or the second red–green pair. The data indicate that performance was not controlled exclusively by either the first or second red–green pair. At a more specific

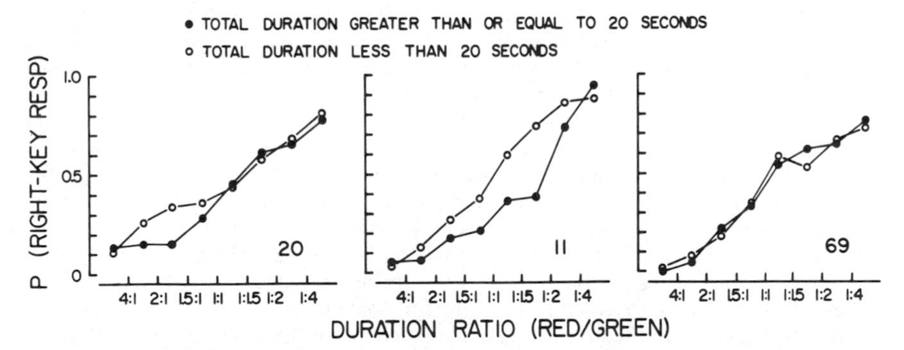

FIGURE 5. Probability of a right-key response as a function of the duration ratio of red to green. Performance is shown separately for problems in which the total time of the four durations was 20 sec or longer (*solid circles*) or less than 20 sec (*open circles*). The data are based on totals of the last five sessions, with roughly half of the problems 20 sec or longer and half shorter than 20 sec.

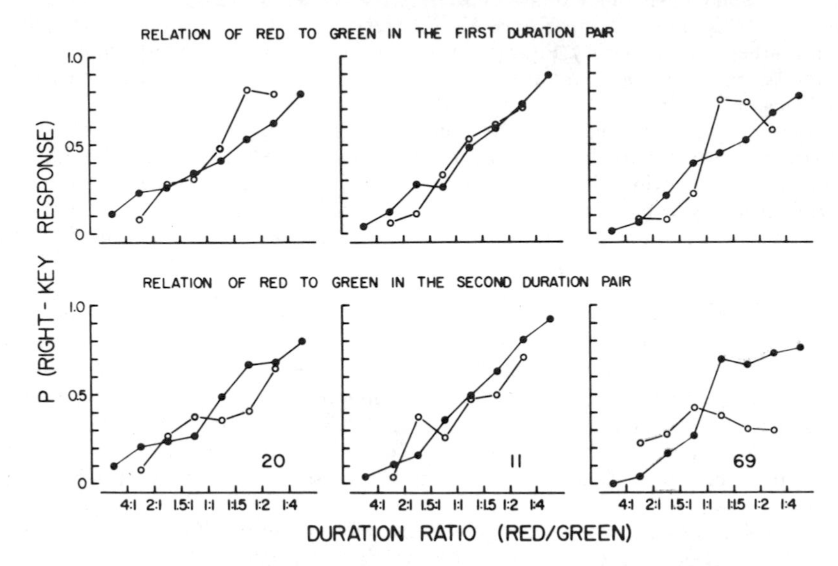

FIGURE 6. Probability of a right-key response as a function of the duration ratio of red to green. Presentation of the data emphasizes the first red–green duration pair (**top**) and the second (**bottom**) regarding whether the relation in that pair was in the same or opposite direction as the total red–green relation. In the **top panels** *solid circles* represent performance when the duration of red (green) was longer than green (red) both in the first pair and also in the total relation. The *open circles* represent performance when duration of red (green) was longer than green (red) in the first pair, but with green (red) longer in the total relation. The **bottom panels** show similar relations computed with respect to the second red–green pair. *Open circles* are not shown at extreme ratios since few instances (generally less than ten in the five sessions) occurred.

level, the data indicate a somewhat greater contribution by the second red–green pair. The data of pigeon 69 are instructive since this bird showed greatest control by the second red–green pair. The bottom panel for this pigeon indicates an increasing function when the second pair relation was in the same direction as the total, but a much flatter function when the two relations were opposite in direction. The small change in performance and generally lower accuracy when the red–green relation in the second pair differed from that of the total indicate greater control by the second red–green pair. Comparison of the data for pigeons 20 and 11 shows that the second-pair functions were more comparable, with overlap in the two sets of points. However, accuracy measures derived from the comparable points indicate that even for these pigeons accuracy was higher when the second-pair relation was in the same direction as the total (68% versus 62% for pigeon 20 and 72% versus 69% for pigeon 11). The data indicate a relatively greater weighting of the second red–green pair in determining choice that ranged from slight (pigeon 11) to moderate (pigeon 20) to strong (pigeon 69). For the first red–green pair (shown in the top panels of FIGURE 6), the functions are roughly similar. If there is any difference, the difference is one of higher accuracy when the relation in the first red–green pair was opposite that of the total. The reason for this result has to do with the interrelation of the first and second red–green pairs as they affected the total. If, for example, red was longer than green in the first pair, then green would have to be even longer in the second pair for the total green duration to be longer than red. The greater difference in the second pair (whether due to a more extreme red–green ratio or to longer durations in the second pair) would enhance accuracy, especially in light of the greater weighting given to the second red–green pair. The data both from the top and bottom portions of FIGURE 6 indicate somewhat greater control by the second red–green pair.

One bird was exposed to a condition in which the durations were increased four-fold. Only one bird was used because of the increased session length of approximately 8 hours. FIGURE 7 shows results like those in the previous figures. The left panel shows performance for all durations and also data from the prior condition for comparison. Performance was similar for the two conditions. The middle panel shows that performance was comparable whether the total duration was more or less than 80 sec. The right panel shows performance computed with respect to the second red–green pair and whether the red–green relation was the same or opposite the total red–green relation; performance was similar to that in FIGURE 6. Basically the results corroborate those of the previous condition while extending them to a longer set of durations.

The results demonstrate that pigeons can integrate and discriminate the duration of events in the task, that accuracy is not markedly affected by the total time of the four durations, and that performance is based on the combination of more than two of the four durations. The results are in general agreement with previous results in which responses depended on the duration of a single stimulus or on the comparison of two durations.[15,16] The different results agree that discrimination accuracy is a function of relative rather than absolute differences in durations and that performance is comparable across different ranges of durations. The most obvious difference in the results is lower accuracy in the present task (for example, the comparison in FIGURE 4). That accuracy was lower is not surprising since the task was more difficult, involving the integration and comparison of several durations.

Performance was controlled by all four durations, but the evidence suggests that there was a greater weight given to the earlier durations and, at the same time, a

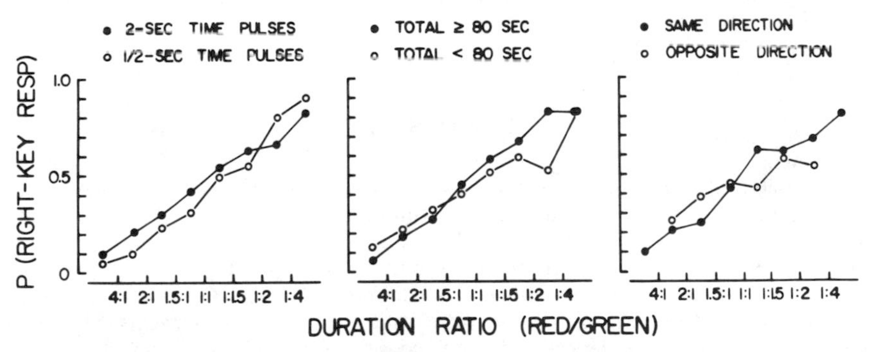

FIGURE 7. Probability of a right-key response ("green longer than red") as a function of the duration ratio of red to green. The **left panel** shows performance on all problems when the time base controlling duration was 2 sec (*solid circles*) and from FIGURE 4 when it was 0.5 sec (*open circles*). The **center panel** shows performance for problems with a total duration of 80 sec or greater (*solid circles*) or less than 80 sec (*open circles*). The **right panel** shows performance computed with respect to the second red–green pair when the relation in that pair was in the same direction as the total (*solid circles*) or the opposite direction (*open circles*). Data points are based on totals of the last five sessions.

greater weight given to the second red–green pair. Evidence that early durations were weighted more heavily comes from findings in FIGURE 4 that accuracy was higher in cases in which red was longer than green as opposed to comparable cases in which green was longer than red (for example, 2:1 versus 1:2). Also, the birds responded as if red was longer in most cases when red and green durations were equal. These results could be due to a left-key bias that was independent of the task, but this possibility seems rather unlikely since performance was similar in all three birds and since the reinforcement contingencies, which fixed relative reinforcement rate, would minimize such a bias. It is more likely that something about the task produced the difference in performance. The results suggest that the early durations might be perceived as being relatively longer, for when red and green durations were equal, the animals more often responded as if red was longer. Perhaps the proper scale for comparison should not be the physical scale based on clock time but instead a psychological scale. Use of a psychological scale that is related to clock time by a power function (with an exponent less than one) produces similar levels of accuracy when the ratios of red to green and green to red are comparable in terms of this scale. The approach of using the psychological scale is reasonable, and previous work has modified scales and durations in light of performance.[16] While an approach that uses a psychological scale as opposed to a physical scale may be reasonable, some caution is necessary. Different experiments have suggested different scales so the present suggestion of a power function conflicts with other suggestions of linear and logarithmic scales.[17] Hasty conclusions should not be made on the basis of one experiment.

Findings of greater control by the second red–green pair are consistent with research involving memory for sequences of events.[18–20] When events occur in sequence, performance measures show greatest influence by events at the end of the series. While our results are consistent with animal memory research in some ways, there is one way in which they are not. Research with other tasks as well as duration tasks indicates a lowering of accuracy as the time separating a stimulus and choice increases.[21] These results imply that accuracy in our task would decrease as the total time of the duration series increased since increases would produce a longer delay between the early durations and a choice. However, findings of similar levels of accuracy over different ranges of durations raise questions about the way in which memory should be considered in this situation.[15]

DISCUSSION

At the outset of this paper, a distinction was made between sensory and perceptual frameworks for considering temporal discriminations in animals. Two experiments were reported which grew out of considerations of temporal discrimination from a perceptual approach. The experiments are only preliminary steps; more research is necessary. As one example, pointing in a different direction, Dreyfus and Stubbs trained pigeons to discriminate fixed from variable sequences of events when one choice was reinforced if preceded by a fixed sequence while a second choice was reinforced if preceded by a variable sequence.[22] The events were response-produced stimuli rather than time-dependent stimuli, but the results indicate that it should be possible to train pigeons to discriminate different patterns of temporal events.[23] Collectively the experiments argue for a research program rooted in the perceptual framework. Different stimuli should be used: not just lights, sounds, shocks and food, but stimuli that change in a variety of ways. Different tasks should be used that tap the various processes like those uncovered with human time perception research. And, experiments should be performed that focus not just on the discrimination of simple duration, but on the discrimination of events with different temporal structure and in different temporal contexts. The study of animal learning has long ignored the findings of perception. Animal time perception provides an opportunity for redressing this neglect by taking advantage of knowledge gained from perception.

REFERENCES

1. RICHELLE, M. & H. LEJEUNE. 1980. Time in Animal Behavior. Pergamon. New York, NY.
2. STUBBS, D. A. 1979. Temporal discrimination and psychophysics. *In* Advances in the Analysis of Behavior: Reinforcement and the Organization of Behavior, Vol 1. M. D. Zeiler & P. Harzem, Ed.: 341–369. Wiley. Chichester, England.
3. GIBBON, J. 1977. Scalar expectancy theory and Weber's Law in animal timing. Psychol. Rev. **84:** 279–325.
4. PLATT, J. R. 1979. Temporal differentiation and the psychophysics of time. *In* Advances in the Analysis of Behavior: Reinforcement and the Organization of Behavior, Vol 1. M. D. Zeiler & P. Harzem, Eds.: 1–29. John Wiley. Chichester, England.
5. CHURCH, R. M. 1978. The internal clock. *In* Cognitive Processes in Animal Behavior. S. H. Hulse, H. Fowler & W. K. Honig, Eds.: 277–310. Lawrence Erlbaum. Hillsdale, NJ.
6. ROBERTS, S. 1981. Isolation of an internal clock. J. Exp. Psychol. Animal Behav. Proc. **7:** 242–268.
7. GIBSON, J. J. 1979. The Ecological Approach to Visual Perception. Houghton Mifflin. Boston, MA.
8. GIBSON, J. J. 1975. Events are perceivable but time is not. *In* The Study of Time: II. J. T. Fraser & N. Lawrence, Eds.: 295–310. Springer-Verlag. New York, NY.
9. GOLDSTONE, S. & W. T. LHAMON. 1971. Levels of cognitive functioning and the auditory visual differences in human timing behavior. *In* Adaptation Level Theory: A Symposium. M. H. Appley, Ed.: 263–280. Academic Press. New York, NY.
10. STUBBS, D. A. 1976. Scaling of stimulus duration by pigeons. J. Exp. Anal. Behav. **2:** 15–25.
11. GOLDSTONE, S. & W. T. LHAMON. 1974. Studies of the auditory-visual differences in human time judgments: I. Sounds are judged longer than lights. Percept. Mot. Skills **39:** 63–82.
12. MANTANUS, H. 1981. Empty and filled interval discrimination by pigeons. Behav. Anal. Let. **1:** 217–224.
13. SPETCH, M. L. & D. M. WILKIE. 1981. Duration discrimination is better with food access as the signal than with light as the signal. Learn. Motiv. **12:** 40–64.
14. STUBBS, D. A., S. J. VAUTIN, H. M. REID & D. L. DELEHANTY. 1978. Discriminative functions of schedule stimuli and memory: A combination of schedule and choice procedures. J. Exper. Anal. Behav. **29:** 167–180.
15. FETTERMAN, J. G. & L. R. DREYFUS. Duration discrimination: A pair-comparison procedure. *In* Quantitative Analyses of Behavior: The Effect of Delay and Intervening Events on Reinforcement Value. M. L. Commons, J. A. Nevin & H. Rachlin, Eds. Ballinger. Cambridge, MA. In preparation.
16. STUBBS, A. 1968. The discrimination of stimulus duration by pigeons. J. Exp. Anal. Behav. **11:** 223–258.
17. GIBBON, J. 1981. Two kinds of ambiguity in the study of psychological time. *In* Quantitative Analyses of Behavior: Discriminative Properties of Reinforcement Schedules, Vol. 1. M. L. Commons & J. A. Nevin, Eds.: 157–189.
18. SHIMP, C. P. 1976. Short-term memory in the pigeon: Relative recency. J. Exp. Anal. Behav. **25:** 55–61.
19. WEISMAN, R. G., E. A. WASSERMAN, P. W. DODD & M. B. LAREW. 1980. Representation and retention of two-event sequences in pigeon. J. Exper. Psychol. Anim. Behav. Proc **6:** 300–313.

20. HONIG, W. K. 1981. Working memory and the temporal map. *In* Information Processing in Animals: Memory Mechanisms. N. E. Spear & R. R. Miller, Eds. Lawrence Erlbaum. Hillsdale, NJ.
21. CHURCH, R. M. 1980. Short-term memory for time intervals. Learn. Motiv. **11:** 208–219.
22. DREYFUS, L. R. & D. A. STUBBS. 1979. Discrimination of a dynamic stimulus. Paper presented at the American Psychological Association, New York.
23. HULSE, S. H., J. HUMPAL & J. CYNX. 1984. Processing of rhythmic sound structures by birds. This volume.

Subjective Duration in Rats: The Psychophysical Function[a]

HANNES EISLER

Department of Psychology
University of Stockholm
S-106 91 Stockholm, Sweden

Rats are like humans, particularly with respect to time perception. Their psychophysical function is likewise a power function of the form

$$\psi = \alpha(\phi - \phi_0)^\beta, \quad (1)$$

where ψ refers to subjective and ϕ to physical duration.

Consider FIGURE 1. The two upper panels describe subjective duration as functions of physical duration. One derives from an experiment with a human observer, the other from rat No. 3. The agreement of the black points with the curves indicates the goodness of fit. Both psychophysical functions build on the parallel-clock model and are derived from duration reproduction data.

Let me briefly recapitulate the parallel-clock model. A thorough description will be found in Eisler,[1] with further clarifications in a subsequent paper.[2] In the present duration reproduction task, the observer is presented with the standard duration, which is indicated by a sound. After a short interruption, the sound starts again and is terminated by a microswitch (or a lever) being pressed when he, she, or it experiences the second duration as equal to the first. According to the parallel-clock model, the observer deals with two subjective quantities, namely, the subjective durations corresponding to (1) the total duration from the start of the standard to the end of the reproduction ($\phi_T = \phi_S + \phi_V$), and (2) the reproduction (ϕ_V), that is, the variable response duration determined by the observer (ϕ denotes physical time in sec, and the subscripts S, V, and T denote standard, variable, and total, respectively). These subjective durations are assumed to be accumulated in two separate sensory registers ("clocks"), hence the name parallel-clock model. The second duration is experienced as equal to the first when the difference between the total subjective duration ψ_T and the variable subjective duration ψ_V equals the variable subjective duration:

$$\psi_T - \psi_V = \psi_V. \quad (2)$$

FIGURE 2 should make this clear.

Rearranging Equation 2 shows that $\psi_V = \frac{1}{2}\psi_T$, and inserting Equation 1 yields the following linear relation between ϕ_V and ϕ_T:

$$\phi_V = (\tfrac{1}{2})^{1/\beta}\phi_T + [1 - (\tfrac{1}{2})^{1/\beta}]\phi_0. \quad (3)$$

With enough points (standard durations) to fit the straight line, Equation 3 allows computation of the parameters of the power function from the slope and intercept. In

[a]This work was supported by the Swedish Council for Research in the Humanities and Social Sciences. The data treatment was carried out at the Computer Center of the University of Leiden, the Netherlands, during my stay as a fellow at the Netherlands Institute for Advanced Study in the Humanities and Social Sciences at Wassenaar, the Netherlands.

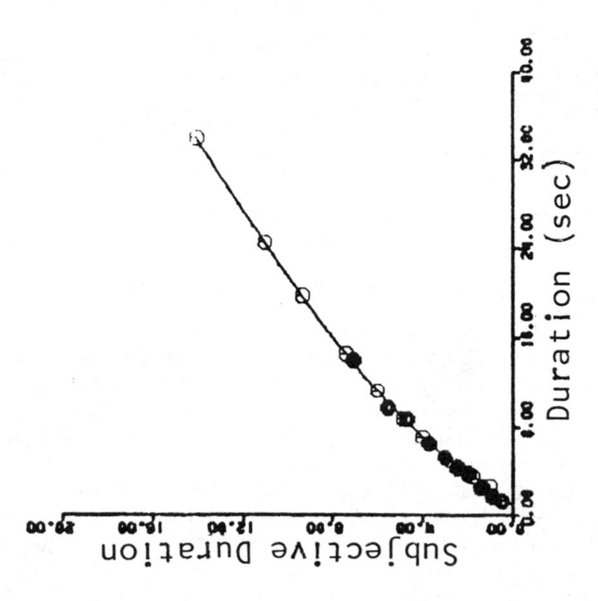
Subjective Duration
0.00
4.00
8.00
12.00
16.00
20.00
Duration (sec)
0.00
8.00
16.00
24.00
32.00
40.00

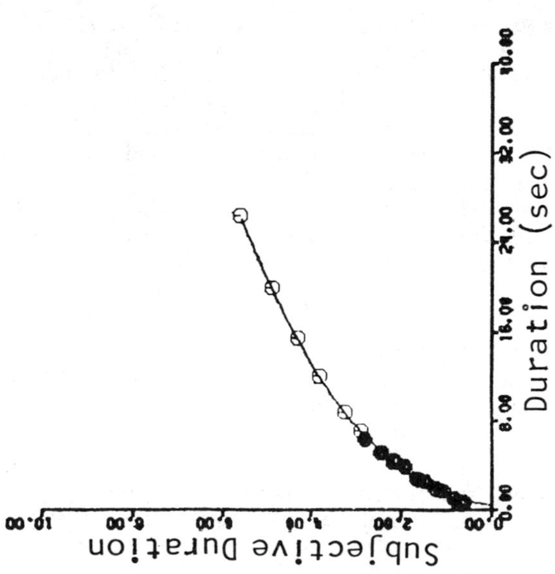
Subjective Duration
0.00
2.00
4.00
6.00
8.00
10.00
Duration (sec)
0.00
8.00
16.00
24.00
32.00
40.00

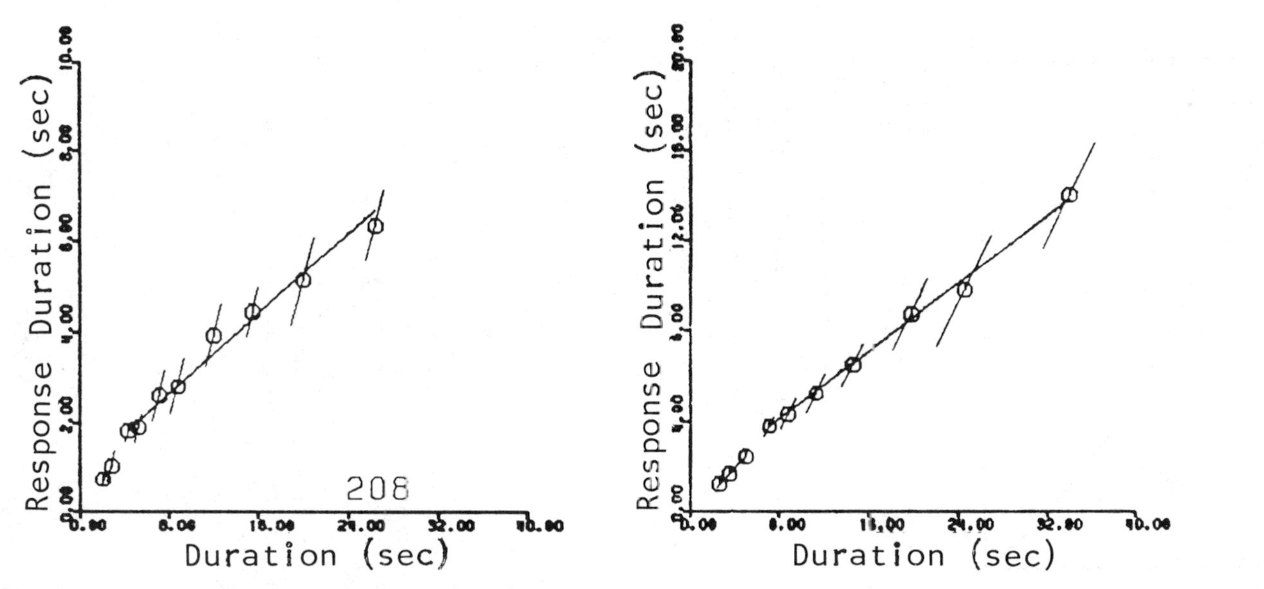

FIGURE 1. Duration reproductions from a rat (*left*) and a human observer (*right*). The *upper panels* describe the psychophysical power function, subjective versus physical duration. *Black dots* indicate goodness of fit. *Lower panels* show plots of response duration ϕ_V versus total duration ϕ_T, together with the fitted straight lines. The *slanted lines* around the points are standard deviations.

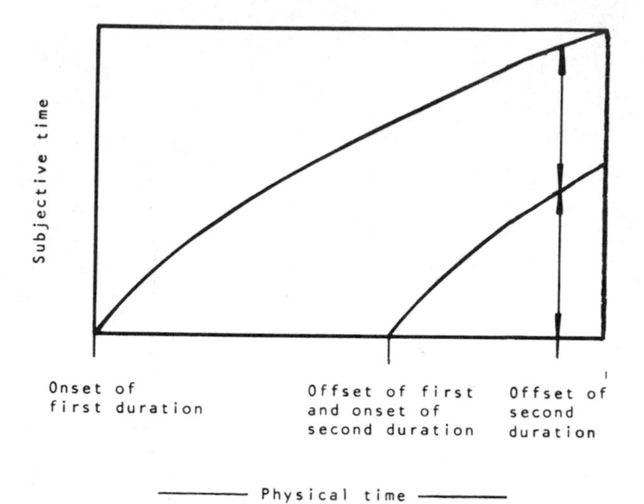

FIGURE 2. Duration reproduction according to the parallel-clock model. Subjective duration is plotted versus physical duration for the total duration (*larger curve*) and the response duration (*smaller curve*). At the point in time when the difference between these two subjective durations (*upper arrow*) equals the subjective response duration (*lower arrow*), the observer reports equality between standard and response duration by pressing the lever (rats) or the microswitch (humans).

this way, application of the parallel-clock model allows one to determine, for example, the exponent β from duration reproduction data. Later I shall describe the data treatment in more detail.

Communication with students of psychology, the data from one of whom I just showed, is comparatively easy, at least compared with communication with rats. But before I go into the details of the experiment, describing the learning procedure that has to replace verbal communication with humans, I want to mention two pitfalls, the avoidance of which probably was a necessary condition for the comparatively successful outcome of my experiment, as opposed to those of Reynolds[3] or Mandell and Atak,[4] for example. The first pitfall is the coupling of two temporal requirements, as when reinforcement after a temporal task is given by a variable-interval schedule instead of using continuous reinforcement (compare the viewpoints of Stubbs[7]).

The other pitfall is the ambition that rats should reproduce time veridically, that is, reinforcing responding within a small band of durations around the standard duration. We know from experiments with humans that reproductions often lie far below the standards: Why should those of rats be closer?

METHODS

Subjects

The subjects were eight male blackhooded rats, obtained 51 days after birth and aged about 1 year and 4 months at the termination of the experiment. They were kept at 80% of their free-feeding weight (compared with a control group of four animals of the same age) in individual plastic cages in a room without windows in which the day–night light cycle was reversed, so that they were taken to the experimental chamber from a dark room.

Apparatus

The experiment took place in a Skinner box, endowed with a lever and a small loudspeaker, the distorted tone of which indicated the durations. Reinforcement consisted of 5-sec access to sugar-sweetened diluted condensed milk delivered by means of a small movable spoon. I do not want to describe the apparatus in detail here; its function will become clear in the following descriptions of the procedure. Stimuli and contingencies were controlled by means of a punched Mylar tape closed to a loop; the rats' behavior was registered both on paper tape and counters. Here we are concerned only with the response durations ϕ_V.

Learning to Reproduce Durations

The experiment consisted of three phases. The first was just magazine training. In the third phase the subject was presented with a standard duration, followed by a short pause of 300 msec, after which the tone started again until it was terminated by a lever press. The object of the second phase was to teach the rats to attend to this short pause, that is, to an interruption of the tone. Before going into phase 2 in detail, I want to describe procedural features common to phases 2 and 3. Sessions were daily (weekdays and holidays alike) and mostly lasted 1 hour. Each session consisted of cycles, and each cycle of four or five parts (FIG. 3). A cycle begins with a long pause of 30 sec, with no tone and during which a lever press is of no consequence. The next part is the standard duration, indicated by the tone, and thereafter the short pause of 300 msec. Then the tone comes on again and is terminated by the rat's lever press. If the rat's previous

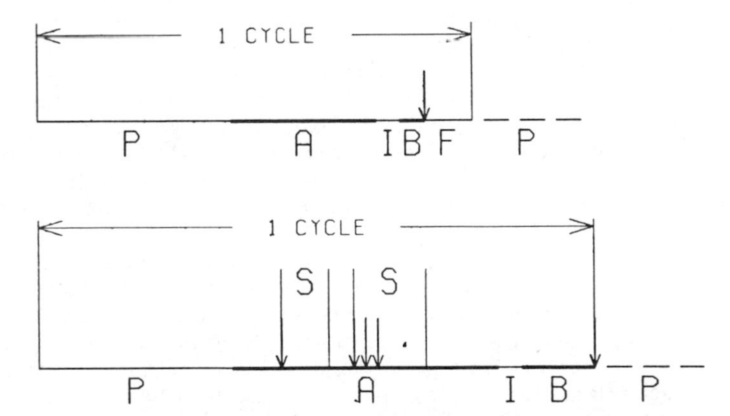

FIGURE 3. Cycle with correct (*upper panel*) and incorrect behavior (*lower panel*) during phase 2. The *thick lines* indicate the ongoing tone, and *vertical arrows* denote lever presses. P = the long pause; A = the presented standard duration, in the *lower panel* lengthened by the punishment time, S; I = the interruption of the tone; B = the latency between the offset of the interruption and the following lever press; F = the feeding period.

behavior during the cycle has been correct (see below), the cycle ends with the feeding period, before the long pause of the next cycle starts. Otherwise the long pause follows directly upon the lever press that terminates the tone. There were ten different standard durations, logarithmically spaced between 1.3 and 20 sec, presented in a pseudo-random series in a block of 30. That is, after 30 cycles, during which every standard duration occurred three times, the whole series was repeated by means of the closed loop of the punched tape. The ten standard durations are found in the second row of TABLE 1.

In phase 2 the rats had to learn to attend to the interruption of the tone. This was achieved by rewarding them for not pressing the lever during the tone before the interruption. A lever press during the standard A (the first tone) lengthened that tone by 4 sec (to avoid superstitious chaining) and reward was withheld, so that the lever press during the second tone was followed by the long pause immediately. If no lever presses occurred during the first tone, the lever press that terminated the second tone was followed by the feeding period before the long pause of the next cycle. Phase 2 was concluded after 68 sessions, at which time the performance of the rats had stabilized at a satisfactory level. There was only a single completely error-free session, and this was achieved by rat No. 3.

In phase 3 the contingencies of lever pressing during the first tone were removed. Instead, the animals were rewarded for pressing the lever during the second tone if the duration of this tone was between bounds that depended on the particular standard duration, irrespective of possible lever presses during the first tone. I made use of two

TABLE 1. The Ten Standard Durations with Lower and Upper Bounds (sec)

Lower bounds	0.35	0.65	1.00	1.40	1.90	2.85	3.85	4.85	6.20	7.50
Standard durations	1.3	1.8	2.5	3.3	4.5	6.0	8.1	11.0	14.8	20.0
Upper bounds	1.30	1.80	2.80	3.80	4.80	6.10	7.40	11.50	15.20	21.00

sets of bounds. The first set consisted of half the standard duration as the lower and 1.5 times the standard duration as the upper bound, and was used for four of the eight rats. The second set, listed in TABLE 1, was used for the remaining four. For these latter bounds, the data obtained from the duration reproduction experiment with humans[1] were used as a guideline. (Originally, I regarded the experiment with humans as a pilot study for the rat experiment.) As the lower bound a value was chosen close to the shortest reproduction given in any of the six trials by any of the 12 observers for that particular standard. The upper bound was chosen similarly. For both bounds, but particularly for the upper, restrictions had to be imposed in order to minimize the overlap between adjacent bands.

Naturally, learning to reproduce durations had to proceed slowly, step by step. Thus, the animals were not exposed to all ten standard durations from the very start of phase 3. The first standard duration used was the second shortest (that is, 1.8 sec), for which a correct reproduction, in accordance with what was said above, had to lie either between 0.9 and 2.7, or between 0.65 and 1.8 sec. These bands have a small overlap with the latencies (B in FIGURE 3) from phase 2, so that the longer of these latencies were reinforced. After a number of sessions with 1.8 sec, the procedure was repeated with 6 and 5 sec, respectively. Then the demands were stepped up further by presenting both 1.8- and 6- (or 5-) sec durations in random order in the same session. Before all ten durations were used, the rats had to work with 20-sec durations, either alone, or in combination with other durations. Which stimuli to choose for a session was to a certain extent a trial-and-error procedure and depended on the individual rat's performance. Accordingly, the rats had different learning histories.

The experiment was terminated when the performance of the rats deteriorated. I have three possible explanations for this deterioration: (1) The clock controlling the light in the animal housing room stopped working, so that it was dark there at all times. This was discovered only after the termination of the experiment. (2) The rats may have reverted to more "normal" behavior (compare the work of Breland and Breland[5]). (3) Senility may have occurred in the rats.

Data Treatment

The data from each session were dealt with separately. Every standard duration was presented between six and nine times during a session (a result of using a block of 30 for the ten standards). The means of the reproductions for every standard duration were calculated.

However, rats are no better than humans. Rats are occasionally subject to mistakes in their reproductions, which could be ascribed, by analogy with those in humans, to anticipatory errors and slackened attention. That is, there were a few lever presses almost directly after the short pause, and a few very late ones, compared to the rest, for a given standard duration. The rule of thumb I used was to remove reproduction values whenever, after rank ordering, the ratio between the second lowest and the lowest value exceeded 1.7, and in like manner for the highest and second highest. In Session 208 of rat No. 3, for instance, which is the one shown in FIGURE 1, of the 79 reproduced durations, three were excluded from the data treatment. Somewhat surprisingly, the scatter, measured as standard deviation (S.D.), is roughly the same for the human and animal subjects, as indicated by the sloped lines through the data points in the lower panels of FIGURE 1. (Their inclination is due to having fallible values on both axes.)

The rest of the calculations are somewhat more complicated than indicated by Equation 3. Both human and animal data show a break or discontinuity in their psychophysical function, entailing more than one straight line when ϕ_V is plotted against ϕ_T (see FIGURE 1, lower panels). These breaks are described in detail in Eisler,[1] and it is interesting to find them again in the rat data. A break indicates a change in one or two of the parameters in Equation 1, namely, in α, ϕ_0, or both.

The lines were fitted by the method of weighted least squares, with the reciprocals of the variances of the points as weights. This was recommendable because of lack of homoscedasticity, one cause of which is a tendency for the standard deviations to increase with the ϕ values, that is, to conform to Weber's law, which seems to hold approximately. The other cause is the division of the psychophysical function into two segments on either side of the break. Near the break the psychophysical function sometimes is not defined, so that the subjective duration corresponding to the abscissa at the break can oscillate between the two segments. This shows up as reproduction values that sometimes seem to belong to one, and sometimes to the other straight line when ϕ is close to the transition point. The implication is that the mean is misleading. On the other hand, assigning single values to the one or the other straight line according to appearance is too arbitrary. When using weighted least squares, however, the impact of such a point on the fit is strongly reduced by its large variance.

Unfortunately, often the data are not so clear-cut that one can be completely certain where the break lies. Therefore I carried out the fitting for every data set with the break positioned at all possible values and chose the parameter set that yielded the least sum of squared deviations. About 20 such runs were carried out for every data set.

RESULTS

Of the eight experimental animals, only two "made it," both belonging to the group with bounds based on the experiment with humans, the group with lowest lower bounds. By "made it" I mean that there was at least one session that fulfilled the following two criteria: (1) 50 "correct" reproductions (a session was shut off after 50 feedings) and (2) at least one "correct" reproduction of every one of the ten durations. The reason for the first requirement was two-fold: good performance (about 60–80% correct cycles) and enough reproduction values for each standard duration to obtain a fairly stable mean. The second requirement excluded the "cheating" animals, who tended to reproduce only the shorter standards correctly and press the level much too soon after the longer ones.

All in all, this result can hardly be considered brilliant, with only two out of eight rats furnishing usable data, but in view of the communicative difficulties, it may be regarded as acceptable. Data from 16 sessions of rat No. 3 and from two sessions of rat No. 10 could be used.

By now I have investigated six sessions with rat No. 3 and one with rat No. 10. The exponents β for the first rat varied between 0.43 and 0.70, with a mean of 0.55 and a median of 0.52. The exponent for rat No. 10 was 0.45.

DISCUSSION

As mentioned before, six of the eight experimental rats never learned their task and one of the remaining two produced only two acceptable sessions. The most obvious reason is that communication failed. The data for most of the rats seem to indicate that they, rather than reproducing durations, categorized the standards into a few classes, each eliciting a corresponding response duration. An exponent as low as 0.5 implies that, for a given time interval of middle or long duration, most of the temporal experience is crammed into its first portion,[6] so that the category "long durations" becomes quite wide. From this it follows that the same response duration is given over a wide range of the higher standards. This interpretation merges with the description of the "cheating" animals given above. It might be worth noting that rat No. 3, the successful one, also was best in phase 2 of the experiment, suggesting individual differences in the sort of ability required for good performance in the present experiment. On the other hand, rat No. 3 was the only one trained during phase 3 with a tape comprising the standards 1.8, 5, and 20 sec, in which the block of 30 consisted of 16 presentations of 20 sec and only seven of the other two standards. So I might just have hit upon a particularly efficient training procedure.

Several conclusions can be drawn from the study reported here. I would like to separate them into three problem areas: (i) procedures for the study of psychophysics in animals, (ii) experienced duration in rats, and (iii) scaling.

Procedures in Animal Psychophysics

In many experiments it has been demonstrated that animals can be brought under the control of temporal stimuli. For an overview see, for example, the work of Stubbs.[7] However, unlike the present experiment, only one or two durations or classes of durations have been used. Furthermore, the present experiment combines stimulus discrimination and response differentiation in the same task, which has been more the exception than the rule previously. It is quite clear that a rather finely graded discrimation, as well as differentiation, has been achieved. However, the task was extremely difficult, as can be concluded from the fact that even the best rat's behavior never stabilized in the sense that performance was reliable from session to session. The 16 sessions with data fit for use were interspersed between sessions 183 and 252 of phase 3.

Experienced Duration in Rats

While I obtained a β value of about 0.5, previous attempts to scale subjective time in animals in terms of power functions resulted in exponents that also characterize results in human adults,[6] that is, a value close to unity.[8] The reason for this latter value of the exponent probably lies in theoretical confusion regarding the data obtained and the internal representation of duration, as explained in detail by Platt.[8] Considering that the β value for children is much lower than for adult humans,[6] the present result gains in plausibility. It is likely that for children, as well as for rats, what happens "now" is more important than what happens later on, that is, that for a given longer time interval, experienced duration is concentrated to its first part (see above).

Scaling

From the scaling aspect it is worth emphasizing that the parallel-clock model seems to work as well with animal as with human duration reproduction data. It is remarkable that the data are so similar, as FIGURE 1 demonstrates, including the break in the psychophysical function. Since this is the third data set accommodating the parallel-clock model (the second are duration discrimination data[9]), the model can be considered well supported. The outcome of the present investigation strengthens the evidence for Stevens' power law (Equation 1) as an internal representation for duration and generalizes it, to my mind convincingly, to animals, or, at least, rats.

I stated at the outset that rats are like humans, at least regarding time perception. Perhaps it would have been more correct to claim that they are like children in that respect.

REFERENCES

1. EISLER, H. 1975. Subjective duration and psychophysics. Psychol. Rev. **82:** 429–450.
2. EISLER, H. 1981. The parallel-clock model: Replies to critics and criticisms. Percept. Psychophys. **29:** 516–520.
3. REYNOLDS, G. S. 1966. Discrimination and emission of temporal intervals by pigeons. J. Exp. Anal. Behav. **9:** 65–68.
4. MANDELL, C. & J. R. ATAK. 1982. Temporal reproduction in the rat. Behav. Anal. Lett. **2:** 141–151.
5. BRELAND, K. & M. BRELAND. 1961. The misbehavior of organisms. Am. Psychol. **17:** 681–684.
6. EISLER, H. 1976. Experiments on subjective duration 1868–1975: A collection of power function exponents. Psychol. Bull. **83:** 1154–1171.
7. STUBBS, D. A. 1979. Temporal discrimination and psychophysics. *In* Advances in Analysis of Behaviour. Reinforcement and the Organization of Behaviour. M. D. Zeiler & P. Harzem, Eds. Vol. **1:** 341–369. Wiley. New York, NY.
8. PLATT, J. R. 1979. Temporal differentiation and the psychophysics of time. *In* Advances in Analysis of Behaviour. Reinforcement and the Organization of Behaviour. M. D. Zeiler & P. Harzem, Eds. Vol. **1:** 1–29. Wiley. New York, NY.
9. EISLER, H. 1981. Applicability of the parallel-clock model to duration discrimination. Percept. Psychophys. **29:** 225–233.

Scalar Timing in Memory

JOHN GIBBON

New York State Psychiatric Institute
New York, New York 10032; and
Department of Psychology
Columbia University
New York, New York 10027

RUSSELL M. CHURCH AND WARREN H. MECK

Department of Psychology
Brown University
Providence, Rhode Island 02912

INTRODUCTION

A recent report of ours[1] proposed an information-processing account of temporal generalization. The account posited a clock process, which was the basic time measurement device, and working and reference memory for storing the output of the clock either temporarily or relatively permanently. Records of time intervals in working and reference memory were then compared using a binary decision process, which dictated responding or not responding. The analysis concentrated on a relativistic Weber's law property of the data from temporal generalization, and the constraints this property imposed on sources of variance in the information-processing stages. Our purpose here is to summarize that work and generalize the model in two ways: First we consider several sources of variance operating simultaneously. The original analysis demonstrated that if only one source of variance is present, it must be a scalar source, that is, it must result in a variable memory for which variance increases with the square of the mean.[2] In the generalized account proposed here, we will develop the conclusion that scalar sources dominate in some time ranges, while other sources may dominate in others. These ideas are then applied to two additional timing tasks with different characteristics.

TEMPORAL GENERALIZATION

The reference experiment for temporal generalization is a straightforward task with rats that was reported earlier.[3] The procedure is simple: The houselight in an operant chamber is turned off for a duration of time, *T*. Then a retractable lever is inserted into the chamber. After a 5-sec opportunity to respond, the lever is withdrawn and a 30-sec intertrial interval begins. If the light-off stimulus lasts for the "correct" duration, say, 4 sec, a response is reinforced with a pellet of food. If it is not 4 sec, no reinforcement is forthcoming.

The typical result is that rats come to respond with high probability when the duration is 4 sec, and with decreasing probability for stimuli either longer or shorter than 4 sec.

FIGURE 1 shows how this decline changes with changes in the size of the reinforced duration ($S+$). The probability of responding on the lever for three groups is shown as a function of signal duration when the rewarded duration was either 2 sec (top panel),

4 sec (second panel), or 8 sec (third panel). In each case the maximum response probability is near $S+$, and the spread of responsiveness around $S+$ increases considerably with increases in the $S+$ value.

The way in which the spread is related to the size of $S+$ is shown in the bottom panel, which plots these data on a relative time scale in which signal durations are taken as proportions of the $S+$ duration. The data from the different groups roughly superpose in this metric. This property, which we have called the scalar property, is ubiquitous in animal timing work in the seconds-to-minutes range.[2] The scalar property we will see exerts strong constraints on admissible sources of variance and on admissible comparison rules for the decision whether to respond or not.

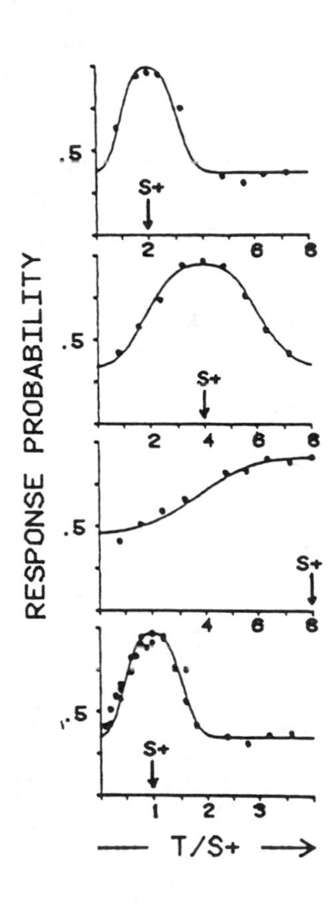

FIGURE 1. Response probability gradients for three groups of rats studied with three different placements of the positive signal. *Data points* represent median response probability and are replotted in the *bottom panel* on a relative time scale. (After Church and Gibbon.[2])

Our information-processing model for this situation is shown in FIGURE 2. The top row shows the clock process, which includes a pacemaker and a switch for gating pulses to an accumulator in working memory. The pacemaker generates pulses at a mean rate (Λ) that we assume is high relative to the time values (seconds to minutes) that we use in these experiments. The switch, after appropriate training (instructions), gates pulses for a mean duration (D_T) to an accumulator in working memory (second row) when the timing signal is present. The accumulator records and stores the number of pulses (mean of M_T). When, at the end of a given trial, a response is made and reinforced ($T = S+$), the time value recorded in working memory on that trial is stored in a more permanent reference memory for reinforced values (mean of M^*_{S+}). The third row shows the decision process. A response occurs when a comparator yields a judgment that the current record in working memory for this trial is "close enough" to the reference memory for the reinforced duration to warrant a response.

In FIGURE 3 we show the clock process in more detail. Imagine a pacemaker generating pulses with interpulse intervals, τ, with mean rate $\Lambda = \lambda$. In this figure, the pulses are evenly spaced to indicate no variance. In later analyses we consider variance in these parameters.[a]

The pulses are switched into the accumulator by the switch indicated in the middle box. The switch (SW) is assumed to have some latency to close (t_1) after the signal goes on, and some latency to open (t_2) after the signal goes off. Thus, the mean time during which pulses are gated into the accumulator is $D_T = T - T_0$, where T_0 is the expected

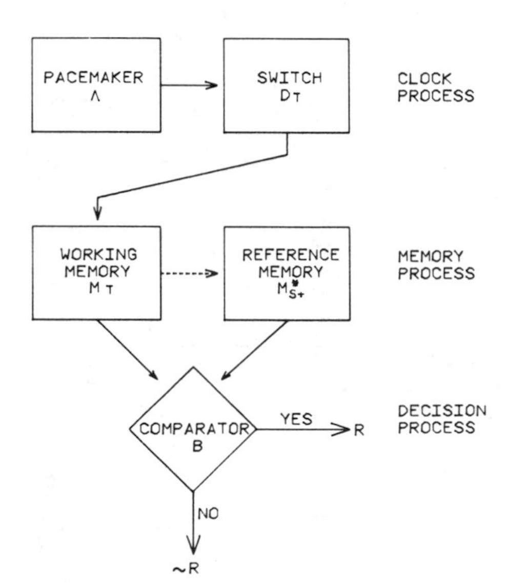

FIGURE 2. Information-processing model. The *top row* describes a clock process with a pacemaker generating pulses that are switched into an accumulator for working memory (*middle row*). After reinforcement, working memory contents are stored in a reference memory for later comparison on subsequent trials with the current working memory value.

difference between the latencies to close and open the switch. In principle, T_0 may be negative as well as positive. The graph above the switch shows this linear relation between D_T and T. The intercept is the minimum signal duration below which counts do not register in the accumulator. For signal durations less than this minimum, the switch is opened before it has closed so no pulses are switched into the accumulator. Conversely, if the latency to reopen the switch exceeds the initial latency to close it ($t_2 > t_1$), even a very short signal duration suffices to allow counts to register during the t_2 latency. If t_1 and t_2 were precisely the same, the switch duration would mimic exactly the signal duration. In general, this situation is unlikely, however, and these two latencies constitute one of the sources of variance that we will consider later.

The accumulator in this scheme simply records the number of impulses gated to it. The mean accumulated number, M_T, therefore, is just the impulse rate times the

[a]We generally refer to random variables with lowercase letters and their expectations with the corresponding uppercase letter (for example, $D_T = E(d_T)$). An exception is λ, which can play a dual role in the Poisson pacemaker.

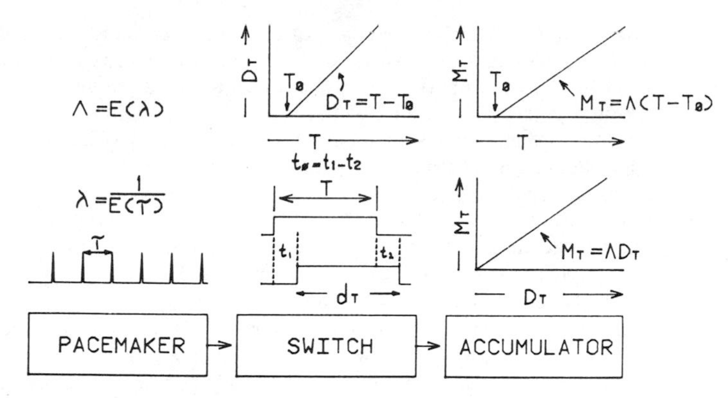

FIGURE 3. Clock process. The pacemaker generates impulses at intervals of τ and these impulses are gated into an accumulator by the switch. The switch has a latency to close at the beginning of the stimulus and to open at the end of the stimulus. The accumulator records the number of impulses gated to it during the interval that the switch is closed. (After Gibbon and Church.[1])

duration that the gate is closed, as shown in the lower graph directly above the accumulator. The upper graph shows the accumulation value as a function of signal duration. The switch introduces the non-zero intercept.

The memory system that we propose here is much simpler than is probably realistic for memory models (for example, that of Heinemann[4,5]). However, it is quite complicated enough to analyze, even in this oversimplified form. We propose that the working memory directly reflects the accumulated count as shown in the proportional plot above working memory in FIGURE 4 (and in FIGURE 3). When, for a given

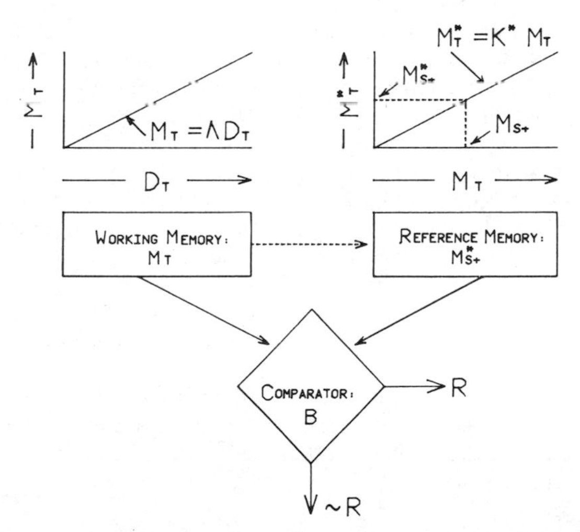

FIGURE 4. Memory process. The accumulator in working memory directly reflects counts gated to it from the pacemaker. On reinforced trials the working memory record for that trial is stored in a more permanent reference memory. (After Gibbon and Church.[1])

condition (series of trials), $S+$ is fixed at some value, and when on a particular trial $S+$ is presented ($T = S+$) and a response is made and reinforced, the value stored in reference memory, m^*_{S+}, is a proportional representation of the value on that trial recorded in working memory (m_{S+}). The comparison, then, on subsequent trials, is between the current value in working memory, m_T, and a stored value in reference memory, m^*_{S+}. The judgment whether or not to respond is based on some kind of comparison between these two values.

Several major features of this account are present in an early proposal of Treisman,[6] in which a pacemaker, counter, store (memory), and comparator occur. To our knowledge, that was the first model to use a form of scalar timing (equation 11) explicitly in a timing system. Creelman[7] some years ago also proposed a timing model in which a pacemaker and a counter were involved, but with a Poisson rather than scalar form of variance. These distinctions will come in for more discussion later.

In our earlier analysis[1] we discuss briefly the need for separate lines of evidence to establish the existence or utility of each of the processes in this account.[8] In the present article we will concentrate on features of the account that may be applied in a similar manner to similar and dissimilar timing tasks.

First, however, a brief summary of our earlier analysis is necessary. We first examined a case in which there was no variance in the system whatever, so that the values in the accumulator in working memory and in reference memory were all one-to-one with the appropriate signal durations.

Two response rules were examined, one of which compared the current time value in working memory with the remembered time value in reference memory, and dictated a response when these two values were "close enough," that is, when the absolute discrepancy between them lay below a certain threshold. The absolute rule for the no-variance case is shown in FIGURE 5 in the top row. The tics on the abscissa in the upper left panel indicate respectively, T_0, S, and $2S$, corresponding to the minimum signal duration, a given $S+$ duration, and a second $S+$ duration set at double the first. These might be appropriate, say, to the 2-sec and 4-sec $S+$ conditions in FIGURE 1. The absolute difference between M^*_S and M_T is shown decreasing to 0 at S and increasing beyond it as the solid line function. The positive diagonal hatching shows the acceptance region within which responses are required when the absolute discrepancy falls below the threshold value indicated by the horizontal line, B. When the $S+$ value is doubled (at $2S$), the dashed line shows the discrepancy function for this case, and the negative diagonal hatching indicates the acceptance region.

In the middle panel of the top row the response consequences are plotted in real time. In the first case for $S+ = S$, responding occurs whenever T is within an absolute window of S, and the same rule for $S+ = 2S$ produces the same spread around the reinforced $S+$ value. In the upper right panel these step functions for response probability are replotted in relative time, as a function of $T/S+$. The efficiency of the absolute rule is seen to increase considerably in relative time as $S+$ is increased. The spread in relative time around $S+ = 2S$ is much smaller than that around $S+ = S$ in this plot. This plot is comparable to that in the bottom panel of FIGURE 1. It shows that an absolute comparison rule is untenable under these assumptions.

In the bottom row of FIGURE 5, we show an alternative response rule, which compares the absolute discrepancy between the current time and the remembered reinforced time, to the remembered reinforced time. This relative rule requires a response whenever

$$\left| \frac{M^*_{S+} - M_T}{M^*_{S+}} \right| < B. \tag{1}$$

The subjective discrepancy is taken as a proportion of the reference memory value for reinforcement. In the panel in the lower left, the solid line function for $S+ = S$ is shown decreasing from 1.0 down to 0 and back up for $T > S$ as in the panel above. Now, however, the threshold, B, is a proportion. When the discrepancy is less than this proportion, responding is dictated. Again the acceptance region is indicated by positive diagonal hatching.

The dashed function shows the relative subjective discrepancy for $S+ = 2S$. When $S+$ is doubled, the window size nearly doubles also, as the negative diagonal hatching shows. In the middle panel the response consequences plotted in absolute time show the increasing spread around $S+ = 2S$, and in the right panel the two acceptance regions

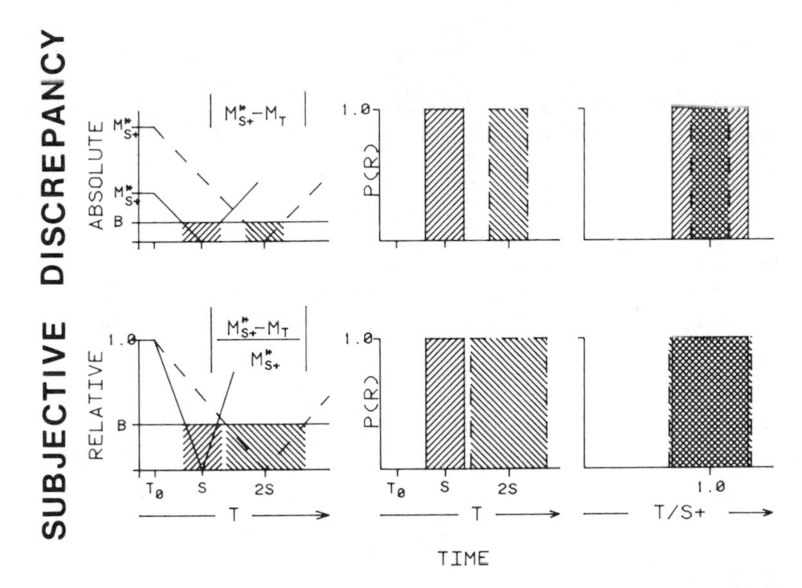

FIGURE 5. Two comparison rules. The *left column* shows subjective discrepancies between working memory and reference memory as a function of signal duration with two different placements of $S+$. The *right two columns* shows response probability. The *top row* describes an absolute discrepancy rule, in which the absolute difference between working and reference memory values forms the basis of the response decision. The *bottom row* describes a relative discrepancy rule in which the absolute discrepancy is taken as a proportion of the representation of $S+$. (After Gibbon and Church.[1])

are nearly equivalent when plotted in relative time. They are not precisely equivalent because the T_0 intercept in the accumulation of subjective time counts more heavily when $S+$ is smaller than when it is larger. However, even for T_0 relatively large, the difference is not great. In this example T_0 has been chosen equal to $S/4$.

Thus, even when there is no variance in the timekeeping and mnemonic system, a relative response rule is needed to accommodate the superposition of generalization gradients in relative time.

Of course, the real response gradients are not square waves, and so a no-variance account will not suffice as a realistic description. In our earlier analysis we studied several sources of variance which would generate smooth shoulders on the gradients, but we found that in no case was the absolute discrepancy rule tenable. Even when variation increased substantially with $S+$, the absolute discrepancy rule is too efficient to accommodate a realistic description of the data as $S+$ is increased.

With the relative discrepancy rule, however, a number of different sources of variance were compatible with both the superposition of the data in relative time and with smooth shoulders around $S+$ (FIG. 1). The analysis showed that at least four distinct sources of variation in perceiving, remembering, and discriminating time intervals were possible. Each of these sources was scalar, that is, multiplicative with the memory for time. Each was associated with different components of the information-processing scheme. Two additional sources of variation in the system are also potential contributors to variability, but these sources were shown to be not feasible as a sole source of variation, were there but one.

One of our current goals is to deepen this analysis by allowing these latter two sources of variation to be imbedded in variation from multiple points in the system. The earlier analysis is summarized with respect to the relative discrepancy rule only. The way in which the two nonscalar sources of variation violate the superposition requirement will be described briefly first.

Switch: Constant

The simplest of these two sources of variation is the variability introduced by varying latency to open and close the switch gating pulses into short-term memory. It is likely that even in highly practiced subjects there is some variation in the perceptual

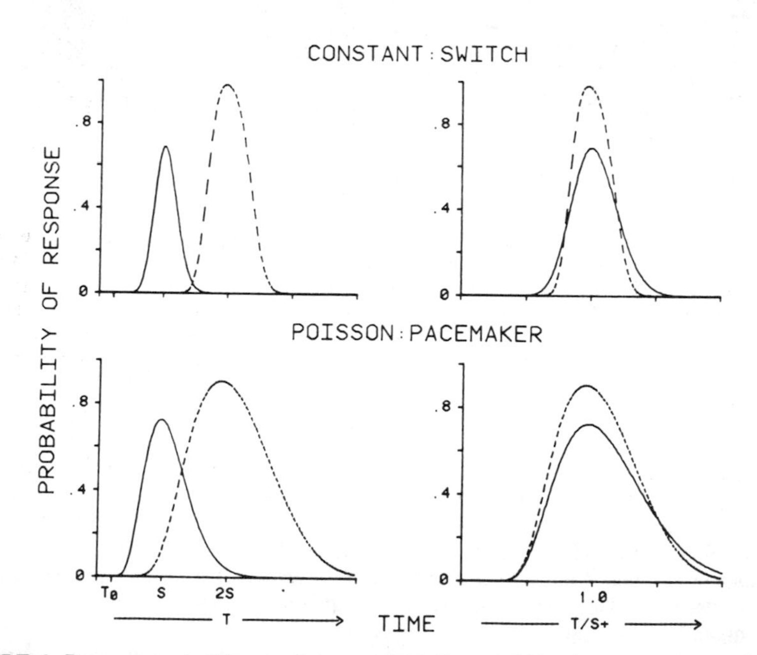

FIGURE 6. Response probability gradients resulting from variation in the switch only (*top*) or from Poisson variation in the pacemaker only (*bottom*). At the *left* gradients are plotted in real time and at the *right* in relative time.

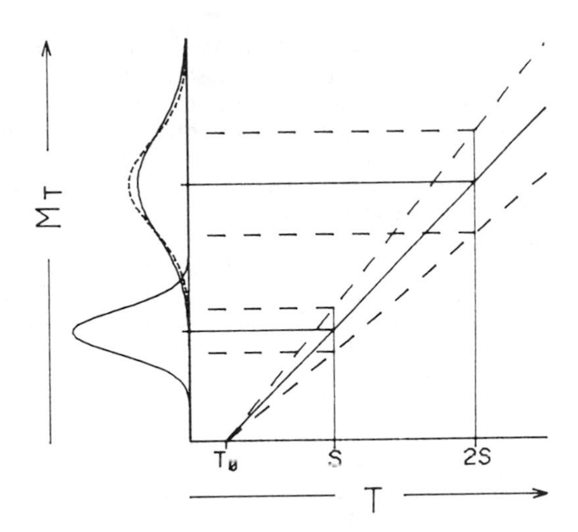

FIGURE 7. Working memory accumulator. Distributions of counts associated with two different $S+$ durations are shown when the only source of variance is the scalar fluctuation in pacemaker rate. (After Gibbon and Church.[1])

response to external stimulation. A gating mechanism introducing variance of this sort is similar to an early proposal of Kristofferson and his colleagues which argued that in some ranges the timing system might reflect only constant, low variance.[9,10,11] A realistic form for this latency difference distribution might be most likely back-to-back exponentials (Laplace distribution). A similar alternative is the triangular distribution analyzed by Kristofferson as the result of a difference between two rectangular variates. If the switch is complex, however, and requires the execution of several components in order to open and close, it might approach a normal form. The niceties of distinctions between forms will not concern us in the present account, since we will rely on rather grosser distinctions which are relatively insensitive to distribution form. We assume a normal form in what follows.

The top row of FIGURE 6 shows the effect on response probability of introducing variance in switch latency only. In the upper left panel response gradients are plotted against absolute time for $S+ = S$ and $S+ = 2S$. Increasing $S+$ results both in greater accuracy at $S+$, and in a broader spread around $S+$. In the panel in the upper right, these effects counterbalance each other somewhat in relative time. There the largest discrepancy is at $S+$, while the increase in the spread with increasing $S+$ is seen to be not sufficient to produce superposition in the wings. Constant variance, which does not change with increasing size of the interval being judged, results in too efficient a system at large values of $S+$. Accuracy is increased at the target duration, and the spread around it does not increase sufficiently to accommodate superposition.

Pacemaker: Poisson

A second source of variation in this system is one that has received classical attention in modeling perceptual systems, namely Poisson variation in the discharge of the pacemaker. From a variety of considerations, variance in the interpulse interval in a neural pacemaker ought to follow the Poisson law.[6,7,12–14] Interpulse intervals are exponentially distributed in a steadily varying stream with intensity $\lambda = \Lambda$. The accumulator in memory is then a Poisson counter, with a variance that increases directly with the mean.

In the lower row of FIGURE 6 response gradients for the Poisson pacemaker source are shown. The gradients in absolute time in the lower left show an increase in accuracy at $S+$ when $S+$ is doubled. However, this increase is not as large as that for the constant variance case, and the increase in spread at $S+ = 2S$ is somewhat broader. Nevertheless, when plotted in relative time on the right, the gradients still deviate substantially from superposition. Thus, this system, which does increase variance with increasing $S+$, does not do so fast enough to accommodate superposition, particularly near $S+$.

These two kinds of variability introduced by different stages of information processing are to be contrasted with two other kinds of variability, both of which are scalar and induce the approximate superposition seen in the data. These are discussed next.

Pacemaker: Scalar

An alternative source of pacemaker variance is a drifting rate. For simplicity we imagine that the time between pulses, τ, is fixed on any trial, but that from trial to trial the pulse rate, $\lambda = 1/\tau$, varies normally around a mean, Λ.[b]

In FIGURE 7 the result for the accumulated count in memory is shown for $S+ = S$ and $S+ = 2S$. The rising solid line reflects the mean value of the rate, Λ, comparable to the rising line in the upper right panel in FIGURE 3. The dashed rays emanating from T_0 indicate plus and minus 1 standard deviation of λ. When $S+$ is doubled, the distribution of accumulated counts is nearly a scale transform of the distribution associated with S, hence the name scalar timing. The solid function distribution for $2S$ is generated by our information-processing system with a fixed T_0. The dashed function which is nearly identical to it is the simple scale transform of the S distribution that would result from strict proportionality, an axis multiplier of 2. The small discrepancy reflects the role of T_0. Again, T_0 was chosen rather large, equal to $S/4$. Even with an intercept of this size, the multiplicative property dominates the result.

This kind of multiplicative variance in combination with the relative discrepancy rule results in the response gradients shown in the top row of FIGURE 8. In absolute time (on the left), it is clear that accuracy when $S+ = S$ and when it equals $2S$ remains the same, but variance increases considerably. In the proportional plot in the upper right panel, the functions now show near superposition, like the data. Thus, if the only source of variance were a drifting rate in the pacemaker, the smooth shoulders, constant discriminability at the positive value, and rough superposition are accommodated by this scalar source.

Memory: Scalar

In our earlier analysis we also studied the possibility of additional noise introduced when the accumulated counts from the pacemaker are stored in working memory, and again when these records are transferred to relatively permanent storage in reference memory. If both of these storage mechanisms involve a proportional transform, then they too may introduce variability which has the scalar form shown for pacemaker rate

[b] A more realistic version might allow both kinds of variation simultaneously. Our later analysis (APPENDIX) allows Poisson variance within trials at a given intensity, λ, which in turn is a random variable that drifts over trials.

variance in the top row of FIGURE 8. The first of these potential sources, that involved in translation to a short-term storage in working memory, will be ignored in what follows here, since with the ratio discrimination rules that we employ, the mean value of the translation process is cancelled, and its variance may be absorbed in the pacemaker variance term. Storage in reference memory, however, plays an important role, and may contribute variance as well as distortion between remembered and current working memory values. Qualitatively, however, these two sources of variability show similar results for the response gradients as long as the reference memory translation is unbiassed. Hence we do not treat them separately here.

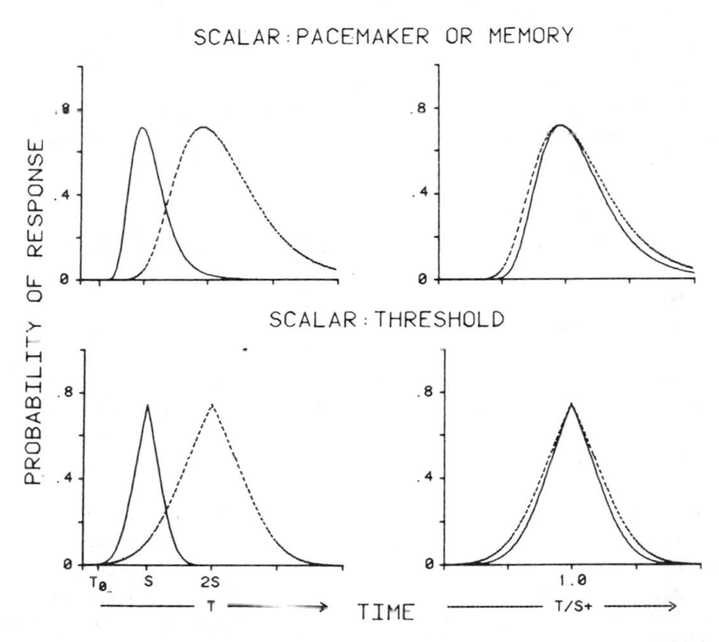

FIGURE 8. Response probability gradients for two scalar sources. The *top row* shows the results of variation in either the pacemaker or the memory constant. The *bottom row* shows the results of variation in the threshold.

Threshold: Scalar

Still another source of variance is realistic for our processing system. This is variance in the relative proximity of T to $S+$ that subjects deem "close enough." One can readily imagine momentary fluctuations in this level. Multiplying through by the norming value in Equation 1, variation in b induces the scalar property on remembered time. The manner in which this property is expressed, however, differs from that for clock or memory variance.

In the bottom row of FIGURE 8 response gradients are plotted assuming a normally distributed threshold about some positive value b. The gradients in the lower left panel are strictly symmetric, have identical accuracy for $S+ = S$ and $S+ = 2S$, and show a prominent discontinuity at the positive value. In the lower right panel the relative time gradients approximately superpose. Thus, this source of variance, like the scalar source for clock or memory in the upper row, induces superposition in relative time, as the data require. The manner in which it is accomplished, however, differs in two ways: First the upper gradients have a slight, positive skew. The lower gradients in contrast are strictly symmetric. Second, the upper gradients are smooth, roughly bell-shaped around $S+$, while the lower gradients are discontinuous at $T = S+$. The discontinuity results from a feature of our assumptions that is open to question here, namely, a normal form for the threshold distribution that includes negative values. In practical terms this assumes that on some trials the threshold is so conservative that not even a comparison of identical values (no subjective discrepancy) would warrant a response. The defect may be remedied by truncating threshold distributions to be always positive, but the result is equally difficult for theory if this is the only source of variation. This modification induces perfect accuracy at $T = S+$ for all $S+$ values, and the data show clearly otherwise.

These features of the threshold variance account might prove troublesome for a description involving only this source of variance. However, in combination with the clock and memory variability we will see that the symmetry in threshold variability is an important feature of our account.

Simultaneous Sources of Variation: General Case

From the analysis thus far, we were able to conclude that were one forced to pick a single source of variation, it would have to be a scalar source, and probably would have to be located in pacemaker or memory. However, while the smooth character of these gradients looks like that of our temporal generalization gradients, and while our temporal generalization gradients have, as do these, some slight skew with a higher right than left wing, the skew in theory is generally larger than the skew in reality. The data fall somewhere between strict symmetry around $S+$, and the scalar forms for clock or memory variance alone.

This discrepancy leads us to analyze the mixture generated by allowing all of these potential sources of variability to operate at once. In the APPENDIX we pursue this analysis, obtaining an approximation for the underlying random variables in working and reference memory, and their discrimination via a variable threshold in a comparator. The result may be summarized as follows: Allowing random variation in constant, Poisson, and scalar sources results, not surprisingly, in constant, Poisson, and scalar components of variance in the composite variables underlying the spread of the response probability gradients. For reasonable choices of parameter values for the constant and Poisson variance sources, however, the scalar sources dominate variance in the seconds to minutes ranges studied here. This may be seen in FIGURE 9. In the APPENDIX the argument is developed showing that under our assumptions the response probability gradients reflect two random variables, one corresponding to each edge of the acceptance region for our comparison statistics. The standard deviations of the window edge variables grow (nearly) linearly with T and $S+$. In FIGURE 9 the standard deviation of the upper window edge is shown as a function of increasing $T = S+$ signal durations for a variety of choices of pacemaker rate, $\Lambda = 5, 10, 15$, and 20 per sec. The figure assumes a switch with a mean and variance of $T_0 = \sigma_0 = 0.5$ sec, variance in the threshold of $\sigma_b = 0.1$, and a scalar pacemaker or memory coefficient of variation of $\gamma = 0.2$. The point of the figure is that the scalar sources of variance rapidly come to dominate the standard deviation of the window edge. These functions are approximately linear in $S+$ for $S+$ values above about 5 sec (dashed line). Since the ranges we study here are usually well above this level, we cannot say whether significant contributions from a Poisson source and a constant source might not be present, but masked. The domination by the scalar sources in these mixtures may be used in the future to specify more precisely the range of parameter values for pacemaker rate and switch variance that are possible and yet still accommodate our data. It is clear from the work presented in this volume that the temporal ranges that interface between milliseconds and hours will be important in the future.

PEAK

A procedure that is very similar to temporal generalization in conception and results, but quite different in the manner of assessing response strength, is the "peak" procedure developed by Seth Roberts[15] and studied by him and Church and Meck.[16] It is a modified discrete-trial free-operant fixed-interval schedule. At the onset of a signal, usually a light or a noise, rats may respond on a lever, which occasionally will "pay off" for the next response after $S+$ sec have elapsed from the beginning of the

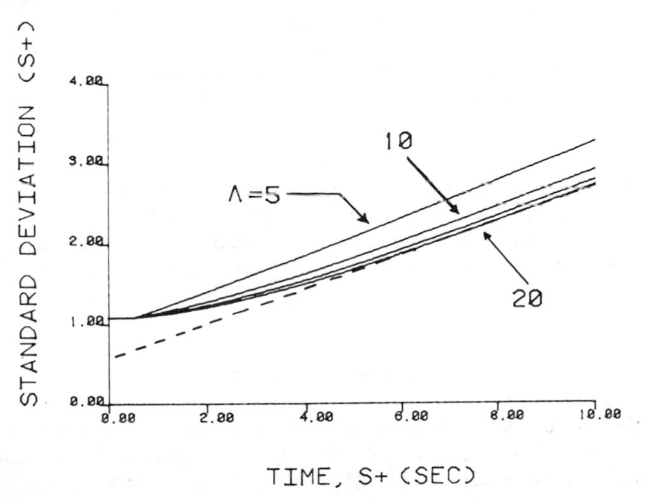

FIGURE 9. Standard deviation of a window-edge decision statistic as a function of the size of $S+$. The parameter is pacemaker rate.

signal. On some trials, however, responding is not reinforced and the signal remains on for a long time. Subjects come to anticipate completion of the $S+$ interval. On trials in which reinforcement is not programmed, they show maximum responding close to that time. Typical results are shown in FIGURE 10. The top row shows responding of two pigeons studied under the peak procedure using a color change on a response key as the $S+$ duration signal. On half of the trials the first response after 15 sec was reinforced. Response rate increases up to a maximum near 15 sec and decreases thereafter.

In the second row, data from a rat studied by Roberts is shown for two different conditions. In the first condition on 50% of the trials the rat was reinforced for responding after 20 sec had elapsed since the onset of an auditory signal. The leftmost function shows the performance. Response rate peaks a little beyond 20 sec and declines at longer times. The broader function on the right was obtained from the same rat under a later condition with $S+ = 40$ sec. These data are quite similar to the

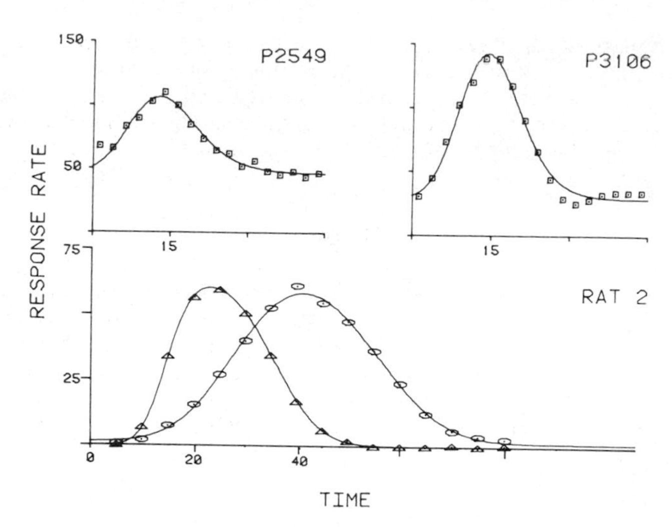

FIGURE 10. Response rate gradients from two pigeons (*top row*) and one rat (*bottom row*) studied under the peak procedure.

temporal generalization data in two respects: Accuracy at the peak time is about the same in both functions, and the spread around the larger $S+$ is approximately proportionally increased. Indeed, the peak procedure might be thought of as a generalization procedure which delivers all possible durations on every unreinforced trial.

Memory Distortion

All of the above cases show some slight displacement of the peak time from the nominal $S+$ value. This is not unusual with the peak procedure. The peak for bird No. 3106 is about 13 or 14 sec and the peak for rat No. 2 studied under $S+ = 20$ sec occurs near 24 sec.

In FIGURE 11 in the left hand column the effect of deviation from $S+$ is shown for the data of rat 2. The top functions show the peak rate as a function of absolute time as in FIGURE 10. In the middle panel, these functions are shown plotted in relative time, $T/S+$, as in FIGURE 1 for temporal generalization. Now superposition is not achieved. The $S+ = 20$ sec curve shows a broader right wing in relative time.

In the model, the information-processing account allows for distortion of the mean in the translation constant between working and reference memory.[17] Working memory is assumed to reflect pacemaker accumulation directly, while the storage into reference memory may introduce both additional variance of its own, and systematic distortion in the mean (upwards or downwards). The translation constant, K^*, required for these data is somewhat above 1.0, hence the failure of superposition.

In the lower left panel, a fit to the peak data has been accomplished and the data are now plotted relative to K^*S+, that is, relative to the mean of the memory distribution for the reinforced time. When adjusted in this manner, superposition succeeds.

A more extreme example of memory distortion which nevertheless preserves superposition in the appropriate metric is obtained from an experiment studying two peak procedures in rats simultaneously. A peak procedure value of 10 sec in the presence of a light and 30 sec in the presence of a noise were studied on randomly intermixed trials. Median performance of a group of ten rats is shown in the right hand column of FIGURE 11. In the upper right panel the two peak functions plotted in real time are seen to be centered close to 10 sec and 30 sec, as would be expected from the preceding data. The functions differ somewhat from those of the individual subjects in FIGURE 10. Both functions are fairly flat near the peak, leaving some ambiguity as to just which time value corresponds to the "true maximum." However, even with this flatness it is clear that the 10-sec data lie mostly above 10 sec, while the 30-sec data lie

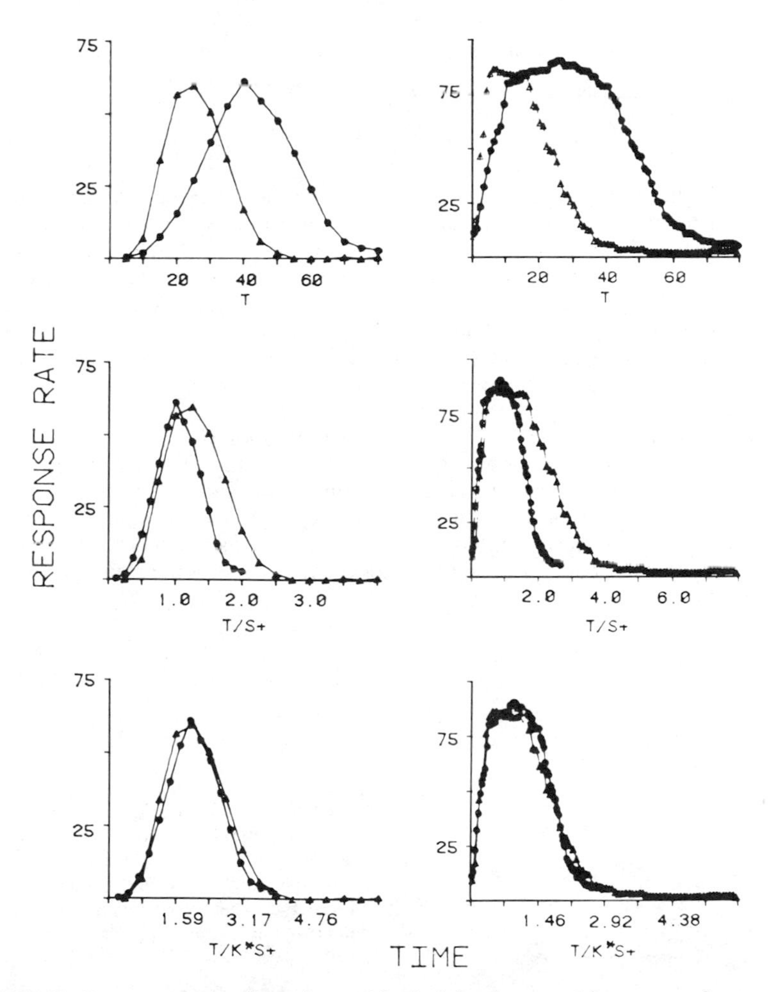

FIGURE 11. Response rate gradients for rat 2 in the left column, and for a group of rats studied under two different $S+$ values on the right. Gradients in the top row are plotted in real time, in the middle row in time relative to $S+$ time, and in the bottom row in time relative to the empirically determined memory representation of $S+$.

mostly below 30 sec. It is as though the subjective representations of the 10-sec and 30-sec $S+$ experiences have become mixed or perhaps mutually attracted in memory.

In the middle panel of FIGURE 11 these data are plotted in relative time, and the systematic distortion is again evident. The 10-sec curve appears broader than the 30-sec curve in relative time, because its "true" $S+$ time is larger than 10 sec and the "true" 30-sec $S+$ time is less than 30 sec. This is similar to the deviation on the left, but more extreme. In the bottom panel, the results of fitting these data, allowing for memory distortion, show clear superposition when plotted in time relative to the memory time (T/K^*S+).

Thus, the peak procedure provides temporal gradients with a maximum near the time of reinforcement, but deviations appear both idiosyncratically and because of potential interactions in reference memory when more than one time value must be retained. The results of this analysis imply that this sort of distortion is a multiplicative one. The scalar property applies to the memory for time, not necessarily to real time.

TIME LEFT

Our final application of these ideas is to a procedure that differs considerably from the temporal generalization or peak procedures. In a previous report, Gibbon and Church[18] studied choice procedures with rats and pigeons; these procedures were designed to reveal preference for the subjectively shorter of two delays to food, when one of these delays was elapsing. The aim of that work was to examine these choices parametrically at a variety of delays, since parametric data should reveal curvature or linearity in subjective time. The linearity result will not be reviewed here for that implication, but rather some new data from the same procedure under several conditions will be examined for the superposition property.

The procedure studied with pigeons is shown in FIGURE 12. At the beginning of the trial (initial link), two keys are lit with different colors, say white and red, and birds distribute responding across the two keys as time elapses during the trial. At some point (T), the next response produces mutually exclusive (terminal link) consequences on either key. If the next response is the white key, the red key is extinguished, and responding on the white key may continue for the remainder of the white-key interval,

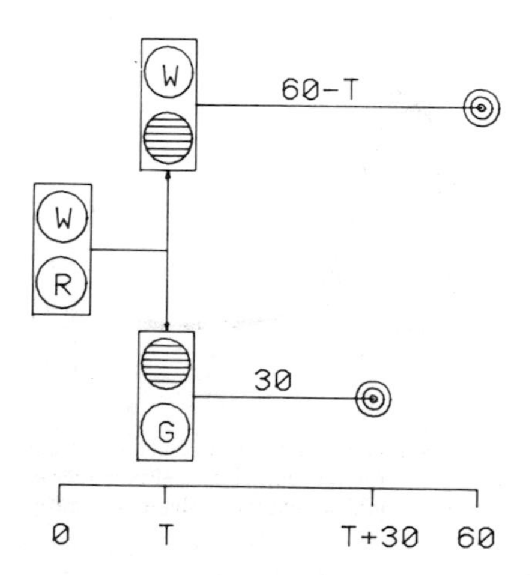

FIGURE 12. "Time left" procedure. Two keys are lit with different colors (white or red) at the beginning of the trial. At some randomly chosen point, T, the next response results in one or the other of two mutually exclusive delays to reinforcement (bull's-eyes).

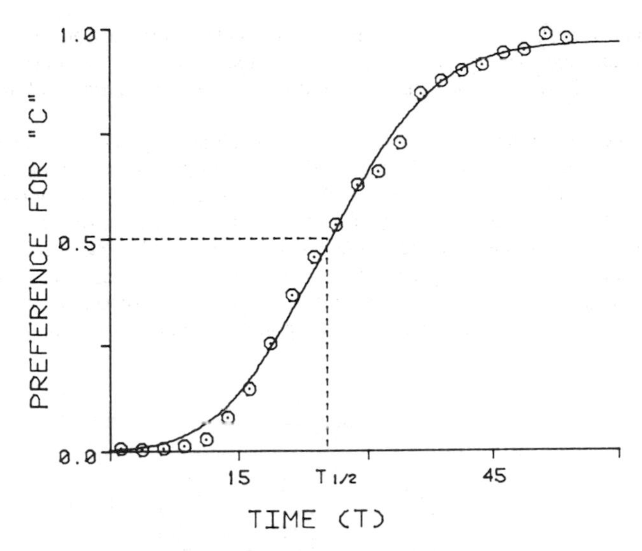

FIGURE 13. Psychometric preference function. Preference, the proportion of "*C*" choices at successive points during the choice period, increases from favoring "*S*" at the start of the trial to favoring "*C*" as the trial elapses. The point of indifference is indicated by the dashed line over $T_{1/2}$.

in this example 60 sec, when reinforcement is made available. Thus, on the white key, reinforcement is available after a total of 60 sec from the beginning of the trial.

If, at the choice point, the next response is to the red key, the white key is extinguished, the red key changes color to green, and responding will be reinforced for pecking the green key after 30 sec have elapsed since the color change. Thus, this delay, called the standard (S), is fixed at 30 sec. The entry times (T) vary from trial to trial.

Early in the trial, it behooves subjects to respond to the red key since the delay to food there is shorter than on the comparison 60-sec interval. However, as the choice period elapses, the time left on the comparison ($C = 60$ sec) side becomes shorter than the 30-sec standard, and now it behooves subjects to respond on the white key. The typical result is shown in FIGURE 13, in which choice responding begins heavily favoring the standard, and at some point in the trial switches over to favor the comparison time-left side as that delay becomes more favorable. The point at which the switch is made, $T_{1/2}$, is a datum of primary interest. If subjective time is linear and subjects are unbiassed in their appreciation of this time, then they should switch over to preferring the time-left side of the choice at precisely $C - S$ sec into a C-sec comparison interval—at 30 sec into the 60 sec interval in the above example. In fact, it is common for subjects to switch somewhat before the midpoint value for $C = 2S$, and the bias in favor of the elapsing interval may be due to a preference for a key color paired with primary reinforcement. In the analysis that follows, the crossover point, $T_{1/2}$, plays a fundamental role.

The information-processing account for this procedure has three distinct conditions: the choice period, and each of the mutually exclusive terminal link consequences. The scheme we use to model choice responding is shown in FIGURE 14. When the choice period begins, the switch gates pacemaker pulses into a working memory accumulator for trial time, T. The difference between the memory for current time and the memory for the comparison (C) interval represents the subjective time left on the comparison side. This remaining time is compared in the comparator with memory for the time "left" on the standard side—the standard delay to food were the terminal link to come on immediately.

As time accumulates during the trial in working memory, the remaining time on the comparison side is continually updated, and compared with memory for the standard delay. At the outset of the trial, the standard delay appears shorter than the remaining time on the comparison delay and hence preference for the comparison interval is low. At some point subjects cross over and prefer the remaining time on the comparison side, when their subjective assessment of this delay makes it appear more favorable, by a potentially biased threshold. Note that this threshold, b', is distinct from the proportional threshold defining a response window in the temporal general-

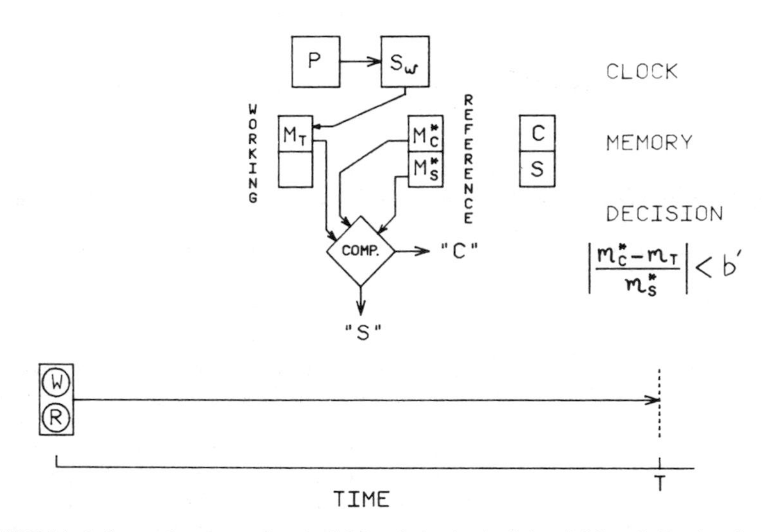

FIGURE 14. Information-processing model for choice in the "time left" task. During the choice period the switch gates pacemaker pulses to working memory for elapsed trial time, T. A comparison is made between the discrepancy between C and the current time—time left to food on the comparison side—and the standard delay to food, S. Preference is for the shorter of these two delays.

ization case. Here b' indicates a bias in favor of the time-left or standard side, with an unbiased mean value represented by $B' = 1.0$. We expect, however, that variance associated with the decision rule for both cases may be quite comparable.

The two terminal link conditions establish the memory values for the comparison and standard delays. In FIGURE 15, processing of the time-left terminal link is shown. The switch continues to gate pulses into the working memory for T, until reinforcement occurs when $T = C$ at the end of the trial. During this terminal link subjects are in a go/no-go situation precisely comparable to the left wing of a peak procedure (a fixed interval schedule). Hence their decision to respond or not to the white key is assumed to be based upon the time left to reinforcement normalized by the overall time to reinforcement, as shown in the decision rule in the lower right. After reinforcement, the reinforced value from working memory (m_C) is stored in reference memory (m_C^*) for the comparison interval.

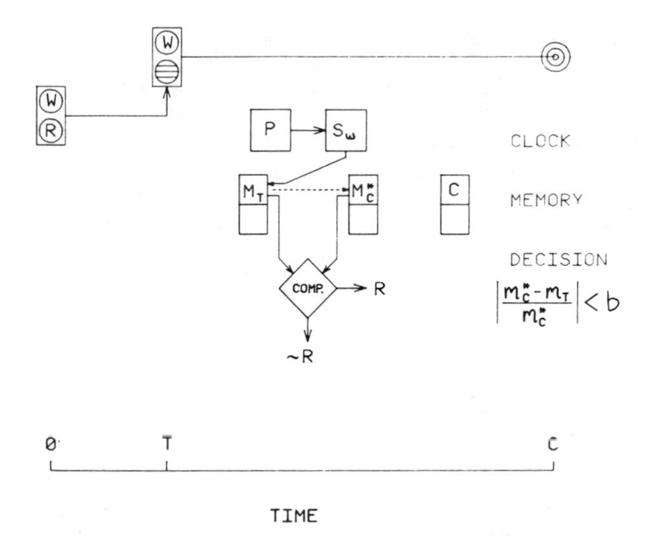

FIGURE 15. Information-processing model for the time-left terminal link. The switch continues to gate pulses to the trial-time accumulator, and the comparison is a go/no-go decision, as in the peak procedure.

When on other trials the choice at the entry point is in favor of the standard, the standard terminal link is entered. FIGURE 16 shows the switch gating pulses into a new working memory accumulator (m_t) as subjects begin timing the standard interval. Go/no-go responding is again assumed to be comparable to the left wing of a peak procedure, and the decision rule reflects the relative proximity to food. After reinforcement, the value in the working memory store (m_S) is transferred to a reference memory for the standard interval (m_S^*). In this way the memories for C and S are built

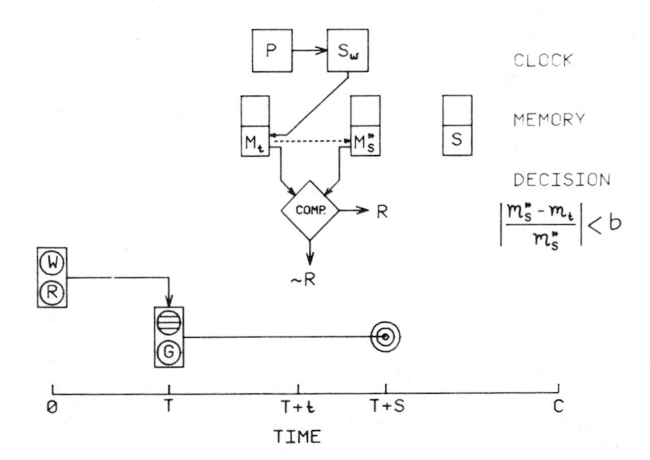

FIGURE 16. Information-processing model for the standard terminal link delay. During this delay the switch gates pulses into an accumulator for the standard delay in working memory, and the comparison is again a go/no-go decision, as in a peak procedure.

up over training and then become available for sampling during the choice period of later trials. In principle it should be possible to predict on individual trials the details of responding in the terminal links from features of responding in the choice link. We will pursue this line in later work. However, in what follows, we concentrate on analysis of preference, pooled over the choice period.

The time, $T_{1/2}$, at which subjects cross over during a sufficiently lengthy choice period from preferring S to preferring time-left is revealing. The decision rule implies that at this point the delays are subjectively equal. On average,

$$M_C^* - M_{T_{1/2}} = B'M_S^*. \quad \mathbf{(2)}$$

Translating these mean values according to Equation 1 gives us

$$T_{1/2} = \left(\frac{C}{S} - B'\right) K^*S + T_0[1 + K^*(B' - 1)]. \quad \mathbf{(3)}$$

Indifference points should be linear in S, for constant ratios of C to S, as described previously.[15] Note that for $T_0 = 0$ and no bias ($B' = 1$) or memory distortion ($K^* = 1$), $T_{1/2} = C - S$, when the actual time left, $C - T_{1/2}$, equals S. Deviations from physical equivalence may be introduced by bias and memory distortion. $T_{1/2}$ may be thought of as a set-point at which these factors have been subjectively equalized.

The performance of three subjects studied under three combinations of $S = C/2$ (7.5, 15; 15, 30; 30, 60) are shown in the three panels in FIGURE 17. The S value is

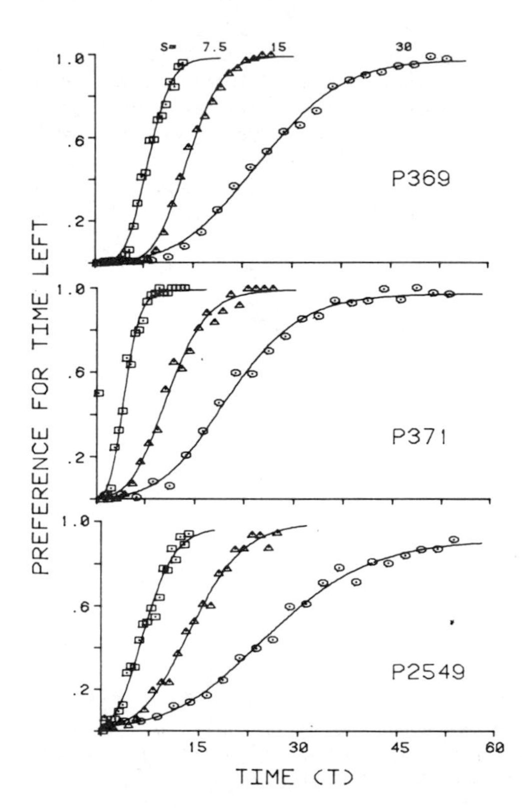

FIGURE 17. Psychometric preference functions for time left for three subjects at three different pairs of $S = C/2$ values. The functions are plotted against real time in the trial. The S parameter values are indicated in the *top panel*.

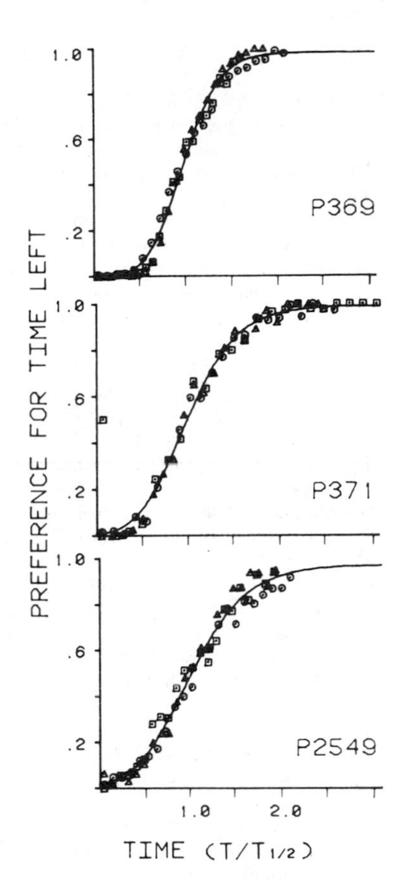

FIGURE 18. Preference functions for these three subjects replotted from FIGURE 17 as a function of time relative to the indifference point, $T_{1/2}$. The symbol code is the same as in FIGURE 17.

indicated above each curve in the top panel. These preference functions, like those published earlier,[15] show a shallower rise as $S = C/2$ is increased, and $T_{1/2}$ values are approximately linear in S, as expected (Equation 3).

Our interest centers on a superposition property of the account, which is developed in the APPENDIX. There it is shown that if these curves are normalized by their $T_{1/2}$ values, the effects of changing bias ($B' \neq 1$) and memory encoding ($K^* \neq 1$) are cancelled in much the same way as the effects of changing K^* may be cancelled in the peak procedure through normalizing by K^*S+. Here normalization by $T_{1/2}$ entails superposition as long as the scalar variance sources dominate. This will be true if the time ranges are large relative to T_0 and the Poisson rate parameter is greater than about 10 per sec. Plotting the preference functions at $T/T_{1/2}$ results in the superposition of these functions shown in FIGURE 18. Thus, again a Weber's law property is revealed, but with the important proviso that it operate on subjective, not objective time. Distortions in memory and innate preferences operate to rescale objective time.

ACKNOWLEDGMENTS

We gratefully acknowledge the assistance of Stephen Fairhurst and Amy Waring in all of the data and theory analysis presented here.

REFERENCES

1. GIBBON, J. & R. M. CHURCH. Sources of variance in an information processing theory of timing. Presented at the Harry Frank Guggenheim Conference on Animal Cognition, Columbia University, June 2–4, 1982. In press.
2. GIBBON, J. 1977. Scalar expectancy theory and Weber's law in animal timing. Psychol. Rev. **84:** 279–325.
3. CHURCH, R. M. & J. GIBBON. 1982. Temporal generalization. J. Exp. Psychol. Animal Behav. Processes. **8:** 165–186.
4. HEINEMANN, E. G. 1982. A memory model for decision processes in pigeons. *In* Quantitative Analyses of Behavior. Vol. 3: Acquisition. M. Commons, R. J. Herrnstein, and A. R. Wagner, Eds. Ballinger. Cambridge, MA.
5. HEINEMANN, E. G. 1982. The presolution period and the detection of statistical associations. *In* Quantitative Analyses of Behavior. Vol. 4: Discrimination Processes. M. Commons, R. J. Herrnstein, and A. R. Wagner, Eds. Ballinger. Cambridge, MA.
6. TREISMAN, M. 1963. Temporal discrimination and the indifference interval: Implications for a model of the "internal clock." Psychol. Monogr. **77:** 13.
7. CREELMAN, C. D. 1962. Human discrimination of auditory duration. J. Acoust. Soc. Am. **34:** 582–593.
8. CHURCH, R. M. 1984. Properties of the internal clock. This volume.
9. KRISTOFFERSON, A. B. 1967. Attention and psychophysical time. Acta Psychol. **27:** 93–100.
10. ALLAN, L. G., A. B. KRISTOFFERSON & E. W. WIENS. 1971. Duration discrimination of brief light flashes. Percept. Psychophys. **9:** 324–327.
11. WING, A. M. & A. B. KRISTOFFERSON. 1973. Response delays and the timing of discrete motor responses. Percept. Psychophys. **14:** 5–12.
12. GETTY, D. J. 1976. Counting processes in human timing. Percept. Psychophys. **20:** 191–197.
13. McGILL, W. J. 1967. Neural counting mechanisms and energy detection in audition. J. Math. Psychol. **4:** 351–376.
14. GREEN, D. M. & R. D. LUCE. 1974. Counting and timing mechanisms in auditory discrimination and reaction time. *In* Contemporary Developments in Mathematical Psychology. Vol. 2: Measurement, Psychophysics, and Neural Information Processing. D. H. Krantz, R. C. Atkinson, R. D. Luce & P. Suppes, Eds. W. H. Freeman. San Francisco, CA.
15. ROBERTS, S. 1981. Isolation of an internal clock. J. Exp. Psychol. Animal Behav. Processes **7:** 242–268.
16. MECK, W. H. & R. M. CHURCH. 1981. Simultaneous temporal processing. J. Exp. Psychol. Animal Behav. Processes **7:** 242–268.
17. MECK, W. H. 1981. Selective adjustment of the speed of internal clock and memory processes. J. Exp. Psychol. Animal Behav. Processes **7:** 242–268.
18. GIBBON, J. & R. M. CHURCH. 1981. Time left: Linear versus logarithmic subjective time. J. Exp. Psychol. Animal Behav. Processes **7:** 87–108.

APPENDIX

Temporal Generalization and Peak

We require the total probability of the event in Equation 1, the decision rule, when all stages in the information-processing scheme may contribute variability. That is, we require the total probability of

$$\left| \frac{m^*_{S+} - m_T}{m^*_{S+}} \right| < b, \tag{A1}$$

where m^*_{S+}, m_T, and b are each independent random variables sampled from reference memory, working memory, and from a threshold distribution on each trial. We develop this calculation by obtaining the mean and variance of the memory random variables, when these receive variance contributions from each of the postulated sources of variance in FIGURES 2 through 4. Then the analysis follows that of Church and Gibbon.[3] At several points in the development normal approximations for distribution forms that are demonstrably slightly skewed will be used. We have in each case run computer simulations to satisfy ourselves that the approximations are reasonable. Our strategy is to add successive sources of variance moving through the information-processing chain from the pacemaker to the comparator.

The pacemaker generates pulses with Poisson variability at an intensity, λ. Imagine first that the effective switch closure time, d_T, is fixed at the mean value, D_T.[c]

The number of counts accumulated in working memory associated with this switch closure time is then a Poisson-distributed variate with a mean and variance

$$E(m) = \mathrm{Var}(m) = \lambda D$$

We now allow the intensity parameter to vary. The moment generating function for the Poisson variate with fixed intensity is

$$M_{m_\lambda}(\theta) = \exp[\lambda D(e^\theta - 1)], \qquad \textbf{(A2)}$$

where we have replaced the subscript T by λ to indicate dependence on λ at any fixed T. If we now let λ vary around a mean Λ with standard deviation $\gamma\Lambda$, the moment generating function of the mixture then becomes

$$M_m(\theta) = \int_{-\infty}^{+\infty} M_{m_\lambda}(\theta) f(\lambda) d\lambda. \qquad \textbf{(A3)}$$

Assuming normality of f, we may complete the square under the integral and factor so that λ is integrated out of **(A3)**. With some rearrangement the moment generating function may be shown to be

$$M_m(\theta) = \exp\,[\Lambda D(e^\theta - 1) + 1/2\,(\gamma\Lambda D)^2(e^\theta - 1)^2]. \qquad \textbf{(A4)}$$

Taking first and second derivatives and setting $\theta = 0$ gives the first and second moments as

$$\mu_1' \equiv E(m) = \Lambda D,$$

$$\mu_2' = \Lambda D + (\Lambda D)^2(1 + \gamma^2), \qquad \textbf{(A5a)}$$

so that the variance of counts in working memory under both Poisson and scalar sources of variation is given by

$$\mathrm{Var}(m) = \mu_2' - \mu_1'^2 = \Lambda D + (\gamma\Lambda D)^2. \qquad \textbf{(A5b)}$$

The distribution form corresponding to the moment-generating function for the mixture **(A4)** is not normal, but approaches normality rapidly as T grows. In the ranges we will discuss, the forms are reasonably well approximated by normal distributions with mean and variance given by Equations A5a and 5b.

[c]Subscripting will be dropped henceforth where the parameter dependence is obvious, and picked up again where needed. In the present case we assume the stimulus is on for an arbitrary but fixed period of time, T sec, and thus dependence on T will be suppressed.

We now wish to allow variability in the latency to open and close the switch. That is, we wish to obtain the mean and variance of the mixture that results from allowing d to vary normally about a mean, $D = T - T_0$ with standard deviation σ_0. The conditional rules for mean and variance of x conditioned on y^d may be written

$$E(x) = E[E(x_y)], \quad \textbf{(A6a)}$$

$$\mathrm{Var}(x) = E[\mathrm{Var}(x_y)] + \mathrm{Var}[E(x_y)]. \quad \textbf{(A6b)}$$

For the present case, allowing switch variance results in a mean and variance for working memory given by

$$E(m) \equiv M = \Lambda D, \quad \textbf{(A7a)}$$

$$\mathrm{Var}(m) \equiv \sigma_m^2 = a(\Lambda\sigma_0)^2 + \Lambda D + (a - 1)(\Lambda D)^2, \quad \textbf{(A7b)}$$

where

$$a = 1 + \gamma^2. \quad \textbf{(A7c)}$$

Working memory may be construed as involving a proportional translation for short-term storage of the count in the accumulator. Such a translation might add another source of variance. This source was considered in our earlier piece[1] and is not discriminable in form from variance induced by pacemaker rate variability, and hence we do not analyze it separately here.

Translation to reference memory occurs only on a subset of the trials, but after long training we assume the reference memory has built up a distribution of remembered, reinforced working memory values via the proportional translation indicated in FIGURE 4. The conditional rule (**A6a**) gives, for $m^*_{k^*} = k^* m_{S+}$,

$$E(m^*) = E[E(m^*_{k^*})] \equiv M^*_{S+} = K^* \Lambda D_{S+}. \quad \textbf{(A8a)}$$

And a lengthy but straightfoward application of (**A6b**) gives, for $\sigma_{k^*} = \gamma^* K^*$,

$$\mathrm{Var}(m^*) = E[\mathrm{Var}(m^*_{k^*})] + \mathrm{Var}[E(m^*_{k^*})],$$

or

$$\sigma^2_{m^*} = K^{*2}[aa^*(\Lambda\sigma_0)^2 + a^*\Lambda D_{S+} + (aa^* - 1)(\Lambda D_{S+})^2], \quad \textbf{(A8b)}$$

where

$$a^* = 1 + \gamma^{*2}. \quad \textbf{(A8c)}$$

We are now in a position to analyze the decision rule, assuming independent, normally distributed memory variates.[e]

If we assume that the three variables in (**A1**) are independent and (essentially)

[d] Feller, W. 1966. An Introduction to Probability Theory and Its Applications. Vol. 1: 164. Wiley. New York, NY.

[e] The assumption of normality in the (product) memory variates violates somewhat the true character of these distributions, which have some positive skew. Again, however, in the range of seconds to minutes, this skew is not substantial and normal approximations are reasonable.

positive, the decision rule may be written

$$1 - b < \frac{m}{m^*} < 1 + b. \quad \textbf{(A9)}$$

The correlated upper and lower limits (window edges) do not pose special difficulties here and may be treated as independent variables (compare the Appendix in Church and Gibbon[3]).

Defining $w_i = 1 + (-1)^i b$ with mean and variance,

$$E(w_i) \equiv W_i = 1 + (-1)^i B,$$

$$\mathrm{Var}(w_i) = \sigma_b^2, \qquad i = 1, 2,$$

Equation A9 becomes

$$m^* w_1 < m < m^* w_2. \quad \textbf{(A10)}$$

Letting $x_i = m - m^* w_i$,

$$E(x_i) = M - M^*_{S+} W_i. \quad \textbf{(A11a)}$$

Repeated application of Equation A6b shows

$$\mathrm{Var}(x_i) = \sigma_m^2 + \sigma_{m^*}^2(\sigma_b^2 + W_i^2) + M^{*2}_{S+}\, \sigma_b^2, \quad \textbf{(A11b)}$$

where σ_m^2 and $\sigma_{m^*}^2$ are given by Equations A7b and A8b. Assuming the x_i are normal, we have

$$P(R \mid T) = \Phi(Z_2) - \Phi(Z_1), \quad \textbf{(A12)}$$

where

$$Z_i = \frac{-E(x_i)}{\sqrt{\mathrm{Var}(x_i)}}, \qquad i = 1, 2,$$

and Φ is the unit normal distribution function. It is convenient to divide numerator and denominator of the Z_i by M^*_{S+}, giving

$$Z_i = \frac{W_i - \dfrac{D_T}{K^* D_{S+}}}{\sqrt{\left(\dfrac{\sigma_m}{M^*_{S+}}\right)^2 + \left(\dfrac{\sigma_{m^*}}{M^*_{S+}}\right)^2 (\sigma_b^2 + W_i^2) + \sigma_b^2}}, \qquad i = 1, 2. \quad \textbf{(A13)}$$

Some algebraic rearrangement in Equations A7 and A8 gives

$$\left(\frac{\sigma_m}{M^*_{S+}}\right)^2 = \frac{1}{(K^* D_{S+})^2}\left[a\sigma_0^2 + \frac{D_T}{\Lambda} + (a - 1)D_T^2\right] \quad \textbf{(A14)}$$

and

$$\left(\frac{\sigma_{m^*}}{M^*_{S+}}\right) = \frac{1}{D_{S+}^2}\left[a^* a\sigma_0^2 + a^* \frac{D_{S+}}{\Lambda} + (a^* a - 1)D_{S+}^2\right], \quad \textbf{(A15)}$$

where a^* and a are as defined in Equations A7 and A8. The denominator of Equation A13 with $i = 2$ was used for the window-edge standard deviation in FIGURE 9. The form (**A12**), with definitions (**A13**), (**A14**), and (**A15**), was used to plot the generalization gradients in the text. The peak procedure gradients require the additional scale factor $\overline{R}_{max}$, translating response probability into response rate,

$$\overline{R}_T = \overline{R}_{max} P(R \mid T). \quad \textbf{(A16)}$$

Time-Left Procedure

The time-left procedure is analyzed in a similar fashion, but with two memory variates, m_S^* and m_C^*, corresponding to reinforced exposure to the S and C delays. The memories are built up in the same way as in temporal generalization and (**A8**) and (**A15**) hold for $S+ = S, C$.

The decision rule for choice for time-left may be written:

$$m_C^* - m_T < b' m_S^*. \quad \textbf{(A17)}$$

Defining

$$x = m_C^* - m_T - b' m_S^*,$$

$$P(R \mid T) = \Phi(Z), \quad \textbf{(A18)}$$

where

$$Z = \frac{-E(x)}{\mathrm{Var}(x)}, \quad \textbf{(A19a)}$$

and

$$E(x) = M_C^* - M_T^* - B' M_S^*, \quad \textbf{(A19b)}$$

$$\mathrm{Var}(x) = \sigma_{m_C^*}^2 + \sigma_{m_T}^2 + \sigma_{m_S^*}^2 (\sigma_{b'}^2 + B'^2) + M_S^{*2} \sigma_{b'}^2. \quad \textbf{(A19c)}$$

Again, it is convenient to divide numerator and denominator of (**A19a**) by M_S^*, giving

$$Z = \frac{B' - \dfrac{D_C}{D_S} + \dfrac{D_T}{K^* D_S}}{\sqrt{\left(\dfrac{\sigma_{m_C^*}}{M_S^*}\right)^2 + \left(\dfrac{\sigma_m}{M_S^*}\right)^2 + \left(\dfrac{\sigma_{m_S^*}}{M_S^*}\right)^2 (\sigma_{b'}^2 + B'^2) + \sigma_{b'}^2}}, \quad \textbf{(A20)}$$

where $(\sigma_{m_S^*}/M_S^*)^2$ is defined by (A15) and

$$\left(\frac{\sigma_{m_C^*}}{M_S^*}\right)^2 = \left(\frac{1}{D_S}\right)^2 \left[a^* a \sigma_0^2 + a^* \frac{D_C}{\Lambda} + (a^* a - 1) D_C^2\right]. \quad \textbf{(A21)}$$

This form (**A18, A20,** and **A21**) was used in fitting the psychometric functions in figures in the text.

To see why near superposition is achieved when $P(\text{“}C\text{”})$ is normalized by $T_{1/2}$, we consider only large S, C, with S/C constant, so that T_0 may be neglected. From (**A17**) $T_{1/2}$ is given by

$$T_{1/2} = \left(\frac{C}{S} - B'\right) \mathrm{K}^*S.$$

The numerator of (**A20**) becomes:

$$\left(\frac{C}{S} - B'\right)(T/T_{1/2} - 1),$$

which is constant at $T/T_{1/2}$.

The denominator terms may be analyzed separately. As S, C becomes large, the constant and Poisson components of variance become negligible and, from (**A21**)

$$\left(\frac{\sigma_{m_C^*}}{M_S^*}\right)^2 \rightarrow (a^*a - 1)\left(\frac{C}{S}\right)^2.$$

From (**A15**)

$$\left(\frac{\sigma_{m_S^*}}{M_S^*}\right)^2 \rightarrow (a^*a - 1),$$

and from (**A14**),

$$\left(\frac{\sigma_{m_T}}{M_S^*}\right)^2 \rightarrow (a - 1)\left(\frac{C}{S} - B'\right)(T/T_{1/2}).$$

Thus, the variance term approaches constancy for constant $T/T_{1/2}$ also, resulting in superposition in (**A18**).

Time Perception: Discussion Paper

R. DUNCAN LUCE

Department of Psychology and Social Relations
Harvard University
Cambridge, Massachusetts 02138

Study of the papers in this session reveals at least three major topics, each of which arises in three or more of the papers. They are: the quality of timing performance, clock models aimed at accounting for the variability in the behavior, and discovery of the scale for the subjective perception of time. I organize my remarks accordingly.

QUALITY OF TIMING PERFORMANCE

Perhaps the most obvious contrast in quality of timing is that between human and animal performance. In both cases considerable evidence is provided that mean times are quite accurate and that, to a first approximation anyhow, all distributions of normalized responses are the same. However, substantial differences exist between the animal and human data in the magnitude of the relative variability: the Weber fraction for the human data runs at about 5% and for the animals nearer to 50%.[1-3] However, in at least two respects these two classes of data are not comparable. First, the ranges over which they have been studied do not overlap, being between tens of milliseconds and a few seconds for the humans and from seconds to tens of seconds for the animals. Second, the pressure in the human experiments has been for precision of performance and it is far from clear that the animal studies have been designed with that in mind. The consquences for an animal who does not exhibit exact timing are really not very severe, being nothing worse than some unrewarded responding. Perhaps it would be useful for someone doing animal studies to attempt to establish the limits of their performance, which we have no reason to expect to be worse by an order of magnitude than that of people. And in the other camp, perhaps it would be useful to determine whether the 5% figure continues to hold into the region of tens of seconds. I do not underestimate the difficulties and effort required in each case, but both questions seem important.

All of us have been astonished by the precision of timing that Kristofferson and his associates have managed to achieve, and until he demonstrated it 15 years ago few of us would have anticipated that the variability in timing would remain constant or nearly so over any substantial region. What is new in the present data, and even more surprising, is the series of plateaus in estimates of the period of the clock (see below), which are spaced by factors of two over intervals that increase by factors of two. This, it seems to me, has important implications for modeling, to which I now turn.

CLOCK MODELS

A number of our authors envisage timing behavior as based upon a clock of some sort. Three of the papers[1,2,4] postulate a real-time digital device with timing arising from a count of the number of events. Hopkins and Kristofferson admit no variability in the clock itself, whereas Gibbon *et al.* explore various possibilities, rejecting as a primary source of the observed variability Poisson noise in the clock and favoring some form of scalar variability either in the clock or in the memory. (I should make clear that their finding that Poisson variability plays little role in timing in no way bears on the physiologically well justified Poisson representation of sensory intensity.) Assuming the period of the clock to be q, everyone agrees that the arrival of a signal will be random relative to the pulse train defining the clock, which introduces a uniformly distributed random variable over the interval $(0,q)$. Wing and Kristofferson[5] suggested that this is just one of three sources of variability leading to the observed variability, the other two being another, but independent, uniformly distributed one also on $(0,q)$, and the third an independent, normally distributed one associated with the response process. Hopkins has shown us that this model gives an almost perfect fit to his data; however, one would also like to see how well the data can be fit by other highly peaked distributions, such as the Laplace, instead of the triangular.

The major problem of that model, it seems to me, is this: Where does the second uniform distribution over $(0,q)$ come from? Hopkins attempted an argument along the following lines: After the count is achieved, the system exits the clock and initiates a response mechanism which is delayed in starting in much the same way the clock is, presumably because it cycles in a clock-like fashion. To fit the data, the two rates must be nearly the same, but to achieve approximate independence he assumed slightly different rates. This argument does not seem very persuasive to me, and I fear that it may run into difficulties with Kristofferson's findings about the plateaus.

Consider how the plateaus of variance may come about. One possibility is that the counter applied to the pulses of the clock has a maximum count, and when a time is wanted that exceeds the capacity of the counter, the system in essence counts every other pulse. This could be achieved by cells that are activated whenever two pulses occur within q time units but are refractory for considerably less than q time units, where we recall q is estimated to be about 12 msec. Such a model produces one uniformly distributed random variable on the interval $(0,2q)$, but I really don't see where the second one is to come from since there is no reason for the quantal character of the response process also to change scale. Because the second random variable seems to arise from exiting the clock and initiating the response process, its distribution should be controlled by the statistics of the response mechanism, not that of the timer. Once that dilemma is solved, then estimating even longer intervals simply involves repeated applications of the same type of cell that responds to every other pulse, but with even broader periods of integration. Such a mechanism generates the factors of two which Kristofferson has found. One cannot but wonder how many of these filtering cells can be arranged in series; presumably that can be estimated by extending Kristofferson's methods to appreciably longer times. It seems important to me that the distributions for $2q$ and $3q$ be studied with the same care the Hopkins has given q to see whether the fit of the convolution of a normal with two identical, independent uniform distributions continues to be equally satisfactory. For the reasons given above, I wonder if an asymmetry will not begin to evidence itself. If it does not, the second uniform distribution on $(0,2q)$ is an interesting theoretical challenge.

Before I turn to my last topic, let me say how pleased I am to find growing evidence for the existence of both good mental clocks and accurate mental counters, which some years ago David Green and I[6] suggested would provide a parsimonious account of some psychophysical speed–accuracy data.

SUBJECTIVE PERCEPTION OF TIME

When we turn to the subjective aspects of the perception of time, the only phrase that comes to my mind is "a can of worms." It is a familiar can to those who, like myself, have fished in psychophysical waters.

With the exception of Eisler (see below), those who have spoken of clock models have postulated periodicity in real time, and to the degree that the models are successful, which is considerable, that can be taken as *prima facie* evidence that at a certain level the perception of time is proportional to physical time. In this view, subjective scales are no more than useful constructs in a theory, and certainly many important constructs of physics—energy, momentum, entropy, and force—gained their status only via theory. However, as psychologists we have, in addition, strong intuitions about the lively existence of subjective attributes that cannot possibly be linear with the usual physical measures as well as the added knowledge that when we ask human subjects, more or less directly, about these attributes, we usually obtain results that are far from linear with physical measures. That makes suspect, but by no means rules out, the proportionality of subjective to physical time which is posited, with success, in these models.

Some[5] observe that the distributional data are describable as arising from a single distribution through scale changes, which is what Weber's law amounts to, and suggest that this in essence determines the needed transformation of time—which transformation is located in memory and not in the clock. This is the original strategy of Fechner, one that postulates a solution which, at least in psychophysics, has been found wanting an empirical basis.

So, one says, almost reflexively, let's decide the matter empirically. It is perhaps well to begin with the blunt admission that psychophysicists have never evolved a way to do so that has commanded wide assent. The direct scaling methods of S.S. Stevens[7] to which Eisler made reference, rest upon a mode of communication that is entirely language based; in fact, these methods rely upon the instruction to the subjects that the numbers they assign to stimuli shall preserve subjective ratios. In whatever way our subjects understand this instruction, they do give consistent, repeatable data; nonetheless, whatever the instruction does mean, we do not have the slightest idea how to communicate it nonverbally. Moreover, through the work of King and Lockhead,[8] we know that magnitude estimates are highly malleable, and quite different functions can be obtained by altering the feedback subjects receive. In brief, we simply do not know how to do scaling experiments with subjects who do not speak our language. Yet, that is exactly what two, and perhaps three, of our authors have claimed to be doing.[1,3,] Do I misunderstand and have they solved the century-old dilemma of the psychophysics of big differences, of what I call global psychophysics? I think not.

So far as I can tell, the researchers working with animals are doing temporal discrimination studies which, just like the discrimination studies of psychophysics, do not tell us much about the overall apprehension of an attribute. The fact that the indifference point between a variable and a standard time interval sometimes is approximately at the standard in no way implies that a linear scale in involved, and the fact that Weber's law holds does not dictate a particular nonlinear transformation. Eisler is quite aware that neither tack will do, but I believe he has slipped into two other traps. First, he has avoided reducing the problem to one of simple discrimination by assuming that the subject selects the second interval not to be equal to the first one, but to be subjectively one-half of the total interval. The motivation for this bit of indirection on the part of the animal, although not the author, escapes me. How does the animal know to use 1/2 rather than any other fraction? Second, and rather more serious, he has used the human data to establish a region within which he reinforced the animals' responses in what amounts to a discrimination study with 10 discriminative stimuli, and the animals—at least two of eight rats—quite reasonably took into account their own variability and stayed well within the reinforced region, thereby nearly reproducing the human behavior. Since we know from years of operant work that animals are quite sensitive to temporal reinforcement and from human work that magnitude estimation scales are malleable, these results persuade me of nothing whatsoever about temporal perception in animals.

I do not wish to disparage efforts toward finding objective ways of eliciting information about internal states, which is what I believe a subjective scale to be, but it is surely going to require a more complex idea than either just discrimination or just reinforcement. For the moment we may have to be satisfied with models of the sort that Kristofferson and his students and Gibbon and Church have been working on to account for these highly regular and, I believe, important temporal discrimination and timing data.

REFERENCES

1. GIBBON, J., CHURCH, R. M. & W. H. MECK. 1984. Clock, memory, and decision processes in different timing. This volume.
2. KRISTOFFERSON, A. B. 1984. Quantal and deterministic timing in human duration discrimination. This volume.
3. STUBBS, D. A., L. R. DREYFUS & J. G. FETTERMAN. 1984. The perception of temporal events. This volume.
4. HOPKINS, G. W. 1984. Ultrastable stimulus–response latencies: Towards a model of response-stimulus synchronization. This volume.
5. WING, A. M. & A. B. KRISTOFFERSON. 1973. Percept. Psychophys. **14:** 5.
6. GREEN, D. M. & R. D. LUCE. 1973. In Attention and Performance: IV. S. Kornblum, Ed.: 547–569. Academic Press. New York, NY.
7. STEVENS, S. S. 1975. Psychophysics. Wiley. New York, NY.
8. KING, M. C. & G. R. LOCKHEAD. 1981. Response scales and sequential effects in judgment. Percept. Psychophys. **30:** 559–603.
9. EISLER, H. 1984. Subjective duration in rats: The psychophysical function. This volume.

The Representation of Mental Activities in Critical Path Networks

RICHARD J. SCHWEICKERT

Department of Psychological Sciences
Purdue University
West Lafayette, Indiana 47907

Suppose a subject executes a number of mental processes to perform a task. We cannot observe the processes directly, but we can observe the time required to complete the task under various conditions. The problem is to determine, as far as possible, how the processes are organized and how much time each process takes. I will summarize an approach to this problem based on the theory of scheduling and on Sternberg's[1] additive factor method. Then this approach will be used to locate the decision process in several cognitive tasks.

The theory of scheduling deals with the optimal organization of processes in computers, factories, and so on.[2–5] One common type of process organization is the critical path network. Each process in the task is represented by an arrow in the network, and if process *a* must be completed before process *b* can start, then the arrow corresponding to *a* precedes the arrow corresponding to *b* (Fig. 1). No arrow is allowed to precede itself, and the network is said to be acyclic.

Processes in a critical path network are related in one of two ways. Those joined by a directed path, such as *a* and *b* in FIGURE 1, must be executed in order and are called *sequential*. Those not joined by a directed path, such as *a* and *c* in FIGURE 1, are called *concurrent* because they can be executed simultaneously. Note that processes in series and in parallel[6] are special cases of sequential and concurrent processes, respectively. Two sequential processes are in *series* if a process is on a path with one of them if and only if it is on a path with the other. Two processes are in *parallel* if they have the same starting and terminating points. For example, *a* and *c* are in parallel in FIGURE 1. We assume that no process can begin until all those preceding it are finished. Every process has a duration, and the duration of the task, the response time, is the sum of the durations of the processes on the longest path through the network, called the *critical path*.

As the critical path method is ordinarily used, the network is given, and the time required to complete the task is to be calculated. We have the opposite problem: We know the time required to complete the task under various conditions and want to construct the unknown network. The key to constructing the network is to use the idea from Sternberg's additive factor method of prolonging processes. The effects of such prolongations are surprisingly informative about the network.

The assumptions here differ from those of the additive factor method in that we allow for the possibility of two processes being executed concurrently. The assumptions are also different from those of McClelland's[7] cascade model. In his model, more than one process is in execution at a time, but the processes are sequential in the sense that information is passed continuously, from one process to the next, in order.

A further assumption will be made: that the process durations are fixed quantities and do not vary from trial to trial. Equations based on this assumption are only approximations to the correct stochastic equations, but with long prolongations and a large number of trials the approximations are not too bad.[8] Without this assumption, the problems are formidable, although some progress has been made.[9–11]

With Sternberg's[1] additive factor method, two experimental factors which increase reaction time are manipulated. Suppose that (a) all the processes of a task are executed in a sequence and that (b) each factor prolongs a different process. Then the effect on reaction time of prolonging both processes will be the sum of the effects of prolonging them individually. A violation of additivity has usually been interpreted in the framework of the method as indicating that (b) is false. But nonadditivity might indicate instead that (a) is false, and that processing is not entirely sequential. Nonadditivity turns out to be likely when separate processes are prolonged in a task involving concurrent processing. For such a task, of course, there is no reason to expect the principles of the additive factor method to apply without modification.

Latent Network Theory

The increase in response time produced by prolonging two processes depends on whether they are concurrent or sequential. Let T denote the response time when all the processes are at the shortest durations used in the experiment. Let $\Delta T(\Delta x, 0)$ denote the increase in T produced by prolonging process x by Δx, leaving y unchanged; other

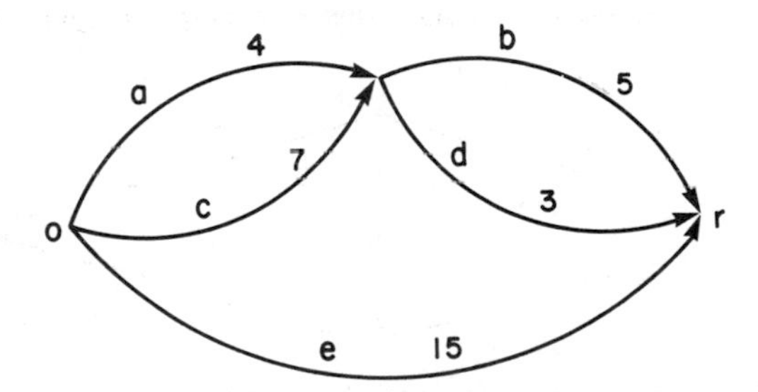

FIGURE 1. Each arrow represents a mental process that must be executed to perform a task. The numbers on the arrows are the process durations.

increases are denoted analogously. It can be shown that if x and y are concurrent, then

$$\Delta T(\Delta x, \Delta y) = \max\,[\Delta T(\Delta x, 0), \Delta T(0, \Delta y)]. \tag{1}$$

(All the equations in this section were derived in an earlier paper.[12])

The situation is more complicated if x and y are sequential. Suppose x precedes y. The amount of time by which x can be prolonged without making y start late is called the *slack* from x to y, written $s(xy)$. Similarly, the amount of time by which x can be prolonged without delaying the response, r, and thereby increasing the response time is called the *total slack* for x, written $s(xr)$. A process is on a critical path if and only if its total slack is zero. If all the processes are in a sequence, there is only one path, necessarily critical, so every process has zero total slack.

Slack is important when two sequential processes are prolonged. Suppose x precedes y on a path. If the prolongations Δx and Δy are not too small, then it can be shown that

$$\Delta T(\Delta x, \Delta y) = \Delta T(\Delta x, 0) + \Delta T(0, \Delta y) + k(xy), \tag{2}$$

where $k(xy) = s(xr) - s(xy)$ is called the *coupled slack* from x to y.

The magnitude of $k(xy)$ does not depend on the magnitudes of Δx and Δy. This fact provides a strong test of whether a network analysis applies to a given set of data: All values of Δx and Δy large enough for Equation 2 to hold should yield the same value for $k(xy)$, the interaction term.

If all the processes are sequential, $k(xy) = 0$ for every pair x and y, and Equation 2 becomes the additive relationship of the additive factor method. In general, however, Equations 1 and 2 indicate that when two separate processes in a network are prolonged, their effects will interact.

The Wheatstone Bridge

A negative value of $k(xy)$ is very informative. If x precedes y and $k(xy) < 0$, then the task network must have a subnetwork in the shape illustrated in FIGURE 2, called a Wheatstone bridge. Moreover, certain relationships hold among the path durations, although these are not relevant here. See Schweickert[12] for the details and proof.

A peculiarity of processes x and y arranged in a Wheatstone bridge with $k(xy) < 0$ is that small prolongations of x and y will result in the holding of Equation 1 rather than 2. That is, for small prolongations, x and y will mimic concurrent processes. This can only occur with a Wheatstone bridge. This mimicking of concurrent processes by sequential ones is the analog of the nonidentifiability discovered by Townsend,[6] that serial and parallel processes often cannot be distinguished on the basis of their completion time distributions.

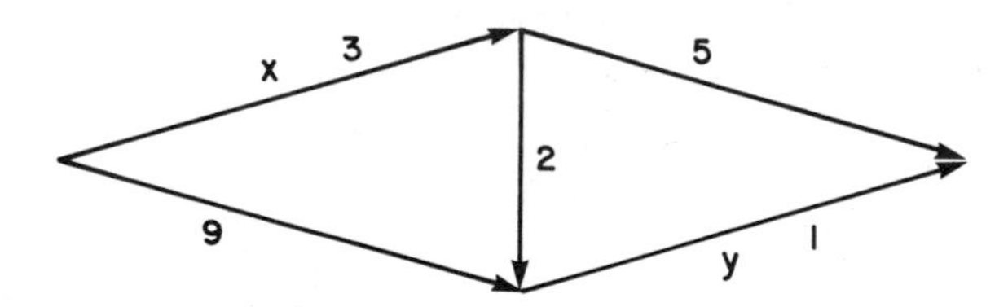

FIGURE 2. A Wheatstone bridge. If the coupled slack for x and y is negative, then the task network has a subnetwork of this shape.

Determining Processing Order

Experiments in which the subject makes two responses on every trial are also informative. Let the two response times be T_1 and T_2, both measured from the same point. Each response time considered alone will satisfy Equation 1 or 2 under the appropriate conditions. Furthermore, T_1 and T_2 are related. Suppose x precedes y, which precedes both responses, and the prolongations Δx and Δy are not too small. Then

$$\Delta T_1(\Delta x, \Delta y) - \Delta T_1(0, \Delta y) = \Delta T_2(\Delta x, \Delta y) - \Delta T_2(0, \Delta y). \tag{3}$$

If, instead, y precedes x, then $\Delta T_1(\Delta x, 0)$ and $\Delta T_2(\Delta x, 0)$ are required in Equation 3 in place of $\Delta T_1(0, \Delta y)$ and $\Delta T_2(0, \Delta y)$. The order of x and y is revealed, then, if one version of Equation 3 holds but not the other. If neither version holds, a network model is invalid.

There is another way to find the order of execution of processes. Suppose x precedes y which precedes z. If the prolongations Δx, Δy, and Δz are not too small, then the combined effect of prolonging all three processes is

$$\Delta T(\Delta x, \Delta y, \Delta z) = \Delta T(\Delta x, 0, 0) + k(xy) + \Delta T(0, \Delta y, 0) + k(yz) + \Delta T(0, 0, \Delta z). \quad (4)$$

This equation is useful for two reasons. First, since all the parameters in it can be determined by prolonging the processes individually and in pairs, the equation makes a prediction which can be tested. Second, the equation gives information about order: y is executed between x and z. To see this, note that for the three processes there are three coupled slacks, $k(xy)$, $k(yz)$ and $k(xz)$, but $k(xz)$, the one corresponding to the first and last of the three processes, is missing in the above equation. If the order were, say x, then z, then y, the missing term would be $k(xy)$.

The method summarized here has three major advantages over most other methods for analyzing reaction times: (1) It can be used even if the processes are not all in series. Furthermore, one can usually determine whether two mental processes are sequential or concurrent. (2) If processes are sequential, one can often determine their order of execution. (3) Information about the durations of processes is provided by the magnitudes of the coupled slacks, $k(xy)$ (see below).

APPLICATION: THE LOCATION OF DECISIONS IN COGNITIVE TASKS

The procedure just described will be applied to four information-processing tasks, each of which involves a decision, that is, a process prolonged by increasing the information in the stimulus.[13,14]

Digit-Naming

Two digit naming experiments will be discussed, one by Sternberg[20] (p. 296) and a closely related one by Blackman.[15] In Sternberg's experiment, subjects were presented visually with a digit and responded with a spoken digit. Three factors were manipulated: (i) Stimulus quality was degraded by superimposing a checkerboard pattern over the digit on some trials. Let E be the process prolonged by this procedure. (*ii*) The number of alternative digits was two in some blocks of trials and eight in others. We will call the process prolonged by increasing the number of alternatives the decision, D. (iii) On some trials the subject named the digit, and on other trials he named the digit plus one. Let Q be the process prolonged when the subject adds one to the digit.

The data are in TABLE 1. The effects of degrading the stimulus and of adding one are additive, indicating that each of those manipulations affects a separate process; the two processes are sequential. The effects of degrading the stimulus and increasing the number of alternatives interact, as do the effects of increasing the number of alternatives and adding one to the digit.

The additive factor explanation of these results assumes that the task is composed of a sequence of processes and degrading the stimulus prolongs one of them, E, while adding one to the digit prolongs another, Q. If increasing the number of alternatives prolonged a third process in the sequence, then the effects of this factor should be additive with the others. The explanation, according to the additive factor method, of the interactions which were found instead is that increasing the number of alternatives

TABLE 1. Changes in Response Times in Sternberg's Digit-Naming Experiment[a]

Factor Levels				
Stimulus Quality	Number of Alternatives	Transformation	Prolongations	ΔT
Degraded	8	$x + 1$	$\Delta E\ \Delta D\ \Delta Q$	197
Intact	8	$x + 1$	$\Delta D\ \Delta Q$	144
Degraded	2	$x + 1$	$\Delta E\ \ \Delta Q$	45
Intact	2	$x + 1$	ΔQ	18
Degraded	8	x	$\Delta E\ \ \Delta D$	97
Intact	8	x	ΔD	43
Degraded	2	x	ΔE	30
Intact	2	x	Baseline	0

NOTE: The baseline reaction time was 328 msec.
[a]Data are from Figure 8 of Sternberg.[1]

affects both E and Q. If we relax the assumption that all the processes are in a sequence, however, the interactions can be explained if each factor affects a process in a critical path network.

Suppose the processes are arranged in a network. The data indicate that Equation 2 holds for every pair of the processes E, D, and Q. The values of the coupled slacks are calculated from TABLE 1.

$$k(EQ) = \Delta T(\Delta E, 0, \Delta Q) - \Delta T(\Delta E, 0, 0) - \Delta T(0, 0, \Delta Q) = -3,$$
$$k(ED) = \Delta T(\Delta E, \Delta D, 0) - \Delta T(\Delta E, 0, 0) - \Delta T(0, \Delta D, 0) = 24,$$
$$k(DQ) = \Delta T(0, \Delta D, \Delta Q) - \Delta T(0, \Delta D, 0) - \Delta T(0, 0, \Delta Q) = 83.$$

These equations indicate that processes E, D, and Q are all on a path together. What is their order? Common sense suggests that E, the process affected by stimulus degradation, would come first. In his discussion of the experiment, Sternberg says that degrading the stimulus and having the subject add one to the digit probably affect processes that are widely separated. This suggests that D is in the middle, and the order is E, then D, and then Q. This order can be tested by Equation 4, which should hold in the following form if the order E, D, Q is correct:

$$\Delta T(\Delta E, \Delta D, \Delta Q) = \Delta T(\Delta E, 0, 0) + k(ED) + \Delta T(0, \Delta D, 0) + k(DQ) + \Delta T(0, 0, \Delta Q).$$

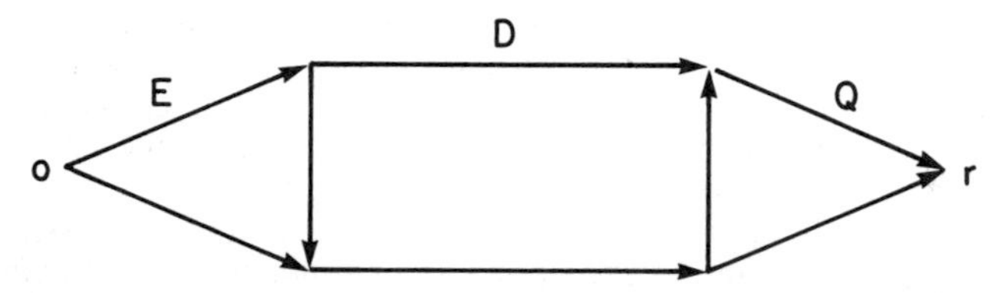

FIGURE 3. A network for the digit-naming experiments of Sternberg and Blackman. The onset of the digit is at o and the response is made at r. Process E is prolonged by stimulus degradation, D is the decision, and Q is prolonged when the subject adds one to the digit.

From TABLE 1,

$$197 \simeq 30 + 24 + 43 + 83 + 18.$$

Since the right hand side is 198, the error is 1 and the equation holds.

An equally good fit would be given by the order Q, D, E, although this order is unintuitive. Every other order does worse. The fact that Equation 4 holds for some order is, of course, support for a network model. A network representing the mental processes in this task is given in FIGURE 3. Details about its construction and about the process durations are in the APPENDIX.

A Related Digit-Naming Task by Blackman

Blackman[15] performed an experiment which is a replication of the one by Sternberg, with one change. Sternberg increased the number of alternative digits from 2 to 8 and thereby prolonged a process we called D. Blackman always presented three stimuli, but varied their probabilities. The most probable stimulus was presented 70% of the time and each of the other two was presented 15% of the time.

TABLE 2. Changes in Response Times in Blackman's Digit-Naming Experiment[15]

Factor Levels				
Stimulus Quality	Stimulus Probability	Transformation	Prolongations	ΔT
Degraded	.15	$x + 1$	$\Delta E\ \Delta D\ \Delta Q$	307
Intact	.15	$x + 1$	$\Delta D\ \Delta Q$	146
Degraded	.70	$x + 1$	$\Delta E\ \Delta Q$	137
Intact	.70	$x + 1$	ΔQ	31
Degraded	.15	x	$\Delta E\ \Delta D$	227
Intact	.15	x	ΔD	76
Degraded	.70	x	ΔE	97
Intact	.70	x	Baseline	0

NOTE: The baseline reaction time was 445 msec.

We will suppose that the process affected by stimulus probability is the same process as that affected by the number of alternatives, since manipulating the latter is a special case of manipulating the former, and we will call this process the decision, D.

The data for Blackman's experiment are given in TABLE 2. The coupled slacks have the following values:

$$k(EQ) = \Delta T(\Delta E, 0, \Delta Q) - \Delta T(\Delta E, 0, 0) - \Delta T(0, 0, \Delta Q) = 9,$$

$$k(ED) = \Delta T(\Delta E, \Delta D, 0) - \Delta T(\Delta E, 0, 0) - \Delta T(0, \Delta D, 0) = 54,$$

$$k(DQ) = \Delta T(0, \Delta D, \Delta Q) - \Delta T(0, \Delta D, 0) - \Delta T(0, 0, \Delta Q) = 39.$$

Since Equation 2 holds for each pair of processes, E, D, and Q are all on a path together. To be consistent with the results found in Sternberg's experiment, their order should be E, then D, then Q. Equation 4 should then hold in the form

$$\Delta T(\Delta E, \Delta D, \Delta Q) = \Delta T(\Delta E, 0, 0) + k(ED) + \Delta T(0, \Delta D, 0) + k(DQ) + \Delta T(0, 0, \Delta Q).$$

From TABLE 2,

$$307 \simeq 97 + 54 + 76 + 39 + 31,$$

the error is 10, and the equation holds.

Since the experiments of Sternberg and Blackman are so similar, one would expect that the networks representing them would be similar, and in fact, the network in FIGURE 3 represents both; the details are in the APPENDIX.

A Memory-Scanning Task

The experiments by Sternberg and Blackman support, through Equation 4, the idea that stimulus quality, stimulus probability and stimulus–response compatibility

TABLE 3. Response Times in the Experiment of Miller and Anbar[16]

	Stimulus Probability		
Degradation	Baseline .31	$\Delta_1 D$ .15	$\Delta_2 D$ .04
	Unexpected		
$\Delta_2 E$ (dim)	849	894	925
$\Delta_1 E$ (dots)	709	786	859
Baseline (normal)	648	659	731
	Expected		
$\Delta_2 E$ (dim)	757	780	823
$\Delta_1 E$ (dots)	622	643	685
Baseline (normal)	547	551	585

NOTE: Data published in Figures 4 and 5 of Miller and Anbar.[16] Numerical values from Miller, J. Personal communication, May 14, 1982.

affect processes arranged in a critical path network. Further support, this time through Equation 2, comes from an experiment by Miller and Anbar[16] (experiment II).

The task was memory scanning using letters as stimuli. Three letters were in the positive set. On each trial, the subject was presented with a probe letter and indicated, by pressing a button, whether the probe was an element of the positive set or not. Stimulus probability and stimulus quality were manipulated. As above, we denote the process affected by stimulus quality as E and that affected by stimulus probability as D. A third factor, expectancy, was also manipulated.

The effects of decreasing expectancy do not agree with the equations we would expect to hold if it were prolonging a single process in a critical path network, although a critical path network describes the effects of the other two factors. The effects of expectancy were somewhat irregular, and I will not discuss them further, although the data are given in TABLE 3 for the interested reader. The data discussed here are for the condition in which the stimuli were expected.

Three levels of stimulus probability were used, .31, .15 and .04. Stimulus encoding was affected in two different ways, by decreasing contrast with a filter over the screen and by superimposing dots over the stimuli. The data are consistent with the idea that changing the contrast and superimposing the dots are two levels of a factor that prolongs a single process, E. It will be convenient for us to accept this idea, although an analysis of the results carried out without this assumption would also support a network model.

The data are in TABLE 3. Note that when the letters are presented normally, there is little effect of changing stimulus probability from .31 to .15, but there is an effect of changing stimulus probability from .31 to .04. The network model explanation is that D, the process prolonged by decreasing stimulus probability, has slack. A small prolongation of D has no effect on response time, while a larger prolongation has an effect.

Let $\Delta_1 D$ be the amount by which D is prolonged when stimulus probability is decreased from .31 to .15, and let $\Delta_2 D$ be the amount by which D is prolonged when stimulus probability is decreased from .31 to .04. The combined effects of changing stimulus quality and stimulus probability are described by Equation 2. That is,

$$T(\Delta_1 E, \Delta_2 D) - T(\Delta_1 E, 0) - T(0, \Delta_2 D) + T(0, 0) = k(ED)$$
$$685 - 622 - 585 + 547 = 25.$$

Furthermore, in accordance with the model, the larger level of prolongation of E, $\Delta_2 E$, leads to the same value for $k(ED)$,

$$T(\Delta_2 E, \Delta_2 D) - T(\Delta_2 E, 0) - T(0, \Delta_2 D) + T(0, 0) = k(ED)$$
$$823 - 757 - 585 + 547 = 28.$$

Since the two values of $k(ED)$ are about the same ($25 \simeq 28$), the equations above support the idea that E and D are sequential processes in a critical path network.

At the normal level of stimulus quality, changing stimulus probability from .31 to .15 had little effect on response time. But TABLE 3 shows that beyond the normal level, the effects of changing stimulus quality and of decreasing stimulus probability have additive effects on response time. The additivity is found for decreasing stimulus probability from .31 to .15, and also for decreasing it from .31 to .04. In other words, the coupled slack between E and D is 0, and in accordance with Equation 2 this value does not depend on the level of prolongation of D. It is likely that at these levels for the duration of E and D, E and D are critical processes. The coupled slack is zero because factors prolonging critical processes are additive factors.

A Dual Task

I will now discuss the location of a decision in a more complex task.[17] In a dual tone and digit identification task, Becker[18] investigated the joint effects of the size of the interstimulus interval and the difficulty of the tone decision. The digit 1 or 2 was presented visually, and after an interstimulus interval of either 90 or 190 msec a high- or low-frequency tone was presented. With his left hand the subject pressed one of two buttons to indicate which digit occurred. With his right hand he responded to the tone. The decision about the tone was manipulated by requiring either one or two alternative responses to the tone. In the one alternative condition, the subject pressed the same button when either tone occurred, and in the two alternative condition he pressed one of two buttons to indicate which tone occurred.

Let I be the interstimulus interval and let N be the process prolonged by increasing the number of alternatives for the tone response. We will call N the decision about the tone. When there are two responses, it is customary to measure the reaction time for the second response from the onset of the second stimulus. For our purposes, it is convenient to measure all the times with respect to the same point, namely the onset of the first stimulus, the digit. Let o_d denote this point. Let $T_d(0, 0) = 346$ be the baseline response time to respond to the digit and let $T_n(0, 0) = 475$ be the baseline response time to respond to the tone, both time intervals measured with respect to o_d.

The time to respond to the digit when I and N are prolonged will be denoted $T_d(\Delta I, \Delta N)$. The increase in response time to the digit when I and N are prolonged will be denoted $\Delta T_d(\Delta I, \Delta N) = T_d(\Delta I, \Delta N) - T_d(0, 0)$, and so on. (In each condition T_d is the same as RT1 in Table 1 of Becker's article, and T_n is RT2 plus the appropriate ISI.) TABLE 4 gives the response times and changes in response times in the various conditions.

Additive Factors

The two experimental factors have additive effects on response time. For the digit responses (see TABLE 4),

$$\Delta T_d(\Delta I, \Delta N) = \Delta T_d(\Delta I, 0) + \Delta T_d(0, \Delta N),$$
$$113 \simeq 21 + 87,$$

with an error of 5. A similar equation holds for the tone responses

$$182 = 37 + 145,$$

with an error of 0.

The additivity is evidence that the interstimulus interval and the decision about the tone are sequential processes. If all the processes in the task were sequential, though, then when I is prolonged, each response time should increase by however much I increases. But when I is prolonged by 100, T_d is only increased by 21 and T_n by 37. What happened to the rest of the time by which I was prolonged?

A solution is to assume that some process, or sequence of processes, is executed concurrently with I and takes longer than I, as illustrated in FIGURE 4, where U indicates this sequence. In the figure, o_d and o_n are the onsets of the digit and tone, respectively. If the lower path takes longer than the upper path by about 79, then when I is prolonged by 100, the first 79 are expended in simply making the upper and lower paths equal in duration, and the remaining 21 yield the observed increase in T_d.

The lower path U of processes concurrent with I is drawn as preceding N in FIGURE

TABLE 4. Changes in Response Times in Becker's Digit and Tone Identification Task[18]

Factor Levels				
ΔI	Tone Responses	Prolongations	ΔT_d	ΔT_n
190	2	$(\Delta I, \Delta N)$	113	182
90	2	$(0, \Delta N)$	87	145
190	1	$(\Delta I, 0)$	21	37
90	1	Baseline	0	0

NOTE: The baseline digit response time was 346, and the baseline tone response time was 475.

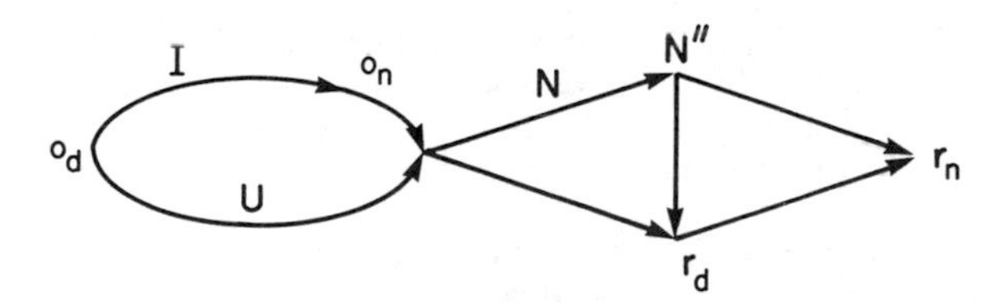

FIGURE 4. A network for Becker's digit and tone identification task. The digit is presented at o_d and the tone at o_n. The interstimulus interval is I and the decision about the tone is N. Responses to the digit and tone are made at r_d and r_n, respectively.

4 for the following reason: Since the effects of prolonging I and N are additive for both reaction times, $k_d(\text{IN})$ and $k_n(\text{IN})$ are both zero, that is, the slack for I with respect to each response. The network in FIGURE 4 is the simplest way of achieving these equalities.

Since U begins before the tone has been presented, U is probably involved in processing the digit. A role for U is suggested by Welford's[19,20] single-channel theory, which says that when a subject is presented with two stimuli, he can only make a decision about one of them at a time, although peripheral processing of the two stimuli might overlap. A clear explication of this theory is provided in an excellent review by Kerr.[21] It may be, then, that U involves the decision about the digit, and must be completed before the decision about the tone can start.

The Order of I and N

Since I and N are on a path together, Equation 3 can be used to establish their order. Common sense suggests that I precedes N, since the decision about the tone can only occur after I has ended and the tone is presented. If, contrary to this idea, N precedes I, we would expect Equation 3 to hold in the form

$$\Delta T_d(\Delta I, \Delta N) - \Delta T_d(\Delta I, 0) = \Delta T_n(\Delta I, \Delta N) - \Delta T_n(\Delta I, 0),$$

but the error would be 53 (see TABLE 4), so N does not precede I.

On the other hand, if I precedes N we would expect Equation 3 to hold in the form

$$\Delta T_d(\Delta I, \Delta N) - \Delta T_d(0, \Delta N) = \Delta T_n(\Delta I, \Delta N) - \Delta T_n(0, \Delta N),$$

and from TABLE 4,

$$113 - 87 \simeq 182 - 145.$$

Here the error is only 11, so the evidence supports the common sense order I then N.

Many more details about the network can be determined; in particular, several relationships between the durations of paths in the network can be derived, the details of which are in the appendix.

DISCUSSION

Factorial experiments on reaction time are commonplace due to the influence of Sternberg's[1] additive factor method. But it is rare for an investigator to use more than two levels of a factor or to use three or more factors. Each level of a factor which is included doubles the size of the experiment, so naturally such experiments are unattractive unless one is looking for something specific. Such experiments are important for testing network models because of the very specific predictions made in Equations 1–4.

The first three experiments discussed were similar in that each manipulated a factor affecting encoding and a factor affecting the information in the stimulus. The results from all three support the idea that these factors prolong different mental processes, provided that we allow for the possibility that more than one process can be executed at a time. In each of the experiments at least one of the network equations was tested and found to hold. Furthermore, two similar experiments, the one by Blackman and the one by Sternberg, led to the same network representation.

More experiments are needed to test the network models, and to determine the principles underlying their forms. One principle is the single-channel hypothesis that a subject can make only one decision at a time.[19,20,23] There may be other limitations at work in the peripheral processing. For instance, Fisher[22] has provided evidence that in search tasks there are a limited number of processors available—perhaps four—and hence there may be a limit on the number of encoding processes that can be executed concurrently. Discovering the arrangements of processes in cognitive tasks is a challenging problem, and discovering the principles underlying these arrangements will be even more so.

REFERENCES

1. STERNBERG, S. 1969. The discovery of processing stages: Extensions of Donders' method. *In* Attention and Performance: II. W. G. Koster, Ed. North-Holland. Amsterdam.
2. COFFMAN, E. G., Ed. 1976. Introduction to deterministic scheduling theory. *In* Computer and Job-Shop Scheduling Theory. John Wiley. New York, NY.
3. CONWAY, R. W., W. L. MAXWELL & L. W. MILLER. 1967. Theory of Scheduling. Addison-Wesley. Reading, MA.
4. KELLEY, J. E. & M. R. WALKER. 1959. Critical path planning and scheduling. *In* Proceedings of the Eastern Joint Computer Conference, Boston, MA: 160–173.
5. MALCOLM, D. G., J. H. ROSEBOOM, C. E. CLARK & W. FAZAR. 1959. Applications of a technique for research and development program evaluation. Operations Res. **7:** 646–669.
6. TOWNSEND, J. T. 1972. Some results concerning the identifiability of parallel and serial processes. Br. J. Math. Statis. Pyschol. **25:** 168–199.
7. MCCLELLAND, J. L. 1979. On the time relations of mental processes: An examination of processes in cascade. Psychol. Rev. **86:** 287–330.
8. SCHWEICKERT, R. 1982. The bias of an estimate of coupled slack in stochastic PERT networks. J. Math. Psychol. **26:** 1–12.
9. CHRISTIE, L. S. & R. D. LUCE. 1956. Decision structure and time relations in simple choice behavior. Bull. Math. Biophys. **18:** 89–112.
10. FISHER, D. L. & W. M. GOLDSTEIN. Stochastic PERT networks as models of cognition: Derivation of the mean, variance and distribution of reaction time using order-of-processing (OP) diagrams. J. Math. Psychol. In press.
11. MCGILL, W. J. & J. GIBBON. 1965. The general-gamma distribution and reaction times. J. Math. Psychol. **2:** 1–18.
12. SCHWEICKERT, R. 1978. A critical path generalization of the additive factor method: Analysis of a Stroop task. J. Math. Psychol. **18:** 105–139.
13. DONDERS, F. C. 1868. [Over de snelheid van psychische processen. Onderzoekingen gedaan in het Physiologisch Laboratorium der Utrechtsche Hoogeschool. Tweede Reeks.] II: 92–120. *In* Attention and Performance: II (1969). Translated by W. G. Koster (Ed.). North-Holland. Amsterdam.

14. HICK, W. E. & A. T. WELFORD. 1956. Comments on "Central inhibition: Some refractory observations" by A. Elithorn and C. Lawrence. Q. J. Exp. Psychol. **8:** 39–41.
15. BLACKMAN, A. R. 1975. Test of the additive factor method of choice reaction time analysis. Percep. Motor Skills **41:** 607–613.
16. MILLER, J. & R. ANBAR. 1981. Expectancy and frequency effects on perceptual and motor systems in choice reaction time. Memory Cognition **9:** 631–641.
17. SCHWEICKERT, R. 1980. Critical path scheduling of mental processes in a dual task. Science **209:** 704–706.
18. BECKER, C. A. 1976. Allocation of attention during visual word recognition. J. Exp. Psychol. Human Percept. Perform. **2:** 556–566.
19. WELFORD, A. T. 1952. The "psychological refractory period" and the timing of high-speed performance—a review and a theory. Br. J. Psychol. **43:** 2–19.
20. WELFORD, A. T. 1967. Single-channel operation in the brain. Acta Psychol. **27:** 5–22.
21. KERR, B. 1973. Processing demands during mental operations. Memory Cognition **1:** 401–412.
22. FISHER, D. L. 1982. Limited channel models of automatic detection: Capacity and scanning in visual search. Psychol. Rev. **89:** 662–692.
23. SCHWEICKERT, R. 1983. Latent network theory: Scheduling of processes in sentence verification and the Stroop effect. J. Exp. Psychol. Learn. Memory Cognition. **9:** 353–383.

APPENDIX

Details about the Experiments of Sternberg and Blackman

Digit-Naming Network

Let x' and x'' denote the starting point and terminating point, respectively, of a process x. There are three properties required by any critical path network representing the data in TABLE 1. (a) None of the processes E, D, or Q is on the critical path. (b) The longest path from E'' to the response at r does not contain D'' or Q'. (c) The longest path from the stimulus presentation at o to Q' does not contain E'' or D'.

To see that (a) is true, consider process E. Since $k(ED) = s(Er) - s(ED) = 24$ is positive, the total slack for E is positive, and E cannot be critical. Similar arguments show that D and Q are not critical.

Part (b) follows from the following equation.[12] Suppose process x precedes process y. Then

$$k(xy) = \delta(or) - \delta(oy') - \delta(x''r) + \delta(x''y'), \tag{5}$$

where $\delta(uv)$ is the duration of the longest path between points u and v.

If Q' were on the longest path from E'' to r, then

$$\delta(E''r) = \delta(E''Q') + \delta(Q'r).$$

Then, by Equation 5, it is easy to show that $k(EQ) - k(DQ) \geq 0$. However,

$$k(EQ) - k(DQ) = -3 - 83 < 0,$$

contradicting the idea that Q' is on the longest path from E'' to r. Similar arguments complete the demonstration of (b) and (c).

Path Durations

Since the reaction time is $\delta(or) = 328$, Equation 5 provides information about the durations of paths. For example, since $k(ED) = 24$,

$$328 - k(ED) = 304 = \delta(oD') + \delta(E''r) - \delta(E''D').$$

The other two coupled slacks lead to similar equations.

More Information about the Network in Becker's Tone and Digit Experiment

Two new equations will be introduced here. First, if process x is prolonged by Δx, then if Δx is smaller than $s(xr)$, no increase in the reaction time occurs. Otherwise, the increase in reaction time is $\Delta x - s(x, r)$. That is,

$$\Delta T(\Delta x) = \max\{0, \Delta x - s(x, r)\}. \quad (6)$$

Second, the total slack for x can be written in terms of path durations as follows:

$$s(xr) = \delta(or) - \delta(ox') - \delta(x) - \delta(x''r). \quad (7)$$

Estimates of Slack

In Becker's experiment the size of the interstimulus interval, I, is known because it was directly controlled by the experimenter. With this information, Equation 6 can be used to estimate the size of the total slack for I. Let r_d *and* r_n be the points at which the responses to the digit and tone, respectively, are made. By Equation 6,

$$\Delta T_d(\Delta I, 0) = \Delta I - S(I, r_d).$$

Because of the experimental procedure, $\Delta I = 100$. Therefore, $s(I, r_d) = 79$. Similarly, $s(I, r_n) = 63$. Since these two estimates are so close, we will assume the slacks are equal, and use the average, 71, as the estimate,

$$s(I, r_d) = s(I, r_n) \simeq 71.$$

Location of the Two Responses

Process N precedes both r_d and r_n because prolonging N prolongs both reaction times. Furthermore, r_d is not on the longest path from N'', the terminus of N, to r_n. To see this, note that

$$\begin{aligned}\Delta T_d(0, \Delta N) - \Delta T_n(0, \Delta N) &= \Delta N - s(N, r_d) - \Delta N + s(N, r_n)\\ &= \delta(o_d, r_n) - \delta(N'', r_n) - \delta(o_d, r_d) + \delta(N'', r_d),\end{aligned}$$

by Equations 6 and 7. If r_d were on the longest path from N'' to r_n, then

$$\delta(N'', r_n) = \delta(N'', r_d) + \delta(r_d, r_n),$$

so

$$\Delta T_d(0, \Delta N) - \Delta T_n(0, \Delta N) = \delta(o_d, r_n) - \delta(o_d, r_d) - \delta(r_d, r_n) \geq 0.$$

But $\Delta T_d(0, \Delta N) - \Delta T_n(0, \Delta N) = 87 - 145 = -58 < 0$, contradicting the idea that r_d is on the longest path from N'' to r_n. Therefore, there is a path from the end of N to r_n, not containing r_d (see FIGURE 4).

Since the subjects were instructed to respond to the digit first, then the tone, I indicate that r_d precedes r_n; this is not necessary to explain the data, however.

I will now show that (a) the longest path from o_d to r_d does not contain N, and (b) the longest path from o_n to r_d does not contain N. I have shown in the equations above that $s(N, r_d) - s(N, r_n) = 58$, so $s(N, r_d)$ is greater than zero, and there must exist a path from o_d to r_d not containing N. Hence, (a) is true. I have also shown above that $s(I, N') = s(I, r_d)$ and by Equation 6 it can be shown that

$$\delta(o_n, r_d) - \delta(o_n, N') = \delta(o_d, r_d) - \delta(o_d, N').$$

Substituting the left hand side of the above expression into the following expression (Equation 7),

$$s(N, r_d) = \delta(o_d, r_d) - \delta(o_d, N') - \delta(N) - \delta(N'', r_d) > 0,$$

we obtain

$$\delta(o_n, r_d) - \delta(o_n, N') - \delta(N) - \delta(N'', r_d) > 0,$$

so (b) is truc. Perhaps the simplest way to incorporate propositions (a) and (b) into the model is with the network of FIGURE 4.

In order to account for all the time elapsing between the onset of the first stimulus and the onset of the last response, one would like to know the duration of each process in the network. It is not possible with these data to completely determine the durations, but we can come close. The duration of the longest path between every two points in the above network can be expressed in terms of four parameters, $\delta(o_n, N')$, $\delta(N)$, $\delta(N'', r_d)$ and $\delta(r_d, r_n)$, and an upper bound can be found for each parameter. The details are left to the reader.

Timing Perturbations with Complex Auditory Stimuli[a]

DONALD G. JAMIESON, ELZBIETA SLAWINSKA,
MARGARET F. CHEESMAN, AND
BLAS ESPINOZA-VARAS

Department of Psychology
The University of Calgary
Calgary, Alberta, T2N 1N4 Canada

In a duration-discrimination task, the order in which stimuli are presented can have a substantial effect on accuracy. For example, subjects are more often able to select the longer of a pair of durations, say 340 msec followed by 300 msec, when the stimuli are presented in that order (that is, longer followed by shorter) than in the reverse order (300 followed by 340).[1,2] Such a presentation order effect is called a positive time-order error (TOE). Time-order errors are not unique to duration comparisons, of course, since they have been reported with a number of stimulus modalities, including brightness and loudness, but they are particularly sizable with durations. For example, for brief-duration stimuli, such as those described above, the time-order error may produce a difference in the proportion of correct responses between the two presentation orders of 0.3, or more.[3]

Positive TOEs are thus well documented for the case of judgments of the overall duration of pairs of stimuli.[2,4] In other situations, when the overall set of stimuli covers a substantial range, it is well established that another form of presentation order effect—the assimilation effect—is likely to occur when subjects judge overall duration.[1,2,5] In a number of interesting perceptual situations, however, the listener must judge the duration of components or segments of complex stimuli. In one class of such complex stimuli, speech sounds, the duration of an acoustic event may provide an important cue to phonetic class. In fact, one such distinction—voice-onset time (VOT)—is arguably the most commonly studied variable in speech perception research.[6-8] VOT discrimination between /ba/ and /pa/ requires, among other things, attention to the duration of events occurring within the first 80 to 100 msec of a syllable that is 300 to 400 msec in length. It is therefore relevant to ask (a) whether TOE and/or assimilation effects occur for such "duration judgments," and (b) how they may compare with the effects known to occur with judgments of overall stimulus duration.

The three experiments reported in this study sought to answer these questions. The first experiment examined judgments with synthetic /ba/–/pa/ syllables differing in VOT. The second experiment examined judgments with pure-tone stimuli consisting of a frequency glide plus a steady-state tone. Here the variable of interest was the duration of the initial transition. The third experiment examined judgments of the identity of two tones containing differences in the duration and/or in the frequency of one tone of the pair. These experiments also allow us to examine the influence of the temporal and spectral complexity of the stimuli on the TOE effects.

[a]This work was supported by grants from the Natural Sciences and Engineering Research Council and the Alberta Heritage Foundation for Medical Research (to D.G.J.)

EXPERIMENT 1: PRESENTATION ORDER EFFECTS WITH SPEECH STIMULI VARYING IN VOICE-ONSET TIME

In articulatory terms, voice-onset time is the interval of time separating the release of the stop from the onset of vocal cord vibration. Acoustically, the release is indicated by a brief noise burst, and vocal cord vibration is indicated by periodicity. Synthetic consonant vowel syllables varying in VOT differ in terms of the delay between onset of the burst and the onset of the periodicity. For "voiceless" consonant–vowel syllables (CVs), such as /pa/ and /ta/, voice-onset time values may reach 80–100 msec; for "voiced" CVs, such as /ba/ and /da/, voice-onset time values are less than 20 msec.

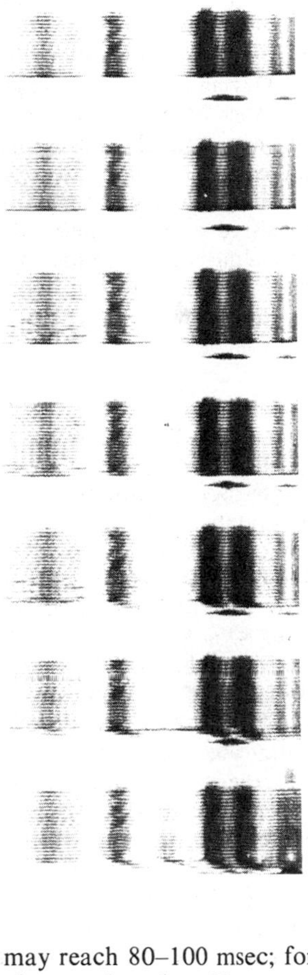

FIGURE 1. The seven synthesized speech stimuli used in Experiment 1. Stimuli towards the *left* have brief VOT values, and are heard as /ba/. Stimuli towards the *right* have longer VOT values, and are heard as /pa/.

Stimuli

A seven-stimulus voice-onset time continuum, ranging from /ba/ to /pa/, was synthesized using a digital speech synthesizer[9] implemented on a PDP11/34. The resulting stimuli are presented in FIGURE 1.

Method

Our procedure involved a "same–different" task, in which a standard stimulus was selected from the continuum, then followed by a comparison stimulus, also from the continuum. On 50% of the trials, the standard and comparison stimuli were physically identical; on the remaining trials, the two stimuli differed by one step of 10 msec. The 10-msec difference could consist of either an increment or a decrement in VOT. Thus, on one-half of the different trials, the comparison stimulus was longer than the standard and on the other half, it was shorter than the standard. The task required the subject to indicate whether the two stimuli presented on any trial were "the same," or whether they were "different," by depressing one of two labeled response buttons. Stimulus presentation, response collection, and all aspects of timing and sequencing were controlled by a PDP11/34 computer.

During testing, the subject was seated comfortably in an IAC sound-attenuating chamber. Stimuli were presented to the right ear at 72 dB SPL, measured by a Bruel and Kjaer type 2118 sound-level meter, coupled to a type 4152 artificial ear. All three subjects had normal hearing, and previous experience listening to synthetic speech.

Subjects served for a total of four 140-trial experimental sessions. Each trial consisted of the followed sequence of events: an intertrial interval of 1000 msec preceding the first stimulus; the presentation of the first stimulus; a silent, 400-msec interstimulus interval; the presentation of the second (comparison) stimulus; a response interval of 4 seconds.

Results and Discussion

The traditional measure of the time-order effects which occur in duration comparison situations is obtained by comparing discrimination scores in the two presentation orders: "shorter, longer" (that is, when the shorter duration precedes the longer), denoted (S, L), and "longer, shorter," denoted (L, S). The statistic, $\text{TOE} = P(C|L, S) - P(C|S, L)$, is zero when there is no time-order error. It is greater than zero when the time-order error is "positive," and less than zero when the time-order error is "negative."

Since our interest is in the possibility of presentation order effects with these stimuli, we will confine our analysis to the time-order error index, $\text{TOE} = P(C|L, S) - P(C|S, L)$, where L indicates the longer VOT interval (that is, more pa-like stimulus), S indicates the shorter VOT interval (that is, more ba-like stimulus), and C indicates that the correct discrimination (different) was made. Thus, $P(C|L, S)$ indexes the proportion of correct responses observed when the longer duration of a stimulus pair was the first presented.

FIGURE 2 presents the value of this index as a function of the duration of the longer VOT of the pair for a typical subject. A clear trend is apparent in the data: the TOE index is uniformly positive for durations at the /ba/ end of the continuum. That is, a comparison VOT, 10 msec shorter than the standard is discriminated better than a comparison VOT 10 msec longer than the standard. This result is consistent with the notion that positive time-order errors should occur when brief durations are compared. However, the TOE index decreases regularly with increasing VOT duration, becoming negative with the longer VOT duration at the /pa/end of the continuum. This latter result is more consistent with the occurrence of assimilation effects[2,5] than with positive TOEs.

It is thus clear that the two fundamental presentation order effects observed when overall stimulus duration is compared are also observed with comparisons of only a portion of a complex acoustical stimulus. First, time-order errors occur even when the perceptual judgment refers to the duration of the initial events of a longer acoustic complex. Second, assimilation effects clearly occur in such a situation. As well, these results indicate that speech stimuli are susceptible to two of the "nuisance" variables of psychophysics, extending a result established with a selective adaptation speech paradigm.[7]

EXPERIMENT 2: DO PRESENTATION ORDER EFFECTS OCCUR WITH ANALOGUES TO SPEECH FORMANT TRANSITIONS?

The results of the previous experiment were obtained with stimuli that were both acoustically complex and capable of inducing a phonetic percept. Experiment 2

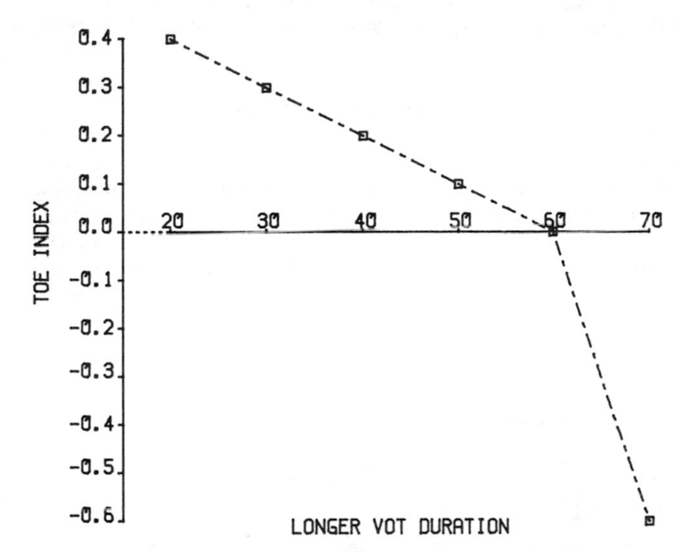

FIGURE 2. The time-order error index, TOE = $[P(C|L, S) - P(C|S, L)]$, for a typical subject, as a function of the longer VOT duration. In each case stimuli are adjacent speech stimuli on a synthesized /ba/–/pa/, VOT continuum. Thus, the longer duration always exceeds the shorter by 10 msec.

eliminated the phonetic component, while retaining some of the complex, time-varying properties of speech sounds. To accomplish this, we used stimuli consisting of an initial frequency glide followed by a steady-state portion, which resembled the variations in first-formant center frequency of a /ba/ sound, but were not heard as speech. These stimuli also have potential importance because such frequency transitions cue important phonetic distinctions such as place of articulation in stop consonants.[6] As well, researchers seeking to study the auditory precursors to speech perception have seized on the importance of sensitivity to frequency transitions, offering evidence that sensitivity to frequency change exists in distinct "frequency channels."[11]

Stimuli

The stimuli consisted of an initial frequency glide ranging from 400 to 700 Hz, followed by a steady-state tone of 550 Hz. The stimuli differed in terms of the rate of the frequency change of the initial glide or, equivalently, in transition duration. The 300-Hz frequency increment occurred over 20 msec for the fastest change, and over 80 msec for the most gradual change. The seven stimuli covered this range in 10-msec steps (that is, in 20, 30, 40, 50, 60, 70, and 80 msec). In each case, the steady-state portion of the stimulus was adjusted to fix overall stimulus duration at 200 msec. Listeners described these stimuli as "chirps" or "whistles"; they were not heard as speech-like by any listener.

Method

The testing procedure used the same–different paradigm, described for Experiment 1. All aspects of the testing situation were as described for that experiment.

Results and Discussion

The TOE index, TOE = $P(C|L, S) - P(C|S, L)$, where L indicates the stimulus with the longer glide duration and S indicates the shorter glide duration, was computed for each stimulus pair and plotted as a function of the duration of the longer glide. FIGURE 3 shows that this index is positive for the three briefest glide durations, but it becomes negative for the two longer glide durations. Improved performance is seen

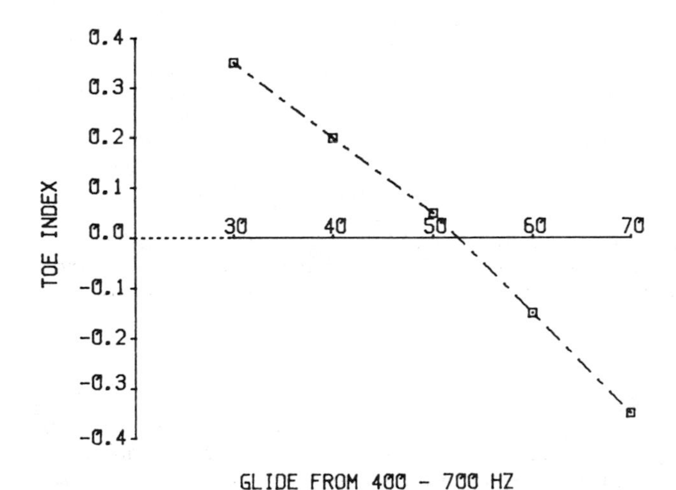

FIGURE 3. The time-order error index, TOE = $[P(C|L, S) - P(C|S, L)]$, for a typical subject, as a function of the longer glide duration. In each case stimuli are pairs of complex sounds, each consisting of a frequency glide from 400 to 700 Hz, followed by a steady-state pure tone at 550 Hz. In each pair, the longer glide duration exceeds the shorter by 10 msec. Overall stimulus duration was fixed at 200 msec in each case; stimuli towards the *left* of the figure have more rapid glides, while stimuli towards the *right* have slower glides.

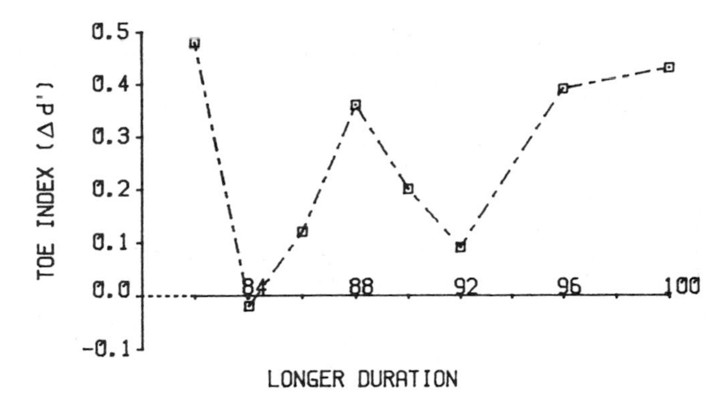

FIGURE 4. The time-order error index, TOE = $[(d' \mid L, S) - (d' \mid S, L)]$, where L is the longer duration of the pair, S is the shorter duration, and d' is the familiar measure of discriminability. Note that the shorter duration is fixed at 80 msec in each case.

when the second-presented stimulus was the more extreme of the pair and not routinely when the longer duration was the first presented. There is thus little or no evidence for a dominant, or even important positive time-order error. Rather, assimilation appears to be the major effect in this experiment.

The failure to find a substantial positive TOE in Experiment 2 limits the generality of the conclusions of Experiment 1. Positive TOEs may thus occur under two types of conditions: in comparisons of the overall duration of simple stimuli, and in comparisons of the duration of cues occupying a portion of sufficiently complex stimuli. In this view, the stimuli of Experiment 2 would be insufficiently complex; a further experiment using still simpler "complex" stimuli would not show positive TOEs.

The consistent occurrence of a substantial assimilation effect in Experiments 1 and 2 is also of interest, of course. Analyses of performance with speech stimuli and with speech-analogue stimuli rarely examine the possibility of such effects. As seen here, however, these effects can be substantial in magnitude, resulting in differences in the proportion of correct responses of as much as 40% between presentation orders. Assimilation effects accompany the use of a substantial stimulus range, however. To eliminate such effects in Experiment 3, we restricted the overall stimulus range substantially, relative to the preceding experiments.

EXPERIMENT 3: PRESENTATION ORDER EFFECTS WITH SPEECH ANALOGUES TO SPECTRAL/TEMPORAL FUSION

Stimuli

This experiment used pairs of pure-tone stimuli which (a) were identical 80-msec 1500-Hz sounds or (b) differed in the frequency of one tone and/or in the duration of the other tone. Stimuli were generated by analogue equipment and presented at 75 dB over matched and calibrated TDH-49 headphones.

Method

The experimental procedure was essentially as described for previous experiments, except that the interstimulus interval was 60 msec and the two conditions in which (a) the first tone of the pair could be incremented in duration and (b) the second tone of the pair could be incremented in duration were run sequentially as different experiments.

Results and Discussion

Performance for each stimulus in the two experimental conditions was first summarized by the d' statistic. The TOE index, TOE = $(d' \mid L, S) - (d' \mid S, L)$, where L indicates the longer duration of the pair and S indicates the shorter duration of the pair (which was fixed at 80 msec), and d' is the familiar discrimination measure of signal detection theory, is presented in FIGURE 4. It is clear from this figure that discrimination scores tend to be higher when the first-presented duration is the longer, consistent with the expectation based on explicit duration-comparison experiments.

GENERAL DISCUSSION

The present experiments show that positive time-order errors and/or assimilation effects occur when brief-duration stimuli are compared, even when duration is but a cue in part of a complex auditory stimulus. As such, these results have potential implications for experimenters whose major interest is far removed from the study of timing and duration perception. Of course, since these results suggest that such phenomena are by no means unique to the explicit duration-comparison situation, they may be considered encouraging to traditional duration-perception research.

REFERENCES

1. ALLAN, L. 1977. The time-order error in judgments of duration. Can. J. Psychol. **31:** 24–31.
2. JAMIESON, D. 1977. Two presentation order effects. Can. J. Psychol. **31:** 184–194.
3. JAMIESON, D. & W. PETRUSIC. 1976. On a bias induced by the provision of feedback in psychophysical situations. Acta Psychol **40:** 199–206.
4. JAMIESON, D. & W. PETRUSIC. 1975. Presentation order effects in duration discrimination. Percept. Psychophys. **17:** 197–202.
5. HELSON, H. 1964. Adaptation-level theory. Harper & Row. New York, NY.
6. ODEN, G. & D. MASSARO. 1978. Integration of featural information in speech perception. Psychol. Rev. **85:** 172–191.
7. DIEHL, R., M. LANG & E. PARKER. 1980. A further parallel between selective adaptation and contrast. J. Exp. Psychol. Human Percept. Perf. **6:** 24–44.
8. WOOD, C. 1976. Discriminability, response bias and phonemic categories in the discrimination of voice onset time. J. Acoust. Soc. Am. **60:** 1381–1388.
9. KLATT, D. 1980. Software for a cascade/parallel formant synthesizer. J. Acoust. Soc. Am. **67:** 971–995.
10. REPP, B., A. HEALEY & R. CROWDER. 1979. Categories and context in the perception of steady state vowels. J. Exp. Psychol. Human Percept. Perf. **5:** 129–145.
11. REGAN, D. & B. TANSLEY. 1979. Selective adaptation to frequency-modulated tones. J. Acoust. Soc. Am. **65:** 1249–1257.

Temporal Order and Duration: Their Discrimination and Retention by Pigeons

EDWARD A. WASSERMAN, ROBERT E. DELONG, AND MARK B. LAREW

Department of Psychology
The University of Iowa
Iowa City, Iowa 52242

The external environment contains a vast variety of energies which organisms may detect with specialized receptor organs. Beyond these highly specific sensory systems, however, organisms may perceive properties of single and multiple stimuli that are in some measure independent of the particular form of energy that is acting upon the sensory receptor.[1,2]

For instance, consider a patch of colored light projected on a nickel-sized disk. A pigeon may readily discriminate the hue, brightness, and shape of the patch. It may also discriminate how long the light has remained on. To the extent that duration is a discriminable property of the visual stimulus for the bird, we have reason to believe that the pigeon would similarly discriminate the duration of nonvisual stimuli, such as sounds, odors, and tastes. In short, duration is a property of every sensory stimulus. What we learn about duration discrimination in one modality ought to hold in another modality (for a possible exception see Spetch and Wilkie[3]).

Now, consider the successive presentation of a pair of stimuli. The two stimuli may either be the same or different. And, if they are different, they may occur in one of two temporal orders. Of course, to discriminate whether two stimuli are the same or different as well as to discriminate the temporal order of two nonidentical stimuli, sensory receptors must be stimulated. But here again, we expect that same–different and temporal-order discriminations ought to occur in all remaining sensory modalities, given successful demonstration in one. In fact, same–different and temporal order discriminations are so general that they ought to hold *across* as well as *within* all sensory systems.

Our interest in duration and temporal order discrimination has prompted a series of experiments using operant conditioning techniques and pigeon subjects. This work has persuaded us that pigeons very ably discriminate these features of visual stimuli. In addition, by inserting delays between the discriminative stimuli and the performance test, we have traced the retention of these discriminations over the course of several seconds. Duration and temporal order are thus both discriminable and rememberable stimulus properties for our avian subjects.

TEMPORAL ORDER

Our investigations into temporal-order discrimination[4–6] were initially inspired by the aim of definitively demonstrating control over behavior by this aspect of two nonidentical stimuli. As we pondered the problem, several possible paradigms fell short of our goal. Suppose, for example, that as stimulus pairs we were to use colored key lights ordered orange–green and green–orange. Were we to reinforce the pigeon's key pecks to a vertical-line test stimulus after orange–green pairs but not to do so after green–orange pairs, we would expect test-key pecking in the first case to exceed that in the second. But would we have clearly shown that stimulus order was the critical aspect of the stimulus pairs that was the basis for the bird's discrimination? No. The problem here is that selectively attending to either the first or to the second color of the pair would be sufficient for the pigeon to respond discriminately on orange–green and green–orange trials: for instance, to peck the vertical line if the second color was green, but not if it was orange.

One solution to this difficulty would be to intermix orange–orange and green-green pairs with trials involving orange–green and green–orange pairs. As before, pecks to the vertical-line test stimulus would result in food only after orange–green pairs. But now, selective attention to either the first or to the second color of the pair would fail to provide the bird with a basis for accurate discriminative responding.

Even with these modifications, pigeons can learn to respond at high rates on orange–green tests and at low rates on green–orange, orange–orange, and green–green tests (see experiments 1 an 2 in Weisman *et al.*[6] for such data with slightly different visual stimuli). This performance clearly indicates that a sequence of two colors presented in a specific temporal order can acquire discriminative control over a pigeon's behavior. Yet there is still a means by which the pigeon could respond discriminatively during the test periods and have no direct access to the identities of the prior color stimuli or to their temporal order. The bird might simply peck the key if the first color were orange and continue to peck into the test if the second color were green; if the first color were green or the second color were orange, the bird might become unresponsive or direct its behaviors away from the response key.[7] Stepwise discrimination of this sort may well afford the pigeon a simple mode of solving this complex task.

A basic problem with the second discrimination task is then that, because only the orange–green sequence is associated with reinforcement, the reversely ordered green–orange sequence can potentially be discriminated as involving nonreinforcement as soon as green is presented as the first color. Therefore, there is no assurance that the pigeon attends to both colors in a sequence and then responds differentially to the test stimulus on the basis of the temporal order of the preceding color pair.

Our solution to this problem was to devise the task outlined in TABLE 1. This task is a variant of a delayed, go/no-go conditional discrimination.[8–11] Here, orange–green and green–orange stimulus orders are associated with either reinforcement or nonreinforcement, depending upon which of two line orientations, vertical or horizontal, is

TABLE 1. Procedure Used in the Study of Temporal Order Discrimination

	Trial Sequence						
Trial Type	Color 1 (2.0 sec)	Interitem Interval (0.5 sec)	Color 2 (2.0 sec)	Retention Interval (0.5 sec)	Test Stimulus (5.0 sec)	Trial Outcome (2.5 sec)	Intertrial Interval (15.0 sec)
AA	Orange	—	Orange	—	Vertical	Blackout	—
BB	Orange	—	Orange	—	Horizontal	Blackout	—
AB	Orange	—	Green	—	Vertical	Food	—
BA	Orange	—	Green	—	Horizontal	Blackout	—
BA	Green	—	Orange	—	Vertical	Blackout	—
AB	Green	—	Orange	—	Horizontal	Food	—
BB	Green	—	Green	—	Vertical	Blackout	—
AA	Green	—	Green	—	Horizontal	Blackout	—

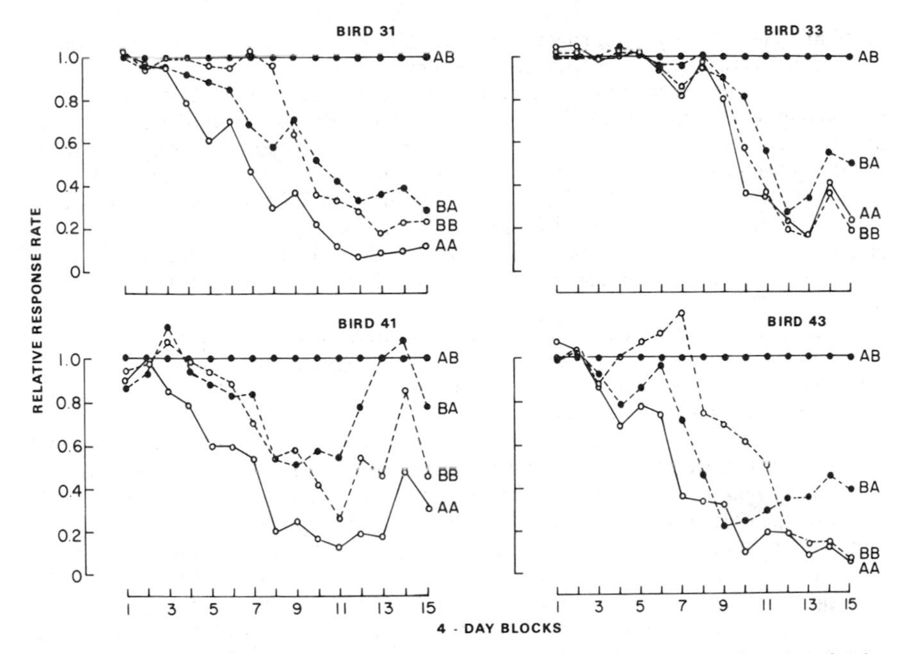

FIGURE 1. Relative rates of responding on *AB*, *BA*, *AA*, and *BB* test trials as a function of 4-day blocks of training. Relative response rates were calculated by dividing a given test rate of response by that obtained on *AB* test trials. (Modified from Weisman *et al.*[6])

presented as the test stimulus. As can be seen in TABLE 1, when an orange–green pair is followed by a vertical line or a green–orange pair is followed by a horizontal line (designated as *AB* trials in the first column), pecking to the test stimulus is reinforced by food. Alternatively, when an orange–green pair is followed by a horizontal line or a green–orange pair is followed by a vertical line (designated as *BA* trials), pecking to the test stimulus is followed by blackout. Thus, the order of color pairs associated with reinforcement is conditional upon the test stimulus. This conditionality sought to require the bird to remember the order of the colors until the test stimulus was presented, because a particular order of the two nonidentical colors could not be discriminated as reinforced or nonreinforced until that time. Note, however, that the conditionality held only for trials involving nonidentical colors; trials involving pairs of identical colors never entailed reinforcement. Here, *AA* and *BB* refer to trials with identical colors in which the prevailing color corresponds, respectively, to the first or to the second item of *AB* trials for a particular test stimulus.

Discrimination

To examine the acquisition of discriminative responding under the procedures of TABLE 1, four pigeons were trained for 60 days, with key pecking separately recorded on *AB*, *BA*, *AA*, and *BB* tests. Daily sessions comprised 80 trials, with each of the eight possible sequences occurring randomly and equiprobably. FIGURE 1 (from experiment 3, replication 1 of Weisman *et al.*[6]) plots responding on all trial types relative to responding on *AB* trials, over successive 4-day blocks of training. Such a depiction, of course, means that scores on *AB* trials always equal 1.0.

In general, discrimination training resulted in a gradual reduction in erroneous responding on nonreinforced test trials (*BA*, *AA*, and *BB*), although the performance of bird 41 deteriorated from its prior high level near the end of training. As to the relative rates of discriminating reinforced from nonreinforced trial types, *AA* trials were most readily discriminated from *AB* trials; *BA* and *BB* trials were discriminated from *AB* trials with greater difficulty. Furthermore, whereas early in training erroneous responding on *BB* tests equalled and sometimes exceeded responding on *BA* tests, later in training erroneous responding on *BA* tests usually exceeded that on *BB* tests.

Clearly, even under the demanding procedure that we devised, pigeons were able to discriminate a recently presented pair of nonoverlapping, nonidentical color stimuli from (a) the same two colors presented in the opposite temporal order and (b) stimulus pairs comprising two identical colors.

The pattern of differential responding over the course of training suggests that final performance followed mastery of two discriminations. First, responding appeared to be based mainly on the second color of the pair and its relationship to the line orientation test stimulus. Although both *A* and *B* colors consistently preceded the test stimuli on reinforced trials, *B*-test associations exerted stronger control over differential responding than did *A*-test associations; thus, response rates on *AB* and *BB* tests initially exceeded those on *BA* and *AA* tests. Second, the birds successfully distinguished *AB* trials from all others. Correlated with this discrimination was a tendency for erroneous responding on *BA* tests to exceed that on *AA* and *BB* tests. This result may be due to the subjects' discriminating trials with identical colors from those with nonidentical colors; only the latter could eventuate in food reinforcement.

Retention

Given clear evidence of stimulus-order discrimination, we were interested in seeing how long such information could be retained. Studies of short-term memory in the pigeon using either choice[12] or go/no-go[13] procedures typically disclose that discriminative performance is a negative function of the sample-test retention interval. Here, we were concerned with determining whether a similar loss of discriminative control would also be observed in our delayed temporal-order discrimination. In particular, the possibility of the pigeon's discriminating the nonoccurrence of reinforcement after pairs of identical colors made it likely that a rather different retention function would be observed following these color pairs than after pairs of nonidentical colors, in which either reinforcement or nonreinforcement were equiprobable outcomes.

Three of the pigeons (birds 31, 33, and 41) continued in this investigation. Daily sessions comprised 96 trials, 12 randomized blocks of the eight possible trial sequences shown in TABLE 1. The first 16 trials in each session were warm-up trials, and the retention interval here was always 0.5 sec; the duration of the retention interval for the final 80 trials was varied between sessions. Retention intervals of 0.5, 1.0, 2.0 and 4.0 sec were randomly presented within 4-day blocks for 16 days. Bird 41 subsequently received 16 additional daily sessions with retention intervals of 0.5, 4.0, 8.0, and 12.0 sec.

The top portion of FIGURE 2 shows that, at the shortest retention interval of 0.5 sec, all three pigeons discriminated reinforced from nonreinforced trials; response rates on *BA*, *BB*, and *AA* tests were all less than half the rate on *AB* tests. Again, response rates were generally ordered $AB > BA > BB > AA$. This ordering is consistent with the birds' discriminating both identical from nonidentical color pairs ($AB + BA > BB + AA$) and earlier from later items in reinforced color pairs ($AB + BB > BA + AA$). Increasing the retention interval from 0.5 to 4.0 sec substantially disrupted discriminative performance on *BA* trials; however, on *BB* and *AA* tests relative response rates remained at or below 0.4 of the rate on *AB* tests.

The generally poorer discrimination on *BA* than on *BB* and *AA* trials at the shortest retention interval plus the greater loss of stimulus control on *BA* than on *BB* and *AA* trials as the retention interval was lengthened are results consistent with different discriminative and retentive processes operating on trials with identical and nonidentical color pairs. Further support for this proposal is given in TABLE 2. This table shows the rate of key pecking of each pigeon to both color items during each of the eight

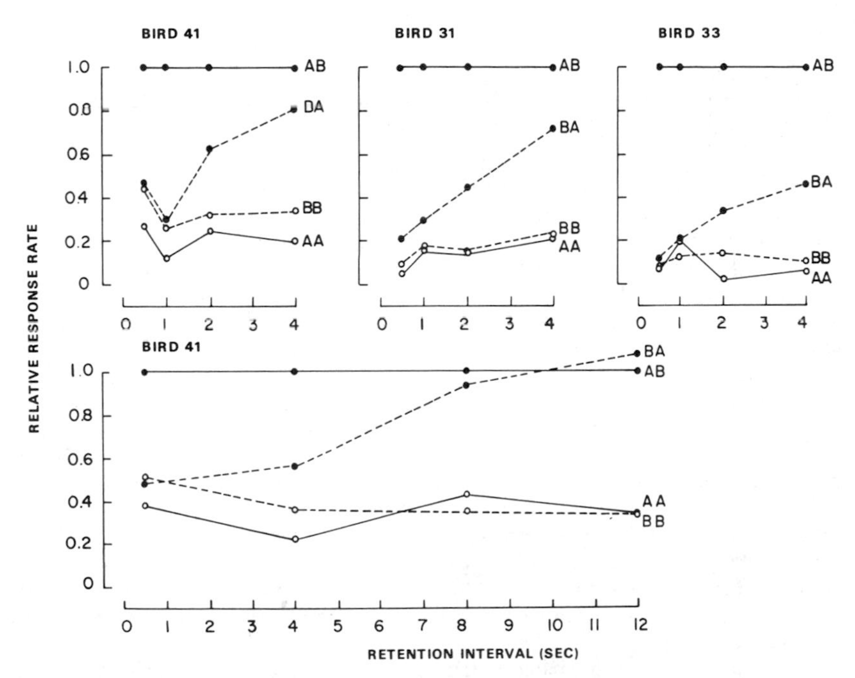

FIGURE 2. Relative rates of responding on *AB*, *BA*, *AA*, and *BB* test trials as a function of the retention interval.

possible color pair-test sequences at the 0.5-sec retention interval. Two trends can be discerned. First, on trials with nonidentical colors, responding to the second item exceeded that to the first. Second, and more pertinent to the present issue, response rates during the second color were much lower when it was the same as the first than when the two colors were different. Thus, the pigeons did indeed discriminate identical from nonidentical colors during presentation of the second item of the pair.

Discrimination of impending nonreinforcement during the second color on *AA* and *BB* trials could be of considerable importance in understanding why little or no loss in discriminative control was observed here compared with that of *BA* trials. Were it easier for pigeons to remember the expectancy of nonreinforcement (on *BB* and *AA*

TABLE 2. Mean Rate of Key Pecking (Pecks per Sec) during Color 1 and Color 2 of Each Color Pair–Line Combination at the 0.5-Sec Retention Interval for Birds 31, 33, and 41

	Color Pair–Line Combinations							
	O-O-V	O-G-V	G-O-V	G-G-V	O-O-H	O-G-H	G-O-H	G-G-H
Bird 31								
Color 1	2.16	2.08	2.09	2.03	2.25	2.20	2.08	2.24
Color 2	0.82	*2.67*	*2.78*	0.72	0.62	*2.73*	*2.25*	0.37
Bird 33								
Color 1	1.43	1.13	1.33	1.73	1.59	1.79	1.38	1.79
Color 2	0.19	*2.08*	*2.28*	0.39	0.30	*2.14*	*2.25*	0.48
Bird 41								
Color 1	0.57	0.45	0.60	0.54	0.63	0.23	0.67	0.47
Color 2	1.04	*1.65*	*1.40*	1.02	1.18	*1.64*	*1.63*	0.93

NOTE: Italicized rates denote responding to color 2 on trials involving nonidentical stimuli.
ABBREVIATIONS: O = orange; G = green; V = vertical; H = horizontal.

trials) than the order of presentation of earlier, nonidentical colors (on *AB* and *BA* trials), then the present results would have a plausible explanation.[14]

Recall that bird 41 received an extra test series with 0.5-, 4.0-, 8.0-, and 12.0-sec retention intervals. The bottom portion of FIGURE 2 again shows: fairly accurate discrimination of stimulus order at the shortest 0.5-sec retention interval; loss of discriminative control on *BA* trials when the retention interval reached 8.0 and 12.0 sec; and no loss of discriminative control on *BB* and *AA* trials, even at the 12.0-sec retention interval.

We now knew that increasing the retention interval between the second color item and the test stimulus impaired the discrimination between prior *AB* and *BA* orders, but had little effect upon the discrimination of prior *AB* from *AA* and *BB* orders. If such different retention results are indeed due to different memory processes operating *after* the birds discriminate identical from nonidentical color pairs, then varying the interitem interval—which elapses *before* this same–different discrimination can take place—should not differentially affect performance on nonreinforced trials involving identical and nonidentical color pairs. Our final experiment in this series examined the issue by systematically varying both the interitem and the retention interval in a within-sessions design.

Two other pigeons with earlier experience in a related temporal-order task[6] were first pretrained on the problem shown in TABLE 1. The experimental phase of training was conducted in 3-day blocks: one session of baseline, one session of interitem interval manipulation, and one session of retention interval manipulation. The baseline session was the first day of each block; the interitem interval and retention interval manipulation sessions were presented on the second and third days of each block with the order of their occurrence reversed in each successive 3-day block. The experimental phase lasted 24 days.

Each session of experimental training comprised 96 trials. Baseline sessions entailed 12 randomized blocks of the eight sequences outlined in TABLE 1, with the interitem and retention intervals set at 0.5 sec. The first 32 (warm-up) trials of interitem interval manipulation sessions and retention interval manipulation sessions were identical to baseline sessions. The final 64 trials comprised two randomized blocks of 32 trials, with each of the eight stimulus sequences combined with each of four values of the interitem interval or the retention interval: 0.5, 1.0, 2.0, and 4.0 sec.

The primary behavioral measure was the rate of key pecking during the test stimuli, recorded separately for each main trial type (*AB*, *BA*, *AA*, and *BB*). For the final 64 trials of interitem interval manipulation sessions and retention interval manipulation sessions, response rates were also calculated separately for each interval value. Mean response rates were obtained over the eight sessions of each of the two interval manipulations, and relative rates were calculated by dividing rates on all trial types by rates on *AB* trials. (Data from the baseline sessions were not scored since each session of experimental training included trials on which the interitem and retention intervals were both set equal to 0.5 sec.)

Relative response rates of both birds on all trial types are shown in FIGURE 3 as a function of the interitem interval (top portion) and the retention interval (bottom portion). Increasing the retention interval again led to a notable loss of discriminative control on *BA* trials and to a much smaller loss on *AA* and *BB* trials. Bird 12 showed no difference in responding after *AB* and *BA* orders at retention intervals of 1.0 sec or more, whereas bird 42 discriminated between prior *AB* and *BA* orders out to a 2.0-sec retention interval.

Unlike the differential effects of retention interval manipulation on discriminative responding, increasing the interitem interval over the same range of values led to a similar loss of discrimination on *BA*, *AA*, and *BB* trials. For neither bird did

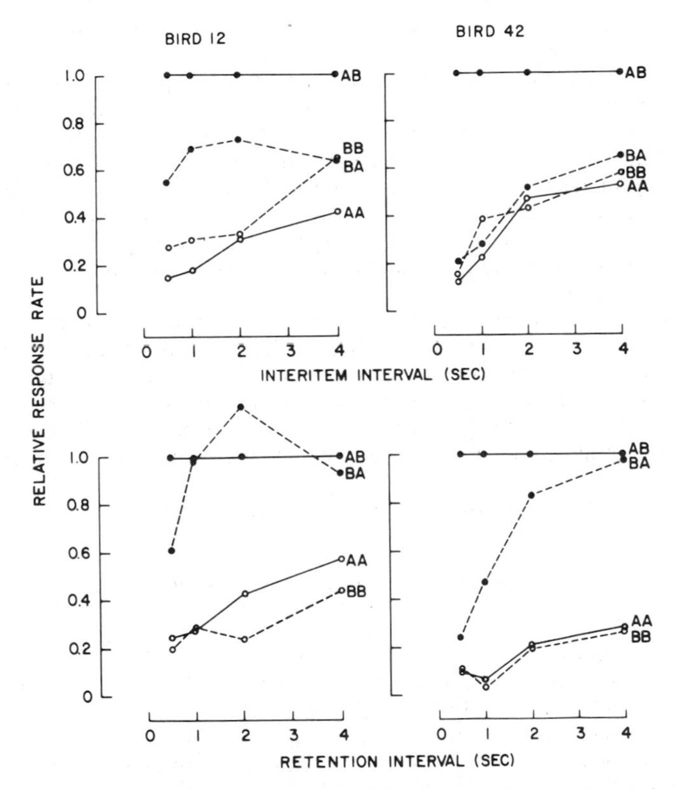

FIGURE 3. Relative rates of responding on *AB*, *BA*, *AA*, and *BB* test trials as a function of the interitem interval (*top*) and the retention interval (*bottom*).

performance on *BA* trials worsen more than on *AA* and *BB* trials. These results thus confirm our suspicion that only *after* presentation of the second color item do different memory processes mediate performance on same-color and different-color trials when the retention interval is lengthened. Because our pigeons could not know in advance whether the second color item would be the same as the first, increasing the interitem interval had equivalent effects on all nonreinforced trial types.

DURATION

That animal behavior can be controlled by the passage of time is obvious when one considers the patterns of responding that are supported by temporal schedules of reinforcement.[15] As noted by Catania,[16] however, such evidence fails to isolate the temporal interval being discriminated from the subject's ongoing stream of behavior. In order to be completely confident in the subject's discriminating two or more stimulus durations, it is necessary to separate the discrimination phase of performance from the report phase.

One rather popular procedure adopting this line of attack proceeds as follows: After pecking a white warning signal, the pigeon is exposed to one of two different durations of a red-lighted key. Following key light offset, two side keys are made available to the pigeon, with left key choices reinforced after one duration and right key choices reinforced after the other duration. If the pigeon were to respond discriminatively after the two stimulus durations, we would have evidence that duration is functioning as a discriminative dimension of the pigeon's environment.

But just how good is this evidence? At least one behavioral strategy can solve this problem without requiring the pigeon to discriminate *each* of the programmed stimulus durations. Suppose that, after pecking the warning signal, the bird stands in front of the choice key affiliated with the shorter stimulus duration. If the stimulus light offsets before some critical time value is reached, the pigeon will peck that key. If, however, the critical time value is reached before the light offsets, the pigeon moves to the other choice key, thus predisposing the bird to making a correct response *before* the longer time value has even elapsed.

One way to remedy this shortcoming is to present two visually distinctive stimuli on the choice keys and to vary their spatial location from trial to trial. If selection of one test stimulus were correct after the shorter interval and selection of the other test stimulus were correct after the longer interval, then accurate test performance would require the subject to attend to both stimulus durations—in their entirety. This technique was employed by Stubbs[17]; it supported orderly discriminative performance by pigeons.

Our own[18] method of separating temporal discrimination from the performance

TABLE 3. Procedure Used in the Study of Duration Discrimination

	Trial Sequence				
Trial Type	Warning Stimulus (FR 1)	Duration Stimulus (red)	Test Stimulus (5.0 sec)	Trial Outcome (3.0 sec)	Intertrial Interval (20.0 sec)
1	White	2 sec	Vertical	Food	—
2	White	2 sec	Slanted	Blackout	—
3	White	16 sec	Vertical	Blackout	—
4	White	16 sec	Slanted	Food	—

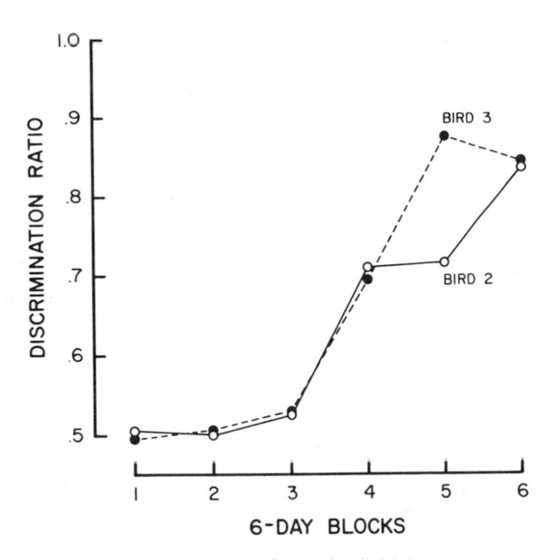

FIGURE 4. Overall temporal discrimination ratios as a function of 6-day blocks of training. Discrimination ratios were calculated by dividing the rate of response on reinforced trials by the combined rate of response on reinforced and nonreinforced trials.

test used the task shown in TABLE 3. Here, pecking responses to a vertical line test stimulus were reinforced after 2.0 sec of red key illumination, but not to a line test stimulus slanted 60 degrees from vertical; conversely, responses to the slanted-line test stimulus were reinforced after 16.0 sec of red key illumination, but not to a vertical line test stimulus. Here, too, discriminative test responding would have to be based upon the subject's discriminating the *full* durations of *both* temporal stimuli.

Discrimination

Two birds were given 36 days of training using the method outlined in TABLE 3. Each session comprised a total of 40 trials, 10 of each type presented in random blocks. From the rates of responding on reinforced and nonreinforced test trials, an overall discrimination ratio was computed that divided the rate of response on reinforced trials by the combined rate of response on reinforced and nonreinforced trials. These ratios are depicted over 6-day blocks in FIGURE 4. It is readily apparent that discriminative performance rose from a chance level of 0.5, ultimately exceeding a ratio of 0.8 for both birds. As in our earlier work on the temporal-order problem, discrimination learning was disclosed by a decrease in erroneous responding on nonreinforced trials.

In order to get a fuller picture of the pigeons' discriminative capabilities, the initial temporal problem was expanded to include a range of short durations (2.0, 4.0, 6.0, and 8.0 sec) and a range of long durations (10.0, 12.0, 14.0, and 16.0 sec).[17] Reinforcement was available on short–vertical and long–slanted trials; reinforcement was not available on short–slanted and long–vertical trials. Training lasted 96 days, with each daily session comprising a total of 80 trials, 5 of each type (8 durations × 2 tests) presented in random blocks.

FIGURE 5 shows discriminative performance portrayed as a function of the duration of the temporal stimulus over the final 24 days of training. Discrimination ratios again compared responding on reinforced trials to responding on both reinforced and nonreinforced trials, but here at each of the eight stimulus durations. Discrimination ratio was a direct function of distance from the reinforcement cutoff of 9 sec.

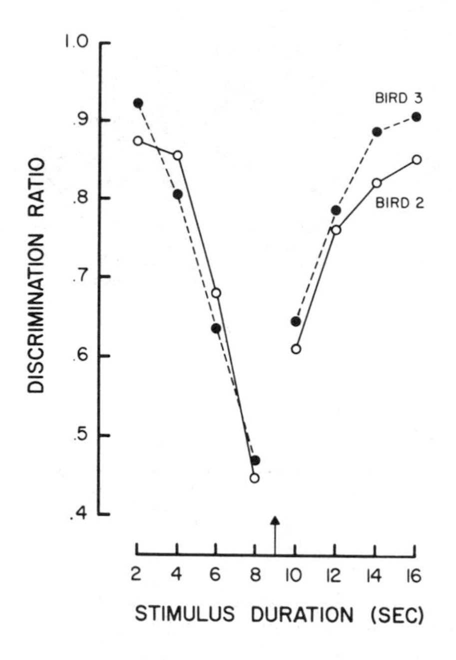

FIGURE 5. Temporal discrimination ratios as a function of the duration of the temporal stimulus. The duration cutoff was 9 sec.

Increasingly different time values were thus easier for pigeons to discriminate than were increasingly similar time values. Additionally, each bird also showed a trend for the short segment of the discriminability function to begin at a lower level and to rise more rapidly than did the long segment. This result is consistent with the birds' overestimating intermediate durations of the key light stimulus: behaving in accord with go/no-go dispositions appropriate to durations a bit longer than those actually given.

Retention

Having determined with our conditional discrimination procedure that pigeons discriminate the durations of key light stimuli and do so with increasing accuracy the more dissimilar the time values, we were interested in ascertaining the pigeons' short-term memory of stimulus durations. To accomplish that goal, we simply inserted retention intervals between offset of the red key light and onset of the vertical or slanted-line test stimulus, in a within-sessions design. If pigeons do forget the prior duration of the red stimulus, we would expect discrimination ratios to fall as the retention interval is lengthened.

Birds 2 and 3 continued as subjects. Prior to the phases of training depicted in FIGURE 6, the birds received extensive training with a new set of duration values: the short stimuli were 1.0, 2.0, 3.0, and 4.0 sec and the long stimuli were 5.0, 6.0, 7.0, and 8.0 sec. Reinforcement was available only on short–vertical and long–slanted trials; the other trial types entailed nonreinforcement. In addition, bird 2 received training with 1.0-, 2.0-, and 4.0-sec retention intervals, and bird 3 received training with 1.0-, 2.0-, 4.0-, and 8.0-sec retention intervals (see DeLong[18] for details).

During the phases depicted in FIGURE 6, birds 2 and 3 received 120 days of training. For bird 2, daily sessions involved 48 trials, 1 of each possible combination (3 retention intervals × 8 durations × 2 tests). For bird 3, daily sessions involved 64 trials, 1 of each possible combination (4 retention intervals × 8 durations × 2 tests).

Discriminative performance from the last 24 days in each of these phases is shown in FIGURE 6. Although the overall performance of bird 2 was poorer than that of bird 3, the collective pattern of results was quite similar. As in FIGURE 5, discrimination improved as more extreme stimulus durations were given. Also as in FIGURE 5, the short segment of the discriminability function started at a lower level and was steeper than the long segment. Finally, and most germane to the issue of short-term memory, discrimination was a strong negative function of the retention interval separating the duration to be discriminated and the subsequent performance test. Thus, like other sensory information, the duration of a stimulus becomes increasingly unlikely to control discriminative performance the longer it has been since stimulus offset.[19] Yet even at a retention interval of 16.0 sec, we found measurable control over behavior by stimulus duration.

A final issue to be considered here concerns the way that increases in the retention interval might lead to a worsening of control by prior stimulus duration.[19] One means involves the subject's forgetting the duration of the red key light as the time since stimulus *offset* and the performance test is lengthened. Another possibility is that subjects are basing their test responding on the passage of time since stimulus *onset;* lengthening the retention interval ought then to lead to poorer discrimination and responding more appropriate to long stimulus durations. A third possibility is that subjective estimates of stimulus duration decrease as the retention interval is increased[20]; subjective shortening will then lower discriminative performance and support responding more appropriate to short stimulus durations. Scrutiny of FIGURE 6 reveals that lengthening the retention interval reduced discrimination scores, but did

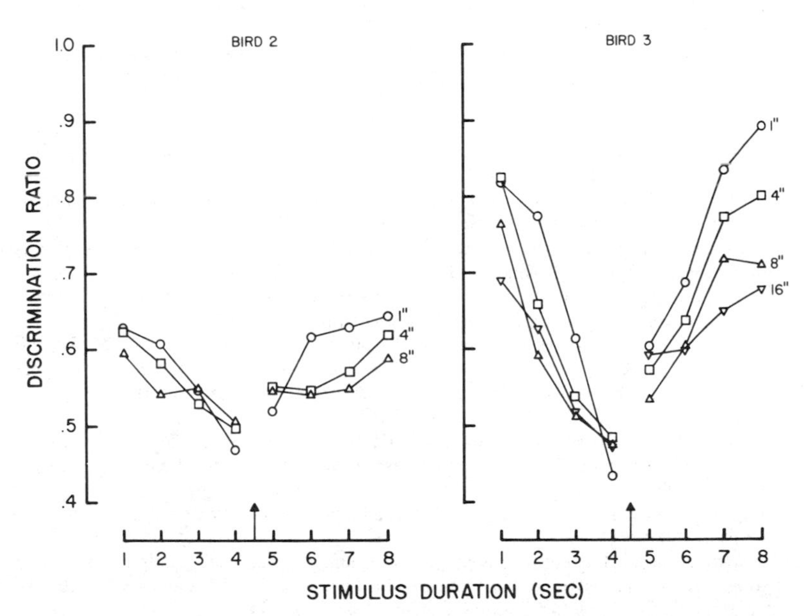

FIGURE 6. Temporal discrimination ratios as functions of the duration of the temporal stimulus and the retention interval. The duration cutoff was 4.5 sec.

not bias performance in any systematic manner. Thus, the latter two interpretations seem not to have been supported by our temporal discrimination data.

CONCLUSIONS

Conditional discrimination procedures reveal that temporal order and duration can serve as discriminative stimuli for pigeon subjects. Not only do pigeons respond discriminatively after different orders of two nonidentical stimuli and different durations of a single stimulus, but they forget this information if the performance test is delayed for several seconds. Discrimination and retention of temporal order and duration suggest that pigeons are able to abstract nonvisual and relational properties from simple visual stimuli.

REFERENCES

1. Gibson, E. J. 1969. Principles of Perceptual Learning and Development. Appleton-Century-Crofts. New York, NY.
2. Marks, L. E. 1978. The Unity of the Senses. Academic Press. New York, NY.
3. Spetch, M. L. & D. M. Wilkie. 1981. Duration discrimination is better with food access as the signal than with light as the signal. Learn. Motiv. **12:** 40–64.
4. Larew, M. B. 1979. Discrimination and Retention of Stimulus Order by Pigeons. Honors thesis, University of Iowa, Iowa City, IA.
5. Wasserman, E. A., K. R. Nelson & M. B. Larew. 1980. Memory for sequences of stimuli and responses. J. Exp. Anal. Behav. **34:** 49–59.
6. Weisman, R. G., E. A. Wasserman, P. W. D. Dodd & M. B. Larew. 1980. Representation and retention of two-event sequences in pigeons. J. Exp. Psychol. Animal Behav. Processes **6:** 312–325.
7. Wasserman, E. A., S. R. Franklin & E. Hearst. 1974. Pavlovian appetitive contingencies and approach versus withdrawal to conditioned stimuli in pigeons. J. Comp. Physiol. Psychol. **86:** 616–627.
8. DeLong, R. E. & E. A. Wasserman. 1981. Effects of differential reinforcement expectancies on successive matching-to-sample performance in pigeons. J. Exp. Psychol. Animal Behav. Processes **7:** 394–412.
9. Konorski, J. A. 1959. A new method of physiological investigation of recent memory in animals. Bull. Acad. Polon. Sci. Ser. Sci. Biol. **7:** 115–117.
10. Nelson, K. R. & E. A. Wasserman. 1981. Stimulus asymmetry in the pigeon's successive matching-to-sample performance. Bull. Psychon. Soc. **18:** 343–346.
11. Wasserman, E. A. 1976. Successive matching-to-sample in the pigeon: Variations on a theme by Konorski. Behav. Res. Methods Instrum. **8:** 278–282.
12. Blough, D. S. 1959. Delayed matching in the pigeon. J. Exp. Anal. Behav. **2:** 151–160.
13. Nelson, K. R. & E. A. Wasserman. 1978. Temporal factors influencing the pigeon's successive matching-to-sample performance: Sample duration, intertrial interval, and retention interval. J. Exp. Anal. Behav. **30:** 153–162.
14. Honig, W. K. & E. A. Wasserman. 1981. Performance of pigeons on delayed simple and conditional discriminations under equivalent training procedures. Learn. Motiv. **12:** 149–170.
15. Ferster, C. B. & B. F. Skinner. 1957. Schedules of Reinforcement. Appleton-Century-Crofts. New York, NY.
16. Catania, A. C. 1970. Reinforcement schedules and psychophysical judgments: A study of some temporal properties of behavior. *In* The Theory of Reinforcement Schedules. W. N. Schoenfeld, Ed. Appleton-Century-Crofts. New York, NY.
17. Stubbs, A. 1968. The discrimination of stimulus duration by pigeons. J. Exp. Anal. Behav. **11:** 223–238.

18. DeLong, R. E. 1983. Control of Responding by Stimulus Duration. Ph.D. dissertation, University of Iowa, Iowa City, IA.
19. Church, R. M. 1980. Short-term memory for time intervals. Learn. Motiv. **11:** 208–219.
20. Spetch, M. L. & D. M. Wilkie. 1983. Subjective shortening: A model of pigeons' memory for event duration. J. Exp. Psychol. Anim. Behav. Processes **9:** 14–30.

Contingent Aftereffects in Duration Judgments[a]

LORRAINE G. ALLAN

Department of Psychology
McMaster University
Hamilton, Ontario, Canada L8S 4K1

Exposure to a repeating event of a constant duration has been shown to influence the judged duration of subsequently presented test events. For example, adaptation to a 4-sec tone results in the shortening of the judged duration of a subsequently present 2-sec test tone.[1] This duration aftereffect is negative in that the judged duration of the test tone is in a direction away from the value of the adaptation duration. Recently, negative duration aftereffects which are contingent on the pitch of the tone or on the temporal order of presentation of two events have been reported.[2,3] These contingent duration aftereffects are reminiscent of the contingent aftereffect discovered by McCollough.[4] Her subjects inspected two visual patterns that alternated every few seconds. Under one condition, one adaptation figure consisted of black vertical bars on an orange background and the other adaptation figure consisted of black horizontal bars on a blue background. During the test, the subject viewed achromatic patterns and reported that the achromatic background of the vertical bars appeared bluish and the achromatic background of the horizontal bars appeared orangeish. McCollough[4] argued that this orientation-contingent negative aftereffect could be understood in terms of color adaptation of orientation-specific edge detectors and she interpreted her result as psychophysical evidence for neural units that are both color- and orientation-specific. While selective adaptation of neural detectors is still the favored explanation,[5] there are data that are supportive of an associative account.[6,7]

In the reports[2,3] of contingent duration aftereffects an attempt was made to explain the data using the two accounts that have been applied to the McCollough effect. Both selective adaptation of neural detectors and associative mechanisms are far removed from the usual theoretical concepts found in the time perception literature. The research to be reported questions the assumption that the contingent duration aftereffects are related to the McCollough effect.

There is considerable evidence in the time perception literature of systematic "biases" in judgments of perceived duration.[8] Duration judgements are influenced by such nontemporal characteristics of the duration marker as energy, complexity, modality, and whether the interval is filled or empty. It is possible that the contingent duration aftereffects are yet another example of judged duration's being influenced by variables in addition to physical duration.

The experiments to be reported compare contingent duration aftereffects with the McCollough effect. A number of studies have shown that McCollough-type contingent aftereffects last for days and even weeks.[5] One purpose was to study the persistence of the contingent duration aftereffects. In one experiment a delay was introduced between adaptation and test and its effect on the size of the order-contingent duration

[a]This research was supported by the Natural Sciences and Engineering Research Council of Canada. Some of the data reported in Experiments 3 and 4 were presented at the 1980 meeting of the Psychonomic Society.

aftereffect was examined. In addition, some subjects ran multiple sessions under both adaptation conditions so that the persistence of the aftereffects over sessions could be studied and the aftereffects for individual subjects could be examined.

For McCollough-type contingent aftereffects, whenever the perception of feature *A* has been found to be contingent on feature *B*, the perception of *B* has also been shown to be contingent on *A*. For example, judgment of color is contingent on line orientation, on spatial frequency, and on motion direction; and judgment of line orientation, spatial frequency, and motion direction is contingent on color.[5] If judgments of duration can be contingent on pitch, one might expect judgments of pitch to be contingent on duration. Another purpose was to determine whether a negative pitch aftereffect contingent on duration could be demonstrated.

A magnitude estimation task [2] and a reproduction task[3] have been used in the studies of contingent duration aftereffects. To extend the generality of results, another psychophysical procedure, forced-choice discrimination, is tried.

PROCEDURE

One hundred twenty-five undergraduates at McMaster University participated as part of a course requirement. In addition, data were obtained from six paid subjects who ran multiple sessions.

Stimulus presentation, recording of responses, and timing were controlled either by a PDP 8/E computer with a video terminal or by a SuperPET. Auditory signals were presented binaurally over headphones. A Wavetek Function Generator was used with the PDP 8/E to produce a 70-dB pure tone signal. Otherwise, the SuperPET's tone generator was used.

In all experiments a session was made up of a preadaptation phase and three adaptation-test sequences. Preadaptation and test consisted of discrimination trials. A trial began with READY displayed on the screen. This was followed by a pair of tones. To insure that subjects make a comparative judgment on each trial, a roving standard design was used. At the end of the second tone, RESPOND was displayed on the screen. The subject indicated his judgment by typing a *1* or a *2* on the keyboard. During adaptation the subject simply listened to a series of tones.

EXPERIMENT 1: DURATION CONTINGENT ON TEMPORAL ORDER[b]

Method

Forty subjects were randomly divided into four groups of 10 subjects each. Preadaptation consisted of 96 duration discrimination trials. READY was displayed for 500 msec and 1 sec later a pair of 364-Hz tones, separated by 400 msec, was presented. The two tones differed in duration by 30 msec. There were four pairs of duration values: 330/360, 360/390, 390/420, and 420/450. Each pair was presented in two orders, short followed by long (SL) or long followed by short (LS). Twenty subjects were asked whether the longer tone occurred first or second ("long instruction") and 20 were asked whether the shorter tone occurred first or second ("short instruction").

[b]Parts of Experiments 1 and 2 are described in a B.Sc. thesis by A. M. Majerovich[9] at McMaster University.

The subject responded by typing *1* (for first) or *2* (for second) on the computer keyboard. The intertrial interval was 1.4 sec.

During adaptation the subject listened to 50 pairs of tones. The two tones in a pair were separated by 400 msec and pairs were separated by 1.4 sec. One tone in a pair was 200 msec, the other was 600 msec. For 10 subjects under each instruction, the first tone was the short (short–long adaptation); for the remaining 10 subjects under each instruction, the first tone was longer (long–short adaptation). Thirty-two duration discrimination test trials immediately followed adaptation. The adaptation–test sequence occurred three times for a total of 96 test trials.

Four subjects ran repeat sessions under similar conditions. The difference in duration was 20 msec and only the long instruction was used. The four pairs of duration values were 370/390, 390/410, 410/430, 430/450. Two subjects (M.G. and H.B.) ran under long–short adaptation and then under short–long; the other two subjects (L.P. and D.K.) received the adaptation conditions in the reverse order. The number of sessions under each adaptation condition is shown in TABLE 2.

Results and Discussion

The probability of a correct duration judgment during preadaptation and during test is present in TABLE 1 for each of the four groups. The data are summarized by the difference between two conditional probabilities, known as the time-order error (TOE):

$$\text{TOE} = P(R_{LS}|LS) - P(R_{SL}|SL)$$

where $P(R_{LS}|LS)$ is the probability of a correct response to the LS order and $P(R_{SL}|SL)$ is the probability of a correct response to the SL order. Most research has indicated that for brief durations TOE is positive and for longer durations it is negative.[8] TABLE 1 shows that for all four groups in this experiment, TOE is positive during preadaptation. While the mean size of preadaptation TOE is smaller under the short instruction, instruction is not a significant variable according to the Mann–Whitney U test ($U = 159.5$, $p > .05$). This result is consistent with other findings.[10–12]

If a negative duration aftereffect can be made contingent on temporal order, then after short–long adaptation, the first test tone presented should appear longer compared to preadaptation judgments and the second presented test tone should appear shorter, resulting in a more positive TOE in test than preadaptation. After

TABLE 1. $P(R_{LS}|LS)$, $P(R_{SL}|SL)$ and TOE: Group Data from Experiments 1 and 2

Instruction	Adaptation Condition	Preadaptation $P(R_{LS}\|LS)$	Preadaptation $P(R_{SL}\|SL)$	Preadaptation TOE	Test $P(R_{LS}\|LS)$	Test $P(R_{SL}\|SL)$	Test TOE
Experiment 1							
Long	Long–short	.73	.62	.11	.69	.71	−.02
	Short–long	.76	.59	.17	.81	.42	.39
Short	Long–short	.66	.62	.04	.56	.70	−.14
	Short–long	.67	.59	.08	.74	.45	.29
Experiment 2							
Long	Long–short	.69	.54	.15	.73	.57	.16
	Short–long	.76	.63	.13	.79	.66	.13

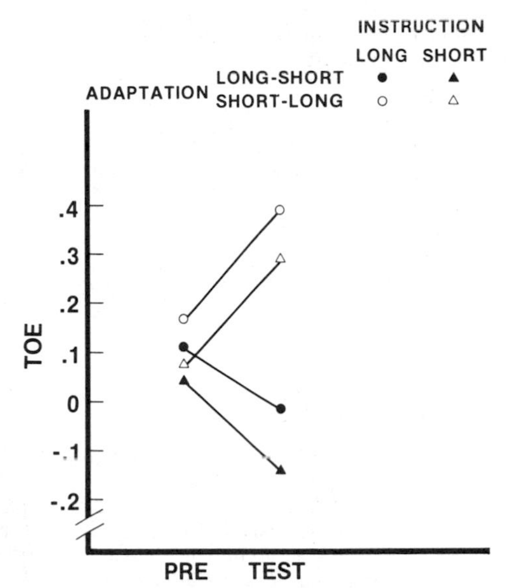

FIGURE 1. Preadaptation and test TOE in Experiment 1 as a function of adaptation condition. The data are shown separately for each instruction.

long–short adaptation, the first presented test tone should appear shorter compared to preadaptation judgments and the second presented test tone should appear longer, resulting in a less positive TOE during test than during preadaptation. TABLE 1 and FIGURE 1 show these predictions to be supported by the data. For 30 of the individual subjects, the change in TOE was in the expected direction: 19 in the short–long groups and 11 in the long–short groups. A χ^2 test indicates that the proportion of the subjects showing a change in TOE in the expected duration was significant ($\chi_1^2 = 10.0$, $p < .005$). These data demonstrate that judged duration is influenced by the pair of stimuli presented during adaptation. One adaptation condition increases the size of the preadaptation positive TOE; the other changes the preadaptation positive TOE to a negative one.

$P(R_{LS}|LS)$, $P(R_{SL}|SL)$, and TOE, averaged over all sessions, are shown in TABLE 2 for each of the four repeat-session subjects. In FIGURE 2, preadaptation and test TOE are plotted for each adaptation condition. D.K. is the only subject who shows no effect of adaptation. For the other three subjects, long–short adaptation results in a more negative TOE during test than during preadaptation and short–long adaptation in a more positive TOE. Thus, for an individual subject, TOE can be made more positive or more negative depending upon whether the subject is exposed to short–long or long–short pairs of durations.

The overall probability of a correct duration response, regardless of the presentation order, is also shown in TABLE 2, where

$$P(C) = P(LS)P(R_{LS}|LS) + P(SL)P(R_{SL}|SL).$$

For subject H.B., $P(C)$ increases from preadaptation to test under both adaptation conditions, whereas for the other three subjects, $P(C)$ decreases from preadaptation to test under both adaptation conditions. Thus, while adaptation has a differential effect on reports about perceived duration, here measured by TOE, adaptation does not differentially affect discriminability, as measured by $P(C)$. This result is consistent

TABLE 2. $P(R_{LS}|LS)$, $P(R_{SL}|SL)$, $P(C)$ and TOE: Repeat-Session Subject Data from Experiment 1

Adaptation Condition and Subject	Preadaptation				Test			
	$P(R_{LS}\|LS)$	$P(R_{SL}\|SL)$	$P(C)$	TOE	$P(R_{LS}\|LS)$	$P(R_{SL}\|SL)$	$P(C)$	TOE
Long–Short								
M.G. (8)[a]	.70	.77	.73	−.07	.62	.82	.72	−.20
H.B. (8)[a]	.64	.55	.60	.09	.57	.66	.61	−.09
L.P. (8)[b]	.66	.59	.63	.07	.48	.69	.58	−.21
D.K. (7)[b]	.64	.66	.65	−.02	.61	.63	.62	−.02
Mean				.02				−.13
Short–Long								
M.G. (8)	.80	.71	.76	.09	.85	.55	.70	.30
H.B. (7)	.56	.59	.57	−.03	.63	.60	.61	.03
L.P. (6)	.59	.55	.57	.04	.67	.45	.56	.23
D.K. (10)	.52	.66	.59	−.14	.50	.64	.57	−.14
Mean				−.01				.10

NOTE: Number of sessions under each adaptation condition is shown in parentheses.
[a]Received the long–short condition first.
[b]Received the short–long condition first.

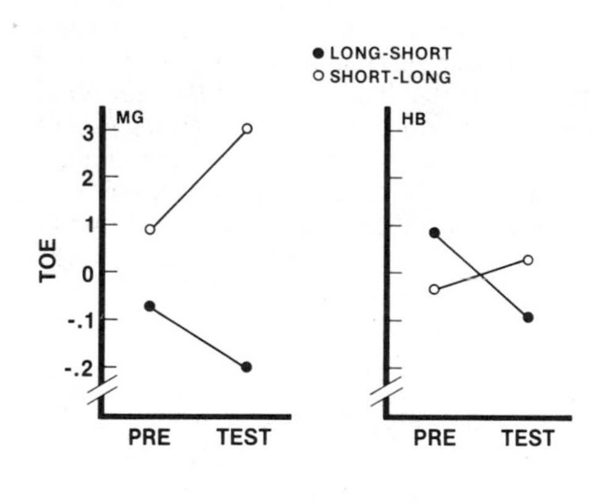

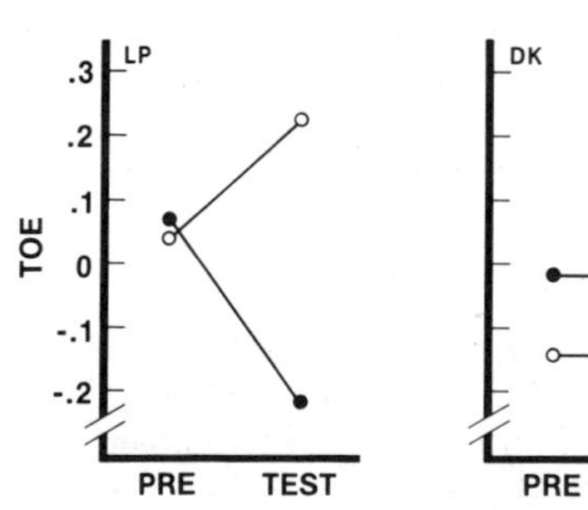

FIGURE 2. Preadaptation and test TOE in Experiment 1 as a function of adaptation condition. The data are shown separately for each repeat-session subject.

with other findings in the literature that variables that influence reports about perceived duration often have no differential effect on discriminability.[8]

Preadaptation TOE is shown in FIGURE 3, separately for sessions 1 and 2 and for sessions (N-1) and N (the last two sessions). If the order-contingent duration aftereffect showed persistence over sessions, one would expect preadaptation TOE to become more negative over sessions under long–short adaptation and more positive over sessions under short–long adaptation. Subject L.P. shows this pattern. Subjects H.B. and D.K. show little change in preadaptation TOE over sessions, and M.G. shows

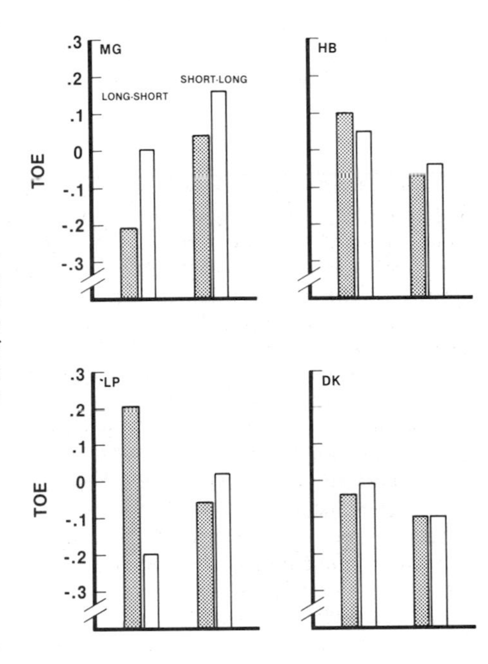

FIGURE 3. Preadaptation TOE in Experiment 1 shown separately for early and late sessions. *Stippled bars* represent mean TOE during sessions 1 and 2 and *open bars* during sessions (N-1) and N. The pair of bars on the *left* are for the long–short condition and the pair on the *right* are for the short–long condition.

the reverse trend under the long–short condition. These data suggest that the adaptation effect does not persist over sessions.

EXPERIMENT 2: PERSISTENCE OF THE ORDER-CONTINGENT DURATION AFTEREFFECT

Method

Twenty subjects were randomly divided into two groups of 10 subjects each. The method was similar to that used in Experiment 1, except that a 25-min delay was introduced between the end of each adaptation period and the beginning of test. The subjects remained in the experimental room and were instructed to keep the headphones on. Only the long instruction was used.

Results and Discussion

The probability of a correct duration judgment during preadaptation and during test and the value of TOE are presented in TABLE 1 for each adaptation group. As in Experiment 1, TOE is positive during preadaptation. Overall, there is no change in TOE from preadaptation to test following either adaptation condition. In the long–short group only five subjects showed the expected change in TOE (more negative) and in the short–long group only four subjects showed the expected change in TOE (more positive). These data suggest that the influence of the adaptation condition has disappeared by the end of a 25-min delay period.

EXPERIMENT 3: DURATION CONTINGENT ON PITCH

Method

Thirty-four subjects were randomly divided into two groups of 17 subjects each. Preadaptation consisted of 96 duration discrimination trials. READY was displayed for 500 msec and 1 sec later a pair of tones, separated by 400 msec, was presented. One tone was 600 Hz, the other 900 Hz. The two tones also differed in duration by 35 msec. There were four pairs of duration values: 330/365, 365/400, 400/435, and 435/470. Two duration orders and two frequency orders resulted in four events, each event occurring 24 times in the 96 trials. The subject was required to indicate which tone, first or second, was longer by typing *1* or *2* on the terminal keyboard. The intertrial interval was 2 sec.

During adaptation the subject listened to two alternating tones, with a 400-msec interval between successive tones. Each tone was presented 100 times. For the *Lo/L–Hi/S* group, one tone was presented at 600 Hz for 600 msec and the other at 900 Hz for 200 msec; for the *Lo/S–Hi/L* group, one tone was presented at 600 Hz for 200 msec and the other at 900 Hz for 600 msec. Sixteen discrimination test trials immediately followed adaptation. The adaptation–test sequence occurred three times for a total of 48 test trials.

Six subjects ran repeat sessions under similar conditions. The difference in duration between the two tones was 20 msec and the four pairs of duration values were 370/390, 390/410, 410/430, 430/450. The intertone interval and the intertrial interval were 1 sec. During adaptation each tone was presented 50 times with a 1-sec interval between tones. Thirty-two discrimination trials followed adaptation. The three adaptation–test sequences resulted in 96 test trials. Three subjects (K.C., D.K., and I.S.) ran under *Lo/L–Hi/S* adaptaton and then under *Lo/S–Hi/L*; the other three subjects (H.B., M.G., and L.P.) received the adaptation conditions in the reverse order. The number of sessions under each adaptation condition is shown in TABLE 4.

Results and Discussion

The probability of a correct duration judgment during preadaptation and during test is presented in TABLE 3 for each adaptation group. The data are organized in matrices where the rows represent duration order, short followed by long (SL) or long followed by short (LS), and the columns represent frequency order, low followed by high (LH) or high followed by low (HL). The negative diagonal of the preadaptation matrices was larger than the positive for 22 of the 34 subjects and was the same for 4 subjects. The Wilcoxon T test shows the sum of the entries along the negative diagonal to be significantly larger than the sum along the positive diagonal ($T = 116.5$, n = 30, $p < .02$). These data indicate that short low-pitched tones are judged shorter than short high-pitched tones and that long high-pitched tones are judged longer than long low-pitched tones. This dependence of duration judgments on frequency is referred to as auditory *kappa* in the literature.[13,14] In the remainder of this paper *kappa* will be defined as:

$$[P(R_{SL} \mid SL \text{ and } LH) + P(R_{LS} \mid LS \text{ and } HL)] - [P(R_{SL} \mid SL \text{ and } HL) + P(R_{SL} \mid LS \text{ and } LH)]$$

If a negative duration aftereffect can be made contingent on pitch, then after *Lo/S–Hi/L* adaptation, low-pitched tones should appear shorter and high-pitched tones should appear longer during test than during preadaptation. That is, *kappa* should be larger after adaptation. The data in TABLE 3 show this to be the case in that the difference between the negative and positive diagonals is greater during test (.43) than during preadaptation (.17). For the *Lo/S–Hi/L* group, low-pitched tones should

TABLE 3. Duration Discrimination Data from Experiment 3

Adaptation Condition	Preadaptation Duration Order	Preadaptation Pitch Order LH	Preadaptation Pitch Order HL	Test Duration Order	Test Pitch Order LH	Test Pitch Order HL
Lo/L–Hi/S						
	SL	.74	.63	*SL*	.80	.58
	LS	.66	.72	*LS*	.53	.74
Lo/S–Hi/L						
	SL	.72	.64	*SL*	.60	.62
	LS	.61	.71	*LS*	.77	.78

NOTE: Each entry is the probability of a correct duration response.

appear longer and high-pitched tones should appear shorter during test than during preadaptation. That is, *kappa* should be attenuated. The data in TABLE 3 show this to be the case in that the difference between the two diagonals during preadaptation (.18) is eliminated during test (−.01). Of the 34 subjects, 22 showed performance during test, relative to preadaptation, that was in the expected direction (12 with larger *kappa* after *Lo/L–Hi/S* and 10 with reduced *kappa* after *Lo/S–Hi/L*) and one subject showed no change. According to the Wilcoxon T test these data indicate that the adaptation conditions had a significant differential effect on performance ($T = 142.5$, n = 33, $p < .02$). One adaptation condition increases the preadaptation tendency of judging low-pitched tones as relatively short and high-pitched tones as relatively long; the other adaptation condition eliminates this tendency.

The probability of a correct duration judgment, averaged over all sessions under a particular adaptation condition, and the value of *kappa* are shown in TABLE 4 for each of the six repeat-session subjects. For five subjects *kappa* was increased after *Lo/L–Hi/S* adaptation and for five subjects *kappa* was decreased after *Lo/S–Hi/L* adaptation. The data averaged over the six subjects show the same patterns as do the group data in TABLE 3. There is a *kappa* effect during preadaptation. *Kappa* is increased from .18 to .30 by *Lo/L–Hi/S* and *kappa* is decreased from .12 to –.06 by *Lo/S-Hi/L*.

The overall probability of a correct response, regardless of presentation order and pitch, is shown in TABLE 5 for each of the six repeat-session subjects. Adaptation condition does not have a systematic effect on $P(C)$ across subjects. As in Experiment 1, while adaptation condition has a differential effect on reports about perceived duration, here measured by *kappa*, adaptation condition does not differentially influence discriminability, as measured by $P(C)$. Again, these results are consistent with other findings that variables that affect reports about perceived duration often do not influence discriminability.[8]

Preadaptation *kappa* is shown in FIGURE 4, separately for sessions 1 and 2 and sessions (N-1) and N. If the pitch-contingent duration aftereffect showed persistence

TABLE 4. Probability of a Correct Duration Response for Each of the Four Stimulus Events and the Value of Kappa for the Six Repeat-Session Subjects in Experiment 3

Adaptation Condition and Subject	Preadaptation					Test				
	LH		*HL*			*LH*		*HL*		
	SL	*LS*	*SL*	*LS*	Kappa	*SL*	*LS*	*SL*	*LS*	Kappa
Lo/L-Hi/S										
H.B. (6)[a]	.77	.39	.73	.55	.20	.76	.44	.61	.62	.32
L.P. (8)[a]	.63	.56	.45	.69	.31	.59	.51	.34	.78	.52
M.G. (8)[a]	.63	.81	.76	.69	–.26	.60	.81	.68	.80	–.08
K.C. (7)[b]	.70	.37	.78	.42	–.03	.70	.26	.62	.49	.31
D.K. (7)[b]	.87	.54	.68	.76	.41	.81	.48	.63	.76	.47
I.S. (7)[b]	.74	.43	.53	.67	.45	.70	.39	.67	.63	.26
Mean					.18					.30
Lo/S-Hi/L										
H.B. (8)	.64	.53	.65	.52	–.02	.60	.58	.67	.52	–.14
L.P. (9)	.68	.56	.51	.71	.32	.57	.67	.55	.72	.07
M.G. (7)	.51	.90	.74	.65	–.48	.46	.86	.73	.69	–.45
K.C. (8)	.70	.44	.56	.65	.35	.61	.44	.57	.57	.17
D.K. (6)	.82	.51	.65	.75	.41	.76	.73	.79	.65	–.10
I.S. (7)	.82	.46	.73	.52	.14	.76	.51	.70	.55	.10
Mean					.12					–.06

NOTE: Number of sessions under each adaptation condition is shown in parentheses.
[a]Received the *Lo/S–Hi/L* condition first.
[b]Received the *Lo/L–Hi/S* condition first.

over sessions, one would expect preadaptation *kappa* to increase over sessions under *Lo/L–Hi/S* and to decrease over sessions under *Lo/S–Hi/L*. A consistent pattern across subjects is not seen. These data, like the data from the earlier experiments, suggest that the influence of adaptation does not persist over sessions.

EXPERIMENT 4: PITCH CONTINGENT ON DURATION

Method

Thirty-one subjects were randomly divided into two groups of 16 subjects and 15 subjects. Preadaptation consisted of 96 pitch discrimination trials. READY was

TABLE 5. Overall Probability of a Correct Duration Response, $P(C)$, in Experiment 3

Condition and Subject	$P(C)$	
	Preadaptation	Test
Lo/L-Hi/S		
H.B.	.61	.61
L.P.	.58	.56
M.G.	.72	.72
K.C.	.57	.52
D.K.	.71	.67
I.S.	.59	.59
Mean	.63	.61
Lo/S-Hi/L		
H.B.	.58	.59
L.P.	.62	.63
M.G.	.70	.68
K.C.	.59	.55
D.K.	.68	.74
I.S.	.63	.63
Mean	.63	.64

displayed for 500 msec and 1 sec later a pair of tones, separated by 400 msec, was presented. The two tones differed in duration (200 msec or 600 msec). The two tones also differed in frequency by 8 Hz. There were four pairs of frequency values: 732/740, 740/748, 748/756, and 756/764. The subject was required to indicate which tone, first or second, was higher in pitch by typing *1* or *2* on the terminal keyboard.

The adaptation conditions were identical to those in Experiment 3. The subject listened to two alternating tones, with a 400-msec interval between successive tones. Each tone was presented 100 times. There were 16 subjects in the *Lo/L–Hi/S* group

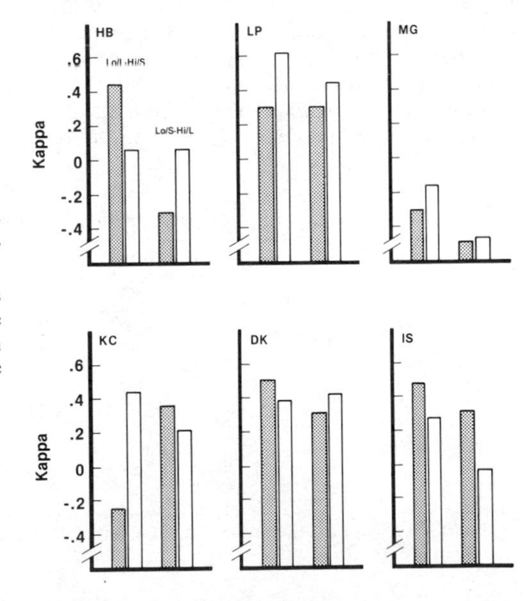

FIGURE 4. Preadaptation kappa in Experiment 3 shown separately for early and late sessions. *Stippled bars* represent mean kappa during sessions 1 and 2 and *open bars* during sessions (N-1) and N. The pair of bars on the *left* are for the *Lo/L–Hi/S* condition and the pair on the *right* are for the *Lo/S–Hi/L* condition.

and 15 in the *Lo/S–Hi/L* group. Sixteen pitch discrimination test trials immediately followed each adaptation period. The adaptation–test sequence occurred three times for a total of 48 test trials.

Six subjects ran repeat sessions under similar conditions. The difference in pitch between the two tones was 9 Hz and the four pitch pairs were 726/735, 735/744, 744/753, 753/762. The intertone interval and intertrial interval were 1 sec. During adaptation each tone was presented 50 times with a 1-sec interval between tones. Thirty-two discrimination trials followed adaptation. The three adaptation–test sequences resulted in 96 test trials. Three subjects (D.K., L.P., and K.C.) ran under *Lo/L–Hi/S* adaptation and then under *Lo/S–Hi/L*; the other three subjects (H.B., M.G., and I.S.) received the adaptation conditions in the reverse order. The number of sessions under each adaptation condition is shown in TABLE 7.

Results and Discussion

The data in TABLE 6 are presented in a similar manner to those in TABLE 3, except that the entries represent the probability of a correct pitch judgment. The negative diagonal of the preadaptation matrices was larger than the positive for 21 of the 31 subjects. The Wilcoxon T test shows the sum of the entries along the negative diagonal to be significantly larger than the sum along the positive diagonal ($T = 103.5$, $n = 31$, $p < .01$). These data show that low-frequency short tones are judged lower in pitch than low-frequency long tones and that high-frequency long tones are judged higher in pitch than high-frequency short tones. The dependence of pitch judgments on duration is referred to as auditory *tau* in the literature.[13,15] In the remainder of this paper *tau* will be defined as:

$$[P(R_{LH} \mid LH \text{ and } SL) + P(R_{HL} \mid HL \text{ and } LS)] - [P(R_{LH} \mid LH \text{ and } LS) + P(R_{HL} \mid HL \text{ and } SL)]$$

For the *Lo/L–Hi/S* group, short tones should appear lower in pitch and long tones higher in pitch during test than during preadaptation. That is, *tau* should be larger after adaptation. The data in TABLE 6 indicate that there is little change in *tau* from preadaptation (.32) to test (.29). If anything, *tau* is reduced. For the *Lo/S–Hi/L* group, short tones should appear higher in pitch and long tones lower in pitch during test than during preadaptation. That is, *tau* should be attenuated. The data in TABLE 6 show this to be the case in that the difference between the negative and positive

TABLE 6. Pitch Discrimination Data from Experiment 4

Adaptation Condition	Preadaptation Duration Order	Preadaptation Pitch Order *LH*	Preadaptation Pitch Order *HL*	Test Duration Order	Test Pitch Order *LH*	Test Pitch Order *HL*
Lo/L–Hi/S						
	SL	.82	.66	*SL*	.89	.67
	LS	.66	.82	*LS*	.77	.84
Lo/S–Hi/L						
	SL	.86	.60	*SL*	.81	.70
	LS	.64	.77	*LS*	.68	.81

NOTE: Each entry is the probability of a correct pitch response.

TABLE 7. Probability of a Correct Pitch Response for Each of the Four Stimulus Events and the Value of Tau for Each of the Repeat-Session Subjects in Experiment 4

Condition and Subject	Preadaptation					Test				
	SL		*LS*			*SL*		*LS*		
	LH	*HL*	*LH*	*HL*	Tau	*LH*	*HL*	*LH*	*HL*	Tau
Lo/L–Hi/S										
H.B. (8)[a]	.69	.38	.59	.51	.22	.68	.40	.61	.42	.09
L.P. (9)[b]	.81	.63	.82	.58	−.06	.87	.60	.82	.56	.00
M.G. (8)[a]	.97	.88	.94	.96	.12	.99	.78	.93	.95	.23
K.C. (7)[b]	.83	.35	.54	.56	.50	.79	.35	.51	.48	.42
D.K. (8)[b]	.79	.79	.70	.79	.09	.81	.69	.68	.77	.21
I.S. (8)[a]	.96	.89	.99	.96	.05	.99	.85	1.00	.94	.07
Mean					.15					.17
Lo/S–Hi/L										
H.B. (8)	.65	.50	.65	.53	.02	.68	.35	.61	.46	.18
L.P. (10)	.71	.84	.78	.74	−.17	.60	.85	.70	.78	−.16
M.G. (8)	.95	.90	.91	.93	.06	.96	.84	.95	.94	.10
K.C. (6)	.73	.42	.57	.52	.26	.78	.37	.58	.43	.26
D.K. (8)	.84	.87	.85	.81	−.07	.79	.84	.90	.78	−.17
I.S. (8)	.81	.84	.88	.90	−.01	.83	.88	.86	.91	−.01
Mean					.02					.03

NOTE: The number of sessions under each adaptation condition is shown in parentheses.
[a]Received the *Lo/S–Hi/L* condition first.
[b]Received the *Lo/L–Hi/S* condition first.

diagonals is less during test (.24) than during preadaptation (.39). Of the 31 subjects, 19 showed performance during test, relative to preadaptation, that was in the expected direction (7 with larger *tau* after *Lo/L–Hi/S* and 12 with reduced *tau* after *Lo/S–Hi/L*), and one showed no change. According to the Wilcoxon T test these data indicate that the adaptation conditions did not have a significant differential effect on performance ($T = 200$, $n = 30$, $p > .05$). Judged pitch does not appear to be differentially influenced by the pair of stimuli presented during adaptation. Both adaptation conditions tended to reduce *tau*.

The probability of a correct pitch judgment, averaged over all sessions under a particular adaptation condition, and the value of *tau* are shown in TABLE 7 for each of the six repeat-session subjects. Only one subject (D.K.) showed the expected result: *tau* increased after *Lo/L–Hi/S* and decreased after *Lo/S–Hi/L*. For the other subjects adaptation did not have a consistent effect. Averaged over the six subjects there is virtually no change in *tau* under either condition. The data from these subjects are in agreement with the group data in TABLE 6. In neither case were we able to establish a negative pitch aftereffect contingent on duration.

DISCUSSION

The data show that judged duration lengthens with increases in pitch and judged pitch increases with duration. While others have reported demonstrations of auditory *tau* and of auditory *kappa*, the effects in the present experiments are clearer and more consistent. The data also provide strong support for the claim that TOE for relatively brief durations is positive.[8,16]

The adaptation conditions had a differential effect on TOE in Experiment 1 and on *kappa* in Experiment 3. This can be taken as evidence for order-contingent and pitch-contingent duration aftereffects respectively. In neither of these experiments did the adaptation conditions differentially influence $P(C)$. Thus, adaptation influences reports about perceived duration, but not discriminability.

If duration aftereffects are variants of the McCollough effect, one would expect to be able to demonstrate a duration-contingent pitch aftereffect. We were unable to do so in Experiment 4. The data from the repeat-session subjects suggest that neither the order-contingent nor the pitch-contingent duration aftereffect persists over sessions. Furthermore, the order-contingent duration aftereffect was not observed when a relatively short delay of 25 min was introduced after adaptation. In contrast, the McCollough effect and its variants persist for days, even weeks. Our data suggest fundamental differences between the duration aftereffects and the McCollough effect.

The favored account of the McCollough effect is the selective adaptation of feature detectors. For duration aftereffects, this account would imply duration detectors selectively tuned to temporal order or pitch. We know of no physiological evidence for such detectors.

TOE is a well-documented finding in the time perception literature. A satisfactory account of the phenomenon does not exist. Woodrow[16] suggested that the perceptual duration of the first presented event "gravitated" towards a remote standard, the indifference interval. This would result in a positive TOE for durations less than the indifference interval and a negative TOE for durations greater than the indifference interval. Aside from defining the indifference interval, this account has difficulty with the finding that TOE, whether positive or negative, tends toward zero as the interstimulus interval is increased.[17,18]

Hellström[11,12] has presented a model that postulates that the two internal durations on a trial are modified by the adaptation level or mean subjective duration, and the differential weights are then applied to the modified internal values. The reponse is based on a comparison of these weighted internal values with a criterion internal value. The differential weights associated with the first and second duration values are identified with retroactive and proactive interference effects respectively. This interpretation of the weight parameters is consistent with the finding that TOE tends toward zero as interstimulus interval is increased. Interference effects should decrease with interstimulus interval and therefore so should TOE. In order to account for a positive TOE with brief durations and a negative TOE with longer durations, Hellström has to assume that proactive interference is the greater for short durations and retroactive is the greater for longer durations. Why this should be the case is not developed in the model.

Allan[19] presented a model for TOE that provided a qualitative account of her data. However, Jamieson[20] showed that the model provided a poor quantitative fit to the data.

The data from the present experiments do not account for the finding that the sign of TOE depends upon duration. However, they do provide information about the differential effect of the two adaptation conditions on the preadaptation positive TOE. As was noted above, a positive TOE, as found during preadaptation in the present studies, indicates that R_{LS} responses are made more frequently than R_{SL} responses. The effect of adaptation on this distribution of responses can be explained as contrast in perceived duration between test and adaptation. The difference between the perceived durations of the two tones during long–short adaptation is large relative to the test difference. Therefore, only when the subject is certain during test that the first stimulus is the longer would the trial be labeled R_{LS}. This would result in fewer R_{LS} responses during test than during preadaptation and a more negative TOE. The reverse is the case for short–long adaptation. Again, the difference between the perceived durations of the two tones during adaptation is noticeable relative to the test difference. Now only when the subject is certain during test that the first stimulus is the shorter would the trial be labeled R_{SL}. This would result in more R_{LS} responses during test than during preadaptation and a more positive TOE. In decision theory language, the effect of adaptation is on the criterion. The criterion value increases after long–short adaptation and decreases after short–long adaptation.

Kappa and *tau* describe the finding that high-frequency tones and long tones "belong" together as do low-frequency tones and short tones. Thus, when tones are presented in the HL order, R_{LS} is more likely than R_{SL}; and when tones are presented in the LH order, R_{SL} is more likely than R_{LS}. This is the *kappa* effect. Similarly, R_{HL} is more likely than R_{LH} when tones are presented in the LS order; and R_{LH} is more likely than R_{HL} when tones are presented in the SL order. This is *tau*.

The contrast hypothesis, used to explain the differential effects of adaptation on TOE, can be applied to the differential effects of adaptation on *kappa* as well. During preadaptation, the criterion on HL trials is lower than the criterion on LH trials, resulting in more R_{LS} responses on HL trials than on LH trials and a *kappa* effect. Under the *Lo/S–Hi/L* adaptation condition, the high tone is always longer than the low tone and the difference between the perceived durations of the two tones is large relative to the test difference. Therefore, on HL test trials, only when the subject is certain that the first tone is the longer is the trial labeled R_{LS}, resulting in a decrease in R_{LS} on HL test trials relative to preadaptation; on LH test trials, only when the subject is certain that the first tone is the shorter is the trial labeled R_{SL}, resulting in a decrease in R_{SL} on LH test trials relative to preadaptation. That is, the criterion for R_{LS} is increased on HL trials and is lowered on LH trials, resulting in a reduced *kappa*. Now consider the *Lo/L–Hi/S* adaptation condition. The high tone is always shorter than the low tone. On HL test trials, only when the subject is certain that the first tone is the shorter is the trial labeled R_{SL}, resulting in a decrease in R_{SL} on HL test trials; on LH test trials, only when the subject is certain that the first tone is the longer is the trial labeled R_{LS}, resulting in a decrease in R_{LS} on LH test trials. That is, the criterion for R_{LS} is lowered on HL trials and is increased on LH trials, resulting in an increased *kappa*.

The contrast hypothesis explains the effect of adaptation through criterion shifts and therefore predicts that that discriminability will not be influenced by adaptation. This is what was found. One would not expect the effect of contrast to be long-lasting. Again, this is in accord with the data.

The data presented in this paper suggest that contingent duration aftereffects are different in kind from the McCollough effect. The contrast hypothesis is presented as a plausible account for the effect of adaptation. The hypothesis is compatible with existing models in the time perception literature. A more formal presentation of the hypothesis needs to be developed so that a quantitative evaluation can be undertaken.

SUMMARY

It has been shown that judged duration of tones depends on pitch (the *kappa* effect), on order of presentation (the time-order error), and on repetition (a negative duration aftereffect). Recently, a duration aftereffect contingent on pitch and a duration aftereffect contingent on order of presentation have been described. Our results suggest that these contingent duration aftereffects differ from the McCollough effect, a color aftereffect contingent on orientation, in two ways. They have a relatively short life and they are not symmetrical, in that while a pitch-contingent duration aftereffect could be established, a duration-contingent pitch aftereffect could not. In contrast, the McCollough effect persists for days and both an orientation-contingent color aftereffect and a color-contingent orientation aftereffect have been reported. A decision theory account for contingent duration aftereffects is outlined.

REFERENCES

1. Huppert, F. & G. Singer. 1967. An aftereffect in judgment of auditory duration. Percept. Psychophys. **2:** 544–546.
2. Walker, J. T. & A. L. Irion. 1979. Two new contingent aftereffects: Perceived auditory duration contingent on pitch and on temporal order. Percept. Psychophys. **26:** 241–244.
3. Walker, J. T., A. L. Irion & D. G. Gordon. 1981. Simple and contingent aftereffects of perceived duration in vision and audition. Percept. Psychophys. **29:** 475–486.
4. McCollough, C. 1965. Color adaptation of edge-detectors in the human visual system. Science **149:** 1115–1116.
5. Stromeyer, C. F. 1978. Form-color aftereffects in human vision. *In* Handbook of Sensory Physiology: (Vol. 8): Perception. R. Held, H. Leibowitz & H. L. Teuber, Eds. Springer-Verlag. Heidelberg, West Germany.
6. Murch, G. M. 1976. Classical conditioning of the McCollough effect: Temporal parameters. Vision Res. **16:** 615–619.
7. Siegel, S. & L. G. Allan. Associative bases of orientation-contingent colour aftereffects. Submitted for publication.
8. Allen, L. G. 1979. The perception of time. Percept. Psychophys. **26:** 340–354.
9. Majerovich, A. M. 1983. A Duration Aftereffect Contingent on Temporal Order. Unpublished B.Sc. thesis, McMaster University, Hamilton, Ontario.
10. Jamieson, D. G., & W. M. Petrusic. 1975. Presentation order effects in duration discrimination. Percept. Psychophys. **17:** 197–202.
11. Hellström, A. 1977. Time errors are perceptual. Psychol. Res. **39:** 345–388.
12. Hellström, A. 1977. On the nature of the time-error. Report from the Department of Psychology, University of Stockholm, Supplement 38.
13. Cohen, J., C. E. M. Hansel & J. C. Sylvester. 1955. Interdependence of judgments of space, time and movement. Acta Psychol. **11:** 360–372.
14. Yoblick, D. A. & G. Salvendy. 1970. Influence of frequency on the estimation of time for auditory, visual, and tactile modalities: The kappa effects. J. Exp. Psychol. **86:** 157–164.
15. Christensen, I. P. & Y. L. Huang. 1979. The auditory tau effect and memory for pitch. Percept. Psychophys. **26:** 489–494.
16. Woodrow, H. 1935. The effect of practice upon time-order errors in the comparison of temporal intervals. Psychol. Rev. **42:** 127–152.
17. Jamieson, D. G. & W. M. Petrusic. 1976. On a bias induced by the provision of feedback in psychophysical experiments. Acta Psychol. **40:** 199–206.
18. Jamieson, D. G. & W. M. Petrusic. 1978. Feedback versus an illusion in time. Perception **7:** 91–96.
19. Allan, L. G. 1977. The time-order error in judgments of duration. Can. J. Psychol. **31:** 24–31.
20. Jamieson, D. G. 1977. Two presentation order effects. Can. J. Psychol. **31:** 184–194.

Contingent Aftereffects and Situationally Coded Criteria: Discussion Paper

MICHEL TREISMAN

Department of Experimental Psychology
University of Oxford
Oxford OX1 3UD, England

It is a pleasure to comment on this series of interesting and instructive papers on time perception.

Richard Schweickert's[1] paper on the application of critical path analysis to information-processing models takes the major advance in interpreting such experiments, the additive factor method first put forward by Saul Sternberg,[2] a step forward. Dr. Schweickert's techniques may make it possible to unravel not only simple sequential models, but also those that include concurrent parallel processes.

Drs. Wasserman, DeLong, and Larew[3] examined the problem of order perception in pigeons. The study of perception in animals, as in infants, has started to take off in recent years, both because of an increased willingness to suppose that complex abilities may exist, and because of improvements in techniques for demonstrating them. This paper clearly demonstrates that pigeons make and retain discriminations of order.

Turning to the perception of temporal order in humans, George Sperling[4] has given us a fine example of his ability for lucid analysis and incisive experimental investigation. His results are of great interest and deserve consideration at leisure.

Finally, both Donald Jamieson[5] and Lorraine Allan[6] have presented us with interesting observations on time-order effects. Dr. Allan has made a careful investigation of contingent aftereffects in duration judgments, giving us a great deal of information. She concludes by rejecting a parallel with the color-contingent aftereffect and suggests that the explanation for the effects she found may lie in contrast mediated by decision processes. I believe that Dr. Allan is on the right track and I would like to take this argument a little further.

"Contrast," like "assimilation," is a familiar term and is often taken for an explanatory term. But is it? I suggest that it is no more than a label for effects lying in a certain direction, and requires explanation itself. Terms like "kappa" and "tau," unfortunately, are not even descriptive.

A theory of criterion-setting on which I have recently been working[7] offers explanations for sequential effects, of which contrast is an example, and it may be of interest to see how well such a theory can account for the present results. The theory derives sequential effects from the operation of mechanisms responsible for establishing the value of the decision criterion from trial to trial.

The theory of criterion-setting has been applied to Thurstonian models of detection, identification, absolute judgment, and magnitude estimation. These models assume that repeated presentation of a stimulus produces a normal distribution of sensory effects on a central scale, and that judgments are made by the use of decision criteria.

Let me briefly summarize the main assumptions of the theory. It addresses the question "Why does a response criterion have one value rather than another?" and it offers the answer that there are three main mechanisms (excluding consideration of feedback) that determine the momentary values of criteria. The first, a global mechanism, establishes a long-term reference value for each criterion. This is the process implicitly assumed by signal-detection theory. It uses past experience and knowledge of global parameters, such as the overall probability of the signal and the payoff values, to determine the best reference criterion.

However, the reference criterion is not maintained without change from trial to trial, save for that produced by random noise. The reason for this is that each trial contributes short-lived information to the observer which he must apply there and then if he is to perform optimally. Consider an observer attempting to detect the presence of a stimulus. On trial *i* he concludes that the stimulus is present. In daily experience, objects that are truly present tend to persist and show the same or similar sensory characteristics, at least for short periods of time. Thus, the observer's criterion on the next few trials should take account of this information, which is equivalent to a short-lived increase in the probability of the stimulus. Accordingly, the observer should lower his criterion for a limited period, which will make repetition of the positive response more likely. (This corresponds to "assimilation.") By lowering his criterion in this case he "tracks" the current state of the world. To meet this requirement, criterion-setting theory proposes that each response sets up a short-lived memory trace or "indicator trace" which indicates a direction and magnitude of shift required for the criterion, and that this trace decays with time. The working criterion at any time is given by the sum of the reference criterion plus the currently nonzero indicator traces.

The tracking procedure may be put in force when the subject makes a decision, covert or overt, categorizing the input. But whether or not he does so, the sensory inputs themselves constitute a third source of information which may be used to stabilize the criterion in relation to the sensory inflow. They indicate whether it is properly centered in relation to the flux of incoming sensory inputs, or is away to one side, in which case little information will be conveyed by the responses. Thus, if the sensory input on trial *i* is greater than the criterion, this indicates that the criterion should shift to the right on the next trial; if the sensory input is less, the criterion should shift to the left. The extent of the shift should increase with the disparity. The overall effect of such shifts will be to center the criterion. It follows that if the sensory input is below the detection criterion on one trial, inducing a shift to the left, the probability of a detection response will be increased on the next. This constitutes "contrast." FIGURE 1 shows a reanalysis of data obtained by Tanner *et al.*[8] in which the subject had to identify a tone as LOUD or SOFT. These data illustrate both tracking and stabilization in operation in the same experiment: if the probability of a LOUD response, P(LOUD), is plotted against previous responses, it is greater after a preceding LOUD response than after a SOFT one. If we plot it against the preceding stimulus magnitude, it is greater following a *soft* stimulus than following a *loud* one.

Tracking is assumed to have relatively short-lived effects, since the relevance of previous judgments may soon decline, but stabilization may produce longer-lived traces, centering the criterion in relation to a larger sample of the sensory flux.

This theory can be applied to provide explanations for the *kappa* effect, the *tau* effect, and at least some contingent aftereffects. For this purpose we add the following two assumptions: First, criteria may be situationally coded. That is, a criterion is defined not only for a given judgment, but also for particular conditions under which this judgment may be made. Then contributions to the reference criterion, and indicator traces, produced under a given condition are labeled as applying to the criterion only when employed under the same condition. Thus, the decision criterion may have different values for different sets of circumstances. For example, if a discriminative stimulus indicates that the probability of the signal is high, this should cause the reference criterion to be lowered when that stimulus is present. This is a natural extension of the assumption that the global criterion-setting mechanism takes account of the prior probability of the signal.

The second assumption is that if a decision takes place in stages, then the evidence provided by an earlier stage may be preserved as a covert quantitative value to be employed at a later stage. But the magnitude of this quantitative value is determined

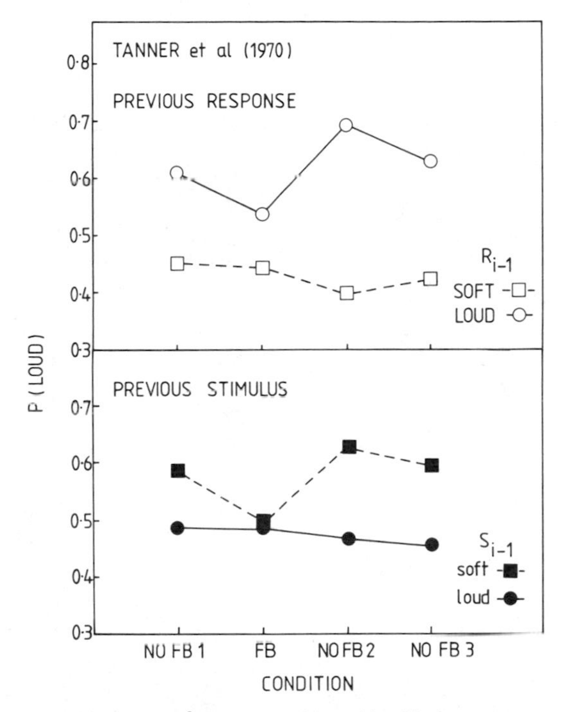

FIGURE 1. Data from Tanner *et al.*[8] in which subjects identified a tone as LOUD or SOFT. In the *upper panel* the probability of the response LOUD is shown as a function of the response on the preceding trial (R_{i-1}); in the *lower panel* it is shown as a function of the preceding stimulus (s_{i-1}). The *upper panel* illustrates a positive sequential effect ("assimilation") of the preceding response; the *lower panel* shows a negative dependency ("contrast") on the preceding stimulus.

by a criterion whose function is to set the origin of the scale on which this value is measured. This assumption will be illustrated below.

We now examine how these assumptions may be used to build models for contingent effects.

THE KAPPA EFFECT

Cohen *et al.*[9] exposed subjects to three lights set in order along a line and flashing in a repeated cycle. Subjects adjusted the time of the central flash to make t_l, the interval between lights L_1 and L_2 (separated by distance d_1) equal to the interval t_2 between the flashes L_2 and L_3 (separated by distance d_2). For equal spatial separations approximately equal time intervals were produced. But if d_1 was greater than d_2, then t_1 was less than t_2. For example, if $d_1/d_2 = 3/1$, the mean settings gave $t_1 = 0.68$ and $t_2 = 0.82$ sec for a total cycle of 1.5 sec.

The criterion-setting model for this experiment is shown in FIGURE 2. When engaged in adjusting the time of L_2, subjects make covert judgments of t_1 and t_2, deciding whether each of them is too short or too long. For each judgment they use a

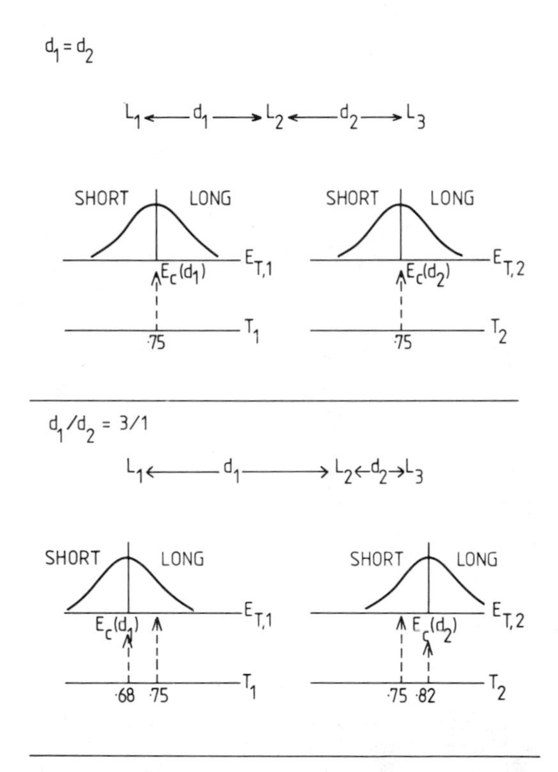

FIGURE 2. The model for the *kappa* effect[9] given by criterion-setting theory. The *upper panel* shows the cycle of three flashing lights, L_1, L_2 and L_3, separated by equal distances d_1 and d_2. The subject adjusts L_2 to flash at half the total duration. The judgments for t_1, represented on the time dimension T_1, and for t_2, represented on T_2, are shown separately. Each duration produces a distribution of central effects on the corresponding central decision axis, E_T, shown for convenience as two axes, $E_{T,1}$ and $E_{T,2}$. After each adjustment the subject decides whether the first interval is too long or too short using a criterion coded for the first distance, $E_c(d_1)$. He makes a similar covert judgment about second interval. Adjustments continue until a setting is found such that the sensory effects of the first interval are approximately equally distributed about $E_c(d_1)$, and those of the second interval about $E_c(d_2)$. Since $d_1 = d_2$, the criteria should be similar and the final judgment unbiased.

The *lower panel* applies when d_1 is three-quarters of the whole distance. The criterion associated with the long distance, d_1, is shifted to the left, and the criterion associated with d_2 is shifted to the right. Then 0.75 seconds will seem too long for the first interval, and too short for the second. Adjustment will find a shorter interval to match $E_c(d_1)$ and a longer interval for $E_c(d_2)$.

situationally coded criterion, that is, a criterion coded by association with the first distance $E_c(d_1)$ or by association with the second $E_c(d_2)$. When $d_l = d_2$, the criteria should be the same. But when the distances are unequal, the global mechanism will take note of this in setting the reference positions for the two criteria. In the real world movements are usually smooth: if an object moves successively through distances d_l and d_2, our experience is almost always that the time taken is monotonically related to the distances, especially when the movement is horizontal. Thus, for a long distance, the global mechanism should expect that the probability of a long interval will be increased and should set the criterion for detecting long intervals low; the reverse will apply when the distance is short. This is analogous to the effect of signal probability on a signal-detection criterion.

The lower panel applies the model when d_1 is long and d_2 short. The first criterion shifts to the left and the second to the right. Since the method of adjustment finds the stimulus whose mean central effect best matches the criterion, it will select t_1 less than 0.75 sec, and t_2 greater. Remember that the stabilization process tends to shift each criterion in the direction of recent sensory inputs; thus, if both start too low (giving covert LONG responses for both intervals) or too high, this will automatically be corrected.

THE TAU EFFECT

We now consider the auditory *tau* effect described by Cohen *et al.*[10] These investigators presented three brief tones in succession, with an interval t_1 between the first two tones and t_2 between the second and third, and required the subject to adjust the middle tone to be intermediate in pitch. With an initial tone of 1000 Hz and a final tone of 3000 Hz, given in a 1.5-sec cycle, the middle tone was set to 2068 Hz for $t_1 = 0.5$ sec, and to 1676 Hz for $t_1 = 1.0$ sec. When the initial and final anchor tones were 3000 Hz and 1000 Hz, respectively, the middle tone was set to 1725 Hz for $t_l = 0.5$ sec, and to 1921 Hz for $t_l = 1.0$ sec.

This effect is formally similar to the *kappa* effect and could be explained by a similar model, although this would rest on the weaker assumption that we expect successive tones to change frequency at a regular rate. But a simpler explanation is suggested if we note that in this experiment time may function in two ways: as a processed sensory input, as the authors assume, or simply as the familiar independent variable present in all experiments. The latter is all that is required for a simpler model in which the results follow directly from the features of the tracking criterion-setting process, without requiring criteria coded for the two time intervals. This model is illustrated in FIGURE 3.

The figure shows the central scale for frequency, E_F, on which presentations of the anchor tones, 1000 and 3000 Hz, produce sensory distributions. It also shows the criterion E_c which would be used in identifying them. When an anchor tone is presented, the subject needs to identify it in order to perform the task; this categorization sets up a tracking adjustment to the criterion. Thus, in the figure, the categorization of each anchor stimulus sets up a corresponding indicator trace. At any moment the criterion is given by the sum of its initial value and the existing indicator traces; as the latter decay, it returns to its initial value. The method of adjustment hunts this criterion. It follows that when this method is used with $t_l = 0.5$ sec, it will find the bisecting stimulus value further removed from the immediately preceding anchor stimulus than when $t_1 = 1.0$ sec. This is sufficient to account for the auditory *tau* effect.

CONTINGENT AFTEREFFECTS: DURATION CONTINGENT ON ORDER

Contingent aftereffects are found when judgments are preceded by an adaptation period during which the subject is passively exposed to a series of stimulus presentations which require no perceptual decisions on his part. In the absence of perceptual categorization, tracking may not occur or, since it is short-lived, its effects will be outweighed by the accumulation of longer-lived stabilization indicator traces. The passive registration of sensory inputs lends itself to the manifestation of stabilization effects. These provide the basis for a model of the contingent aftereffects described by Dr. Allan.

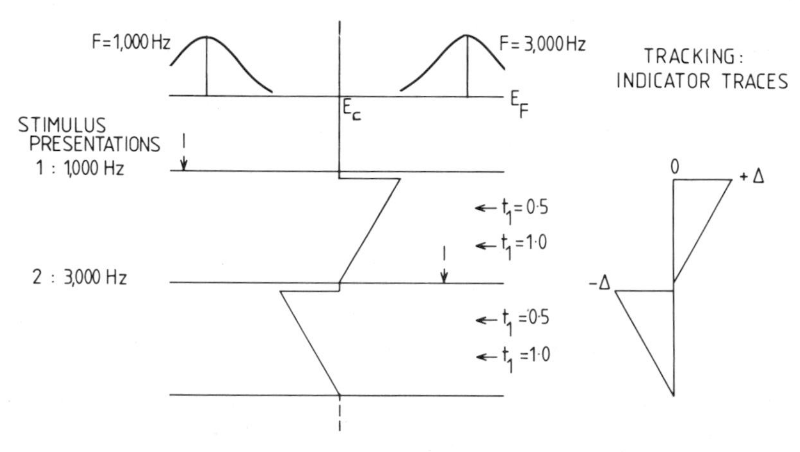

FIGURE 3. The auditory *tau* effect: an expression of the tracking mechanism. Repeated presentations of two anchor frequencies, 1000 and 3000 Hz, produce central effects distributed on a central scale, E_F, on which there is also an intermediate criterion, E_c. We see the effect of a presentation of the anchor stimulus, 1000 Hz. (Time proceeds downward in the figure, and the decision axis, E_F, is redrawn for each stimulus presentation.) The sensory input falls below E_c and is categorized as "1000." The tracking mechanism produces a corresponding adjustment in criterion: an indicator trace of value $+ \Delta$ is set up, and the value of this trace is added to the initial value of the criterion to give its current value. As the indicator trace decays, the current value of the criterion correspondingly returns to its initial value. In this illustration the indicator trace disappears completely before the next stimulus presentation. A second anchor stimulus is then presented, is classified as "3000" and sets up an indicator trace of magnitude $- \Delta$. This is again added to the criterion. For interpresentation intervals of 1.5 seconds, the position of the criterion when $t_1 = 0.5$, or when $t_1 = 1.0$ seconds, is indicated in each case.

The upper panel of FIGURE 4 illustrates the preadaptation stage. The subject must decide which of two successive tones is the longer. This may be done in two stages. In the first, the stimuli are assessed as they occur, their strength being represented by quantitative outputs taken from the criterion as origin, separate coded criteria being used for interval 1 and interval 2. These outputs are compared in the second stage and determine the response: if the difference between the second measure and the first is positive, the second interval is chosen as longer, otherwise the first is selected. The upper panel assumes, for simplicity, that there is no initial bias. Then the criteria coded for the first and second intervals are the same, so that the mean sensory magnitude assigned to a given stimulus will be the same in either interval. The subject's

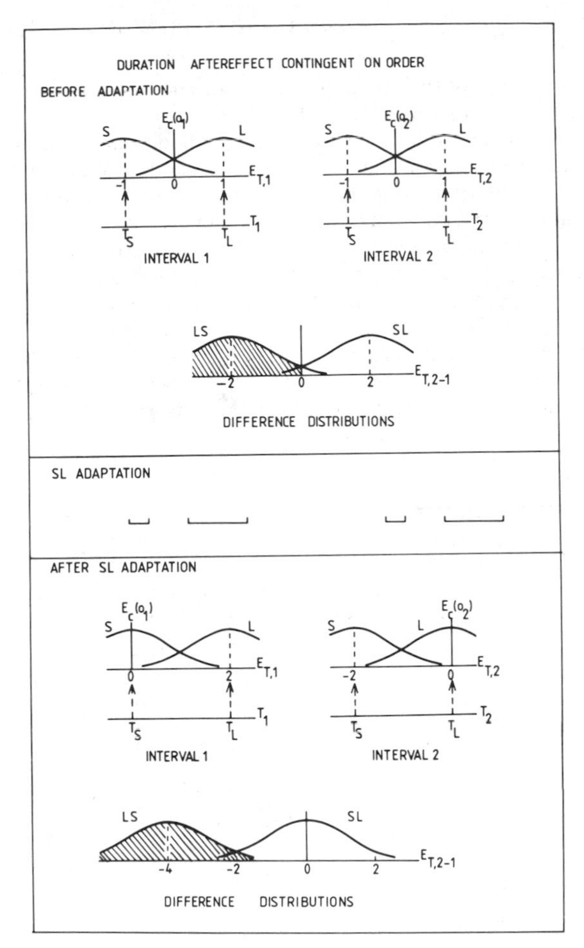

FIGURE 4. Order-contingent duration aftereffect: the model given by criterion-setting theory. Pairs of durations are presented and the subject makes a forced-choice response. The time dimension and the central effects of presented intervals are represented separately for the first and second intervals. Assuming there is no preexisting bias, the two coded criteria, $E_c(o_1)$ and $E_c(o_2)$, are identically placed. Each determines the origin on the scale of central effects, E_T, for intervals in the corresponding position. Since the origins are initially the same, the short interval (T_S) and the long interval (T_L) produce distributions of central effects with means -1 and 1, respectively, whether these intervals occur first or second. A quantitative first-stage output is found for each interval, given by the measure of the sensory effect registered on the E_T scale for that interval. The difference between the two measures on a given trial, $E_{T,2} - E_{T,1}$, may be represented on a difference axis, $E_{T,2-1}$. The difference distributions resulting from LONG-SHORT (*LS*) presentations and from SHORT-LONG (*SL*) presentations are shown on this axis. For *SL* pairs the mean of the corresponding distribution is 2, and for *LS* pairs the mean is -2. The probability of a correct response to *LS* pairs, $P(R_{LS}|LS)$ is the area of the *LS* distribution to the left of the origin (*stippled area*) and the corresponding probability for *SL* pairs is the area of the *SL* distribution to the right of the origin. Thus, the time-order error as defined by Dr. Allan,[6] TOE = $P(R_{LS}|LS) - P(R_{SL}|SL)$, is zero.

The effects of *SL* adaptation, to repeated SHORT-LONG pairs, on later judgments is illustrated in the *lower panel*. The criterion for the first interval is shifted to the left, taking the origin with it, and $E_c(o_2)$ is shifted to the right. The sensory inputs produced by the two stimuli are unchanged. But the mean sensory magnitude assigned to a presentation of T_L in the first interval is now 2, not 1, because of the criterion shift to the left, and in the second interval it is now 0, not 1, because the origin has moved to the right. Thus, the means of the difference distributions are changed for each order of presentation. Both are shifted to the left and the *TOE* is increased.

forced-choice decisions are determined by the difference distributions shown on the difference decision axis. (A roving standard design was used but we may disregard this since a difference distribution is calculated.) Dr. Allan defines the time-order error (TOE) as the proportion of the *LS* distribution to the left of the origin, reduced by the proportion of the *SL* distribution to the right of the origin. In the figure this is zero. In fact, for her results we should assume that the first criterion is displaced to the left or the second to the right, to account for the initial small positive error.

The effects of adaptation to short–long pairs are shown in the lower panel. During adaptation $E_c(o_1)$, the criterion associated with the first interval, is subject to the stabilization effect of the short intervals repeatedly presented in position 1. Their sensory inputs will lie well to the left of $E_c(o_1)$ and their stabilizing indicator traces will produce a shift of this criterion to the left. Similarly, the long intervals given in position 2 during adaptation will shift the criterion for this interval to the right. The sensory

BEFORE ADAPTATION

LOW-HIGH SEQUENCES

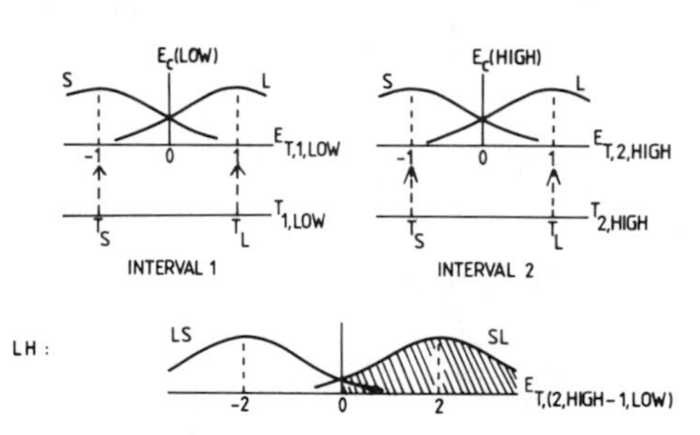

HIGH-LOW SEQUENCES

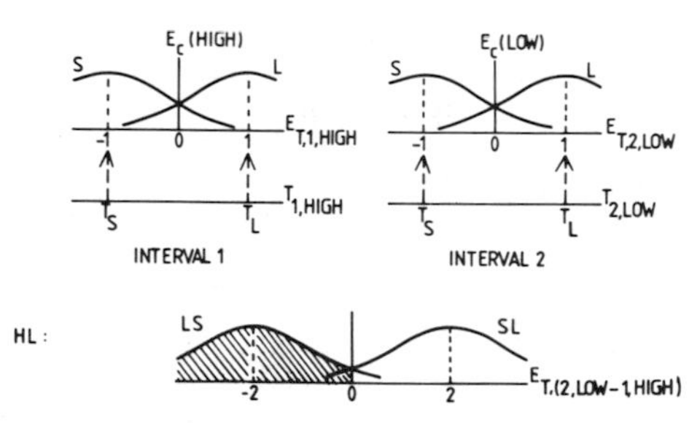

FIGURE 5A. Duration aftereffect contingent on pitch: judgments made prior to adaptation, assuming that there is no initial bias. Whether the first tone is low and the second high, or the first high and the second low, the two coded criteria, $E_c(LOW)$ and $E_c(HIGH)$, are the same in each case. Thus, the mean quantitative measure assigned to the sensory effects of a given duration, T_S or T_L, is the same whether it is the first or the second interval, and presented as a low or a high tone. Consequently, the difference distributions for the *SL* and *LS* orders will be the same and symmetrical for each tone order. *Kappa* is given by the proportion of the *SL* distribution to the right of the origin, for the low–high tone order, plus the proportion of the *LS* distribution to the left of the origin for the high–low tone order (both shown stippled) reduced by the proportion of *LS* to the left of the origin in the first case and the proportion of *SL* to the right in the second. In this case, it is zero.

inputs given by T_S and T_L are unchanged. But the quantitative measures they give rise to in each interval are shifted because of the shifts in origin (the order-coded criteria) on the E_T scale for each interval, and in consequence the *LS* and *SL* difference distributions are each moved to the left. This will manifest as an increase in *TOE*. In the same way, *LS* adaptation would produce a decrease in *TOE*.

DURATION AFTEREFFECT CONTINGENT ON PITCH

In her fourth and fifth experiments Dr. Allan adapted subjects to short and long intervals (200 and 600 msec) given as low and high tones (600 and 900 Hz). The

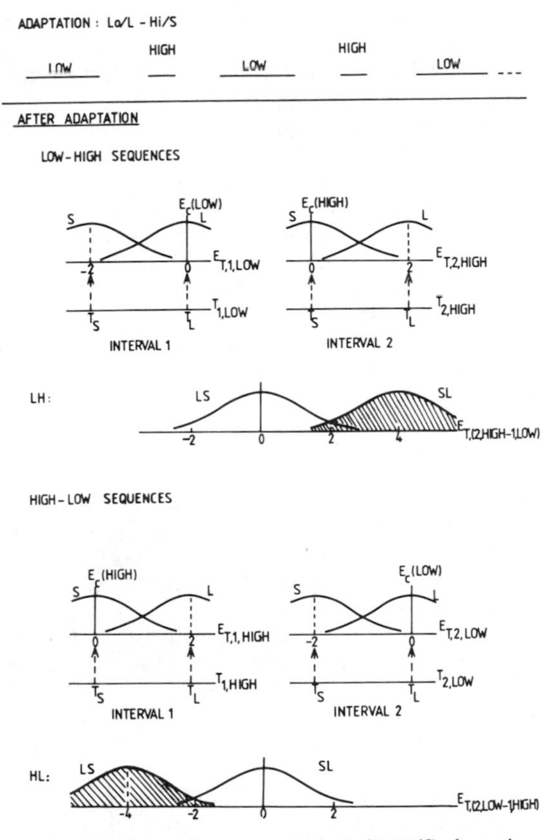

FIGURE 5B. Duration aftereffect contingent on pitch: *Lo/L-Hi/S* adaptation and the state of the criteria after such adaptation. The criterion coded by association with the low tone, $E_c(LOW)$, has been exposed to long intervals during adaptation. It has shifted to the right, so that in this illustration the origin now coincides with the mean of the T_L distribution. This holds for either tone sequence during judgment. Similarly, $E_c(HIGH)$ is shifted to the left in both cases. Although the sensory inputs produced by T_L and T_S have not changed, they will now both be assigned lower values when presented as low tones and higher values when presented as high tones. Consequently both difference distributions are shifted to the right for the low–high tone sequences and to the left for the high–low sequences. Consequently, *kappa* is increased.

subject was subsequently required to decide which of two tones (one at 600, the other at 900 Hz) was the longer. An effect was found which Dr. Allan defines as

$$\text{kappa} = [P(R_{SL} | \text{SL and LH}) + P(R_{LS} | \text{LS and HL})] - [P(R_{SL} | \text{SL and HL}) + P(R_{LS} | \text{LS and LH})].$$

Adaptation to low-frequency long tones and high-frequency short tones increased *kappa*; the reverse adaptation reduced it. These results can also be explained by stabilization effects on coded criteria, the criteria in this case being coded by association with the low or high frequencies. The model is illustrated in FIGURES 5A and B.

Prior to adaptation, and assuming no initial bias, the criteria coded by pitch are the same, giving symmetrical difference distributions and a zero value for *kappa*. If low-frequency long-duration intervals are given during adaptation, these will act only on the criterion coded for association with low frequency. Since they are long, they will cause stabilization shifts of $E_c(LOW)$ to the right. Equivalent effects will be produced by other adaptation combinations. The figure demonstrates how *Lo/L-Hi/S* adaptation shifts the criteria selectively and consequently causes the difference distributions to shift to the right for low-frequency–high-frequency sequences and to the left for high–low sequences. This gives a positive value of *kappa*. Similarly, it can be shown that *Lo/S–Hi/L* adaptation will cause a reduction in *kappa*.

Prior to adaptation, Dr. Allan's subjects showed a positive value of *kappa*. This may possibly reflect a bias in the environmental noises we experience, low-pitched rumbles tending to be longer than higher-pitched squeaks.

It appears that criterion-setting theory, extended by the hypothesis of situational coding of criteria, can account for the major intersensory effects involving time. This result has a number of interesting implications.

1. If criterion-setting theory can account for the present contingent aftereffects, it is possible that it may also account for many of the contingent aftereffects found in other modalities, such as the color-contingent aftereffect.[11] This would relate such phenomena directly to sequential dependencies in general[7] and so provide a more parsimonious account than do theories that require the assumption that neural feature detectors are the site of adaptation or theories that postulate special types of learning.

2. Whereas the color-contingent aftereffect may endure days or weeks,[12] Dr. Allan found that her order-contingent aftereffect did not last 25 minutes. The explanation for this difference may lie in situational coding: if a coded criterion is modified only in the situation it is intended for, then its life may be a function of the frequency with which it encounters that specific situation and is modified in it. The natural environment frequently presents us with sequences of sounds of varying durations, which may expose criteria coded for order to noisy influences, whereas the bar and color patterns used in inducing the color-contingent aftereffects are rarely seen under natural conditions.

3. In her final two experiments, Dr. Allan failed to find an effect on pitch contingent on the duration of adapting stimuli. Thus, this experiment provides no evidence that pitch criteria may be coded by temporal duration. We have also seen above that the auditory *tau* effect of Cohen *et al.*[10] does not require coded criteria for its explanation. These observations raise the possibility that duration cannot readily be used to code pitch criteria and perhaps criteria on other modalities. The presence or absence of contingent aftereffects may provide us with evidence on relations between modalities which allow inputs in one to affect sensory decisions in another.

4. The present analysis provides a basis for distinguishing between different types

of cross-sensory effects on decision. Thus we can contrast Cohen *et al.*'s[9] *kappa*, which we have attributed to the global criterion-setting process, and the finding described by Dr. Allan, which may be mediated by stabilization of coded criteria.

REFERENCES

1. SCHWEICKERT, R. J. 1984. The representation of mental activities in critical path networks. This volume.
2. STERNBERG, S. 1969. The discovery of processing stages: Extensions of Donders' method. *In* Attention and Performance: II. W. G. Koster, Ed. North-Holland. Amsterdam.
3. WASSERMAN, E. A., R. E. DELONG & M. B. LAREW. 1984. Temporal order and duration: Their discrimination and retention by pigeons. This volume.
4. SPERLING, G. Temporal order of brief visual events. Presented at this conference, but not submitted for publication.
5. JAMIESON, D. G., E. SLAWINSKA, M. F. CHEESMAN & B. E. VARAS. 1984. Timing perturbations with complex auditory stimuli. This volume.
6. ALLAN, L. G. 1984. Contingent aftereffects in duration judgments. This volume.
7. TREISMAN, M. 1984. A theory of criterion setting with an application to sequential effects. Psychol. Rev. **91:** 68–111.
8. TANNER, T. A., J. A. RAUK & R. C. ATKINSON. 1967. Signal recognition as influenced by information feedback. J. Math. Psychol. **7:** 259–274.
9. COHEN, J., C. E. M. HANSEL & J. D. SYLVESTER. 1955. Interdependence in judgments of space, time and movement. Acta Psychol. **11:** 360–372.
10. COHEN, J., C. E. M. HANSEL & J. D. SYLVESTER. 1954. Interdependence of temporal and auditory judgments. Nature (London) **174:** 642–644.
11. MCCOLLOUGH, C. 1965. Color adaptation of edge-detectors in the human visual system. Science **149:** 1115–1116.
12. STROMEYER, C. F. 1978. Form-color aftereffects in human vision. *In* Handbook of Sensory Physiology, Vol. 8. R. Held, H. Leibowitz & H. L. Teuber, Eds. Springer-Verlag, Heidelberg.

Introduction

ALAN M. WING

Medical Research Council
Applied Psychology Unit
Cambridge CB2 2EF, England

Of the following four papers, three—those of Stelmach and Requin and Semjen, and my own—are concerned with the temporal patterns produced in sequences of movements, while the other, that of Jones, describes a pitch-recognition task in which the stimuli are presented in different temporal patterns. All four papers use human subjects, while my own, with Keele and Margolin, takes advantage of deficits in performance resulting from a neurologic disorder to explore a theoretical model developed on the basis of the more usual "normal" population of college students.

The papers by Jones and Requin and Semjen return to a theme taken up in other research reported at this conference: Physically distinguishable dimensions of the stimuli (the interstimulus interval of Schweickert) or of the movements (the interresponse interval of Eisler) interact in their effects on performance. The internal representation of a series of movements and the consequences of timing of the movements is addressed by three of the papers: in the paper by Stelmach the timing is incidental to the production of handwriting, but in the work of Requin and Semjen and Wing *et al.* the subjects have to produce well-regulated interresponse intervals. All three papers examine the intervals between excessive movements, either in terms of means (Stelmach and Requin and Semjen) or in terms of variances (Wing *et al.*). In addition, the papers by Stelmach and Requin and Semjen attempt to characterize the nature of separation for movements using the latency with which the sequences are initiated.

Motor Programming and Temporal Patterns in Handwriting[a]

GEORGE E. STELMACH,[b] PATRICIA A. MULLINS,[c] AND
HANS-LEO TEULINGS[d]

[b]*Motor Behavior Laboratory*
University of Wisconsin
Madison, Wisconsin 53706

[c]*Committee on Cognition and Communication*
University of Chicago
Chicago, Illinois 60637

[d]*Psychological Laboratory*
University of Nijmegen
Nijmegen, the Netherlands

Over the past decade there has been considerable interest in the process by which motor commands pass from some abstract representation in the brain to a functional code at the muscular level.[1-4] The term most frequently used in conjunction with this aspect of movement specification is that of a motor program. Henry and Rogers[3] and Keele[4] purported the motor program to be a specification of parameters of action such as force, velocity, duration, and sequencing of the involved muscles. Over the years this idea has changed so that today a motor program is thought of as an abstract non-muscle-specific representation of motor acts.

Recently, an increasing number of studies have investigated motor programs in complex motor tasks, such as the production of words,[7,8] typewriting,[7,10] and piano playing,[9] in an attempt to describe the characteristics of these programs. Handwriting, involving the coordination of orthogonal muscle systems of the forearm, hand, and fingers, with intricate timing relationships, is another skill that is proving to be useful for studying the nature of motor programming.

As described from a motor control perspective, handwriting can be viewed as the production of flexion and extension movements of the thumb, index, middle, and ring fingers, and abduction and adduction of the hand around the wrist joint. For the most part, vertical strokes of letters are produced by thumb–finger flexions and extensions, and horizontal strokes are produced by hand abduction and adductions. A steady rotation about the shoulder produces left-to-right progression. Despite the rich theoretical motor control aspects of handwiting, most research until recently has concentrated on handwriting characteristics and pathologic abnormalities.[11,12] However, the widely quoted description of the handwriting simulated by Hollerbach,[13] the expanding research effort from the Psychological Laboratory at the University of Nijmegen,[14-16] and the time–space constancy data by Viviani and Terzuolo[17] have all increased interest in handwriting as a skill for gaining insight into the representation of complex actions and the structure and organization of motor commands.

[a]This research was conducted at the University of Nijmegen and was partially sponsored by the Graduate School Research Committee of the University of Wisconsin; by the Scientific Affairs Division of NATO; by the Senior Fulbright Scholar Program; and by the Netherlands Organization for the Advancement of Pure Research.

Recent research from Nijmegen[14–16,20] has clearly shown that handwriting movements, often considered to be continous for letters and words, can be decomposed into individual strokes. The resulting dynamic description of size, velocity, and duration parameters of the writing trace is used to help understand the control parameters of the motor output. Using such measures, Viviani and Terzuolo[17] found in handwriting an invariant structure for space–time parameters, leading them to conclude that the basic unit of organization for motor output corresponds to the sequence that is specified rather than letter-to-letter transitions. The homotetic behavior in the time domain for handwriting is captured by the tangential velocity when subjects are required to intentionally modify their writing speed while keeping constant the size of the letter. When the total duration of the writing movement changes, instantaneous values of the velocity change proportionally in such a way as to leave invariant the ratios among the times of occurrence of the major features of the trajectory profile. Similar trajectory profiles are also observed when changes in writing size are required.[17,23] This observation led Viviani and Terzuolo to conclude that the observed homotetic behavior in the time domain is a general organizational principle of learned movements.

Along a different dimension, Raibert[6] has shown that handwriting possesses striking similarities when writing with the left or right hands, right arm, pen taped to the right foot, or pen held in the mouth. The observation suggests that there is a common memorial representation for all of the writing behaviors that is implemented regardless of limb and muscles involved.[5] Keele,[1] attempting to relate these data to Hollerbach's[13] spring model, suggests that if, in each of Raibert's handwriting examples the muscles are organized into relationships that operate at right angles to each other, then the same time patterns could be applied and the output, except for size, would be very much the same.

These studies suggest that cursive handwriting may result from the relative timing of sets of muscles in a generalized force–time relationship. A program for a letter could then be thought of as a specification of the phasing between coupled oscillations in the horizontal and vertical direction with transitions from one letter to another occurring when the phases are changed by some underlying timing component. The unit that is specified would be in this sense a complete cycle at one phase setting of the agonist–antagonist muscle pairs.[1]

In search of a unifying principle underpinning these various hypotheses, two fundamental issues are investigated herein, namely, the generality of the invariant characteristics of the relative timing of a handwriting sequence, and the nature and specification of a unit of the motor program for handwriting. The first experiment to be reported examines the susceptibility to modification of the widely acclaimed space–time invariance reported by Viviani and Terzuolo,[17] in a prepared handwriting sequence. Stroking characteristics in situations where a given sequence is prepared and executed are compared to those where the prepared sequence must be modified at the time of an imperative response signal. Modifications of a prepared allograph required the subjects to either increase handwriting size (3 cm) or decrease it (1.5 cm). Preparation state was manipulated by expectancy information that created a bias toward preparing one of the two handwriting sizes. This type of paradigm was first used by La Berge, van Gelder and Yellott[18] and has been recently employed by Larish and Stelmach[19] and Stelmach and Teulings[20] to examine the restructuring aspects of motor programming processes.

At issue in the present experiment is whether the space–time invariance phenomenon operates differentially under advance-planning and parameter-restructuring situations. Because size is the only variant between probability conditions in the experiment, the main difference between conditions is the "reparameterizing" or readjustment of force–time parameters required for proper execution.[21,22] Heretofore, the space–time invariance has only been observed in situations where the subject could fully prepare the execution of writing different sizes. One of the issues to be explored in this experiment is whether the effect of change in size prescription is limited to the initial stroke or whether the context of the size preparation will alter the duration of the stroke characteristics throughout the allograph. If learned handwriting movements are represented in some abstract code and retrieved as a unit, it may be posited that in the restructuring situation the subject will have difficulty adjusting size parameters throughout the handwriting sequence.

EXPERIMENT 1

Method

Subjects

Subjects were 13 right-handed male and female psychology students from the University of Nijmegen. They were either paid or given class credit for participation.

Apparatus

The writing movements were recorded by a computer–controlled digitizer (Vector General Data Tablet DTI). The position of the tip of the electronic pen, expressed in horizontal and vertical coordinates with a combined RMS error better than 0.2 mm, was sampled at a rate of 200 Hz.[20] The pen tip was an ordinary ballpoint refill and the subject wrote on a sheet of paper fixed on the digitizer surface. The digitizer was positioned such that the subject's individual writing slope was parallel to the horizontal axis of the digitizer. Direct vision of the writing hand was eliminated by the placement of a shield above the writing surface.

A display (Vector General Graphics Display Series 3 Model 2DS) was positioned at a distance of 125 cm directly in front of the subject at eye level and it allowed the tachistoscopic presentation of stimuli (the stimuli were built up within 1 msec).

Procedure

Each trial began with a buzzer and after a 1-second delay, a writing stimulus was displayed for 100 msec. One allograph was presented (*hye*) and depending on the probability condition, the subjects were required to write it in one of two sizes (large or small). The subject's task was to initiate a response to the writing stimulus as fast as possible with minimal error. All pen movements were recorded and stored and 1 second later the recorded writing trace, the stimulus, and the reaction time were displayed to the subject for 2 sec. The fed-back writing trace served also as a means to instruct the subjects in the correct writing size. As such, the traces of the letters were fitted between two horizontal lines representing the correct writing size of 1.5 or 3 cm. These lines were also useful to inform the subject that s/he was writing horizontally and correctly.

The subjects were familiarized with the writing task and given sufficient practice (100 trials of each allograph size) until they had no difficulty. Then each subject participated in a block of 55 trials where the type of allograph displayed and written altered between *hye* large (3 cm) or small (1.5 cm). As an aid to preparation, the highly probable allograph size was continually visible to the subject during a block of trials and when the subjects heard the buzzer (1 sec prior to "go" signal), they were to prepare the displayed handwriting reponse and write it as fast as possible when the allograph was flashed on the display. The distinguishing feature between blocks was the designation of which allograph size appeared most frequently. Within a block of trials, 40 trials consisted of the flashed allograph matching the displayed allograph (80%), 10 where it did not (20%) and 5 that were catch trials (no target signal). Depending on the block, the subjects participated in sessions where they either executed *hye* large 80% of the time, but on 20% of the trials had to switch to *hye* small, or else they executed *hye* small 80% of the time, but had to switch to *hye* large on 20% of the trials. Thus, for each allograph size there was a highly prepared execution (80% trials) and one that was executed only after it was restructured (20% trials).

It was stressed that all up and down movements had to be made clearly, that at the start of a trial the pen had to be in contact with paper, and that the pen should not be lifted during the writing movement. The subjects were instructed to perform the writing task as fast and as accurately as possible. It was stressed that the subject should make optimal use of the probability information by preparing the most probable allograph on every trial and switching to the alternative sequence only if the stimulus display required it. Presentation order of allographs and probability levels were balanced over subjects. The first five trials were exclusively the most probable stimulus, and no two successive catch trials with the least probable grapheme occurred.

Response Analysis

The vertical coordinate as a function of time was differentiated and filtered at 16 Hz, yielding the vertical velocity, and time marks were determined where the vertical velocity changed sign (that is, a downward movement is passed into an upward movement or vice versa). The allographs were chosen such that the movement from one time mark to the following could be regarded as one stroke.[13] In addition to the intervals between these time marks, the velocity, the net length, and the direction of the strokes were also determined for the first eight strokes of the allographs. The time marks, as calculated by a computer algorithm, were made visible so that the analysis of each trial could be verified.[16] If the algorithm did not work perfectly, the time marks could be readjusted by hand, using a moving cursor controlled by a 10-turn potentiometer.[20] If the response obviously contained an execution error, the trial was excluded from analysis.

Results

Reaction Time

The mean reaction times (RTs) within each handwriting size and probability level were determined and then compared, and these reaction times for the four conditions are reported in Table 1. Inspection of the table reveals that for both handwriting sizes the RTs are considerably faster in the high-expectancy conditions (80%) compared to those of low expectancy (20%). By employing a one-tail sign test and using subjects as uncorrelated variables, we observed that both handwriting sizes were initiated

TABLE 1. Mean Reaction Times (msec) for *hye* Written Small or Large under Two Probability Levels

	Small	Large
hye–20% condition	383	362
hye–80% condition	317	301
Mean difference	66	61

significantly faster (large, 301 versus 362 msec, $p < .05$; small, 317 versus 383 msec, $p < .01$) in the advanced-planning conditions, indicating that the subjects were indeed using the probability information to prepare the expected handwriting sequence.

It is also worthy of note that when comparisons in RTs are made between handwriting sizes, the RTs for the smaller-sized writing are consistently longer, regardless of expectancy level. For the 80% condition, the difference was 16 msec, and for the 20% condition it was 21 msec. While the exact locus of this effect is unknown, it most likely due to differences in the force–time prescription at the muscular level. Handwriting consists of a set of orthogonal muscle control systems with intricate timing relationships, and in this case, when large amounts of force are required, the wrist joint and its constituent muscle groups probably play a larger role in the initial stroke than in the smaller-execution size, where finger movement dominates. Similar findings were reported by Bonnet, Requin and Stelmach,[22] who found that specification of short dorsal-flexion foot movements took longer to initiate than substantially longer ones.

Stroke Characteristics

The average length per stroke for the production of the two allograph sizes is reported in TABLE 2. It is easily seen that the execution of the two allograph sizes produces substantially different lengths per stroke and that, in general, having to switch to a different size allograph had little effect on stroke length. This indicates that regardless of the type of preparation, the subjects are able to accomplish the switch to producc the required lengths, with the differences being consistent throughout the first eight handwriting strokes. However, close inspection of the first stroke for each probability condition reveals that the type of preparation alters the produced length of

TABLE 2. Mean Lengths and Velocities per Stroke for the *hye* Allograph Written Small and Large for each Probability Level

	Stroke Number							
	1	2	3	4	5	6	7	8
Length (mm)								
hye-large 20%	1.47	1.53	0.61	0.42	0.56	0.31	0.55	1.21
hye-small 20%	.97	.81	.34	.23	.31	.16	.32	.68
hye-large 80%	1.62	1.58	.66	.45	.58	.31	.58	1.21
hye-small 80%	.84	.74	.34	.23	.31	.16	.30	.70
Velocity (cm/sec)								
hye-large 20%	12.16	18.71	8.14	6.88	6.55	5.57	6.12	12.97
hye-small 20%	9.05	10.41	4.45	3.97	3.78	6.10	3.58	8.07
hye-large 80%	14.52	20.21	9.17	7.76	7.28	5.83	6.61	13.92
hye-small 80%	7.58	9.97	4.47	4.08	3.64	3.01	3.33	8.08

the first stroke. Under conditions where the subject prepared for a small stroke, but then had to execute a large one, the length of the first stroke is shorter than when he or she prepared to write a large stroke and then executed it. A similar context effect, but in the opposite direction, is found for the first stroke when subjects prepared to write large, but had to switch to execute a small stroke. Both of these context effects are significant ($p < .05$).

Velocities per stroke are also reported in TABLE 2 and it is apparent that with an increase in length per stroke there is an associated increase in velocity, the longer stroke lengths possessing large velocities. These increases in velocity remain fairly constant across the eight strokes regardless of the type of preparation. There also seems to be a proportional increase in velocity for the longer strokes within an allograph, as is seen by making comparisons between length and velocity on a stroke-by-stroke basis. Moreover, when we examine the lengths of the first strokes, we see a context effect in the first stroke for velocity. When the subjects prepared for a large stroke and then executed it, the did it faster than when they had to switch to execute a small stroke, and vice-versa.

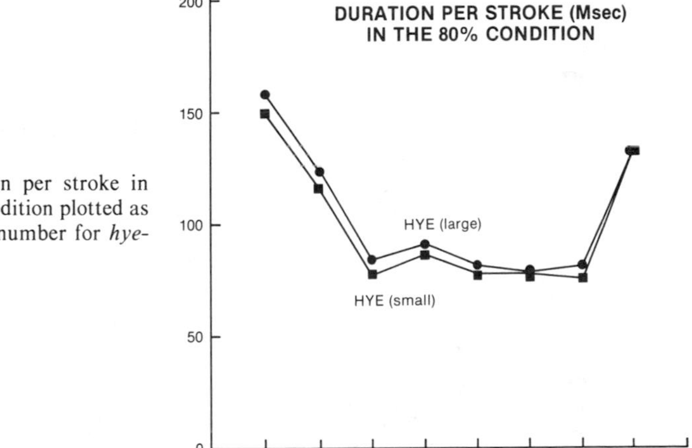

FIGURE 1. Duration per stroke in msec for the 80% condition plotted as a function of stroke number for *hye*-large and *hye*-small.

While the observed stroke parameters for length and velocity are along the lines suggested by a space–time invariance,[17] the data on durations per stroke are rather informative. FIGURE 1 plots the durations per stroke for small versus large in the 80% condition, where the subject executed the prepared size. Statistical comparisons between the durations per stroke revealed that there are no differences between small and large sizes, supporting the well-known time-constancy invariance often observed in handwriting.[17] FIGURE 2 plots the durations per stroke for the "reparameterizing" condition (20%) as a function of large and small sizes. In contrast to the results in FIGURE 1, there are substantial differences between writing size for stroke durations across the eight strokes plotted. All of these differences are significant at the $p < .01$ level and suggest that the force–time phasing is not normally specified. Yet it can be seen by the shape of the curve that the relative ratios among the times of occurrence of the individual strokes is invariant.

In comparing across figures it can be seen that most of the duration effect is due to differences in the execution of the large allograph. The duration for large strokes in the 20% condition is considerably longer, whereas there is little difference in the 80% condition. This implies that it is more difficult to increase the force required when the preparation is for a small size, than when a decrease in size is required. Remarkably, this duration effect remains for the entire eight strokes, suggesting that there is an overall force–time prescription that modulates the entire allograph.

Discussion

The foregoing data suggest that the type of preparation (80% versus 20%) within a given size affects the execution of the first stroke of an allograph, as seen in the significant differences observed in velocity and length parameters in TABLE 2. This

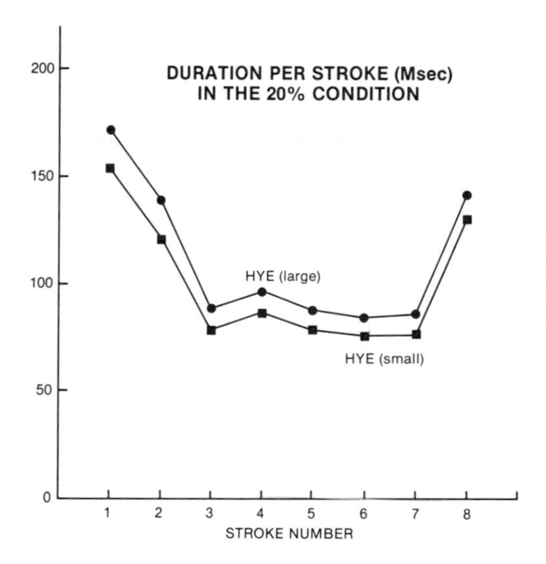

FIGURE 2. Duration per stroke in msec for the 20% condition plotted as a function of stroke number for *hye*-large and *hye*-small.

indicates that not all effects of programming are indexed by the reaction time. The approximate cost of "reparameterizing" the size prescription for both writing sizes was 60 msec. Apparently, this additional time was not sufficient to get the force–time relationship organized so that the sequence could be normally executed. An examination of advance-planning effects as seen by comparing length and velocity parameters within a handwriting size reveals that the observed preparation effects are limted to the first stroke. One interpretation of this part of the data is that the motor programming in these handwriting tasks is limited to the first stroke, with the remainder of the programming done on line. This observation is similar to that proposed by Hulstijn and van Galen,[15] who found that interdigit intervals appear not to be influenced by the sequence length, thereby reflecting a local programming process for the execution of each individual digit. Our data, interpreted as such, are somewhat at odds with those of Sternberg *et al.*,[7] who proposed that an entire sequence of words (speech) or characters (typewriting) are being programmed prior to the initiation of the first response and that a global search process is necessary for the execution of each individual response in the sequence.

A slightly different picture of motor programming is obtained when one compares large versus small writing sizes at each of the expectancy conditions (80% versus 20%). Under the "reparameterization" condition (that is, when a movement sequence is prepared under a specific force–time relationship and this preparation has to be restructured) it was found that the original preparation interferes with the new one. The surprising aspect of this finding is that the lack of adequately changing the durations in the handwriting sequence for the 20% condition remains throughout the execution of the allograph and the timing ratios of successive elements are preserved. As for the debate mentioned earlier, over the nature of motor programming and the type of units involved, these data suggest that motor programming for handwriting may be characterized as a series of horizontal and vertical movements which are modulated by some underlying timing component involving the entire sequence. A few years ago, Wing[23] came to essentially the same conclusion from his work on durations between stroke segments and suggested that the basic program unit being timed was not a single stroke, but that an underlying metronomic process was involved.

As can be seen from the evidence presented thus far, a programming unit for handwriting has been described variously as an agonist–antagonist phase cycle,[13] a metronomic process,[23] word,[17] and a letter.[15] Further, since it has been found that increasing the number of line segments in the production of a grapheme increases the reaction time,[14] there has also been some speculation that individual strokes may be represented as programming units. If we are to interpret work on motor programming in handwriting, it is clear that what constitutes an elementary unit of a program must be defined. To this end, two experiments were designed to uncover whether the most fundamental strokes, straight or curved, could function as programming units in a paradigm similar to that used by Sternberg *et al.*[7,8] The number of strokes was manipulated in order to test whether the length of the sequence had an effect upon the RT interval. A linear increase in RT and a quadratic increase in duration would then indicate advance planning of a sequence of strokes, with the unit being the individual stroke.

The second and third experiments employed a modified simple reaction-time paradigm with a handwriting response. Figures consisting of 1–5 line segments (\ V N W M) and 1–6 line–curve permutations (\ u u w uu uw) were presented individually in random order. The subjects' task was to reproduce the prespecified figure, completing it as quickly as possible after receipt of an imperative signal. If the unit of programming in handwriting is not the stroke, but is one complete cycle at a particular phase setting,[1,13] with letters emerging as modulations of underlying oscillation patterns, or if it is a letter, then differential programming effects should occur not only among items in one experiment, but also between items when comparing both experiments.

EXPERIMENT 2

Methods

Subjects

Subjects were four right-handed male and female psychology students drawn from the same population as Experiment 1.

Apparatus

The apparatus and experimental configurations were basically the same as in Experiment 1, except that direct vision of the hand was possible, and that the vertical coordinates were filtered and differentiated at 32 Hz.

Procedure

A typical trial proceeded as follows: A line-segment figure (for example, и) appeared on the display screen for 500 msec and then went off. After a 2500-msec delay, which subjects were encouraged to use for response preparation, two 100-msec warning beeps (1000 Hz) occurred, 500 msec apart. They were followed 500 msec later by a third 100-msec beep having a higher frequency (2000 Hz) tone. This was the signal to write the figure, completing it as rapidly and accurately as possible. All movements of the pen were recorded and stored and 500 msec later a feedback trace of the response was displayed and the next trial began automatically. In order to avoid anticipation, on 16% of the trials the third tone did not sound, indicating a catch trial on which the subject was not to respond.

Summary information was given at the end of each block in terms of the number of errors out of the total number of real trials, the average movement time in milliseconds between the signal and the end of the writing response, and a score. The score was based on time, errors, and standard deviation, encouraging the subjects to optimize the speed–accuracy trade-off. The subjects were familiarized with the writing task and practiced 375 trials, attaining a high level of performance. They were then presented with four blocks of 75 trials, 12 repetitions of each stimulus item per block plus catch trials. Each block was pseudorandomized with a constraint of no two identical successive items. The subjects were encouraged to write the figures in a consistent size with all up and down movements being made clearly and precisely. They were informed that the pen had to be in contact with the paper at the start of a trial and could not be lifted during the writing movement.

EXPERIMENT 3

The same subjects participated in this experiment as Experiment 2, with the order of presentation counterbalanced. The stimulus items were the only change in this experiment. Six different figures were used here (\ υ u w uu uw). After 450 practice trials, subjects were presented with four blocks of 90 trials each. This experiment was identical in all other respects to Experiment 2.

Results

Reaction Time

The mean reaction time for each sequence length, averaged over subjects and over sessions, is shown in FIGURE 3. Results for the stimulus items in Experiment 2 and Experiment 3 are considered both separately and in relation to each other.

Visual inspection reveals some similarity in the shape of functions, with the most striking effect being the increase in RT for the first stroke in Experiment 2 and the first

2 strokes in Experiment 3. This finding has been demonstrated in previous work on handwriting,[15] where a possible explanation was put forth that for short movements the stop (which for handwriting measurements must be very precise) is programmed before the movement starts. Sternberg *et al.*[7] discuss the element invariance requirement in which the equivalence of single, beginning, interior, and terminal elements must be assumed in order to interpret the latency function for sequences starting with $n = 1$. If this requirement is not met, and indeed it may not be in handwriting, given the bidirectional phase shifts, our attention must be restricted to performance functions for $n \geq 2$ (or $n \geq 3$, in Experiment 3, where *u* may be considered a single element).

The best-fitting straight line in Experiment 2 was calculated for the remaining points and a trend analysis on the linear component was performed. This analysis

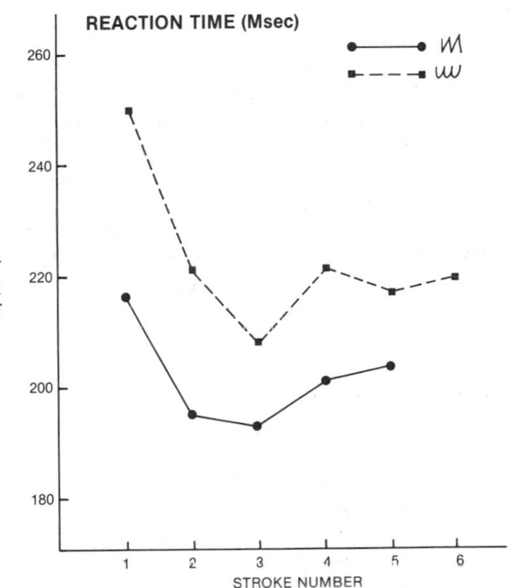

FIGURE 3. Reaction time in msec for each of the line and line–curve segment experiments plotted as a function of stroke number.

yielded a slope of 5.2 msec of the fitted latency function $F(1, 6) = 7.84$ $p < .05$. Deviations from linearity were not significant ($p < .05$). This trend toward linearity gives a weak indication of the individual stroke's being the unit of programming in this context. However, there was no linear trend for RT in Experiment 3 $F(1, 9) = 1.74$. Since all of the line-segment tasks in these experiments consisted of connected strokes, there is some degree of "co-ography" here, more so in Experiment 3 than in Experiment 2. One suspects that the effective linearity may strengthen with isolated, iterative strokes.

Comparing the items in Experiment 2 with those in Experiment 3, the RTs were, on average, 21.6 msec faster for the former, with differences ranging from 14 to 33 msec. The differences for lengths 1–5 are all highly significant ($p < .002$). There is also a significant effect of items for both experiments ($p < .002$). This provides some indication that subjects treated line segments and curves differently.

If the unit of programming in handwriting is a complete cycle at one phase setting of agonist–antagonist muscles, as argued by the spring model,[13] then the latency functions should increase from stoke 2 to stroke 4 in Experiment 2 and from strokes 2 to 4 to 6 in Experiment 3, since these are the items with complete phase cycles. A significant increase was found for Experiment 2 ($p < .002$); however, no appreciable RT difference was detected for Experiment 3. Thus, phase cycles could be a significant factor in programming connected, alternating straight-line segments, but do not seem to be a determinant in planning handwriting sequences with continuous curved lines.

The letter has also been advanced as a possible unit of programming. Hulstijn and van Galen[15] found a linear increase in RT with a sequence length of 1 to 4 letters, although the slope of this latency function decreased rapidly with practice. Considering that *v* and *w* are each single letters, if the unit were this large, one would expect a flat latency function. The difference in the mean latencies is significant, however ($p < .002$). The same finding holds true for comparison of the letters *u* and *w* in Experiment 3 ($p < .01$). Thus, it would seem that in this context the letter was not an integrated unit.

Another possibility to be considered here, and one that seems most promising, is that the unit by which the program for handwriting output is built up is the downstroke. Taking into account just the number of downstrokes in Experiment 2, we find that RT increases from 196 msec for one downstroke to 198 msec for two to 204 msec for three. Similarly, in Experiment 3, RT increases from 209 msec for *u* to 218 msec for *uu*. An efficient way to mentally or verbally code the task sequences is by using a form of rhythmic notation, that is, "1-and" = *v*; "1-and-2" = *N*, *u*; "1-and-2-and" = *W*, *uu*; "1-and-2-and-3" = *M*. An analogy can be drawn here to the idea of Sternberg *et al.* that the stress group is a unit of programming in speech where a primary stress can be followed by one or two unstressed syllables or words to form a unit.

Movement Time

FIGURE 4 shows the mean duration or total movement times (that is, the time from the first movement of the pen until the end of the last stroke) for each sequence length in Experiments 2 and 3. Data from both experiments are well described by linear functions, indicating that mean duration increased approximately linearly with the length of the sequence. Using a least-square method for linear regression, a slope of 112 msec was obtained in Experiment 2, while a slope of 95 msec was obtained in Experiment 3. The quadratic component predicted by the model of Sternberg *et al.*[7,8] was not evident here. Rather, each stroke added a constant amount of time to the total duration. However, the slopes of the fitted linear functions differ significantly between experiments, with durations for the items in Experiment 2 being 18 msec slower on average. Thus, although the writing of these items is initiated reliably faster, it is carried out more slowly. Worthy of mention is the highly significant ($p < .002$) difference between the means for $n = 3$ and $n = 5$ in the two experiments. It can be seen that the items written more slowly are those in Experiment 2, which end in a downstroke with an incomplete phase resolution. This suggests that a similar mechanism may be effective for RT and MT and contributes further evidence that the phase cycle and the downstroke may delimit a unit of programming in handwriting. It may be argued that the items for $n = 3$ and $n = 5$ in Experiment 3 also end in a downstroke with an incompete phase cycle, but show no significant departure from linearity. One possible explanation is that subjects indeed treated the items in the two experiments differently, with the downstroke of a curved segment not being as well defined. Admittedly, this interpretation is speculative.

Discussion

The results of these experiments seem to indicate that the unit of programming in handwriting may be smaller than the letter, at least in certain contexts. There is some evidence that the downstroke may be a significant factor in determining the boundaries of a unit. The number of complete phase cycles may play a role in distinguishing units in Experiment 2, but there is clearly no difference in RT among items with differing numbers of complete phase cycles in Experiment 3. It may be that, in the case of temporally overlapping curved segments, once a particular phase is set any number of cycles may be run off. In other words, there are no differential effects of interactive phases evident in the reaction-time interval. Considering the MT data, items with complete phase cycles are more alike in the two experiments than the other items, indicating a possible significance of this factor.

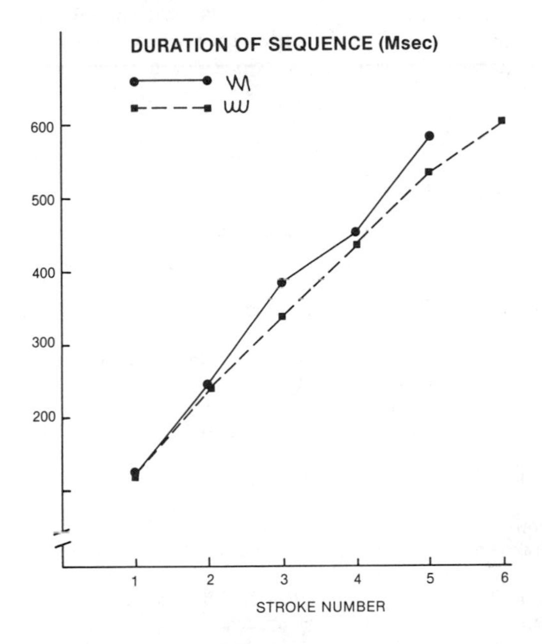

FIGURE 4. Duration of sequence in msec for each of the line and line-curve segment experiments plotted as a function of stroke number.

GENERAL DISCUSSION

The main result obtained in the first experiment, namely, an effect of reducing the opportunity for advance planning, upon the stroking characteristics of the first letter of the sequence, seems to indicate that subjects began to write before they were fully prepared, then adjusted their motor program on-line. In this "reparameterization" condition they may not have taken time to complete programming prior to movement; therefore, the process of adjusting the program may have lingered over into the execution of the first stroke. Thus, size prescription (force–time relationship) of a handwriting sequence seems to be an integral aspect of a motor program.

The notion of space–time invariance[17] is supported by similarities in stroke parameters for length and velocity. The shape of the duration curve also indicates the presence of an invariant timing characteristic. Thus, the ability to program or plan a handwriting sequence in advance can be seen to affect the RT interval, and to carry over into the characteristics of the movement itself, while the ratios among the strokes remain relatively intact. Time and space, letter-size in this case, are then not necessarily consequences of each other, but may be general organizational principles underlying handwriting.

Information about the elements of organization in handwriting is addressed by focusing on the results of the last two experiments. Tentative evidence is provided for the downstroke, possibly integrated in some way with the phase cycle, to be a unit of programming in these handwriting tasks. Incidental support is found for Sternberg *et al.*'s[7,8] idea of a length-dependent increase in RT, although a quadratic component for MT was not evident. The general indication seems to be in agreement with Keele[1] and Wing[23] that timing is not only an integral part of motor sequencing, but also aids in the specification of programming units. It has been demonstrated in these experiments that some aspect of relative timing is invariant in the stroking characteristics for a handwriting sequence and that particular details of timing within the sequence, such as rhythmical patterning perhaps, are related to the preparatory time interval in terms of unit specification and total time.

REFERENCES

1. KEELE, S. W. 1981. Behavioral analysis of movement. *In* Handbook of Physiology. Vol. II: Motor Control, part 2. V. B. Brooks, Ed. American Physiological Society. Baltimore, MD.
2. STELMACH, G. E. & V. A. DIGGLES. 1982. Control theories in motor behavior. Acta Psychol. **50:** 83–105.
3. HENRY, F. M. & D. E. ROGERS. 1960. Increased response latency for complicated movements and a "memory drum" theory of neuromotor reaction. Res. Quart. **31:** 448–458.
4. KEELE, S. W. 1968. Movement control in skilled motor performance. Psychol. Bull. **70:** 387–403.
5. MERTON, P. A. 1972. How we control the contraction of our muscles. Sci. Am. **226:** 30–37.
6. RAIBERT, M. H. 1977. Motor control and learning by the state-space model. Technical report AI-TR-439. Artifical Intelligence Laboratory, Massachusetts Institute of Technology, Cambridge, MA
7. STERNBERG, S., S. MONSELL, R. L. KNOLL & C. E. WRIGHT. 1978. The latency and duration of rapid movement sequences: Comparison of speech and typewriting. *In* Information Processing in Motor Control and Learning. G. E. Stelmach, Ed.: 118–150, Academic Press. New York, NY.
8. STERNBERG, S., C. E. WRIGHT, R. L. KNOLL & S. MONSELL. 1980. Motor programming and rapid speech: Additional evidence. *In* Perception and Production of Fluent Speech. R. Cole, Ed.: 508–533. Lawrence Erlbaum. Hillsdale, NJ.
9. SHAFFER, L. H. 1980. Analysing piano performance: A study of concert pianists. *In* Tutorials in Motor Behavior. G. E. Stelmach, Ed.: 443–456. North-Holland. Amsterdam.
10. SHAFFER, L. H. 1978. Timing in the motor programming of typing. Quart. J. Exp. Psychol. **30:** 333–345.
11. VREDENBREGT, J. & W. G. KOSTER. 1971. Analysis and synthesis of handwriting. Philips Tech. Rev. **32:** 73–78.
12. ELLIS, A. W., 1982. Spelling and writing. *In* Normality and Pathology in Cognitive Functions. A. W. Ellis, Ed. Academic Press. London.
13. HOLLERBACH, J. M. 1981. An oscillation theory of handwriting. Biol. Cybernet. **39:** 139–156.

14. Van Galen, G. P. 1980. Handwriting and drawing: A two-stage model of complex motor behavior. *In* Tutorials in Motor Behavior. G. E. Stelmach & J. Requin, Eds.: 567–578. North-Holland. Amsterdam.
15. Hulstijn, W. & G. P. Van Galen. 1983. Programming in handwriting: Reaction time and movement time as a function of sequence length. Acta Psychol. **54:** in press.
16. Teulings, J. L. H. M. & A. J. W. M. Thomassen. 1979. Computer-aided analysis of handwriting movements. Visible Lang. **13:** 219–231.
17. Viviani, P. & V. Terzuolo. 1980. Space-time invariance in learned motor skills. *In* Tutorials in Motor Behavior. G. E. Stelmach & J. Requin, Eds.: 525–539. North-Holland. Amsterdam.
18. La Berge, D. H., P. Van Gelder & J. Yellott. 1980. A cueing technique in choice reaction time. Percept. Psychophys.: 51–67.
19. Larish, D. & G. E. Stelmach. 1982. Preprogramming, programming and reprogramming of aimed hand movements as a function of age. J. Motor Behav. **14:** 322–340.
20. Stelmach, G. E. & J. L. H. M. Teulings. 1983. Motor programming and temporal characteristics in handwriting. Acta Psychol. **54:** in press.
21. Denir Van Der Gon, J. J. & J. T. Thuring. 1965. The guiding of human writing movements. Kybernetik **2:** 145–148.
22. Bonnet, M., J. Requin & G. E. Stelmach. 1982. Specification of direction and extent in motor programming. Bull. Psychonom. Soc. **19:** 31–34.
23. Wing, A. M. 1978. Response timing in handwriting. *In* Information Processing in Motor Control and Learning. G. E. Stelmach, Ed.: 153–172. Academic Press. New York, NY.

The Patterning of Time and Its Effects on Perceiving

MARI RIESS JONES

Department of Psychology
The Ohio State University
Columbus, Ohio 43210

The problem addressed here involves the patterning of events in time and the way this patterning influences a person's attention to and perception of pitch relationships that are embedded within simple music-like tonal sequences. This paper reports the outcome of the first of several experiments on this problem that I have been conducting at Ohio State with the invaluable assistance of Beth Marshburn and Gary Kidd. These studies all consider the special effects that might arise from some larger rhythmic context, one that is set up within the experimental session itself, upon a person's ability to recognize melodic relationships within individual auditory patterns.

The general idea involves manipulation of the rhythmic context afforded by certain patterns and pairs of patterns in a melodic recognition task. A melody recognition task is a task wherein, on any given trial, in a multiple-trial experimental session, a person hears two auditory patterns, the first one being a standard pattern and the second being a comparison pattern. The listener's task is to decide whether the second melody is the same or different from its standard, when, in fact, half of the time it contains some melodic changes of interest. In the present case we are interested in a change involving the pitch of a single note in the comparison pattern. Effectively this changes two pitch intervals in that pattern, where pitch intervals refers to the pitch distance between a pair of notes.

FIGURE 1 shows a typical recognition trial in one of our studies: a listener hears a warning tone; then, after a pause, the standard pattern of a pair occurs, and it is followed then by the comparison pattern; finally, a response period of 6 seconds is presented. The patterns are temporal patterns, each involving 12 square-wave tones with fundamental frequencies corresponding to tones of the C major scale. FIGURE 1 also shows a typical standard and comparison melody based on these notes. Comparison melodies, if they differed from the standard, would always preserve its basic shape and the changed note would still be consistent with the key of C major. In this example, a changed comparison pattern is shown: the pitch of the note occurring at the 11th serial position is shifted up one scale step, changing its relative distance from immediately neighboring notes. A deviation, if present, could occur at one of three different serial positions.

We created many such melodies using some systematic rules for arranging pitch intervals and contour. And since all patterns were necessarily temporal in nature, the melodies so constructed also had a rhythm. Indeed, rhythm was the variable of major interest here. The pair of melodies presented on any given trial always shared the same rhythm, but rhythm could differ between pairs of patterns encountered on different trials in an experimental session. A given pair of patterns was developed either in an isochronous rhythm or in what is termed here a duple rhythm. These two rhythms are shown in FIGURE 2. The isochronous rhythm is based on a recurrent stimulus–onset–asynchrony (SOA) of 300 msec, whereas the duple rhythm consists of a long–short time pattern with alternating SOAs of 300 msec and 200 msec, respec-

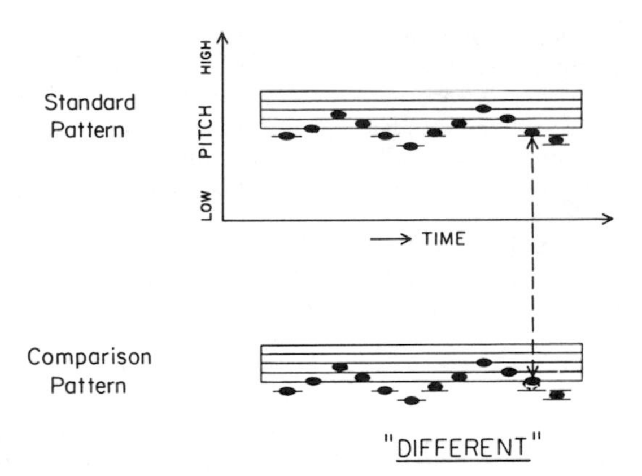

FIGURE 1. A single trial presented a listener with a standard melodic pattern (*top*) and then a comparison (*bottom*) when the comparison contained (half of the time) a changed tone (*broken line*).

tively. (In both rhythms, these SOAs reflect primarily tonal durations; off-times between tones were constant at 5 msec.) Obviously, the difference between the two rhythms is that the duple rhythm introduces *agogic* (that is, time-based) accents through relatively lengthened events. These accents fall on the first tone in a pattern and on all odd-numbered serial locations. Musically speaking, it is possible to indicate the duple rhythm as |♩·♩| and the isochronous as |♩· ♩·| or |♩· ♩· ♩·|, and so forth, where ♩ = 200 msec. Both rhythms, however, hold constant (at 300 msec) the duration of a changed note (at serial positions 5, 7, and 11, shown by arrows in FIGURE 2) when such occurs in a comparison pattern.

Our interests in the effects of rhythm upon melody recognition have arisen for several reasons. One reason involves purely experimental considerations. In a number of earlier studies we uncovered some intriguing effects of rhythmic context upon judgments about temporal aspects of pattern structure such as temporal order and tone

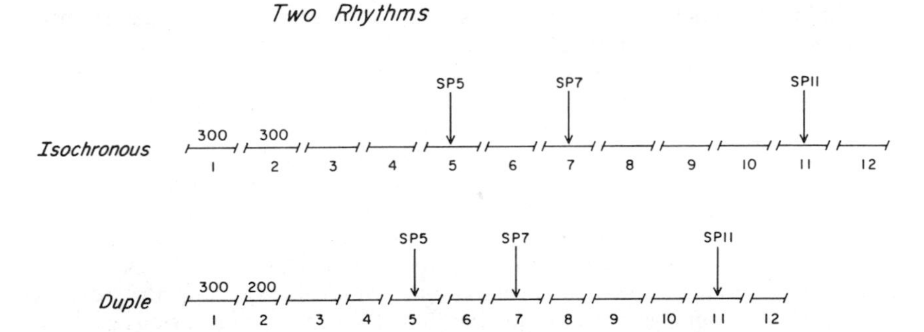

FIGURE 2. Two rhythms used in the present study: The isochronous rhythm (*top*) has an SOA of 300 msec throughout, while the duple (*bottom*) has alternation of SOAs of 300 msec and 200 msec.

duration. In these studies, there was some suggestion that when rhythm is varied as a within-subject so that a listener encounters several different rhythms within a single experimental session, overall performance is lowered relative to the case where rhythm is treated as a between-subject variable.[1,2] In the present series of studies we extend this investigation to the examination of recognition judgments about the nontemporal structure of melodies, namely to perception of pitch relationships embedded within a melody.

Our interest in choosing to extend this investigation specifically to perception of melodic relationships highlights another motivation for this research. There is also a theoretical reason for this inquiry. Briefly, this relates to ideas, which I have developed elsewhere, that attending to dynamic patterns is itself a dynamic activity responsive to relative time relationships within unfolding patterns.[3,4] This means that the rhythms of patterns encountered by a listener can guide attending so as to prepare the listener for "when" certain future serial events will occur. In this view, perceiving the musical relationship between a pair of tones will be more accurate if attentional energy is targeted to the temporal occurrence of this tonal interval. The idea is that a person abstracts rhythmic invariants and uses these to guide attending. This, in turn, immediately raises the issue of the internal "representation" of rhythm. Without belaboring this issue, let me simply state that I find it most parsimonious to subscribe to a biological model and to merely assume that all organisms are essentially rhythmic. Others who have considered the timing and coordination of motor productions have made similar assumptions.[5,6] From the perspective of an attentional theory, this means that when one encounters temporally patterned energy in the environment, one is capable of responding in kind with one's own physical, attentional energy in a sort of "act of synchrony." Through this act, internal organismic rhythmicities are awakened, and I assume that these help to guide future attending over time. Indeed, when we consider that rhythm is defined in terms of relationships between time periods, it is not unreasonable to assume that abstracted rhythmic invariants guide attending over several time periods simultaneously where relatively smaller time intervals are contained in (nested) relatively longer ones.[3,7]

In the present study, the rhythms involved can be described in terms of invariant time ratios that obtain between neighboring and remote events. For example, the ratio of SOAs between neighboring events in the isochronous rhythm is 1/1 whereas for the duple rhythm it is 3/2. Such ratios specify temporal grouping (or lack of it). That is, because the isochronous rhythm has a ratio of 1/1, for a sequence of 12 tones, it lends itself to several potential groupings with accents on every other note, for example, or on every third, or every fourth note. The duple rhythm, however, tends to suggest lower-order groups of two notes with the first element accented. At the higher temporal level, both rhythms permit equal spacing of accents in time, although the actual periods between accents differ for the two rhythmic types. For example, even if both patterns are grouped in twos, the higher–order time period operating between accents in the isochronous rhythm spans 600 msec, whereas that of the duple spans 500 msec.

Thus, if temporal contexts can direct attending "toward" or "away from" certain temporal locations, this may enhance or retard pickup of related melodic information. Targets of temporally guided attending are accents. By manipulating the rhythmic context to induce accenting within individual melodic sequences, we recently found evidence to support this prediction.[8] In this study, we held constant the melodic and temporal relationships between three central tones within nine-tone auditory sequences while varying the timing of the surrounding tones. When contextual timing induced attentional targeting to the temporal location of a possible pitch change in the central tone group of a comparison pattern, persons were much more accurate at identifying it as different from the standard.

In the present study we go farther and manipulate the rhythmic environment around a pair of melodies. From the data afforded by the study of Jones, Boltz, and Kidd[8] discussed above, we determined that often the regularity of isochronous rhythms is as effective as certain agogic-accent patterns in directing attending to critical serial positions within a sequence of events. Thus, in the present study we anticipated that the two rhythms we selected might be of comparable effectiveness in guiding attending within a pattern or over a pair of patterns sharing that rhythm on any given trial, all other things equal. But we are interested in effects that might emerge over trials if these two rhythms were combined in a single session. Is it possible that the temporal structure of surrounding pattern pairs (that is, those occurring on other trials) influences the course of attending locally within sequences of a given standard–comparison pair? Since accent locations follow different respective trajectories in the isochronous and duple rhythms, when both occur in a single session could rhythmic carryover effects result that might establish an occasional inappropriate dynamic set? If so, a listener would target attending to the wrong temporal location and thereby miss a critical pitch interval in the standard or comparison melody. Thus, experimentally, the question of interest becomes: Does it matter if persons experience only one rhythm within an experimental session or if they experience two?

The experimental design addressed to answering this question is outlined in TABLE 1. Subjects participated in 24 trials on one of the two rhythms (isochronous or duple) in a rhythmic adaptation phase and then, without a break, they participated in a test phase consisting of a total of 84 more trials. In the test phase, two kinds of trials

TABLE 1. Experimental Design

	Test Phase	
Adaptation Phase	Context Trials	Test Trials
Isochronous	Isochronous	Isochronous
Rhythm	Isochronous	Duple
Duple	Duple	Isochronous
Rhythm	Duple	Duple

occurred: (1) context trials, which occurred in runs of two and three (equally often), and in which medolic pairs always preserved the adaptation rhythm; and (2) test trials, which always occurred singly and in which the two melodies of a pair *either* both shared the adaptation rhythm or they both were switched to a new test rhythm. The basic idea is simple: in the adaptation phase persons get used to anticipating future events on the basis of rhythmic invariants in a given adaptation rhythm. This rhythm continues into the test phase as the context rhythm. If in the test phase this context rhythm is supplemented on test trials with the same rhythm, then performance here will be relatively good. On the other hand, if in the test phase a person encounters a novel rhythm in addition to the context rhythm, there now is some basis for interference. On some trials attentional rhythmicities excited by one rhythm will inappropriately carry over to the next trial and thereby reduce the effectiveness of temporal targeting of attention. In terms of the relative timing of points where deviant pitch relationships could occur, the two rhythms differ and so abstraction of the rhythmic invariants of one should inappropriately guide attending in the other. Thus, when text and context rhythmicities differ, performance in melody recognition should suffer.

To summarize, two different rhythms were used to study the effects of larger temporal context in inducing attentional targeting over time. In an adaptation phase, where listeners experienced only one of the rhythms, large differences in melodic recognition were not expected. In a test phase where listeners could, in some conditions, experience a combination of both rhythms, melody recognition performance was expected to be poorer than in the case when listeners encountered the same rhythm on all trials. In short, in the test phase we predicted a context rhythm by test rhythm interaction wherein performance was poorer whenever context and test rhythms differ.

METHOD

Subjects

Sixty subjects participated in the experiment in return for credit in an introductory psychology course. Twelve subjects were eliminated on the basis of failure to differ significantly from chance performance during adaptation. Questionnaires indicated that subjects had a background of musical study ranging from 0 years to 13 years. The median number of years of music study was 3.5 years for the 48 individuals whose data were analyzed.

Apparatus

All tones were square waves generated by a Wavetek Model 159 waveform generator controlled by a Cromemco Z-2 microcomputer. A 20-msec rise and a 20-msec fall time was imposed on each tone. Stimulus sequences were tape-recorded on a Nakamichi LX-3 cassette recorder. Prior to recording, all tones were equated for subjective loudness according to criteria of three judges.

The prerecorded tapes were played on a Tandberg Model TCD 310 MK II cassette recorder at a comfortable listening level over AKG K-240 headphones.

Materials

Melodic manipulations involved a total of 18 standard patterns. Each pattern consisted of 12 tones arranged serially according to rules applied to the C major diatonic scale. Patterns were all constructed using the following guidelines: Regardless of a pattern's initial tone (which was randomly selected), the scale distances between note pairs spanning serial locations 1–2, 3–4, 5–6, 7–8, 9–10, and 11–12 were constant within a pattern. These distances could realize either ± 1, ± 2, or ± 3 scale steps (that is, next N^j, rules where $j = \pm 1, \pm 2, \pm 3$). This constraint provided a basis for grouping notes into pairs within a sequence. Contour was also systematically varied so as to provide an alternative basis for grouping involving clusters of three tones having a common pitch trajectory (for example + + + or − −). This was realized through constraints imposed upon the pitch interval connecting successive pairs of tones within a sequence: the direction of this interval, whether up (+) or down (−) in pitch, was selected so as to induce sets of at least three ascending or descending tones. Thus, pattern contours could be of the form + + − − − + + + − − − or − − + + + − − − + − −, and so forth. The result of these two kinds of rules was a large set of melodies that afforded grouping into pairs or into triplets of tones. The underlying rationale was to build a set of melodies potentially compatible with grouping properties of a range of rhythms. From this set a number of resulting melodies were eliminated using the following criteria: (1) Melodies containing two temporally adjacent occurrences of a given tone; (2) melodies containing successive tone groups that were either identical or were exact reversals (for example, A_4 G_4 E_4 G_4 A_4 within a sequence was not allowed); (3) melodies with notes below G_3 or above G_5.

From each of the 18 remaining standard melodies, four types of comparison melodies were constructed; (1) same comparison in which the comparison pattern was identical to the standard; (2) deviant comparison with a changed note at the fifth serial position (SP5); (3) deviant comparison with a changed note at the seventh serial position (SP7); (4) deviant comparision with a changed note at the eleventh serial position (SP11). Any deviant note, if it occurred, never occurred at an extreme point (that is, highest of lowest note of pattern) in the pattern and it was always an in-key contour-preserving note change of either ± 1 or ± 2 semitones (ST) (with the majority involving a 2-ST shift).

Pairs of standard and comparison patterns were constructed and arranged in two different random orders. In each order, half of the time the unchanged comparison patterns were paired with the standard and half of the time a deviant comparison was paired with the standard (with deviant comparisons involving SP4, SP7, and SP11 occurring equally often). The ordering of pattern pairs was constrained according to experimental design requirements that a given session contain an adaptation phase, in which pattern pairs occurred only in a context rhythm, and a test phase, in which pattern pairs could appear in either the pre-established context rhythm or in a test rhythm. The adaptation phase consisted of 24 trials of randomly ordered pattern pairs arising from four different standard melodies. The test phase consisted of 84 trials of which 60 were termed context-rhythm trials and 24 were termed test-rhythm trials. The constraint in the test phase was that a single test-rhythm trial was programmed to occur (equally often) after either two or three context-rhythm trials. Within both the context-rhythm and test-rhythm trials sets, equal numbers of "same" and "different" standard comparison pairs occurred. Melodies in the test phase were based on the remaining 14 standard patterns not used in the adaptation phase.

Rhythmic manipulations involved imposing upon these melodies one of two rhythms, depending upon the experimental conditions. The rhythms, isochronous and duple, are shown in TABLE 1. The isochronous rhythm was based upon stimulus–onset–asynchrony (SOA) of 300 msec; the duple rhythm was a long–short recurrent pattern, wherein the 300-msec tone was now relatively long and the 200-msec tone was shorter, thus placing agogic accents on serial positions 1, 3, 5, 7, 9, and 11, assuming that the longer events would be heard as accented. (This was, in fact, the case according to several independent judges.)

Design

The design consisted of an adaptation phase and a test phase and data from the two phases were analyzed separately. In the adaptation phase, context rhythm (isochronous, duple) and test rhythm (isochronous, duple) conditions yielded a 2×2 between-subjects' factorial design. In the test phase, a $2 \times 2 \times 2$ factorial design involved context rhythm (isochronous, duple), test rhythm (isochronous, duple) and trial type (context, test). Only trial type was a within-subjects' variable. Twelve subjects were randomly assigned to each of the four context $\times$ test conditions.

Procedure

Subjects were instructed, via prerecorded tapes, that they had to judge whether the second melody of a pair contained a single changed note (they received an illustrative diagram) on each of a series of trials. They were told that half the time the second melody would, in fact, be different from the first and furthermore that a change would never violate the contour, or shape, of the melody, only the pitch distance. They responded to each pair in writing by checking "same" or "different," plus indicating certainty in their judgment, using a score of 1–5 (1 = very certain, 5 = uncertain).

Participants listened to the melodies over AKG (Model K-240) headphones and responded to a total of 108 trials plus two practice trials. A high-pitched 1-sec warning tone (3,500 Hz) signaled the onset of each trial. This was followed by a 2-sec silence and then by the pair of patterns separated from one another by a 3-sec pause. A 6-sec response interval intervened between the offset of a comparison pattern and the onset of a warning tone signaling the next trial.

Subjects were tested in groups of two to four persons, but they were individually separated from one another by partitions, within a sound-deadened room.

RESULTS

Scoring

Both *PC*, proportion correct, and A_g, a nonparametric ROC measure of recognition accuracy, were computed for each subject. The A_g score is computed by the trapezoidal rule[9,10] based on a program developed by Davison and Jagacinski.[11] An A_g score estimates unbiased recognition accuracy in the two-choice care where chance is .50. An A_g score of 1.00 reflects perfect discrimination and .50 is random guessing. Analyses were carried out on both dependent measures with results essentially equivalent for both. For this reason, only the results of A_g are reported here.

Adaptation Phase

The four different groups of subjects who were assigned to context–test combinations in the test phase were compared with respect to their performance on the first 24 adaptation trials. No significant differences among the four groups was observed $F(1, 44) < 1$ for the context-by-test interaction, indicating that these subjects were performing indistinguishably from one another prior to the test phase. It is also important to note that melody recognition accuracy was not significantly affected by the context rhythm (isochronous, duple) in adaptation. These rhythms were equally effective in determining recognition accuracy, $F(1, 44) < 1.00$.

Test Phase

Means and standard deviations for A_g scores as a function of context–test combinations for these two trial types are in TABLES 2 and 3. In all conditions performance was significantly above chance levels on this measure ($p < .05$). The data in TABLES 2 and 3 show that in both test and context trials subjects were better in those conditions where the rhythms remained the same across context and test trials than

TABLE 2. Means and Standard Deviations (A_g) on Context Trials in Test Phase

Context Rhythm	Test Rhythm	
	Isochronous	Duple
Isochronous		
Mean	.68	.60
S.D.	.10	.10
Duple		
Mean	.59	.66
S.D.	.08	.09

when the rhythms differed. Overall, the context-by-text rhythm interaction was statistically significant, $F(1, 44) = 4.40, p < .05$. A significant three-way interaction of trial type with context-by-test rhythm did not emerge, indicating that introduction of a novel test rhythm during the test phase has a pervasive effect. Subjects are not merely unable to lock into this new rhythm, but there is also a pronounced "boomerang" effect which lowers performance with the contextual, or adapted-to, rhythm when a novel test rhythm is suddenly inserted into the experimental session.

DISCUSSION

The data are consistent with the hypothesis that the larger rhythmic context within an experiment can influence attending to serial relationships that are not primarily temporal. Poorer performance observed in the test phase for listeners who experienced two rhythms suggests the importance of temporal patterning in guiding attending.

However, even in conditions where performance is best, namely, in the unchanged rhythm conditions, melody recognition does not appear, at least superficially, to be exceptionally good (that is, A_g scores of .66–.72). But this is deceptive. If we weigh this performance against what is known about melody recognition in comparable situations, this study actually reflects fairly good performance. That is, the pattern recognition task was difficult for at least three reasons: (a) all pitch interval deviations reflected in-key violations; and (b) all pitch interval deviations did not break contour; and (c) all patterns were relatively long (12 tones). The work of Lola Cuddy[12,13] and Jay Dowling[14–16] has shown repeatedly that the first two features lead to very poor melody recognition. And my own research, as well as general expectations, suggests pattern recognition difficulty increases with pattern length. When these facts are considered, it turns out that performance in the present task is relatively good.

Thus, it appears that when fairly subtle pitch changes in a melody were introduced,

TABLE 3. Means and Standard Deviations (A_g) on Test Trials in Test Phase

Context Rhythm	Test Rhythm	
	Isochronous	Duple
Isochronous		
Mean	.64	.63
S.D.	.16	.13
Duple		
Mean	.66	.73
S.D.	.11	.11

it helps to be able to anticipate "when" in time these changes will transpire. I should mention that these findings are not encompassed by proposals that emerge from recent research on temporal coding such as Povel's[17] or Handel's.[18] This research is related to the present study in that it reveals the psychological salience of time ratios on performance in tapping tasks. But the formal descriptions of performance in tapping tasks are concerned primarily with the effects of *temporal* structure on *temporal* reproduction. Here, we find effects of *temporal* structure on *melody* recognition. The present data provide evidence for the general influence of higher-order temporal context upon attending and perceiving in auditory sequences.

There are some pragmatic implications of these findings. The most obvious one is that the decision to treat rhythm as a between- or within-subject variable should not be made lightly. Furthermore, we must weigh cautiously the generality of conclusions about listeners' performance levels in tasks that have exposed participants to a range of different rhythms and that have failed to take account of possible contextual effects.

Finally, consider some theoretical implications of these findings. One tempting means of conceiving of these effects at a more general level involves the idea of temporal uncertainty. That is, one could attempt to devise some metric that would reflect the fact that addition of a novel rhythm in certain conditions merely adds to the uncertainty about a pattern's rhythmic scheme. But is important to be careful about using this term "uncertainty" as a complete explanation of these effects. In the first place, in conditions with two rhythms there is, in fact, no uncertainty on many trials about *which* rhythm will occur on that trial because of the trial-to-trial contingencies of context and test trials. These trials were not randomly arranged and furthermore there were more context trials than test trials. And, of course, because standard and comparison melodies always share the same rhythm, if this contingency is considered, there is no uncertainty about the rhythm of a given comparison pattern. When these kinds of redundancies are considered, it quickly becomes clear that some information metric based on uncertainty will be a complex one subject to all the problems investigators encountered in the 1960s using these metrics to describe higher-order contingencies in pattern structure.[19]

There are other reasons why a simple rhythmic-uncertainty explanation seems incomplete. It is more descriptive than explanatory. Why should uncertainty in rhythmic structure affect sensitivity to melodic structure? Melodies remain the same across all four rhythm conditions. One might be tempted to rely here for an explanation upon some version of temporal parsing under uncertainty. However, it seems that when we finally become quite specific about what "parsing" means and how it is determined in this context, it is quite difficult to distinguish it experimentally from the simpler interpretation, which simply states that attending follows relative time rules that afford grouping. To be sure, uncertainty in temporal structure is in some way important to perception of nontemporal structure. My suggestion is that greater uncertainty, in a loose, descriptive sense, is detrimental because it affects the reliability with which the listener abstracts and uses appropriate rhythmic invariants to guide responding. On a number of trials, a listener may experience conflicting rhythmic expectations or may simply rely upon the wrong one.

REFERENCES

1. JONES, M. R., G. KIDD & R. WETZEL. 1981. Evidence for rhythmic attention. J. Exp. Psychol. **7:** 1059–1073.
2. JONES, M. R. & M. BOLTZ. 1981. Temporal context: What is it and what does it do for you? Paper presented at Psychonomic Society meeting, Philadelphia.

3. JONES, M. R. 1976. Time, our lost dimension: Toward a new theory of perception, attention, and memory. Psychol. Rev. **83:** 323–355.
4. JONES, M. R. 1981. A tutorial on some issues and methods in serial pattern research. Percept. Psychophys. **30:** 492–504.
5. KELSO, J. A. S., D. L. SOUTHARD & D. GOODMAN. 1979. The nature of human interlimb coordination. Science **203:** 1029–1031.
6. KELSO, J. A. S., B. TULLER & K. S. HARRIS. A 'dynamic pattern' perspective on the control and coordination of movement. *In* The Production of Speech. P. MacNeilage, Ed. Springer-Verlag. New York, NY.
7. YESTON, M. 1976. The Stratification of Musical Rhythm. Yale University Press. New Haven, CN.
8. JONES, M. R., M. BOLTZ & G. KIDD. 1982. Controlled attending as a function of melodic and temporal context. Percept. Psychophys. **32:** 211–218.
9. BAMBER, D. 1975. The area above the ordinal dominance graph and the area below the receiver operating characteristic graph. J. Math. Psychol. **12:** 387–415.
10. POLLACK, I., D. A. NORMAN & E. GALANTER. 1964. An efficient nonparametric analysis of recognition memory. Psychonomic Sci. **1:** 327–328.
11. DAVISON, T. C. B. & R. JAGACINSKI. 1977. Nonparametric analysis of signal detection confidence ratings. Behav. Res. Methods Instrum. **9:** 545–546.
12. CUDDY, L. L. & A. J. COHEN. 1976. Recognition of transposed melodic sequences. Q. J. Exp. Psychol. **28:** 255–270.
13. CUDDY, L. L., A. J. COHEN & D. J. K. MEWHORT. 1981. The perception of structure in short melodic sequences. J. Exp. Psychol. **7:** 869–883.
14. DOWLING, W. J. 1972. Recognition of melodic transformations: Inversion, retrograde, and retrograde inversion, Percept. Psychophys. **12:** 417–421.
15. DOWLING, W. J. 1978. Scale and contour: Two components of a theory of memory for melodies. Psychol. Rev. **85:** 341–354.
16. DOWLING, W. J. & D. S. FUJITANI. 1971. Contour, interval, and pitch recognition in memory for melodies. J. Acoust. Soc. Am. **49:** 524–531.
17. POVEL, D. J. 1981. Internal representation of simple temporal patterns. J. Exp. Psychol. **7:** 3–18.
18. HANDEL, S. & J. S. OSHINSKY. 1981. The meter of syncopated auditory polyrhythms. Percept. Psychophys. **30:** 1–9.
19. JONES, M. R. 1974. Cognitive representations of serial patterns. *In* Human Information Processing: Tutorials in Performance Cognition. B. Kantowitz, Ed. Erlbaum. Potomac, MD.

On Controlling Force and Time in Rhythmic Movement Sequences: The Effect of Stress Location

ANDRAS SEMJEN,[a] ADELA GARCIA-COLERA, AND JEAN REQUIN

Department of Experimental Psychobiology
Institute of Neurophysiology and Psychophysiology
National Center for Scientific Research
Marseille, France

INTRODUCTION

The work that we shall report has two facets: it relates to the rather broad problem of advance planning of rapid movement sequences; and it relates to the question of how the mechanisms that control timing and intensity of the component movements are coordinated in such rapid sequences.

The assumption that rapid movement sequences are planned in advance derives from at least two kinds of observations: those pertaining to structural properties of such sequences as they are executed, and those pertaining to the time it takes to prepare their execution.

The first approach is exemplified by Vince's experiment in which subjects drew vertical lines by continuously moving a pencil between two horizontal marks. When these movements were executed at a high speed, aiming errors were not corrected between individual movements, but rather between groups of movements (for example, between a first group in which all movements fell short of the mark, and a second group in which all movements attained the mark). From this observation the inference was drawn that, under the speeded-up condition of movement execution, decisions concerning movement extent were made for groups of movements, rather than for individual movements. Hence, the group was "planned" in advance as a whole.[1]

The second approach is exemplified by studies from a number of investigators who, following Henry and Rogers' pioneer work,[2] attempted to relate the time cost of programming a movement sequence to its "complexity." Here the basic idea is that if time to initiate a rapid movement sequence (after a "go" signal) depends on the number of elements contained in the sequence, and/or on the specific relationships between these elements, this implies treatment of the sequence as a whole before its first element is emitted. "Treatment" may mean that a motor program is generated or activated for the execution of the whole sequence, much as Henry and Rogers' original memory-drum theory suggested. However, the minimum requirement for the advance planning notion is that the movement sequence be somehow identified as a particular response.

The reaction time (RT) analysis of how rapid movement sequences are planned and executed has been extensively applied to speech and typewriting.[3,4] It has been

[a]Address for correspondence: C.N.R.S.–I.N.P. 03, 31, Chemin Joseph-Aiguier, 13402 Marseille Cedex 9, France.

amply demonstrated that the RT of a spoken or written sequence increases as a function of the number of elements in the sequence (that is, the number of words or the number of keystrokes). Furthermore, it has been shown that this factor also affects the timing of the sequence, that is, the intervals between the component units.

The notion of advance planning of rapid movement sequences is clearly supported by the variations in RT provoked by the manipulation of the number of elements. However, the question can be raised as to whether more intrinsic properties of a sequence, such as the time distribution of motor intensities over the elements it contains, are also taken into account before the execution of the sequence starts. This question seemed to us of particular interest, since force, along with time, might constitute a basic dimension in the control of movements.

In a series of previous experiments, the results of which will be reported elsewhere, we asked subjects to produce rapid sequences of cadenced finger taps and to accentuate one of them. The sequence always contained the same number of taps (usually five), but the location of the stress varied within the sequence. The main finding from these experiments was that when the serial position of the element to be stressed was indicated to the subject by the go signal (a choice RT condition), the RT of the sequence was considerably longer than when the subject received such information prior to the go signal (a simple RT condition).

One possible interpretation of this "stress position uncertainty" effect is that stress location differentiates the whole sequence as a particular response. RT would then increase because, prior to its initiation, the required response must be identified, and perhaps programmed in accordance with the time and force distribution properties of the sequence. Since RT is known to increase as a function of the number of equally likely response alternatives,[5] the obvious prediction from the above interpretation is the following: RT of rapid sequences of cadenced finger taps should increase as a function of the number of equally likely locations of the stressed tap within the sequence. The first aim of the experiment to be reported here was to test this hypothesis.

The second aim of the experiment was to elucidate the origin of the effect that stressing a tap had on the timing of the sequence. From our previous experiments it was established that stressing a tap lengthened the intervals that immediately preceded and followed the stress. While the variations in tapping cadence may reflect what Ostry[4] calls "real-time organization of the sequence," they could also be caused by purely peripheral, biomechanical factors. For example, it can be argued that in order to produce stress, that is, more force, the subject moves his finger farther away from the key, and that the lengthened finger trajectory causes a lengthened tapping interval. While such an explanation could be valid for the interval that precedes the stress, it cannot hold true for the interval that follows it. Variations in the tapping cadence seem to depend on processes that control variations in the force level, rather than on processes linked to the force generation itself. In order to assess more carefully this possibility, we included in the present experiment a condition under which subjects were asked to tap one element of the sequence less forcefully than all the others. Even though under this "inverted stressing" condition the force patterns should be quite different from those observed under the ordinary stressing condition, we predicted that the variations in the tapping intervals before and after the critical ("negatively stressed") element would be approximately the same as under the ordinary stressing condition.

No explicit prediction was made as to the RT effect of "inverted stressing." However, since under this condition subjects were required to produce rather unusual force patterns, it was logical to expect longer RTs insofar as response organization concerned the sequence as a whole. Under the alternative hypothesis, namely, that only the very first element of the sequence is prepared during the RT period, the subject would have to make the same binary choice between high or low level of force, independent of the stress condition. Hence the reaction time should remain constant.

METHOD

Task

Subjects were instructed to execute a sequence of four taps on a key with the index or middle finger of their preferred hand, reproducing a temporal model that was presented to them on each trial. This model consisted of four clicks of equal intensity, separated by 180-msec intervals. Subjects were also required to make one of the taps either stronger (stress + condition) or weaker (stress – condition) than the other three. These two conditions were run in separate series. The tapping response was to be started as quickly as possible after the display of a digit which indicated the serial position of the tap to be stressed in the ongoing trial.

The experiment took place in a noise-insulated chamber. The subject was seated facing the key, and a digital display unit. A trial began with a warning signal consisting of a "0" that was displayed for 500 msec. Two seconds later, the temporal model was delivered to the subject via headphones. After a 1-sec period, the display of a digit indicating the location of the stress signaled the onset of the movement sequence. The intertrial interval lasted 10 sec.

Measurements

All experimental events and measurements were automatically controlled by a PDP-12 computer (Digital Equipment Co.). The RT, measured as the delay between presentation of the digit used as response signal and onset of the first tap, and the duration of the intervals between the four taps were recorded for each trial. The force of each tap was measured in arbitrary units as the output voltage of a strain-gauge incorporated in the key.

A test was automatically performed on the measurements to determine whether a sequence had been executed as required. The trials in which the response was not correctly performed were repeated once at the end of each block. A response was classified as an error: (1) if the number of taps was other than four; (2) if the stress was not placed on the required tap; or (3) if the intervals exceeded certain limits that were fixed in advance. According to the temporal model, the second, third, and fourth elements of the movement sequence were to be tapped 180, 360, and 540 msec, respectively, after the first one. A time error was registered if any tap was executed more than 90 msec before or after its required delay.

Design

All subjects served in two experimental sessions. Half of the subjects received the stress+ condition in the first session and the stress– condition in the second one, while the order was reversed for the other half.

The experimental session was divided into three parts. The purpose of the first part was to acquaint the subject with the task. It was designed as a simple RT situation in which the subject was instructed to stress a given tap before the beginning of a series of trials. Each possible stress location was practiced in a block of 10 trials.

The second and third parts of the session were conducted according to a choice RT paradigm. The subject was ignorant of the serial position of the stress until he was informed by the digit used as response signal. There was a four-choice RT condition and a 2-choice RT condition which were balanced across parts two and three. Under the four-choice RT condition, the four positions were presented as equally likely alternatives on all trials. This condition was run in three 32-trial blocks in which the four possible digits appeared in random order. Under the two-choice RT condition, there were only two positions in which the stress could be located. These two alternative positions were equiprobable. They were indicated to the subject before a series of trials began. All possible pairwise combinations between the four different stress locations were used. The six combinations (pairings) were tested in six blocks of 16 trials each. The order of presentation of these combinations was balanced across subjects.

Subject Selection

Twenty-four paid volunteers participated in a selection procedure which comprised the execution of five-tap sequences under simple and five-choice RT conditions. The task was performed in the same manner as the stress+ condition described above. The twelve participants with a higher proportion of correctly performed responses and faster RTs were selected as subjects for the experiment.

RESULTS

Errors

After its completion, each response sequence was tested for number of taps, stress location, and time intervals produced. Unlike testing for number of taps and stress location, testing for time intervals is arbitrary. No *a priori* definition can be given of the amount of discrepancy between the temporal model provided and actual response timing, which would unequivocally qualify a response as a time error. Nevertheless, we introduced this test in order to detect systematic departures from the required cadence and, most importantly, any strategy based on a partitioning of the sequence into well-separated subunits. Such a strategy on the part of the subjects would have compromised our attempt to utilize RT as an index of advance planning of the sequence as a whole.

The overall time-error rate was 5.9% and it was roughly the same under the stress+ and stress− conditions. Given the arbitrary character of the interval-testing procedure, individual mean scores for each of the response parameters recorded (that is, RT, interval duration, and tapping force) were computed in two ways: either excluding from or including in the computation the response sequences qualified as time errors. Manner of computation produced virtually no difference between group means for either RT or force. Interval mean and standard deviation were slightly different, depending on the way the computations were made. However, analyses of variance applied to these data showed the same factors to be significant (or nonsignificant), regardless of the manner in which individual scores were computed. Therefore, it was decided to include the "time errors" together with the correct responses in the presentation of the results.

TABLE 1 shows the combined error rates for number of taps and stress location, averaged over subjects. A greater number of errors was produced under the stress−

TABLE 1. Average Error Rates

		Stress Location				
		1	2	3	4	Mean
Stress +	2 C	7.7	13.6	14.7	3.8	9.9
	4 C	13.6	14.8	16.9	11.8	14.2
Stress −	2 C	25.7	39.1	20.5	27.9	28.3
	4 C	26.0	32.4	24.4	22.6	26.3

than under the stress+ condition. The effect of number of choices and stress location depended on the type of stress (+ or −). The number of choices caused a slight increase in error rate only under the stress+ condition. Concerning the effect of stress location, the highest error rate was observed under the stress− condition when the stress was required on the second tap; the lowest error rate was observed under the stress+ condition when the stress was required on the fourth tap. The comparison between error rate and RT data indicates no tendency for speed–accuracy trade-off.

Force Patterns

The force data are shown in FIGURE 1. Each point in the figure represents the mean force of a tap. The points have been grouped by four; these groups represent the four successive taps in a sequence. Stress locations are indicated by arrows. The force measurements recorded under the two-choice and four-choice conditions were pooled since there was no difference between the two conditions.

Stressed (either + or −) and nonstressed taps are sharply differentiated. This differentiation remains stable across all stress locations. However, the change in force

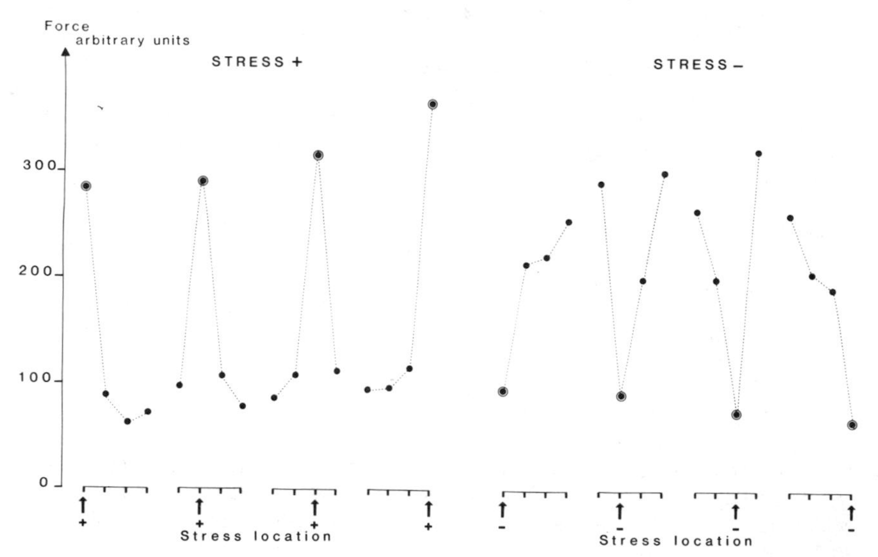

FIGURE 1. Mean force of the four successive taps in a sequence, plotted as a function of stress location (*arrows*) and stress type (+ and −).

level (from a nonstressed to a stressed tap and vice versa) seems to be more abrupt under the stress+ than under the stress− condition. As a consequence, the force level of the nonstressed taps presents much less variability under the former than under the latter condition.

The force of the stressed tap increases for stress+ and decreases for stress− as the stress moves from the first to the fourth position. Trend analysis confirmed that these effects were significant (for stress+: $F_{LIN} = 15.72$, d.f. = 1,11, $p < .01$ and $F_{QUA} = 6.72$, d.f. = 1,11, $p < .05$; for stress−: $F_{LIN} = 42.29$, d.f. = 1,11, $p < .001$).

The force level of the nonstressed taps shows a similar increase under the stress+ condition ($F_{LIN} = 21.61$, d.f. = 1,11, $p < .001$), whereas it shows a more complicated, curvilinear evolution under the stress− condition ($F_{QUA} = 19.99$, d.f. = 1,11, $p < .001$).

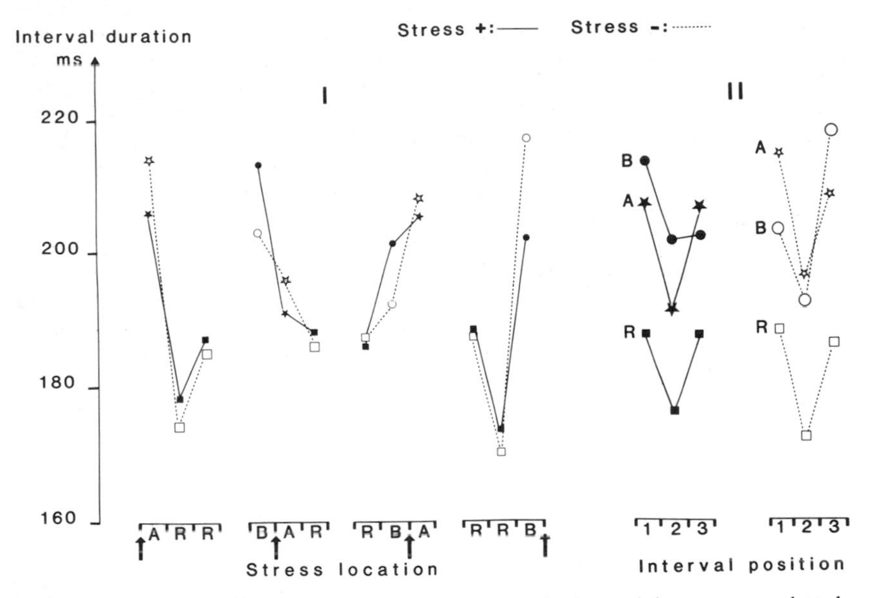

FIGURE 2. Part I (*left*): Mean duration of the three successive intervals in a sequence, plotted as a function of stress location (*arrows*) and stress type. Part II (*right*): Same data as in Part I, replotted according to interval position in the sequence (1, 2, 3), and interval position with respect to the stressed tap (B = before stress; A = after stress; R = remote). Remote intervals sharing the same position in the different sequences were averaged.

Sequence Timing

The mean duration of the successive intervals in a sequence (points grouped by three) is shown on the left (I) of FIGURE 2. Interval duration varied as a function of the interval's position (first, second, third) within the sequence and the location of the stressed tap. This was confirmed by a four-way analysis of variance (ANOVA), which yielded significant effects for stress location (F = 10.89, d.f. = 3,33, $p < .001$), interval position (F = 24.33, d.f. = 2.22, $p < .001$), and the interaction of these two factors (F = 37.48, d.f. = 6,66, $p < .001$). Neither the number of choices as a main factor, nor its interaction with the other factors, was significant. Stress type was not significant, but the three-way interaction between interval position, stress location, and stress type reached significance ($F = 2.50$, d.f. = 6,66, $p < .05$). In other words, the joint effect of interval position and stress location was slightly modulated by stress type. However, the timing of the sequence was fundamentally the same for stress+ and stress−, as can be seen in FIGURE 2 (I).

The time structure underlying all these interval configurations can be further clarified by regrouping the intervals according to their position in the sequence (first, second, third) on the one hand, and to their position with respect to the stressed tap (before, after, remote) on the other. Regrouping of the intervals according to this double classification leads to the time patterns shown on the right (II) of FIGURE 2. The intervals that are located before and after a stressed tap are longer than those located in other (remote) positions. In addition, the first and last intervals are longer than the second one.

Interval standard deviations (S.D.), averaged over subjects, are presented in FIGURE 3 (I). A four-way ANOVA yielded significant effects for stress location ($F = 6.14$, d.f. = 3,33, $p < .05$) and interval position ($F = 8.04$, d.f. = 2.22, $p < .01$), while number of choices and stress type were nonsignificant. The following interactions were significant: stress location by stress type ($F = 5.96$, d.f. = 3,33, $p < .05$), stress location by interval position ($F = 12.80$, d.f. = 6,66, $p < .001$), and stress location by interval position by stress type ($F = 7.94$, d.f. = 6,66, $p < .001$). The latter three-way interaction indicates that the effect of the stress on the interval that immediately precedes and follows the stressed tap strongly depends on the stress type.

On the right part (II) of FIGURE 3, interval S.D.s have been regrouped according to the double classification principle explained above. It can be observed that interval variability (that is, the regularity of timing) was more strongly influenced by "inverted stressing" than by ordinary stressing. It also appears that, unlike interval duration, interval variability does not present a unique configuration linked to the interval's position within the sequence.

One of the criteria for regrouping the intervals (on the right side of FIGURES 2 and

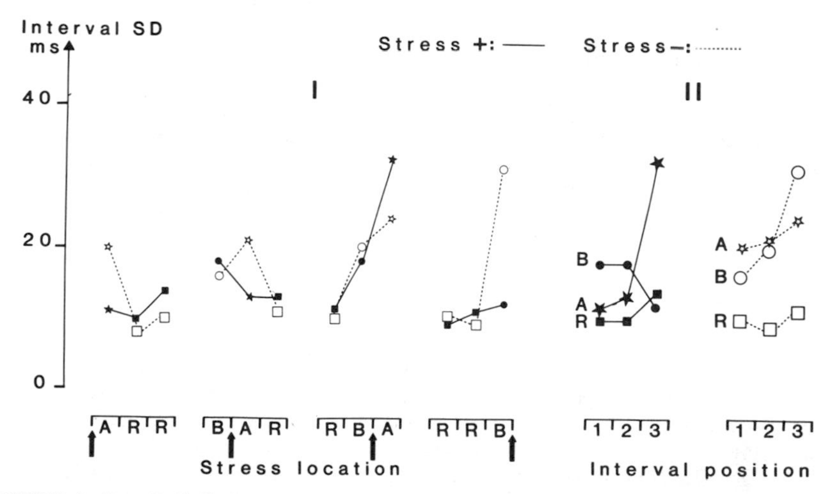

FIGURE 3. Part I (*left*): Mean standard deviations (SD) of the successive intervals in a sequence, as a function of stress location (*arrows*) and stress type. Part II (*right*): Same data as in Part I, replotted as in FIGURE 2, Part II.

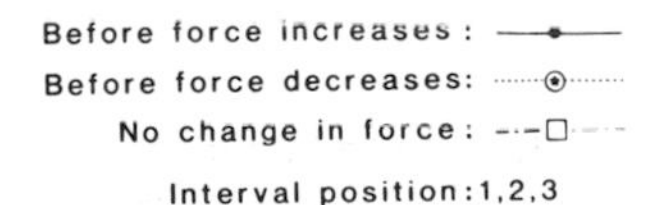

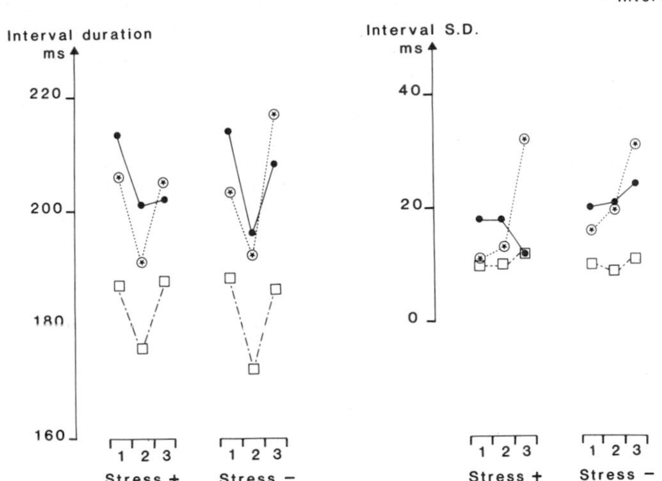

FIGURE 4. Interval mean duration and standard deviation (SD) plotted as a function of stress type (+ and −), interval position in the sequence (1, 2, 3), and interval position with respect to the force variations associated with stress production (before force increases, before force decreases, when no force change occurs). This figure is based on the same data as in FIGURES 2 and 3.

3) was their position with respect to stress location. It should be noted, however, that the direction of force change is opposite for the two types of stress. The interval produced before the stress precedes a force increase under the stress+ condition and a force decrease under the stress− condition. Conversely, the interval produced after the stress precedes a force decrease under the stress+ condition and a force increase under the stress− condition (see FIGURE 1).

In FIGURE 4, interval means and S.D.s are plotted according to the position of the interval in the sequence and with respect to the required variations in force level (before force increases, before force decreases, and when no force change occurs). The figure shows most clearly that variations in force level have the same lengthening effect on the duration of the intervals that precede these variations, regardless of the stress type. The interval data were subjected to a four-way ANOVA. Interval position within the sequence produced a significant main effect ($F = 25.08$, d.f. $= 2,22$, $p < .001$), whereas number of choices and stress type were nonsignificant. Position of the interval with respect to force change produced a significant main effect ($F = 29.91$, d.f. $= 2,22$, $p < .001$). Subsequent partial comparisons showed that intervals that preceded an increase and those that preceded a decrease in force did not differ significantly. However, the interaction of this position factor with interval position in the sequence was significant ($F = 9.41$, d.f. $= 2,22$, $p < .01$). This interaction signifies that, at the beginning of the sequence, a force increase had a greater lengthening effect on the interval than a force decrease, whereas at the end of the sequence this relationship was reversed. A similar interaction appeared also in the S.D. data, as can be seen in the right part of FIGURE 4.

Reaction Times

Mean individual RTs were calculated for each stress location, each stress type and each level of choice. These means were subjected to a three-way ANOVA, which yielded significant main effects for stress type ($F = 23.49$, d.f. = 1,11, $p < .001$) and number of choices ($F = 62.76$; d.f. = 1,11, $p < .001$), while the interaction between these two factors was nonsignificant. As shown in the left part of FIGURE 5, RT increased as the number of equally likely stress locations increased from 2 to 4, and it was longer under the stress– condition than under the stress+ condition. There was no significant main effect for stress location, but a highly significant stress location by stress type interaction ($F = 8.59$, d.f. = 3,33, $p < .001$). The origin of this interaction clearly appears in the middle part of FIGURE 5, where RTs are shown as a function of

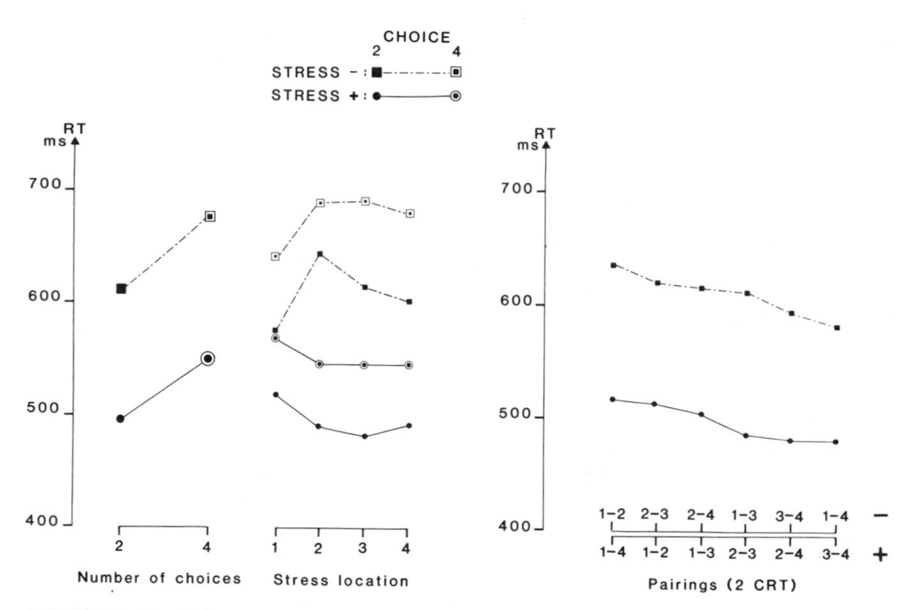

FIGURE 5. (*Left*) Mean reaction time (RT) as a function of number of choices and stress type. (*Middle*) Mean RT as a function of stress location, number of choices and stress type. (*Right*) Mean RT for the pairings in the two-choice condition. Pairings are ranked in decreasing order for each stress type.

stress location. It can be seen that stressing the first tap provoked the longest RT within the stress+ condition, and the shortest RT within the stress– condition. It must be noted, however, that for a given level of choice, RT was always longer for the stress– than for the stress+ condition.

Subsequent partial comparisons showed that stress location was a significant factor within each stress-type condition (for stress+: $F = 3.39$, d.f. = 3,33, $p < .05$; for stress–: $F = 6.36$, d.f. = 3,33, $p < .01$). There was also a significant stress location by number of choices interaction within the stress– condition. This interaction was mainly due to RT being particularly long under the two-choice condition when the stress was located on the second tap.

RTs obtained under the two-choice condition were also analyzed in order to determine whether pairing of the sequences was by itself a significant source of variance in the two-choice RT. It should be recalled that the six possible pairwise combinations between the four different stress locations were used. Mean individual RTs obtained for the six pairings were subjected to a one-way ANOVA, which showed a significant treatment effect both for the stress+ and stress− condition (the F ratios were, respectively, 3.27 and 2.87, d.f. = 5,55, $p < .05$).

The variations in RT as a function of sequence pairings are shown in the right part of FIGURE 5. RTs were ranked in decreasing order. Only some of these values differed significantly from each other in pairwise *post hoc* comparisons. Under the stress+ condition, the 1–4 pairing differed significantly from the 2–3, 2–4, and 3–4 pairings. Also, the 1–2 pairing differed significantly from the 2–4 pairing. Under the stress− condition, the 1–2 pairing differed significantly from the 1–3, 3–4, and 1–4 pairings. Also, the 2–3 pairing differed significantly from the 3–4 pairing. All other differences were nonsignificant. It is important to note that the ordering of the pairings was clearly different between the stress+ and stress− conditions.

DISCUSSION

This section shall involve two main parts. First, RT data will be evaluated with regard to the advance planning problem. Second, the results concerning the timing of the sequence will be considered.

The Advance Planning Problem

Number of Alternative Stress Locations

As predicted, RT increased as a function of the number of equiprobable locations of the stress within the sequence of four finger taps. Since RT is known to increase when the number of response alternatives increases,[5,6] a simple interpretation of our results is that stress location differentiates the whole sequence as a particular response. Therefore, under the four-choice condition RT lengthens as compared to the two-choice condition because in the former case the subject must identify the appropriate response within a larger set of response alternatives than in the latter case. Although *sequence identification* does not imply that a detailed program is also established for the *execution* of the sequence, it appears nevertheless as a prerequisite for generating such a program.

Our interpretation assumes that response identification was the main contributing factor to the observed variations in RT, whereas the time needed for stimulus discrimination changed little, if any, as a consequence of the increase in the number of equiprobable *stimuli*. This assumption is based on known evidence that manipulation of the number of equiprobable stimulus–response pairs produces sizable choice RT effects; however, this is true only if the conditions of the experimental procedure make it difficult to identify the responses assigned to the different stimuli.[5,7] When response identification is easy and direct, as in numeral-naming tasks, where subjects are simply asked to read aloud (to name) visually presented digit stimuli, the response latencies do not show much variation as a function of the number of equiprobable digits. When the same digit stimuli are associated each with the movement of a different finger (button press), the movement latencies increase sharply as a function of the number of equally likely stimulus alternatives.[5,8]

The assumption that in our experiment the manipulation of the number of alternative stress locations primarily affected response identification rather than stimulus discrimination was corroborated by the results of a control experiment in which our subjects performed a numeral-naming task. Although naming latencies turned out to be longer under the four-choice condition than under the two-choice condition, the difference was only one-third of that found when sequences of finger movements were associated with the same digits.[b]

The sequence identification hypothesis states that before the subject starts to execute such rapid movement sequences, he identifies the whole sequence as the required response. An alternative hypothesis would be that, during the RT period, only the first element of the sequence is selected for execution. Under the first-item identification hypothesis RT should depend on the predictability of that first element.

Under the two-choice condition the predictability of the first element of the sequence (that is, whether a high- or a low-force tap) varied as a function of pairings. For the pairings in which the stress was never located on the first tap (2–3, 2–4, 3–4), the first element was fully predictable. For the other pairings (1–2, 1–3, 1–4), the probability that the first tap was a high- or a low-force tap equaled .50. Under the first-item identification hypothesis one would predict sizable RT differences between the pairings in which the first tap was fully predictable and those in which a high- or a low-force tap was equally probable. Moreover, identical RT would be predicted for pairings belonging to the same category. Our data lent no support for either of these predictions.

Under the four-choice condition, the probability that the first tap was a high- or low-force one was .25 and .75 under the stress+ condition, and .75 and .25 under the stress− condition, respectively. The first-item identification hypothesis would predict that sequences starting with a .75 probability tap (under the four-choice condition) would have shorter RT than those in which a low- or high-force tap is equally probable (pairings 1–2, 1–3, 1–4 under the two-choice condition). *Post hoc* analysis of the data lent no support for this prediction; RT under the four-choice condition was longer than under the two-choice condition regardless of the probability of the first tap.

The predictability of the first element of the sequence varied according to pairings and choice conditions, but not according to type of stress. The first-item identification hypothesis would predict identical RT for stress+ and stress− sequences since in either case the subject would have to make the same binary choice between a low- and high-force tap. This hypothesis was not supported by the experimental data.

Since none of the predictions of the first-item identification hypothesis received

[b] In an additional session, subjects were presented with the same digital stimuli used in the main experiment. The task consisted of naming a digit as quickly as possible after its presentation. Simple, two-choice, and four-choice RT conditions were run as described in the METHOD section. Mean naming latencies, together with the RTs for the tapping task, were subjected to an ANOVA. Significant effects were obtained for type of task (naming or tapping) ($F = 28.38$, d.f. = 1,11, $p < .001$) and number of choices ($F = 69.20$, d.f. = 1,11, $p < .001$). The interaction between these two factors was also significant ($F = 11.26$, d.f. = 1,11, $p < .01$). On the average, the four-choice RT was 60.6 msec longer than the two-choice RT for the tapping task, while there was only a 19-msec difference for the naming task.

support from the data, it is safe to conclude that the sequence as a whole, rather than its first element, was identified as the required response during the RT.[c]

Type of Stress

As expected, it took more time for the subject to initiate a sequence in which a tap was weaker than all the others, than to initiate a sequence in which one tap was stronger than all the others. Furthermore, the effect of type of stress and that of the number of alternative stress locations combined additively. From the perspective of Sternberg's additive factor method for analyzing RT,[9] number of alternative stress locations and type of stress could be regarded as two factors that relate to separate processes in response organization. The first factor may be linked to a sequence identification stage, whereas the second one may be linked to a stage of planning or programming of the response execution.

From the present results, it is difficult to determine the kind of information and the degree of detail contained in the plan for response execution. Timing of the sequence, as revealed by the duration of the intervals between the successive taps, did not depend on type of stress. Recently, Klapp contended that response timing is the main factor that provokes changes in response program complexity and hence in programming time.[10] However, our experimental data indicate that timing was not a critical factor for the lengthening of RT under the stress− condition. The time distribution of motor intensities (forces) over the component elements was very different according to the type of stress. Positive stress involved the generation of a high-force element on the background of a stable low-level force selected for nonstressed taps. It may be that under this condition planning concerned only the force and the time location of the stressed tap, whereas the background low-level force was regulated by more automatic control mechanisms. In contrast, negative stress involved the generation of a low-force element on the background of an unstable high-level force selected for nonstressed taps. Here the planning may have concerned not only the force and time location of the negatively "stressed" element, but also the maintenance of the background force on a unusual high level. This more complicated planning would result then in a longer RT.

Effect of Stress Location

The RT of a sequence beginning with a stressed element was different from that of a sequence beginning with a nonstressed element. Under the stress+ condition, the stressed tap was of high force and the nonstressed taps of low force. The sequences beginning with a high-force element had longer RT than those beginning with a

[c]How is the response sequence identified? It has been suggested[5] that response identification involves search for the abstract name code of the response. However, such an abstract representation of the response may be of little use for the motor system. If so, the abstract response code must be translated into more concrete representations. We feel that the ultimate identifier of the response is nothing more than the plan or program that permits its execution. It seems likely that in the course of response organization intermediate response representations are generated between the abstract response code and the plan for response execution. The sensory or proprioceptive image of the response could be such a representation.[12] We are presently investigating the hypothesis that the rhythmic response sequences are identified in terms of the acoustic feedback images that are normally associated with the production of such sequences.

low-force element. Under the stress− condition, the stressed tap was of low force and the nonstressed taps of high force. Here again, the sequences beginning with a high-force tap had longer RT than the sequences beginning with a low-force tap. From these results one can conclude that programming a high-force initial element required more time than programming a low-force initial element. This conclusion is in accordance with our previous suggestion that low force was regulated in a more automatic way than high force. However, it must be emphasized again that type of stress effect does not simply derive from the decision concerning the level of force of the initial tap. Sequences that started with a high-force tap under the stress− condition had RTs 100 msec longer than sequences beginning with a high-force tap under the stress+ condition. The same relationship held true for sequences that started with a low-force tap. Thus, type of stress effect would be related to the planning of the sequence as a whole, while the observed stress location effect would be related to the programming of the initial element.

In conclusion, one can suggest that the short rhythmic sequences studied in this experiment are planned in advance of their execution. Planning involves, first, the identification of the sequence as a particular response, and second, the generation of a time–force plan that concerns the sequences as a whole. From the stress location effect we can infer that, in addition to the general time–force plan of the sequence, an executive program is generated that specifies the appropriate force level for the first element of the sequence.

The Timing of the Sequence

In our experiment the subjects had to reproduce a temporal model in which the duration of the intervals was the same. What was actually produced were strings of finger taps in which the intervals showed lawful variations with respect to the model. The structure of these variations could not be apprehended in the movement sequences as they were produced. It became apparent only when the time intervals produced were sorted according to a double classification principle based on position of the interval in the sequence and on position of the interval with respect to stress (or to change in force). This classification of the intervals permitted us to uncover: (1) a time structure linked to interval position in the sequence; and (2) the modulation of that basic structure by the stress.

Interval Modulation by Its Position in the Sequences

The time structure underlying the movement sequences was characterized by the fact that the first and last intervals tended to be equal, while both were considerably longer than the intermediate one. This time structure was force-independent in the sense that it was basically the same whether at the end of the intervals a force increase, a force decrease, or no change in force should occur. In previous experiments the same relationship between intervals was observed in five-tap strings as was demonstrated in the present experiment in four-tap strings. These time configurations are not easily accounted for by models in which an internal clock sends trigger pulses directly to the motor system for producing cadenced finger taps. Although the aim of the present work was not to test any formal model of sequence timing, our results might lend some support for a model in which the role of the internal clock is to provide a temporal reference for movement execution, rather than to trigger each component movement.[11]

Interval Modulation by Stress

The intervals that immediately preceded and followed an inverted stress were lengthened similarly to those that preceded and followed an ordinary stress. This indicates that the lengthening of the intervals is related to a change in force level, regardless of the direction of that change. Consequently, the effect of the stress on interval duration cannot be explained by the intervention of peripheral (biomechanical) factors responsible for the generation of higher level of force. Instead, the lengthening of the intervals that precede and follow a stressed element seems to be related to central processes that control the appropriate change in force level. This suggests that, in addition to the plan established during the RT, real-time control processes are also involved in the execution of the sequence. These processes could comprise a memory search for the target force level and/or updating the executive program with specification of the new level of force.

SUMMARY

Accentuation involves modulation of motor intensity. It differentiates a movement from others within a motor sequence. Does the serial position of the accent characterize the whole sequence as a particular response? How are the control of time and force coordinated in the motor sequence? Subjects produced sequences of four fingertaps on a key. Time of onset and force of each tap were recorded. Tapping rate was imposed by a string of four clicks delivered at 180-msec intervals before each trial. A flashed digit served as go signal. It indicated to the subject which of the four taps had to be tapped stronger (stress+) or weaker (stress−) than all the others. These conditions were run in separate series. Reaction time (RT) of the sequence increased when the number of equally likely locations of the stress increased from 2 to 4. RT was also longer under the stress− than under the stress+ condition. Tapping intervals were longer before and after the stressed tap than elsewhere in the series. The first and last intervals tended to be longer than the second one. These effects were the same under both stress conditions. The RT data indicate that the motor sequence is identified as a particular response before it starts. Timing is partly force-independent, but is modulated by central processes that control force.

REFERENCES

1. VINCE, M. A. 1948. Corrective movements in a pursuit task. Q. J. Exp. Psychol. **1:** 85–103.
2. HENRY, F. M. & D. E. ROGERS. 1960. Increased latency for complicated movements and a "memory drum" theory of neuromotor reaction. Res. Q. **31:** 448–458.
3. STERNBERG, S., S. MONSELL, R. L. KNOLL & C. E. WRIGHT. 1978. The latency and duration of rapid movement sequences: Comparisons of speech and typewriting. *In* Information Processing in Motor Control and Learning. G. E. Stelmach, Ed.: 117–152. Academic Press. New York, NY.
4. OSTRY, D. J. 1980. Execution-time movement control. *In* Tutorials in Motor Behavior. G. E. Stelmach & J. Requin, Eds.: 457–468. North-Holland. Amsterdam.
5. THEIOS, J. 1975. The components of response latency in simple human information processing tasks. *In* Attention and Performance. P. M. A. Rabbitt & S. Dornic, Eds.: 418–440. Academic Press. London and New York.
6. HICK, W. E. 1952. On the rate of gain of information. Q. J. Exp. Psychol. **4:** 11–26.

7. BRAINARD, R. W., T. S. IRBY, P. M. FITTS & E. A. ALLUISI. 1962. Some variables influencing the rate of gain of information. J. Exp. Psychol. **63:** 105–110.
8. THEIOS, J. 1973. Reaction time measurement in the study of memory processes: Theory and data. *In* The Psychology of Learning and Motivation: Advances in Research and Theory. G. H. Bower, Ed. Vol. 7: 43–85. Academic Press. New York and London.
9. STERNBERG, S. 1969. The discovery of processing stages: Extension of Donder's method (Attention and Performance: II). Acta Psychol. **30:** 276–315.
10. KLAPP, S. T. 1977. Reaction time analyses of programmed control. Exercise Sport Sci. Rev. **5:** 231–253.
11. SHAFFER, L. H. 1981. Performances of Chopin, Bach and Bartok: Studies in motor programming. Cognitive Psychol. **13:** 326–376.
12. GREENWALD, A. G. 1970. Sensory feedback mechanisms in performance control with special reference to the ideo-motor mechanisms. Psychol. Rev. **77:** 73–99.

Motor Disorder and the Timing of Repetitive Movements[a]

ALAN M. WING

Medical Research Council
Applied Psychology Unit
Cambridge CB2 2EF, England

STEVEN KEELE AND DAVID I. MARGOLIN

Cognitive Neuropsychology Laboratory
Good Samaritan Hospital and Medical Center
Portland, Oregon 97210

INTRODUCTION

In psychology, there are a number of instances in which accounts of human information processing have been advanced by the study of pathologic states. In the case of language, for example, the study of aphasia resulting from brain disease not only provided the first links toward understanding language and brain anatomy, but has also contributed to the psychological understanding of language in terms of the functional mechanisms involved. This is true because language breakdown is often selective; it is quite common to observe patients with some aspects of language function severely disrupted, while other aspects remain relatively intact. An example is provided by the contrasting selective deficits of Broca's and Wernicke's aphasia with damage to anterior and posterior areas of the brain.

In the present paper, we consider a theoretical model of the timing of repetitive movements that was developed by Wing and Kristofferson[1] as an account of data from normal subjects. The model postulates the existence of two separate processes contributing to the variability observed in the intervals between a stream of supposedly regular responses. We ask: In motor disorders of the central nervous system can a selective deficit be demonstrated in one or other of these processes? We address this question by presenting a detailed study of finger tapping by a patient with Parkinson's disease.

Parkinson's disease is a degenerative disease affecting the dopaminergic pathways of the basal ganglia. Symptoms of the disease affecting arm movements can include rest tremor (a tremor seen when the arm is held in a given posture, of larger amplitude and lower frequency than normal physiological tremor), rigidity (heightened resistance of the arm to passive manipulation), and bradykinesia (slowness of voluntary movement). The basal ganglia are bilateral structures of the midbrain involved in control of movement in the contralateral side of the body, but cases of predominantly unilateral symptoms have been reported.[2–4] We describe a case of hemi-parkinsonism in which a primary symptom was bradykinesia of the dominant (right) hand, which

[a]This research was supported in part by Grant NSF BNS 8119274 from the National Science Foundation (to S.K.), by Individual Research Award F32 NSO 6788 from the National Institute for Neurological and Communicative Disorders and Stroke (to D.I.M.), and by a grant from the American Parkinson Research Foundation.

had deleterious effects on the timing of movements in that hand. In this paper we compare the application of the Wing and Kristofferson model to timing data from each hand.

Consider a task in which the subject is instructed to produce a series of responses, such as tapping the index finger of one hand, in synchrony with a periodic auditory stimulus. After a few taps to get into phase with the pacing stimulus (an account of the processes involved may be found in Michon[5]), the auditory pacer is discontinued and the subject is asked to continue to tap at the same rate. Under these conditions subjects are able to produce long series of "free" responses with stable interresponse intervals (IRIs) with little drift in mean.[1]

Wing and Kristofferson's model of self-paced responding[1] (FIG. 1) proposes that departures from periodic responding may arise from imprecision in a hypothetical timekeeper and from temporal "noise" in the execution of responses triggered by the timekeeper. At time intervals, C, the timekeeper emits pulses, each of which initiates a motor response. A motor output delay, D, intervenes between the initiation and occurrence of the overt response, R. Each interresponse interval is thus the sum of a timekeeper interval plus the difference in motor delays associated with the initiating and terminating responses:

$$I_j = C_j + D_j - D_{j-1}. \quad (1)$$

Provided timekeeper intervals and response delays are both sequences of independent random variables and are mutually independent, it can be shown[6] that the IRI autocovariance function may be expressed as:

$$\text{autocov}_I(0) = \text{var}(I) = \text{var}(C) + 2\,\text{var}(D) \quad (2)$$

$$\text{autocov}_I(1) = -\text{var}(D) \quad (3)$$

$$\text{autocov}_I(k) = 0, \quad k > 1. \quad (4)$$

Thus, the model predicts negative covariation between immediately successive IRIs ($\text{autocov}_I(1) < 0$). Chance variation in any particular delay, D_j, about the mean motor output delay will tend to produce deviations of opposite signs in I_j and I_{j+1} about the mean IRI. The covariance between adjacent intervals, involving the products of such deviations, thus tends to be negative. Intervals that are separated by at least one intervening IRI and so do not share a common boundary should have zero covariance.

In previous experimental work,[6] various lines of research have provided evidence in support of the two-process model. First, estimates of the covariance between adjacent IRIs are usually less than zero.[1] Second, Wing[7] found that changes in the mean interval between responses only affect timekeeper variance (which may be estimated by solving for var[C] in Equations 2 and 3). Third, Wing[8,9] has found that, with changes of effector, timekeeper variance is more stable than motor delay variance (however, another parameter was added to the model, a point we will return to in the DISCUSSION).

If timing of the left and right hands of a patient with symptoms of parkinsonism affecting one side more than the other are compared, the theoretically interesting question is whether left–right differences in the variance estimates of the hypothesized underlying processes provide evidence that the processes are dissociated. If there is greater timing variability in the tapping of one hand compared with that of the other, will this be attributable to elevated motor-delay variance (larger, more negative lag-one variance) or to elevated timekeeper variance (larger IRI variance with no difference in lag-one covariance)? Or, will there be changes in both?

THE PATIENT

Clinical Report

The patient (M.F.), a right-handed woman aged 44 at the time of first testing for the present study, had had a 4-year history of predominantly unilateral parkinsonism. About 4 years prior to first testing she noticed that handwriting required more effort than normal and she sought medical attention some 18 months later. A computerized tomography scan of the brain was within normal limits, and she was begun on anti-parkinsonian medication. She did not receive much benefit from trihexyphenidyl or amantadine, and Sinemet was begun 1 year prior to testing. The prescribed dosage was low (10/100 mg three times a day) and, furthermore, she has never taken this medication consistently.

When first seen by us at the start of testing she showed mild, generalized symptoms of parkinsonism, including a minimal decrease in spontaneous trunkal and facial movements, stooped posture, and slight loss of postural reflexes. The only extremity involved was the right arm. Walking appeared normal except for decreased associated movement of the right arm. There was mild rigidity at the wrist and bradykinesia of voluntary finger movements. There was no apparent rest tremor. Mental status was found to be preserved on clinical examination.

During the period of testing (1 year) there was little if any increase in the axial signs of parkinsonism. However, the right hand became more rigid and bradykinetic. The right leg became noticeably stiff, sometimes causing the leg to drag when walking.

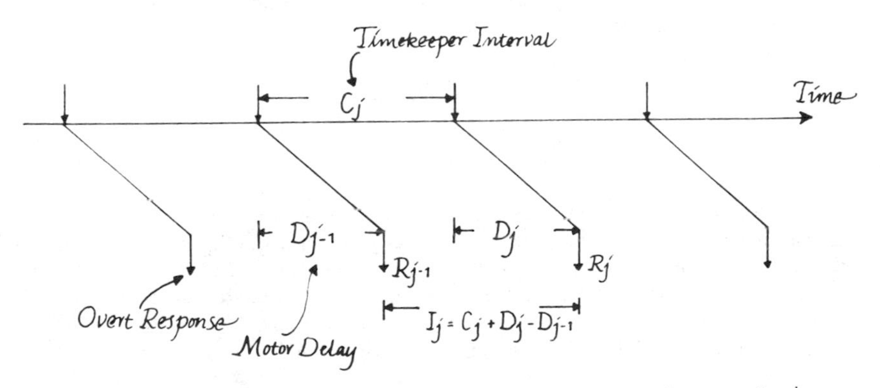

FIGURE 1. Wing and Kristofferson's model of self-paced periodic responding.[1]

Assessment of Function of Upper Limbs

In addition to the paced responding task described in the next section, a number of tests were used to document various aspects of the slowness of M.F.'s upper limb movements:

1. *Handwriting.* Because of her difficulty with writing using the preferred (right) hand, M.F. had been doing some writing with the left hand. She was asked to write a series of the cursive letters *e* and *l*. The average velocity for the production of the first and tenth occurrence of the letters was computed.[10]

The average velocity of the left hand was greater than that of the right. In the left hand the velocity showed a slight increase from the first (1.15 m/sec) to the tenth letter, (1.36 m/sec). In contrast there was a decrease in velocity in the right hand from the first (0.66 m/sec) to the tenth letter (0.42 m/sec).

2. *Reciprocal aimed tapping.* The patient was asked to use a pencil in the right or the left hand to tap alternately in pairs of targets 5 mm or 20 mm wide, 51 mm or 205 mm apart. The number of taps made within a 10-sec trial was recorded for each of the four combinations of width and distance. Over eight trials for each combination the time per tap for the left hand was reliably less than that for the right hand (TABLE 1). There was a reliable hand-by-width interaction; the right hand failed to speed up movements as much as the left hand to capitalize on the easiness of the wide targets.

3. *Purdue pegboard.* A task requiring perceptual–motor coordination similar to the previous task is the Purdue pegboard (Science Research Associates). Small cylindrical pegs must be picked up and placed in a row of drilled holes. From the published norms for this test, it was established that M.F.'s performance was below the first percentile for the right hand. But even with her left hand, her score was only at the fifth percentile. Although M.F. is used as her own control in the experiment described below, it is important to recognize that left-hand performance is not unimpaired. It might be noted that studies have also indicated that changes associated with Parkinson's disease tend to be bilateral even if markedly asymmetric.[3]

4. *Reaction time.* In a task developed for studying possible effects of line-drawing movement complexity on reaction time, it was found that M.F.'s simple and choice reaction were slower than normal controls. The median latency of the right hand was some 16 msec longer than that of the left hand.

5. *Maximum rate of tapping.* The maximum tapping rate was slower in the right hand (315 msec per tap) than in the left hand (215 msec per tap). Both values are longer than the normal range.[11]

EXPERIMENT

Task

The subject was seated with the arm used for tapping resting on a table. A touch plate was placed in front of the subject who placed her hand on it, palm down. Tapping the index finger on the touch plate completed a low-voltage electrical circuit, which provided a pulse to an Apple II computer, allowing time of contact to be measured to the nearest millisecond.

On each trial a series of 20 brief, clearly audible tones were presented at regular intervals under computer control. The standard intervals used were 450 and 550 msec. The subject was instructed to use movements at the carpo-metacarpophalangeal joint to tap in synchrony with these tones and to continue tapping at the same rate when the

TABLE 1. Reciprocal Tapping Task: Time per Tap (msec) as a Function of Target Distance and Width for Patient M.F.

	Left Hand		Right Hand	
	Wide	Narrow	Wide	Narrow
Near	328	414	416	451
Far	439	549	545	623

TABLE 2. Frequency of IRIs Excluded from Analysis

	Left Hand		Right Hand	
	450 msec	550 msec	450 msec	550 msec
Number of trials	43	45	40	48
Number of excessively long IRIs excluded	0	0	4	3
Number of very short IRIs excluded	1	1	30	12

tones ceased sounding. After approximately 30 such unpaced taps had occurred, the subject was asked to stop tapping. There was an interval of about 30 seconds between trials. A session comprised up to 24 trials; 6 trials with each hand at each of the two intervals. There were seven sessions spread over a period of a year; the first four were separated from the last three by 7 months.

Results

Observations made while M.F. was tapping revealed that movements of her more affected right hand were of smaller amplitude than those of her left hand and sometimes evidenced unwanted tremor between taps. On occasion the unwanted movement resulted in a response's being recorded by the computer. Each sequence of IRIs was therefore checked for the presence of intervals shorter than 250 msec and such intervals were dropped from the sequence. The immediately preceding and following intervals, one of which, when combined with the very short interval, might approximate the target interval, were also taken out of the sequence. Less common than unwanted responses were "missed" responses, in which the finger accidentally failed to contact the touch plate. These were identified by setting a criterion IRI of twice the required period less 150 msec. Intervals longer than this were removed from the sequence. TABLE 2 indicates the frequency of occurrence of such events. Missed and unwanted responses occurred mainly in the right hand. If they had not been removed, the variance estimates of the right hand would have been considerably larger.

The autocovariance function for lags zero through five of each sequence was estimated using

$$A_k(I) = \sum_{j=1}^{N-k} (I_j - \bar{I})\,(I_{j+k} - \bar{I})/(N - k - 1),\; k = 0{,}5,$$

where $\bar{I}$ is the sequence mean. In presenting the results we have employed the signed square root modulus of the autocovariance, which, in the case of the lag-zero autocovariance, is the same as the standard deviation of the IRIs. In FIGURE 2 the averaged functions at the standard intervals of 450 msec and 550 msec are plotted separately for the left and right hands. Two standard errors on either side of zero are indicated for lags two through five. The autocovariance function does not depart reliably from zero, except at lags zero and one (where the standard errors of estimate are, in fact, smaller).

Recall that, according to the Wing and Kristofferson two-process model, motor-delay variability is estimated from the lag-one autocovariance. Estimates of the square root of the autocovariance at lag zero (that is, the standard deviation) and at lag one

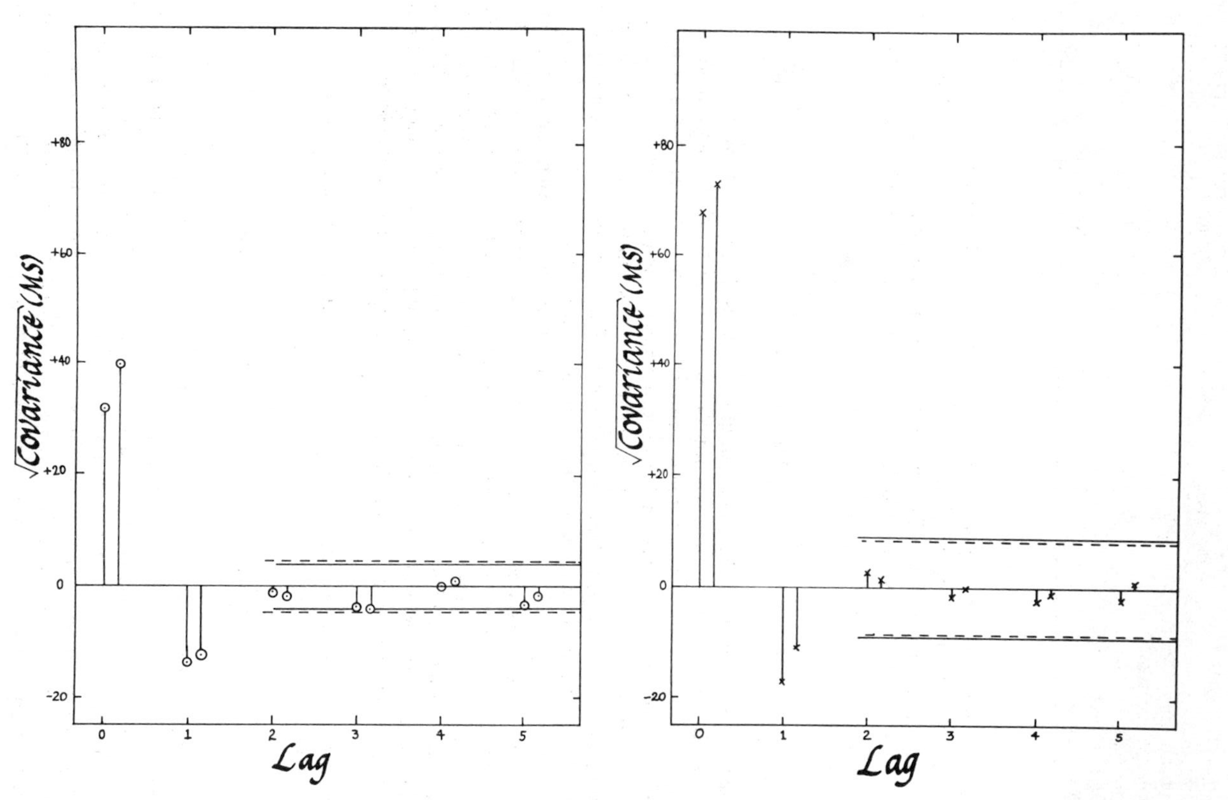

FIGURE 2. Signed square root modulus of IRI autocovariance at lags 0 through 5 plotted separately for the left (*circles*) and right (*crosses*) hands.

are shown in FIGURE 3 plotted against mean IRI. Data for the first group of four sessions are shown on the left; data for the second group of three are seen on the right. A three-factor analysis of variance on the standard deviations revealed significant main effects of session [$F(1,168) = 56.41$, $p < 0.01$], hand [$F(1,168) = 215.53$, $p < 0.01$], and interval [$F(1,168) = 9.57$, $p < 0.01$]. Although the second session variance levels were higher, and the variance at 550 msec was greater than at 450 msec, there was no session-by-interval interaction. The two-way session-by-hand interaction resulting from the increase in the right-hand variability in the second session relative to the left hand was significant [$F(1,168) = 44.89$, $p < 0.01$]. A separate analysis of variance on the signed square root modulus of the lag-one autocovariance indicated that there were no significant effects. In particular, there was no difference between the hands [$F(1,168) = 0.09$].

DISCUSSION

The tapping data from both hands of the patient (M.F.) with hemi-parkinsonism described in this paper show clear negative lag-one autocovariance in the IRIs, as predicted by the two-process model of Wing and Kristofferson.[1] Estimates of the covariance of IRIs separated by one or more intervening intervals (that is, the autocovariance at lag two or higher) did not differ significantly from zero. This is also consistent with the two-process model. The variability of the right hand, which was more affected by parkinsonism, was greater than that of the left hand. However, since the lag-one autocovariance was the same in the two hands, the inference from the model is that the elevated IRI variability of the right hand should be attributed to the timekeeper intervals and not the motor delays.

The data reported in this paper were collected from a single subject. Until similar results are obtained from other patients with hemi-parkinsonism, conclusions about processes in response timing or about the role of the basal ganglia are necessarily limited in their generality. Possible alternatives to studying patients with hemi-parkinsonism that retain the design philosophy of using subjects as their own controls include the evaluation of deterioration in performance with disease progression or the assessment of performance changes associated with dopaminergic medication. In the case of M.F. we looked for, but did not find, differences in tapping performance as a function of medication. This, perhaps, buttresses M.F.'s own observation that her use of medication is intermittent since it does not provide reliable symptomatic relief. However, in the 7-month interval separating the two groups of sessions, the performance decline in the right hand provided further evidence of dissociation of timekeeper intervals and motor delays.

The pattern of M.F.'s results indicates that the timekeeper variability is higher in the right hand than in the left. Should this be taken as evidence for two separate timekeepers? While we have no definitive data to refute this possibility, we are inclined against this view on the basis of a preliminary analysis of a new experiment conducted with M.F. In this study we used two force transducers mounted 70 mm apart with the points of finger contact 10 mm above the surface on which the hand rested. The transducers were interfaced through an analog–digital converter to an LSI-II computer. The response times of the index finger of the left hand tapping alone, the right hand tapping alone, or both hands tapping together were recorded together with the peak forces developed. On each trial, after 10 intervals of approximately 500 msec measured by a metronome, sampling was initiated for a period of 30 sec. Eight trials were run with each of the left and right hands alone and with the left and right hands together.

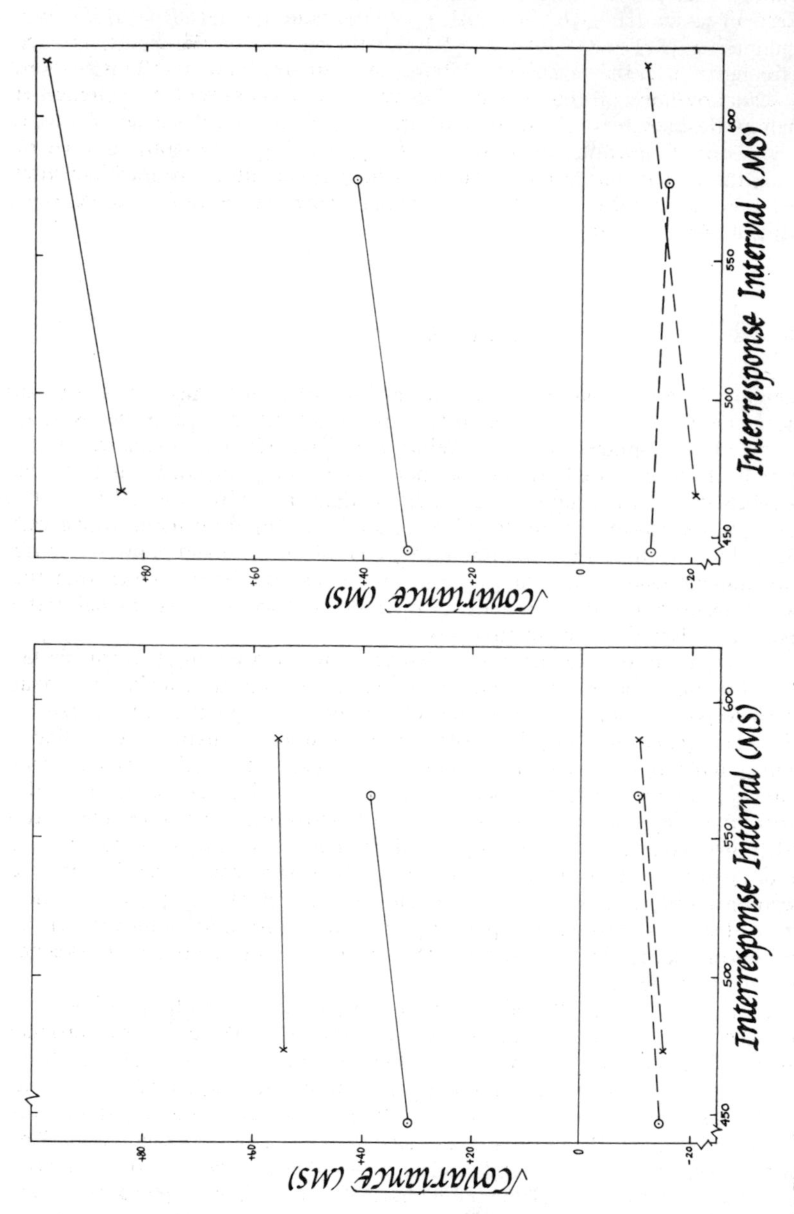

FIGURE 3. Signed square root modulus of IRI autocovariance at lag 0 (*solid lines*) and lag 1 (*dashed line*) for left (*circles*) and right (*crosses*) hands as a function of mean IRI.

The overall average interresponse interval was 523 msec. The average peak force per tap developed by the right hand was 18 g, just over half the 30 g developed by the left hand. A video recording showed that the amplitude of the right index finger excursion was one-third that of the left hand. In the hands-together condition, the left hand led the right hand on an average by 14 msec. The average standard deviation of asynchrony was 32 msec. An analysis of variance with hand (left versus right) and movement condition (tapping alone versus tapping together) as factors was performed on the standard deviation of the interresponse intervals of each sequence (see TABLE 3). The main effect of hand was significant [$F(1,28) = 24.39, p < 0.01$], but there was no reliable main effect of movement condition. Particularly interesting is the reliable hand-by-movement condition interaction [$F(1,28) = 9.57, p < 0.01$].

The lag-one autocovariance was only reliably less than zero in the right-hand hands-together condition. If, nonetheless, we assume constancy of motor-delay variance, the finding that the IRI S.D. of the left and right hands is the same is consistent with a single source of timekeeper variability when the hands tap together. But the effect of tapping with the hand more affected by parkinsonism (the right hand) indicates a sensitivity of the timekeeper to the conditions of movement that is not represented in the account of timing embodied in FIGURE 1.

TABLE 3. Interresponse Interval: S.D. (msec) of Left and Right Hands as Function of Whether Hands are Tapped Alone or Together

	Alone	Together
Left	26.2	40.9
Right	54.4	47.4

Relations between estimates of timekeeper variance and type of movement have been observed by Wing.[8,9] In that study it was found that, for example, tapping using finger flexion and extension was more variable than tapping using extension and flexion at the elbow. Because there was not a corresponding difference in lag-one autocovariance, the interpretation based on Wing and Kristofferson's two-process model[1] was that timekeeper variance had changed. However, it was noted that the estimated autocovariance at lag two was not zero, and this was taken as justification for adding to the model an extra parameter pertaining to dependence in the motor delays. In the present study no differences between the hands are seen in the autocovariance function at lag one, and lags greater than one are not different from zero. Thus, there are no grounds for supposing that there is motor-delay dependence and we are left with the conclusion that parkinsonism increases variability in the triggering of movement by the timekeeper.

In conclusion this case study has caused us to look afresh at the two-process timing model. Since it is a single case it will be very important to assess the generality of our ideas with follow-up research on available patients. Of special interest would be a patient with a contrasting disorder, in which motor-delay rather than timekeeper-interval variability was elevated.

SUMMARY

This paper is concerned with the timing of regular repetitive movements. The two-process model of Wing and Kristofferson[1] attributes variability in self-paced interresponse intervals to imprecision in a timekeeper and to temporal noise in the execution of motor responses triggered by the timekeeper. Assuming independence of timekeeper intervals and motor delays, the variance of each may be estimated from interresponse-interval statistics. Comparison of changes in timing performance associated with alterations in motor-system functioning offer the possibility of a new approach to investigation of this model. Illustrative data are presented from a case study of a patient with Parkinson's disease whose lesions affecting the dopaminergic pathways of the basal ganglia have given rise to asymmetric symptoms, including differences in timing performance of the two hands. Analysis of interresponse-interval variability according to the two-process model indicates that the elevated variability of the side more greatly affected by parkinsonism is attributable to the timekeeper intervals rather than the motor delays.

REFERENCES

1. WING, A. M. & A. B. KRISTOFFERSON. 1973. Response delays and the timing of discrete motor responses. Percept. Psychophys. **14:** 5–12.
2. SCOTT, R. M. & J. A. BRODY. 1971. Benign early onset of Parkinson's disease: A syndrome distinct from classic postencephalitic parkinsonism. Neurology **21:** 366.
3. MARTINEZ, A. J. & R. A. UTTERBACK. 1973. Unilateral Parkinson's disease. Clinical and neuropathologic findings. Neurology **23:** 164–170.
4. GILBERT, G. J. 1976. A pseudohemiparetic form of Parkinson's disease. Lancet (August 28).
5. MICHON, J. A. 1967. Timing in Temporal Tracking. Institute for Perception RVO-TNO. Soesterberg, the Netherlands.
6. WING, A. M. 1980. The long and short of timing in response sequences. *In* Tutorials in Motor Behavior. G. E. Stelmach & J. Requin, North-Holland. Amsterdam.
7. WING, A. M. 1973. The Timing of Interresponse Intervals by Human Subjects: Experiment 2. Ph.D. thesis, McMaster University, Hamilton, Ontario, Canada.
8. WING, A. M. 1977. Effects of type of movement on the temporal precision of response sequences. Br. J. Math. Stat. Psychol. **30:** 60–72.
9. WING, A. M. 1979. A note on the estimation of the autocovariance function in the analysis of timing of repetitive responses. Br. J. Math. Stat. Psychol. **32:** 143–145.
10. MARGOLIN, D. I. & A. M. WING. Agraphia and micrographia: Clinical manifestations of motor programming disorders. Acta Psychol. In press.
11. KEELE, S. W. Motor control. *In* The Handbook of Perception and Performance. K. F. Boff, Ed. John Wiley. New York, NY. In press.

Timing of Motor Programs and Temporal Patterns: Discussion Paper

EUGENE GALANTER

Psychophysics Laboratory
Columbia University
New York, New York 10027

The opening papers of this volume address the role of time in psychology from the point of view of stimulus description and its psychic representation in animals and humans. The idea is that the sensation scale values of this ostensible sensory continuum can be used to characterize responses to stimulus changes by reference to the scaled perceptual effects. The intuitive plausability of this approach gives it life even in the face of its consistent analytical failure. Over and over we seem to rediscover that stimulus measurement does not generalize to form a coherent representation of response. Studies of the psychophysics of temporal perception can, in my view, fare no better. Such research will not lead to a rational theory of time as a sensation. Even the recent successes of the study of shape and form, motion, and depth perception, which depend upon a multidimensional representation of the environment, have failed to convince us that a one-dimensional analysis of the stimulus is an unlikely path to understanding.

But now the topic has changed. In this section of this book we move away from a single stimulus representation to the idea that time acts as a behavioral constraint and a perceptual entity. As a perceptual entity, time serves as a contour. Jones' trochaic rhythms demonstrate the value of a psychophysics of syncopation. In her experiments time is introduced as a part or feature of a multidimensional display. Indeed, the Ohio researchers may not have gone far enough. I would suggest introducing dynamics (that is, amplitude variations) into the stimulus mix to allow even more detailed analysis of the dependence of response on the perceptual display.

We have already seen the rough outlines of the effects of time emerge as a contour of perception in the extensive studies of speech perception. I agree wholeheartedly with Jones that we would lose the structure of any reasonable models if we embraced some concept of information as an explanatory construct for the analysis of rhythmic patterns. Such a statistical tool is of little use in the analysis of configurations that are highly repetitive (and therefore information-poor, as rhythms are) and yet generate a variety of distinguishable patterns of behavior.

The second, and perhaps most important function of time in psychology, as evidenced by several papers in this section, is as a matrix for embedding the behavior stream. The major impetus to this effort is, or course, the primary finding of B. F. Skinner that time patterns of continuous activity may be construed as the fundamental information available for an analysis of behavior. His enormous influence on psychology lies, I think, in recognizing this important concept and in coupling it to the central theoretical notion in psychology, associationism, and its primary mechanism, contiguity.

It was not until Lashley's comments made in 1951, as many of our contributors have pointed out, that even when connected to configural perceptual mechanisms, associationism may not be sufficient as a theory of human and animal action. His reference to glissandos and other such motor programs provided little more than a complaint about the limits of simple associations.

The point was made well enough by Lashley, however, to disenchant many with prevailing stimulus–response associationism. But servo-models, structured hierarchically to avoid a mere replacement of the *S–R* chain, were also criticized, for example by Suppes.[1] Similar arguments had been voiced much earlier by Spence and Hull about the attempts by Tolman and his collaborators to replace the *S–R* bond with cognitive maps. Suppes's telling criticisms force an examination of the intrinsically learned or cultural aspects of planned behavior. "Totes,"[2] and their evolution into plans,[3] are nothing more or less than habits with external control over their execution. The conclusion, after 20 years of "cognitive science," is that chaining models of behavior are still the most powerful that we have. Even though they may fail to satisfy our intuitions about continuous behavior, they are hard to fault as a basis for summarizing observations.

These comments are meant to give voice to my concerns about servo-models as a solution to the behavior pattern problem. There is no doubt that servo-systems operate in the control of many aspects of behavior, but such schemes are not sufficiently constrained to serve as real theories of observable behavior. One alternative is that ballistic acts are a major feature of ongoing behavior. Although modeled on the saccade, my own view is that these behavioral "thrusts" can be found both in the small and in the larger features of behavior. To counter criticisms of observational invisibility, ballistic behavior can be identified by behavioral criteria. There are three main identifiers of ballistic acts and a fourth observation that supports the concept. They are: (1) Speed between component acts is less than stimulus–response RT. (2) Errors of components are cumulative. (3) Components are not segmentable. The

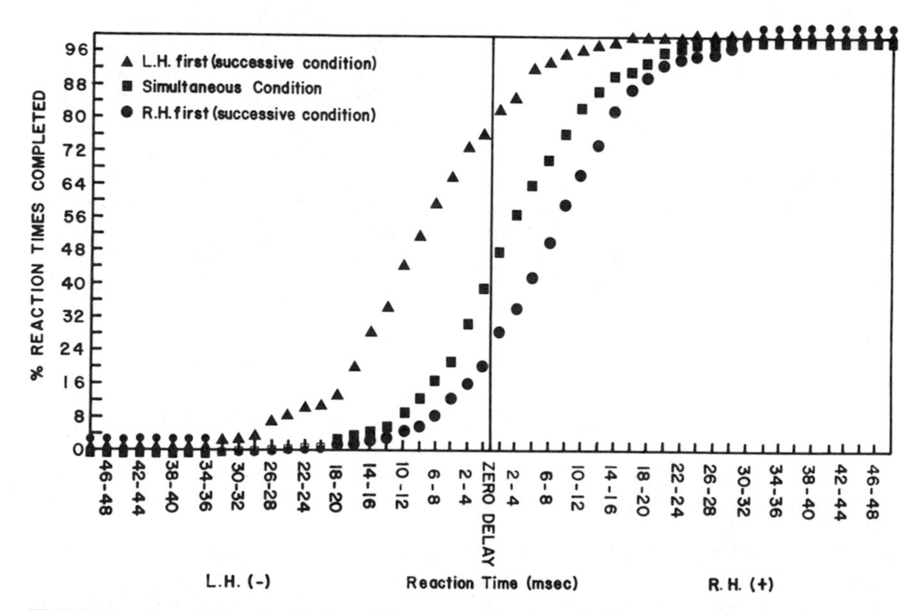

FIGURE 1. Cumulative distributions of interresponses times for lifting the left and then the right hand: *triangle* = left hand first; *square* = simultaneously; *circle* = right hand first.

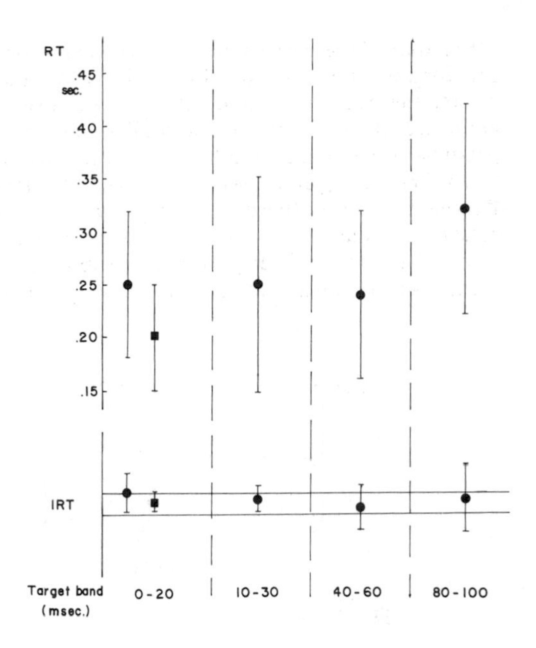

FIGURE 2. *Upper panel:* Response times (RT) in msec before initial response of lifting the left and right hand successively. *Lower panel:* target interresponse times (IRT) for positive payoffs. Subject adjusted the interval between lifting the two hands to match the target time. The "whiskers" represent 1 standard deviation.

fourth concept is the observable time required to organize an unanticipated ballistic act.

Let me demonstrate the first point based on work in our laboratory conducted by Rochelle Frankel. In these experiments subjects were required to lift their left (or right) forefinger from a key, followed as quickly as possible by their right (or left). A control consisted of simultaneous lifting of both. FIGURE 1 shows these data as cumulative frequencies of IRTs for the three conditions. Right hand first for right-handed subjects shows a median IRT of 7 msec. This goes to 10 msec for the left hand. Simultaneous response shows a slight (1.5 msec) bias toward the preferred hand. These speeds are consonent with component speeds of handwriting responses as reported in this volume by Stelmach, Mullins and Teulings.

Their work also demonstrates the importance of preplanning, and the potential for fine-tuning that this preplanning provides. This speaks to my fourth observation, and is well supported by our own work, as shown in FIGURE 2, which demonstrates control of IRTs between the two hands as evidenced by differential payoffs for IRT delays that fell in various narrow time (20-msec) bands.

Other evidence of the need for preplanning, and some support for the second point above can be seen in the paper by Semjen and Requin. In their report I also note that certain variables, such as the force of movement, play a less central role than might be thought in the ongoing behavior stream. That force differences have little effect on, say, the RT can be seen in an experiment from our lab by Bernice Rogowitz. FIGURE 3 shows the essential identity of simple RT distributions for 12-g, 100-g, and 1360-g "hold-down" forces for a single subject.

The final paper in this section, by Wing, Keele, and Margolin, attempts to investigate the two-process account (of Wing and Kristofferson) of response timing (tapping) by reference to response data from a patient with physiological damage. I am out of my depth in commenting on the clinical data, but I do want to sound a warning that such research arouses in me. Illness, physiological deficit, and drug-induced modulation of the nervous substrate may and probably do involve extensive psychological reorganization. It is therefore unlikely that the analysis of behavior from people in these states can lead to useful information concerning normal function. This position is not inviolable. For example, the studies made by Graham *et al.*[4] of the color vision of patients with asymmetric color deficits are inherently immune to this criticism.

In Wing's paper we see a similar application in a clinical case, a patient with Parkinson's disease. Because of the asymmetry of the symptoms in this patient, the researchers could use the patient as her own control, and thereby test the two-process model. This model, which partitions the observed behavioral variability of tapping into a motor-delay component and a timekeeper component, suggests that Parkinson's

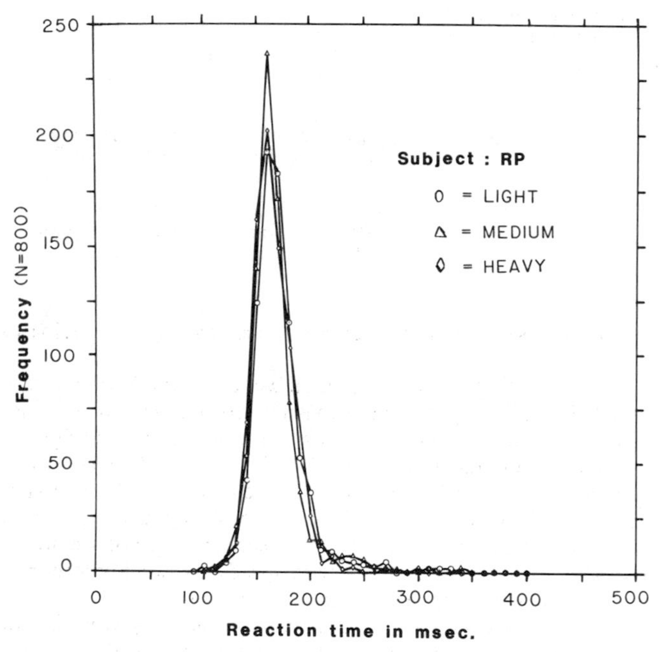

FIGURE 3. Distributions of simple key-release RT to 1000-Hz tone at 82 dB SPL for hold-down forces of about 10, 100, and 1400 g.

disease insults the timekeeping aspect of the model. The authors suggest that such behavioral data may shed light on the locus of neural control of response timing. But even without follow-up studies to support the physiological psychology, the vitality of the two-process motor-timing theory is enhanced.

Taken together, the work presented in this section of this volume demonstrates the vigor of research on temporal processes in behavior and the central importance of temporal perceptual and motor patterns. Some theoretical insights and potentially practical information have been imparted. We must still press for a more encompassing and coherent theory that will allow us to synthesize as well as to analyze the observations and experiments that result from some of the new techniques presented in this section.

REFERENCES

1. SUPPES, P. 1969. Stimulus–response theory of finite automata. J. Math. Psychol. **6:** 327–355.
2. GALANTER, E. & G. A. MILLER. 1960. Some comments on stochastic models and psychological theories. *In* Mathematical Methods in the Social Sciences: 1959. K. J. Arrow, S. Karlin & P. Suppes, Eds. Stanford University Press. Stanford, CA.
3. MILLER, G. A., E. GALANTER & K. PRIBRAM. 1960. Plans and the Structure of Behavior. Holt, Rinehart & Winston. New York, NY.
4. GRAHAM, C. H., H. G. SPERLING, Y. HSIA & A. H. COULSON. 1961. The determination of some visual functions of a unilaterally color-blind subject: Methods and results. J. Psychol. **51:** 3–32.

Introduction

HERBERT S. TERRACE

Department of Psychology
Columbia University
New York, New York 10027

An organism's unlearned sensitivity to time has long been the focus of research on such diverse phenomena as circadian rhythms and migratory cycles. After Pavlov's discovery of the conditioned reflex and Thorndike's formulation of the law of effect, temporal parameters such as the interval between the conditional and unconditional stimuli (CS and US), delay of reinforcement, interresponse time, rate of responding, and postreinforcement-pause have been the subject matter of countless experiments.

While the role of timing in animal learning is a common theme of the papers in this section, the various ways in which timing enters into contemporary analyses of learned behavior is revealing of the many advances that have occurred in this area. One of the most obvious examples is the demonstration by Rescorla and others that temporal contiguity per se between the CS and the US is not sufficient to produce conditioning. Jenkins' paper provides an incisive analysis of the strengths and weaknesses of three alternative theories of conditioning: his own "relative waiting time" model, the Rescorla–Wagner model (referred to as a "contiguity-plus-competition model"), and Gibbon's scalar expectancy theory (referred to as a "contiguity-plus-comparison model"). Balsam's paper provides an illuminating discussion of how scalar expectancy theory can be applied to trace-conditioning. Taken together, both papers show the importance for the learning process of an animal's assessment of the time that elapses between successive presentations of the US.

An animal's ability to integrate and compare relative rates of reinforcement provided by different schedules of reinforcement (as demonstrated by Herrnstein and others) has served as another focus of much recent research on animal learning. The papers by Krebs and Lea relate this line of research to recent investigations of foraging. Both papers show how variables such as choice of schedule type and the consequences of shifting from one schedule to another have important analogues in the study of foraging. They also demonstrate the influence exerted on an animal's foraging behavior by the rate of food intake in various locations of its environment, and the cost and risk of shifting from one location to another. At the same time they provide a good example of the fruitfulness of relating classic problems of ethology to laboratory studies of "matching" and "maximizing" and vice versa.

Before one can generalize about the role of timing in animal learning, one must have a broad base of information regarding the ability of various species to learn to resolve temporal intervals under conditions that minimize the contributions of nontemporal factors such as motivation and response biases. The papers by Platt, Richelle and Roberts summarize our knowledge of these issues clearly and also point to a variety of interesting questions that need to be addressed in future research.

Motivational and Response Factors in Temporal Differentiation[a]

JOHN R. PLATT

Department of Psychology
McMaster University
Hamilton, Ontario, Canada L8S 4K1

Temporal differentiation paradigms provide reinforcing or punishing stimuli contingent on a specified temporal property of a response. Beginning with Catania,[1] a number of investigators[2–6] have treated temporal differentiation paradigms as animal analogues to various paradigms used to scale subjective time in humans.[7] Such animal analogue studies universally find mean produced values of the differentiated temporal property to be described by a fractional-exponent power function of the minimum values required for reinforcement. This result has sometimes been interpreted as suggesting that the animal's subjective representation of time bears a power relationship to physical time.[3]

More recently, I have argued that the proposed analogy between temporal differentiation in animals and various human scaling paradigms is inappropriate.[8] Other authors have also criticized the inference of subjective time scales from temporal differentiation data.[9,10] An unfortunate byproduct of this demise in temporal differentiation as a basis for inferring subjective time scales seems to be a general loss of interest in temporal differentiation in the recent literature in animal research. Even if temporal differentiation paradigms are useless for scaling, they embody a fundamental operant process involved in many reinforcement schedule effects.[11] Temporal differentiation also clearly involves timing mechanisms that need to be understood and related to timing mechanisms inferred from other paradigms. The present experiment represents an attempt to identify some of the processes, in addition to a timing mechanism, which determine performance in temporal differentiation.

Shimp has provided evidence of a preference for shorter response values in a temporal differentiation paradigm that reinforced all values equally.[12] I have suggested that this preference can be conceptualized as an incentive motivational process favoring immediacy of reinforcement.[8] This process would oppose production of longer response values by temporal differentiation contingencies, and could easily be integrated into an incentive-based timing position such as scalar expectancy theory.[13] Such a hybrid theory is potentially capable of accounting for the extant data from temporal differentiation paradigms, including the fractional-exponent power relationship between mean temporal response value and minimum value required for reinforcement. It thus seems timely to provide a firmer empirical basis for an incentive motivational process favoring shorter response values in temporal differentiation. It would also be useful to know whether this process is independent of the particular behaviors defining the temporal interval, or whether more specific response factors must also be considered.

One way to investigate these proposed motivational and response factors is simply to compare differentiation of various temporal properties of several responses. On the

[a]This research was supported by Grant A8269 from the National Sciences and Engineering Research Council of Canada.

working hypothesis that various temporal properties of different responses share a common timing mechanism, resulting differences in the temporal response values produced could reasonably be attributed to the kinds of motivational and response factors under consideration. In addition, an incentive motivational process favoring shorter temporal values should be independently accessible in terms of a decreasing willingness to engage in a response as one of its temporal properties is lengthened by differential reinforcement, and a reversion to shorter values if reinforcement is subsequently presented independently of that property.

A comparative analysis of various temporal properties of different responses cannot be achieved using common temporal differentiation paradigms that reinforce responses exceeding some fixed criterion value of a specified temporal property. Various temporal properties of different responses yield different distributions of values prior to differential reinforcement. Under these circumstances, introduction of a fixed temporal criterion for reinforcement will produce different probabilities of reinforcement, depending on the property being differentiated. In the extreme, some temporal properties may not even contact a particular fixed criterion in the sense that all responses may have temporal values falling on one side of the criterion. There is thus no way to tell whether resulting differences in performance reflect differences in effects of differential reinforcement on various properties, or differences in contact with the reinforcement criterion. This confounding is further potentiated when changes in performance produced by differential reinforcement lead to even larger differences in contact with the reinforcement criterion. In addition, fixed reinforcement criteria limit the usefulness of amount of change in response values as a dependent variable, because the criterion imposes an upper limit on the amount of change to be expected.

The problems associated with a comparative analysis based on fixed temporal criteria for reinforcement can be obviated by use of a percentile reinforcement paradigm which holds constant an organism's contact with a differential reinforcement contingency.[14] Such a paradigm reinforces a response if the value of its relevant property is longer than those for a specified proportion of the organism's most recent responses. With an appropriate selection of parameters, such a procedure has the effect of differentially reinforcing a constant proportion of the organism's responses that have the longest values of the relevant temporal property at the time the responses are emitted.[14]

The current experiment applied percentile reinforcement to four temporal properties of two different responses. The four temporal properties were those most commonly used in the temporal differentiation literature—namely, latency, interresponse time (IRT), changeover IRT, and duration. One of the two responses was lever pressing, a commonly used operant with rats. The other response involved rearing on the hind legs and extending the snout into an opening in the ceiling of the chamber. This is a common response in the rat's "natural" repertoire, involving postural orientation rather than manipulation of an environmental object such as a lever.

METHOD

Subjects

Thirty-two experimentally naive male Wistar rats, approximately 120 days old, were maintained at 85% of their free-feeding weights by supplemental rations of Purina Lab Chow administered in the home cages after each experimental session.

Apparatus

Four identical Lehigh Valley rodent test chambers 30.5-cm long by 24-cm wide by 27-cm high were contained in sound-attenuating enclosures. A 3.8-cm circular aperture in the front wall allowed access to a food tray into which standard 45-mg Noyes Precision Food Pellets were delivered. Chamber illumination was provided by a 2.8-W incandescent bulb mounted 17.5 cm above the food tray. Lehigh Valley retractable levers were mounted 6.5 cm to either side of the food tray, 5.0 cm above the floor. The levers could be extended or retracted in 1.9 sec and required a force of .22 N to record a response.

A Plexiglas false ceiling could be inserted into the chamber, reducing the chamber height to 18.5 cm. Centered in the false ceiling was a 5-cm square aperture into a 5-cm-deep box, which will be referred to as the "chimney." The rear wall of the chimney was hinged at the false ceiling and could be rotated 90 degrees by a small motor so as to close the aperture. A photobeam 1.5 cm into the chimney was used to detect the presence of a rat's snout.

Experimental control and data acquisition was accomplished by a Digital Equipment Corporation PDP 8/e computer. The computer was programmed to resolve real time to the nearest .02 sec.

Procedure

All rats were trained to approach and eat from the food tray over three sessions. Each session consisted of 30 response-independent presentations of two food pellets on a variable time schedule with a mean value of 30 sec. The rats were then randomly and equally divided into two groups. One group was trained to press the right lever (the "lever animals"), while the other group of rats was trained to rear on their hind legs and insert their snouts into the chimney (the "chimney animals"). This training was carried out in a single session consisting of 100 food presentations. Every response produced two food pellets and initiated an 8-sec intertrial interval (ITI), during which the chamber was dark, the levers retracted and the chimney closed.

The lever and the chimney animals were then each divided randomly and equally into four subgroups which differed with respect to the temporal property of the appropriate response subsequently differentiated. These properties were the latency of a single response (latency), the duration of a single response (duration), the time between two offsets of the same response (IRT), and the time between offsets of two spatially distinct responses (changeover IRT). For lever animals, a changeover IRT was from the left to the right lever, while for chimney animals it was from the chimney to the right lever. For the remainder of the experiment a trial consisted of one response for latency and duration animals, two identical responses for IRT animals, and two different responses for Changeover IRT animals. Each trial was followed by an 8-sec ITI, and sessions consisted of 100 trials or 50 min, whichever came first.

All rats were given two sessions in which all trials terminated with food presentation. Percentile reinforcement of the appropriate temporal property was then instituted. In the first session a trial terminated with food presentation if its relevant temporal property was longer than it had been on at least 8 of the last 13 trials. In the next session this criterion was increased to 10 of the last 13. Thirty sessions were then run in which food presentation depended on the relevant temporal property being longer than it had been for at least 12 of the last 13 trials. This last condition reinforces the longest 14.3% of temporal values emitted by the rat at any point in training.[14]

After percentile reinforcement, an additional 12 sessions were conducted in which the contingency between temporal response value and food presentation was removed without changing the probability of food presentation. This was accomplished by using a pseudorandom number generator to present food on 14.3% of the trials, without regard for the temporal response value. The purpose of this procedure was to examine maintenance of temporal differentiation in the absence of differential reinforcement.

One rat in the lever duration group died during the experiment. This condition thus contained three rats; all others contained four.

RESULTS

When percentile reinforcement of any temporal response property was instituted, the mean value of that property increased in a negatively accelerated manner and became asymptotic within 25 sessions. FIGURE 1 shows the mean produced value of the

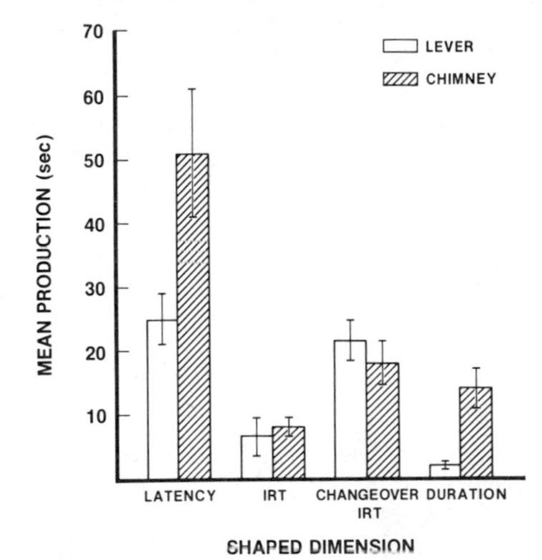

FIGURE 1. Means and standard errors of the various differentiated temporal response properties over the last three sessions of percentile reinforcement.

differentiated temporal property for each group over the last three sessions of percentile reinforcement.

Despite the comparability of differential reinforcement across responses and temporal properties achieved by the percentile reinforcement procedure, asymptotic production values varied extensively from a mean of 1.9 sec for lever duration to a mean of 50.9 sec for chimney latency. The chimney response yielded longer values than the lever response with respect to latency and duration, but quite similar values for IRT and changeover IRT. Across the two responses, latency yielded longer values than changeover IRT, which in turn yielded longer values than IRT and duration. There was a strong interaction between responses and temporal properties for the latter two properties. With the lever response, IRT yielded longer values than duration, while the reverse was true for the chimney response. Indeed, lever duration increased only slightly and one of the three rats in this group showed no increase over the entire course of percentile reinforcement.

Produced response values also varied considerably across temporal properties when differential reinforcement was subsequently replaced by nondifferential reinforcement. During the 12 sessions of nondifferential reinforcement, each previously differentiated temporal property gradually shorten in value. TABLE 1 shows the group means of each rat's lowest session mean value of the previously differentiated temporal property during subsequent nondifferential reinforcement. There was very little difference in these maintained temporal values between the two responses, but considerable difference between temporal properties. Latency and changeover IRT means were in excess of 5 sec, while IRT means were on either side of 1 sec and duration means were less than 0.5 sec. The longer maintained values for latency and changeover IRT might account for the larger increases obtained for these properties during percentile reinforcement. Otherwise there was very little relationship between maintained response values (TABLE 1) and differentiated values (FIG. 1).

Differences between responses and temporal properties in asymptotic values during both differential and nondifferential reinforcement renders absolute values useless for comparing maintenance of the various temporal differentiations when reinforcement was subsequently made nondifferential. Instead, a relative maintenance measure was devised on the basis of Anderson's shape function method[15] for comparing resistance to extinction across populations with different acquisition and extinction performance

TABLE 1. Mean and Standard Error of Lowest Session Mean Production (sec) for the Various Groups During Nondifferential Reinforcement

Shaped Dimension	Response	
	Lever	Chimney
Latency	5.5 ± 0.97	5.0 ± 0.92
IRT	1.1 ± 0.04	0.6 ± 0.13
Changeover IRT	5.2 ± 1.20	8.1 ± 1.53
Duration	0.4 ± 0.15	0.4 ± 0.09

asymptotes. A relative maintenance function was constructed for each rat according to the expression

$$M_n = \frac{\overline{T}_n - T_{MIN}}{T_\infty - T_{MIN}},$$

where M_n is the maintenance index for session n, $\overline{T}_n$ is the mean response value of the previously differentiated temporal property in session n, T_∞ is the mean value of the differentiated property over the last three sessions of percentile reinforcement and T_{MIN} is the lowest session mean value of the previously differentiated temporal property during nondifferential reinforcement. This computation assigns a value of 1.0 to asymptotic performance during temporal differentiation and a value of 0.0 to the lowest performance during nondifferential reinforcement. Performance during nondifferential reinforcement sessions is scaled relative to these two reference values.

FIGURE 2 shows the group means of the individual relative maintenance functions over the 12 sessions of nondifferential reinforcement. A negatively accelerated decrease in the previously differentiated temporal response property is clearly evident, although chimney changeover IRT showed an initial increase. Overall maintenance comparisons can more easily be made in FIGURE 3, which shows the means of the relative maintenance functions over all 12 sessions of nondifferential reinforcement. The two responses employed made little difference for maintenance of the temporal differentiation, except that chimney latency was maintained slightly better than lever

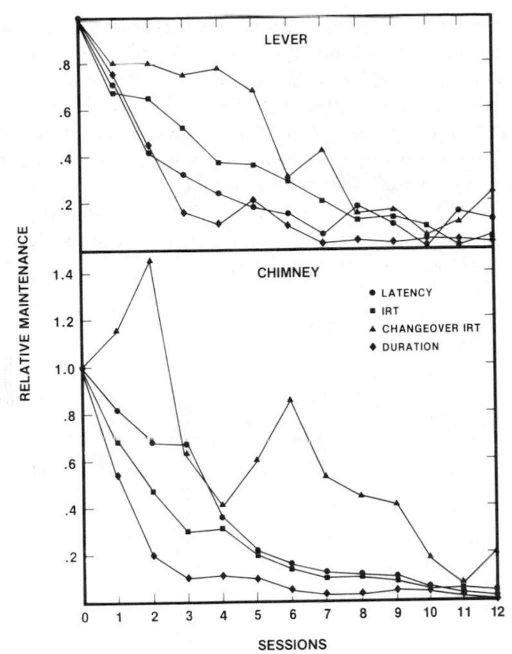

FIGURE 2. Mean relative maintenance functions for the various temporal differentiations over 12 sessions of subsequent nondifferential reinforcement.

latency. On the other hand, the temporal property differentiated produced considerable variation in maintenance. Changeover IRT was maintained most and duration least, while IRT and Latency were intermediate and quite similar to each other.

Yet another way to examine a possible preference for shorter values of temporal response properties is to look for a decreased willingness to engage in a response when the value of one of its temporal properties is increased by differential reinforcement.

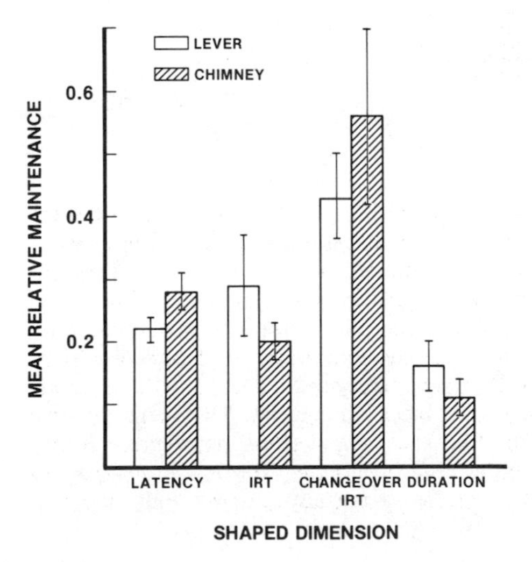

FIGURE 3. Means and standard errors of overall relative maintenance for the various temporal differentiations during subsequent nondifferential reinforcement.

Except when latency itself is being differentially reinforced, this willingness might be measured in terms of response latency. FIGURE 4 shows mean response latencies for each group. The first bar of each pair represents mean latency over the last three sessions of percentile reinforcement, the second represents the same measure over the last three sessions of subsequent nondifferential reinforcement. Data are included for latency groups to provide reference values reflecting the effect of explicit differential reinforcement of response latency.

When reinforcement was nondifferential, all groups except changeover IRT showed relatively short and similar response latencies in the range of 2.5 to 7.8 sec. With the exception of the lever changeover IRT group, response latencies were

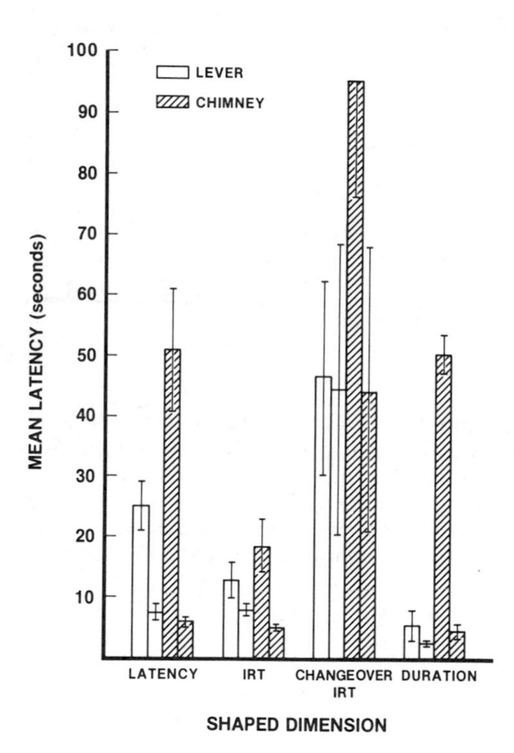

FIGURE 4. Means and standard errors of response latencies for the various groups. The *first bar* for each group represents the last three sessions of differential percentile reinforcement. The *second bar* represents the last three sessions of subsequent nondifferential reinforcement.

considerably longer when values of any temporal property of the response were increased by percentile reinforcement. In some cases this latency increase was as great or greater than that produced by differential reinforcement of latency itself. Furthermore, the magnitude of increase in response latency resulting from differential reinforcement of any temporal property was generally related to the amount of increase in the differentiated property (FIG. 1), except in the case of changeover IRT. Thus, except in the case of changeover IRT, response latency does seem to be reflecting a preference for shorter temporal response values. Changeover IRT required two different responses in a prescribed order. Subjects would often perseverate in the second of these responses before initiating the IRT by emitting the first response, thus greatly inflating the latency measure. This probably accounts for the long latencies (about 45 sec) in changeover IRT groups, even when differential reinforcement was not in effect.

DISCUSSION

The primary purpose of the present experiment was to examine possible preferences for shorter response values in temporal differentiation. Such a preference might be conceptualized in terms of an incentive motivational process favoring immediacy of reinforcement, and would serve to oppose the acquisition and maintenance of longer temporal response values. A second purpose was to ask if this process is independent of the particular behaviors involved, or if more specific response factors are also relevant.

The present results provide several sources of support for the proposed incentive motivational process. When longer values of any temporal property of either response were differentially reinforced, the mean value of that property increased in a negatively accelerated manner and became asymptotic within 25 sessions. In a conventional fixed-criterion temporal differentiation paradigm, the asymptotic nature of these acquisition functions would be expected as more and more of an organism's responses exceeded the criterion for reinforcement. However, the percentile reinforcement schedule employed here is a bit like riding a donkey while dangling a carrot on a stick in front of it—the donkey can endlessly approach the carrot, but never gets any closer to it. With percentile reinforcement, any shift in a response distribution towards longer reinforced temporal values is met by a corresponding increase in the values required for subsequent reinforcement. Under these circumstances there is no reason to expect increases in the value of the relevant temporal property to show negative acceleration toward a limit unless processes other than differential reinforcement oppose such increases. The proposed incentive motivational process favoring immediacy of reinforcement would certainly provide such opposition.

Another feature of the present results supporting the proposed incentive motivational process is the finding that all differentiated temporal properties of both responses rapidly fell to shorter values when reinforcement was subsequently made nondifferential. This result might be expected simply on the basis of more shorter response values being reinforced under the latter condition. However, it has recently been shown that nondifferential reinforcement following differential percentile reinforcement of a spatial response property often produces no change in subsequent response distributions.[16] This suggests that the increased variability in reinforced response values when reinforcement is changed from differential to nondifferential is not in general sufficient to change the response distribution. The rapid shortening of previously differentiated *temporal* response values in the present nondifferential reinforcement condition is readily explained, however, in terms of an incentive motivational process favoring immediacy of reinforcement, and hence shorter temporal response values.

Possibly the best independent evidence for the proposed incentive motivational process is provided by changes in response latencies when some other response property underwent temporal differentiation. When longer mean values of any temporal property of either response were produced by differential percentile reinforcement of that property, response latency also increased. When mean response values decreased during subsequent nondifferential reinforcement, response latencies declined in all cases except lever changeover IRT, were latencies were likely contaminated by perseveration of the terminal response. Furthermore, differences in response latencies between differential and nondifferential reinforcement conditions were roughly related to differences in mean values of the differentiated temporal property between these conditions. The greater the increase in the differentiated property, the greater the increase in latency. These results suggest a decreasing willingness to engage in a response as one of its temporal properties is lengthened by differential reinforcement and is exactly what would be expected from an incentive motivational process favoring immediacy of reinforcement.

An alternative explanation of the systematic relationship between response latency and differentiation of other temporal properties of the response might be that the temporal differentiation generalized to the latency dimension. Several considerations make this alternative account implausible. First, it would be very surprising if the various temporal properties involved in the present experiment did not differ considerably in their similarity to latency, and hence produce very different amounts of generalization of the temporal differentiation to the latency dimension. Yet latency increases with differentiation of another temporal property appeared to be determined by the amount of increase in the differentiated property, and not by which property was differentiated.

An even stronger argument against the generalization account is provided by response durations which were reported here only for groups in which that property was differentiated, but which were collected for all groups. As already noted, response latencies increased when duration was differentiated. However, there were no systematic changes in response duration when latency was differentiated. If latency and duration were sufficiently similar to produce generalization of temporal differentiation from duration to latency, similar generalization should have been observed from latency to duration.

Although the present results provide strong support for the existence of the proposed incentive motivational process and its assumed role in temporal differentiation, they also indicate the importance of more specific response factors. A preference for shorter response values based on immediacy of reinforcement should oppose differentiation and maintenance of longer values equally for all temporal properties and responses. The amount of opposition to a particular response would simply depend on its temporal value, and not on the nature of the response. Nevertheless, the various temporal properties and responses examined in the present study differed considerably in their asymptotic levels of differentiation and the rate of loss of this differentiation during subsequent nondifferential reinforcement. If such differences are also to be conceptualized in terms of an incentive motivational process, this process would have to depend on response factors in addition to time. One such possibility would be response effort. The main problem with such an approach is the lack of a suitable way to provide *a priori* specification of the effort involved in various responses. Perhaps a better approach would be to use choice between various reinforcement requirements for different temporal response properties to achieve a scaling of the incentive motivational function opposing longer response values in any particular case. Of course such a scaling exercise would be extremely laborious.

Yet another possibility is that differences in asymptotic levels of differentiation and loss of differentiated responding during nondifferential reinforcement may result directly from the mechanics of the various response systems, rather than through a resulting incentive structure. The various temporal properties examined in the present experiment did, for instance, differ in their values in the absence of differential reinforcement. However, these differences did not appear to be generally related to subsequent differences in asymptotic differentiation or its maintenance during nondifferential reinforcement. The present experiment does not provide an adequate basis for speculation as to relevant response mechanics, and such an approach seems unlikely to lead to a parsimonious behavioral theory.

The obvious role of specific response factors in the present results makes findings of quantitatively similar functional relationships for temporal differentiation of different responses and response properties all the more surprising. With fixed-criterion temporal differentiation paradigms, asymptotic mean value of the differentiated property ($\overline{T}$) has repeatedly been found to be related to the minimum value required for reinforcement (t) by a power function of the form

$$\overline{T} = kt^{\text{n}}$$

where k and n are empirical constants.[8] Furthermore, k invariably takes on values greater than 1.0, while n takes on values slightly less than 1.0. A power function with these parameter values indicates a departure from proportionality between $\overline{T}$ and t such that $\overline{T} > t$ for small values of t but $\overline{T} < t$ for larger values of t. As indicated previously, some have interpreted this power function in terms of a relationship between subjective and physical time,[3] while others have argued against this interpretation.[8-10]

Support in the present experiment for an incentive motivational process favoring immediacy of reinforcement provides an alternative interpretation for the power relation between $\overline{T}$ and t, and also makes clear the reason for the parameter values $n < 1.0 < k$. This process provides increasing opposition to the production of temporal response values as those values become longer, and the subject must balance this incentive factor against that resulting from the fact that values shorter than t are not reinforced. With short values of t there is very little opposition based on immediacy of reinforcement so the subject emits most of its responses with values in excess of t and $\overline{T} > t$. As t becomes larger, this opposition systematically increases so that subjects emit fewer and fewer responses with values in excess of t and $\overline{T}$ becomes progressively less than t. It is thus clear that, in general, $n < 1.0 < k$ will hold. However, within this restriction it would be expected to find variations in n and k with specific response factors and such variations are commonly reported.[8]

In summary, the present results indicate that future theoretical treatments of temporal differentiation should include a preference for shorter response values, such as is provided by the proposed incentive motivational process favoring immediacy of reinforcement. In addition, such theories should also provide for differences in the differentiability of various temporal properties and responses. Anyone still believing that temporal differentiation provides a direct window into the nature of subjective time or the workings of an internal clock should find these results very disconcerting.

REFERENCES

1. Catania, A. C. 1970. Reinforcement schedules and psychophysical judgements: A study of some temporal properties of behavior. *In* The Theory of Reinforcement Schedules. W. N. Schoenfeld, Ed. Appleton-Century-Crofts. New York, NY.
2. DeCasper, A. J. & M. D. Zeiler. 1974. Time limits for completing fixed ratios: III. Stimulus variables. J. Exp. Anal. Behav. **22:** 285–300.
3. DeCasper, A. J. & M. D. Zeiler. 1977. Time limits for completing fixed ratios: IV. Components of the ratio. J. Exp. Anal. Behav. **27:** 235–244.
4. Kuch, D. O. 1974. Differentiation of press durations with upper and lower limits on reinforced values. J. Exp. Anal. Behav. **22:** 275–283.
5. Platt, J. R., D. O. Kuch & S. C. Bitgood. 1973. Rats' lever-press durations as psychophysical judgements of time. J. Exp. Anal. Behav. **19:** 239–250.

6. RICHARDSON, W. K. & T. E. LOUGHEAD. 1974. Behavior under large values of the differential-reinforcement-of-low-rate schedule. J. Exp. Anal. Behav. **22:** 121–129.
7. TREISMAN, M. 1963. Temporal discrimination and the indifference interval: Implications for a model of the "internal clock." Psychol. Monogr. **77** (13, Whole No. 576).
8. PLATT, J. R. 1979. Temporal differentiation and the psychophysics of time. *In* Advances in Analysis of Behavior: Vol. 1. Reinforcement and the Organization of Behavior. M. D. Zeiler & P. Harzem, Eds. John Wiley. Chichester, England.
9. CHURCH, R. M. & M. Z. DELUTY. 1977. Bisection of temporal intervals. J. Exp. Psychol. Animal Behav. Processes **3:** 216–228.
10. GIBBON, J. 1981. Two kinds of ambiguity in the study of psychological time. *In* Quantitative Analysis of Behavior: Vol. 1. Discriminative Properties of Reinforcement. M. L. Commons & J. A. Nevin, Eds. Harper & Row. Cambridge, England.
11. PLATT, J. R. 1979. Interresponse-time shaping by variable-interval-like interresponse-time reinforcement contingencies. J. Exp. Anal. Behav. **31:** 3–14.
12. SHIMP, C. P. 1973. Synthetic variable-interval schedules of reinforcement. J. Exp. Anal. Behav. **19:** 311–330.
13. GIBBON, J. 1977. Scalar expectancy theory and Weber's law in animal timing. Psychol. Rev. **84:** 279–325.
14. PLATT, J. R. 1973. Percentile reinforcement: Paradigms for experimental analysis of response shaping. *In* The Psychology of Learning and Motivation, Vol. 7. G. Bower, Ed. Academic Press. New York, NY.
15. ANDERSON, N. H. 1963. Comparison of different populations: Resistance to extinction and transfer. Psychol. Rev. **70:** 162–179.
16. DAVIS, E. R. & J. R. PLATT. 1984. Contiguity and contingency in the acquisition and maintenance of an operant. Learn. Motiv. In press.

Relative Time in Trace Conditioning

PETER BALSAM

Department of Psychology
Barnard College
Columbia University
New York, New York 10027

Time and timing are an integral part of our analyses of associative learning. Time underlies our construction of both dependent and independent variables and much progress in understanding associative learning parallels our progress in characterizing the temporal dimensions of behavior and environments. Even the most primitive characterizations of the conditioning process require temporal order information. Conditioning is inferred to the extent that behavior occurs prior to motivationally significant events and studying the temporal order of events provided some of the earliest experimental manipulations and associative typologies.[1,2] Subsequently, demonstrations of orderly changes in behavior as a function of temporal manipulations provided the foundations of conditioning theory.

Early experiments in associative learning clearly demonstrated that temporal contiguity was a powerful modulator of associative strength. Consider the two experimental procedures shown in FIGURE 1. The top line shows a delay conditioning procedure in which the conditioned stimulus (CS) is turned on periodically and remains present until an unconditioned stimulus (US) is presented. The second row of the figure shows a trace conditioning procedure that is identical to the delay procedure in all respects except that the CS begins and ends earlier such that there is gap between the offset of the CS and the onset of the US. As this gap is lengthened, performance gradually deteriorates until there is no evidence of excitatory conditioning. This type of result confirms that the temporal contiguity of events is an important determinant of performance. This result tells us further that contiguity is not an all-or-none matter. Subjects show a quantitative appreciation of the temporal relationship between CS and US. This chapter is about which events are timed in classical conditioning procedures and how the duration of these events interact with each other to determine the strength of a conditioned response (CR).

The primary data for the analyses reported here comes from autoshaping experiments.[3] In these experiments pigeons are exposed to CSs consisting of the illumination of a keylight and USs consisting of a few seconds of access to grain. After a number of keylight–grain pairings, pigeons approach and peck the keylight. We assume that the strength of excitatory CS–US associations is monotonically reflected in the strength of the tendency to approach and peck the keylight.

TIMING OF EVENTS IN CLASSICAL CONDITIONING

The classical conditioning procedures illustrated in FIGURE 1 shows five commonly studied intervals between stimuli that can conceivably influence associative learning. They are the interstimulus interval (ISI) or trial duration (T), CS (S) duration, intertrial interval (ITI), interreinforcement (IRI) or cycle (C) time, and the trace interval (TI) or gap (G) duration. Notice that although we have identified five intervals in these procedures, a full specification of all of the temporal relationships

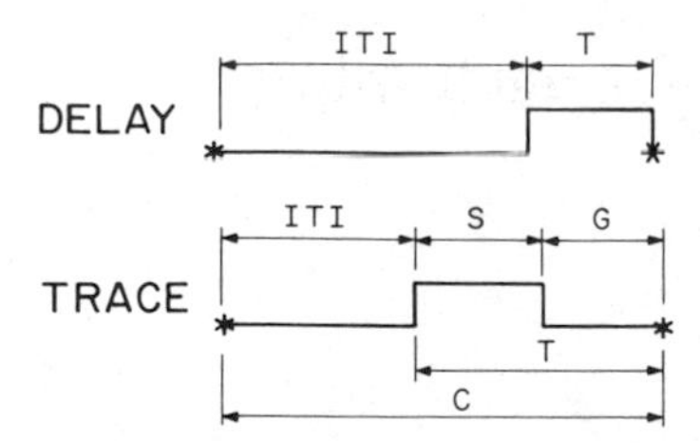

FIGURE 1. Schematic representation of the temporal intervals in classical conditioning. C is the duration of the cycle or interreinforcement interval. T is the interstimulus interval. G is the trace interval duration. The ITI is the intertrial interval and S is the duration of the CS.

does not require that all intervals be described. For example, if T and G are given, then S is fixed. Likewise if C and T are given, then the ITI is known. Hence, variation in some intervals necessarily produces changes in other intervals and our treatment outlined below reflects our attempts to characterize the conditioning process in terms of the smallest number of temporal intervals that still specify all of the relevant variables. Specifically, we have chosen to frame our analysis in terms of the cycle time, C; the interval from CS onset to US onset, T; and the trace interval duration, G, with an occasional reference to the CS duration, S. The effects of variation in each of these intervals is described below.

Trial Duration

There is an inverse relation between the time from CS onset to US onset and the speed of acquisition in autoshaping. FIGURE 2 shows the effects of varying T in both delay and trace conditioning. These data were taken from an autoshaping experiment in which grain was presented every 96 seconds and keylight illumination preceded each food delivery. In the delay groups the CS remained on throughout the trial period, whereas in the trace groups the CS was fixed at 4 seconds. The interval from CS onset to US onset was varied from 4 to 16 seconds. In both cases the effect of increasing T is to slow the speed of acquisition. In all cases the trace conditioning groups are slower to keypeck than is the delay group with the comparable T. There is, thus, an effect of trace conditioning procedure in addition to the effects of changing T.

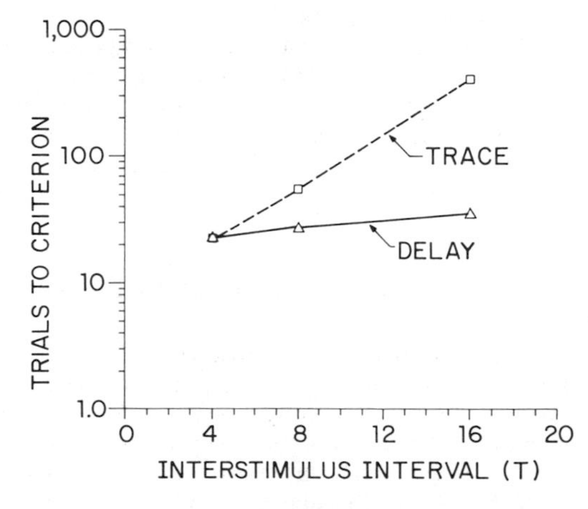

FIGURE 2. The median number of trials to the fifth trial with at least one response is shown as a function of T in both delay and trace conditioning.

CS Duration

In delay conditioning, there is no interval from CS offset to US onset, hence the effects of varying S, the CS duration, are equivalent to variations in T. In the trace procedure, changing the CS duration necessarily means that either the trace interval is filled with different CS durations when T is held constant or that T changes with CS duration when G is held constant. FIGURE 3 shows the effects of both of these kinds of changes in CS duration separately. These two ways of increasing CS duration do not affect behavior in similar ways over the range of values that we have studied. The effect of filling a given T with more CS is facilitative, whereas the effect of lengthening CS duration by altering its onset time is detrimental.

Thus far there are two generalizations that we can make about the effects of varying stimulus durations in autoshaping. Performance is inversely related to T, the time from CS onset to US onset, and similarly for a given T, it is inversely related to G, the time from CS offset to US onset. Hence, the trace procedure produces worse conditioning than the delay procedure shown in FIGURE 1 for two reasons. Conditioning is diminished both because T is longer and because there is a gap from CS offset to US onset.

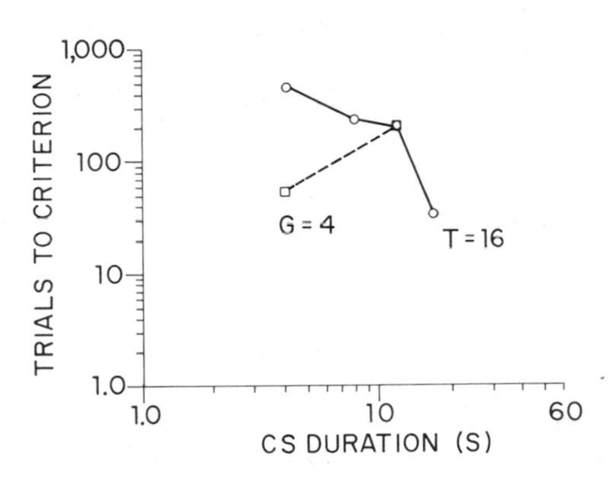

FIGURE 3. Acquisition speed is shown as a function of CS duration (S) in procedures in which T is either fixed or varied across groups. In the fixed groups (*solid line*) T was set at 16 sec. In the varied groups (*dashed line*) G was fixed at 4 sec.

Cycle Time

The effects of varying C are well known and described elaborately for delay conditioning.[4–7] Acquisition is faster with spaced than with massed trials. Likewise in trace conditioning, increases in C facilitate acquisition. FIGURE 4 shows how increases in C affect the rate of acquisition. The points in the figure are taken from studies which allowed comparisons of different IRIs for groups of subjects exposed to identical interstimulus intervals (T) and trace interval durations (G). All groups were exposed to a 16-sec trial and they differed in the durations of G and C. Acquisition is facilitated by increases in the cycle time at all of the trace intervals.

In summary, we have described all of the effects of varying the temporal intervals between events in classical conditioning. A simple CS–US contiguity account is not adequate in accounting for these findings. The absolute contiguity between CS and US

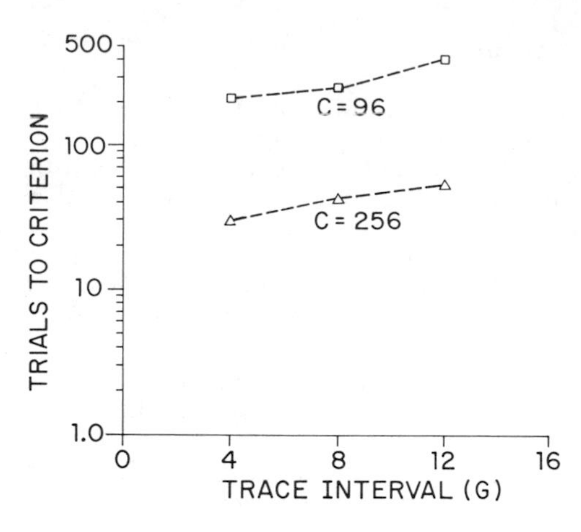

FIGURE 4. Acquisition speed is shown as a function of the trace interval for groups of subjects exposed to 16-sec T durations and either 96-sec or 256-sec cycle times.

can be held constant, but performance is drastically altered by changing the cycle time. Specifically, across the range of intervals that we have studied, performance is directly related to the time between reinforcements, C, and inversely related to the time from CS onset to US onset, T. Finally, performance is enhanced the briefer the duration of the trace interval, G, during the trial period. The next section describes how these different intervals interact with one another to determine performance. First the interaction of C and T is described for delay conditioning and then generalized to the trace conditioning case.

RELATIVE TIME AND PERFORMANCE

The results of studies that have varied temporal parameters in delay procedures are best summarized in terms of the ratio of C to T.[5,6] In other words, acquisition speed does not depend on the absolute values of C and T; rather, the relative time to reinforcement is the controlling variable. The independent variable has been characterized as relative delays to reinforcement,[5] relative wait time,[6] and delay reduction.[8] While there are differences in these accounts, what they all have in common is the notion that behavior is sensitive to the proportion of the total interreinforcement interval that the CS occupies. I will reiterate briefly the account of delay conditioning that John Gibbon and I developed and then extend that account to an analysis of trace procedures.

Scalar Expectancy Theory

In 1981 Gibbon and Balsam proposed a model based on an application of scalar expectancy theory (SET)[9] to an analysis of acquisition in autoshaping. The central features of this account for delay conditioning in which a US immediately follows a CS are shown in the top row of FIGURE 5. Expectancy is shown on the ordinate and time is shown on the abscissa. Expectancy (H) may be thought of as the excitatory strength or value associated with a particular, unconditioned stimulus. A reinforcer supports a fixed amount of expectancy (H), which is distributed uniformly and independently over both the CS and background or contextual cues. The expectancy level associated with a particular stimulus is inversely proportional to the estimated delay to reinforcement in that stimulus. Specifically, the asymptotic heights of the trial and background expectancies in FIGURE 4 are:

$$h_T = H/T \qquad \text{(1a)}$$

and

$$h_C = H/C \qquad \text{(1b)}$$

respectively, where T and C refer to the durations of trial and IRI, respectively. We assumed that as conditioning progresses, subjects compare these values during CS presentation by taking their ratio,

$$r = h_T/h_C = C/T. \qquad \text{(2)}$$

If this ratio exceeds some threshold criterion, b, terminal responses, such as the keypeck, will be educed.

This formulation specifies how temporal parameters exert control over acquisition speed. Specifically, speed of acquisition should be directly related to the ratio of C to T duration, and previous results of autoshaping research on the effects of varying CS and IRI durations have been consistent with this expectation.[4-6]

Trace Conditioning

Let us now consider how to deal with a trace conditioning procedure in these terms. Theoretically, there are two ways in which a trace procedure might produce worse conditioning than delay procedures. The presence of a gap between CS offset and US

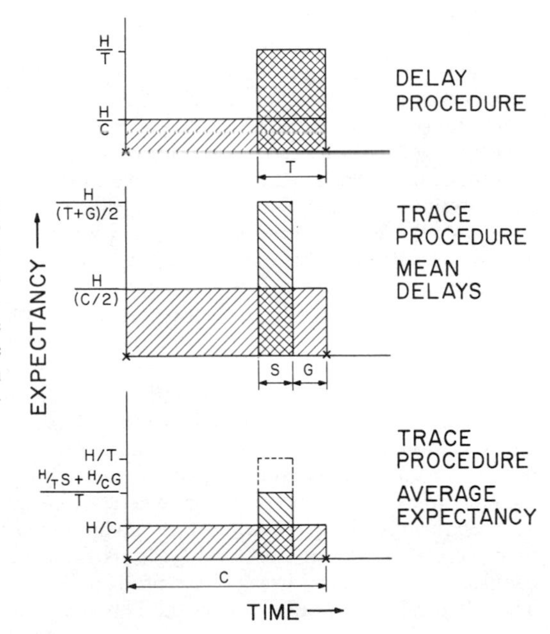

FIGURE 5. Expectancy diagrams for delay and trace procedures. *Top row* shows the basic model for delay conditioning. The *middle row* shows the expectancy for a trace conditioning procedure based on the mean delay to reinforcement. The *bottom row* shows expectancy in trace conditioning based on an average expectancy in the ISI.

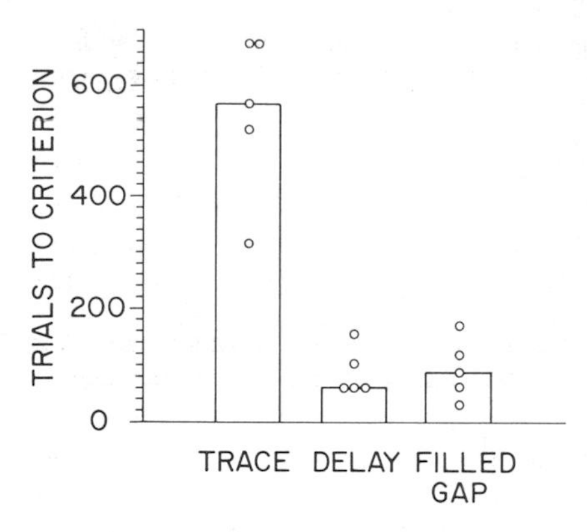

FIGURE 6. Acquisition scores for delay, trace, and serial conditioning groups with identical T values.

onset might decrease the associative value of the CS and/or increase the associative value of the context.[5,10] In either case there will be a decrement in performance with trace procedures relative to the performance generated by delay conditioning. In our first approach to this question[11] we reasoned that if some of the decrement produced by the trace procedure is the result of an increase in the context value, then the introduction of a different stimulus during the trace interval should attenuate the trace deficit.

The acquisition scores for three of the groups in the study by Balsam and Gibbon[11] are shown in FIGURE 6. All subjects were exposed to response-independent food presentations spaced 96 sec apart. In the delay group a keylight CS remained on for the 16 sec prior to each feeder operation. In the trace group, the CS remained on for only 8 sec and the ITI conditions were restored for the remaining 8 sec prior to food. The filled trace group was exposed to a procedure that was identical to the trace procedure except that a tone was presented during the trace interval. As FIGURE 6 shows, the presence of a tone during the gap completely alleviated the trace decrement. This finding, while consistent with the idea that increases in background value may be responsible for the trace decrement, is open to a number of other interpretations.[11–15] Part of the problem in interpreting these results is that the keypeck measure is used simultaneously as a measure of both CS and context value. We cannot really say whether the introduction of the tone in the trace interval has prevented some context conditioning or whether it has introduced a way of increasing CS value. In either case, keypecking would be enhanced. We have, therefore, subsequently analyzed the effects of the trace procedure with techniques that allow us to independently measure the associative values of cues and contexts. One of the techniques for doing this employs a blocking test of associative value.[16] A cue previously associated with a US can interfere with the conditioning of a novel cue with which it is compounded and paired with reinforcement. Hence, the capacity of a cue to interfere with subsequent conditioning of other cues can be used as an assay of strength of its association with the US.

In the first experiment in this series we replicated the basic trace effect, but employed tones as the conditioned stimuli. Two groups of subjects were given 125 pairings of a tone with grain every 48 sec. In a delay group the tone was presented for 8 sec prior to each grain presentation, whereas in a trace group there was an 8-sec gap between the offset of the tone and each food presentation. The associative values of the tones were then assayed in a blocking test. All subjects were given 25 daily pairings of a compound stimulus consisting of the pretrained tone and a novel keylight with grain. The left panel of FIGURE 7 shows the keypeck acquisition scores for both groups of subjects. The trace tone formed a less effective association with the US than did the delay tone, as evidenced by its diminished effectiveness in blocking keypeck acquisition. The basic difference between trace and delay conditioning was thus replicated with the tone CSs.

In the next experiments, we examined the source of the trace decrement by independently analyzing CS and context conditioning in trace and delay procedures. Our first technique for independently evaluating context conditioning comes from the fact that keypeck acquisition speed is inversely related to the associative value of the context in which it is tested. Acquisition is slower in contexts previously associated with reinforcers than in contexts not previously associated with reinforcers,[17,18] and acquisition speed is inversely related to the number of prior pretraining trials in that context.[17,19,20] Thus the acquisition of keypecking can serve as a sensitive assay of previously established context–US associations.

Two groups of subjects were pretrained in procedures that were identical to the ones employed in the preceding experiment. Subjects received either trace or delay presentations of the tone. The associative value of the context was then evaluated in all of the groups with a keylight acquisition test. Any differences between groups in context conditioning during training with the tone should be reflected in the speed of subsequent keypeck acquisition. Twenty five times a session the keylight was illuminated 8 sec prior to each of the grain presentations, which were spaced 48 sec apart. The middle panel of FIGURE 7 shows that the two groups acquired the keypeck response at approximately the same rate. Thus, there appears to be no difference between the contextual conditioning induced by trace and delay procedures. By default, the difference between trace and delay procedures must be in the degree of excitatory strength controlled by the CS. We wished to confirm this more directly by evaluating the associative value of trace and delay CSs in contexts that we were sure were of equal associative value. We achieved this in the next experiment by interposing a context extinction phase between the training of the CSs and the subsequent test of their associative value.

Two groups of subjects were pretrained with tone CSs as in the trace and delay groups of the preceding experiments. All subjects were then exposed to five sessions of nonreinforced exposure to the context. Prior work has shown that five such sessions would be adequate for extinguishing the context conditioning established by hundreds

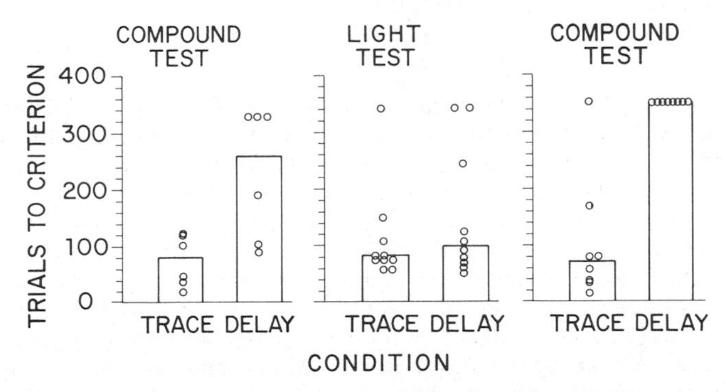

FIGURE 7. All panels show the median trials to acquisition as a function of pretraining with either trace or delay pairings of a tone with grain. The *left panel* shows acquisition to a keylight paired with grain. The *center panel* shows acquisition to a tone–keylight compound. The *right panel* shows acquisition to a tone–keylight compound after intervening context extinction.

or even thousands of pretraining trials.[18,19,21] Additionally, general activity was monitored throughout the course of this experiment, which provides us with another measure of context conditioning.[19,22,23] General locomotor activity is a response that gets conditioned to contextual cues. It is inversely related to C,[22,24] extinguishes,[19,22,23] shows spontaneous recovery,[19,22] and it is under the specific control of stimuli associated with the US.[19,22,23]

There was no significant difference between the general activity engendered by the trace and delay procedures during any part of the experiment. This result provides further support for our conclusion that trace and delay conditioning do not produce differential context conditioning. General activity declined over the course of the context extinction sessions until very low levels of activity were recorded by the fifth day. We were thus successful in providing both groups of subjects with equivalent and low-valued contexts in which to test the value of the tones.

In the final phase of the experiment the tone values were tested in a compound blocking test. All subjects received pairings of a tone–keylight compound with grain. The rightmost panel of FIGURE 7 shows the keypeck acquisition scores for both groups of subjects. The data show that even when the stimuli are evaluated in contexts that we know to be of equivalent associative value, the delay tone still shows evidence of having accrued a greater association with the US than the trace tone.

The results of this series of experiments tells us that the source of the trace decrement is that in a trace procedure there is a less effective CS–US association formed than in a delay procedure. How to characterize this decline in effectiveness is explored in the next section of this paper.

Relative Time in Trace Conditioning

The relative delay to reinforcement hypothesis can easily be extended to trace procedures by assuming that subjects are sensitive to the mean reinforcement delay associated with a stimulus. The second row of FIGURE 5 shows how expectancy would be distributed in a trace procedure if subjects were sensitive to the mean delay to reinforcement. Equation 3 shows that the mean delay (D) in the CS increases with increases in T and G, and thus provides a potential account of the difference between trace and delay procedures:

$$D = \frac{T + G}{2}. \tag{3}$$

Notice, though, that a given increase in either T or G produces the same effect on the mean delay. In other words, increasing the interstimulus interval by x seconds should have the same effect as increasing the trace interval duration by x seconds. We can therefore evaluate this model by examining whether the rate of change in performance as a result of increasing T in delay procedures (when G is fixed) is the same as the rate at which performance changes as a function of trace interval duration when T is fixed. Some of the data from FIGURES 1 and 2 have been replotted in FIGURE 8 to allow this comparison. It is clear from the figure that the trace manipulation is far more detrimental to performance than the delay manipulation. The trace procedure, therefore, involves more than alterations in the mean delay to reinforcement.

A clue to what might be responsible for the trace decrement comes from our filled trace study. It would appear that the failure to discriminate the trace interval from the background stimuli is an essential contributor to the trace effect.[11,13,25] When a different CS fills the gap between CS offset and US onset, conditioning is facilitated. One interpretation of this finding is that the associative value of the stimuli that follow the trace signal influence the level of expectation controlled by the CS itself. A return to the low-valued background stimuli in trace procedures decrements CS expectancy, while the change to a second CS in the filled trace procedure does not. Expectancy in the CS might be a mixture of the expectancies associated with all of the stimuli present between CS onset and US onset. As a first approximation, we propose that the expectancy associated with a particular ISI is an average of the expectancies appropriate to the delays signaled by the stimuli within the ISI.

The third row of FIGURE 5 shows this sort of analysis for trace conditioning. When the CS comes on, there is an expectancy level appropriate to, T, H/T. S seconds later

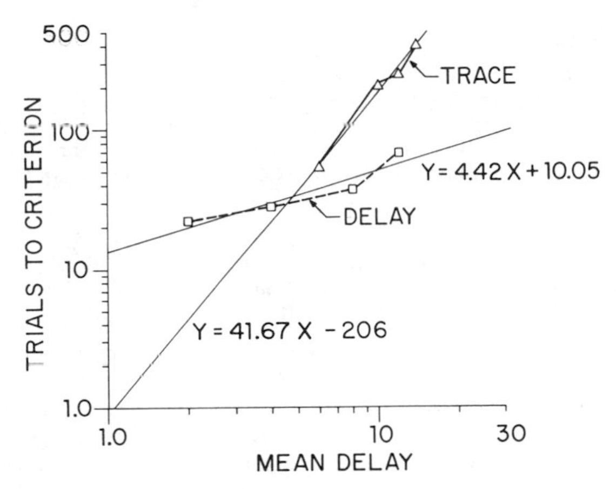

FIGURE 8. Acquisition speed is shown as a function of the mean delay to reinforcement (D) for delay and trace conditioning groups exposed to a 96-sec cycle.

the CS is turned off and expectancy reverts to the background level, H/C. The mean expectancy associated with T, h(T) is

$$h(T) = \frac{\frac{H}{T}S + \frac{H}{C}G}{T}. \tag{4}$$

Since S = T − G, the expectancy ratio, r = h(T)/h(C) is:

$$r = \frac{C}{T} - \frac{G}{T}\left(\frac{C}{T} - 1\right). \tag{5}$$

This expectancy ratio has some interesting properties. First, notice that in delay procedures where G = 0, this reduces to the expectancy ratio for delay procedures given in Equation 2. Secondly, as G ⟶ T, R ⟶ 1.

FIGURE 9 shows the acquisition data presented earlier as well as data from other published studies replotted as a function of this ratio. As a first approximation, the data are well described by the model in the sense that across an order of magnitude of

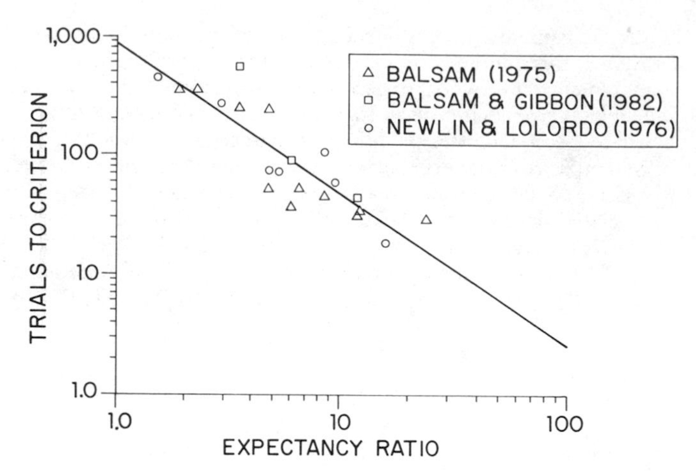

FIGURE 9. Acquisition scores are shown as a function of the expectancy ratio described in Equation 5. (Replotted from Balsam[43]; Balsam and Gibbon[11]; and Newlin and Lolordo.[45])

changes in the ratio, there is a generally monotonic change of nearly two orders of magnitude in acquisition scores. The regression line, $Y = 857.2\ X^{-1.3}$, accounts for about 80% of the variance. A similar fit for the rate of responding at the end of training is shown in FIGURE 10. Again, the expectancy ratio is monotonically related to the dependent measure. In this case, the regression accounts for 68% of the variance.

This model, then, does a fairly good job of predicting acquisition speed and response rates in the autoshaping preparation. We will explore the generality of these effects below but first some interesting implications of the model will be described.

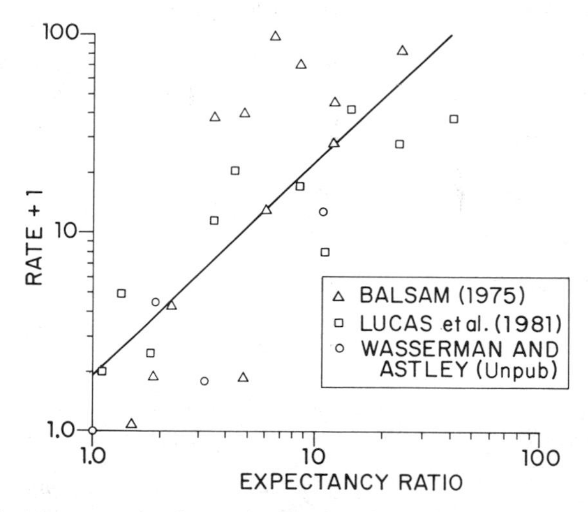

FIGURE 10. Response rates are shown as a function of the expectancy ratio. (Replotted from Balsam[43]; Lucas *et al.*[44]; and Wasserman and Astley.[47])

EXCITATION AND INHIBITION

FIGURE 11 shows how the expectancy ratio changes as a CS of fixed duration is placed at different points within a reinforcement cycle. The effects of such manipulations have only been studied in restricted ranges of this continuum until quite recently. The points on the left side of the continuum represent trace procedures, CSs in the middle are called explicitly unpaired, and those farthest to the right represent backwards pairing procedures. The general finding has been that while trace procedures are excitatory, explicitly unpaired and backward procedures generally result in inhibitory associations. The form of the expectancy ratio, however, allows some predictions about how changes in CS placement in this dimension would affect the strength of these associations. If we assume that the threshold ratio for excitation is about 2.0, as our data suggest, and that ratios below this value might be inhibitory, then this manipulation has a much greater effect on excitation than on inhibition, assuming that both large and small ratios are translated into performance with the same rules. We predict that there will be steep changes in excitation in above-threshold trace procedures, while grading in the strength of inhibition for ratios below threshold will be difficult to detect.

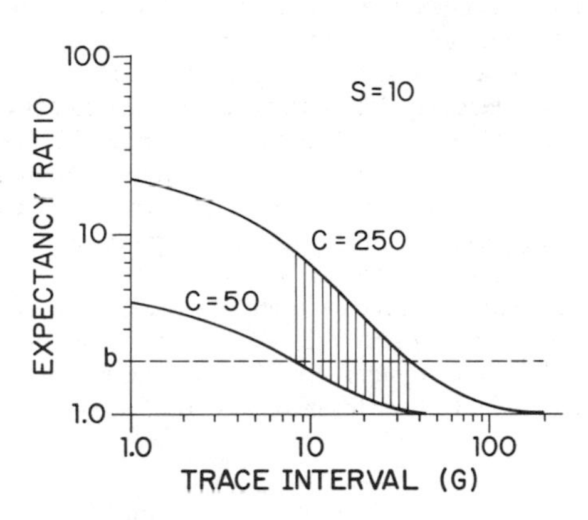

FIGURE 11. Theoretical functions showing how the expectancy ratio changes as a function of CS placement within the cycle (G/T ratio). The function is shown for two different cycle (c) durations. The hatched area shows the range over which inhibition will be produced at the short cycle while excitation is produced at the long cycle.

A second implication of this formulation is that whether or not a stimulus will be excitatory or inhibitory will depend on the ratio of C/T as well as the ratio of G/T. In other words, it should be possible to fix CS and gap duration and make a stimulus either a conditioned excitor or inhibitor by varying C. This is illustrated by comparing the two functions shown in FIGURE 11.

Fortunately for us, Peter Kaplan[26] has recently conducted some experiments that allow us to test these implications of the model. In these experiments both the placement of the CS was varied within the reinforcement cycle and T and G were fixed while C varied. The technique for these experiments was developed by Eliot Hearst and his colleagues at the University of Indiana. The experiments employ an autoshaping preparation in which approach and withdrawal from the conditioned stimulus are used as measures of conditioned excitation and inhibition, respectively. Keylight presentations are randomly alternated from one side of the chamber to the other from trial to trial. The amount of time that a pigeon spends on each side of the chamber is recorded on each trial and an approach–withdrawal ratio is computed by dividing the amount of time spent on the side with the lighted keylight by the total trial time. Values above 0.5

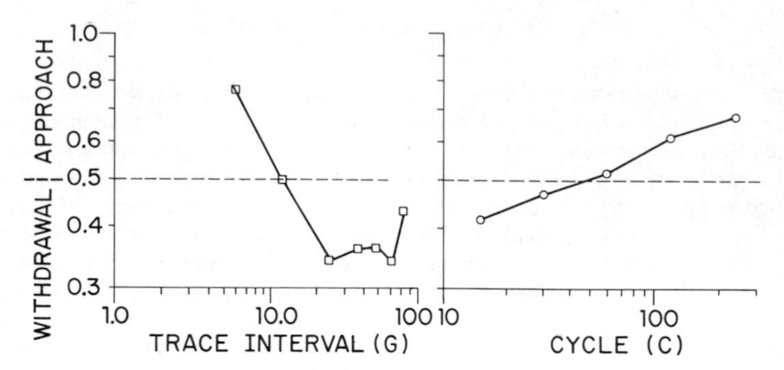

FIGURE 12. Approach–withdrawal ratios are shown as a function of gap size in the *left panel* and cycle time in the *right panel.* (Replotted from Kaplan.[26])

indicate approach whereas values below 0.5 indicate withdrawal or avoidance of the keylight. The withdrawal measure reflects inhibition since it is highly correlated with more traditional measures of inhibitory control, such as summation, generalization, and resistance to reinforcement.[26-29]

I have replotted Kaplan's results in FIGURE 12. In all of these experiments one 12-sec keylight illumination was presented during each cycle. The left panel shows the effects of varying G when C was fixed at 87 sec. The keylight becomes less excitatory as G increases and becomes inhibitory with gaps of at least 24 sec. Notice that excitation declines rapidly and that the strength of inhibition shows only shallow grading as a function of the temporal manipulation. The right panel shows the effects of varying C when T and G were fixed at 24 and 12 sec, respectively. The ratios are once again a graded function of the temporal manipulation and the results clearly show that a given G/T ratio can be either excitatory or inhibitory. Thus, both excitation and inhibition are controlled by relative time and are therefore potentially described by our model. FIGURE 13 shows the data from both of Kaplan's experiments replotted as a

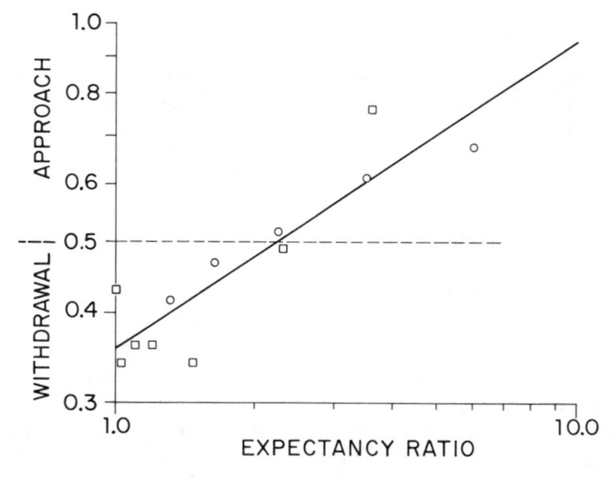

FIGURE 13. The approach–withdrawal ratios from Kaplan[26] are replotted as a function of the expectancy ratio.

function of the expectancy ratio. The ratio is monotonically related to approach–withdrawal ratios and the regression accounts for 81% of the variance. Thus, there is a temporally based continuum that is mapped into a behavioral continuum from inhibition to excitation.

According to our view of how relative time influences performance, it is the value of a trial relative to the context value which determines whether or not a stimulus will be an inhibitor or an excitor. Consistent with this notion, Kaplan and Hearst[30] have trained CSs as conditioned inhibitors and then independently manipulated the associative value of the context before testing the CS. They have found that the expression of inhibition does indeed require the presence of an excitatory background.

We have not yet described how to treat the background itself, which at first blush might seem to be an inhibitory case of the model. If the background were compared to itself, it would have a ratio of 1 and by our argument be an inhibitor. There is ample evidence, however, that contexts are excitatory.[19,31,32] We think this arises from the fact that the ratio comparison is always one of a stimulus to other stimuli in which it is embedded and never to itself. This hierarchical structuring of contexts suggests that the CS is compared to the experimental context and that the experimental context is compared to the laboratory context in which experimental sessions are embedded to determine their respective excitatory strengths.[19] The experimental context, therefore, will generally be excitatory since US presentations outside of the experimental chamber are generally kept to a minimum.

PARAMETRIC AND RESPONSE SYSTEM GENERALITY

A great deal of research has been generated in search of the optimal ISI in classical conditioning. The general finding in these studies has been that conditioning is a bitonic concave downward function of T. At very short Ts, conditioning is poor, but rapidly improves for most preparations as the CS-onset–US-onset interval is lengthened. Beyond this point and across a much broader range, performance declines as T is further increased.[12,33] The poor conditioning at very short T values may reflect processes other than a direct influence of time on associative learning. Indeed, the manipulation of temporal intervals other than the absolute T value exerts little effect on the strength of the CR when T values are below the maximal ISI. For example, Reynolds[34] found no effect of varying C when very short T values were employed in human eyeblink conditioning. Similarly, Schneiderman[35] concludes that there is no difference between trace and delay conditioning for very brief T values in nictitating membrane and heart-rate conditioning. These data suggest that there is an effect of the absolute value of T but no sensitivity to other temporal parameters for values that are below the "optimal ISI."

Beyond very short ISIs, CR strength declines with increases in T in both delay and trace procedures.[12,33,36] Furthermore, trace conditioning is generally considerably weaker than delay conditioning at a given T in preparations such as salivary,[2,37] eyeblink,[38] heart rate,[39] conditioned emotional response (CER),[40–42] and autoshaping.[11,43–45] Similarly, variations in C in other preparations show that across a wide range of values performance is facilitated by spaced trial presentation as compared to massed presentations.[33,36] Hence, there is considerable generality to the absolute temporal features of the account that we have outlined. It appears, though, that the two arms of the bitonic ISI function are modulated by different processes. We suggest that decrements in performance produced by increasing T beyond the optimal ISI are the result of changes in the expectancy ratio, whereas decrements in performance below the optimal ISI may not relate to a direct temporal influence on associative learning but rather be due to an effect on response production or stimulus processing mechanisms.

A full test of the generality of our account of the control by relative time in associative learning, however, requires that both C and T be manipulated in comparable procedures. Such a study was done by Reynolds[34] in human eyeblink conditioning. Subjects were exposed to ITIs of either 15 or 90 sec, the CS duration was fixed at 50 msec, and T varied from 250 to 2250 msec. FIGURE 14 shows Reynold's data, excluding the 250 msec (below the optimal ISI) groups, replotted as a function of the expectancy ratio. Even though the absolute T values are very much briefer than those we have previously studied, the control by relative time is still pronounced, as evidenced by the generally monotonic relationship between the expectancy ratio and response probability. The regression accounts for about 90% of the variance.

Similarly, in the conditioned emotional response (CER) paradigm, performance also appears to be generally controlled by the relative proximity to reinforcement. In this sort of experiment, the suppression of ongoing operant behavior is used as the index

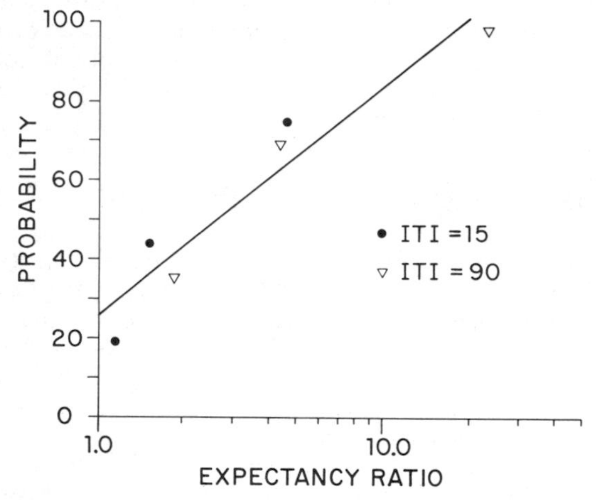

FIGURE 14. Probability of an eyeblink is shown as a function of the expectancy ratio. (Replotted from Reynolds.[34])

of Pavlovian fear conditioning. The level of suppression is directly related to the C/T ratio in delay procedures.[46] Unfortunately, no one has reported parametric effects of varying simultaneously C, T, and G in the CER paradigm.

There is thus great generality to the basic idea that we have proposed. In autoshaping we have shown that both excitation and inhibition are ordered on a dimension defined by the expectancy ratio. In other preparations, the model does well in predicting the effects of C, T, and G on performance. While the specific account that is proposed here may be incomplete in some ways, the general description that it provides of control by relative time is accurate and any inclusive account of classical conditioning must, therefore, incorporate these temporally defined phenomena into the basic conceptualization of the associative learning process.

These examples of how the SET account can help integrate a wide range of data in Pavlovian conditioning are of great theoretical interest. But even more generally, they illustrate the potential power of a temporal analysis of associative learning. This sort of analysis makes clear that by understanding how time is processed by organisms, we will arrive at a deeper understanding of the rules of learning and performance.

REFERENCES

1. EBBINGHAUS, H. 1913. Memory: A Contribution to Experimental Psychology. Translated by H. A. Ruger & C. E. Bussenius. Columbia University Press. New York, N.Y.
2. PAVLOV, I. P. 1927. Conditioned Reflexes. Dover Publications. New York, NY.
3. BROWN, P. L. & H. M. JENKINS. 1968. Autoshaping of the pigeon's keypeck. J. Exp. Anal. Behavior **11:** 1–8.
4. GIBBON, J., M. BALDOCK, C. LOCURTO, L. GOLD & H. TERRACE. 1977. Trial and intertrial durations in autoshaping. J. Exp. Psychol. Anim. Behav. Processes **3:** 264–284.
5. GIBBON, J. & P. D. BALSAM. 1981. The spread of association in time. *In* Autoshaping and Conditioning Theory. C. M. Locurto, H. S. Terrace and J. G. Gibbon, Eds. Academic Press. New York, NY.
6. JENKINS, H. M., R. A. BARNES & F. J. BARERRA. 1981. Why autoshaping depends on trial spacing. *In* Autoshaping and Conditioning Theory. C. M. Locurto, H. S. Terrace and J. G. Gibbon, Eds. Academic Press. New York, NY.
7. TERRACE, H., J. GIBBON, L. FARREL & J. BALDOCK. 1975. Temporal factors influencing the acquisition and maintenance of an autoshaped keypeck. Anim. Learn. Behav. **3**(1): 53–62.
8. FANTINO, E. 1981. Contiguity, response strength, and the delay-reduction hypothesis. *In* Predictability, Correlation, and Contiguity. P. Harzem & M. H. Zeiler, Eds. Wiley. London.
9. GIBBON, J. 1977. Scalar expectancy and Weber's law in animal timing. Psychol. Rev. **84:** 279–325.
10. RESCORLA, R. A. & A. R. WAGNER. 1972. A theory of Pavlovian conditioning: Variations in the effectiveness of reinforcement and non-reinforcement. *In* Classical Conditioning II: Current Theory and Research. A. H. Black & W. F. Prokasy, Eds.: 64–99. Appleton-Century-Crofts. New York, NY.
11. BALSAM, P. D. & J. GIBBON. 1982. Factors underlying trace decrements in autoshaping. Behav. Anal. Lett. **2:** 197–204.
12. GORMEZANO, I. & E. J. KEHOE. 1981. Classical conditioning and the law of contiguity. *In* Predictability, Correlation and Contiguity. P. Harzem & M. H. Zeiler, Eds. Wiley. London.
13. KAPLAN, P. & E. HEARST. 1982. Bridging temporal gaps between CS and US in autoshaping: insertion of other stimuli before, during and after CS. J. Exp. Psychol. Anim. Behav. Processes **8:** 187–203.
14. RASHOTTE, M. E. 1981. Second-order autoshaping: Contributions to the research and theory of Pavlovian reinforcement by conditioned stimuli. *In* Autoshaping and Conditioning Theory. C. M. Locurto, H. S. Terrace and J. G. Gibbon, Eds. Academic Press. New York, NY.
15. RESCORLA, R. A. 1982. Effects of a stimulus intervening between CS and US in autoshaping. J. Exp. Psychol. Anim. Behav. Processes **8:** 131–141.
16. KAMIN, L. J. 1969. Predictability, surprise, attention and conditioning. *In* Punishment and Aversive Behavior. B. A. Campell & R. M. Church, Eds. Appleton-Century-Crofts. New York, NY.
17. BALSAM, P. D. & A. L. SCHWATZ. 1981. Rapid contextual conditioning in autoshaping. J. Exp. Psychol. Anim. Behav. Processes **1:** 382–393.
18. TOMIE, A. 1981. Effects of unpredictable food upon the subsequent acquisition of autoshaping: Analysis of the context blocking hypothesis. *In* Autoshaping and Conditioning Theory. C. M. Locurto, H. S. Terrace & J. G. Gibbon, Eds. Academic Press. New York, NY.
19. BALSAM, P. D. 1982. Bringing the background to the foreground: The role of contextual cues in autoshaping. *In* The Harvard Symposium on the Quantitative Analysis of Behavior: Acquisition Processes. M. Commons, R. Herrnstein & A. Wagner, Eds. Ballinger. Cambridge, MA.

20. BALSAM, P. D. & J. GIBBON. Tones paired with food block keylights but not contexts. In preparation.
21. TOMIE, A. 1976. Interference with autoshaping by prior context conditioning. J. Exp. Psychol. Anim. Behav. Processes **14:** 126–132.
22. BALSAM, P. D. General activity induced by periodic food presentation is a conditioned response. In preparation.
23. DURLACH, P. 1982. Pavlovian learning and performance when CS and US are uncorrelated. *In* The Harvard Symposium on the Quantitative Analysis of Behavior: Acquisition Processes. M. Commons, R. Herrnstein, A. Wagner, Eds. Vol. 3. Ballinger, Cambridge, MA.
24. KILLEEN, P. 1975. On the temporal control of behavior. Psychol. Rev. **82:** 89–115.
25. MOWER, O. H. & R. R. LAMOREAUX. 1951. Conditioning and conditionality (discrimination). Psychol. Rev. **58:** 196–212.
26. KAPLAN, P. The importance of relative temporal parameters in trace autoshaping: From excitation to inhibition. J. Exp. Psychol. Anim. Behav. Processes. In press.
27. BOTTJER, S. W. 1982. Conditioned approach and withdrawal behavior in pigeons: Effects of a novel extraneous stimulus during acquisition and extinction. Learn. Motiv. **13:** 44–67.
28. HEARST, E., S. W. BOTTJER & E. WALKER. 1980. Conditioned approach-withdrawal behavior and some signal-food relations in pigeons: Performance and positive vs. negative "associative strength." Bull. Psych. Soc. **16:** 183–186.
29. HEARST, E. & S. FRANKLIN. 1977. Positive and negative relations between a signal and food: Approach-withdrawal behavior to the signal. J. Exp. Psychol. Anim. Behav. Processes **3:** 37–52.
30. KAPLAN, P. & E. HEARST. 1984. Contextual control and excitatory vs. inhibitory learning: Studies of extinction, reinstatement, and interference. *In* Context and Learning. P. Balsam & A. Tomie, Eds. Erlbaum. Hillsdale, NJ.
31. RESCORLA, R., P. DURLACH & J. GRAU. 1984. Contextual learning in Pavlovian conditioning. *In* Context and Learning. P. Balsam & A. Tomie, Eds. Erlbaum. Hillsdale, NJ.
32. RANDICH, A. 1984. The role of contextual stimuli in mediating the effects of pre and post exposure to the unconditioned stimulus alone on acquisition and retention of conditioned suppression. *In* Context and Learning. P. Balsam & A. Tomie, Eds. Erlbaum. Hillsdale, NJ.
33. MACKINTOSH, N. J. 1975. Theory of attention. Psychol. Rev. **72:** 276–298.
34. REYNOLDS, B. 1945. The acquisition of a trace conditioned response as a function of the magnitude of the stimulus trace. J. Exp. Psychol. **35:** 15–30.
35. SCHNEIDERMAN, N. 1972. Response system divergencies in aversive classical conditioning. *In* Century Psychology Series. Classical Conditioning: II: Current Theory and Research, K. MacCorquodale, G. Lindzey & K. E. Clark, Eds. Appleton-Century-Crofts. New York, NY.
36. GORMEZANO, I. & J. W. MOORE. 1969. Classical conditioning. *In* Learning: Processes. M. Marx, Ed. MacMillan. Toronto.
37. ELLISON, G. D. 1964. Differential salivary conditioning to traces. J. Comp. Physiol. Psychol. **57:** 373–380.
38. WERDEN, D. & L. E. ROSS. 1972. A comparison of the trace and delay classical conditioning performance of normal children. J. Exp. Child Psychol. **14:** 126–132.
39. FITZGERALD, R. D. & T. J. TEYLER. 1970. Trace and delayed heart-rate conditioning in rats as a function of US intensity. J. Comp. Physiol. Psychol. **70:** 242–253.
40. BOLLES, R. C., A. C. COLLIER, M. E. BOUTON & N. A. MARLIN. 1978. Some tricks for ameliorating the trace-conditioning deficit. Bull. Psych. Soc. **11:** 403–406.
41. GRAY, T. 1978. Blocking in the CER: Trace and delay procedures. Can. J. Psych. **32:** 40–42.
42. KAMIN, L. J. 1965. Temporal and intensity characteristics of the conditioned stimulus. *In* Classical Conditioning: A Symposium. W. F. Prokasy, Ed. Appleton-Century-Crofts. New York, NY.
43. BALSAM, P. D. 1975. The effects of varying the interreinforcement interval, inter-stimulus interval, and trace interval duration on acquisition and auto-maintenance in the pigeon. Doctoral dissertation, University of North Carolina.

44. LUCAS, G. A., J. D. DEICH & E. A. WASSERMAN. 1981. Trace autoshaping: Acquisition, maintenance, and path dependence at long trace intervals. J. Exp. Anal. Behav. **36:** 61–74.
45. NEWLIN, R. & V. LOLORDO. 1976. A comparison of pecking generated by serial, delay and trace autoshaping procedures. J. Exp. Anal. Behav. **25:** 227–241.
46. GIBBON, J., R. BERRYMAN & R. L. THOMPSON. 1974. Contingency Spaces and Measures in Classical and Instrumental Conditioning. J. Exp. Anal. Behav. **21:** 585–605.
47. WASSERMAN, E. A. & S. L. ASTLEY. Positive and negative sign-tracking in the pigeon: Temporal versus correlational control. Unpublished manuscript.

The Function of Time Discrimination and Classical Conditioning[a]

SETH ROBERTS AND MARK D. HOLDER

Department of Psychology
University of California, Berkeley
Berkeley, California 94720

Time discrimination and classical conditioning are common and general laboratory results.[1-4] By "time discrimination" we mean arbitrary duration discriminations, such as those seen with fixed-interval schedules, Sidman avoidance schedules, and inhibition-of-delay procedures. The durations are usually seconds or minutes. As an example, FIGURE 1 shows some results from an experiment[5] that used rats pressing a lever to get food. After a random intertrial interval, either the 30-sec signal (light) or the 60-sec signal (sound) began. After the appropriate time—30 sec for light, 60 sec for sound—the next response produced food, and the signal was turned off. On the last day, but not the first day, there was a clear time discrimination. We use the term "classical conditioning" in a broad way; we include any result showing that the animal's response (the conditioned response, or CR) to one event (the conditioned stimulus, or CS) depended on the time between the CS and another event (the unconditioned stimulus, or US). An example is taste-aversion learning; making a rat sick within a few hours after it has tasted a novel flavor will cause it to avoid the flavor. The flavor is the CS; the sickness the US; and the avoidance of the flavor the CR.

Everyone agrees that time discrimination and classical conditioning are worth explaining; their function, in the evolutionary sense,[6] is part of the explanation. Questions about the function of laboratory behavior are of course questions about the function of the mechanism that the behavior reflects—why were animals with the mechanism favored over animals without it?

METHODS OF STUDYING FUNCTION

Zoologists, and a few psychologists, have used three empirical methods to learn about the function of behavior:

One method, the most direct, is to try to measure the effects of the behavior. This is done, for example, by preventing the behavior, undoing its immediate effects, or comparing (within species) the success of animals with high and low amounts of the behavior. A well-known example of this approach is the study of Tinbergen and others,[7] who asked why black-headed gulls retrieve broken fragments of eggs near their nest. They found that scattering shell fragments around the nest—that is, undoing their retrieval—increased predation on the nest, and concluded that the function of shell retrieval was to lower predation. An observational example is the work of Seely and Hamilton,[8] who asked why some desert beetles construct straight ridges about a meter long. Comparing water content of the ridges to water content of the nearby sand, they found that the ridges collected water; measuring the water content of the ridge

[a]This work was supported by grant BNS 79-00829 from the National Science Foundation.

before and after a beetle crossed, they found that the beetles collected water from the ridge. They concluded that the function of ridge-making was to collect water. Sometimes, of course, the effects of a behavior are so clear that no experiment or careful observation is needed; hunting and eating are two examples.

Experiments by Zamble[9] show how this method can be used to study the function of classical conditioning. He compared two groups of food-restricted rats. Both groups got 30-min access to food at unpredictable times: For one group, the feedings were signaled by a 15-min light or sound; for the other group, the feedings were unsignaled. The group with signaled feedings ate more and lost less weight than the other group. This suggests that one function of classical conditioning is to make eating more efficient. We admire these experiments, and hope that this approach will be used more widely. On the other hand, Zamble's approach—or any other use of this method—has important limitations. One is that, to be convincing, it requires a "realistic" situation. Time discrimination and classical conditioning are difficult to observe in nature, and it is hard to judge the realism of a laboratory arrangement. Another limitation is that the method is essentially confirmatory—it can test ideas, but it cannot usually create

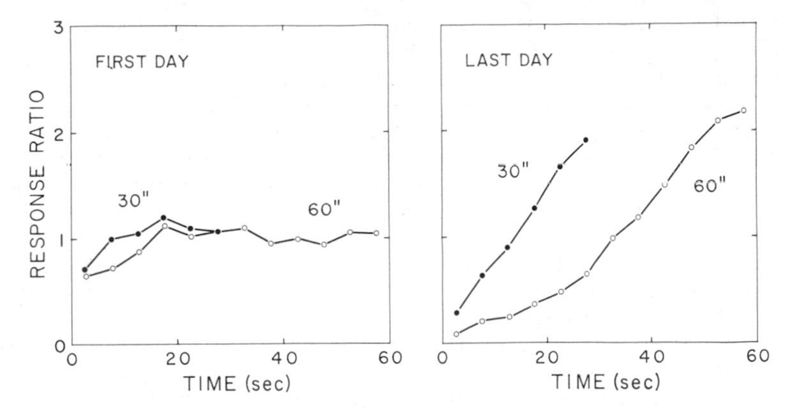

FIGURE 1. An example of a time discrimination. (Response ratio is response rate at the given time divided by overall response rate.) (From Roberts and Church.[5] Reprinted by permission.)

ideas. When we began to study time discrimination, we had no ideas about its function. A third limitation is that it requires that the behavior have a circumscribed, easy-to-measure effect. The within-animal generality of time discrimination and classical conditioning (for example, many different responses can be classically conditioned) makes us doubt that their effects are restricted to a single situation.

A second method of studying function is to make cross-species comparisons to learn what habitats or behaviors are correlated with the behavior of interest. Some correlations are better explained by some functions than others. For example, why do birds defend territories around their nest? One possible reason is that the territory provides a food supply.[6] This idea is supported by the finding that territory size is highly correlated with body weight, and thus food requirements; it is also supported by the finding that birds that eat animals defend larger territories than do birds that eat plants; plants are a denser source of food than animals. The limitation of this approach, for our purposes, is that it requires that closely related species vary substantially in the behavior of interest. (The less related the two species, the harder it is to argue from the observed correlation to causation.) We know of no examples of this for either time discrimination or classical conditioning.

The third method is to make within-species comparisons to learn what changes the expression of the behavior. Some effects are more understandable given some functions than others. This is like using the accessories of a tool to judge the function of the tool; for example, pencils often have erasers, erasers make pencils better for writing but not for punching holes, therefore pencils were more likely designed for writing than for punching holes. After a species has acquired the ability to do behavior x, it can also use a variety of modifiers to optimize the expression of behavior x. One use of this method is by Sherman,[10] who found that squirrels with relatives were more likely to give alarm calls than squirrels without relatives. Presumably, having relatives increased alarm calls. This effect made alarm calls better suited for alerting relatives, but not for a number of other functions, and Sherman took it to support the idea that the function of alarm calls is to alert relatives.

This method is sometimes used in the study of classical conditioning. For example, rats more easily learn to avoid tastes that have been followed by illness than sounds that have been followed by illness. Because illness is more likely to be caused by something with an unusual taste than something with an unusual sound, Dickinson[11] took this finding to support the idea that the mechanism underlying classical conditioning was designed to detect causality. A weakness of this method is the need to assure that the observed effect makes the behavior better for some functions than others; this assumption can rarely be supported with evidence.

This paper uses a fourth method, which, as far as we know, is new. It consists of comparing behaviors (either within or between species). Similarities between behaviors can be due either to (a) similarities in the place where evolution started (similarities in the material, such as neurons, on which evolution worked) or (b) similarities in the place that evolution was trying to reach (the function of the behavior). When similarities between two behaviors are unlikely to be due to similarity of building materials, it is reasonable to conclude that they are due to similarity of function. This method is used in a simple way when we apply the same name to different behaviors and ask about the function of all of them at once—for example, when we ask about the function of classical conditioning rather than about the function of each example of classical conditioning (such as salivation learning or taste-aversion learning). We do so, of course, because the different examples have much in common. Like the other methods, this method has important limitations. One is that similarity is a vague notion; another is that it is hard to judge the likelihood that a similarity is due to chance. The strength of this method, probably, is that its limitations are quite different from the limitations of the other methods. For example, it is less confirmatory than the first method—one does not need ideas about function in order to use it.

We use this method three times. First, we compare "laws" of classical conditioning with a list of rules devised by a philosopher (Hume) for inferring causality. Second, we compare the internal clock used in time discrimination with a stopwatch. Finally, we compare time discrimination and classical conditioning. In each case, similarities of "form" suggest a similarity of function.

CLASSICAL CONDITIONING AND HUME

In Book 1, Section 15 of *A Treatise of Human Nature,* first published in 1739–40, David Hume listed eight "rules by which to judge of causes and effects."[12] At least five of them correspond to what are sometimes called "laws" of classical conditioning.

TABLE 1 shows the correspondence. Most of the factors that influenced Hume's judgment about whether A causes B can have very large effects on the strength of a CR when A is a CS and B is a US. For example, Hume's second rule is "the cause must precede the effect." The corresponding result in classical conditioning is that in most cases the CS must precede the US. In other words, backward conditioning (where the CS follows the US) is rarely effective. Some of the rules (the fifth, the unlisted seventh and eighth, and perhaps the first part of the fourth) have no corresponding result (at least not yet). In addition, it is not always clear to us what Hume means (for example, how is Rule 3 different from Rule 4?). In spite of these problems, the correspondence seems strong enough to be worth explaining.

TABLE 1. Similarities between Hume's Rules and "Laws" of Classical Conditioning

Hume	Classical Conditioning
1. The cause and effect must be contiguous in space and time.	Space: increasing the space between CS and US decreases the CR.[13] Time: increasing the time between CS and US decreases the CR.[3] (Trace conditioning.)
2. The cause must precede the effect.	The CS must usually precede the US to produce a CR.[3] (Backward conditioning.)
3. There must be a constant union between the cause and the effect.	Presenting the CS without the US reduces the CR.[3] (Extinction, partial reinforcement, and latent inhibition.)
4. The same cause always produces the same effect, and the same effect never arises but from the same cause.	Presenting the US without the CS reduces the CR.[3,14] (Contingency, US pre-exposure.)
5. Where several different objects produce the same effect, it must be by means of some quality, which we discover to be common among them.	(No clear analogy.)
6. The difference in the effects of two resembling objects must proceed from that particular in which they differ.	Following A with the US and following AB with nothing makes B reduce the CR.[3] (Conditioned inhibition.) Wagner's relative-validity work.[15a]

[a]By "Wagner's relative-validity work" is meant experiments that measured the CR to a CS after the CS had been followed by the US on half the trials. The CS was conditioned in compound with a second stimulus that took one of two values (for example, a high or low tone). The value of the second stimulus either (a) predicted or (b) did not predict whether the US would occur. When the value of the second stimulus predicted the US, the CR to the CS was reduced.

We cannot rule out the possibility that research on classical conditioning has somehow been constrained by human ideas about causality, so that the two lists are similar because of a similar source. Still, the more likely explanation is that the correspondence is due to a similarity of function. Because the function of Hume's list is to help the reader detect causality, this suggests that the function of classical conditioning resembles the detection of causality.

Others[11,16] have reached the same conclusion for different reasons, and many have reached related conclusions.[17,18] Hollis[4] points out, quite correctly, that this conclusion says nothing about the nature of the responses that are conditioned (such as salivation or taste aversion), and she shows that the responses that can be conditioned are very diverse. She proposes that the function of classical conditioning is to help the animal "optimize interaction with" the US; this, of course, is close to the older idea that the function of the CR is to prepare the animal for the US.[19] There is some direct evidence for Hollis's proposal (for example, Zamble's work, described earlier). A problem, though, is that the notion of "optimizing interaction with" is vague. If the US is food, for example, this idea does not seem to distinguish between the function of classical conditioning and the function of digestion. But we agree with Hollis that detection of causality cannot be the whole function of classical conditioning. Experiments such as Zamble's should show how knowledge gained about causality is used.

THE CLOCK USED IN TIME DISCRIMINATION AND A STOPWATCH

To discriminate time requires some sort of clock—something that changes with time in a regular way. Most of our work (reviewed in Roberts[2]) has tried to determine properties of the clock that rats use to discriminate times on the order of minutes and seconds. One of the earliest conclusions from this work was that there were seven similarities between the rat's clock and an ordinary stopwatch.[5] In later work[20,21] more similarities were found. TABLE 2 lists the properties that we now think the two clocks share. Because the properties, and the supporting evidence, are described in detail elsewhere,[2] this description will be brief.

The shared properties can be divided into three groups. First, there are properties that the rat's clock shares with most or all man-made clocks, but not with all possible clocks. For example, although all man-made clocks have linear scales, other clocks do not (for example, the clock used in carbon dating). The first property in this group is distinctness. This means that the clock can be changed without changing other things, and other things can be changed without changing the clock. For example, time discrimination not only requires a clock but also a stimulus-to-response mapping; apparently the rat's clock can be changed without changing the mapping, and the mapping can be changed without changing the clock. The second property is a linear scale, that is, equal distances on the scale correspond to equal differences in seconds. The third property is that the clock is relatively accurate compared to other operations used in time discrimination. This means that most of the variance in time discrimina-

TABLE 2. Similarities Between the Rat's Clock and a Stopwatch

I. Properties of most or all man-made clocks:
  1. Distinct.
  2. Scale linear with physical time (equal distances on scale correspond to equal differences in seconds).
  3. Relatively accurate (compared to other operations used in measuring time).
  4. Changes along only one dimension.

II. Properties of man-made clocks that measure duration:
  5. Times selectively.
  6. Has internal pacemaker.
  7. Can time stimuli from more than one modality.
  8. Times different intervals using the same rate.

III. Properties of man-made clocks that measure unknown durations:
  9. Can time more than one interval.
  10. Times up (rather than out).
  11. Can be reset quickly.
  12. Can be stopped temporarily.

TABLE 3. Two-way Classification of Man-Made Clocks

	Is Time Known before Measurement?	
Type of Time	Yes	No
Duration	Kitchen timer	Stopwatch
Location in cycle	Alarm clock	Wall clock

tion is not due to variance in the clock; for example, it might be due to variance in the criterion. The fourth property is that the clock changes along only one dimension. This means that the state of the clock can be fully described with one number (the time that it reads).

Man-made clocks vary. A deluxe digital watch (Casio), in May 1983, has four modes, each with a different time: (a) the time of day; (b) the time of day to which an alarm is set; (c) the elapsed time on a stopwatch-like clock; (d) the not-yet-elapsed time on a timing-out clock. TABLE 3 shows that the four modes correspond to the four cells of a two-way (2 × 2) classification of man-made clocks. One division is whether duration or location in cycle is being measured; the other division is whether time is being estimated or produced. Clocks in different cells have different properties. The second group of similarities between the rat's clock and a stopwatch consists of properties possessed by man-made clocks designed for measuring duration (stopwatches or kitchen timers, for example) but not by man-made clocks designed for measuring location in cycle (such as alarm clocks or wall clocks). The first property in this group is that the clock times selectively. This means that the clock is not always running; it does not time all detectable stimuli. The second property is that the clock has an internal pacemaker. In other words, the rate of the clock is determined by something inside the animal (in the case of the rat's clock) or inside the watch (in the case of the stopwatch). Other clocks have pacemakers that are far away (sundials or house-current clocks, for example). The third property is that the clock can time stimuli from more than one modality. For example, the rat's clock can time both lights and sounds. The final property in this category is that the clock times different intervals using the same rate. For example, a stopwatch goes at the same rate whether it is timing a 30-sec interval or a 60-sec interval. In contrast, location-in-cycle clocks must measure different intervals (cycle lengths) using different rates.

The final group of similarities between the rat's clock and a stopwatch are those properties shared by clocks designed for measuring durations not known before the measurement (such as a stopwatch), but not by clocks designed for measuring durations known before the measurement (such as a kitchen timer or hourglass). The first property in this group is that the clock can time more than one interval. A stopwatch, of course, can time many different intervals; an hourglass cannot. The second property is that it times "up" (rather than out). Some clocks, such as a stopwatch, measure intervals of different lengths by starting at the same point (zero) and ending at different points. Other clocks, such as a kitchen timer, measure intervals of different lengths by starting at different points and ending at the same point. The third property is that the clock can be reset quickly. A stopwatch can be reset quickly; an hourglass cannot. The final similarity is that the clock can be stopped temporarily. A stopwatch, of course, can be stopped temporarily, but most kitchen timers cannot.

Some might question one or two of the similarities (for example, does the rat's clock have a linear scale? is a stopwatch relatively accurate?), but this would hardly change the overall picture. The similarities are certainly not due to similarity of building materials. Thus they suggest that the rat's clock and a stopwatch have similar functions; in particular, that the rat's clock, like a stopwatch, was designed for measuring unknown durations. Our procedures, such as fixed-interval schedules, in a sense required the rats to measure unknown durations, and the rats did so, but it was a little surprising that they did so using a clock designed for the task. Fixed-interval schedules do not correspond to any obvious natural problem. That an animal can readily solve a laboratory problem does not mean that it has a mechanism designed to solve the problem. For example, laboratory work shows that rats can detect X rays, and that at some intensities they do this with their olfactory system[22]; yet we can be sure that the olfactory system was not designed to detect X rays.

The comparison of animal clocks and man-made clocks may be useful in the study of other animal clocks. Just as our research has emphasized the differences between kitchen timers and stopwatches, a person studying circadian clocks might emphasize the differences between alarm clocks and wall clocks. Consideration of the difference between alarm clocks and wall clocks shows that a circadian clock might have two different uses: (a) to make sure the animal does certain things at certain fixed-before-birth times (the alarm-clock use); or (b) to learn when different things happen, such as when food is available (the wall-clock use). All animals show innate circadian rhythms (the alarm-clock use); at least some animals, such as rats[23] and bees,[24] can learn the time of day of an event (the wall-clock use). One question that this distinction suggests: Is the same clock used for both? Recent work[25] suggests that the answer in some cases is no. Evidence has been accumulating that there are at least two circadian clocks,[26] but there has been no conceptual basis for the differences between them.

TIME DISCRIMINATION AND CLASSICAL CONDITIONING

In recent work, we have found similarities between time discrimination and classical conditioning. We will describe an experiment illustrating two of the similarities, and then describe other similarities more briefly. Finally, we will give two possible reasons for the similarities.

The experiment asked the following question: When does the clock used in time discrimination time a stimulus? The clock might time all stimuli that are detectable, or it might time only some of them. In other words, the clock might or might not be selective. In this experiment, the stimulus of interest was sound, so that our theoretical question became: When does the clock time sound?

If the clock were selective, how would it select? On the basis of theories of selective attention (such as that of Sutherland and Mackintosh[27]), it was reasonable to assume that the clock might select among stimuli based on their signal value, that is, to assume that stimuli with different signal values might be timed to different extents. Therefore we varied the signal value of the sound. The sound was treated in three ways: (a) *Exposure.* First, the sound was made familiar. It was presented alone for random durations (averaging 20 sec). (b) *Pairing.* Next, the sound was made a signal for food using a classical conditioning procedure. The sound lasted a fixed duration (20 sec) and ended with a pellet of food. (c) *Extinction.* Finally, the sound was extinguished. It lasted a fixed duration (20 sec) and ended without food.

After each treatment, we wanted to know whether the clock timed the sound. To answer this question, we took advantage of earlier results[21] that showed that there could be transfer of a time discrimination from light to sound. These results implied that the same clock timed both light and sound. In this experiment, we trained rats to time light, and then asked whether they timed sound. We made two assumptions. First, we assumed that if the clock timed the sound, there would be transfer from light to sound. Second, we assumed that if the clock did not time the sound, there would not be transfer from light to sound. Justification of these assumptions is beyond the scope of this paper, but it may be helpful to know that experiments that do not make these assumptions suggest the same conclusions as this one.

The time discrimination was trained with a psychophysical choice procedure. The subjects were eight rats; they worked in standard lever boxes with two retractable levers. During intertrial intervals, the levers were retracted. There were three types of trials, which were randomly mixed throughout a session: (a) *Training trials (with light)*. These trials established and maintained the time discrimination with light. They began with light lasting 3 or 12 sec. When the light ended, the two levers were extended into the box, and the rat chose between them. After a choice (one response), the levers were retracted. If the rat had chosen correctly, it was rewarded with a pellet of food. The correct choice depended on the duration of the light: After a 3-sec light, one lever (the "short" lever) was correct; after a 12-sec light, the other lever (the "long" lever)

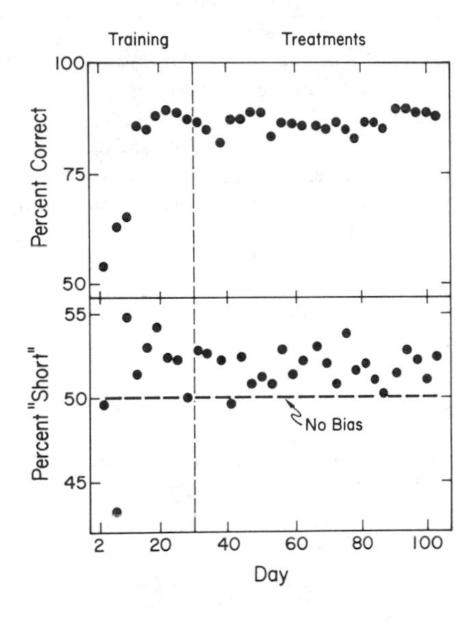

FIGURE 2. Accuracy (*upper panel*) and bias (*lower panel*) with light were roughly constant throughout the testing of sound. (Each point is a mean over 3 days, eight rats, and two durations.)

was correct. Throughout the experiment, training trials were always at least 70% of all trials. (b) *Treatment trials (with sound)*. These trials changed the signal value of the sound. On these trials, the sound was treated as described earlier. The levers remained withdrawn. Treatment trials were 10–20% of all trials. (c) *Probe trials (with sound)*. These trials tested for transfer of the time discrimination. They began with sound lasting either 3 or 12 sec. Then the levers were extended, and the rat chose between them. After a choice, the levers were retracted; neither choice was rewarded. Early in the experiment, no probe trials were given. On the days when they were given, they were 10% of all trials.

FIGURE 2 shows the acquisition of the time discrimination with light. The upper panel shows a measure of accuracy (percent correct); it started at chance (50%) and then rose to about 90%. The lower panel shows a measure of bias (the overall probability of a "short" response). Because short (3-sec) and long (12-sec) trials were equally likely, an unbiased rat would choose the "short" lever on 50% of all trials. The lower panel shows that there was a negligible bias in the "short" direction, on the order of a few percent. Probe trials were used during the phase of the experiment labeled "Treatments"; FIGURE 2 shows that accuracy and bias were roughly constant during this phase.

FIGURE 3 shows the results from probe trials. In general, cross-modal transfer means that the animal responds to stimuli in the new modality as if they were stimuli in the old modality. In this experiment, increasing the duration of the light increased the probability of a "long" response after the light. Thus, in this experiment, cross-modal transfer would be the finding that increasing the duration of the sound increased the probability of a "long" response. FIGURE 3 shows that there was transfer after pairing ($p < .01$), but not after exposure or extinction ($p > .05$).

FIGURE 3 also shows that during exposure and extinction, when responding to the two probes was equal, responding was not at chance (50%). In both cases, the probability of a "long" response after a probe was about 20%, reliably less than 50%. In other work, we have found that after time-discrimination training, if no stimulus at all is presented—if the levers are just extended after the intertrial interval—then the rats will choose the "short" lever about 80% of the time (equivalent to choosing the "long" lever 20% of the time). An absent stimulus can be thought of as having a duration of zero. Thus, during exposure and extinction the rats were treating 3- and 12-sec sounds as if they had durations of zero.

The results from probe trials suggest that the clock timed the sound after pairing, but not after exposure or extinction. We can conclude that the change from exposure to pairing increased the timing of the sound, and the change from pairing to extinction decreased the timing of the sound. Here are two similarities between timing and classical conditioning. Suppose that we had done this experiment using dogs instead of rats, and had measured salivation during the sound. We would certainly have found that the change from exposure to pairing increased salivation during the sound, and that the change from pairing to extinction decreased salivation during the sound.

We have confirmed the conclusions of this experiment in other experiments, some of them not involving cross-modal transfer, and we have done other experiments that asked whether timing and classical conditioning are changed in the same way by various manipulations. TABLE 4 summarizes what we and others have found. There are eight similarities, and they can be divided into two groups. The first group consists of

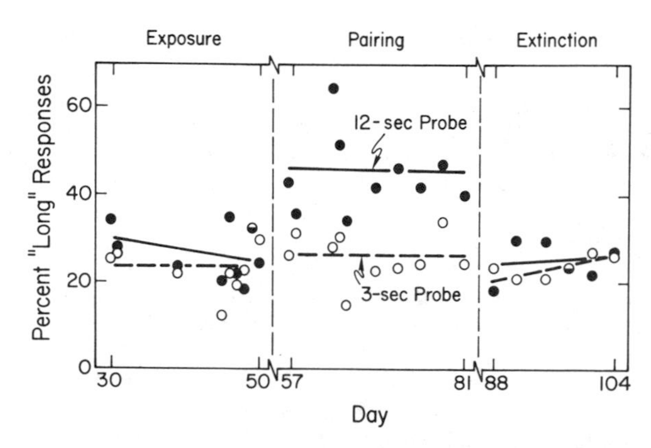

FIGURE 3. Sound was timed when followed by food (pairing), but not when presented alone (exposure and extinction). (Straight lines were fit by least-squares.)

TABLE 4. Changes with Similar Effects on Classical Conditioning and Timing

I. Changes with large effects on classical conditioning and timing:
1. *Probability of US.* After exposure of the CS, following the CS with the US increases CR rate during the CS and increases timing of the CS. (Forward conditioning.)
2. *Probability of US.* After forward conditioning, presenting the CS alone reduces CR rate during the CS and reduces timing of the CS. (Extinction.)
3. *Order of CS and US.* Backward conditioning (CS follows US) is much less effective than forward conditioning (CS precedes US).
4. *Time between CS and US.* Increasing the time between the CS and US reduces CR rate during the CS and reduces timing of the CS. (Trace conditioning.)

II. Changes with little or no effect on classical conditioning and timing:
5. *Information provided by CS duration.* The CS can be a fixed or geometrically distributed duration.
6. *Choice of CS.* Many stimuli can be used as the CS.
7. *Choice of US.* Many events can be used as the US.
8. *Choice of animal.* Found in most or all vertebrates.

experimental changes that have large effects on both timing and classical conditioning; the second group consists of changes that have little or no effect on both. So far, we have not found any changes that affect one but not the other.

The first two similarities involve sequential changes in the probability of the US after the CS. They were illustrated in the experiment just described. After exposure of the CS, an increase in the probability of the US increases both timing and the CR. After the CS and US have been paired, a decrease in the probability of the US decreases both timing and the CR.

The third similarity is the importance of the order of CS and US. With classical conditioning, of course, backward conditioning (CS follows US) usually produces no excitatory conditioning in the same situations in which forward conditioning (CS precedes US) produces considerable excitatory conditioning.[3] We have found that backward conditioning does not produce timing of the CS in a situation in which forward conditioning does produce timing of the CS.

The fourth similarity is the importance of the time between CS and US. With classical conditioning, of course, increasing the time between the end of the CS and the start of the US eventually decreases the CR.[3] It also decreases timing of the CS. We have measured classical conditioning and timing in the same situation; we found that, after pairing, putting a 20-sec gap between the end of the CS and a food pellet eliminated both the timing of the CS and the CR.

The second group of similarities consists of changes with little or no effect on classical conditioning and timing.

The fifth similarity, the first in this group, is the importance of the information provided by CS duration. The contrast is between CSs with fixed durations and CSs with geometrically distributed durations. We have studied the case where the US comes at the end of the CS. When the CS has a fixed duration, timing the CS helps the animal anticipate the US—as time passes during the CS, the US in a sense becomes more likely. But when the CS has a geometrically distributed duration, timing the CS does not help the animal anticipate the US. Some theories of selective attention (such as Sutherland and Mackintosh's[27]) assume that attention to irrelevant dimensions decreases. By analogy, we might expect that a CS will be consistently timed only if timing it helps the animal to anticipate the US. In other words, we might expect the information provided by CS duration to affect timing of the CS. But apparently it does not; we have found that both fixed and geometrically distributed durations can produce consistent timing of the CS. Classical conditioning can be effective whether the CS is a fixed or geometrically distributed duration (for example, see Libby and Church[28]).

The sixth change with similar effects is the choice of CS. A fundamental finding about classical conditioning is that with any US there are always many possible CSs; with food as the US, for example, Pavlov found that lights, sounds, smells, and tactile stimuli were effective CSs[29]. Reviews of the animal-timing literature show that many stimuli can be timed.[1,2]

The seventh change is the choice of US. Classical conditioning can be produced by a wide range of USs, such as food, water, acid, electric shock, and drugs. The same range of events has also been found to produce timing. However, it seems likely that USs given by injection, such as morphine and poison, may produce classical conditioning, but not timing. It is too early to speculate about this possible difference.

The final change is the choice of animal. Both classical conditioning and time discrimination have been found in a wide range of vertebrates.

These similarities could hardly be due to the building materials; for example, evolution could surely manage to restrict time discrimination to only a few vertebrates. Therefore the similarities suggest that the mechanisms that underly classical conditioning and time discrimination have similar functions. What is the similarity of function? At the moment, two possibilities seem consistent with all of the evidence: (a) The two mechanisms have the same function; (b) the two mechanisms have similar but different functions.

Same function. The clock used in time discrimination may also be used in classical conditioning to help the animal detect causality. Then the clock used in time discrimination has the same function as the mechanism producing classical conditioning because it is part of that mechanism. To say that event A causes event B usually implies that soon after or during event A, the rate of event B increases. Thus judgments of causality require comparisions of rates. To measure rates requires timing. Recent work on autoshaping[30-33] has encouraged explicit recognition of the timing that goes on in classical conditioning, and has revealed some features of the timing; however, any theory of classical conditioning that tries to explain graded effects of contingency[34] must assume that the animal measures the rate of the US. The clock used in time discrimination might be used to measure the rate of the US during the CS. If so, then the timing of a CS, unlike the CR to the CS, should not be sensitive to changes in contingency.

Similar but different functions. Another possibility is that the clock used in time discrimination is different from the clock(s) used to measure US rates. In this case, the similarities between time discrimination and classical conditioning would be understandable if the function shared by both of them was prediction of the future: The mechanism behind classical conditioning helps the animal to predict *what* will happen, the quality of future events, while the mechanism behind time discrimination helps the animal predict *when,* the time of future events.

All of the features of timing described in this paper make sense in terms of either possibility. Both functions—measuring the rate of the US during the CS, and prediction of when—require that the clock measure unknown durations, accounting for the similarities between the clock and a stopwatch. Both functions are facilitated if the clock does not time stimuli that signal nothing, in accord with the sensitivity of timing to the signal value of the stimulus. Both functions would be useful in a wide range of situations, accounting for the generality of timing.

There is no direct evidence for the first possibility (measuring the rate of the US during the CS), but there is direct evidence for the second (prediction of when). It is simple: In laboratory situations, animals commonly anticipate (and thus must predict) the time of important events, such as USs. One example is shown in FIGURE 1; on the last day of training, response rates reached a maximum near the times that food had been given on earlier trials. Thus the rats must have used measurements of the time of food on earlier trials to predict the time of food on the current trial. Many other procedures (such as Sidman avoidance or classical conditioning with a fixed-duration CS) show the same thing.[2] In a park, the probability that a bird waiting for flies would leave one perch and find another increased with the time since the last capture attempt.[2,35] This suggests that the prediction of when may help animals forage more efficiently, a point often made in the optimal-foraging literature.[36] It also suggests that time discrimination happens outside of laboratories.

A third possibility is that the function of the clock used in time discrimination is both to measure US rates during the CS and to allow prediction of when. In rats, the eyes are used both to make visual discriminations and to entrain circadian rhythms.[37] The clock used in time discrimination might be used in more than one important way.

SUMMARY

This paper makes three main points:

1. There are many similarities between man-made rules for detecting causality and "laws" of classical conditioning, presumably generated by nature. They suggest that the function of classical conditioning resembles the detection of causality.
2. There are twelve similarities between the clock used by rats to discriminate time and a stopwatch. They suggest that the rat's clock, like a stopwatch, was designed for measuring unknown durations.
3. There are eight similarities between time discrimination and classical conditions. In particular, whether or not a stimulus is timed depends on the signal value of the stimulus. The similarities suggest that time discrimination and classical conditioning—their underlying mechanisms—have similar functions.

ACKNOWLEDGMENTS

We thank Michael Brown, Thelma Rowell, and Sonja Yoerg for their comments.

REFERENCES

1. RICHELLE, M. & H. LEJEUNE. 1980. Time in animal behaviour. Pergamon Press. Oxford.
2. ROBERTS, S. 1983. Properties and function of an internal clock. *In* Animal Cognition and Behavior. R. L. Mellgren, Ed. North-Holland. Amsterdam.
3. MACKINTOSH, N. J. 1974. The Psychology of Animal Learning. Academic Press. London.
4. HOLLIS, K. L. 1982. Pavlovian conditioning of signal-centered action patterns and autonomic behavior: A biological analysis of function. Adv. Study Behav. **12:** 1–64.
5. ROBERTS, S. & R. M. CHURCH. 1978. Control of an internal clock. J. Exp. Psychol. Anim. Behav. Processes **4:** 318–337.
6. CLUTTON-BROCK, T. H. 1982. Function. *In* The Oxford Companion to Animal Behavior. D. McFarland, Ed.: 220–223. Oxford University Press. Oxford.
7. TINBERGEN, N., G. J. BROOEKHUYSEN, F. FEEKES, J. C. W. HOUGHTON, H. KRUUK & E. SZULC. 1962. Egg shell removal by the black-headed gull, *Larus ridibundus* L.: A behaviour component of camouflage. Behaviour **19:** 74–118.
8. SEELY, M. K. & W. J. HAMILTON, III. 1976. Fog catchment sand trenches constructed by tenebrionid beetles, *Lepidochora*, from the Namib desert. Science **1983:** 484–486.

9. ZAMBLE, E. 1973. Augmentation of eating following a signal for feeding in rats. Learn. Motiv. **4:** 138–147.
10. SHERMAN, P. W. 1977. Nepotism and the evolution of alarm calls. Science **1977:** 1246–1253.
11. DICKINSON, A. 1980. Contemporary Animal Learning Theory. Cambridge University Press. Cambridge, England.
12. HUME, D. 1978. A Treatise of Human Nature. 2nd ed. Oxford University Press. Oxford.
13. RESCORLA, R. A. & C. L. CUNNINGHAM. 1979. Spatial contiguity facilitates Pavlovian second-order conditioning. J. Exp. Psychol. Anim. Behav. Processes **5:** 152–166.
14. RANDICH, A. & V. M. LOLORDO. 1979. Associative and non-associative theories of the UCS preexposure phenomenon: Implications for Pavlovian conditioning. Psychol. Bull. **86:** 523–548.
15. WAGNER, A. R. 1969. Stimulus validity and stimulus selection in associative learning. *In* Fundamental Issues in Associative Learning. N. J. Mackintosh & W. K. Honig, Eds. Dalhousie University Press. Halifax.
16. TARPY, R. 1982. Principles of animal learning and motivation. Scott, Foresman. Glenview, IL.
17. MACKINTOSH, N. J. 1977. Conditioning as the perception of causal relations. *In* Foundational Problems in the Special Sciences. R. E. Butts & J. Hintikka, Eds. D. Reidel Publishing Co. Dordrecht, Holland.
18. BOLLES, R. C. 1972. Reinforcement, expectancy, and learning. Psychol. Rev. **79:** 394–409.
19. MILLER, R. R., C. GRECO, M. VIGORITO & N. MARLIN. 1983. Signaled tailshock is perceived as similar to a stronger unsignaled tailshock: Implications for a functional analysis of classical conditioning. J. Exp. Psychol. Anim. Behav. Processes **9:** 105–131.
20. ROBERTS, S. 1981. Isolation of an internal clock. J. Exp. Psychol. Anim. Behav. Processes **7:** 242–268.
21. ROBERTS, S. 1982. Cross-modal use of an internal clock. J. Exp. Psychol. Anim. Behav. Processes **8:** 2–22.
22. GARCIA, J. & R. A. KOELLING. 1971. The use of ionizing rays as a mammalian olfactory stimulus. *In* Handbook of Sensory Physiology. Vol. 4: Chemical Senses. L.M. Beidler, Ed.: 449–464. Springer-Verlag. Berlin.
23. BOLLES, R. C., & L. W. STOKES. 1965. Rat's anticipation of diurnal and a-diurnal feeding. J. Comp. Physiol. Psychol. **60:** 290–294.
24. KOLTERMANN, R. 1974. Periodicity in the activity and learning performance of the honeybee. *In* Experimental Analysis of Insect Behavior. L. B. Browne, Ed. Springer-Verlag. Berlin.
25. BOULOS, Z., A. M. ROSENWASSER & M. TERMAN. 1980. Feeding schedules and the circadian organization of behavior in the rat. Behav. Brain Res. **1:** 39–65.
26. MOORE-EDE, M. C., F. M. SULZMAN & C. A. FULLER. 1982. The Clocks That Time Us. Harvard University Press. Cambridge, MA.
27. SUTHERLAND, N. S. & N. J. MACKINTOSH. 1971. Mechanisms of animal discrimination learning. Academic Press. New York, NY.
28. LIBBY, M. E. & R. M. CHURCH. 1975. Fear gradients as a function of the temporal interval between signal and aversive event in the rat. J. Comp. Physiol. Psychol. **88:** 911–916.
29. PAVLOV, I.P. 1927. Conditioned Reflexes. Oxford University Press. Oxford.
30. GIBBON, J. & P. BALSAM. 1981. Spreading association in time. *In* Autoshaping and Conditioning Theory. C. M. Locurto, H. S. Terrace & J. Gibbon, Eds. Academic Press. New York, NY.
31. JENKINS, H. M., R. A. BARNES & F. J. BARRERA. 1981. Why autoshaping depends on trial spacing. *In* Autoshaping and Conditioning Theory. C. M. Locurto, H. S. Terrace & J. Gibbon, Eds. Academic Press. New York, NY.
32. GIBBON, J., & M. D. BALDOCK, C. LOCURTO, L. GOLD & H. S. TERRACE. 1977. Trial and intertrial durations in autoshaping. J. Exper. Psychol. Anim. Behav. Processes **3:** 264–284.
33. JENKINS, H. M. & D. SHATTUCK. 1981. Contingency in fear conditioning: A reexamination. Bull. Psychon. Soc. **17:** 159–162.

34. Rescorla. R. A. 1968. Probability of shock in the presence and absence of CS in fear conditioning. J. Comp. Physiol. Psychol. **66:** 1–5.
35. Davies, N. B. 1977. Prey selection and the search strategy of the spotted flycatcher (*Muscicapa striata*): A field study on optimal foraging. Anim. Behav. **25:** 1016–1033.
36. Charnov, E. L. 1976. Optimal foraging: The marginal value theorem. Theor. Pop. Biol. **9:** 129–136.
37. Rusak, B. & I. Zucker. 1979. Neural regulation of circadian rhythms. Physiol. Rev. **59:** 449–526.

Time and Contingency in Classical Conditioning

H. M. JENKINS

Department of Psychology
McMaster University
Hamilton, Ontario, Canada L8S 4K1

Pavlov considered the temporal contiguity of conditioned stimulus (CS) and unconditioned stimulus (US) to be a fundamental condition of association in conditioning. The principle of contiguity still serves as starting place for certain contemporary theories of classical conditioning despite a forceful challenge to the adequacy of the contiguity principle from the work of Rescorla.[1,2] He showed, as Prokasy[3] suggested, that with identical conditions of CS–US contiguity, an association might or might not develop, depending on the degree of correlation or contingency between CS and US. A substantial part of the recent history of conditioning theory can be viewed as an attempt to preserve the principle of contiguity in the face of this challenge.

Contingency is reduced when a CS is presented without a US or when a US is presented without a CS. Under certain conditions, which we will later try to identify, CS-alone and US-alone presentations interfere with the ability of a contiguous presentation of CS and US to develop and maintain conditioned responding. Let us consider first the question of how a presentation of US-alone could reduce the effectiveness of a contiguous CS–US presentation.

The Rescorla–Wagner theory[4] provides an account of interference from US-alone presentations which preserves the principle of contiguity. The essential points are these: (1) A CS is always presented in the context of a background which can be treated as any other stimulus. (2) Presentations of the US-alone condition the background. (3) The background then acts as a blocking stimulus, which competes with the CS for a share of the total signal strength that can be supported by the US. The form of the competition between background and CS is specified by the equations of the theory. For the special case of no contingency between CS and US, in which the probability of US presentation per unit time is the same in the presence or absence of the CS, the equations yield the asymptotic prediction of no signal value for the CS; a result that has been frequently, although not invariably, obtained (for a review, see Gormezano and Kehoe[5]).

We may note that in the Rescorla–Wagner theory, conditioning occurs as the result of contiguity unless competition with a concurrent, fully conditioned stimulus prevents contiguity from doing its work. The theory can be characterized as a contiguity-plus-competition theory.

The equations of the theory also imply that exposure to noncontingent reinforcement of the CS after the CS has acquired signal value from contingent reinforcement removes the signal value of the CS. This implication is consistent with the well established observation that previously conditioned responses cease as the result of extended exposure to noncontingency. The explanation rests on the idea that as the background becomes more and more excitatory through US-alone presentations, the total excitatory value of the CS plus the background exceeds the maximum value supportable by the US. When this happens the equations show that both the background and the CS will lose signal value even though the CS continues to occur in temporal conjunction with the US. Because the background regains strength as a consequence of the US-only presentations, while the CS continues to lose strength on CS–US presentations, the asymptotic result is a fully conditioned background and a completely neutralized CS.

A second explanation for interference by US-alone presentations comes from the work of Gibbon and Balsam.[6] Unlike the Rescorla–Wagner theory, this explanation utilizes time as a fundamental variable. Much of the evidence comes from experiments on the acquisition of autoshaped keypecking in pigeons in which a lighted key serves as the CS and food as the US. Over broad limits, the longer the wait between feedings, the

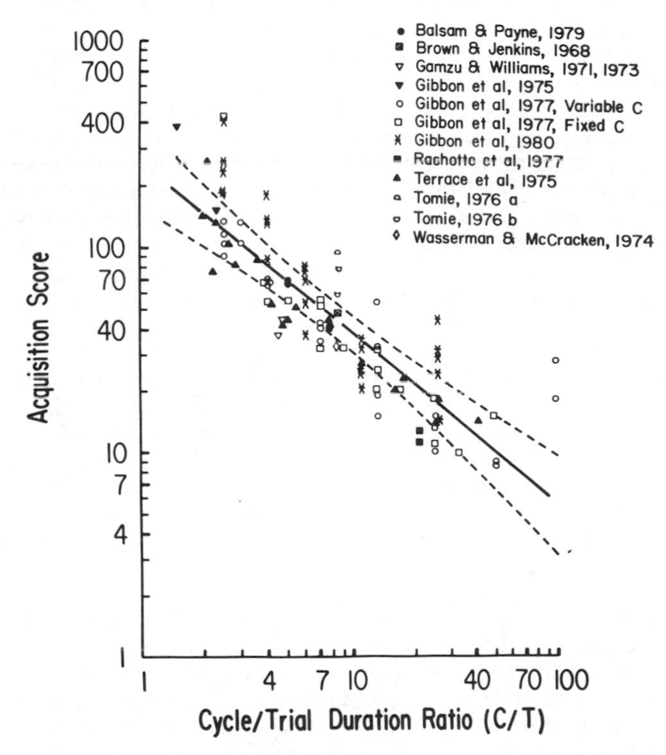

FIGURE 1. Reinforcers to acquisition (n) versus cycle to trial ratios (C/T) on double log coordinates. Data are from the studies indicated on the *right*. The *heavy line* is the least-squares regression, $n = 260.6\ (C/T)^{-.8294}$. The *dashed lines* represent the 95% confidence limits around the regression. (From Gibbon and Balsam.[6] Reprinted by permission.)

more rapid is the acquisition of autoshaped pecking. The time spent waiting in the CS for the delivery of food is also important; the longer the wait in the CS, the poorer is the acquisition of autoshaped pecking. Moreover, there is a trade-off between these two waits. The important variable appears to be the ratio of the average waiting time between feedings to the average waiting time in the CS. Over a considerable range of values, equal ratios yield equal acquisition. Results from many autoshaping experiments by different investigators have been brought together by Gibbon and Balsam[6] in the plot shown in FIGURE 1.

There is evidence for a similar relation in the conditioned emotional response (CER) procedure in rats. Stein, Sidman, and Brady[7] combined different durations of the CS with different durations of the non-CS, or intertrial interval. From their data, asymptotic suppression levels can be obtained for each of 20 different waiting-time ratios. The waiting-time ratios are calculated by dividing the average time between shocks by the CS duration (each CS terminated with a single shock). The correlation of asymptotic suppression with the ratio of waiting times was +.92, which suggests that relative waiting times might play a role in CER conditioning similar to the role they play in autoshaping.

The relation shown in FIGURE 1 is expressed in terms of average durations—the average duration of the wait between feedings and the average duration of the wait in the presence of the CS. The emphasis on averages is warranted by experimental results on autoshaping which show a rather surprising insensitivity of responding to the time between the immediately preceding feeding and the occurrence of the CS. Moreover, there appears to be little difference between a variable wait between feedings and a constant wait at the mean value of the variable waits. Although, as noted below, recent waiting times do exert a stronger effect than distant waiting times (the averaging process is not entirely cumulative), the fact that needs emphasizing is the rather surprising insensitivity of autoshaping to the time between the immediately preceding feeding and the onset of the CS.[6,8]

Data of the kind I have just reviewed led Gibbon and Balsam to adapt the basic ideas of scalar expectancy theory, which Gibbon[9] had developed earlier in connection with studies of animal timing, to the acquisition process in autoshaping. Experiments on autoshaping in my laboratory gave results in accord with the basic implications of scalar expectancy theory. There were, however, certain findings which led me to prefer a description of the data in terms of the ratio of the overall waiting time to the waiting time in the CS, rather than in terms of the ratio of overall food expectation to food expectation in the CS. For my immediate purposes, I wish to put aside the theoretical issue that lies behind this difference in terminology. The empirical generalization is that the effectiveness of a CS–US pairing depends on the ratio of the average US-to-US interval to the average CS-to-US interval. The issue I wish to examine is whether the putative effects of CS–US contingency can be subsumed by this generalization.

Consider a sequence of CS–US pairings occurring at certain times within a conditioning session. The addition of the US-alone trials will, of course, reduce the average US-to-US waiting time. Could the interfering effect of the additional USs be due entirely to this variable? If so, one might expect the same effect from adding CS–US pairings as from adding US-alone trials inasmuch as the addition of the CS–US pairings would result in the same reduction in the US–US waiting time. Otherwise put, if the ratio of waiting times is the critical variable, the addition of signaled or unsignaled USs should have a very similar effect on acquisition.

Although the exploration of this possibility was begun with autoshaping experiments, I turn first to an experiment on classical fear conditioning in rats.[10] Following classical pairings of a CS with shock, fear conditioning was tested by suppression of food-reinforced bar-pressing during presentations of the CS, the so-called CER procedure. In the classical conditioning phase, all groups received shocks at the same rate during the CS-on periods. A control group received 12 2-min trials, 40% of which were reinforced, during each 2-hr conditioning session. Strong fear conditioning was expected in this group. Each of two other groups received about ten added shocks in the 2-hr session. In one group the added shocks occurred only during added CS trials. Since the added shocks were signaled, the CS–US contingency was no less than in the control group. In another group, the additional shocks were unsignaled. Consequently, contingency in this group was substantially reduced.

The test results are shown in FIGURE 2. Adding shocks to the control condition resulted in less fear conditioning to the US as assessed by the suppression ratio. The important result was that interference with fear conditioning was no less for the group that received these shocks unsignaled, with a consequent reduction in contingency, than for the group that received them signaled, with no loss of contingency. It would appear that the effect of reducing contingency through the addition of US-alone presentations might be subsumed as an effect of relative waiting times.

The classic experiment that purported to show the effect of CS–US contingency in fear conditioning was reported by Rescorla.[2] It did not, however, include a control for variations in overall shock rate. As it turns out, his results are correlated with relative waiting times as closely as they are with differences in CS–US contingency or with the

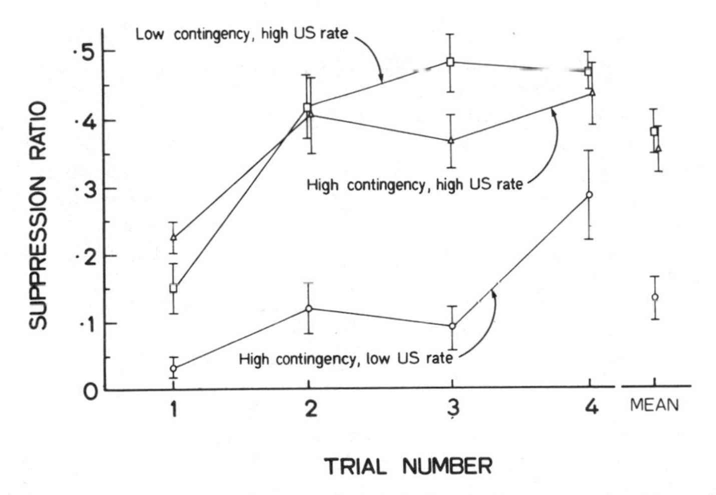

FIGURE 2. Suppression ratio as a function of trials during the first test session. Mean of the four trials for each group is shown at the *right*. **Bars** show 1 standard error of the mean. (From Jenkins and Shattuck.[10] Reprinted by permission.)

orderings of the groups on the basis of the Rescorla–Wagner theory (see Jenkins and Shattuck[10]).

There is reason to believe that relative waiting time, not contingency, also governs acquisition in the autoshaping experiment. As we have seen, an implication of the relative waiting time hypothesis is that the proportion of feedings that is signaled by the autoshaping CS is not important as long as the average waiting time between feedings, and the average waiting time when the CS is present, are unaltered. This implication was tested in an autoshaping experiment in which the control group received a single CS–US pairing at the middle of a 21-min session (Jenkins *et al.*,[8] experiment 13). Each of four other groups received the same mid-session trial, referred to as the reference trial, but they also received 30 additional feedings at the rate of one every 40 sec. An equal number of these occurred before and after the reference trial. In one group, 100% of these extra feedings were preceded by the same CS that appeared on the reference trial; in a second group 50% were signaled; in a third group 30% were signaled; and in a fourth group none of the added feedings was signaled. Since the reference trial itself was signaled, this group had approximately 3% of all the feedings signaled.

Acquisition, measured on the reference trial only, is shown for each group in FIGURE 3 by 10-session blocks. The control group, which received no added feedings, acquired within the first ten trials. At the end of 40 sessions each of the other groups was responding, on average, to about 50% of the trials. As would be expected, the groups with higher percentages showed an initial advantage because by the time they reached the first reference trial they had received many more trials than the groups with lower percentages of feedings signaled. Nevertheless, as training proceeded, the groups showed similar, statistically undistinguishable, results. When acquisition is considered on a per trial basis, plotting the data trial-by-trial for all trials rather than for just the reference trial, the lower percentage groups, which received more widely spaced CS presentations, showed more rapid acquisition than the higher percentage groups.

The results of this experiment are consistent with the waiting-time relation. Wide variations in the degree of CS–US contingency did not show up in the later stages of acquisition. The interference with acquisition due to additional feedings was no less when the feedings were signaled than when they were unsignaled.

The results from the experiments I have just reviewed are not in agreement with the Rescorla–Wagner theory. According to that theory, the addition of US-alone presentations should slow acquisition more than the addition of signaled USs. The US-alone presentations would be expected to condition the background more strongly than would a signaled US presentation because the signal would be expected to block conditioning to the background.

As we have noted, previously conditioned responses can be eliminated when the CS–US contingency is removed through the addition of US-alone presentations. If the CS–US contingency is reduced to zero through the introduction of US-alone presentations, the overall waiting time between USs becomes equal to the waiting time within the CS. The relative waiting-time generalization might, therefore, subsume this effect as well. Again, a critical question is whether the addition of signaled USs would eliminate responding as rapidly as would the removal of contingency through the addition of US-alone trials. This should be the case if the results are to be subsumed by relative waiting times. On the other hand, it should not be the case if the Rescorla–Wagner theory of conditioning to background by US-alone presentations applies.

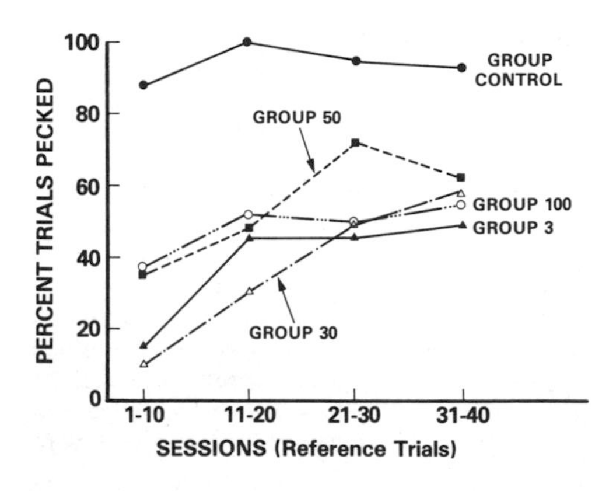

FIGURE 3. Percentage of reference trials pecked over 10-session blocks; experiment 13. Group control received only the reference trial. Other groups received a total of 30 feedings, of which from 100% to 3% were signaled. (From Jenkins, Barnes, and Barrera.[8] Reprinted by permission.)

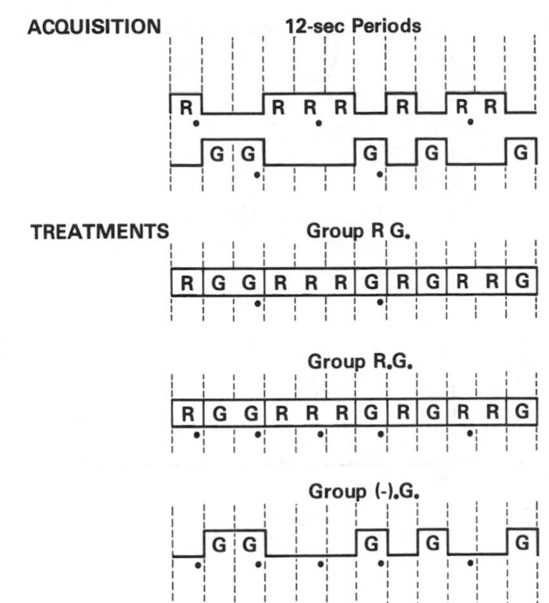

FIGURE 4. Effect of added feedings: signaled versus unsignaled. Design of experiment 1. The location of feedings is shown by dots. R stands for red lighted key, G for green.

The plan of an autoshaping experiment in which this issue was examined (Jenkins and Lambos[11], experiment 1) is shown in FIGURE 4. The purpose of the first phase, in which all subjects were treated alike, was to autoshape the keypeck in alternating sessions to a red keylight and to a green keylight. Training sessions were programmed on the basis of 100 12-sec periods. In the last phase of training, the key was lighted on a random half of these periods (red in one session, green in the other), and 40% of the lighted periods contained a 4-sec feeding in the last 4 sec of the 12-sec period (see the upper panel of FIGURE 4).

After training, subjects were divided into three matched groups for the treatment phase. In the group labeled R.G., all of the events of the separate sessions were combined: the key was always lighted either red or green and 40% of the 12-sec periods in red or green were reinforced. Note that the CS–US contingency is thereby removed, and the ratio of waiting times is 1.0. Feedings continued, however, to be accompanied by the already established signals. Consequently, the Rescorla–Wagner formulation implies continued responding. In group (-).G. one of the previously trained keylight colors was dropped, but the feedings previously associated with it continued. This is the typical noncontingent arrangement. Both the Rescorla–Wagner theory and relative waiting times imply a loss of the previously autoshaped responding. Finally, in group R G. the feedings associated with one of the keylight colors was dropped. The relative waiting time now remains what it was in the separate sessions. On either theory responding to the reinforced keylight is expected to continue.

It is evident from the results shown in FIGURE 5 that the decline in responding for group R.G. and group (-).G. was very similar. The effect of added feedings was no different when they were signaled by a well-established key color or introduced in unlit periods that had never previously been associated with feedings. Group R G. showed a substantial contrast-like effect which is implied by neither relative waiting times nor by

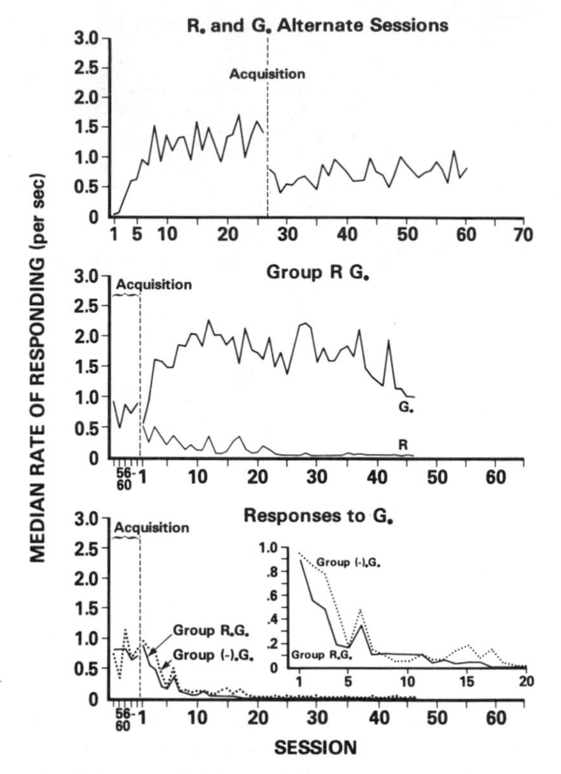

FIGURE 5. Effect of added feedings: signaled versus unsignaled. Rate of responding on trials during acquisition and treatment sessions. *Inset* at lower right replots data in lower panel for first 20 treatment sessions on an expanded scale.

the Rescorla–Wagner theory. The results for this group do serve to establish that the additional feedings, not the change from an unlit to a lighted key, were responsible for the decline in responding in the other groups.

It might be thought that the equivalence of signaled and unsignaled feedings in the elimination of autoshaped responding is peculiar to the circumstance of complete removal of CS–US contingency. In the experiment just described, the red and the green keylights were reinforced alike, and neither sustained responding when they occupied the entire sesion. Would equivalence of signaled and unsignaled feedings still apply when the stimulus used to signal the additional feedings maintains its value as a CS throughout?

In order to examine this question, Murray Goddard and I carried out an experiment (unpublished) on the following plan. We began by establishing each of two key colors as food signals. One color was subsequently used as the reference CS. A single, reinforced, 12-sec presentation of the reference CS occurred per daily session. The other key color was used to signal the occurrence of added feedings. These signals were 8 sec in duration with the feeding in the last 4 sec.

The added feedings were introduced in two well-separated batches, as shown in FIGURE 6. Feedings within a batch were separated by 8 sec. The first batch occurred near the beginning of the session and was followed by a 10-min wait. The second batch occurred immediately prior to the reference trial. The last feeding of this batch occurred just 4 sec before the onset of the reference CS. In group S–U, the feedings of the first batch were signaled, while those of the second batch were unsignaled. In group U–S, the location of the signaled feedings was interchanged so that they came directly before the reference CS. Each group began the experimental phase with no added feedings (reference CS only). In subsequent sessions the number of feedings per batch was increased progressively from 1 to 30.

If the signaled feedings prevent background conditioning, and if the loss of responding that accompanies the added feedings is due to competition between background and the reference stimulus, then the group with signaled feedings immediately prior to the reference trial (group U–S) should maintain a higher level of responding. For this group, the 10-min wait that follows the unsignaled feedings of the first batch would be expected to reduce background conditioning through extinction, and the signaling of the feedings prior to the reference trial would block reconditioning of the background. For group S–U, on the other hand, which received the unsignaled feedings immediately before the reference trial, the background at the time of the reference CS should be more strongly conditioned. This design controls for stimulus generalization between the signaling stimulus and the reference stimulus since each group received the same set of events; only the spacing was varied.

Responding on the reference trial is shown as a function of the number of feedings in each batch in FIGURE 7. The level of responding to the signaling stimulus is not shown, but it remained high (between 2 and 3 responses per sec) throughout the experiment. It is apparent that the decline in responding to the reference trial was no different for the group that received the second batch of feedings signaled than for the group that received the second batch unsignaled.

The last nine sessions of the standard procedure were run with 30 feedings in a batch. For the next five sessions the batch immediately prior to the reference trial was shifted forward in the session so that the 10-min wait occurred immediately before the reference trial in both groups. As shown in FIGURE 7, this produced a substantial recovery of similar size in each group. The recovery suggests that feedings immediately prior to the reference trial were having a greater effect that those that occurred 10

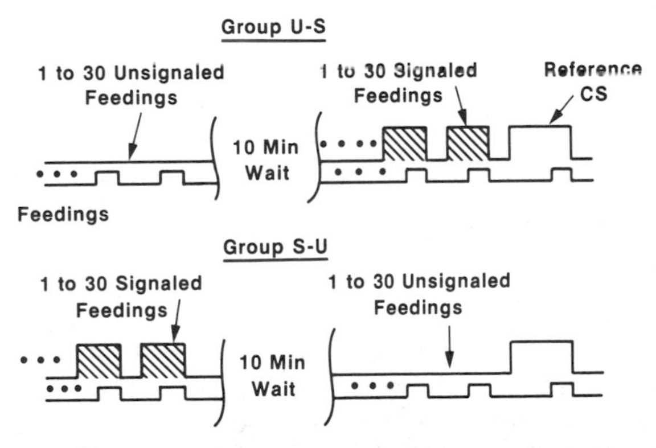

FIGURE 6. Plan of experiment on the effect of the location of signaled and unsignaled feedings on responding to a reference CS. An equal number of feedings occurred before and after the 10-min wait. Starting with 0, the number of feedings was increased progressively using the values 1, 2, 4, 8, 10, 14, 18, 22, 26, and 30. Two sessions were run at each value before increasing the number, except that 10 sessions were run at 10 feedings, and 9 at 30 feedings.

minutes before the reference trial. There appears to be a limit to the averaging of waiting times over a session.

Taken together with the previous experiment, these results provide strong support for the idea that the decline in responding that accompanies the complete removal, or the reduction, of CS–US contingency by US-alone presentations is due to a change in relative waiting time. If the decline does involve conditioning background stimuli, then it appears, contrary to the Rescorla–Wagner theory, that such conditioning occurs to much the same extent for signaled and unsignaled feedings.

There is another fact about the elimination of responding from the removal of contingency that goes against the view that competition for associative value between the background and the CS is the cause. The elimination is not due to a loss in the associative value of the CS since the removal of all food (that is, extinction) results in a recovery of responding to about the level that existed prior to the removal of contingency.[11,12] The reversibility of the decline indicates that it has more to do with performance than with the unlearning of an association.

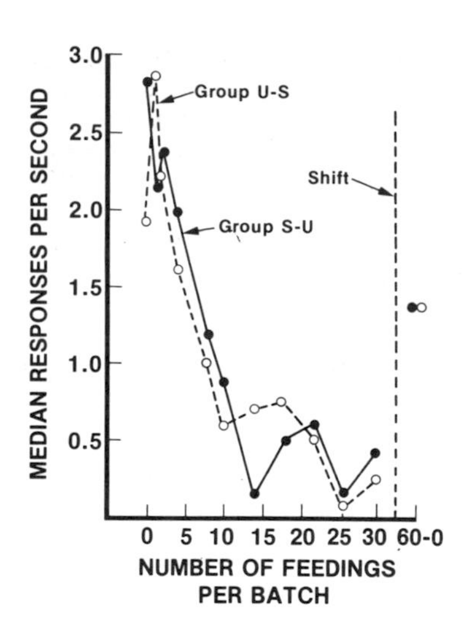

FIGURE 7. Median responses per second on the reference trial as a function of the number of added feedings. Group U-S ($n = 5$) received the signaled feedings immediately before the reference CS, and group S-U ($n = 6$) received the unsignaled feedings immediately before the reference trial. After the shift (see text) 60 feedings (30 signaled and 30 unsignaled) were delivered in one batch followed by a 10-min wait and the reference CS.

There is a troublesome exception to the generalization that signaled and unsignaled USs interfere equally with the acquisition of autoshaped responding. Durlach[13] presented intertrial feedings at a rate sufficient to remove the correlation between the CS (lighted key) and food. When the intertrial feedings were signaled by a previously conditioned tone, a relatively slow but clear acquisition to the CS occurred. There was no acquisition when the intertrial feedings were unsignaled. This result might be due to the blocking of background conditioning by signaling, but there are reasons to doubt that interpretation. No evidence was presented to show that the prior tone–food pairings contributed to acquisition. Moreover, acquisition curves in the signaled and unsignaled groups were indistinguishable for a number of sessions and well below the curve for a control group in which no intertrial feedings were presented—a puzzling finding in view of the extensive prior conditioning to the tone. Although stimulus generalization to a keylight from an auditory food signal, which does not itself engender pecking, is implausible, some form of transfer between these signals is a possibility. It is interesting that when conditioning to background is assessed by general activity, signaled and unsignaled feedings give equivalent results.[13,14]

I have argued that the relative waiting-time generalization can subsume the effect of US-alone presentations. I should like to point out that there is evidence to support the view that relative waiting times might also subsume the effect of CS-alone presentations. The evidence comes from an extensive experiment by Gibbon and his associates[15] in which the proportion of CSs reinforced was varied over a wide range of values. These proportions were factorially combined with different overall waiting times between reinforcers. The interesting result was that the median number of reinforcers to an acquisition criterion was not significantly affected by the proportion of CSs reinforced, although, as expected, it was strongly affected by the time between reinforcers. Since a reduction in reinforcement probability was obtained by deleting reinforcers, not by adding CSs, the ratio of the waiting time in the CS relative to the waiting time between USs was unaffected. Reinforcing half of the CSs, for example, doubles the total time in the CS per reinforcement while also doubling the time between reinforcers.

As Gibbon and his associates point out, the apparent integration of CS durations over separated presentations suggests that trials are not the natural units of classical conditioning. It follows that the probability of reinforcement on a trial is not a fundamental variable. If, for example, reinforcing every 10-sec CS is equivalent to reinforcing 50% of 5-sec CSs, then, to put the matter casually, the animal is not interested in whether or not a CS presentation ends in a reinforcer, but rather, in how long it must wait in the CS for a reinforcer. Temporal durations rather than probabilities appear to be fundamental (see also Gibbon[16]).

I have presented evidence that the relative waiting-time generalization might subsume the effects normally attributed to CS–US contingency on acquisition and maintenance of certain forms of classical conditioning. I have argued that equivalent interference from signaled and unsignaled USs, as well as recovery in extinction from that interference, are not in accord with the Rescorla–Wagner theory. I have not, however, described an alternative theory for a conditioning process that might underlie the empirical generalization about waiting times. I wish now to comment on the alternative theory proposed by Gibbon and Balsam, scalar expectancy theory.

According to scalar expectancy theory, a conditioning regime establishes two noninteracting expectancies for when reinforcement is due. One of these is governed by the overall average time between feedings within the experimental environment. The other is governed by the average duration of the CS per reinforcer. The strength of responding to the CS is governed by a ratio comparison between these independently acquired expectations. The theory could be characterized as a contiguity-plus-comparison theory to contrast it with the Rescorla–Wagner, contiguity-plus-competition theory. The term contiguity is appropriate for scalar expectancy theory because the expectancy in the CS is acquired without reference to any differential between the events within the CS presentation as compared with events that occur before or after the CS. It is governed simply by the time to reinforcement within the CS. Similarly, no reference is made to the possible role of a difference in feeding rates in the experimental apparatus as compared with rates outside the apparatus in establishing the overall food expectation in the apparatus.

There is evidence, however, that a strict contiguity theory is untenable; a differential of some kind between the conditions of reinforcement in CS and nonCS periods appears to be necessary for acquisition. I have noted previously that autoshaped responding acquired under the usual procedure of differential waiting times is eliminated when added feedings equate CS waiting time with overall waiting time. Responding recovers strongly, however, when all feedings are removed in subsequent extinction sessions. In contrast, animals exposed to nondifferential waiting times from the outset do not begin responding when later placed on extinction.[12] If nondifferential reinforcement in CS and nonCS had established a CS-specific food expectation, the removal of food would be expected to result in responding, just as it does when nondifferential reinforcement follows an initial exposure to differential reinforcement. It appears that contiguity is not sufficient to develop expectations specific to a CS.

A closely related problem for scalar expectancy theory is posed by a recent experiment on the elimination of autoshaped responding (see Jenkins and Lambos,[11] experiment 2). It should be possible, according to the theory, to eliminate autoshaped responding by repeated feedings in the presence of the background only; that is, with the CS entirely removed. Sufficient exposure to background-only feedings at an interfeeding interval equal to the average time per feeding in the CS during the previous differential training phase of the experiment should equate the overall background food expectancy to the CS-specific food expectancy. When the CS is reintroduced, while background-only feedings continue, it should now fail to evoke responding.

We carried out an experiment on this plan, but it did not produce the expected result. Despite as many as 21 sessions of background-only feedings, the initial level of responding to the CS when it was reintroduced in the context of these background feedings was no less than it was for subjects that received no sessions of background-only feedings prior to receiving the CS. Once established as a signal, direct exposure to the absence of a differential between alternating CS and nonCS periods appears to be necessary in order to eliminate the ability of the CS to evoke responding. The importance of the presence, and absence, of a differential in reinforcing conditions from CS to nonCS is underestimated by the contiguity assumption in scalar expectancy theory.

The same appears to be true of the Rescorla–Wagner theory. We have seen that when a differential in reinforcing conditions was removed (Jenkins and Lambos,[11] experiment 1), responding declined at the same rate whether or not the conditions were, according to the Rescorla–Wagner theory, such as to prevent or to allow competition between background and CS. The presence or absence of a differential appears to be more fundamental to the maintenance and elimination of responding than is recognized by the contiguity assumption in the Rescorla–Wagner theory.

If the contiguity principle is inadequate, what theoretical alternatives are available? The results we have reviewed do not favor a theory based on CS–US contingency. Nevertheless, the operation of instituting, or removing, a correlation between CS and US has major effects and it is necessary to ask what, if not the contingency itself, might mediate those effects. I would like to suggest the possibility that the CS–US contingency results in a differential, or contrast, between the conditions of reinforcement during the CS and the surrounding nonCS periods. The presence of this more or less local contrast could be the fundamental condition for CS–US association, and its absence could be the fundamental condition for elimination of conditioned responding. The dimensions of the differential are yet to be identified, but the results we have reviewed suggest that a promising candidate is reinforcer waiting times in the CS as compared with the context. Attention to the events in the CS in relation to neighboring nonCS periods might eventually lead to an alternative to CS–US contiguity, and to CS–US contingency, neither of which seem capable of providing a satisfactory account of the associative process in classical conditioning.

REFERENCES

1. Rescorla, R. A. 1967. Pavlovian conditioning and its proper control procedures. Psychol. Rev. **74:** 71–80.
2. Rescorla, R. A. 1968. Probability of shock in the presence and absence of CS in fear conditioning. J. Comp. Physiol. Psychol. **66:** 1–5.
3. Prokasy, W. F. 1965. Classical eyelid conditioning: Experimenter operations, task demands, and response shaping. *In* Classical Conditioning. W. F. Prokasy, Ed. Appleton–Century–Crofts. New York, NY.
4. Rescorla, R. A. & A. R. Wagner. 1972. A theory of Pavlovian conditioning: Variations in the effectiveness of reinforcement and nonreinforcement. *In* Classical Conditioning II: Current Research and Theory. A. H. Black & W. F. Prokasy, Eds. Appleton-Century-Crofts. New York, NY.
5. Gormezano, I. & E. J. Kehoe. 1981. Classical conditioning and the law of contiguity. *In* Predictability, Correlation, and Contiguity. P. Harzem & M. D. Zeiler, Eds. John Wiley. New York, NY.
6. Gibbon, J. & P. D. Balsam. 1981. Spreading association in time. *In* Autoshaping and Conditioning Theory. C. M. Locurto, H. S. Terrace & J. Gibbon, Eds. Academic Press. New York, NY.
7. Stein, L., M. Sidman & J. V. Brady. 1958. Some effects of two temporal variables on conditioned suppression. J. Exp. Anal. Behav. **1:** 153–162.
8. Jenkins, H. M., R. A. Barnes & F. J. Barrera. 1981. Why autoshaping depends on trial spacing. *In* Autoshaping and Conditioning Theory. C. M. Locurto, H. S. Terrace & J. Gibbon, Eds. Academic Press. New York, NY.
9. Gibbon, J. 1977. Scalar expectancy theory and Weber's Law in animal timing. Psychol. Rev. **84:** 279–325.
10. Jenkins, H. M. & D. Shattuck. 1981. Contingency in fear conditioning: A reexamination. Bull. Psychon. Soc. **17:** 159–162.
11. Jenkins, H. M. & W. A. Lambos. 1983. Tests of two explanations of response elimination by noncontingent reinforcement. Anim. Learn. Behav. **11:** 302–308.
12. Lindblom, L. L. & H. M. Jenkins. 1981. Responses eliminated by noncontingent or negatively contingent reinforcement recover in extinction. J. Exp. Psychol. Anim. Behav. Processes. **1:** 175–190.
13. Durlach, P. J. Pavlovian learning and performance when CS and US are uncorrelated. *In* Quantitative Analyses of Behavior: Volume III, Acquisition. M. Commons, R. Herrnstein & A. Wagner, Eds. Ballinger. Cambridge, MA. In press.
14. Balsam, P. D. Bringing the background to the foreground: The role of contextual cues in autoshaping. *In* Quantitative Analyses of Behavior: Volume III, Acquisition. M. Commons, R. Herrnstein & A. Wagner, Eds. Ballinger. Cambridge, MA. In press.
15. Gibbon, J., L. Farrell, C. M. Locurto, H. J. Duncan & H. S. Terrace. 1980. Partial reinforcement in autoshaping with pigeons. Anim. Learn. Behav. **8:** 45–59.
16. Gibbon, J. 1981. The contingency problem in autoshaping. *In* Autoshaping and Conditioning Theory. C. M. Locurto, H. S. Terrace & J. Gibbon, Eds. Academic Press. New York, NY.

Timing Competence and Timing Performance: A Cross-Species Approach

MARC RICHELLE[a] AND HELGA LEJEUNE[b]

Laboratory of Experimental Psychology
University of Liège
Liège, Belgium

INTRODUCTION

How animal behavior adjusts to or is shaped by time has been the central concern in at least two different scientific fields, fields that unfortunately have been kept separate until recently.[1-3] One is chronobiology, where the emphasis has been upon biological rhythms, as evidenced, at the behavioral level, by cyclic properties (circadian, circatidal, and the like) of spontaneous motor activity. The other field is experimental psychology, where the study of temporal regulations of behavior, initiated in Pavlov's laboratory, has received extensive treatment since operant techniques were developed. A wide variety of situations (schedules of reinforcement) are now available to investigate an animal's capacity to regulate its own behavior in time or to estimate the duration of external events. Refined analyses today combine the possibilities of conditioning procedures with sophisticated psychophysical models.[4-14] This approach differs from the chronobiological tradition in two respects: First, it is mainly concerned with adjustment to arbitrarily chosen short durations (of the order of seconds or minutes), while chronobiology is concerned mainly, although not exclusively, with longer natural cycles. Second, chronobiology has been, from its earliest time, comparative and evolutionary, exploring a wide variety of species, plants and animals; while behavioral research on temporal regulation has been characterized by an almost total neglect of cross-species comparisons, which reflects the persistent belief that schedule-controlled behavior is essentially alike across animal species, including man.[15,16] Data are obviously lacking to support such a general claim, and for what is known, it is clear that cross-species differences exist and should not be explained away.

Despite an increased interest in biological constraints on learning, very few systematic studies have been devoted to cross-species comparison of timing behavior and time estimation. Reviewing the literature 4 years ago, we could list only a dozen mammal species (besides the widely used rats and monkeys), half a dozen birds species (besides the much-favored pigeon), and half of a half-dozen of fish species; finally, bees were the only insect studied. Part of these studies are anecdotal, part bear only on one type of situation (fixed interval [FI] or differential reinforcement of low rates [DRL] in most cases), and part were carried out in our laboratory. Four more years have not added much to the record, which is in contrast with the expansion of research in more traditional laboratory species (rats and pigeons).

Other species investigated from 1979 include wood mice (*Apodemus sylvaticus* and *flavicolis*),[17] turtle doves (*Streptopelia rizoria*),[18] fresh-water turtles (*Pseudemys*

[a]Address for correspondence: Université de Liège, Psychologie expérimentale, 5 boulevard du Rectorat (B32), Sart-Tilman, 4000 Liège, par Liège 1, Belgium.

[b]Senior Research Fellow of the Fonds National de la Recherche Scientifique.

scripta elegans),[19] fish (*Tilapia nilotica* and *aurea*),[20] snakes (*Lampropeltis getulus floridana*),[21] and fish crows (*Corvus ossifraguz*).[22]

The problems to be discussed in this paper will be illustrated with data from some of these studies. They are by no means sufficient to provide a clear picture of timing capacities throughout the animal kingdom or to make a choice among the various hypotheses that come to mind to account for cross-species differences. These admittedly speculative hypotheses have been formulated at length elsewhere.[2] They can be summarized as follows:

a. The evolutionary hypothesis: Temporal regulations of behavior become more refined and more efficient as we climb up the phyletic scale; briefly stated, they would parallel increased structural and functional complexity of the nervous system or increased potential for learning.

b. The "egalitarian" or reductionist hypothesis: The capacity for temporal regulations is equally distributed among all species—it would be as primitive and universal as biological rhythms are. In this view, observed differences would be nonessential, reflecting only inappropriate selection of responses, reinforcers, or situations by the experimenter.

c. The ethological hypothesis: Different species would exhibit different capacities for temporal regulations as a function of the particular repertoire evolved under selective pressure, similar demands being found at very different levels of the evolutionary scale. To make the point clear with a somewhat caricature-like example, predators might be thought of as having developed watching behavior—an advantage in temporal conditioning—while prey (at least those reacting to threat by running away) would not have developed such capacity.

Experimental strategies to be applied if one wishes to progress in the understanding of these problems are obvious, as are the difficulties involved. A first step, but it will be a long step, is to multiply the number of species studied, and to explore new species in different phylae of the zoological tree, which means bringing under experimental control wild species not used to living in the laboratory or in man's company. We cannot hope to study all created species, but we cannot dispense with the investigation of a reasonably diversified sample.

A second strategy consists in comparing species closely related on the phyletic scale. Will they exhibit similar performances or show differences, possibly more striking than those observed between more distant species?

A third strategy addresses a classical problem in cross-species study of behavior: Are we sure that the species compared are tested in equivalent situations? Has lever-pressing the same status for a rat, a cat, a monkey or a pigeon? What if we cannot bring a turtle to press with its paw? Will we use head-pushing as an equivalent unit of behavior? And still more puzzling for our concern: How can we know whether the delays and durations used are equivalent? Exploring various responses, reinforcers, or various ranges of delays in the same species is, of course, the only, if fastidious, way to answer such questions.

It is clear that data obtained using each of these three categories are to be interpreted in the light of data obtained with the others. The next sections provide illustrations of research with each strategy, kept apart for the sake of clarity.

I. COMPARING SPECIES AT VARIOUS LEVELS OF THE PHYLETIC SCALE

The experimental condition that has been most extensively used in exploring temporal regulations in various species is the fixed-interval (FI) schedule of reinforcement.

Under these contingencies, a reinforcement is delivered as a consequence of a subject's response, only after a fixed interval of time has elapsed since the last reinforced response. The periodicity of reinforcement entrains a regular alternation of postreinforcement pauses and operant activity, which is not the condition for reinforcement, but eventually develops spontaneously.

A dozen species have been submitted to FI schedules. Besides traditional laboratory or domestic animals such as rats, mice, cats, and pigeons, species less familiar to experimenters were tested, including turtledoves, wood mice, hapalemurs, tilapias and fresh-water turtles (more properly named terrapins[23]). Conditioning new species raises in each case the problem of selecting an appropriate response (see section III later) and reinforcer. It also implies that the animal can adjust to the experimental situation. In some cases, for example, with hapalemurs and wood mice, subjects were living in seminatural conditions, with the conditioning chamber directly connected to their living quarters, so that they would present themselves spontaneously for the experimental sessions. These precautions do not solve all the problems involved in comparing performances under conditions that the experimenter might consider fairly similar, but that animals might view quite differently.

Behavior under FI schedules is amenable to analysis in terms of rate of responding or in terms of some index of temporal regulation. Rate of responding will be left out here, since it is known not to correlate with timing.[24] Among the various measures of temporal regulation that have been proposed (for a review of these see Richelle and Lejeune[2]), the curvature index (CI) of Fry *et al.*[25] has been used. It is computed from the distribution of responses in successive fractions of the interval, its maximal value being dependent on the number of subdivisions. Since the number of subdivisions was not always the same in different experiments, because of technical reasons, the value of the curvature index has been expressed as a ratio of the maximum value. Performances for those species in which curvature indices were available[c] are shown in FIGURE 1. The interval values ranged from 20 sec to 10 min, but not all intervals were explored in all species.

Those species which rank highest are mammals, although not all mammals appear at the top. Cats, rats and mice (both laboratory mice and wood mice) show values around .70. Incidentally, the suggestion that the status of prey or predator might make a difference does not hold, at least in this particular experimental situation. Hapalemurs do not perform very well compared with other mammals: their curvature index ratio (CIR) is between .40 and .50. It might be that these very wild animals are not at their best in the experimental situation despite our efforts to make it as acceptable to them as possible. This might be the case also for the single representative of another prosimian species, *Perodicticus potto edwarsi,* which shows a very poor curvature index.

Second on our scale, right after mammals (with the reservations just formulated for prosimians), but with some overlapping, are birds—pigeons and turtledoves. Starlings, for which a curvature index is lacking, would plausibly join in the picture. Differences observed between pigeons and turtledoves will be discussed below.

Next come fishes (tilapias) and finally fresh-water turtles, which do not develop anything like the classical scalloping or break-and-run patterns of FI cumulative records.

As a first approximation, these comparative results suggest significant differences that might relate to the evolutionary status. The sample is, however, very limited and

[c]For a few species (golden hamsters, guinea pigs, starlings), tested with minimum technical facilities, only cumulative curves and total number of responses were available. They are not included here.

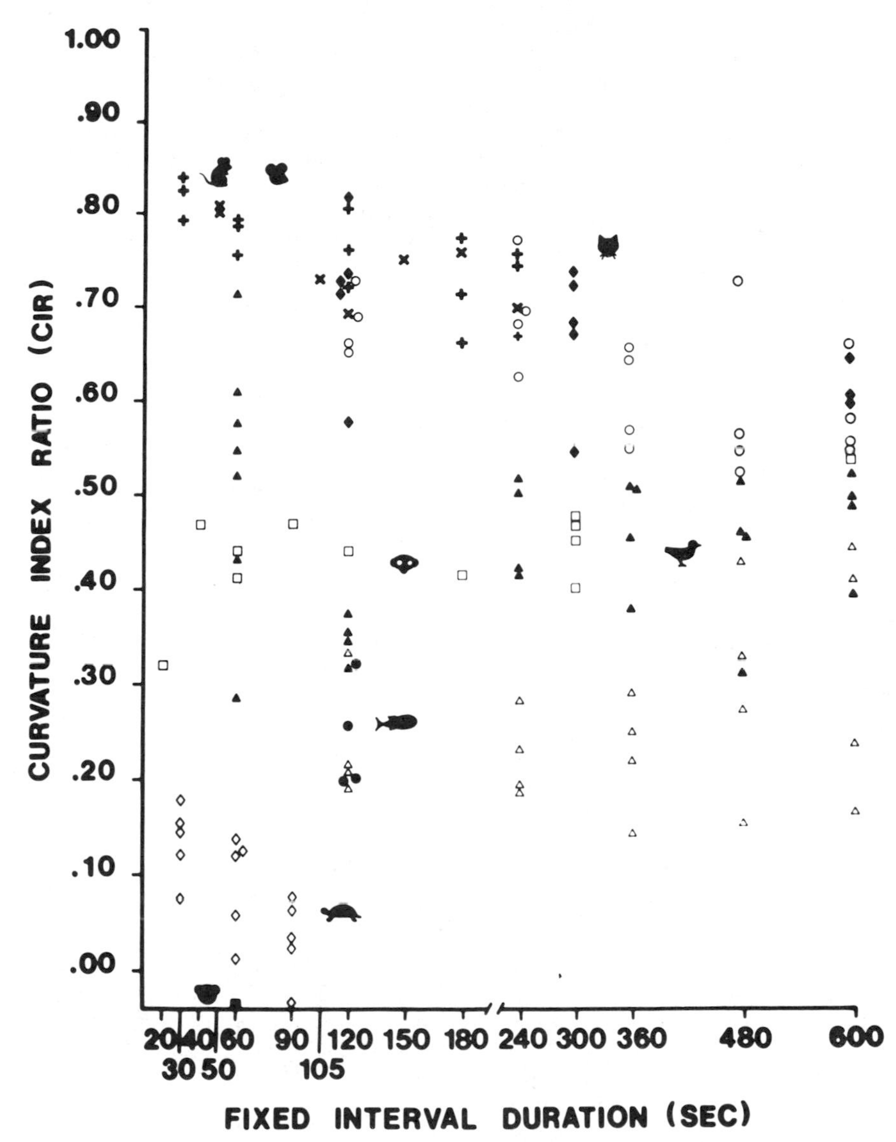

FIGURE 1. Curvature index ratio as a function of interval duration in fixed-interval schedule of reinforcement for individual subjects belonging to ten different species. Points plotted correspond to values averaged from a number of sessions of stabilized performance. Species are identified according to the following code: × = wood mice; + = laboratory mice (NMRI); ♦ = cats; ○ = rats; ▲ = pigeons; □ = hapalemurs; ● = tropical fish (tilapias); △ = turtledoves; ◇ = freshwater turtles; ■ = *Perodicticus potto*.

fragmentary; nothing conclusive could reasonably be stated until many more species are tested. FIGURE 1 might suggest four qualitative levels, rather than a continuous progression, but at this stage such considerations are but speculative. Grouping of species, such as mammals at the upper level, might indeed result from the fact that the measure used leaves out more subtle differences in their respective competence for temporal regulations. The CIR values plotted on the graph were averaged from a number of successive sessions after behavior had "stabilized"—they correspond, it is hoped, to asymptotic behavior with respect to temporal regulation. It can be argued that this is exactly what we want to know. However, species might differ, not in their final adjustment to arbitrary (as opposed to natural) periodicities, but in the speed of

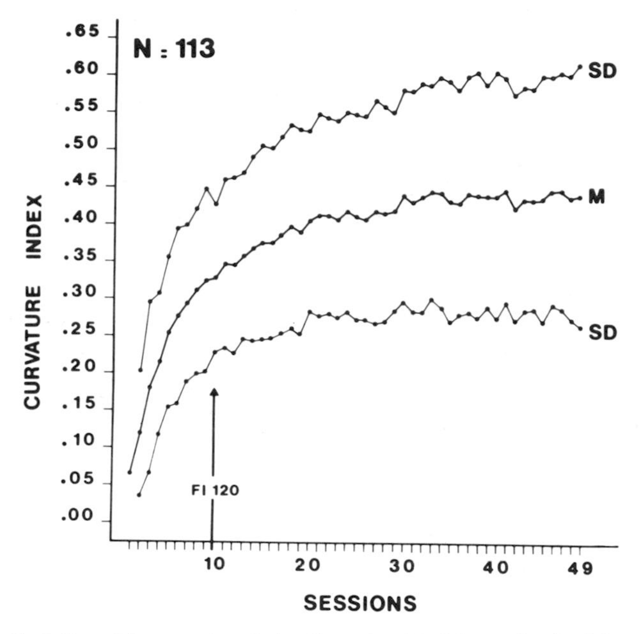

FIGURE 2. Evolution of the curvature index throughout training under fixed-interval schedule of reinforcement in rats. Results plotted are averaged from 113 animals. The interval was brought to 120 sec after 10 sessions. Because it is computed from a partition of the interval into four segments, the curvature index can take a maximal value of .75. The line in the middle represents the means; the lines above and below represent the standard deviations.

acquisition. Cross-species performances should be compared, for instance, in terms of the number of sessions to stabilized behavior. Given within-species interindividual differences, that type of comparison would require larger samples of subjects than are usually available in experiments that extend over several months. An exceptionally large sample of rats has been studied by one of us.[26] FIGURE 2 shows the average "acquisition curve" from 113 subjects run on FI 120 sec. After progressive increase of the interval to its final value over 10 sessions, it took about 20 sessions on 120 sec to reach stabilization (that is, asymptotic behavior).

Interindividual differences deserve further attention. Frequently noted by experimenters in temporal schedules, these differences have not prompted systematic inquiries into the variables responsible for them. While the mean CI remained fairly constant from the 29th session on, the standard deviation increased throughout sessions 20 to 49. Thus, individual differences rather than leveling out became more marked as learning progressed. Rank correlations (Kendall's *tau*) computed between adjacent sessions, and between early sessions and later sessions were in the range of .30 to .50 (statistically significant at $p \leq .001$). This indicates that subjects who ranked high in CI early in training tended to remain good, and vice versa. These data encourage a behavioral genetics approach, with classical problems of chronobiology in mind, that is, the genetic aspects of biological (and for that matter, behavioral) clock(s). However, before starting a systematic selection of good and bad "temporal regulators," the behavioral scientist first must make decisions as to the criteria to be used in such categorizing and as to the relevant schedule. Things would be easy if subjects ranked similarly in different experimental conditions, so that, for instance, FI performances correlated with DRL performances, or with discrimination of response duration, and so forth. Preliminary results comparing FI and DRL in the same subjects showed that this is not the case, so that one has to make an arbitrary choice. Moreover, order effects are important. For example, FI performances are much influenced by previous exposure to DRL schedules.[27]

The schedule of differential reinforcement of low rates (DRL) is, after FI, the most widely investigated experimental situation. Temporal regulation is, in DRL, the condition for reinforcement. An often-noted paradox, which remains to be elucidated, is that animals who pause spontaneously after FI reinforcement for periods of one or several minutes seem unable to refrain from responding for delays beyond a few seconds in DRL. This has been repeatedly described for pigeons, using the traditional key-peck response. While pigeons would pause for more than 1 minute in FI 120 sec (and longer in FI with longer intervals), they do not adjust efficiently to DRL beyond 12 or 15 sec, showing a IRT median value farther and farther away from the critical reinforced value.[28,29] The same relation holds, if sometimes less dramatically, for other species submitted until now to DRL schedules.

FIGURE 3 shows individual results in DRL (5 to 60 sec) for different species studied in our laboratory. They are fewer in number than for the FI schedule, because it is both logical and technically easier to start with FI.

Incomplete as it is, the picture is no less complicated than for FI. Mammals rank highest, with cats and rats first, together with wild mice. Hapalemurs are slightly below, but are clearly higher than birds. Laboratory mice perform poorly. Subjects from three different strains (NMRI, C57BL/6J and Balb/c) have median IRTs ranging from 5.6 to 18.8 sec for critical delays of 20 sec. This is in contrast with their wild brothers, whose median IRTs match the critical value up to 20 sec (and up to 30 sec for one individual).

Pigeons and turtledoves perform poorly, with pigeons slightly better, as in FI schedule.

In interpreting these results, one should take into account the fact that the longest delay explored was dictated by the limits of performance observed in a given species. Exploring values beyond 10 to 15 sec in pigeons, turtledoves or laboratory mice was revealed to be useless.

Median IRT is only one of several possible ways of characterizing timing performance under DRL contingencies. It recommends itself as a practical, simple and plausibly valid index, useful at this still crude stage of cross-species comparisons. It might very well be that more refined statistics will be needed when more subtle differences between species will have to be examined. Dispersion of IRT distribution, for instance, might reveal a more discriminative tool.

The same remark just formulated about the use of stabilized performance under FI

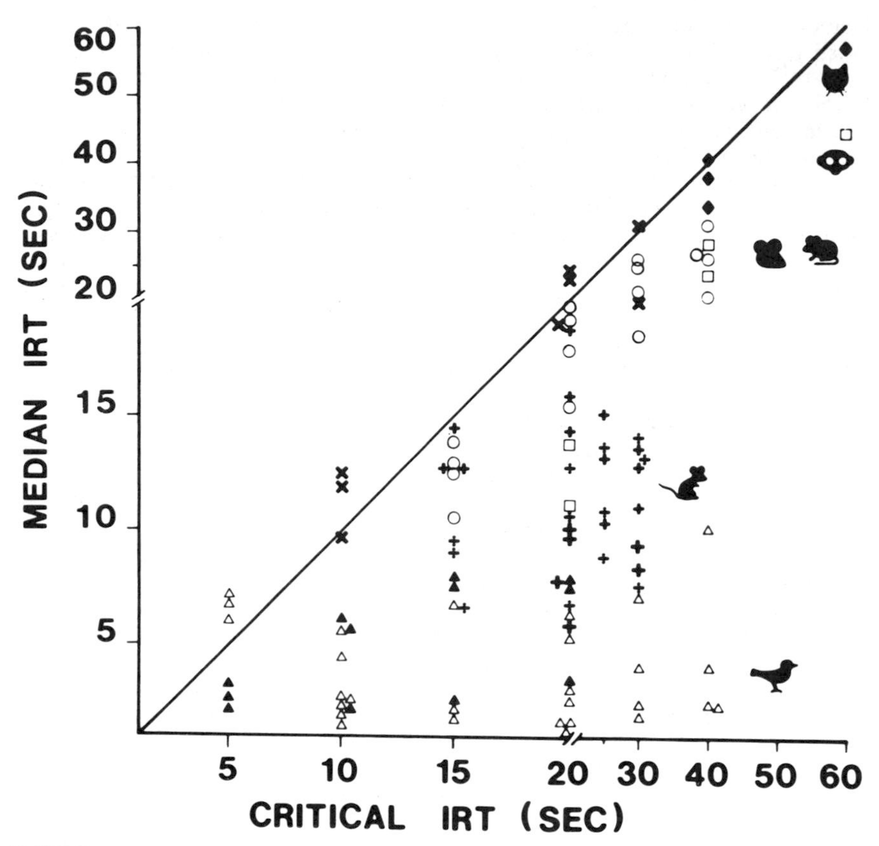

FIGURE 3. Median interresponse time (IRT) as a function of the critical delay in a schedule of differential reinforcement of low rates (DRL) in individual subjects belonging to seven different species. The diagonal line corresponds to a perfect match of median IRT with the critical delay. Results plotted correspond to values averaged from a number of sessions of stabilized behavior. Species are identified according to the following code: × = wood mice; + = Balb/c mice; + = NMRI mice; ÷ = C57BL/6J laboratory mice; ♦ = cats; ○ = rats; ▲ = pigeons; △ = turtledoves; □ = hapalemurs.

schedules holds true for DRL: Species might differ in their speed of acquisition—length of exposure to contingencies needed to reach asymptotic performance—rather than in their stabilized temporal regulation.

Comparing the results summarized in FIGURES 1 and 3 suggests a few additional comments.

The position of hapalemurs in contrast with birds (pigeons and turtledoves) is very different in FI and DRL: Our prosimians generally performed less well than birds in FI, while they seem much better integrated among the mammal group under DRL. However, one should not speculate about these differences until one considers the issues discussed in section III.

More interesting is the evolution of temporal regulation of behavior as a function of interval or delay under FI and under DRL. In the first case, the CIR sometimes increases (improved performance), sometimes decreases (impaired performance), or remains unchanged, depending upon the species and/or the individual. Of course, the observed trends should not be extrapolated to longer intervals. However, it is worth noting that performance (in terms of CIR) sometimes improves as the interval increases within the limited range of intervals explored.

Nothing similar is observed under DRL. At best, performance comes close to the diagonal line of the median IRT/critical delay plot, up to a certain delay, and then it drifts away more or less abruptly. This might reflect an important difference between what we have called *spontaneous* temporal regulation of behavior versus *required* temporal regulation.[2]

II. COMPARING CLOSELY RELATED SPECIES

Before drawing any tentative conclusion from the cross-species comparison, we must take a closer look at the differences between the two pairs of closely related species that have been deliberately introduced in our studies.

A first observation is that closely related species may rank differently in different situations. Laboratory mice and wild mice perform similarly under FI, but the former are distinctly below the latter in DRL. That the difference goes in that direction is altogether intriguing and reassuring—intriguing because one would somewhat anthropomorphically expect that laboratory animals would be better prepared to laboratory situations, and reassuring because this indicates that our efforts to make the experimental situation acceptable to wild animals have been fairly successful.

Secondly, although they are drawn from two situations only, these data indicate that it would be unadvisable to characterize timing competence from behavior observed in one single situation. Which of FI or DRL (and of many other conditions one might wish to explore) is the best indicator of such competence is, of course, undecidable.

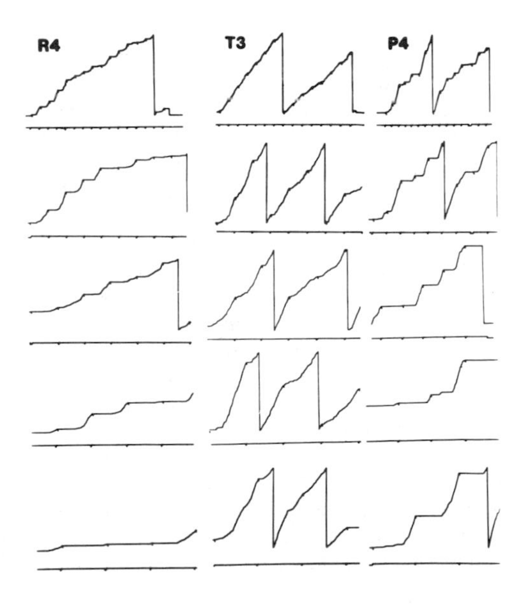

FIGURE 4. Typical cumulative records from one turtledove (T3), one pigeon (P4), and one rat (R4) trained under a fixed-interval schedule, with interval values (*top to bottom*) of 2, 4, 6, 8 and 10 minutes. Availability of reinforcement is signaled by deflection of the pen tracing the horizontal line under the cumulative curve. The cumulative pen sets back after 360 responses.

Under FI contingencies, pigeons differ from turtledoves both in rate of responding (which is higher in turtledoves) and in curvature index (higher in pigeons). Details of the comparison between the two species, and between each of them and rats, for FI delays from 2 to 10 min can be found elsewhere.[30] Subjects in this experiment experienced long exposures (30 to 40 sessions) to each value of the interval. Turtledoves did not exhibit, except occasionally, the typical postreinforcement pause, which has usually been taken as a universal pattern, as can be seen from the cumulative records in FIGURE 4. It had been observed previously that this pattern was not as marked in pigeons as it is in rats.[31] Its almost total absence in turtledoves' behavior is all the more striking since they are zoologically close to pigeons and since a response of identical topography was used.

These comparisons between closely related species throw some doubt on the plausibility of the eco-ethologic hypothesis: How can we account for these differences in terms of differences in selective pressure? We throw out this challenge to our ethologist colleagues.

This comparison also throws some doubt on the evolutionary hypothesis, at least if we want to give it a refined form and are not satisfied with a crude three or four-level scale. Differences between cats and pigeons should not be given too much weight if differences almost as large are observed between pigeons and turtledoves.

III. COMPARING RESPONSES WITHIN SPECIES

Before speculating on the position of a species on an ideal timing competence scale, one should ask another critical question: Are the experimental conditions adequate to reveal the real competence of our subjects? Just as psychologists studying intelligence years ago asked whether IQ tests are culture-free or culture-fair tests, so too have behavioral scientists been questioning for some 10 or 15 years the validity of the selected responses or stimuli. (Credit should be given, however, to a few early pioneers; see Bitterman.[32]) The early belief in the arbitrariness of operant responses (with its corollary: the assumed cross-species equivalence of similar or even nonsimilar motor units) has been shaken under the influence of ethology after a number of experimental findings drew the attention of experimeters to what is now familiar under the label *biological constraints on learning,* or *species-specific status of response or stimuli.*[33–37] Pigeons' key-pecks have been submitted to special scrutiny in this respect. The status of that particular response in the natural food-searching and eating repertoire of the species and the fact that it is amenable to Pavlovian conditioning (as seen in the famous autoshaping paradigm[38,39]) have led us to wonder whether the legendary mediocrity of pigeons under DRL might not be response-bound. It is well known that IRT distributions are biased by the presence of response bursts that load the shortest IRT class, and that these particular responses, usually of extremely brief duration themselves, are not amenable to contingency control.[40,41] These responses are observed in FI schedules as well and their presence in the first part of the interval accounts for part of interindividual variability in CI. Data from pigeons in FI 60 sec plotted in FIGURE 5 have been treated both ways: including and excluding responses occurring in the first tenth part of the interval. It is clear that the second procedure considerably reduces variability.

Several investigators have resorted to a response less involved in the alimentary behavior of pigeons, that is, treadle-pressing.[42] Using this response, pigeons perform better in DRL than they do with key-pecking. Lejeune and Jasselette[43] have compared treadle-pressing and key-pecking in the same individual subjects, both in FI and in fixed-time (FT) schedules—an equivalent of Pavlov's periodic conditioning, where the grain reinforcement is presented at regular intervals independently of the subject's behavior. Subjects were trained to press a treadle under FI 60 sec for 90 sessions, then under FT 60 sec for 35 sessions, and were exposed to FI 60 again for 20 sessions. Then they were submitted to a similar program with key-pecking. Their data confirm previous results[44] showing that response rate is higher for key-pecking than for treadle-pressing. There is, however, no significant difference between CI values for the two responses, whether or not responses in the first segment are taken into account for computation. Suppression of response-reinforcement contingencies (FT schedule) reduces overall response output, but does not significantly affect temporal regulation, until responding eventually is extinguished. This is true, in this experiment, for key-pecking as well as for treadle-pressing contrary to other authors' findings[31]

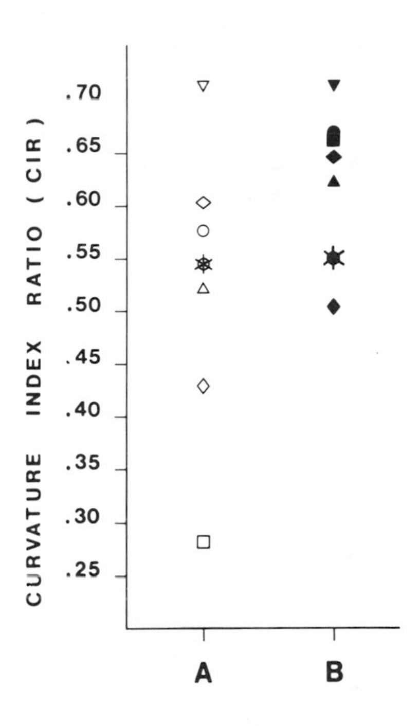

FIGURE 5. Values of the curvature index ratio obtained in pigeons when responses emitted in the first tenth of the interval are included or left out for computation. Individual results computed from the last ten sessions of a 90-session run under fixed interval 60 sec. Left (**A**, *open symbols*): first segment responses included; right (**B**, *solid symbols*): first segment responses left out.

showing that key-peck responses are maintained at high levels under FT 60. When FI contingencies are reinstated, the key-pecking response rate quickly returns to previous FI level while the treadle-pressing rate remains much lower.

The idea of training pigeons to press a treadle with their foot is, obviously, an unimaginative transfer on the part of experimenters used to train rats to lever press. In most persons' eyes—not only in ethologists'—this might appear as a strange and difficult movement for those birds. Looking for a more natural operant, plausibly more appropriate to reveal timing competence in birds, we designed an experimental situation where perching was the response. Pigeons, and later turtledoves, were trained to jump on a perch and sit on it for a minimal period of time (defined as the *critical response duration*) in order to have access to grain. This schedule is more appropriately labeled *differential reinforcement of response duration* (DRRD) rather than DRL. Keep in mind that pigeons do not succeed in spacing their key-pecks to match critical delays longer than 12 or 15 sec or so in DRL. Subjects were trained both in DRRD (perching) and in DRL (key-pecking). Critical delays in DRRD were 10, 20, 30, 40, 50 sec; they were 5, 10, 15 and 20 sec in DRL (exploring longer delays was meaningless, since performance was already very poor at these values). Details of the procedures and results for pigeons can be found in Lejeune and Richelle.[45] Results obtained with turtledoves are still unpublished. Main findings for both species are plotted in FIGURE 6, showing median response durations and median IRTs for idividual subjects as a function of *critical response duration* and of critical DRL delay, respectively. Both species are able to estimate duration of their own perching behavior much better than they are able to control IRTs in key-pecking DRL. Median response duration for pigeons is very close to the critical value up to 40 or 50 sec. Turtledoves compare with pigeons up to 20 sec; beyond that value they fall a bit lower, except for one individual subject that ranks first at 50 sec. Relative frequency distributions of response duration clearly reflect the quality of temporal regulation as contrasted with IRT distribution in DRL. FIGURE 7 shows a typical individual example for both responses in the two species.

In both species, the symmetric distributions of perching durations, with a mode close to the critical value, contrast with asymmetry of IRT distributions, whether or not the first time bin is taken into account.

One might argue that the two situations differ in other respects than the response used. For instance, in the perching situation, the subject initiates a trial by jumping on the perch, while in classical DRL (that is, in our key-pecking situation) a key-peck altogether terminates the "trial" (IRT) and initiates the next one, making the time estimation task a continuous one. Consuming the reinforcer is always part of the

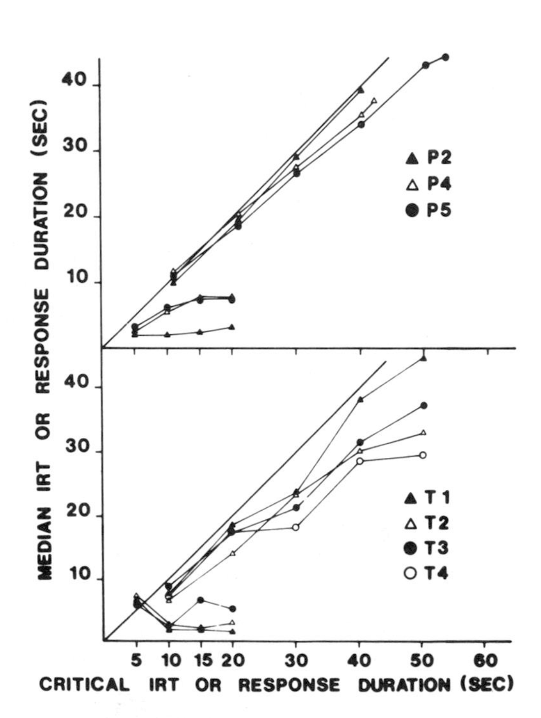

FIGURE 6. Comparison in pigeons and in turtledoves between performance under DRL schedule/key-pecking response and performance under a schedule of differential reinforcement of response duration (DRRD)/perching response. *Ordinate:* median IRT (for DRL/pecking) or response duration. *Abscissa:* critical delay or response duration. *Top:* pigeons (P); *bottom:* turtledoves (T). Individual data were averaged from a number of sessions of stabilized behavior.

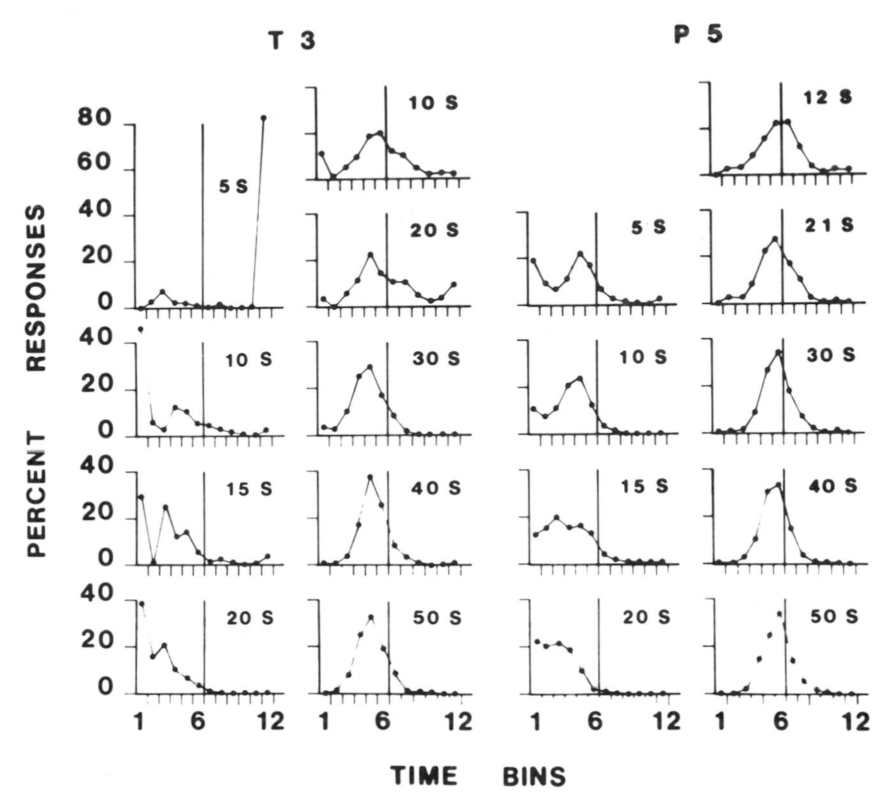

FIGURE 7. Relative frequency distribution of IRTs and of response durations obtained under DRL/key-pecking and DRRD/perching schedules, respectively, for increasing values of the critical delay or duration (*top to bottom*). Data presented are for two typical individuals: one pigeon (*right columns*) and one turtledove (*left columns*). Each time bin on the abscissa is equal to one-sixth of the critical delay or duration. IRTs or response durations falling in the seventh bin or above (that is, right of the vertical fine line) were reinforced. Note that the critical perching duration has been extended up to 50 sec with excellent performance, while with key-pecking, critical delays beyond 20 sec were not judged worth exploring.

postreinforcement IRTs in DRL, while it is not included in response duration in perching DRRD. These differences might be easily eliminated by slight modifications in procedures (see Lejeune and Richelle[45] for suggestions). These differences, however, are not crucial for our present concern. Whatever the variables accounting for contrasting performances in the two situations, it is clear that using another response and a slighty different schedule reveals in birds a timing competence that went unsuspected when the most popular operant behavior was used.

DISCUSSION

Experimental data presented in this paper illustrate the extreme intricacies of any attempt to build a coherent cross-species characterization of temporal regulation of behavior. They indicate how difficult it is to appraise the very competence of various animal species in adjusting their own behavior to arbitrary durations as defined by experimenters. It might very well be that this competence is indeed inaccessible, or even uninferable, or that no such thing exists at all. Performances would be the only things to look at, and all that could be compared, at best, would be the best performances obtained in each species studied.

This implies that experimenters not only explore various species, but look for "species-fair" tests that reveal the best of each species' timing capacities. In the examples above, the nature of the response has been emphasized. The response has been, no doubt, the privileged variable in the recent awareness-raising movement concerning species-specific constraints among conditioning psychologists. Stimuli and reinforcers deserve similar attention.

The use of alimentary reinforcers (food or water) raises a number of problems when the experimenter deals with species whose eating and drinking habits are simply not known or are not as familiar to him or her as those of traditional laboratory animals. An ethological description of alimentary behaviors and a physiological study of metabolic processes typical of a species are prerequisites before any interpretation of the performances observed can be seriously proposed, especially when performances show limitations. In view of the poor performance obtained from turtles in our laboratory, and of the failure to obtain in snakes anything like the classical FI "scallops," as reported earlier by Kleinginna and Currie,[20] it is tempting to conclude that reptiles are poor "timers." Besides the fact that two species can hardly be taken as representative of a whole zoological class, one should ask whether daily sessions with intermittent food or water reinforcers, using arbitrarily selected delays (usually after the experimenter's habit with rodents or birds) are really adequate conditions to test timing capacities in these organisms.

In the particular case of temporal regulations of behavior, which are our concern here, it would seem essential to pay special attention to species-specific constraints on the delay variable. Such constraints might be thought of as bound to the natural temporal spacing of feeding behavior, as just mentioned above, but also to the rhythm of motor activity, the sleep cycles, sexual cycles, and the like. Rather than using arbitrary delays, intervals, or stimulus durations, experimenters should take into account species-specific biological cycles and match their temporal parameters to natural periodicities as experienced or exhibited by the organism under study. There are some, unfortunately few, indications that species such as bees can be trained to forage for food at specific intervals of time approximating their circadian cycle, which is in contrast to the absence of conditioned periodicity when they are exposed to FI contingencies up to 90 sec.[46–48] The interactions and interdependencies between "natural" biological rhythms, as studied in the field of chronobiology, and temporal regulations of behavior, as analyzed in the laboratory by psychologists, offer a fascinating area of research, still almost totally unexplored. Such an integrated approach is the condition for testing some of the hypotheses that come to mind concerning the timing competence of various animal species. One hypothesis we would favor for future research might be phrased as an adapted version of Kurt Richter's view: Through biological evolution, species would exhibit increased capacities to free themselves—when useful for survival in some way—from basic biological periodicities and to adjust to "arbitrary" delays or durations, that is, delays without relation to biological rhythms.

REFERENCES

1. RICHELLE, M. 1968. Notions modernes de rythmes biologiques et régulations temporelles acquises. *In* Cycles Biologiques et Psychiatrie J. de Ajuriaguerra, Ed.: 233–255. Georg et Cie, Genève; Masson, Paris.

2. RICHELLE, M. & H. LEJEUNE. 1980. Time in Animal Behaviour. Pergamon Press. Oxford.
3. ELSMORE, T. F. & S. R. HURSH. 1982. Circadian rhythms in operant behavior of animals under laboratory conditions. *In* Rhythmic Aspects of Behavior, F. M. Brown & R. C. Graeber, Eds.: 273–310. Lawrence Erlbaum Associates.
4. GIBBON, J. 1977. Scalar expectancy theory and Weber's law in animal timing. Psychol. Rev. **84:** 279–325.
5. GIBBON, J. 1979. Timing the stimulus and the response in aversive control. *In* Advances in Analysis of Behaviour (Vol. 1): Reinforcement and the Organization of Behaviour. M. D. Zeiler & P. Harzem, Eds. Wiley. Chichester and New York.
6. GIBBON, J. & R. M. CHURCH. 1981. Time left: Linear versus logarithmic subjective time. J. Exp. Psychol. Anim. Behav. Processes **7:** 87–108.
7. PLATT, J. R. 1979. Temporal differentiation and the psychophysics of time. *In* Advances in Analysis of Behavior (Vol. 1): Reinforcement and the Organization of Behaviour. M. D. Zeiler & P. Harzem, Eds. Wiley. Chichester and New York.
8. ZEILER, M. D., E. R. DAVIS & A. J. DE CASPER. 1980. Psychophysics of key-peck duration in the pigeon. J. Exp. Anal. Behav. **34:** 23–33.
9. ROBERTS, S. 1981. Isolation of an internal clock. J. Exp. Psychol. Anim. Behav. Processes **7:** 242–268.
10. ROBERTS, S. 1982. Cross-modal use of an internal clock. J. Exp. Psychol. Anim. Behav. Processes **8:** 2–22.
11. MECK, W. H. & R. M. CHURCH. 1982. Abstraction of temporal attributes. J. Exp. Psychol. Anim. Behav. Processes **8:** 226–243.
12. STUBBS, D. A. 1980. Temporal discrimination and a free-operant psychophysical procedure. J. Exp. Anal. Behav. **33:** 167–185.
13. CHURCH, R. M. & J. GIBBON. 1982. Temporal generalization. J. Exp. Psychol. Anim. Behav. Processes **8:** 165–186.
14. PLATT, J. R. & E. R. DAVIS. 1983. Bisection of temporal intervals by pigeons. J. Exp. Psychol. Anim. Behav. Processes **9:** 160–170.
15. SKINNER, B. F. 1965. A case history in scientific method. Am. Psychol. **11:** 221–233.
16. LOWE, C. F. 1983. Radical behaviorism and human psychology. *In* Animal Models of Human Behavior. G. C. L. Davey, Ed. J. Wiley. New York, NY.
17. LEJEUNE A. & H. LEJEUNE. 1983. Fixed interval and differential reinforcement of low rate performances in wood mice (*Apodemus sylvaticus and flavicollis*). Unpublished report. Laboratory of Experimental Psychology, University of Liège.
18. LEJEUNE, H. 1983. Differential reinforcement of perching duration in the turtle dove: A comparison with differential-reinforcement-of-low-rate key-pecking. Unpublished report, Laboratory of Experimental Psychology, University of Liège.
19. LAURENT, E. 1983. Régulation temporelle acquise chez la tortue semi-aquatique *Pseudemys scripta elegans (Wied)*. Unpublished Master's thesis, University of Liège.
20. GRAILET, J. M., JEANGILLE, L., MICHAUX, L., QUEVRIN, A., LEJEUNE, A. & H. LEJEUNE. 1981. Etude de la régulation temporelle en FI 120 chez les *Tilapia nilotica et aurea*. Unpublished report, Laboratory of Experimental Psychology, University of Liège.
21. KLEINGINNA, P. R. & J. A. CURRIE. 1979. Effects of intermittent reinforcement in the Florida kingsnake (*Lampropeltis getulus floridana*). J. Biol. Psychol. **21:** 14–16.
22. POWELL, R. W. & W. A. KELLY. 1979. Crows learn not to respond under response-independent reinforcement. Bull. Psychonom. Soc. **13:** 397–400.
23. CAGLE, F. R. & A. H. CHANEY. 1950. Turtle population in Louisiana. Am. Midl. Nat. **43:** 383–388.
24. DUKICH, T. D. & A. E. LEE. 1973. A comparison of measures of responding under fixed-interval schedules. J. Exp. Anal. Behav. **20:** 281–290.
25. FRY, W., R. T. KELLEHER & L. COOK. 1960. A mathematical index of performance on fixed-interval schedule of reinforcement. J. Exp. Anal. Behav. **3:** 193–199.
26. LEJEUNE, H. 1983. Inter-individual differences in FI performance in rats: A large group study (N = 113). Unpublished report, Laboratory of Experimental Psychology, University of Liège.
27. FRANSOLET, R. 1981. Contribution à l'étude des régulations temporelles FI et DRL chez le rat. Unpublished Master's thesis, University of Liège.
28. DEWS, P. B. 1970. The theory of fixed-interval responding. *In* The Theory of Reinforcement Schedules. W. N. Schoenfeld, Ed.: 43–62. Appleton-Century-Crofts. New York, NY.

29. STADDON, J. E. R. 1965. Some properties of spaced responding in pigeons. J. Exp. Anal. Behav. **8:** 19–27.
30. LEJEUNE, H. & M. RICHELLE. 1982. Fixed interval performance in turtle doves: A comparison with pigeons and rats. Behav. Anal. Lett. **2:** 87–95.
31. LOWE, F. & P. HARZEM. 1977. Species differences in temporal regulation of behavior. J. Exp. Anal. Behav. **28:** 189–201.
32. BITTERMAN, M. E. 1975. The comparative analysis of learning. Science **188:** 699–709.
33. BRELAND, K. & M. BRELAND. 1961. The misbehavior of organisms. Am. Psychol. **16:** 661–664.
34. SHETTLEWORTH, S. J. 1972. Constraints on learning. *In* Advances in the Study of Behavior, Vol. 4. D. S. Lehrman, R. A. Hinde & E. Shaw, Eds.:1–68. Academic Press. New York, NY.
35. HINDE, R. A. & J. STEVENSON-HINDE. 1973. Constraints on Learning. Academic Press. London and New York.
36. SELIGMAN, M. E. P. & J. L. HAGER, Eds. 1972. Biological Boundaries of Learning. Appleton-Century-Crofts. New York, NY.
37. SCHWARTZ, B. 1978. Psychology of Learning and Motivation. W. W. Norton. New York, NY.
38. BROWN, P. L. & H. M. JENKINS. 1968. Auto-shaping of the pigeon's keypeck. J. Exp. Anal. Behav. **11:** 1–8.
39. LOCURTO, C. M., H. S. TERRACE & J. GIBBON, Eds. 1981. Autoshaping and Conditioning Theory. Academic Press. New York, NY.
40. BLOUGH, D. S. 1966. Reinforcement of least-frequent interresponse times. J. Exp. Anal. Behav. 1966. **9:** 581–591.
41. MILLENSON, J. R. 1966. Probability of response and probability of reinforcement in a response-defined analogue of an interval schedule. J. Exp. Anal. Behav. **9:** 87–94.
42. HEMMES, N. S. 1975. Pigeon's performances under differential reinforcement of low rate schedule depends upon the operant. Learn. Motiv. **6:** 344–357.
43. LEJEUNE, H. & P. JASSELETTE. 1983. Fixed interval and fixed time treadle pressing in the pigeon: A comparison with FI and FT keypecking in the same subjects. Unpublished report, Laboratory of Experimental Psychology, University of Liège.
44. RICHARDSON, W. K. & N. RAINWATER. 1979. A comparison of the keypeck and treadlepress operants in the pigeon: Fixed interval schedule of reinforcement. Bull. Psychon. Soc. **14:** 333–336.
45. LEJEUNE, H. & M. RICHELLE. 1982. Differential reinforcement of perching duration in the pigeon: A comparison with differential-reinforcement-of-low-rate keypecking. Behav. Anal. Lett. **2:** 49–57.
46. BELING, I. 1929. Z. Vgl. Physiol. **9:** 259–338.
47. GROSSMAN, K. E. 1973. Continuous, fixed-ratio and fixed-interval reinforcement in honey bees. J. Exp. Anal. Behav. **20:** 105–109.
48. PALMER, J. D. 1976. An Introduction to Biological Rhythms. Academic Press. New York and San Francisco.

The Integration of Reinforcements over Time

S. E. G. LEA[a]

Department of Psychology
University of Exeter
Exeter, England

S. M. DOW

Department of Zoology
University of Bristol
Bristol, England

INTEGRATION OVER TIME: A NEGLECTED PROBLEM

The title of this paper could as well have been "The Integration of Prey over Time." We are concerned with a problem that is common to the experimental analysis of animal behavior as it occurs in Skinner boxes, and to the empirical investigation of foraging as it occurs in the animal's natural environment. It is the problem of how a series of discrete events, deliveries of reinforcers or captures of prey, are transformed into an estimate of the density of reinforcements available on a schedule, or the density of prey present in a patch.

All popular accounts both of foraging behavior and of behavior under schedules of reinforcement make use of the results of such an integration. Herrnstein's[1,2] matching law and "quantitative law of effect" predict that behavior under schedules of reinforcement will be a function of the reinforcement rates experienced under the schedules. Charnov's[3] marginal value theorem predicts that natural foraging behavior in a patchy environment will be a function of the rates at which prey can be captured within each patch and within the environment as a whole. The functions concerned are different, but the problem is common to both.

You or I, faced with the problem of estimating the rate at which discrete events occur, would doubtless solve it by counting them, then dividing by the time that had elapsed, or the number of responses we had made. We have no evidence that animals can use this solution. So far as we can tell, the capacities of even the largest-brained birds to discriminate between numbers are quite modest.[4] (Other kinds of "counting" behavior, which appear to be much more accurate,[5,6] turn out to be no different from the kind of noncounting mechanism we propose below.) We have no evidence at all that animals can manipulate numbers in the kind of way required for division. Besides, this solution ignores an additional, fundamental aspect of the problem: the density that is being estimated is liable to change.

In the laboratory we do, of course, change the schedules of reinforcement that our subjects experience. But, except in the case of the transition to extinction, we very rarely study the effects of these transitions. Instead, we maintain the same conditions in force until we are sure we have achieved "steady-state" behavior, and look only at

[a]Address for correspondence: S. E. G. Lea, University of Exeter, Department of Psychology, Washington Singer Laboratories, Exeter EX4 4QG, England.

the asymptotic performance. In so doing, we ignore some of the most interesting problems in the integration of reinforcements over time. Very possibly, too, we constrain our subjects not to show us the full range of choice behavior of which they are capable.

These problems have been much more obvious to the behavioral ecologists. In nature, the prey density an animal faces is always likely to change—most obviously, because of depletion due to the forager's own efforts, or the efforts of the flock to which it belongs.[7] The whole essence of Charnov's marginal value theorem is to predict how a forager should react to the depletion it produces in whatever patch it is currently using.

The existence of depletion, however, is only part of the problem posed by varying prey densities. If the world is liable to change, the forager must always pursue two simultaneous goals, which may not be wholly compatible.[8–10] It must try both to extract as much prey as possible from the environment, per unit time spent foraging, and also to obtain the information on which estimates of prey densities can be based—not just the density of the patch of prey currently being exploited, either, but also the densities of other accessible patches.

THE "COMMON MODEL"

The literature contains several attempts to solve the fundamental problem of integration over time. Surprisingly, all the solutions offered are essentially the same. We shall refer to this unanimous proposal as the "common model" of information integration. Of course, what the various proposed answers have in common is only their core ideas—the details vary greatly. The purpose of our paper is to use some empirical results to restrict somewhat the variants of the "common model" we need to consider.

The essence of the common model is as follows. For each "patch" in the environment, the subject is supposed to maintain a current estimate of the prey density (per unit time or per unit response). Obviously, what is good for each patch in a forager's environment can equally be said for each schedule in an operant conditioning experiment, but from now on we shall stick with the ecological terminology. Ordinarily, this estimate remains constant from moment to moment, but at certain points in time, the subject updates it. It is the updating process that constitutes the common core of the model. It involves multiplying the old estimate by a weighting factor less than unity (let us call it a), and adding to it a quantity that depends on whether or not a prey has just been captured, multiplied by the complement of the weighting allocated to the old estimate, that is, by $1 - a$. Usually this current-capture quantity will simply be either 0 or 1, but it could be the reciprocal of the number of responses required to catch the latest prey, or the time spend doing so. Mathematical learning theorists will recognize that our common model is a variant of the simple "linear operator" learning model of Bush and Mosteller.[11] In order to complete the specification of our model, we have to specify the rule by which the estimated prey density is transformed into behavior.

The nearer the weighting factor, a, is to unity, the more the subject is influenced by the past rather than the present. Another way of putting this is to say that high a values imply a long memory. In fact, the quantity $1/(1 - a)$ is an estimate of how far in the past a previous prey capture has to be before it has effectively no influence on current behavior: it is the temporal width of the "memory window."

With the description just given, we can claim as cases of the "common model" the proposals of Bobisud and Voxman,[12] Myerson and Miezin,[13] Commons *et al.*,[14] and unpublished work by us and by several others with whom we have talked. Of course, while having a common skeleton, these models vary greatly in the flesh they put on it. The following are the major ways in which they can vary:

1. The nature of the parameter being estimated. The major possibilities are prey density per response, and prey density per unit time. In the latter case, we might have a local estimate or a global one—prey found in the patch per time in the patch, or prey found in the patch per total time spent foraging.

2. The event that triggers updating of the density estimate. The major possibilities are the passage of a certain period of time, the occurrence of a response, and the capture of prey. There are also compound possibilities: for example, updating might occur upon capture of the prey—or whenever a certain period of time had passed without a capture.

3. The value of the past *versus* present weighting parameter, a, and whether it is a fixed property of the organism or free to vary. If the weighting is variable, it might vary from day to day, but have the same value for every patch on a given day; or it might have different values for each discriminably different patch.

4. The rule translating the estimated prey density into behavior. For example, is the translation of prey density into behavior absolute or stochastic? That is, does the subject always goes to the patch with the highest current density estimate? The obvious alternative is for time allocation to be a continuous function of estimated density, so that all patches are used, although the one with the highest estimated density is used most. But the subject might simply stay in any patch which yields enough prey to provide his necessary daily intake within the time available for foraging.

5. The value given to the estimated prey density before the beginning of a foraging bout or experimental session. Obvious possibilities for initial estimates of prey density include: zero; the average found in similar environments in the past; and the value needed to ensure necessary daily intake within the time available for foraging.

Our major interest in the present paper is to say something about question 3 in the above list—the value and the variability of a, the past-versus-present weighting factor. Except in the simplest possible cases, the kind of model we produce does not lend itself to analytical treatment. If we want to compare the predictions of any fully specified variant of the common model with observed behavioral data, we have to obtain our theoretical predictions by running computer simulations. This means that, as well as stipulating answers to all five questions given above, we shall need to be able to supply plausible estimates for any free parameters our models may contain. The biggest problem in this kind of theory-testing is retaining generality. For this reason, the focus in the present paper lies, wherever possible, on qualitative trends rather than quantitative fits between data and theory.

EXPERIMENTAL PROCEDURES

Although we have stated the models mainly in ecological language, our data come from fairly conventional operant conditioning apparatus. It has been argued repeatedly that there are close parallels between the preoccupations of reinforcement theory and of foraging ecology.[15–20] Our experiments made use of that analogy. Our subjects (pigeons), our equipment (conventional Skinner boxes controlled by computers), and the outlines of our procedure come from the operant-conditioning tradition. But the detailed conditions that we have examined draw their inspiration from ecological analyses. In particular, as you might expect given the way this paper began, we are concerned with conditions where reinforcement frequencies (or should we say prey densities?) change rapidly and frequently.

Pigeons were tested under the kind of concurrent schedules of reinforcement introduced by Findley.[21] This means that the pigeon was faced with two pecking keys, the right key and the center key. The center key was always lit, from behind, with white light. The right key could be lit with either red, green or yellow light. If the pigeon pecked the right key, it would sometimes (not often) be rewarded by the presentation of a hopper full of preferred grain: the pigeon could then eat for a few seconds. Pecks on the center key were never reinforced with food, but they did cause the color of the side key to change. Different colors on the right key were correlated with different schedules of food reinforcement, that is, different rules determining whether or not a peck would be reinforced. These schedules were all designed to simulate, in their statistical properties, one or another of the kinds of patch that a forager might encounter in its natural environment.

The basic design of the schedules/patches was as follows. At the beginning of an experimental session, each patch consisted of a large number of cells. A few (2–22%, typically no more than 10%) of these cells were designated as containing "prey." Every time the pigeon pecked the right (colored) key, one of the cells was chosen at random. If, and only if, the chosen cell contained a prey item, the pigeon was rewarded. Many variations on this basic patch-sampling procedure are possible. Some of those we used were the following:

1. *Replacement of sampled cells.* Once a cell had been sampled, it could either remain available for future sampling, or it could be made inaccessible. In the latter case, the effective prey density would not change much as a function of the predator's behavior, since the ratio of prey-containing to empty cells remaining available would be roughly constant. The situation would be much like that of a natural forager working systematically through a field of prey that do not move about, remembering where it had already been and being careful not to retrace its steps. But if sampled cells remained accessible, the effective prey density might be reduced over time, depending on the decision taken on the second kind of variation.

2. *Replacement of prey "eaten."* In the case where cells were replaced once sampled, two possibilities arose when the pigeon "found" a "prey-containing cell." The prey might be removed from the cell for future samples, or it might be replaced with a new one. If prey taken were not replaced, the pigeon would face a slowly declining "prey density," a situation analogous to that faced by many natural predators, feeding off any nonrenewing food source where either the predator cannot remember its route so as not to retrace it, or where the prey move around within the patch.

3. *Refreshment of the patch.* After the pigeon had left it, the patch might either stay as it was, or it might revert to its starting condition. In the former case, a particular key color would correspond to a particular geographical patch, which a forager might visit, leave, and then return to, finding the prey just as depeleted as ever on a revisit. In the latter case, a key color would correspond to a generic type of patch, with many instances of that type present in the environment; consequently, when the forager leaves a patch of that type, if it finds another of the same type, it will have the same density as the original did before depletion.

4. *The size of the patch.* If a patch contained more cells than the pigeon was likely to make pecks to it, there was never any question of the patch becoming completely exhausted. Usually, we set up patches of 1000 cells, and arranged for the session to terminate as soon as the pigeon had made 960 pecks, so that patch exhaustion was virtually impossible. But in other conditions, we set up patches of 320 or even 200 cells, so that the pigeon could remove all the prey in a patch. A particularly interesting case is that where the predator faces a roughly constant prey density until the last prey has been taken, at which point density suddenly falls to zero. We call such conditions "sudden death" patches; in operant-conditioning jargon, they correspond to a switch to extinction conditions, a situation that has recently been considered from an optimal foraging point of view.[22]

Typically, in our experiments, one color of the right key has been associated with one type of patch, and the other color either with a different type, or with the same type but a different prey density. The patch types in use, and their correlations with particular right-key colors, have remained constant from day to day: we wanted to give the pigeons a chance to adapt their behavior to the patch types in use. But the prey densities in the patches have varied from day to day, so that the pigeons had to form a new estimate of prey density in every session.

SELECTED RESULTS

In this paper, we want draw out a selection of results which have particular implications for the five variations on the common model which we considered above.

Session Length and the Variability of the Weighting Factor

The first result is described in more detail by Dow.[23] The method used two patches of the nondepleting type (cells once found were made inaccessible). Prey densities varied from 1/10 to 1/35 prey per cell in whichever patch was better for a given day, and from 1/20 to 1/50 in the worse patch. The experimental variable was the length of the session: either 256 or 1024 pecks to colored keys were allowed. Session length remained constant for a block of 15 sessions, while prey densities, as usual, varied every day. The experiment was following up an idea of Krebs *et al.*,[9] who claimed that the optimal sampler, faced with two patches of unknown density, should start by visiting them both equally, and then after a certain time exploit, exclusively, the one that has yielded more prey so far. It is quite clear that, for optimal performance, the longer the session, the longer sampling should continue. It follows that overall reinforcement rates should be lower at the start of a long foraging bout or session than in the corresponding period of a shorter one—a surprising prediction since it involves "delay of gratification," something pigeons are ordinarily very bad at.[24] Nonetheless, the prediction was confirmed; in 256-peck sessions, pigeons obtained one or two more reinforcements than they did during the first 256 pecks of 1024-peck sessions. This difference, although small, was significant.

The common model can only explain this result if the weighting parameter, a, is allowed to vary between long and short sessions. We have to suppose that in short sessions, the past is weighted less heavily than in long ones, so that the subject is more easily "captured" by a current reinforcement. Put another way, the memory window was shorter when there was less time available for foraging. Krebs and Kacelnik[10] reach a similar conclusion using data from great tits.

Choice between Two Depleting, Nonrefreshing Patches

Consider next the very simple case where both patches have the property that prey, once taken, are not replaced, while cells, once found, remain available. The prey densities in both patches will slowly decline throughout the session. The pigeons adapted to this condition fairly well: with initial densities of 1/15 to 1/50 prey per cell in both patches, they made an average of 4.3 changeovers in a 960-peck session, enough to let them spend most of their time in the more advantageous patch.

It is easy to show that there is one combination of model properties that will never give this result. If the subject estimates prey density as reinforcements per unit of session time (rather than per response, or per unit of time in the relevant patch), and if it chooses which patch to visit absolutely rather than stochastically, then the model says that the subject will stay in the patch it is currently visiting, regardless of its depleting density. The reason for this is that, although the subject is not finding any prey in the current patch, so that its estimated density is depleting, it is not finding any prey in the other patch either (obviously!), so the estimate of prey density there is also depleting just as fast.

It seems absurd for the forager to evaluate patch densities per unit of session time. But, as Herrnstein and Loveland[25] have shown, evaluation per response or per unit of time in the patch will not, with stochastic choice between patches, lead to consistent choice of the better of two nondepleting patches. Yet Herrnstein and Loveland showed that such consistent choice did occur (they called their patches concurrent fixed-ratio schedules of reinforcement, but as Krebs[26] has noted, the principle holds). We got the same result in the long/short session experiments with nondepleting patches, mentioned above. Thus, given the common model, the data are consistent with either absolute choice or with computation of prey densities per unit of session time, but not with both.

Choices Involving Refreshing Patches

The problem of two depleting patches becomes more acute when one or both of them "refreshes," that is, returns to its original value if it is left and then re-entered. Under these conditions, the common model seems likely to do exceedingly badly compared with a more complex, "cognitive" model, which could be given insight into the refreshment condition, and so would take as its estimate of the prey density of patches other than the current one, the best estimate so far of initial density.

Faced with a choice between two depleting, refreshing patches, our pigeons showed almost twice as much switching between patches as when either or both of the patches failed to refresh. This is, of course, a "sensible" response—the optimal performance for an omniscient forager being to take one prey item from the patch of higher density, make one response in the lower-density patch, and continue alternating in the same fashion. The common model cannot do as well as that, but nor did the pigeons. If the weighting factor a is set to a rather low value, then more frequent switching will be expected, regardless of the rule for translating estimated prey densities into behavior. Thus the pigeons' success in adapting to the situation where both patches depleted is further evidence of their ability to vary the weighting factor in accordance with the necessities of the situation.

Patches That Run Out of Prey

Finally, consider the situation where one of the patches had a constant prey density, but could completely run out of prey, while the other had a steadily reducing prey density that never reached zero (neither of them refreshed). "Sensible" behavior in this situation involves being more willing to switch away from the "sudden death" patch than the patch that depletes gradually, because the past of a sudden death patch is an unreliable guide to its future. Sure enough, with broadly comparable prey densities overall, we found that our pigeons stayed less time in a "sudden death" patch without finding prey than in a depleting patch, before switching to the other patch (though this only held so long as the sudden death patch was still delivering some prey). In contrast, if an ordinary nondepleting patch was paired with a depleting patch, "giving-up times" were longer in the nondepleting patch. Willingness to abandon the patch that might run out must, therefore, be a function of its "sudden death" quality.

If this result is to be accommodated within the common model, the value of *a*, the factor that weights the past against the present, must be allowed to be different for the different patches. There are two reasons for hesitating over this step. First, to allow *a* to vary between patches introduces extra parameters into the model. Secondly, our way of conceiving of the parameter *a* has always stressed its alternative interpretation as a "memory window." If it is allowed to vary between patches, we may have to drop this view; can an animal look into its past through two different memories at once?

CONCLUDING REMARKS

There are many other questions, relating to the "common model" and extending beyond it, which we could consider with the aid of data from these experiments, but space does not permit that here. There are three general issues, though, on which we must comment before closing.

First, it would be wrong to leave the reader with the impression that we actually favor the common model, or that our pigeons' behavior is normally consistent with simulations based on it. The model is certainly imperfect. Using it, we have never been able to find simulation parameters that will get the number of changeovers between patches even approximately right, no matter what rule we use for translating estimated prey densities into behavior. Absolute preference rules give too few changeovers; stochastic ones, such as Herrnstein's matching law, give too many. We are impressed by the way that absolute preference models tend to get "trapped" in suboptimal patches (a very general tendency, which has been explored by Bovet[27]), and we believe that Herrnstein's approach comes closer to realism. But we suspect that it will have to be supplemented with a parameter to implement reluctance to shift from the current patch. Even so supplemented, though, we suspect that the common model will ultimately fail. Our pigeons consistently managed to respond appropriately to the general conditions of the patches we set up—depletion, refreshment, sudden death, and so forth. They were also able to respond to factors such as the number of different parameter values we had in use. We suspect that ultimately a model that allows the subject to "know" such "structural" properties of its environment will have to replace the common model, which only allows the subject to learn parameters. Krebs and Kacelnik[10] mention further evidence that leads in the same direction.

Secondly, all our experiments drive us to the conclusion that our pigeons were "satisficers"[28] rather than "maximisers." Typically, they left patches only when they had good evidence that the prey density had fallen below the sort of value that would give them a full crop by the end of the session; if they found a reinforcement rate that was satisfactory in that sense, they did not leave it, even though they had plenty of experience of higher rates. Consistent with this idea, optimizing models fitted our pigeons' behavior much more closely at low overall prey densities than in a generally "richer" environment.

Finally, our most successful simulations have always been those that have based their estimates of present prey density not on the outcome of a single response, or a single unit of time, but on the time or effort required to procure the last prey, or the time elapsed since prey was last found. This is consistent with much data on steady-state choice between schedules of reinforcement,[29] where average delay to reinforcement seems a much more useful parameter than average reinforcement frequency (though the delays may need to be transformed).

None of these remarks undermines the principle with which we started this paper: that foragers, whether in nature or in Skinner boxes, need a mechanism by which to integrate discrete events into a prey density or a reinforcement rate. Nor do they undermine the conclusions about that process that we have reached here. We remain convinced that pigeons are able to vary the weighting they give to past experience, relative to the present, both from day to day, and from patch to patch within a foraging bout.

REFERENCES

1. Herrnstein, R. J. 1961. Relative and absolute strength of response as a function of frequency of reinforcement. J. Exp. Anal. Behav. **4:** 267–272.
2. Herrnstein, R. J. 1970. On the law of effect. J. Exp. Anal. Behav. **13:** 243–266.
3. Charnov, E. L. 1976. Optimal foraging: The marginal value theorem. Theoret. Pop. Biol. **9:** 129–136.
4. Simons, D. 1976. "Zahl"-Versuche mit Kolkraben anhand der Methodik der Musterwahl—ein Beitrag zum Verständnis von Problem-Lösungs-Verhalten bei höheren Tieren. Z. Tierpsychol. **41:** 1–33.
5. Church, R. M. & W. H. Meck. The numerical attribute of stimuli. Paper read at the Harry P. Guggenheim conference on Animal Cognition, New York, June, 1982.
6. Rilling, M. 1967. Number of responses as a stimulus in fixed interval and fixed ratio schedules. J. Comp. Physiol. Psychol. **63:** 60–65.
7. Cody, M. L. 1971. Finch flocks in the Mohave desert. Theoret. Pop. Biol. **2:** 142–158.
8. Oaten, A. 1977. Optimal foraging in patches: a case for stochasticity. Theoret. Pop. Biol. **12:** 263–285.
9. Krebs, J. R., A. Kacelnik & P. Taylor. 1978. Test of the optimal sampling by foraging great tits. Nature **275:** 27–31.
10. Krebs, J. R. & A. Kacelnik. Time horizons of foraging animals. This volume.
11. Bush, R. R. & F. Mosteller. 1951. A mathematical model for simple learning. Psychol. Rev. **68:** 313–323.
12. Bobisud, L. E. & W. L. Voxman. 1979. Predator response to variation of prey density in a patchy environment: A model. Am. Nat. **114:** 63–75.
13. Myerson, J. & F. M. Miezin. 1980. The kinetics of choice: An operant systems analysis. Psychol. Rev. **87:** 160–174.
14. Commons, M. L., M. Woodford, J. R. Ducheny & J. R. Peck. The acquisition of performance during shifts between terminal links. Paper read at the Fourth Harvard Symposium on the Quantitative Analysis of Behavior, Cambridge, Massachusetts, June, 1981.
15. Collier, G. H. & C. K. Rovee-Collier. 1981. A comparative analysis of optimal foraging behavior: Laboratory simulations. *In* Foraging Behavior. A. C. Kamil & T. D. Sargent, Eds.: 39–76. Garland, New York, NY.
16. Killeen, P. R., J. P. Smith & S. J. Hanson. 1981. Central place foraging in *Rattus norvegicus*. Anim. Behav. **29:** 64–70.
17. Lea, S. E. G. 1979. Foraging and reinforcement schedules in the pigeon: Optimal and non-optimal aspects of choice. Anim. Behav. **27:** 875–886.
18. Lea, S. E. G. 1981. Correlation and contiguity in foraging behaviour. *In* Advances in Analysis of Behaviour, Vol. 2: Predictability, Correlation and Contiguity. P. Harzem & M. D. Zeiler, Eds: 355–406. Wiley. Chichester.
19. Lea, S. E. G. 1982. The mechanism of optimality in foraging. *In* Quantitative Analysis of Behavior, Vol. 2: Matching and Maximizing Accounts. M. D. Commons, R. J. Herrnstein, & H. Rachlin, Eds: 169–188. Ballinger. Cambridge, MA.
20. Mellgren, R. L. 1982. Foraging in a simulated natural environment: There's a rat loose in the lab. J. Exp. Anal. Behav. **38:** 93–100.
21. Findley, J. D. 1958. Preference and switching under concurrent scheduling. J. Exp. Anal. Behav. **1:** 123–144.

22. McNamara, J. & A. Houston. 1980. The application of statistical decision theory to animal behaviour. J. Theoret. Biol. **85:** 673–690.
23. Dow, S. M. Foraging in the lab: Effects of session length. This volume.
24. Ainslie, G. 1975. Specious reward: A behavioral theory of impulsiveness and impulse control. Psychol. Bull. **82:** 463–496.
25. Herrnstein, R. J. & D. H. Loveland. 1975. Maximizing and matching on concurrent ratio schedules. J. Exp. Anal. Behav. **24:** 107–116.
26. Krebs, J. R. 1978. Optimal foraging: Decision rules for predators. *In* Behavioural Ecology. J. R. Krebs & N. B. Davies, Eds: 23–63. Blackwell. Oxford.
27. Bovet, P. 1980. La valeur adaptive des comportements aléatoires. Année Psychol. **79:** 505–525.
28. Simon, H. A. 1957. Models of Man. Wiley. New York, NY.
29. Killeen, P. 1968. On the measurement of reinforcement frequency in the study of preference. J. Exp. Anal. Behav. **11:** 263–269.

Time Horizons of Foraging Animals[a]

JOHN R. KREBS AND ALEJANDRO KACELNIK

Edward Grey Institute of Field Ornithology
Oxford OX1 3PS, England

INTRODUCTION

Time as a Cost

The universal importance of time for living organisms is underlined by the fact that many evolutionary models of behavior, life history strategies, mating, and other aspects of design, measure benefits and costs as *rates*.[1,2] In models of foraging behavior, benefit is often expressed as intake of food or calories per unit time[3]; in models of mating, benefit is in terms of eggs fertilized per unit of time[4,5]; and in models of reproduction, it is expressed as eggs laid or young produced per unit of time.[2,6,7] These evolutionary models are generally concerned with comparing the "success" of alternative hypothetical strategies: eating one kind of prey only compared with eating all kinds encountered, making a new nest compared with taking over an existing one, and excluding territorial rivals compared with letting them in. In such models time is essential in comparing the success of alternatives because if two alternatives yielded the same quantity of benefit (for example, number of eggs fertilized), but one took twice as long as the other, natural selection would generally favor the one taking the shorter time, in other words, the one yielding a higher rate. The reason for this is perhaps most easily illustrated by referring to reproductive rates. If two genotypes produce ten offspring, but one takes 1 year to do it, while the other takes 5 years, the rapid genotype will soon come to predominate in the population at the expense of the slow one: selection will tend to maximize reproductive rate. It is generally assumed that selection will also tend to maximize the rates of other activities, such as nest building or feeding, in part because these activities contribute eventually to reproductive rate, and in part because time spent in one activity cannot be spent in another. As we shall see later, however, in models of foraging behavior maximizing rate is not the only criterion of "success" that has been considered.

We are referring to time as a *cost*. The benefit for a male dungfly (*Scatophaga stercoraria*) of guarding a newly fertilized female, for example, is that he prevents other males from displacing his sperm,[8] while the cost is the time the male loses from searching for another female.[4] In fact, if guarding could be done at the same time as, and with no encumbrance to, searching, then the male would always be able to guard until after the eggs have been laid. However, given that the two activities cannot be done at the same time, calculation of the strategy that maximizes total rate of fertilizing of eggs has to take into account the time cost of guarding and it does not necessarily pay the male to guard for as long as possible. In a similar way time enters as a cost into foraging models: time spent handling a prey item, like time spent guarding a female, cannot be used for searching for the next item, and time spent foraging in one patch cannot be used for traveling to the next one.

Time, while a pervasive and important cost in models of behavior, is not the only

[a]This work was supported by the Natural Environment Research Council and the Netherlands Foundation for Pure Research (Z.W.O.)

possible one. There may be other costs associated with particular activities, such as wear and tear on the body, risk of succumbing to a predator, or, as we shall discuss later, consumption of energy. In these cases computations of rate of gain using simply time as a cost may be misleading. If the male dungfly referred to earlier was more vulnerable to predator attack while searching than while guarding, he may persist in guarding for longer than would be predicted from models of maximizing number of eggs fertilized per unit time.

Time in Foraging Models

Optimal foraging theory is an attempt to understand the decision rules of foraging animals from a functional (adaptive) standpoint.[3,9–12] For the reasons outlined above, it is often assumed in optimal foraging models that animals are designed by natural selection to maximize rate of food intake. From this premise it is possible to formulate models predicting how animals ought to behave in response to particular foraging problems. Such models have been quite successful in accounting for observed decisions about choice of food items and persistence in foraging in a patch or site (these are reviewed in Krebs *et al.*[13]). Other foraging models, particularly some models of foraging in stochastic environments, do not assume that the animal's activity is designed to maximize rate of gain. If the animal lives in an environment where food supplies fluctuate in an unpredictable way (and this must hold for many natural environments), the animal might be adapted to adjust to the stochastic changes. There are two aspects to models of stochasticity: risk sensitivity and information.[14] The former refers to the possibility that animals are senstitive to variance (the risk of doing well or badly) as well as to the mean rate of intake, while the latter refers to models in which the animal uses its experience of the environment to estimate expected future payoffs. Although the models are not concerned with rate maximizing, time enters into models of both risk and information because the time scale over which the animal makes its decisions (the "time horizon")[15–17] has a critical effect on the optimal policy (see section 2).

Our aim in this paper is to illustrate, by means of results from recent work in our group, some of the ways in which time enters into models of foraging behavior. In particular, we will distinguish between short- and long-term time scales in foraging decisions. We will have nothing to say about mechanisms of time measurement (optimal foraging models are primarily about the consequences of behavior rather than the mechanisms) and will present no direct evidence for time measurement abilities. Instead, we present some functional arguments about why animals *ought* to have certain time-measuring abilities. The other point we will make is that time alone may be an insufficient measure of cost for some foraging models; as we have already mentioned, animals may be sensitive to other costs such as energy expenditure. This is discussed further in section 3.

SHORT-TERM TIMING IN FORAGING MODELS

Involvement Time

Many optimal foraging models assume that animals behave as if their foraging decisions depend on moment-to-moment timing of "involvement times" and intercapture intervals. In the classical model of prey selection,[18] the predator is assumed to be capable of ranking different kinds of prey according to their profitability, defined as energy gain (E) divided by involvement time (h) (the time required to detect, capture and consume the prey item, as distinct from search time). Since some animals do, in fact, prefer the prey of highest rank when offered a sequential choice,[19–22] they behave as if they are measuring E and h. Sometimes, however, further analysis may reveal that the animal uses a correlated cue such as size to judge the profitability of a prey[20] and cannot respond appropriately to independent manipulations of E and h. In other examples animals switch their preferences when E is held constant and h is changed. Erichsen *et al.*[23] were able to reverse the prey preferences of great tits by changing the profitability rankings of two prey types. The change was achieved by increasing the involvement time of the formerly more profitable prey to make it less profitable than the alternative; the involvement time in this case included time spent discriminating the prey from its background. It is worth noting here a contrast between the results of "naturalistic" studies of prey choice and the results of operant simulations such as those of Lea[24] and Snyderman.[25] The operant studies indicate that pigeons in a prey-choice experiment fail to show self-control by waiting for a big reward after a long delay instead of taking a small reward after a much shorter delay, even if the latter has a higher profitability. On the other hand, naturalistic laboratory and field studies have repeatedly shown that animals prefer larger prey items even if their involvement times are longer than for small ones, as long as their overall profitability is higher.[21,22,26] Perhaps an important difference between the two types of study is that in operant studies "involvement time" is simply a delay after the animal chooses a reward before it can eat it, while in natural foraging, although most of the reward is obtained at the end of involvement time, during the involvement the animal gets continual secondary reinforcement due, for example, to the kernel of the nut gradually emerging from its case.

So far we have assumed, in common with the classical foraging models, that involvement time is a fixed constraint, although this is known not to be strictly true.[16,27] In an unpublished experiment, Krebs, Dawkins and Erichsen[28] have investigated the ability of great tits to adaptively modify one component of involvement time, namely, the time spent scanning for a cryptic prey (see also Kamil and Yoerg[29]). The birds were presented with the opportunity to choose between scanning for cryptic prey and searching for them. The prey were presented to caged individuals on a moving belt similar to that used by Krebs *et al.*[19] The prey were pupae of houseflies and each one was presented in a shallow pot with six dried-out pupa cases (dummies). The birds do not eat the cases, so the dried pupae serve as a "background" for the cryptic real prey. The prey and dummies, which were almost indistinguishable to the human eye, could be inspected by the bird as they passed a 4-cm slit in the cover on the belt. The bird could stop the belt to scan the pot by putting its head down to break a photocell beam just above the belt; and by raising its head, it could stop scanning and let the belt continue, equivalent to "searching" for the next prey.

In one experiment, the birds' scanning was interrupted by inserting a variable timer to cause the belt to restart at a set time after the bird had stopped it to inspect a pot. This allowed the construction of a "constraint curve" describing the relationship between duration of scanning time and the probability of detecting the prey (the latter measured by whether or not the bird picked up the prey) (FIG. 1A).

In a second experiment, the average waiting time until the next pot was manipulated as the independent variable and the dependent variable was the amount of time spent scanning each pot: the birds could choose their scanning times by the length of time they interrupted the photocell beam.

This problem bears a superficial resemblance to that represented in the marginal value model[30] with a negatively accelerating gain function within each patch and the prediction of a relationship between waiting (– traveling) time and scanning (– patch) time.[31–34] It is, however, a different kind of foraging problem, since the marginal value model assumes perfect knowledge of patch quality, while in our experiment the animal acquires information about the patch (whether or not it contains a prey) while scanning.[36] If every pot is known to contain a prey, and if the expected probability of discovery within a patch remains constant with time as in the exponential curve of FIGURE 1A, the animal should never leave a patch until it has found the prey, since it

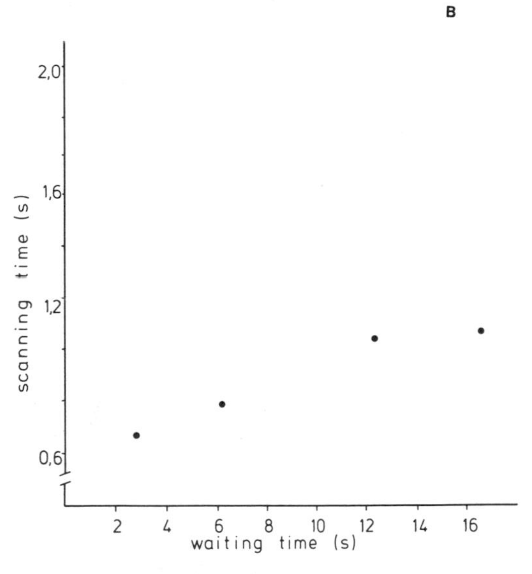

FIGURE 1. (**A**) The probabilty of detecting a cryptic prey as a function of scanning time. The equation $p = 0.91\ (1 - e^{-0.69}\tau)$ explains 95% of the variance. The points are means for four birds. (**B**) The relationship between scanning time and waiting time.

could do no better in the next patch than it is doing at the moment. If, however, a proportion of pots contains no prey or prey that are hard to locate, the optimal policy predicts a positive relationship between waiting and scanning time.[36] In FIGURE 1A, the detection curve asymptotes at 0.9, indicating that the birds' experience is that 10% of pots contain no prey. Therefore a relationship between scanning time and waiting time, as shown by the data in FIGURE 1B, is predicted.

Thus the birds can modify their involvement time in a way that is qualitatively consistent with the optimal policy. Further analysis is required to test whether or not the change is quantitatively as predicted. The adjustment of involvement time in this experiment allows the animal to respond to uncertainty about patch quality.

Intercapture Intervals

One frequently studied foraging problem is that of a predator exploiting discrete patches in which food availability declines with foraging time, the decline being referred to as "resource depression."[37] The animal chooses where to give up on a curve of diminishing returns. If the animal exploits discrete prey items, however, the curve is actually a series of steps, and there has been considerable discussion in the literature as to how an animal feeding on discrete prey items might estimate its position in this curve.[38,39] As McNair[40] has pointed out, the giving-up time rule does not follow directly from the marginal-value model, as assumed by some authors,[41,42] since the model refers only to the case of continuous resource depression curves. Under some circumstances, the giving-up time rule will, however, perform better in terms of the intake rate it yields, than will other simple rules, such as staying in each patch for a fixed time.[40]

Recent evidence consistent with a giving-up time rule comes from the work of Ydenberg[43] on great tits. He simulated in the laboratory the problem of foraging on patches with resource depression by presenting the birds with an operant patch which offered rewards on a stochastic progressive ratio schedule. On each successive response the probability of reward declined according to a square-root function. At any point the bird could choose to fly to a "reset perch" at the other side of the aviary and reset the schedule to the beginning. Ydenberg considered three possible decision rules the birds might use for leaving the patch to fly to the reset perch: leave after a set time, leave after a set number of rewards, and leave after a set time of unsuccessful responding. The three rules make different predictions about the distribution of unsuccessful responses before leaving. The fixed-time rule predicts that the birds should depart at random with respect to the stochastic sequence of rewards just prior to leaving; the fixed-number rule predicts that the birds should leave after a capture; and the giving-up time rule predicts that departures should be preceded by longer than random runs of bad luck (ROBLs). As FIGURE 2 shows, the birds' behavior fits most closely to the predictions of the third hypothesis: departures are not at random with respect to rewards, but tend to follow a run of about three unsuccessful responses, which is longer than expected from random. A complementary analysis showed that the probability of departure increases steeply as a function of ROBL length. These analyses, while consistent with a giving-up time rule, were actually done in terms of responses. While the two are equivalent in the experiment since the animals work at a constant rate, it would be of interest to vary them independently of one another.

Further experiments by Ydenberg involved manipulating the stochastic schedule either by *interrupting* the sequence of rewards (introducing a ROBL) or by *interlaying* extra rewards to exclude ROBLs longer than the hypothesized giving-up time of three responses. Interrupted, interlaid, and control schedules were alternated randomly in the course of an experimental session. The two kinds of manipulated schedule were arranged so that the giving-up time hypothesis predicted that the animals would stay for 9.00 responses longer in the interlaid than in the interrupted schedule. The mean difference between treatments for four birds was 9.85, although there was a great deal of variation between individuals.

LONG-TERM TIMING IN FORAGING MODELS

Estimating Reward Rates

We refer to timing of events outside the immediate context of ongoing behavior as "long-term" timing. Most foraging models, for example, assume that the animal has some estimate of its long-term average rate of intake, or at least that it behaves as though it has such an estimate. In both prey- and patch-choice models, current gain is compared with the expected gain from the environment as a whole, which could be obtained by projecting forward from past experience. To fully review the ways in which past experience might influence estimates of future reward rate is beyond the scope of this article and would involve surveying an extensive literature on animal learning (see also Lea & Dow[44]). Instead we simply raise some questions which relate to foraging models in particular. The first is the question of the time scale over which the estimating is done. At one extreme, Elner and Hughes[26] report that shore crabs

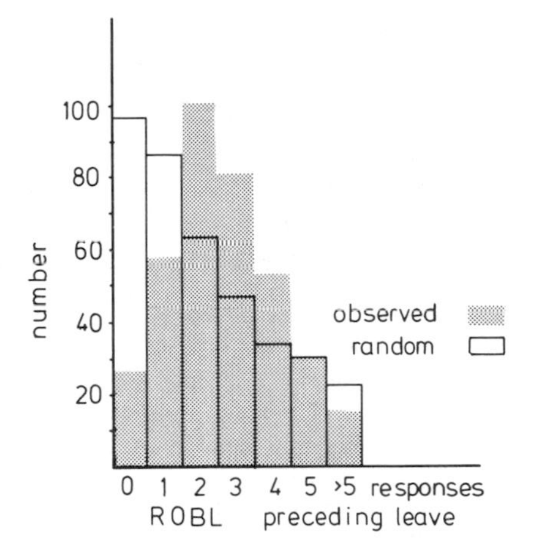

FIGURE 2. The frequency distribution of runs of bad luck experienced by great tit before leaving a patch. (From Ydenberg.[43] Reprinted by permission.)

(*Carcinus maenas*) are sensitive in their choice of mussels (*Mytilus edulis*) to very short sequences of encounters: the probability of accepting a mussel of low profitability increases markedly after even three or four encounters in a row without finding a high-ranking prey.[45] On the other hand Getty and Krebs[46] found that prey choice in great tits feeding off the conveyor belt referred to earlier was hardly influenced by short runs of good or bad luck. Past experience did, however, influence choice and the birds tended to make choices closer to the optimum predicted from the previous experimental session than from the present session: they lagged behind changes in the environment. This brings us to the second point, namely, whether or not the time scale over which animals average is adjustable in relation to the time scale of environmental changes. It would seem reasonable to expect individuals that experience frequent fluctuations in the environment to average over shorter time periods than those living in more stable conditions, but as far as we know this has not been studied in detail. The third point, an extension of the second, refers to the *kind* of changes an animal experiences. Kacelnik and Krebs[47] found that the response of starlings to a stepwise decrease in reward probability from one of two alternative feeding sites could not be accounted for by models that have been successful in describing responses to a more gradual change. In the step-change experiment, the animals appear to be estimating both the reward rate and the probability that a change had occurred. The general point here is that there may be different rules for responding to different types of environmental change.

Time Horizons

The time horizon of a forager is the interval available to the animal for foraging. We use the term here to refer to a property of the environment, leaving aside the important question of how the animal perceives it. Examples of time horizons are the duration of an experimental session, the time left until the end of the day, or the time until the tide comes in for an animal feeding on intertidal mudflats. Houston *et al.*[17] on whose work much of the ensuing discussion is based, point out that animals could be viewed as responding in one of three ways with respect to time horizons: they could behave as if the horizon is infinite, they could behave as though there is a fixed horizon, or they could discount time into the future, a compromise between the first two. The classical foraging models assume that animals have an infinite time horizon, although one can think of special cases where this assumption is clearly inappropriate. Consider, for example, an animal exploiting patches. According to the classical model, the best point at which to leave a patch is when the gain rate drops to the average expected from the environment. But if the time required to travel to the next patch is greater than the time left in the day, it would clearly not make sense to leave the current patch and travel, say, half way to the next one before having to break off to go to a roosting site.[17]

The influence of time horizon on foraging has been studied in detail in relation to stochastic foraging models of information and of risk. Models of information consider how animals ought to sample an environment where reward probabilities are unknown in advance.[15,17,48–50] The important point for our present discussion is that the optimal sampling policy may depend on the time horizon. In order to acquire information, the animal must sacrifice some short-term gain in intake rate by sampling alternatives that appear currently not to be the most profitable, but that may eventually turn out to be better than current estimates suggest. The potential gain from this sampling can only be realized in the future, that is, it depends on the time horizon.

A simple sampling problem that has been analyzed theoretically and experimentally illustrates this point. The problem is the so-called "two-armed bandit problem," in which the sampler is faced with two alternatives with differing, unknown, reward probabilities.[15–17] By using information acquired during sampling the alternatives, the animal could in theory obtain estimates of the two reward probabilities and use the estimates to guide its future choices. It is easy to see where the time horizon comes in here. If the animal had only two trials (opportunities to try the sites) left, it could sample each one once, but the information gained about the difference in reward probability could not be used for later exploitation, so sampling would not pay. In general, the longer the time horizon (the future for using the information) the more worthwhile it is to sample. In an experiment akin to a concurrent VRVR study, Kacelnik[16] found some indication that great tits spent longer times "sampling" when they were trained to expect long sessions than when accustomed to short experiments (see Houston *et al.*[17] for further details and Dow[51] for a similar result).

A second facet of stochasticity in foraging models is risk. Caraco *et al.*[52] were the first to introduce explicitly this concept into the foraging literature, although it had been alluded to earlier by Thompson *et al.*[53] Caraco *et al.* showed that yellow-eyed juncos are sensitive to variance in reward rate when the mean is held constant. The experiment consisted of a choice between two alternative foraging sites, one of which offered a constant reward p and the other offered $2p$ on half the occasions and 0 on the other half. The tendency to choose high or low variance was related to hunger: very hungry birds were risk-prone (chose high variance), while less hungry birds were risk-averse. The accepted interpretation of this is known as the expected energy budget rule,[54,55] which states that if the expected gain is less than the expected requirements (the expected energy budget is negative), the animal should be risk-prone, otherwise it should be risk-averse. This is because for an animal with negative expected energy budget, the certain option will never provide enough food for survival, while the risky one has some probability of doing so (FIG. 3*a*). This formulation assumes that the animal's goal is to maximize survival probability over a fixed time period (usually taken to be a day) and that there is only one foraging choice within this period. In the

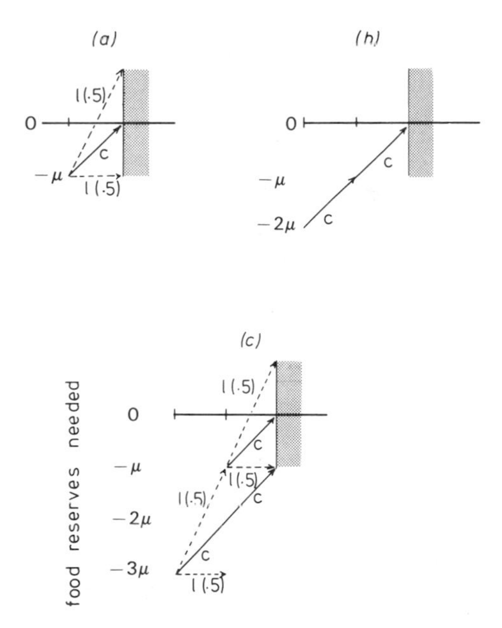

FIGURE 3. Graphic representation of risk sensitivity. The *y*-axis of each graph is the additional food needed to ensure reserves to survive the night. In (*a*) the animal has one choice point: its shortfall at this point (μ) is equal to the gain from choosing the certain option (c); this option therefore guarantees survival. The risky option (l) yields the same mean, but 2 μ on half the occasions and zero on the other half. Thus, it gives a survival probability of 0.5. In (*b*) there are two choice points and the animal can get enough to survive by taking option c on both occasions. In (*c*), however, the animal's requirement at the first point is 3 μ, so only by choosing l at least once can the animal have a survival probability greater than zero. The highest survival probability is given by the sequence l then c (or vice versa).

experiment by Caraco *et al.* the hungry birds were shown by calculations of energy consumption to be in negative expected budget, while the less hungry ones were in positive expected budget. The latter were expected to be risk-averse since the certain option always provides enough to survive, while the risky option may provide more than enough or not enough.

The effect of time horizon on risk sensitivity has been discussed by Houston and McNamara[56] and Stephens[57] and is illustrated in FIGURE 3*b* and *c*. The critical variable is the number of choice points occurring before the end of the time period under consideration, for example, a day. As already mentioned, the expected energy budget rule assumes only a single choice, but now suppose that there are two choice points. FIGURE 3 shows two important features of sequential choice. First, the best policy may be for the animal to change its mind: in FIGURE 3*c* the highest probability of survival comes from choosing the risky option first and then the certain one (or vice versa) and not from choosing always the risky or the certain option. The reason for this is that a risk-prone decision at stage 1 may bring the animal close enough to its daily requirement for it be in positive expected energy budget (and hence risk-averse) at the second choice. Therefore the survival of a risk-prone animal may be higher than would appear possible at first sight. The second point is that the value of a risk-prone decision relative to that of risk-aversion depends on the number of choices left: a risky choice further away from dusk is generally safer than one taken close to dusk. It would be interesting to test whether animals change their preference for risk as dusk or the equivalent end point approaches.

ENERGY VERSUS TIME

As we mentioned in the INTRODUCTION, time is only one of many possible costs that could influence foraging decisions. In this section we describe an experiment with starlings (*Sturnus vulgaris*) carried out by Kacelnik, Tinbergen and Bloem[61] to test whether time or energy costs give a better account of the birds' choice rule. Other studies of foraging which have directly or indirectly demonstrated the importance of energetic costs in decision rules include those of DeBenedictis *et al.*,[58] Williams,[59] Cowie,[31] and Waddington.[60]

The starlings were offered a choice between two food dispensers placed 7 meters apart. Each dispenser consisted of a perch and a reward-delivering machine,[15] where a reward was delivered with a certain probability after the bird hopped on the perch. Before trying either site, the subjects were required to hop at least once on a "choice perch" (C) placed 1 m from one site (patch 1; the "close" side) and 6 m from the other (patch 2; the "distant" side). The reward probabilities were chosen so that the average number of trials (each trial consisting of a hop in the choice perch, a round trip to the patch, and a hop in the feeding perch) required to obtain a reward in patch 1 (r_1) was higher than in patch 2 (r_2). The expected time (T) to obtain one reward in patch i is given by

$$T_i = r_i(t_i + S) + H \tag{1}$$

where t_i is the round-trip travel time between C and patch i, S is the sum of the times required to hop once on the choice perch and in the feeding perch, and H is the handling time required to consume one reward (we assume S and H to be identical for both sites, but this is not necessarily true).

The expected energy expenditure (E) to obtain one reward in patch i is

$$E_i = r_i t_i m_l + (r_i \mathrm{S} + H) m_2 \tag{2}$$

where m_1 and m_2 are the rate of expenditure during travel and "non-travel" activities, respectively.

Using Equations 1 and 2, Kacelnik *et al.*[61] present three putative choice criteria. If the animals ignore energy costs of flight and perching, choice might be based on minimizing *time* to the next reward (that is, maximizing gross yield per unit time). At the other extreme, the animals might ignore time costs and maximize *energetic efficiency* (gain/expenditure), a goal that would make sense if, for example, expending energy had a deleterious effect on body condition. Between these two extremes is the third possibility, that the animals respond to both time and energy costs by maximizing *net rate of energy gain;* this is the hypothesis most frequently used in foraging models.

The methods of calculating the predictions of the three hypotheses are as follows:

(a) Minimize time per reward. The values of r_1, r_2, for which both patches have the same time expectation per reward, can be found by equating T_1 and T_2,

$$r_1(t_1 + S) + H = r_2(t_2 + S) + H,$$

implying that the distant patch will have a lower time expectation when

$$r_2 < \frac{t_1 + S}{t_2 + S} r_1. \tag{3}$$

If time alone is being minimized, patch 2 should be chosen whenever inequality (3) holds true.

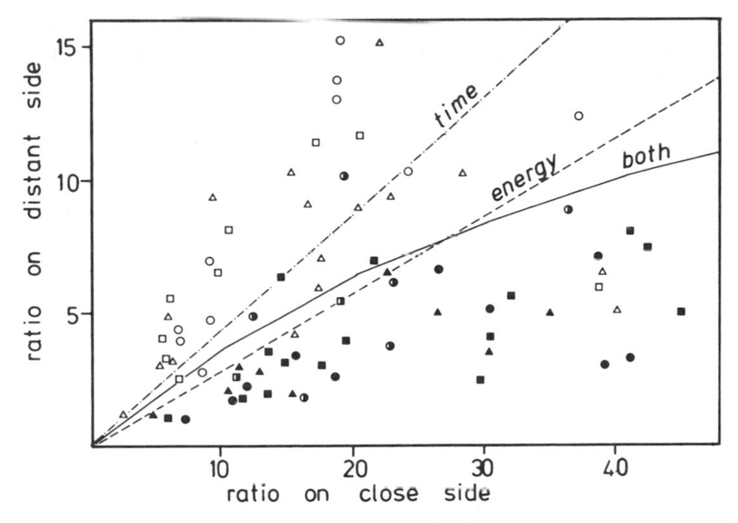

FIGURE 4. Time versus energy. The axes of the graph are the reward ratios at the two sites, one close (x-axis) and one distant (y-axis). *Black symbols* (three sets for three different birds) show the ratio combinations for which the birds preferred the distant site, and *white symbols* are combinations for which the close site was preferred. The *split symbols* are points of no preference. The lines are the predictions of the three models referred to in the text of where the split between black and white points should be. The pure-time model does less well than the other two.

(b) Maximize energetic efficiency. The pair r_1, r_1, for which a reward costs the same amount of energy in either patch, is given by equating Equation 2 for each patch:

$$r_1 t_1 m_1 + (r_1 S + H)m_2 = r_2 t_2 m_1 + (r_2 S + H)m_2,$$

and patch 2 should be preferred when

$$r_2 < \frac{t_1 m_1 + S m_2}{t_2 m_1 + S m_2} r_1. \tag{4}$$

(c) Maximize net rate of gain. Net profitability for patch i is defined as

$$R_{ni} = \frac{V - E_i}{T_i} \tag{5}$$

where V is the assimilable energy contained in one reward. Replacing E_i and T_i by Equations 1 and 2 and solving for r_2, we find the pair r_1, r_2 for which $R_{n1} = R_{n2}$.

$$r_2 = \frac{V(t_1 + S) + Ht_1D}{V(t_2 + S) + Ht_2D + r_1S(t_2 - t_1)D} r_2, \tag{6}$$

where $D = m_2 - m_1$.

In the experiment, three subjects experienced a number of day-long sessions with different pairs r_1, r_2, and a choice for one patch was scored if more than 75% of that day's trials was made in that side. For sessions when neither patch reached this criterion, an "undecided" point was scored. FIGURE 4 shows that the hypothesis that choice may be based on minimizing time per reward alone (maximizing gross yield per unit time) is poor in predicting actual choices, while the predictions of the other two alternatives are reasonably good predictors of choice. In order to find out whether time is not important (hypothesis 2) or whether it is taken into account as a component of the net rate of gain (hypothesis 3), a different set up must be used where the predictions of these last two alternatives are clearly different. A further point of note is that the exact position of the lines for efficiency and net gain depend on assumed costs of flying and perching. Kacelnik *et al.*[61] discuss the relationship between physiological estimates of these costs and the estimates based on the behavioral measures of indifference.

CONCLUSIONS

The main points we have made are as follows. (i) Foraging models imply that animals may have abilities to time events such as intercapture intervals and involvement times lasting in the order of a few seconds or less. Evidence suggests that birds can modify the time spent looking for cryptic prey in an adaptive way and that patch-leaving decisions may depend on intercapture intervals. The timing abilities implicated in these experiments are analogous to those studied in detail by, for example, Gibbon and Church.[62] (ii) Over a longer time scale, of minutes or hours, foraging models refer both to the problem of averaging rewards obtained over time (see Lea and Dow[44]) and to the amount of time left for foraging (the time horizon). While some work has been done on the former problem, the latter has not been investigated in the foraging literature and would make a fruitful area for further work. (iii) Time is not the only cost in foraging models, and in some cases predictions based on time costs alone may be less accurate than those based on other costs, such as energy. The role of energy costs was demonstrated directly in the experiment with starlings, and in other situations costs such as the risk of predator attack may be important. This raises the interesting question, which we have not tried to tackle, of whether psychological models of subjective perception of time might be eventually related to functional models of the significance of time spent in different activities or at different stages of the day.

Finally, there are other aspects of timing in relation to foraging that we have not had space to consider, a notable example being the importance of rhythmicity in foraging decisions. It has been shown that predators can time their hunting activities to coincide with circadian, tidal, and shorter-term activity rhythms of their prey (see the review in Daan and Aschoff[63]). In an extension of this approach Krebs and Kacelnik[64] and Kacelnik[16] have suggested that the timing of daily routines of behavior and daily changes in body weight of small birds might be related adaptively to changing conditions of food availability.

ACKNOWLEDGMENTS

We thank the various investigators who have allowed us to cite their unpublished work.

REFERENCES

1. Krebs, J. R. & N. B. Davies, Eds. 1983. Behavioural Ecology, 2nd ed. Blackwell Scientific Publications. Oxford.
2. Brockmann, H. J., A. Grafen & R. Dawkins. 1979. Evolutionarily stable nesting strategy in a digger wasp. J. Theor. Biol. **77:** 473–496.
3. Pyke, G. H., H. R. Pulliam & E. L. Charnov. 1977. Optimal foraging: A selective review of theory and tests. Q. Rev. Biol. **52:** 137–154.
4. Parker, G. A. & R. A. Stuart. 1976. Animal behaviour as a strategy optimizer: Evolution of resource assessment strategies and optimal emigration thresholds. Am. Nat. **110:** 1055–1076.
5. Davies, N. B. & T. R. Halliday. 1979. Competitive mate searching in male common toads, *Bufo bufo*. Anim. Behav. **27:** 1253–1267.
6. Maynard Smith, J. 1977. Parental investment: A prospective analysis. Anim. Behav. **25:** 1–9.
7. Clutton-Brock, T. H., F. E. Guinness & S. D. Albon. 1982. The Red Deer. Chicago University Press. Chicago, IL.
8. Parker, G. A. 1978. Searching for mates. *In* Behavioural Ecology: An Evolutionary Approach. J. R. Krebs, & N. B. Davies, Eds. Blackwell Scientific Publications. Oxford.
9. Krebs, J. R. 1978. Optimal foraging: Decision rules for predators. *In* Behavioural Ecology: An evolutionary approach. J. R. Krebs & N. B. Davies, Eds.: 23–63. Blackwell Scientific Publications. Oxford.
10. Townsend, C. R. & R. N. Hughes. 1981. Maximising net energy returns from foraging. *In* Physiological Ecology: An Evolutionary Approach to Resource Use. C. R. Townsend & P. Calow, Eds.: 86–108. Blackwell Scientific Publications. Oxford.
11. Krebs, J. R. & R. H. McCleery. 1983. Optimisation in behavioural ecology. *In* Behavioural Ecology, 2nd ed. J. R. Krebs & N. B. Davies, Eds., Blackwell Scientific Publications. Oxford.
12. Kamil, A. C. & T. D. Sargent. 1981. Foraging Behaviour: Ecological and Psychological Approaches. Garland S.T.P.M. Press. New York, NY.
13. Krebs, J. R., D. W. Stephens & W. J. Sutherland. 1983. Perspectives in optimal foraging. *In* Perspectives in Ornithology. G. A. Clark & A. H. Brush, Eds. Cambridge University Press. New York, NY.
14. Stephens, D. W. & E. L. Charnov. 1982. Optimal foraging: Some simple stochastic models. Behav. Ecol. Sociobiol. **10:** 251–263.
15. Krebs J. R., A. Kacelnik & P. Taylor. 1978. Test of optimal sampling by foraging great tits. Nature (London) **275:** 27–31.
16. Kacelnik, A. 1979. Studies of Foraging Behaviour in Great Tits (*Parus major*). Unpublished D. Phil. thesis, Oxford University.

17. Houston, A. I., A. Kacelnik & J. McNamara. 1982. Some learning rules for acquiring information. *In* Functional Ontogeny. D. J. McFarland, ed.: 140–191. Pitman. London.
18. Charnov, E. L. 1976. Optimal foraging: Attack strategy of a mantid. Am. Nat. **110:** 141–151.
19. Krebs, J. R., J. T. Erichsen, M. I. Webber & E. L. Charnov. 1977. Optimal prey selection in the great tit (*Parus major*). Anim. Behav. **25:** 30–38.
20. Barnard, C. J. & C. A. J. Brown. 1981. Prey size selection and competition in the common shrew (*Sorex areneus* L.). Behav. Ecol. Sociobiol. **8:** 239–243.
21. Davies, N. B. 1977. Prey selection and social behaviour in wagtails (Aves: Motacillidae). J. Anim. Ecol. **46:** 37–57.
22. Goss-Custard, J. D. 1977. Optimal foraging and the size selection of worms by redshank *Tringa totanus*. Anim. Behav. **25:** 10–29.
23. Erichsen, J. T., J. R. Krebs & A. I. Houston. 1980. Optimal foraging and cryptic prey. J. Anim. Ecol. **49:** 271–276.
24. Lea, S. E. G. 1979. Foraging and reinforcement schedules in the pigeon: Optimal and nonoptimal aspects of choice. Anim. Behav. **27:** 875–886.
25. Snyderman, M. 1983. Optimal prey selection: Partial selection, delay of reinforcement and self-control. Behav. Anal. Lett. **3:** 131–142.
26. Elner, R. W. & R. N. Hughes. 1978. Energy maximization in the diet of the short crab, *Carcinus maenas*. J. Anim. Ecol. **47:** 103–116.
27. Krebs, J. R. 1980. Optimal foraging, predation risk and territorial defence. Ardea **68:** 83–90.
28. Krebs, J. R., M. Dawkins & A. I. Houston. Unpublished experiment.
29. Kamil, A. C. & S. I. Yoerg. 1982. Learning and foraging behavior. *In* Perspectives in Ethology. P. P. G. Bateson & P. H. Klopfer, Eds. Vol. 5: 325–364. Plenum Press. New York, NY.
30. Charnov, E. L. 1976. Optimal foraging: The marginal value theorem. Theor. Popul. Biol. **9:** 129–136.
31. Cowie, R. J. 1977. Optimal foraging in great tits *Parus major*. Nature (London) **268:** 137–139.
32. Giraldeau, L. A. & D. L. Kramer. 1982. The marginal value theorem: A quantitative test using load size variation in a central place forager, the Eastern chipmunk *Tamias striatus*. Anim. Behav. **30:** 1036–1042.
33. Carlsson, A. & J. Moreno. 1982. The loading effect in central place foraging wheatears *Oenanthe oenanthe* L. Behav. Ecol. Sociobiol. **11:** 173–184.
34. Kasuya, E. 1982. Central place water collection in a Japanese paper wasp, *Polistes chinensis antennalis*. Anim. Behav. **30:** 1010–1014.
35. Kacelnik, A. 1984. Central place foraging in starlings (*Sturnus vulgaris*). I: Patch residence time. J. Anim. Ecol. **53:** 283–300.
36. McNamara, J. & A. Houston. 1984. A simple model of information use in the exploitation of patchily distributed food.
37. Charnov, E. L., G. H. Orians & K. Hyatt. 1976. The ecological implications of resource depression. Am. Nat. **110:** 247–259.
38. Krebs, J. R., J. C. Ryan & E. L. Charnov. 1974. Hunting by expectation or optimal foraging: Study of patch use by chickadees. Anim. Behav. **22:** 953–964.
39. Waage, J. K. 1979. Foraging for patchily-distributed hosts by the parasitoid *Nemeritis canescens*. J. Anim. Ecol. **48:** 353–371.
40. McNair, J. N. 1982. Optimal giving up times and the marginal value theorem. Am. Nat. **119:** 511–529.
41. Bond, A. B. 1981. Giving up as a Poisson process: The departure rule of the green lacewing. Anim. Behav. **29:** 629–630.
42. Townsend, C. R. & A. G. Hildrew. 1980. Foraging in a patchy environment by a predatory net-spinning caddis larva (*Plectrocnemia conspersa*): A test of optimal foraging therory. Oecologia (Berlin) **47:** 219–221.
43. Ydenberg, R. C. 1984. Great tits and giving up times: Decision rules for leaving patches. Behaviour. In press.
44. Lea, S. E. G. & S. M. Dow. This volume.

45. SHETTLEWORTH, S. J. 1983. The ecology of learning. *In* Behavioural Ecology, 2nd ed. J. R. Krebs & N. B. Davies, Eds. Blackwell Scientific Publications. Oxford.
46. GETTY, T. & J. R. KREBS. Lagging partial preferences for cryptic prey: A signal detection analysis of great tit foraging. Am. Nat. In press.
47. KACELNIK, A. & J. R. KREBS. In preparation.
48. GREEN, R. F. 1980. Bayesian birds: A simple example of Oaten's stochastic model of optimal foraging. Theor. Popul. Biol. **18:** 244–256.
49. MCNAMARA, J. M. 1982. Optimal patch use in a stochastic environment. Theor. Popul. Biol. **21:** 269–288.
50. OATEN, A. 1977. Optimal foraging in patches: A case for stochasticity. Theor. Popul. Biol. **12:** 263–285.
51. DOW, S. M. In preparation.
52. CARACO, T., S. MARTINDALE & T. S. WHITHAM. 1980. An empirical demonstration of risk-sensitive foraging preferences. Anim. Behav. **28:** 820–830.
53. THOMPSON, W. A., I. VERTINSKY & J. R. KREBS. 1974. The survival value of flocking: A simulation model. J. Anim. Ecol. **43:** 785–820.
54. STEPHENS, D. W. 1981. The logic of risk-sensitive foraging preferences. Anim. Behav. **29:** 628–629.
55. MCNAMARA, J. & A. I. HOUSTON. 1982. Short term behaviour and lifetime fitness. *In* Functional Ontogeny. D. J. McFarland, Ed.: 60–87. Pitman. London.
56. HOUSTON, A. I. & J. MCNAMARA. 1982. A sequential approach to risk-taking. Anim. Behav. **30:** 1260–1261.
57. STEPHENS, D. W. 1982. Stochasticity in Foraging Theory: Risk and information. Unpublished D.Phil. thesis, Oxford University.
58. DE BENEDICTIS, P. A., F. B. GILL, F. R. HAINSWORTH, G. H. PYKE & L. L. WOLF. 1978. Optimal meal size in hummingbirds. Am. Nat. **112:** 301–316.
59. WILLIAMS, D. 1982. Studies of optimal foraging using operant techniques. Unpublished Ph.D. thesis. Liverpool University.
60. WADDINGTON, K. D. 1982. Honey bee foraging profitability and round dance correlates. J. Comp. Physiol. **148:** 297–301.
61. KACELNIK, A., J. M. TINBERGEN & G. BLOEM. In preparation.
62. GIBBON, J. & R. M. CHURCH. 1981. Time left: Linear vs. logarithmic subjective time. J. Exp. Psychol. Anim. Behav. Proc. **7:** 87–107.
63. DAAN, S. & J. ASCHOFF. 1982. Circadian contributions to survival. *In* Structure and Physiology of Vertebrate Circadian Rhythms. J. Aschoff, Ed. Springer Verlag. Heidelberg.
64. KREBS, J. R. & A. KACELNIK. 1983. The dawn chorus in the great tit (*Parus major*): Proximate and ultimate causes. Behaviour **82:** 287–309.

Timing in Animal Learning and Behavior: Discussion Paper

EDMUND FANTINO

Department of Psychology
University of California, San Diego
La Jolla, California 92093

The seven papers in this section provide us with many and diverse potential areas for discussion. I will emphasize two themes, each of which is relevant to several papers and also to ongoing research in our laboratory. One area concerns the importance of changeover time—the time taken to shift from one option to another, which was discussed by Platt and by Lea and Dow. I am especially interested in the relation of changeover time to foraging. The second area concerns the effects of temporal intervals which immediately precede a stimulus associated with reinforcement on the strength of that stimulus—an issue relevant to several of these papers, especially those of Balsam and Jenkins. Before taking up these themes I will make some quick and, for the most part obvious, specific comments on the papers.

Temporal capacity—stressed in some of Roberts' and Richelle's talks—sets limits on, but in no way insures, temporal control of behavior. Thus, Jenkins has noted "the surprising insensitivity of autoshaping to the time between the immediately preceding feeding and the onset of the CS." This insensitivity is surely not due to temporal incapacity, but to procedural factors. Later we suggest conditions that may eliminate this particular insensitivity. For now I note that Roberts and Holder are providing the data base needed to close the chasm between data on temporal capacity and on the effects of temporal variables on conditioning. Similarly, Richelle and Lejeune are amassing an important data base on cross-species factors in timing in what has been an even more neglected area.

Roberts and Holder made an interesting and potentially important case for similarities between timing and conditioning. One of several similarities concerns the fact that timing does not occur when backward conditioning fails to occur. The case would be strengthened further if they could show that timing does occur when backward conditioning occurs. This is a particularly fertile area since backward conditioning may be demonstrated much more readily than had been realized previously.[1,2] Thus, backward conditioning procedures may permit an informative assessment of the degree of correspondence between timing and conditioning.

Richelle and Lejeune compared pecking and perching responses on temporal schedules and found performance more sensitive to the temporal cues when perching was the response. The authors suggest that the perching response is "a more natural operant, plausibly more appropriate to reveal timing competence in birds." As they also suggest, however, the two situations "differ in other respects than the response used." In order to isolate the effects of the response variable, Richelle and Lejeune should consider studying the two responses (perching and pecking) on both types of temporal schedules (differential reinforcement of low rates, or DRL, and differential reinforcement of response duration, or DRRD) in a two-by-two design. Thus far they have completed only two cells of this design: Pecking on a DRL and perching on a DRRD. Superior performance on the latter may reflect the response factor, the schedule factor, or both. My guess is that perching may be no more sensitive to temporal values than pecking when a DRL schedule is employed.

Among the most interesting results of Lea and Dow's fascinating paper is their support for Krebs' counterintuitive (to me as well as Lea and Dow) prediction that foragers would take more time sampling two patches in longer than in shorter sessions. This result would be more forceful if it were not confounded with the fact that the longer and shorter sessions also differed in that they were immediately preceded by daily sessions with larger and smaller numbers of reinforcers, respectively. Thus, their result may not be entirely free of the "carryover" effects of these differential numbers of prior reinforcers. Given the conterintuitive nature of the results it would be desirable to control more precisely for such carryover effects before the results may be taken as supporting Krebs' view unambiguously.

Moving on to the first more general theme, both Lea and Platt, in very different contexts, reported that data involving changeovers did not behave according to certain expectancies. In Platt's case changeover time (CO IRT) was one of four shaped dimensions, along with latency, IRT and duration. Changeover time had longer latencies with both differential reinforcement and subsequent nondifferential reinforcement for both the lever and chimney responses. Platt noted there was no "suitable way to provide *a priori* specification" of this difference. *After* the fact these results appear sensible, however, as we shall see. Similarly, Lea noted that stochastic preference rules, such as Herrnstein's, give too many changeovers. Finally, there is a third set of data that are potentially troublesome for optimal foraging theories and for the extension of my own delay-reduction hypothesis to foraging. I summarize this hypothesis very briefly, especially since it is also relevant to the second theme of this discussion, before discussing the potentially offending changeover (CO) data, providing a resolution and tying these threads together.

According to the delay-reduction hypothesis, developed in studies of both choice and conditioned reinforcement in the late 1960s,[3] the effectiveness of a stimulus as a conditioned reinforcer may be predicted most accurately by calculating the reduction in length of time to primary reinforcement correlated with the onset of the stimulus in question relative to the length of time to primary reinforcement measured from the onset of the preceding stimulus. The simplest form of the parameter-free delay-reduction hypothesis may be stated as:

$$\text{Reinforcing strength of stimulus } A = f\left(\frac{T - t_A}{T}\right) \quad (1)$$

where t_A is the temporal interval between the onset of stimulus A and primary reinforcement and T is the total time between reinforcer presentations. As Abarca and Fantino[15] have shown, when applied to operant analogs of foraging, the delay-reduction hypothesis, as well as Charnov's version of optimality theory, both require that foraging become *less* selective as search time increases. As Baum has noted, following Krebs' lead, CO time in concurrent schedules is analogous to search time in foraging. Studies from at least three laboratories have explored the relation between CO length and preference in simple concurrent schedules *not* designed—and, I emphasize, *not* intended—to stimulate foraging.[4-7] Data from all three labs are consistent despite marked differences in procedure. As CO time increases, preference becomes *more,* not less, selective. If this relation were obtained with concurrent schedules intended to simulate foraging, then optimality theory and the delay-reduction hypothesis would be in trouble.

Nureya Abarca and I have completed experiments on this question using a design inspired by Stephen Lea's work. In an operant analog to foraging we varied CO time under different deprivation conditions (using both open and two arrangements of closed economies; see Hursh[8]) and with two pairs of feeding schedules (VI 5 sec and VI

20 sec; VI 30 sec and VI 60 sec). The results were entirely consistent with delay-reduction and optimality: As CO time increased, preference became less selective. These results have encouraged us to continue testing delay-reduction and optimality accounts (which prove to be remarkably similar, including the treatment of the travel time predictions noted in Krebs' talk) in a variety of operant analogs to foraging. For the present, however, we face the following question: Why does one get the opposite result with simple concurrent schedules? Perhaps it is for the same reason that Platt obtained long CO times and Lea found a reluctance to shift: When the subject is responding in the preferred schedule with a CO requirement it faces a choice not simply between outcomes A and B but a choice between A and a chain consisting of the CO and then B. But we know that simple (or unsegmented) outcomes are disproportionately preferred to chained (or segmented) ones, disproportionate, that is, compared to preferences based only on temporal factors.[2] From this fact it can be shown that each of the three sets of problematic results follow: longer CO times in Platt's case, reluctance to shift in Lea's case, and increased preference for the preferred schedule with increasing CO times in simple concurrent schedules. An adequate general model of timing and/or foraging, when applied to choice, must carefully consider the role of CO time. Returning briefly to Platt's results, consider his resistance-to-change data: The dimension most closely tied to the response (duration of the response) was least resistant; those involving at least two similar responses (latency and IRT) were intermediate; that involving at least two *different* responses (CO time) was most resistant. These results *may* be related to the increased resistance-to-extinction of the initial links of chain (and other more complex) schedules relative to simpler ones under some conditions.

Moving to the final theme, as I noted (Equation 1), the effectiveness of a stimulus as a conditioned reinforcer may be predicted by calculating the reduction in the length of time to primary reinforcement indicated by the onset of the stimulus in question, relative to the length of time to primary reinforcement measured from the onset of the preceding stimulus. This delay-reduction hypothesis has been extended, with some success, to areas such as: observing,[9-11] self-control,[12,13] three-alternative choice,[14] operant analogs to foraging,[15] and, much more modestly, automaintenance procedures.[16,17]

As has been noted most recently by Dinsmoor,[18] when applied to a standard conditioning paradigm, Equation 1 specifies a simple formula which, in Dinsmoor's terms, is "almost identical in form" to the formulation of Gibbon and Balsam. Dinsmoor also notes that "the parallel is striking" with the "very similar calculation . . . developed by Jenkins." This is not to say that the extension of the delay-reduction hypothesis is identical to the views of either of these more recent formulations—there are differences, as Balsam noted today—but rather to point out that research and theory in at least three laboratories are progressing along similar lines (and there are others, such as Brown and Hemmes' group at Queens College in New York City). It appears that the extensive work by both Balsam and Gibbon (including Balsam's illuminating analysis of trace conditioning today) and Jenkins' group (including Jenkins' powerful analysis of contingency effects and reanalysis of the classic Rescorla–Wagner experiments), coupled with the encouragingly wide applicability the delay-reduction hypothesis has had in more operant paradigms, offers a fair degree of converging evidence and promise for the temporal principles embodied in these formulations.

I conclude by discussing briefly the issue Jenkins raised in his paper of "the surprising insensitivity of autoshaping to the time between the immediately preceding feeding and the onset of the CS." I suggest that the "surprising insensitivity" may be illusory. Using a *within-session and within-subjects* design, we have been able to show clear effects of the immediately preceding interval on *maintained* responding. These differences show up in the first sessions. For example, Ray Preston has conducted a study in my laboratory in which two types of trials are interspersed randomly within a session: Two distinct key lights each precede food by 20 seconds. The contextual cues prior to light onset (that is, during the intertrial interval or ITI) are identical, other than temporal cues. One stimulus follows a 380-sec ITI and is therefore correlated with a 95% reduction in time to reinforcement (380/400); the other stimulus follows a 20-sec ITI and is therefore correlated with only a 50% reduction in time to reinforcement (20/40). We find significantly higher rates of responding to the stimulus correlated with the greater delay reduction. We are extending this work presently (including assessments of whether the same relation holds for acquisition data, and whether the relation depends upon factors identified in Killeen's theory of arousal[19]). While our present data are only suggestive, at this point we suspect that the time between the immediately preceding feeding and CS onset may need to be reckoned with after all. The issue, previously identified by Gibbon, Terrace, Balsam, and associates, is an important one: Does the organism time from the preceding salient event or does it average over many events, demonstrating relative insensitivity to molecular temporal events? I suspect the answer will depend critically on the species and situation under study. Fortunately, the research programs represented in today's papers suggest that before too long we may be in a better position to specify the conditions under which temporal contextual cues affect behavior.

In summary, the present papers suggest progress in the areas of animal timing and temporal control. With respect to the papers on foraging and on conditioning parallel conclusions appear to apply: Just as foraging is a function of the rates at which prey may be captured within each patch and outside the patch, so conditioning is a function of relative times—whether expressed as relative delays or waiting times or delay reductions—to reinforcement in the presence of the CS and in the presence of contextual stimuli in which the CS is embedded.

REFERENCES

1. Spetch, M. L., D. M. Wilkie & J. P. J. Pinel. 1981. Psychol. Bull. **89:** 163–175.
2. Fantino, E. & C. A. Logan. 1979. The Experimental Analysis of Behavior: A Biological Perspective. W. H. Freeman. San Francisco, CA.
3. Fantino, E. 1969. J. Exp. Anal. Behav. **12:** 723–730.
4. Baum, W. M. 1982. J. Exp. Anal. Behav. **38:** 35–49.
5. Dunn, R. M. 1982. J. Exp. Anal. Behav. **38:** 313–319.
6. Pliskoff, S. S., R. Cicerone & T. D. Nelson. 1978. J. Exp. Anal. Behav. **29:** 431–446.
7. Pliskoff, S. S. & J. G. Fetterman. 1981. J. Exp. Anal. Behav. **36:** 21–27.
8. Hursh, S. R. 1980. J. Exp. Anal. Behav. **34:** 219–238.
9. Case, D. A. & E. Fantino. 1981. J. Exp. Anal. Behav. **35:** 93–108.
10. Fantino, E. & D. A. Case. 1983. J. Exp. Anal. Behav. **40:** 193–210.
11. Fantino, E., D. A. Case & D. Altus. 1983. J. Exp. Child Psychol. **36:** 437–452.
12. Navarick, D. J. & E. Fantino. 1976. J. Exp. Psychol. Anim. Behav. Processes **2:** 75–87.
13. Ito, M. & K. Asaki. 1982. J. Exp. Anal. Behav. **37:** 383–392.
14. Fantino, E. & R. Dunn. 1983. J. Exp. Psychol. Anim. Behav. Processes **9:** 132–146.
15. Abarca, N. & E. Fantino. 1982. J. Exp. Anal. Behav. **38:** 117–123.
16. Fantino, E. 1981. *In* Advances in Analysis of Behaviour. Vol. 2: Predictability, Correlation, and Contiguity. P. Harzem & M. D. Zeiler, Eds.: 169–201. John Wiley. Chichester, England.
17. Fantino, E. 1982. Behav. Anal. Lett. **2:** 65–70.
18. Dinsmoor, J. A. 1983. The Behavioral and Brain Sciences. **6:** in press.
19. Killeen, P. R. 1979. *In* Advances in Analysis of Behaviour. Vol. 1: Reinforcement and the Organization of Behaviour. M. D. Zeiler & P. Harzem, Eds. John Wiley. Chichester, England.

Introduction

H. L. ROITBLAT

Department of Psychology
Columbia University
New York, New York 10027

The papers in this section consider the relationship between time and cognitive processes. It is clear that the two are inexorably linked. Cognitive processes take place in time, judge time, and bridge temporal intervals. A prominent feature of these papers (although not the only important feature) is the relationship between time and memory. At least five positions can be identified in the following papers: (1) Memory is that cognitive process which bridges temporal gaps. By this view, time is simply that which separates earlier from later experience. (2) Time is a special attribute of experience. (3) Time is like any other stimulus attribute. (4) Time experience is derived from other properties of memory. (5) Memory is temporal discrimination.

Michon and Jackson argue that temporal properties of events are functionally identical to other stimulus properties. Cognitive processing of the temporal properties, such as encoding and remembering durations, is essentially identical to the active processing of any other stimulus element.

Heinemann and Massaro take a similar position. They both use models developed to explain nontemporal discriminations to account for judgments based on temporal attributes of stimuli.

Staddon, on the other hand, makes the converse argument. In certain memory tasks, the same small set of stimuli is used again and again. In these experiments, the difficulty for the subject lies not in remembering attributes of the stimuli per se but in remembering which of the well-learned stimuli was presented most recently. In such tasks memory could reduce to judgments of relative recency. Staddon argues that discrimination between streams of events is the basis for performance in these working-memory tasks.

Roberts and Kraemer also entertain the possibility that working-memory can be assimilated to temporal judgments, but they reject the notion. Their work, that of Grant, and work from my laboratory indicates that working memory is more active and flexible and does not conform to the patterns that would be predicted by the temporal discrimination hypothesis.

Shimp is also concerned with temporal judgments. In his analysis, however, time judgments are based on an indirect assessment of time. Specifically he assumes that each stimulus presentation activates a set of attributes in memory. As time passes, these attributes return stochastically to an inactive state and the subject uses the proportion still active as a measure of the time passed since the presentation of the stimulus.

Attentional Effort and Cognitive Strategies in the Processing of Temporal Information[a]

JOHN A. MICHON AND JANET L. JACKSON

Institute for Experimental Psychology
University of Groningen
Haren, the Netherlands

1. INTRODUCTION

For many, many years the study of temporal phenomena in human experience has progressed along two virtually disconnected lines: time psychology proper and the study of the temporal organization of memory. Time psychology dealt largely with the psychophysics of duration, with the rate of flow of subjective time and with the experience of the so-called "specious present." Memory research, on the other hand, studied the sequential organization of stimulus events—order and position of items in a series—and serial recall. Only recently do the two trickles appear to be merging into a single stream that may even be heading towards its first rapid.

A burning question—to be answered, perhaps, in this volume—is whether the many and varied phenomena, flow, present, duration and order, may indeed be treated as manifestations of a coherent set of processes or, instead, only as a collection of essentially unrelated processes that entertain only superficial relations to each other. Until further notice we prefer to adhere to the first view, relying, perhaps somewhat naively, on William of Ockham's advice not to multiply explanatory concepts beyond necessity.

In this paper we shall discuss some views and some results that may contribute towards answering the question: How is *conscious time experience connected to memory for lag and order?* Before tracing our main theme, however, we shall outline our general views on the nature, or status, of psychological time.

2. PHYSICAL VERSUS PSYCHOLOGICAL TIME

Humans live in an ever-changing environment. Discounting metaphysical cavils about the *ultimate* reality of Change and Time, we adopt the stance of scientific realism, a position that has shown great merits in physics and biology, for example. Events occur *in reality,* more or less—but not exactly—as and when we observe them occurring. Had it been different, the human race would not have survived so long: a species that systematically misinterprets the ongoings and events in its environment stands no chance in evolution.

In physics relations between events can be specified in terms of *simultaneity* and *succession* (although Einstein has taught us that this is at best a useful parochial view).

[a]This research was supported by a grant (under project number 15–23–15) from the Netherlands Organization for The Advancement of Pure Research.

Before and *after* also have a "real" physical basis: thermodynamics, the expansion of wave fronts and the decay of certain elementary particles prove that "time's arrow" is not an illusion.

Physically, that is about all there is to be said about time.[b] Psychologically, it is not: psychological time has at least two properties that have no representation whatsoever in physics.[12] In the first place, psychological time *streams*, and it does so at an irregular rate: depending on various factors it may drag or fly. In the second place, psychological time has a preferred point: the present or now. In contrast with any definite point in the time of physics the present possesses historical uniqueness. It has moreover a certain width, epitomized by William James[20] as its saddleback quality. Stated more acutely: psychological time is characterized by FLOW and NOW as experiential extras that have no representation as such in the framework of physics.

This immediately specifies the psychologist's task. Ultimately we should be able to provide a coherent description of the experience of FLOW and NOW together with an explanation of these phenomena in terms of underlying processes that comply with the physical concepts and relations of simultaneity and order.

3. MECHANISMS OF TIME AND TIMING

Life distinguishes itself from inorganic nature in its potential for *self-organization,* the ability to develop an internal structure that creates a certain, growing independence from the vicissitudes of the environment. Greater independence is usually achieved by *internalizing* the environment. Thus the early marine organisms developed ways of keeping their internal environment at a comfortable level of salinity by growing an epidermis. This allowed them to survive in, and migrate to, less friendly environments.

At one stage in their evolution it has occurred to most species that one of the most powerful aids for self-organization is the *internalization of change as such.* This allowed them to anticipate certain regular changes in their environment and to return, for instance, to their lair before it became dark in the evening.

There are two fundamental categories of change in nature: cyclical (such as the diurnal rhythm of day and night) and monotonic or progressive (such as radioactive decay, or a stone falling). Not surprisingly evolution has been rather successful in copying both types of change in living organisms and, more specifically, in humans. On the one hand we find a number of endogenous rhythms, some of which seem to have a paleontologically long history, in particular the circadian rhythm.[2,27] On the other hand there are many (more recently developed?) mechanisms for coping with progressive (if not unique) events.[32]

We need not dwell on the various physiological implementations of "internalized change" that chronobiology and chronopsychology have been able to uncover over the past decades. In fact, we agree with Richelle and Lejeune[32] that the construction of a time base (necessary for the timing of a particular behavioral sequence) is a highly idiosyncratic activity from which few if any general insights can be derived that would be pertinent for our understanding of conscious time experience. In their scholarly review, *Time in Animal Behavior,* in which they discuss these various "temporal

[b] As a second intrinsic property of physical time we might consider repetitiveness or recurrence: repeatedly entering into the same state. This however, requires an *external* entity counting the number of repetitions, or remembering the previous occurrence(s) of that state, and therefore it cannot be an intrinsic property of time.

regulations," they repeatedly stipulate that "multiple time bases are continuously constructed in response to the particular requirements of each situation and replaced by others when they become useless."[32] (p. 165)

Therefore, we shall refrain from attempts to account for subjective time experience in terms of physiological clocks or inhibitory neural circuits. We shall leave the rich, but variable organismic tapestry of timekeepers for what it is and, instead, concentrate on the functional aspects of temporal information processing.

4. TIME AS INFORMATION

Quite some time ago one of us proposed that temporal information is essentially *equivalent* to other types of information normally encountered in human information processing. This proposal took the form of what was called the "equivalence postulate":

> Duration appears under a dual aspect. On the one hand it is a physical construct. On the other hand it appears as a "property" of patterns of information in the real world, which we call objects and events. There is no useful way for distinguishing between this latter aspect and the other properties of information such as size, intensity or locus.[24] (p. 254)

If we are to treat information about temporal relations between events as proper information we ought to be able to integrate a model of subjective time experience with the pertinent insights regarding the human information-processing system.

Although we assume that the readers of this volume have a working knowledge of the general characteristics of the type of model that psychologists use to account for what happens to input information from the instant it reaches the sensory systems of the organism, we shall briefly review its major features (FIG. 1).

We suppose that incoming stimuli are buffered in a sensory register, for up to perhaps a second, during which they are transferred to working memory for a progressively deepening analysis. Working memory should not be conceived as an independent component in the system, but rather as that part of our associative memory network which is active at a particular instant. Depending on the task at hand, the process of analyzing input information will in most cases lead to a response, overt or covert. Whether or not a response ensues, in either case the information may leak away, some 20 or 30 seconds after it is no longer being (re)activated, or it may be transferred partially or in its entirety to permanent memory. In other words, it may leave a more or less robust permanent trace, which, at future times, may become reactivated again.

Most current models of human information processing make a distinction between two basic processing modes: *automatic* and *deliberate* (or *controlled*). Automatic processing is supposed to be a fast process that accepts many inputs simultaneously and produces outputs almost instantaneously. It requires no "effort" and there appears to be no upper limit to the structural complexity of the inputs it can handle. Since, by contrast, the deliberate processing mode can only treat inputs in a sequential fashion, it is a comparatively slow process which is limited in its capacity and requires conscious mental effort. Deliberate processing, in other words, requires attention. The focus of attention can be determined by the nature of the task, or by external or internal instructions. Also, if for one reason or another the progress of automatic information processing is interrupted, the deliberate mode will take over in an attempt to reinstate the normal processing routines.

This is a fairly standard description of the basic layout of what we now believe to be the functional composition of the human information-processing system. Details can be found in any introductory text on cognitive psychology or human performance theory (see, for example, Anderson[1]).

5. THE QUEST FOR THE TEMPORAL ATTRIBUTE

If we accept the view that temporal information is proper information, we still have to answer a number of basic questions. The first asks what in a sequential stimulus configuration actually constitutes the functional stimulus or temporal attribute that serves as the input for temporal judgments. Despite a century-long line of empirical

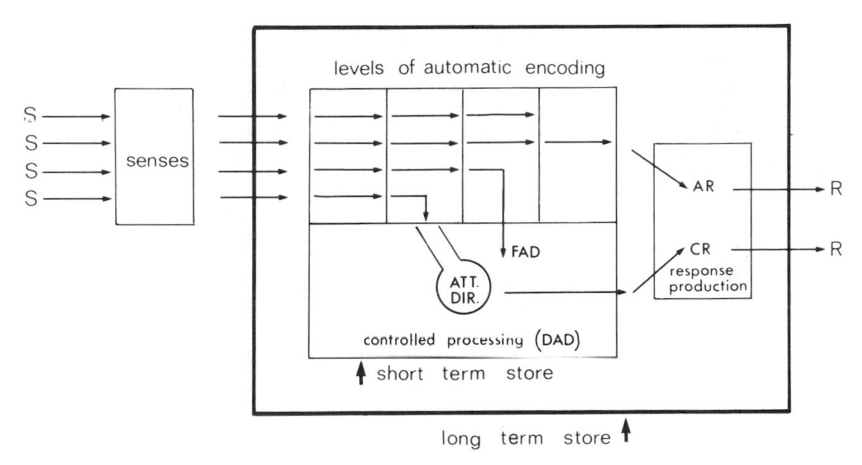

FIGURE 1. A conceptual model of human information processing. Information from the environment (S) reaches the sensory systems, which include a precategorical storage (senses), and then is subjected to a gradually deepening analysis. This is carried out in short-term storage, which is the activated part of long-term or permanent storage. Much of this analysis is carried out automatically, attention being directed (ATT. DIR.) to only a small part of the information being processed. This is the information that is processed under conscious control and that will lead to controlled response production (CR). If attention is occupied, then other stimuli requiring attention for proper processing will be neglected (FAD) as attention capacity is limited. Many stimuli will, however, be processed without conscious "intervention" at all; they may eventually lead to automatic responses (AR). Controlled and automatic responses may be produced simultaneously, as when we talk when walking. (From van Zomeren.[44] Reprinted by permission.)

research and notwithstanding a fair amount of consensus among contemporary investigators that it is in one way or another related to the content and complexity of the associative context, "there remains a prominent lack of understanding of exactly what constitutes the temporal attribute."[14] (p. 525)

The prevalence of descriptive theory in the approach to temporal information processing has thus far rendered few if any explicit aids to better understanding of what constitutes the ultimate temporal attribute. Numerous cognitive and motivational factors have been identified, each of which appears to contribute to the impression of time flow. Surprisingly, however, most can be traced back to the original formulation of a cognitive theory of time experience by the French philosopher,

TABLE 1. Factors Determining Subjective Time Judgments According to Guyau[16] and Their Modern Equivalents

Guyau's Idiom	Modern Idiom
The estimation of duration is dependent on:	
1. The intensity of the represented events (images)	Intensity; S/N ratio; clarity
2. The intensity of the differences between these events	Discriminability
3. The number of events and the number of their differences	Information (bits); event uncertainty
4. The rate of succession of these events	Rate of information (bits/sec)
5. The relations between these events, their intensity, their similarities and distinctions, their temporal distance and their position in time	Structural information, novelty contextual change, sequential uncertainty, grammar
6. The time necessary to "conceive" these events and their relations	Decision time; identification time
7. The intensity of the attention devoted to these events and the feelings of pleasure and effort that accompany them	Selective attention; deliberate processing; mental effort
8. The desires and emotions that accompany these events	Motivation, "set," interest
9. The relations between these events and our attitudes and expectations	Anticipation; expectancies; temporal perspective

Guyau.[16] It seems appropriate to reconsider Guyau's original list of determinants (TABLE 1).

What happened since 1890 amounts to the following. First, empirical results testify that Guyau's original list is fairly complete. Second, we see a continuous process of terminological revisions; with the change of the winds of fashion in psychological research Guyau's determinants are renamed. However, updating of terminology does not automatically imply progress. So, to see what extent progress has indeed been made, let us consider four recent propositions.

1. Event uncertainty. Michon[23] showed that serial production of two second intervals was inversely related to the information content of stimuli. More importantly, he also showed that interval productions were affected only when information was actively processed. When subjects performed the time production task just by looking at stimuli projected on a screen and not reacting to them, no effect on interval productions was found. It soon became clear, however, that "number of bits processed" is too shallow a means of describing cognitive activity, including the estimation of time.

2. Memory storage. A few years later, Ornstein[30] introduced his memory storage metaphor: the amount of time retained is a function of the memory capacity consumed by the information stored during the interval. This point of view quickly gained considerable popularity, partly as a result of the elegance of Ornstein's arguments.

3. Processing effort. Again some years later, Block[4] proposed that it was rather the effort of remembering the contents of the original interval that appeared to determine retrospective time judgments. This view, however, did not settle the matter either.

4. Contextual change. In 1978 Block[5] introduced the somewhat elusive concept of contextual change as the prominent factor determining the subject's time estimations. Contextual change may be taken to provide an incentive to a subject to temporarily pay more attention to factors that are not directly task-related. This will increase the probability of noticing which temporal cues happen to be available in the environment. As such, this process resembles the orientation response.

There is no doubt that these authors have empirically confirmed a number of important aspects of temporal information processing, but neither of the four approaches reveals what constitutes the effective temporal attribute: What information qualifies as temporal information? What aspect of memory storage, of processing effort, or contextual change is indeed interpreted as temporal?

Upon further reflection there is only one candidate immediately available from the considerations offered thus far, namely, sequential *order*. Order, as we have already pointed out in section 2, is physically determinate and it is also directly related to the concept of information. Temporal judgments require an ordered memory representation of the event sequence under concern and this requires at least a partial retrieval of the order of the constituent events.

Tzeng, Lee, and Wetzel[38] were among the first to use the principle of intrinsic order in a process model of temporal information coding. Their argument ran as follows. Subjects who are given a list of words to study for later recall can derive temporal information from the fact that earlier words in the list constitute a background for the word currently being presented. This is due to the fact that they keep these earlier words active in their working memory by continually rehearsing them (study-phase rehearsal). This intrinsic old–new relation constitutes the stuff of which temporal information is made. We agree with this view in principle, but not, as we shall argue later, with the conclusion Tzeng *et al.* draw, namely, that the encoding of the order relation takes place automatically!

Perceptual examples of the intrinsic temporal order between successive events have been given by Collard and Leeuwenberg.[10] Consider, for instance, the two shapes in FIGURE 2. Collard and Leeuwenberg argue as follows. FIGURE 2A has a simpler representation than FIGURE 2B in terms of structural complexity. Therefore, if subjects observe these figures in the order *A–B*, then they are in a more difficult position than if the order is reversed. (The reason is that 2A can be coded as "three identical adjacent triangles" and 2B, less specifically, as "a zig-zag with a slanted line projecting from the left end of the zig-zag"). While 2A can be coded as a special case of 2B, the reverse is by no means the case. This asymmetry specifies a natural order for the two figures since people, when dealing with their environment, tend to stick to a chosen frame of reference as long as possible. If subjects are asked to recall the order of presentation of a pair like 2A and 2B, there is a marked preference for the facile order *B–A*.

In our investigations we have adopted a similar stance. We assume that successive events (more specifically: the items in a list of words) have a number of readily available attributes or cues that enable an item's association to a context. Some of these attributes may serve as temporal cues and, naturally, some will be more "powerful" than others. Thus, for instance, temporal-order information might be inferred from word pairs that are causally related, such as *fall-injury* or from pairs which are logically related, such as *pianist–music*. Even spatial relations may occasionally be taken to represent temporal order as in pairs like horse–cart or head–toe. The more

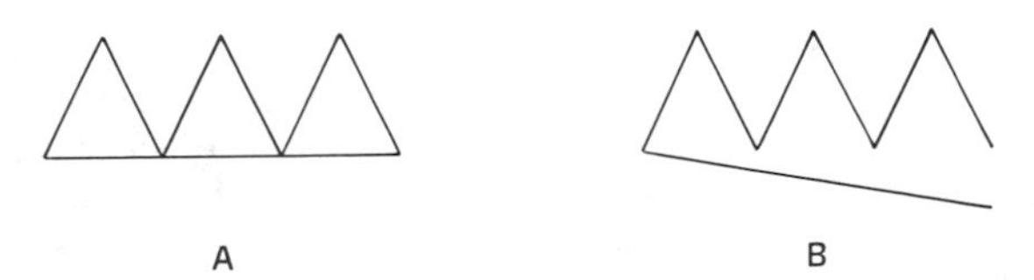

FIGURE 2. Two structurally asymmetric shapes. (After Collard and Leeuwenberg.[10])

powerful cues, that is cues that strongly imply some natural order, may indeed operate more or less automatically in the sense that subjects will find it difficult *not* to notice their presence. If, however, such powerful cues are not available, temporal relations between list items may still be constructed to the extent that the available cues can be given a (quasi-)temporal interpretation.

Temporal information processing then, in our view, means establishing an associative network in relation to both temporal and topical knowledge as it is represented in memory. The quality of temporal information processing will be dependent on factors that orient subjects towards using cues that can be given a temporal (that is, an order) interpretation.

6. IS TEMPORAL INFORMATION PROCESSING DELIBERATE OR AUTOMATIC?

The four theories discussed in the previous section not only fail to specify the temporal attribute, but they are also weak with respect to at least two more crucial issues.

A second question asks whether the temporal attribute, once identified, is encoded upon acquisition or, instead, constructed during retrieval. This is not a trivial matter at all. Those who consider the strength of a memory trace,[18] or its resistance against further decay,[41] as the temporal attribute necessarily embrace a retrieval theory. In contrast, associative theories[13,38] assume that temporal tags are associated with incoming stimuli, and this naturally implies some sort of coding process. Our own theory naturally falls into the latter category, although the idea that topically encoded associative cues may sometimes be interpreted *post hoc* as temporal does not entirely rule out the possibility of temporal information being generated only upon retrieval.

The third important question to ask is whether temporal information is processed as an inevitable automatic by-product of information processing in general, or whether temporal information processing in itself requires a specific cognitive effort. Answering this question will be one major problem addressed in the present paper.

The four theories mentioned in Section 5 make no explicit assumptions about either the coding versus retrieval nature of temporal information processing. Neither do they suggest anything about the strategies that might be available to a subject who needs to cope with a task requiring temporal judgments. As far as these theories are concerned, temporal information processing might well be a simple matter of automatic processing, although that would be quite the opposite of what their proponents intended!

It has recently become popular among certain authors to claim that temporal information is automatically encoded. To what extent is this claim justified? In our view the tendancy to consider temporal information primarily as a free gift, an automatic concomitant of the processing of nontemporal information, is unwarranted. It is hardly, if at all, supported by reliable evidence. We wish to argue instead that subjects cannot organize this material temporally unless they explicitly devote attention to it. We maintain that temporal coding is essentially restricted to the deliberate processing mode.

In part, the claim of automaticity stems from a conceptual confusion. The fact that events that are successive in physical reality are normally perceived in their correct order should not be called automatic processing but a psychophysical necessity. Instead, we should stick to a more appropriate, limited definition of automaticity: automatic are those processing activities that do not require attentional resources and that, consequently, are not subject to interference with other concurrent tasks or task-components. It is from this viewpoint that we shall now consider a number of aspects of temporal information processing, in order to demonstrate that each and any of these aspects implies in a plausible way the need for the deliberate processing of temporal information. Generally speaking, these demonstrations take the form of showing that experimental manipulation of the aspect under concern will effectively influence the quality of temporal information processing. One would not expect such an influence if the processing were automatic.

In a well known paper Hasher and Zacks have analyzed the concepts of automatic and deliberate (effortful) processing, arguing that "when a person attends to an input, some of its attributes are automatically encoded into long-term memory, whereas others require more or less effortful processing to be encoded into a permanent memory trace."[17] (p. 358) They specified several criteria for determining through which information processing mode a certain attribute is actually encoded. In particular varying *instructions, level of practice, state variables, divided attention requirements,* and *developmental trends* ought not to affect task performance if the information is processed in the automatic mode. On the basis of some rather inadequate evidence (in particular, a paper by Zimmerman and Underwood,[43] discussed in Jackson and Michon[19]), Hasher and Zacks[17] reached the conclusion that temporal information is among the attributes of information which are indeed automatically encoded.[c]

Using a line of reasoning similar to that of Hasher and Zacks, we have, in a recent series of experiments, come to a totally opposite conclusion. In our view *temporal information is not encoded unless noticed and not noticed unless meaningful!* Very little temporal information, if indeed any at all, is encoded if no deliberate processing is involved. Our experiments dealt with a number of criteria, some of which were identical to those of Hasher and Zacks:

1. Semantic status of stimulus elements. If temporal information is processed automatically, similar temporal judgments should be expected for words that have different semantic characteristics, such as concrete versus abstract.

2. Instructions. The weight of Hasher and Zacks' conclusion[17] that instructions do not affect temporal information encoding was based on the distinction between incidental and intentional learning instructions. The experimental evidence Hasher and Zacks quoted seem to support the automaticity claim. However, attempts to replicate the effect have failed, and lead to the opposite conclusion that instructions *do* indeed affect temporal information processing.

3. Developmental trends should not be effective if temporal information is encoded automatically, according to Hasher and Zacks.[17] Yet, our results show clearly that developmental stage does have a large effect: 5-year-old children and 11-year-old children behave in very different ways.[47] The developmental criterion under study will not be further discussed in the present paper, although the experimental results are given elsewhere in this volume.[45]

Having arrived at the conclusion that the automaticity hypothesis cannot be upheld and that temporal information processing is very much dependent on deliberate encoding, the next step is to seek for a more detailed insight into the actual processing performed on temporal information. The following two examples illustrate this approach.

4. Sequential structure of the stimulus series. Manipulating the serial associative

[c]The impact of this unwarranted position is already visible in recent introductory texts. In Stephen Reed's *Cognition: Theory and Applications,*[46] the discussion relies heavily on Hasher and Zack's paper. Reed concludes: "Evidence indicates that people can effectively record frequency, spatial and temporal information when they are not trying to learn this information. . . . [These variables] do not interfere with each other and show little developmental change." (p. 52–53)

relations between items in a list should not affect the (automatic) processing of temporal information.

5. *Context.* Contextual cues may help to temporally organize a series of events. If encoding would be automatic no differential effects on temporal judgments would be expected from manipulations of that context.

6. *Levels of processing.* By varing the depth at which topical information is to be processed in a memory task, the trade-off between topical and temporal information may be revealed.[48] This factor will not be treated in this paper, but, again, it appears elsewhere in this volume.[45]

7. *Individual strategies.* At this point, one may start to wonder whether perhaps the conventional experimental paradigms are too simple. When processing deliberately, subjects normally engage in a rich assortment of coding strategies and consequently only a detailed analysis of these strategies will eventually open the gate to better understanding the FLOW and NOW of subjective time. Individual strategies turn out to be prominent features of temporal information processing. They deserve to be studied in the framework of the analysis of verbal protocols since there appears to be no direct way of influencing subjects' choice of strategy. An analysis of individual strategies also reveals a considerable effect of "level of practice," a criterion which, according to Hasher and Zacks,[17] is indicative of the deliberate information-processing mode.

7. SOME EXPERIMENTAL RESULTS

Semantic Status: Concrete versus Abstract Words

Recently, Tzeng, Lee, and Wetzel[38] suggested that temporal-order information is acquired as an automatic by-product of rehearsing prior times while a new item in a list of words is being presented. The items in the set of rehearsed words constitute the context for ordered associations between old and new items, and in their opinion this is a necessary and sufficient condition for temporal coding to occur.

The evidence for the automatic character of temporal information processing stands on very weak ground, however. One important point is that the supporting evidence has been obtained only with highly familiar stimuli: concrete words or pictures. However, in their now classic "time tag" study Yntema and Trask[42] already used both concrete and abstract words—a rarely quoted fact! What is more, they found a considerable difference in precision of temporal judgments between the two word types. This result does suggest that encoding of temporal information is not automatic but instead requires a certain amount of conscious effort. If the relation between items in the rehearsal set and the current items would be a both necessary *and* sufficient condition for temporal information to be encoded, then the distinction between abstract and concrete words should make no difference for the quality and precision of temporal judgments. But if, on the other hand, we do find a difference, then rehearsal as such cannot possibly be a sufficient condition for temporal coding to take place. Instead, something which is related to the concrete–abstract dimension would be involved in temporal coding. This in turn would suggest a role for such deliberate strategies as imagery, elaboration, and categorization.[17]

We recently tested this expectation.[45] On the assumption that temporal information processing is automatic, we would expect no difference between the temporal-order retention of concrete and abstract words, given that the words are indeed in the "rehearsal set." Unfortunately, we cannot know with certainty whether or not a word has actually been in that set during the encoding stage. If a word can be recalled, however, we know it must have been in the rehearsal set (otherwise, its likelihood of being there would be very small, except perhaps for the most recent words in the list). Furthermore we can manipulate the probability of a word entering the rehearsal buffer by explicitly instructing the subject to either remember or forget an item.[6]

In our experiment subjects saw a list of 40 words, either all concrete or all abstract. Half of these words were cued to be forgotten (*F*-cued), the other half to be remembered (*R*-cued). The order of the forgetting cues was random. Immediately after presentation of the entire list, subjects were asked to recall as many *R*-cued words and, completely unexpectedly, also as many *F*-cued words as they could. Five minutes later they were given an unexpected recognition test along with an equally unexpected temporal-order retention test. The experiment was later repeated with slight modifications to clear up some smaller issues. The main outcome of both experiments was identical though. The results are presented in FIGURE 3.

The diagrams show the relation between the true positions (blocked in 8 units of 5 positions each) and the position according to the subject. If temporal judgments had been perfect, all points would fall along the dashed diagonal, a result that was not

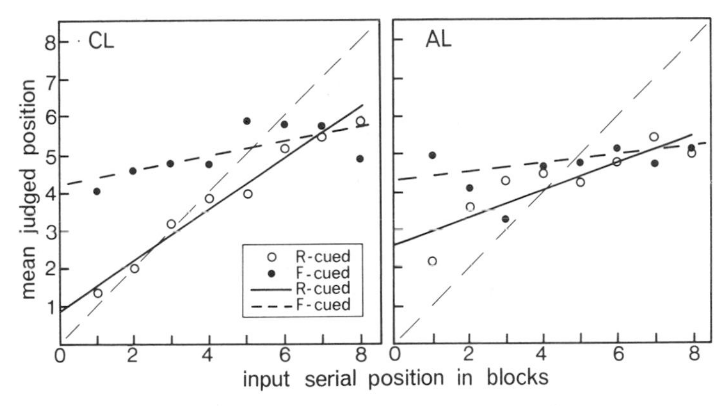

FIGURE 3. Mean judged position for the *R*-cued and the *F*-cued words as a function of their actual input serial position (in blocks); within the concrete (CL) and abstract (AL) word lists. (*R* = word to be remembered; *F* = word to be forgotten.)

obtained. In fact, the words that subjects were instructed to forget carry virtually no temporal information. The little that appears to be there is entirely due to the fact that some of the to-be-forgotten words succeed in being recalled, and these words do indeed carry some temporal information. The words which were to be remembered were, not surprisingly, recalled much better and they also produced better performance on the temporal-order test. Although these results confirm the earlier findings of Tzeng *et al.*,[38] the major difference found in our results is that which exists between concrete and abstract words: for those words that were indeed recalled, concrete words did quite appreciably better than abstract words.

This result is in line with our expectations: being in the rehearsal buffer at encoding time *is not a sufficient condition* for temporal information processing to occur. Something else is needed, something that is related to the encoding strategies used by subjects to meet the requirements set by the memory task—in this case recalling concrete or abstract words. Such mnemonic strategies are, however, according to Hasher and Zacks, for instance, typical for deliberate processing rather than for automatic processing.

Instruction

One very direct way of challenging the automaticity assumption is the incidental learning paradigm. Someone may perform a task and at the same time acquire certain unrelated extra knowledge without effort (showing in the absence of systematic changes in learning time, work load, response latencies, and so forth). In that case, we say that this extra knowledge has been acquired automatically and without conscious control. Conversely, if certain information is being acquired at no extra effort, such information should not have a detrimental effect on the performance of the main task. Both phenomena have been reported in the experimental literature. More specifically, it has been claimed that instructions to pay attention to the temporal organization of words in a list will neither affect later recall of serial position,[37,43] nor later item recall.[43] Unfortunately this evidence is either not very reliable or it can be interpreted in the opposite direction for reasons that we will explain elsewhere. In our opinion the matter is entirely undecided and for that reason we have performed some experiments in which we have looked at the role of instruction on the coding of concrete as well as abstract words.[26]

TABLE 2. Average Recognition, Recall, and Temporal-Order Retention in Concrete and Abstract Lists

	Experiment	Concrete Lists		Abstract Lists		
		RI	TOI	RI	TOI	
1.	Recognition	.77	.48	.63	.27	(% correlated)
	Temporal-order retention	.16	.31	.08	.18	(pooled)
2.	Recognition	.98	.99	.95	.95	(% correlated)
	Temporal-order retention	.44	.67	.04	.18	(pooled)
1.	Recall	.57	.63	.35	.52	(% correlated)
	Temporal-order retention	.17	.68	.03	.60	(pooled)
2.	Recall	.83	.88	.59	.72	(% correlated)
	Temporal-order retention	.42	.79	.08	.27	(pooled)

Abbreviations: RI = recognition instructions; TOI = temporal-order instructions.

When subjects are explictly instructed in advance to pay attention to the temporal order of words in a list for a later temporal-order retention test, they should more readily use the temporal cues available in the coding stage than if they receive instructions for a later recognition test. Temporal information will not facilitate item recognition because an item in a recognition test is presented in isolation ("have you seen this word before?"). It requires matching of the cues that are intrinsic to that particular item rather than matching any of the cues that the item might entertain with other words. The result should be twofold: first, we expect a detrimental result of a temporal-order instruction on recognition scores. Second, words that are recognized will not necessarily be characterized by high temporal retention. In contrast with recognition, recall *does* require an adequate, active reconstruction of the coding context, and temporal cues clearly serve a purpose in that case. Consequently, we expect temporal cues to be effective in a recall task but not in a recognition task.

TABLE 2 shows the relevant results of this experiment and its replication (experiments 1 and 2). The replication had a comparable design, except for a few insignificant improvements (a larger number of subjects and a better balanced order of testing) and, more importantly, shorter lists (30 versus 16 words). The latter factor causes the only systematic difference in the results: the recall and recognition scores are consistently higher in the second experiment because of the shorter lists. There is a slight but not very consistent increase in temporal-order retention as well.[d]

The results of both experiments are in agreement for the temporal scores. In experiment 1, recognition scores suffer from temporal-order instructions as expected. In experiment 2 the lists were much shorter and apparently the recognition scores show ceiling effects in this case. Recall on the other hand does indeed profit from temporal-order instructions: recall scores are consistently higher under temporal-order instructions than under recognition instructions, in both experiments. This gain is much larger for abstract words than it is for concrete words, which suggests that relatively speaking temporal cues constitute a larger part of the total set of available cues in the case of abstract words (even though the total cue set is smaller in an absolute sense).

As far as temporal-order retention is concerned, we see that it suffers considerably under recognition instructions, compared with retention under temporal-order instructions. The latter increase the level of temporal-order retention, in particular when we look at the recalled words. In this case again, the effect is relatively much larger for abstract words.

Consequently, we are forced to disagree with those authors quoted above who claim that incidental and intentional instructions in a temporal information processing context do show the automaticity of this processing. On the contrary, we have added considerable evidence for our view that temporal information is largely a deliberate process which requires attentional resources.

Cognitive Streaming

Time being notoriously one-dimensional we can cope with certain complexities only successively, at least in the deliberate mode of information processing. The way of coping with this difficulty is to perceptually split what is going on in a meaningfully patterned "objectified" foreground, pushing all else into the background. A seeming simultaneity may then be maintained by quick alternation between foreground and background. This can be achieved only if there are indeed several concurrent patterns of information. The phenomenon has been studied extensively in auditory perception, where it is known as streaming.[8,29] It also is known from the macrosphere of daily life, at least among those who are performing quite incompatible roles in their professional or personal life.[33]

The problem, in the present context, is the following. The streaming effect will occur when deliberate processing is required: it reveals that selective attention operates in situations that exceed the capacity of the information-processing system. Stable streaming will occur only if the structural coherence between nonsuccessive elements in a sequential pattern is sufficiently large. In cognitive terms this would mean that categorization (or other topical principles of organization) may overcome the regular temporal organization of simple linear succession. This is true, of course, under the

[d] In our experiments we have used two different performance measures for temporal-position judgments. The first is the amount of variance in a subject's performance that is accounted for by assuming that correct position is retained. It is expressed as r^2, the square of the product moment correlation between time and judged temporal position. The other is a deviation index, which takes better account of partially correct, but wrongly located segments in a list. (For examples, see FIG. 8, subject 2 versus subjects 3 and 4).

assumption that processing of either form of information (temporal and topical) requires effort and that increasing the impact of categorical information will decrease temporal information processing. Evidence for this assumption was found by Murdock:[28] categorization leads to better item recall and to less order retention. Consequently, one would expect the decrement of temporal information retention not to occur indiscriminately, but between rather than within the cognitive streams, in much the same way the temporal relations *between* rather than within auditory streams are *indeterminate*.

In summary, losses of temporal information occur when the subject is dealing with structurally complicated topical information. This is not due to a shortcoming of memory in general, but because of a rather specific mechanism: *cognitive streaming*. The prediction to be derived from this assumption is that within-stream temporal retention remains unaffected (because attentional effort remains more or less the same, as in simpler nonstreaming cases), while between-stream temporal retention will be negatively affected. It should be emphasized though, that categorization as such does not necessarily distract from temporal organization: if categorically similar items in a word list are serially blocked (for example, a number of tools followed by a number

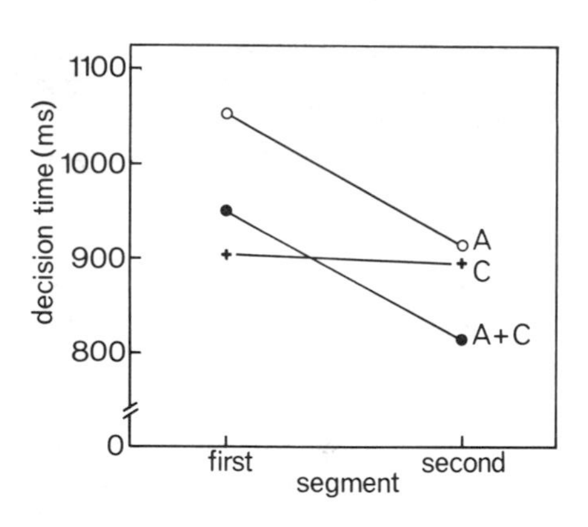

FIGURE 4. Time to decide whether a 6-frame movie fragment was *old, new* or *paraphrase,* for fragments before or after a change in scheme (A), a cut (C), or the combination of the two (*A* + *C*). (After Carroll and Bever.[9])

of animals, and so forth), temporal organization may be facilitated because hierarchical organization principles can be applied.[22,39,40,49] Such clustering, however, will not be considered in the present discussion.

In order to see whether topical organization (categorization) will indeed interfere with the coding of temporal order, we performed some experiments in which subjects were presented with series of 30 simple pictures from a set of 260 designed by Snodgrass and Vanderwart.[36] Pictures were chosen for their concreteness and also because some later experiments were to be carried out with young children.

Two subsets of 30 pictures were selected. The first provided us with the stimuli from which we derived "some-of-everything" or *noncategory* (*NC*) sequences. The other sequences, the *category* (*C*) lists, consisted of permutations of 15 pictures of tools (such as hammer, spanner, and axe) and 15 pictures of common animals. The subject was instructed to remember such a series of 30 pictures for a later memory test. After presentation, a recognition test as well as an unexpected temporal-order test were given.

In order to see whether the temporal coherence of such a sequence could be disrupted we *enhanced* the categorical distinction. Enhancement of a potential disruptive factor is frequently an illuminating operation. A good example of its power

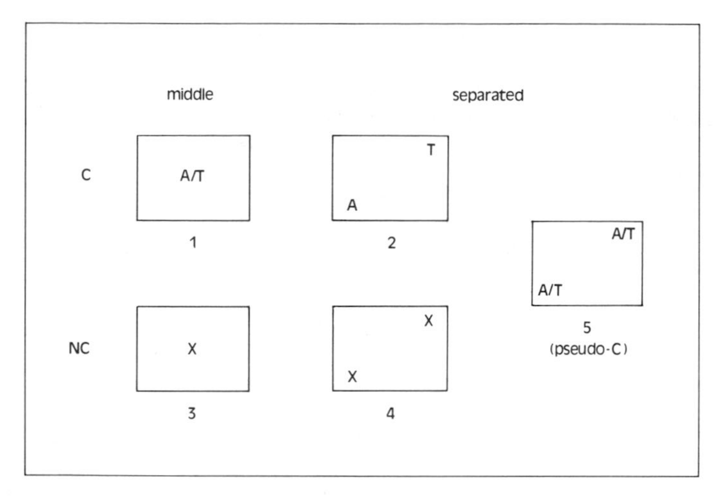

FIGURE 5. Presentation conditions of category (C) and noncategory (NC) series of pictures.

is revealed by Carroll and Bever.[9] They showed a brief movie scene and then tested subjects by showing them either a fragment of the same scene again or a paraphrase, or a new fragment. Each scene contained a "shift," either a change in action (*A*) or a cut (*C*) or a combination of the two (*A* + *C*). The test fragments consisted of the six frames immediately before or immediately after a shift.

The result is shown in FIGURE 4. It reveals that postshift fragments are recognized (or rejected) much faster than are preshift fragments. More important for the present discussion is that action changes have an effect while cuts by themselves do not. Cuts, however, enhance the effect of action changes on recognition quite considerably. These findings suggest that structural factors may have an influence which will become manifest only if mediated by a meaningful stimulus situation.

FIGURE 5 shows the various presentation conditions. In all five conditions pictures were shown at a rate of one per 4 sec. In the first and third conditions the series (*C* or *NC*) were shown in the middle of the screen. In the second condition the category distinction was enhanced in the *C* list by showing the items of one category in the upper right hand corner of the screen, and the items of the other category in the lower left hand corner. The fourth and fifth conditions were "pseudo-enhancements." In condition 4 the items of the *NC* list were randomly assigned to the upper right and lower left corners, while the pictures of the *C* list were assigned in a scrambled order to these corners in condition 5.

TABLE 3. Mean Proportion of Correct Order Judgment as a Function of List Structure and Type of Test Pair

Control Conditions	Within Categories	Between Categories	Within/Between
1. C Middle	.72	.74	.98
2. C Separated	.70	.50	1.40
3. NC Middle	.63	.56	1.13
4. NC Separated	.58	.64	.90
5. PsC Separated	.58	.54	1.05

ABBREVIATIONS: C = Category; NC = noncategory; PsC = pseudo-category.

The results of this experiment are summarized in TABLE 3. It gives the proportion of correct temporal-order judgments separately for each of the five conditions for within-category and between-category picture pairs. The "control" conditions (3, 4 and 5) appear to produce better than chance-order judgments, the average being close to 60% (50% being chance level, and primacy–recency effects being eliminated by analyzing word pairs from the positions 5 to 26 only).

The most interesting comparison is between conditions 1 and 2. These two *C*-conditions show a higher performance (70% or better), except for the between-category pairs that are spatially separated, which is exactly at chance level (50%). We conclude that spatial separation does effectively destroy the temporal information about between-category pairs when separation enhances the categorical information, but not if the separation is not "supporting" such a meaningful distinction in the stimulus material as in condition 1 (*NC*) or in condition 5 (pseduo-*C*). This result ties in with the finding of Carroll and Bever[9] discussed before. The third column in TABLE 2 gives the ratios of the within/between category scores that are proportion-correct. The higher the value of this fraction, the more "streaming" to render the temporal relations between cognitive streams indeterminate.

We conclude that cognitive streaming does indeed take place, but only if the materials can form meaningful streams. In the present case this meaningfulness is supplied by the category distinction between the words.

Context

We have already argued that asymmetric associative structure is an important temporal cue (and perhaps the only one). Even when there is no such asymmetry inherent in the material presented, there is still an opportunity for subjects to derive temporal information from the stimulus sequence. Subjects may be able to associate successive stimulus items to events that are actually part of the experimental context but otherwise unrelated to the stimuli. If this context has an intrinsic temporality—and this will normally be the case—then correct temporal coding of stimulus items will become feasible even when they themselves lack intrinsic temporal cues.

The following experiments were designed to see if, and to what extent, subjects are indeed capable of using such contextual cues. The way in which we achieved this was by systematically providing a context *during* stimulus presentation, adopting a technique that was earlier used by Guenther and Linton.[15] These authors presented a series of pictures against an auditory background of a simple, temporally well-structured story. Subjects were explicitly instructed to associate the pictures with the concurrent line in the story. As was expected, performance on a temporal-position judgment task improved relative to a control group of subjects who were not asked to connect the pictures and the story.

This experiment leaves several questions unanswered. In particular, it does not reveal *how* the structure of the context actually affects performance. In other words, the experiment gives no indication about the extent to which the temporal coding process is controlled by the story. This was the main problem we wanted to tackle in our own experiments.

As stories we used a set of seven simple scripts[34] describing stereotypical activities such as making a trip by train, eating in a restaurant, or visiting a patient in a hospital. There is a strong tendency for people to reproduce a text based on a script in the normal, canonical order of the story actions. By varying the instructions (incidental versus intentional) and the temporal structure of the scripts (by means of flashbacks), it should be possible to highlight the ways in which temporal judgments about a concurrent word sequence are affected by the context provided by the script.

In a first experiment we compared the retention for temporal position as a function of the type of word (concrete versus abstract) and as a function of the type of instruction. In three of four conditions subjects heard a "shopping script." In one of these conditions subjects received explicit instructions to associate the words with the story; in the second they were instructed to count the number of definite articles in the script text; and in the third condition they were informed that they might be able to use the story context, but no further details were given. In the fourth condition, they were only told that it might help them to imagine what they would do when shopping in a supermarket. The results of this experiment may be summarized as follows.

In the first place the explicit instructions for contextual associations of items in a word list can be obeyed. We observed a significant improvement of the temporal coding of abstract words in comparison with concrete words, but also in comparison with abstract words presented without concurrent story, as in the fourth condition. This leads to the conclusion that only words that entertain impoverished associative relations to each other—such as abstract words—profit from context.

A second experiment that is relevant to our present discussion was aimed at studying the effect of changes in the temporal structure of the context by introducing flashbacks and neutral statements into the story. Flashbacks reverse the conventional order of the events in a sequence ("Before they went to the platform they bought a newspaper"), neutral statements refer to events that can happen at any time in a story ("She polished her glasses"). If it is indeed possible to use contextual temporal cues for position encoding we would expect the judged position of items in a concurrent word list to be related to the remembered position of events in the contextual story. Both neutral statements and flashbacks are known to shift positions occasionally,[3] and we may expect the associated list items to go along with these story elements.

Six scripts were constructed, each containing some statements that entertain a natural order relative to one another plus some temporally neutral statements. Versions with and without flashback were then presented to groups of subjects. The concurrent word lists contained either high-associative word pairs (uncle–aunt) or low-associative word pairs (pot–food). Word pairs could either have a symmetric or an asymmetric associative strength (such as the pair *dentist–drill* versus *drill–dentist*). Lists consisted of 48 words, arranged in such a way that words belonging to a pair were never farther than 4 words apart. The design of this experiment is somewhat complicated but essentially amounts to a 2^4 factor between subject design. The factors in the design were context/no context, normal/flashback story, word association level high/low, and finally asymmetric word pairs presented in high/low association order.

The results of the experiment indicate that the influence of context varies with the structure of the word lists as well as with the structure of the context.

A first result is shown in FIGURE 6. This reveals that there is a difference in temporal-order retention between high- and low-associative word lists. The effect is, however, restricted to the first few block-serial positions. In other words, this factor is capable of increasing the primary effect. FIGURE 6 also shows that the presence of the story context is only of advantage toward the end of the list, for block position 5 and 6. A further analysis of this finding indicates that it is entirely due to the presence of flashbacks (FIGURE 7). This suggests that flashbacks do indeed function as prominent attention-getters: their influence may be similar to the contextual change effect reported by Block and Reed.[5]

The effect appears to be independent of the type of word list: both high- and low-associative word pairs show the same results. This is confirmed also by the analysis of what happens to those words that have been associated to the story context in a

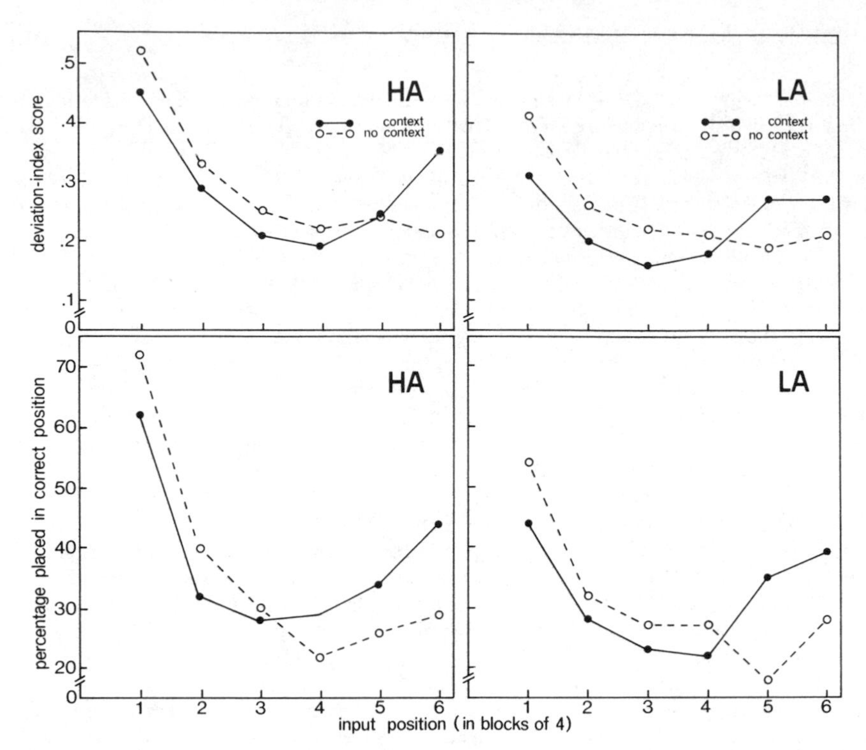

FIGURE 6. Percent of correct judgments and deviation index score for judgments as a function of their actual input serial position (in blocks) and condition (context/no context) within the high-associative (HA) and low-associative (LA) word lists.

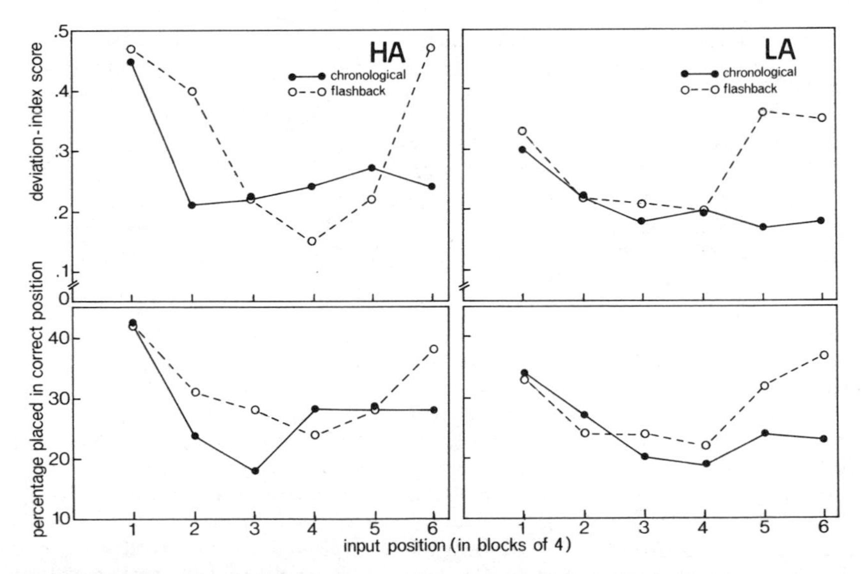

FIGURE 7. Percent of correct judgments and deviation index score for judgments as a function of their actual input serial position (in blocks) and condition (chronological/flashback) within the high-associative (HA) and low-associative (LA) word lists.

flashback version of the story. The expectation is that if the story sentences will, upon recall, be reverted to their canonical positions in the scripts, the corresponding words would go with them. It turns out, however, that very few flashbacks are "undone"; subjects apparently have little difficulty in remembering stories as actually told, a result that is essentially in agreement with Baker's[3] findings.

This is altogether different from what happens to the neutral story sentences. They have no structurally determined position in the story and indeed are found to assume any position in their original story, or even to transgress story boundaries (each session consisted of an uninterrupted sequence of three stories). An analysis of the relation between these neutral sentences and the concurrent words revealed a certain effect which indicates that contextual displacement does indeed have an impact on the temporal-order information retained. This, however, applies only if deliberate attention is drawn to the possibility of explicitly connecting the words from the list and the story. The analysis also showed that the associability of the list items will counteract this displacement effect (TABLE 4). Words that are difficult to associate are more likely to migrate away from their original position than are words that are easily associated. In the latter case the effect is essentially absent.

The general conclusion to be drawn from this experiment is that subjects can use contextual information to improve their temporal-order retention. However, they will do this only when the associability between the events in their main task is inadequate for that purpose, and particularly when the instructions explicitly provide them with

TABLE 4. Correlation between Shifted Neutral Sentences and Word Displacement

Condition	*r*
Low-associative-Low	.36
Low-associative-High	.25
High-associative-Low	.11
High-associative-High	−.05

the appropriate strategy. The effect of contextual cues appears to be rather short-lived as it is found only in the recency part of the performance curve. At longer range it even seems to be detrimental (negative primacy effect), perhaps because interference effects make the retained position information instable, particularly in cases where the intrinsic temporal structure between the list items is weak.

8. STRATEGIES

Not only is the descriptive approach criticized in section 5 of limited usefulness, but also, an approach that is more directed at known or assumed processing stages and mechanisms turns out to be of limited value too. This insight arose while we were working along the lines sketched in the previous section. While we were successful in controlling most of the experimental variables, there was one serious factor that turned out to be extremely difficult to keep under control: the individual subject's choice of strategy. It is easy to show the effect of strategy by looking at individual results of four subjects who were performing a temporal coding task with exactly the same stimulus material, presentation times, instructions, and so on (FIG. 8).

Even if one attempts to control the associability of successive items in a list of words, and even if the type of associative cues is controlled to the best of the investigator's ability, everything that has been put into the experimental trial is turned

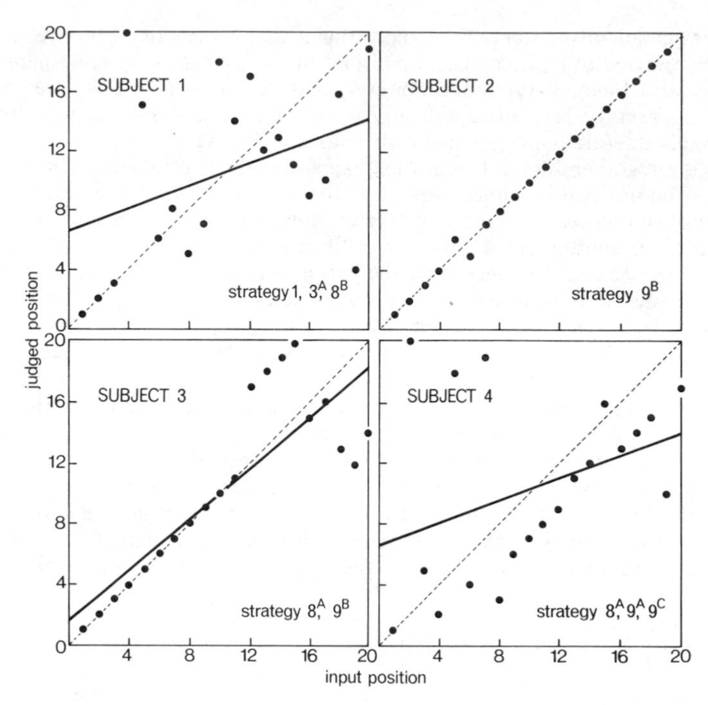

FIGURE 8. Judged position of words as a function of their actual input position: results from four individual subjects.

upside down when the subject happens to observe, for instance, that three successive words begin with the letters B, B and C, respectively, and when he subsequently uses that information for storing the order of these words, paying not the slightest heed to the carefully controlled causal, logical, rhyming or other cues put into the list by the experimenters.

In short, subjects behave in quite varied, unpredictable ways, some of them using, in one form or another, the whole bag of tricks known from mnemonics. Thus, even if data from temporal-judgment experiments produce rather clear-cut results at the aggregate level, as they did in some of our studies, the actual processing going on in the subjects' efforts to cope with their task escaped us.

It became more and more obvious that although we were controlling objective organization (that is, manipulating stimulus characteristics), our instructional variables were not nearly precise enough. Giving global free recall instructions or simply instructions to pay attention to temporal order leaves the whole area of subjective organization untapped. As Reitman[31] has shown, the behavior of experimental subjects in memory experiments often includes efforts to cope with the requirements of the task at hand. In other words, subjects are actively solving problems while learning. From Reitman's protocols it appears that subjects spend much of their time working out and testing various coding and retrieval strategies. Reitman found that at different times people may concentrate on the sounds of words, or on forming associations between consecutive words, on meanings, on forming associations between consecutive words, on classification and categorization procedures, and on imagining physical and conceptual schemes into which words may be inserted as they are presented.

Two experimental roads were open to us. One was to try giving more explicit and precise instructions, taking deliberate measures to encourage subjects to stick to one strategy and the second was to make use of protocols. We knew that performance can indeed be varied by verbal instructions, but we also know that subjects may become dissatisfied with their performance and change to some other way of doing the task. Also, since we are not yet able to predict the efficiency of particular strategies when temporal judgments are required, the second solution offered us the better chance of gaining more insight into both the number and nature of available strategies and into the likelihood of success that comes from using such strategies.

In an experiment to explore the development and stabilization of efficient temporal coding strategies we turned to an in-depth study of four subjects. Subjects were trained to verbalize concurrently with the task, both in the encoding/storage phase and in the retrieval/test phase. With such instructions to verbalize a direct trace of the information in working memory is obtained, and hence indirect evidence for the internal cognitive processing steps. Although our subjects were highly motivated, they were initially untrained in thinking aloud. Consequently, not all of the protocols were as rich as we had hoped for, and—particularly with subject 1—we were forced to add to the information by means of retrospective interviewing. These additions were open-ended and involved asking our subjects immediately after the test phase to describe how they had gone about coding the input information and making their judgments.

The results of the protocol analysis have been summarized in TABLE 5, which shows the various strategies actually adopted by our subjects. The first distinction found was between objective and subjective organization. Objective organization shows that subjects are able to take advantage of any organization provided, deliberately or unknowingly, by the investigator. Subjective organization on the other hand shows subjects imposing their own personal organization on any randomly selected lists. Subjective organization is further divided into simple versus elaborative rehearsal strategies, a breakdown that entertains many similarities with the distinction between maintenance rehearsal and elaborative rehearsal proposed by Craik and Lockhart.[11]

Two of the strategies listed in TABLE 5 (namely, 6 and 10), were never found in coding protocols, but they figured prominently in retrieval protocols and in retrospective reports. The latter—forming a visual, interacting image—does not come as a surprise, since its visual nature precludes explicit verbalization. The former strategy, however, characterized as "beginning–middle–end," is more interesting. It suggests

TABLE 5. Summary of Coding Strategies Obtained from Protocol Analysis

Objective Organization	
1. Physical characteristics	
2. Functional and/or category associations	
Subjective Organization	
A. Simple Rehearsal	B. Elaborative Rehearsal
3. Repetition	7. Number pegs
a. Single words	8. Combining
b. Blocks of words	*a*. Two words into one word
c. Corresponding number	*b*. Two words into one sentence
4. First letter or word	9. Story
5. Before–after	*a*. Elaboration of a word
[a]6. Beginning–middle–end	*b*. One connected story
	c. Two unconnected stories
	d. A number of unconnected stories
	[a]10. Visual interacting

[a]Strategies frequently found in retrieval protocols, but never in coding protocols.

that some sort of implicit *anchoring* takes place at encoding and that these anchor points can be used at retrieval if all else fails.

From our results it appears that the exclusive use of simple rehearsal strategies does not yield correct judgments of temporal position. If we consider subject 1, for instance, who made extensive use of simple rehearsal, we find very little temporal-information retention ensuing. Although the use of elaborative strategies yields much better performance, differences are apparent in this case as well. Subject 2 used strategy 9b, that is, he made one connected story which can later be rerun at retrieval and thereby produces intact temporal judgments. Subject 3, on the other hand, used strategy 9c, a succession of unrelated stories which produce, at retrieval, consistent within-story order effects, but poor between-story judgments. A similar between-story deficit is evident in the results of subject 4, who also made many nonconnected stories. A further strategy used by this subject (9a—the elaboration of a single word) is one which, although appropriate for a conventional free-recall task, does not increase performance on temporal judgment tasks.

9. CONCLUSION

In this paper we have first outlined a programmatic approach to our initial question: How do we connect conscious time experience with memory processes? We are not able to provide a meaningful answer to this question yet, but the results discussed in the preceding sections suggest at least the direction of the answer we are aiming for. Let us summarize.

Psychology's task is in the first place to provide a coherent *description* of the phenomena of time experience. This description must take into account that subjective time differs from physical time at least in the basic experiential properties of FLOW and NOW. In the second place, psychology will have to *explain* these qualities of psychological time experience in terms of underlying functional processes. These processes should ultimately rely only on physically identifiable properties of time, that is, simultaneity and order. These are, in our opinion, the principal stimulus attributes that qualify as temporal *information.*

We consider temporal information equivalent to topical information and we have shown that temporal information draws on the subject's limited processing resources. The assumption that temporal information processing is *deliberate* rather than automatic was tested in several ways and was confirmed in all respects.

This brought us to a conception of the temporal encoding process as the active use of associative cues available in the stimulus sequence and which can be given a temporal interpretation. Some of these cues are intrinsically temporal, that is, they are based on a *structural asymmetry* of before and after. Others can be given a pseudo-temporal interpretation at coding and, to some extent, upon retrieval as well. We have found subjects to use the whole gamut of mnemonic strategies, some of which prove quite effective in encoding temporal information.

Where do we go from here? Expanding from the rather straightforward order and position judgments reported in this paper the most suitable next step would seem to integrate the various types of temporal judgment (order, position, and lag) into one processing model. Presently, we are indeed investigating the degree to which these various judgments do indeed derive from a "common code," a single representation of temporal relation in a series of events. An example of our attempts is described elsewhere in this volume.[45]

From there the next step towards connecting conscious time experience and memory processes would be to attempt an explanation of time estimation (judged duration of time intervals) in terms of the processes studied thus far. Both relative position (lag) and duration are distance judgments; only the units of measurement are different, although perhaps in a fundamental way. From there we will then have to make a jump to the impression of FLOW by measuring the effects of systematic variations of the structural asymmetry of the associative cues in the stimulus sequences. Only then can we hope to establish the connection between conscious time experience and memory processes. In this context we may finally point out that the views expressed in this paper are conceptually consistent with (and implicitly based on) the view that the psychological NOW is an active construction, a sequential testing and updating of hypotheses about the events in progress.[7,21,25] The experience of NOW in this frame of thought is derived from the organism's attempts to predict the immediate future on the bases of hypotheses derived from the immediate past. It is ultimately life's way of coping with the on-line problems of having internalized Time.

ACKNOWLEDGMENTS

Harm Hospers and Egbert Knol assisted in the collection and processing of the data.

REFERENCES

1. ANDERSON, J. R. 1980. Cognitive Psychology and Its Implications. Freeman Publications. San Francisco.
2. ASCHOFF, J. 1984. Circadian timing. This volume.
3. BAKER, L. 1978. Processing temporal relationship in single stories: Effects of input sequence. J. Verb. Learn. Verb. Behav. **17:** 559–572.
4. BLOCK, R. A. 1974. Memory and the experience of duration in retrospect. Mem. Cognit. **2:** 153–160.
5. BLOCK, R. A. & M. A. REED. 1978. Remembered duration: Evidence for a contextual-change hypothesis. J. Exp. Psychol. Hum. Learn. Mem. **4:** 656–665.
6. BJORK, R. A. 1972. Theoretical implications of directed forgetting. *In* Coding Processes in Human Memory. A. W. Melton & E. Martin, Eds. Winston. Washington, DC.
7. BREGMAN, A. S. 1977. Perception and behavior as compositions of ideals. Cognit. Psychol. **9:** 250–292.
8. BREGMAN, A. S. 1978. The formation of auditory streams. *In* Attention and Performance. VIII. J. Requin, Ed. Lawrence Erlbaum. Hillsdale, NJ.
9. CARROLL, J. M. & T. G. BEVER. 1975. Segmentation in cinema perception. Science **191:** 1053–1055.
10. COLLARD, R. F. A. & E. L. J. LEEUWENBERG. 1981. Temporal order and spatial context. Can. J. Psychol. **35:** 323–329.
11. CRAIK, F. I. M. & R. S. LOCKHART. 1972. Levels of processing: A framework for memory research. J. Verb. Learn. Verb. Behav. **11:** 671–684.
12. DAVIES, P. C. W. 1981. Time and reality. *In* Reduction, Time and Reality: Studies in the Philosophy of the Natural Science. Cambridge University Press. Cambridge.
13. FLEXSER, A. J. & G. H. BOWER. 1974. How frequency affects recency judgments: A model for recency discrimination. J. Exp. Psychol. **103:** 706–716.
14. GALBRAITH, R. C. 1976. The effects of frequency and recency on judgments of frequency and recency. Am. J. Psychol. **89:** 515–526.
15. GUENTHER, R. K. & M. LINTON. 1975. Mechanisms of temporal coding. J. Exp. Psychol. Hum. Learn. Mem. **104:** 182–187.
16. GUYAU, M. 1980. La Genèse de l'Idée de Temps. Alcan. Paris.
17. HASHER, L. & R. T. ZACKS. 1979. Automatic and effortful processes in memory. J. Exp. Psychol. Gen. **108:** 356–388.
18. HINRICHS, J. V. 1970. A two-process memory-strength theory for judgment of recency. Psychol. Rev. **77:** 223–233.

19. JACKSON, J. L. & J. A. MICHON. Effects of item concreteness on temporal coding. Acta. Psychol. In press.
20. JAMES, W. 1890. Principles of Psychology (2 vols.). Holt. New York, NY.
21. JONES, M. R. 1976. Time, our last dimension: Toward a new theory of perception, attention and memory. Psychol. Rev. **83:** 323–355.
22. LEE, C. L. & W. K. ESTES. 1981. Item and order information in short-term memory: Evidence for multilevel perturbation processes. J. Exp. Psychol. Hum. Learn. Mem. **7:** 149–169.
23. MICHON, J. A. 1965. Studies on subjective time. II: Subjective time measurement during tasks with different information content. Acta Psychol. **24:** 205–219.
24. MICHON, J. A. 1972. Processing of temporal information and the cognitive theory of time experience. *In* The Study of Time. J. T. Fraser, F. C. Haber & G. H. Mueller, Eds. Springer-Verlag. Heidelberg.
25. MICHON, J. A. 1978. The making of the present: A tutorial review. *In* Attention and Performance. VII. J. Requin, Ed. Lawrence Erlbaum. Hillsdale, NJ.
26. JACKSON, J. L. & J. A. MICHON. Effects of instruction on temporal coding. Submitted for publication.
27. MOORE-EDE, M. C., F. M. SULZMAN & C. A. FULLER. 1982. The Clocks That Time Us: Physiology of the Circadian Timing System. Harvard University Press. Cambridge, MA.
28. MURDOCK, B. B. 1976. Item and order information in short-term serial memory. J. Exp. Psychol. Gen. **105:** 191–216.
29. VAN NOORDEN, L. P. A. S. 1975. Temporal Coherence in the Perception of Tone Sequences. Doctoral dissertation, Technical University of Eindhoven, The Netherlands.
30. ORNSTEIN, R. E. 1969. On the Experience of Time. Penguin. Baltimore, MD.
31. REITMAN, W. 1970. What does it take to remember? *In* Models of Human Memory. D. A. Norman, Ed. Academic Press. New York, NY.
32. RICHELLE, M. & H. LEJEUNE. 1980. Time in Animal Behavior. Pergamon Press. Oxford.
33. RIEGEL, K. F. 1977. Toward a dialectical interpretation of time and change. *In* The Personal Experience of Time. B. S. Gorman & A. E. Wessman, Eds. Plenum Press. New York, NY.
34. SCHANK, R. C. & R. ABELSON. 1977. Scripts, plans, goals, and understanding. Lawrence Erlbaum. Hillsdale, NJ.
35. SHIFFRIN, R. M. & W. SCHNEIDER. 1977. Toward a unitary model for selective attention, memory scanning, and visual search. *In* Attention and performance. VI. S. Dornic, Ed. Lawrence Erlbaum. Hillsdale, NJ.
36. SNODGRASS, J. G. & M. VANDERWART. 1980. A standardized set of 260 pictures: Norms for name agreement, image agreement, familiarity, and visual complexity. J. Exp. Psychol. Hum. Learn. Mem. **6:** 174–215.
37. TOGLIA, M. P. & G. A. KIMBLE. 1976. Recall and use of serial position information. J. Exp. Psychol. Hum. Learn. Mem. **2:** 431–445.
38. TZENG, O. J. L., A. T. LEE & C. D. WETZEL. 1979. Temporal coding in verbal information processing. J. Exp. Psychol. Hum. Learn. Mem. **5:** 52–64.
39. TZENG, O. J. L. & B. COTTON. 1980. A study-phase retrieval model of temporal coding. J. Exp. Psychol. Hum. Learn. Mem. **6:** 705–716.
40. UNDERWOOD, B. J. & R. A. MALMI. 1978. An evaluation of measures used in studying temporal codes for words within a list. J. Verb. Learn. Verb. Behav. **17:** 279–293.
41. WICKELGREN, W. A. 1972. Trace resistance and the decay of long-term memory. J. Math. Psychol. **9:** 418–455.
42. YNTEMA, D. B. & F. P. TRASK. 1963. Recall as a search process. J. Verb. Learn. Verb. Behav. **2:** 65–74.
43. ZIMMERMAN, J. & B. J. UNDERWOOD. 1968. Ordinal position knowledge within and across lists as function of instructions in free-recall learning. J. Gen. Psychol. **79:** 301–307.
44. VAN ZOMEREN, A. H. 1981. Reaction Time and Attention after Closed Head Injury. Doctoral dissertation, University of Groningen, The Netherlands.
45. JACKSON, J. L., J. A. MICHON & A. VERMEEREN. 1984. The processing of temporal information. This volume.
46. REED, S. 1982. Cognition. Theory and Applications. Brooks/Cole. Monterey, CA.

47. VAN SCHAGEN, I., J. L. JACKSON & J. A. MICHON. 1983. De invloed van ontwikkeling op oordelen over temporele positie. [Developmental effects in temporal position judgments.] Internal Report, Institute of Experimental Psychology, University of Groningen, The Netherlands.
48. VAN DER VELDE, J., H. BOONSTRA, J. A. MICHON & J. L. JACKSON. 1983. Effecten van verwerkingsdiepte op codering van temporele informatie. [Effects of level of processing on temporal coding.] Internal Report, Institute of Experimental Psychology, University of Groningen, The Netherlands.
49. MENSINK, G. J. 1982. Gehengenorganisatie en sequentieel leren. [Memory organisation and sequential learning.] Internal Report, Institute of Experimental Psychology, University of Groningen, The Netherlands.

Time and Memory[a]

J. E. R. STADDON

Department of Psychology
Duke University
Durham, North Carolina 27706

There are always two ways to look at behavioral change: in terms of the process by which change occurs, and in terms of the nature of the change itself—learning versus "what is learned." This distinction between dynamic and static approaches, clear in the case of simple discrimination learning, is blurred when what is learned involves time. The change wrought by training in a pigeon that has learned to peck a red key and not a green one can be compactly described. Not so the effect of a brief event that has temporal significance: The effect of a trace-conditioned stimulus, or the food delivery that marks the beginning of a fixed interval, is not constant, but must change with lapse of time, as behavior changes.

How best to characterize the difference between discrimination learning, where time can for many purposes be ignored, and temporal control, where it cannot? The distinction between *reference* and *working* memory is one possibility: Honig[1] and Olton[2] proposed the label "reference memory" for the faculty that animals use for those aspects of the training situation that are fixed, and the term "working memory" for the faculty used to cope with information that changes from trial to trial (the latter usage being slightly different from Baddeley and Hitch's original[3]). Thus, the animal in a delayed-matching-to-sample (DMTS) experiment can rely on the fact that the keys, not the feeder or houselight, are the things to peck. This information is presumed to reside in reference memory. But the animal cannot be sure from trial to trial *which* key is correct. This information is contained in working memory, which must therefore be altered frequently if the bird is to perform correctly.

The locution that reference and working memory "contain" information suggests that they are in some sense physically distinct. Many will vigorously deny this implication, the philosophical sins of quasi-physiological assertions of this sort being now widely recognized. The consensus is likely to be that the locational usage is a harmless convenience. Perhaps, but it does distract attention from an otherwise obvious difference in the *kind* of information represented by these two types of memory: Reference memory contains static information about relations or associations ("red means food"); whereas working memory contains dynamic information about *recency* ("red occurred last"); more will be made of this distinction in a moment. Because working memory always involves time in some way, *event* memory is perhaps a better label: Event memory is memory for *how long ago* something happened, as opposed to its significance in terms of other events.

DMTS is a task in which the animal must identify the stimulus that has occurred most recently.[4] Ability to perform correctly implies sensitivity to the *age* of events. This very same sensitivity is involved in *temporal control,* that is, control of behavior by a time marker. Efficient performance on a fixed-interval schedule, for example, demands that the animal not respond until the age of the most recent time marker (food delivery) has reached a value determined by the interfood interval. It is likely, therefore, that temporal-control experiments, such as fixed-interval schedules, and

[a]This work was supported by grants from the National Science Foundation.

event-memory experiments, such as DMTS, are studying the same capacity. It may be a matter of choice and tradition whether we term this capacity time discrimination or memory.

This paper has three objectives: (a) to support the assimilation of temporal control and event memory by giving examples of proactive and retroactive effects in temporal-control experiments; (b) to argue that similar principles govern the effects on temporal control and event memory of the temporal spacing of events; and (c) to see whether performance differences between tasks such as DMTS, the radial maze, and delayed spatial alternation can be accounted for by these principles. I will conclude that performance differences among these tasks are just about what we would expect from the different requirements they impose: It is not necessary to postulate special processes to account for rats' excellent performance in radial-maze tasks, for example.

TEMPORAL CONTROL: PROACTIVE AND RETROACTIVE EFFECTS

Animals soon learn to pause after food on fixed-interval (FI) schedules, and the length of the pause is systematically related to the FI value.[5,6] Indeed, so regular is this behavior, that it has been attributed to an internal "clock," reset by food delivery.[7] In its simplest form, the clock analogy implies that behavior at a given time depends only on time elapsed since the preceding time marker, which "resets" the clock.

Temporal control certainly involves some kind of internal clock, but performance is not so simply related to the clock as the reset model implies. For example, if feeder duration varies unpredictably from fixed interval to fixed interval, the postfood pause is longest after longer durations, shortest after short[8] (evidently long feeder durations reset the clock more effectively than short). But even this interpretation is incomplete, because the differential effect of feeder duration disappears if the animal experiences only a single duration—the effect depends on intercalation of long and short durations.[9,10] Behavior is affected not just by the most recent time marker, but also by others more remote in time. The same conclusion follows from several experiments that have shown rats and pigeons able to track, with no phase lag, sequences of schedules or interfood intervals that follow a repeating cycle.[5,11–15]

There are other demonstrations. For example, some years ago, Nancy Innis, John Kello and I carried out numerous experiments with fixed-interval schedules in which some food deliveries were omitted, either entirely, or replaced by a brief, "neutral" stimulus such as a key-color change or time out (FI-omission procedures). Pigeons and rats are usually unable to pause after these neutral stimuli as they do after food, even though the stimuli have the same temporal (predictive) properties as food. We concluded that the ineffectiveness of neutral stimuli under these conditions is a proactive effect of preceding food deliveries, which "overshadow" the neutral stimulus and gain control of behavior at its expense.[16]

We found two ways to eliminate this overshadowing effect: Either eliminate the temporal significance of food, or provide a separate context for postfood time and post-neutral-stimulus time. An example of the first method is shown in FIGURE 1.[17] Pigeons were trained on a mixed schedule with two alternating, 2-min components. The components were variable interval (VI 1 min) and fixed interval (FI 2 min). Each FI component comprised a single 2-min interval, initiated by a 3-sec vertical-line stimulus presented on an otherwise-white response key. As the left cumulative record shows, the brief neutral stimulus gained generally good temporal control under these conditions. Evidently the proactive interference from food delivery can be abolished by eliminating the temporal-cue significance of food delivery.

In another experiment[16] we showed that a brief stimulus could achieve good temporal control even on the usual FI-omission procedure, as long as intervals that began with the neutral stimulus were marked off by a key color (context) different from intervals that began with food. Thus, either contextual separation or elimination of the temporal significance of food can mitigate the overshadowing effect of food on neutral trace stimuli.

The right-hand record in FIGURE 1 shows the effect of an apparently trivial modification of the mixed VI–FI procedure. Instead of scheduling the 2-min fixed interval during each 4-min cycle, it was scheduled on only half the cycles. It is unlikely that this change would have had any significant effect by itself as long as the brief vertical-line stimulus continued to provide a reliable temporal cue. However, during those cycles when no fixed interval was scheduled (that is, when the VI 1-min schedule remained in effect), a brief *horizontal-line* stimulus was projected on the response key. The first effect of this change was that the pigeons paused indiscriminately after both horizontal and vertical stimuli. This, of course, led to their sometimes waiting longer than necessary after the horizontal stimulus. Eventually, the animals ceased to pause after either stimulus, as shown in the right-hand record in FIGURE 1.

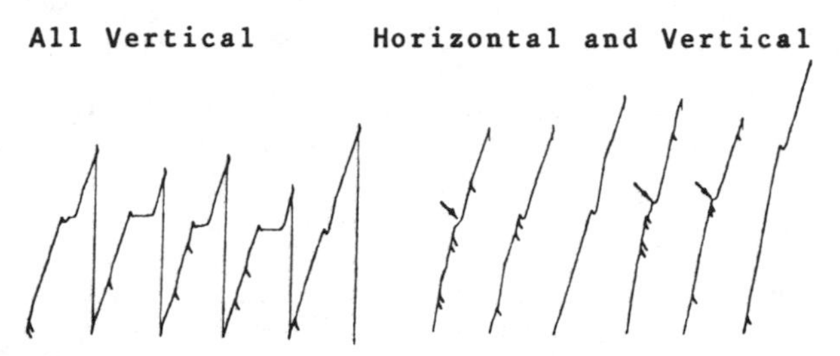

FIGURE 1. Cumulative record of the performance of a single pigeon on a procedure in which a fixed-interval schedule, initiated by a brief neutral stimulus, alternated with a variable-interval schedule. *Left record:* A brief vertical-line stimulus initiated the FI. *Right record:* A brief horizontal-line stimulus with no temporal significance occurred during the half the cycles: Temporal control by the vertical-line stimulus is abolished.[17]

Pigeons have no difficulty in telling vertical from horizontal lines in standard simultaneous or successive discrimination procedures. Their failure to pause differentially after the two stimuli in this experiment does not reflect some kind of perceptual limitation. The problem seems to be that in this experiment the animals were not required to respond *in the presence* of the stimuli. Instead they had to behave differently *after* the stimuli had come and gone, pausing after the vertical lines, not after the horizontal lines. Perhaps because of the similarity of the two stimulus complexes—identical in every way save line orientation—the pigeons were unable to reliably remember which stimulus was most recent, and hence showed erratic temporal control by the vertical lines. Similar effects, showing confusion between two similar samples, have been demonstrated in DMTS experiments.[18,19]

Temporal control, like event memory, is evidently susceptible to proactive interference. I believe that it is also subject to *retroactive* interference, although the evidence here is less clear because the effect is studied under a different name: If a novel stimulus is presented during the pause on a fixed-interval schedule, animals will usually begin responding at once. Although normally termed *disinhibition,* it also fits the definition for retroactive memory interference: a later event (the novel stimulus) impairing recall of the earlier one (food, the time marker).

Performance on temporal schedules is routinely affected by events prior to the most recent event. Sometimes this sensitivity to past history is essential to good performance, as in temporal tracking; sometimes it is harmful, as in some of the fixed-interval examples. But in no case can animals ignore all but the most recent event: Temporal control is subject to the same kinds of proaction, retroaction, and context effects as event memory.

EFFECTS OF EVENT SPACING ON MEMORY AND TEMPORAL CONTROL

It is impossible to summarize in a small space the extensive experimental literature on the effects on memory and temporal control of event spacing. I will emphasize just two well-known effects: the rough proportionality between temporal just-noticeable-difference and absolute time,[20,21] a property of temporal control, and Jost's law,[22] a property of memory.

The relation of event memory to time discrimination is as the relation between perception and psychophysics: Memory is the perception of the past. Just as perceptual principles bear some relation to psychophysical limitations, so memory principles are related to the properties of time discrimination. For example, intensive continua follow Weber's law: Variability in judgment (standard deviation) is proportional to the absolute value judged. This is a psychophysical result. For such continua, the judged magnitude of a given physical change is also inversely related to its base level. This is a perceptual estimate. Obviously the perception is related in some way to the psychophysical constraint, although the exact form of this relation—a problem since the time of Fechner—remains unresolved. A similar relation holds between memory and time discrimination: Time discrimination also follows a Weber-type relation; and changes in the age of events produce smaller and smaller perceptual effects as age increases. Let us look at the relation between time discrimination and the effects of age on the salience of memories.

Models for time discrimination all agree that the accuracy with which the age of a particular event can be estimated is inversely related to age: A difference of 5 seconds is detectable as being between events that occurred 7 and 12 seconds ago, not between events that occurred 1 hour and 1 hour and 5 seconds ago. This relation implies that the salience of a temporally extended event is both directly related to its duration and inversely related to its age.

These effects of age and duration on salience can be interpreted as properties of memory or properties of learning. For example, suppose that instead of considering just a particular event, such as the duration of an individual stimulus, we consider also the duration of a particular *training experience:* the number of trials or sessions devoted to a given procedure. Suppose further that the salience of a particular experience, like the salience of a particular stimulus, is subject to the Weber principle—an experience of longer duration having a greater effect (at a certain "age") than a similar experience of shorter duration, differences in event duration having effects inversely related to absolute duration. Commonplace facts of learning follow at once: The more training an animal receives at something, the better he learns it and the longer he remembers it—the longer a training period, the more salient its effects. And additional increments of training have smaller and smaller effects: Learning curves are negatively accelerated. These truisms follow from common sense, but also from any principle that relates the salience of a past event inversely to its age.

If we give it some quantitative form, this same principle can also predict *changes* in relative salience of different experiences with time. For example, suppose we assume that the perceptual effect of an event is inversely related to its age according to the power relation:

$$E = At^m, \qquad m < 0, \tag{1}$$

where m represents the rapidity with which age diminishes the effect of an event and A is a constant of proportionality. For a temporally extended event, therefore, salience will be given by

$$S = A\,(t_1^m - t_2^m), \tag{2}$$

where t_1 is how long ago the event ended, and t_2 is how long ago it began. It is easy to show that if two experiences have equal effects (saliences) at some time (which implies that the older experience must be of longer duration), then with lapse of time, the older will gain at the expense of the newer. This is just a restatement of Jost's (second) law, a principle from the early days of modern psychology: "Given two associations of the same strength, but of different ages, the older falls off less rapidly in a given length of time" (Ref. 22, p. 649).

Thus, the Weber's law principle, which implies that the salience of a past event is inversely related to its age, can easily be expressed in a way that leads to familiar principles of learning and memory.

This model for the salience of past events is a great simplification, of course. It ignores possible differences of process and in the effects of factors other than temporal spacing, such as differences in the intensity of experiences of similar durations, similarity relations among events, the kind of behavior used to measure the effects of past events, and the details of proaction and retroaction effects. The properties of the perceptual process to which sets of recencies, transformed according to Equation 1, are input are also unspecified. Above all, it ignores the question of *what* is remembered, looking only at the *effectiveness* of a past experience, not its quality. Nevertheless, several apparently puzzling differences between different memory experiments can be understood just by looking at the probable effect of type and spacing of events on their effectiveness as controlling stimuli. In the final section I look at some familiar memory experiments from this point of view.

Reference and Working Memory: Time Scale or Information Type?

There is little reason to doubt that the principles that govern the effectiveness of past experience on present behavior are uniform over a substantial time scale; very short time periods may demand special treatment, but beyond a few seconds the same principles appear to hold. Looked at in these terms, there is no difference between temporal control and memory, or between working and reference memory: The fact that the information in working memory must change while the information in reference memory need not, is really just a difference of time scale: After all, when a procedure changes in some way, which all eventually must, reference memory must alter appropriately. Nothing is forever. Clearly, the working/reference distinction is much more one of procedure than of process.

Transient versus permanent is not the most natural way to dichotomize memory. As the earlier discussion pointed out, the real distinction is based on the *type* of information involved, not its stability: If one wishes to retain the terms, then "reference memory" should be reserved for associative properties, and "working memory" for information about recency. Since every event and experience has a recency, there is no reason to restrict working memory to within-session changes, and perhaps no reason to retain that term, with its inappropriate connotations, at all. If we are concerned with the *relative effectiveness* of past events or experiences on present behavior, then we are studying event memory. If we are concerned with what those effects are, we are studying something else—call it reference memory, if you like. The natural dichotomy is much closer to Tulving's[23] concepts of *semantic* and *episodic* memory than the more recent *working* versus *reference.* Perhaps it is the linguistic connotations of "semantic,"or the reluctance to retain an old term when a new one would do as well, that has prevented Tulving's usage from catching on among animal memorialists.

SOME MEMORY EXPERIMENTS: LEAST- AND MOST-RECENT DISCRIMINATION

Let us look at three common memory paradigms—delayed matching to sample (DMTS), delayed alternation, and the all-arms-baited radial maze—from the point of view of event memory. I will follow the same pattern in each case: first, define the procedure (in particular, define which aspects are typically varied); second, look at the pattern of recencies that must be discriminated by the animal; and, third, reach some conclusions on features of the task that might be expected to make it particularly easy

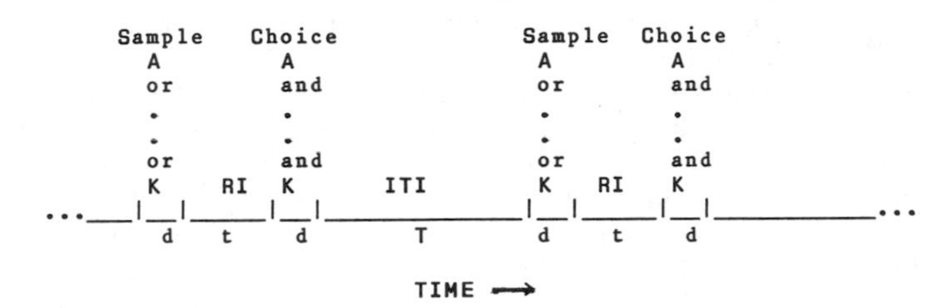

FIGURE 2. Procedure for delayed matching to sample.

or hard. I conclude by comparing the three procedures. My objective throughout is to relate broad differences among procedures to major task differences, rather than to reconcile minor differences between similar experiments.

Delayed Matching to Sample

A common version of the DMTS procedure is illustrated in FIGURE 2: A sample, *A* (selected from a set of K items, K > 1) is presented for a time *d;* after a retention interval *t,* a choice set, consisting of stimulus *A,* plus at least one other alternative, is presented. A response yields either food, no food, or no food plus an added delay. (For simplicity, I assume perfect responding.) After food, a third time interval, *T,* the intertrial interval (ITI), elapses; a further sample (*B* in the figure) is then presented, and the process repeats, for 50 or more cycles in each daily session. Things typically varied are the three times (*t, d,* and *T*), the properties of the stimuli, and the number of stimuli in the choice set.

If events are represented for the animal along the lines I have proposed, then after a few trials, the two-choice DMTS task must appear to him as one of discriminating between two arrays: one in which the most recent stimulus is one sample; the other in which the most recent stimulus is the other sample. For behaviorists, these arrays can be thought of as pictures of the "functional stimulus" for responding on a choice trial[24]; for cognitivists, they can be thought of as memory representations.

Four such arrays are illustrated in FIGURE 3. For simplicity, the arrays show only sample trials (that is, I assume negligible interference from choice trials, but addition of this factor does not alter my conclusions), and sample stimuli *1* and *2* alternate. In each array, time past is represented by vertical distance: the most recent events are at the bottom, the oldest at the top. The horizontal lines correspond to the subjective value of the recencies of sample events *1* and *2*. For ease of comparison, the arrays are lined up at the most recent sample. Distances between stimulus presentations decrease with age, according to Equation 1 with m = −0.5. I assume that semantic information is preserved even for the oldest stimuli, where recency information is lost (that is, the animal "sees" every *1* and *2* in the array, but may be unsure about their relative recency). Stimulus duration is neglected (d = 0).

FIGURE 3 shows four configurations, differing according to intertrial interval (ITI: *T*), retention interval (RI: *t*), and terminal delay (that is, time elapsed since the most recent stimulus, normally equal to the RI). In each case, the DMTS task requires the animal to discriminate the array ending with *1* from the complementary array ending in *2*, choosing *1* in the first case, *2* in the second. (In a real DMTS experiment the task is rather more difficult, namely, to discriminate a member of the set of all possible arrays terminating in *1* from all possible arrays terminating in *2*). The difficulty of this task will be related to the perceptual distinctness of the most recent event (at the bottom of each array in the figure) in each case.

Comparison of the first and second array-pairs shows the effect of absolute time: In both cases, ITI equals RI, but clearly discrimination will be easier in the second case, where RI is 1 sec, than the first one, where it is 2 sec. Array-pairs 3 and 4 show the effect of extra-long delays interpolated after the most recent sample: Longer delays produce increasing confusion about temporal order—compare the vertical separation between the two most recent stimuli in array-pairs 2, 3 and 4.

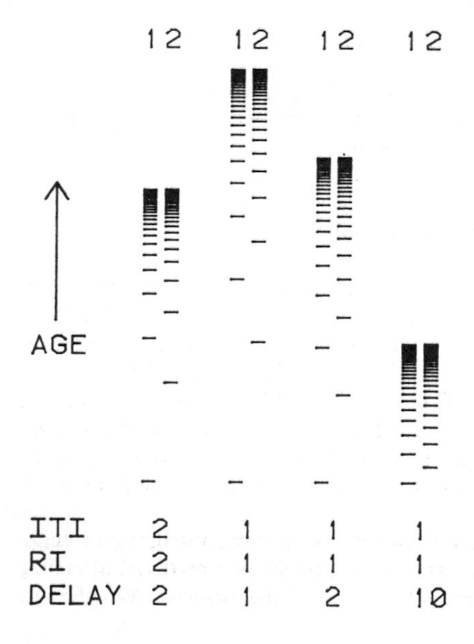

FIGURE 3. Memory arrays for the two-stimulus DMTS procedure with ITI, RI, and delay values indicated. Horizontal lines represent the perceptual effect of past presentations of stimuli *1* and *2*. Spacing of events follows Equation 2, with t_1 equal to the delay, and t_2 equal to the age of the event. Arrays are lined up at the most recent event.

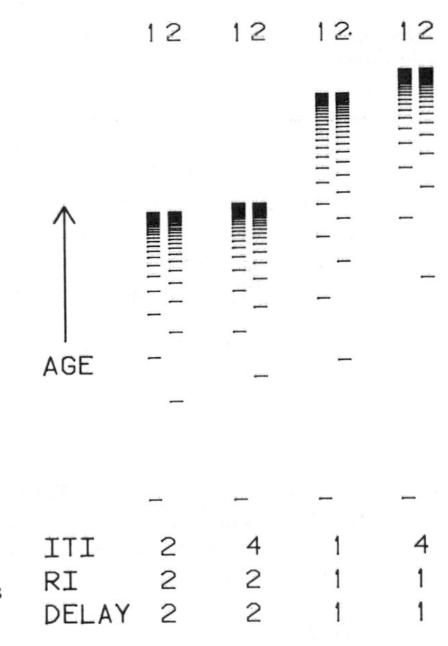

FIGURE 4. Further arrays for two-stimulus DMTS.

If past events appear to the animal as these arrays, then the DMTS task should be quite difficult, because the animal must ignore everything but event spacing. By the hypothesis, semantic information from remote trials is present in the array, and both stimuli appear repeatedly. So, unless prior trials are separated from the most recent one in some way—by time (a long intertrial interval) or by context—these more remote trials will function as distractors. Second, if there is some limit to temporal "acuity," there will be a limit on the number of different events whose recencies can be discriminated. This limit may correspond to the cognitivists' "transfer from short-term to long-term memory."

FIGURE 4 shows the effect on the memory arrays of changing the ITI. Doubling the ITI (compare columns 1 and 2) somewhat increases the discriminability of the most recent event, but halving the absolute time scale has a bigger effect (columns 1 and 3). A fourfold increase in ITI has a larger effect (columns 1 and 2).

FIGURES 3 and 4 imply that increases in ITI should reduce, and increases in retention interval should increase, task difficulty. The retention-interval effect is universal.[25] The ITI effect is not always found.[26–29] It is not clear whether this signifies a weakness in the perceptual approach, either in general or in this version, or whether other factors, such as contextual separation between retention interval and ITI (the two are usually delimited by a salient stimulus such as the presence or absence of a houselight), may sometimes reduce interactions for other reasons. The perceptual approach implies that increases in sample duration should improve performance and data support this.[28,30] A limit on temporal acuity implies the acknowledged limit on the capacity of working memory.

In addition to changes in event spacing and context, FIGURE 3 indicates at least one other way the task might be made easier: by increasing the number of choices. In this way, occurrences of a given sample will on average necessarily be separated by longer intervals of time (since samples normally occur with equal probability on each trial), and hence will appear farther apart in the array. If the different sample stimuli are not (semantically) confusible with one another, then multi-choice DMTS may prove easier than two-choice.

Delayed Alternation

The delayed-alternation task is similar in its temporal structure to DMTS. A spatial arrangement, such as a T- or Y-maze, is used. The animal is constrained to enter one arm; then returned to the beginning of the maze and, after a delay, given a free choice of both arms; reward is given for entry into the previously blocked arm; an ITI follows.

Rats and many other animals are aided in this task by a built-in tendency to alternate between spatially distinct locations (*spontaneous alternation*[31–33]). The memorial information available is as indicated for DMTS in FIGURE 3: The animal's task is to discriminate the array whose most recent entry is "L" from the array whose most recent entry is "R," and then to make the appropriate choice: "R" in the first case, "L" in the second. The task is likely to be difficult for the same reasons as DMTS: each stimulus appears multiple times in each array, and as retention interval increases, separation between trials decreases most rapidly for the most recent trials. Like DMTS, delayed alternation requires the animal to attend to the most recent event.

Radial-Maze-Type Tasks

Both delayed-alternation and DMTS require the animal to identify and choose a particular stimulus on each trial. In recent years a very different class of tasks has come into prominence, of which the radial maze is the typical example. In the radial maze, at the beginning of a trial, food is available in the goal box at the end of every arm. Hence, the rat's task is not to choose any particular arm, but rather to avoid revisiting arms. Krebs and his collaborators[34,35] have studied marsh tits hoarding, and later retrieving, a small number of pieces of food, stored among 100 potential sites. This task improves on the radial maze by allowing measurement of two additional types of error, for a total of three: visiting an empty site, revisiting an emptied site, and failing to visit a full site.

Rats' excellent performance on the radial maze and similar tasks is by now well known.[2,33] Three distinctive properties of these tasks seem to be responsible: They are spatial, hence the choice stimuli are highly distinct; there are many alternatives; and, most important of all, correct performance (not revisiting an arm within a trial) requires only that the animal identify the *least recently chosen* alternative. This third property helps in two ways: the least-recent event will usually be highly salient in the memory array; and (as we have just seen) rats have a built-in tendency to choose the least recently visited location.

The effect of these properties can be seen in the three arrays shown in FIGURE 5. Each arm entry is indicated by a horizontal line, and each array shows eight choices on trial N and seven choices on trial N + 1. The three arrays differ only in the delay, that is, the time after the seventh choice when the eighth is to be made. Because the ITI in radial-maze experiments is always much longer than the inter-arm-choice time (ICT), the last arm entry on day N will be clearly discriminable from the first arm entry on day N + 1. Moreover, because of the properties of Equation 1—or any similar equation—this difference will be very resistant to increases in retention time, which, as we have seen, have their main effect on the discriminability of the *most* recent events. This comparison is evident across all three arrays: The recencies always fall into two clearly separate groups, even at the longest delay. Since (by the hypothesis) semantic information is preserved, the animal should have no difficulty identifying which arm is not present in the most recent group, that is, which arm has not yet been entered on a given trial. Thus, radial maze performance should be exceedingly resistant to lengthy delays.

The longest retention time tried with the radial maze seems to have been in an experiment by Beatty and Shavalia,[36] who allowed rats to choose four arms of an eight-arm maze, then restrained them for periods up to 4 hours, after which they were allowed to choose among all eight arms. Even after this delay, greatly in excess of delays compatible with good DMTS performance, the rats performed well. Nevertheless, if we grant two assumptions—(a) that the memory representation of time differences follows a principle such as Equation 1 that is compatible with Weber's and

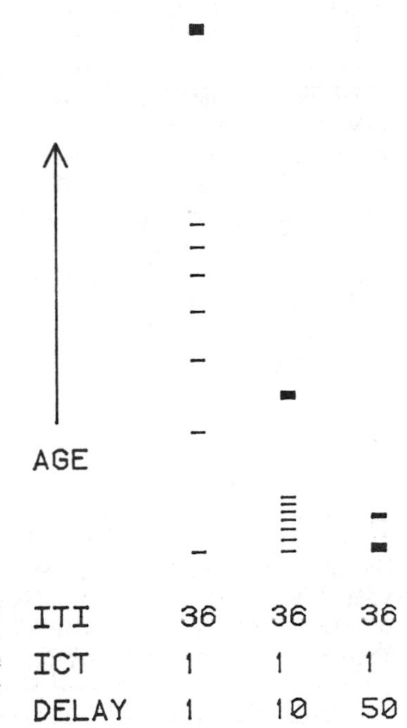

FIGURE 5. Memory arrays for two successive radial-maze trials. Each *bar* represents an arm entry. The three arrays show the effects of increasing delay after the seventh arm entry on the second trial.

Jost's laws and (b) that the salience of events in such a representation follows familiar perceptual principles—then the relative insensitivity to retention delay of radial maze performance and the relative sensitivity of DMTS performance follow.

Delayed-alternation performance is much more sensitive to delay than performance in the radial maze,[37] yet the two tasks are formally identical, differing only in the number of alternatives among which the animal can choose and in details of event spacing. The difference derives from the limit on event memory implied by a perceptual model: In the radial-maze task, after the (typically long) ITI, the recencies of the eight arms visited on the preceding trial will not be discriminable from one another, but will be clearly discriminable from each choice made in the current trial (FIGURE 5). Hence, so long as an arm has not been chosen on a trial, its (most recent) recency will be clearly discriminable from the recencies of already-chosen arms. In delayed-reaction tasks, on the other hand, the ITI is typically quite short and several trials are simultaneously represented in event memory. The array will look much like those in FIGURES 3 and 4. Hence, the correct choice will not stand out as clearly as in the eight-arm-maze experiments.

CONCLUSION

This paper is a speculative attempt to bring together three topics: psychophysical properties of temporal control, the properties of memory for event recency, and the sensitivity to retention delays of performance on memory tasks such as delayed matching to sample and the radial maze. I have argued that if event memory is subject to the same kinds of Weber's law limitation as temporal discrimination, and if past events are "perceived" accordingly, then the broad performance differences between tasks such as DMTS and the radial maze are readily explained. In particular, no new principles are required to account for the great resistance to retention delay of radial maze performance, nor do we need some special account for animals' inability in DMTS to discriminate sample stimuli after quite modest delays. Performance is different because the tasks are different, not because the radial maze draws upon some special "foraging" ability or because rats are "prepared" to do well in naturalistic tasks. Indeed, the truth may be the converse: Rats do well at naturalistic tasks not because these call on special abilities, but because the *general* principles of rat memory are just those needed for efficient performance in natural tasks. That rats do poorly at DMTS is for natural selection a matter of indifference; that they remember which food sites have been visited least recently, hence are most likely to have been replenished, may be vital. Memory mechanisms seem to have evolved accordingly.

SUMMARY

Standard animal memory tasks require judgments of event recency: Delayed matching to sample (DMTS) requires that the animal identify the stimulus seen most recently; radial-maze-type (RM) tasks require that the animal identify the place visited least recently. Delayed-reaction tasks are intermediate. I argue that time discrimination (temporal control) and event memory call on the same processes: Proactive and retroactive effects occur in both, brief events have less effect than protracted events, and increases in event duration have smaller and smaller effects. If the "ages" of past events are represented by animals in a way consistent with Weber's and Jost's laws, and if there is a limit to the number of different recencies that can be discriminated, then the major differences between these three types of memory task can be explained. DMTS performance is poor because the animal must discriminate between two sets of recencies (memory arrays) that differ only in respect of the most recent event; RM performance is good because the recencies of places visited on the current versus earlier trials are always clearly discriminable.

ACKNOWLEDGMENTS

I thank Robert Dale, Nancy Innis and Ken Steele for helpful comments on an earlier version.

REFERENCES

1. Honig, W. K. 1978. Studies of working memory in the pigeon. *In* Cognitive Processes in Animal Behavior. S. H. Hulse, H. Fowler & W. K. Honig, Eds. Erlbaum. Hillsdale, NJ.
2. Olton, D. S. 1978. Characteristics of spatial memory. *In* Cognitive Processes in Animal Behavior. S. H. Hulse, H. Fowler, & W. K. Honig, Eds. Erlbaum. Hillsdale, NJ.
3. Baddeley, A. D., & G. Hitch. 1974. Working memory. *In* The Psychology of Learning and Memory. G. H. Bower, Ed. Vol. 8. Academic Press. New York, NY.
4. D'Amato, M. R. 1973. Delayed matching and short-term memory in monkeys. *In* The Psychology of Learning and Motivation: Advances in Research and Theory. G. H. Bower, Ed. Academic Press. New York, NY.
5. Innis, N. K. & J. E. R. Staddon. 1971. Temporal tracking on cyclic-interval reinforcement schedules. J. Exp. Anal. Behav. **16:** 411–423.
6. Schneider, B. A. 1969. A two-state analysis of fixed-interval responding in the pigeon. J. Exp. Anal. Behav. **12:** 677–687.
7. Church, R. M. 1978. The internal clock. *In* Cognitive Processes in Animal Behavior. S. H. Hulse, H. Fowler, & W. K. Honig, Eds. Erlbaum. Hillsdale, NJ.
8. Staddon, J. E. R. 1970. Effect of reinforcement duration on fixed-interval responding. J. Exp. Anal. Behav. **13:** 9–11.
9. Hatten, J. L. & R. L. Shull. 1983. Pausing on fixed-interval schedules: Effects of the prior feeder duration. Behav. Anal. Lett. **3:** 101–111.
10. Harzem, P., C. F. Lowe & G. C. L. Davey. 1975. After-effects of reinforcement magnitude: Dependence upon context. Q. J. Exp. Psychol. **27:** 579–584.
11. Ettinger, R. H. & J. E. R. Staddon. 1982. Decreased feeding associated with acute hypoxia in rats. Physiol. Behav. **29:** 455–458.
12. Ettinger, R. H. & J. E. R. Staddon. 1983. The operant regulation of feeding: A static analysis. Behav. Neurosci. **97:** 639–653.
13. Johnson, D. F. & H. P. Wheeler. 1982. Comparisons between one-key and two-key versions of the sinewave schedule for pigeons. J. Exp. Anal. Behav. **38:** 101–108.
14. Keller, J. V. 1973. Responding maintained by sinusoidal cyclic-interval schedules of reinforcement: A control-systems approach to operant behavior. (Doctoral dissertation. University of Maryland, 1973). Diss. Abst. Int. **24:** 2901B–2902B.
15. Staddon, J. E. R. 1969. The effect of informative feedback on temporal tracking in the pigeon. J. Exp. Anal. Behav. **12:** 27–38.
16. Staddon, J. E. R. 1974. Temporal control, attention and memory. Psychol. Rev. **81:** 375–391.
17. Staddon, J. E. R. 1975. Limitations on temporal control: Generalization and the effects of context. Br. J. Psychol. **66:** 229–246.
18. Medin, D. L. 1969. Form perception and pattern reproduction by monkeys. J. Comp. Physiol. Psychol. **68:** 412–419.
19. Wilkie, D. M., & R. J. Summers. 1982. Pigeons' spatial memory: Factors affecting delayed matching of key location. J. Exp. Anal. Behav. **37:** 45–56.
20. Gibbon, J. 1977. Scalar expectancy and Weber's law in animal timing. Psychol. Rev. **84:** 279–325.
21. Platt, J. R. Temporal differentiation and the psychophysics of time. *In* Reinforcement and the Organization of Behaviour: Advances in the Analysis of Behaviour. Vol. 1. M. D. Zeiler and P. Harzem, Eds. John Wiley. New York, NY.
22. Hovland, C. I. 1951. Human learning and retention. *In* Handbook of Experimental Psychology. S. S. Stevens, Ed. John Wiley. New York, NY.
23. Tulving, E. 1972. Episodic and semantic memory. *In* Organization of Memory. E. Tulving & W. Donaldson, Eds. Academic Press. New York, NY.
24. Shimp, C. P. 1976. Short-term memory in the pigeon: Relative recency. J. Exp. Anal. Behav. **25:** 55–61.
25. Roberts, W. A. 1972. Spatial separation and visual differentiation of cues as factors influencing short-term memory in the rat. J. Comp. Physiol. Psychol. **78:** 281–291.
26. Hogan, D. E., C. A. Edwards & T. R. Zentall. 1981. Delayed matching in the pigeon:

Interference produced by the prior delayed matching trial. Anim. Learn. Behav. **9:** 395–400.
27. MEDIN, D. L. 1980. Proactive interference in monkeys: Delay and intersample interval effects are noncomparable. Anim. Learn. Behav. **8:** 553–560.
28. RILEY, D. A. & H. L. ROITBLAT. 1978. Selective attention and related processes in pigeons. *In* Cognitive Processes in Animal Behavior. S. H. Hulse, H. Fowler and W. K. Honig, Eds. Lawrence Erlbaum. Hillsdale, NJ.
29. ROBERTS, W. A. 1980. Distribution of trials and intertrial retention in delayed matching to sample with pigeons. J. Exp. Psychol. Anim. Behav. Processes **6:** 217–237.
30. WILKIE, D. M. & M. L. SPETCH. 1978. The effect of sample and comparison ratio schedules on delayed matching to sample in the pigeon. Anim. Learn. Behav. **6:** 273–278.
31. DENNIS, W. 1939. Spontaneous alternation in rats as an indicator of the persistence of stimulus effects. J. Comp. Psychol. **28:** 305–312.
32. DOUGLAS, R. J. 1966. Cues for spontaneous alternation. J. Comp. Physiol. Psychol. **62:** 171–183(a).
33. DALE, R. H. I. & J. E. R. STADDON. A temporal theory of spatial memory. Unpublished manuscript.
34. SHERRY, D. F., J. R. KREBS & R. J. COWIE. 1981. Memory for location of stored food in marsh tits. Anim. Behav. **29:** 1260–1266.
35. SHETTLEWORTH, S. J. 1983. Memory in food-hoarding birds. Scientific American **248**(3): 102–110.
36. BEATTY, W. W. & D. A. SHAVALIA. 1980. Spatial memory in rats: Time course of working memory and effect of anesthetics. Behav. Neural Biol. **28:** 454–462.
37. ROBERTS, W. A. 1974. Spaced repetition facilitates short-term retention in the rat. J. Comp. Physiol. Psychol. **86:** 164–171.

Temporal Variables in Delayed Matching to Sample[a]

WILLIAM A. ROBERTS AND PHILIPP J. KRAEMER

Department of Psychology
The University of Western Ontario
London, Ontario, Canada N6A 5C2

The investigation of short-term memory (STM) in animals has flourished as an area of research in the last 12 years or so, and delayed matching to sample has been the most extensively used procedure for this investigation. In keeping with the emphasis of this volume on timing processes, we would like to present evidence on the effects of temporal variables on delayed matching in pigeons and to discuss how these effects may be understood within a theory of memory.

There are actually several paradigms now available for studying delayed matching, but most of the work we will describe has been done with the choice procedure diagrammed in FIGURE 1. A pigeon is trained to respond to a panel containing a row of three translucent discs or keys, with an aperture containing a grain-filled hopper located below the center key. A typical trial would begin with illumination of the center key with white light, and a single peck on this key would introduce the sample stimulus. The sample stimulus would be a black and white pattern, such as a horizontal or vertical line, or, as in FIGURE 1, a colored field. After the pigeon had been exposed to the sample for a fixed period of time or had made a fixed number of pecks to it, the center key would be turned off, and the subject would remain in the dark for a predetermined delay or retention interval. At the end of the delay, the retention test would be initiated by illumination of the side keys, with the center key now darkened. The side keys contain the comparison stimuli: in the example, a matching red field and a nonmatching green field. If the pigeon now pecks the correct red key, the key is darkened, and it is allowed to eat from the grain hopper for 2 sec; a peck on the nonmatching green key turns off the key, but yields no food. After either one of these two events, the pigeon spends an intertrial interval in darkness, followed by initiation of the next trial by presentation of the white center key. A session may contain as many as 48 trials, with the two stimuli, red and green, alternating randomly as sample stimulus and incorrect comparison stimulus. The left–right position of the matching and nonmatching side keys is varied randomly, so that spatial position cannot provide a cue for accurate choice. Well-trained pigeons perform at a fairly constant level of accuracy on this task, and a variety of experimental variables can be manipulated both within and between sessions. The extent to which performance exceeds the chance level of 50% is taken as an indicator of how well birds remember the sample stimulus.

Within the delayed matching paradigm, several temporal variables can be investigated. Two variables of immediate interest to us were the length of exposure to the sample stimulus and the length of the delay between termination of the sample stimulus and onset of the comparison stimuli. In one experiment,[1] exposure to the sample stimulus was manipulated by requiring pigeons to make 1, 5, or 15 responses to the sample in order to advance to the delay interval. The delay interval was set at

[a]This paper was supported by Grant A7894 to W. A. R. from the National Sciences and Engineering Research Council, Ottawa, Ontario, Canada.

values of 0, 1, 3, and 6 sec. The resulting retention curves, seen in FIGURE 2, showed that accuracy of delayed matching declined progressively as the delay or retention interval was increased. Further, the retention curves for different numbers of pecks or fixed ratios (FRs) declined in parallel to one another, with the height of the curves directly related to the FR. In short, performance improved the longer the pigeon had to peck the sample stimulus and became worse the longer the bird had to wait for the comparison stimuli.

Still another question of interest was how repetition of the sample would affect retention. In this type of experiment, the sample stimulus was presented for a given period of time, the chamber then was darkened, and the sample stimulus then was presented again for some designated period. The sample stimulus could be repeated one or more times in this fashion, with an eventual delay and retention test. Not surprisingly, it has been found that performance improves as the number of repetitions

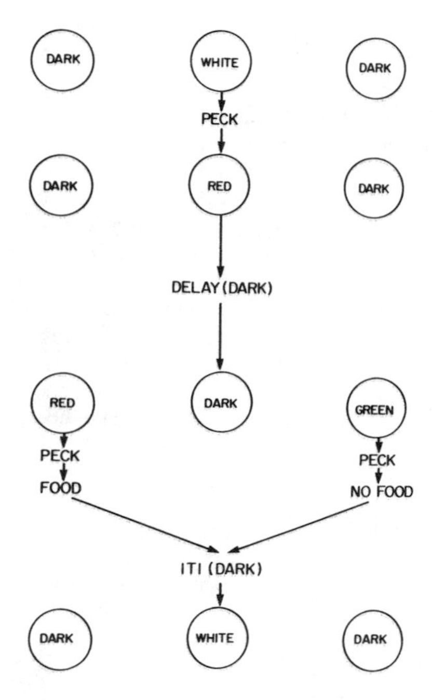

FIGURE 1. The sequences of events occurring on a delayed matching to sample trial. The circles represent rows of three stimulus keys.

increases.[1] A third temporal variable could be examined concurrently with the effects of repetition. This variable was the length of time that elapsed between successive repetitions of the sample stimulus, or the interstimulus interval (ISI). The effect of the ISI was of particular interest, because in studies of human memory for verbal materials memory generally improves as the ISI or degree of item spacing increases.[2,3] This was clearly not the case with pigeons since performance was adversely affected the longer the interval between sample repetitions. The data shown in FIGURE 3 help to summarize the effect of the ISI, as well as the effects of delay and sample presentation time. In this experiment, Roberts and Grant[4] presented the sample stimulus for an initial period (P1) of either 1 or 4 sec. The pigeon then spent an ISI in darkness for 0, 2, or 5 sec, with the 0-sec ISI actually representing no interruption of the sample stimulus. Following the ISI, the sample was presented for a second period (P2), which

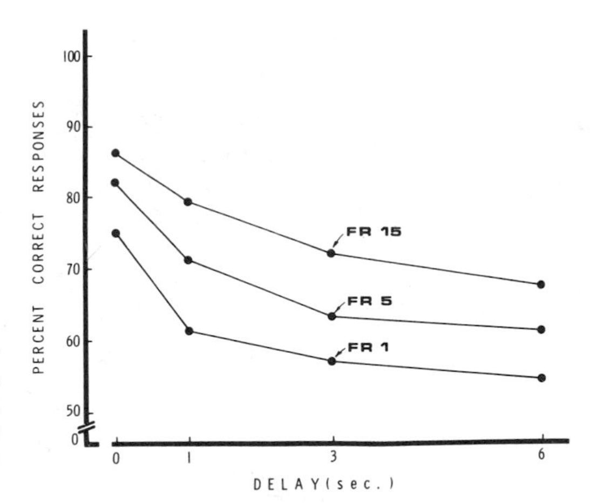

FIGURE 2. Retention curves for pigeons in a delayed matching to sample experiment. FR refers to the number of pecks a pigeon was required to make on the sample stimulus.

lasted for 0, 1, 2 or 5 sec. At the end of P2, retention was tested after delays of 0 or 2 sec. It can be seen that percentage of correct responses improved as both P1 and P2 were made longer and that performance shown in the lower panels, with a 2-sec delay, was inferior to that seen in the upper panels, with a 0-sec delay. The ISI is the parameter in these curves, and level of accuracy clearly dropped as the ISI was increased from 0 to 5 sec. Quite in contrast to the effects of spacing on human retention, spaced repetition was harmful to pigeon STM.

Roberts and Grant[1,4,5] felt that all of these effects of temporal variables upon pigeon STM could be parsimoniously explained within a model of trace strength

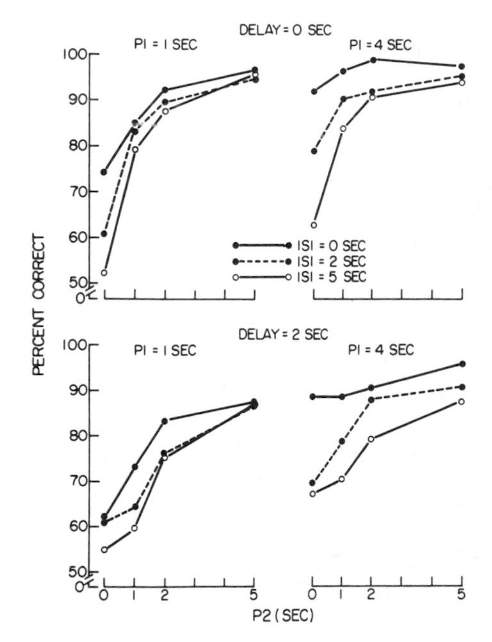

FIGURE 3. Accuracy of delayed matching to sample plotted as a function of P2 length following an initial P1 of 1 or 4 sec and an ISI of 0, 2 or 5 sec. *Top curves* plot data for the 0-sec delay, and *bottom curves* plot data for the 2-sec delay.

growth and decay. The model suggested that trace strength grew in the presence of a sample stimulus and decayed in the absence of the sample stimulus. Further, it was assumed that accuracy expressed in percentage of correct choices was directly related to trace strength. Therefore, accuracy improved with increasing exposure to the sample stimulus because longer exposures produced stronger memory traces. As soon as a delay in darkness was introduced, the trace began to decay, and the falling retention curve tracked that decay process. Similarly, as the ISI was lengthened, more trace strength would be lost between repetitions, and the total trace strength left after the final repetition would be inversely related to ISI.

The Roberts and Grant model of pigeon STM provided a very passive repletion and depletion view of memory. Since the presentation of that model, new findings have led theorists to suggest that STM in pigeons and other animals may involve more active and controlled processes.[6–8] For one thing, it has been proposed that pigeons may not always remember an iconic trace of the sample stimulus. Several theorists have suggested that pigeons may prospectively code sample stimuli into instructions about what response to make at the end of a retention interval.[6,9–11] It has been suggested, further, that the process underlying retention is not passive, but is a more active process of rehearsal. Grant[9] has suggested that sample stimuli activate response codes residing in long-term memory and that retention is based on the extent to which a response code is rehearsed during a retention interval. Unlike decay, the process of rehearsal is not held to follow an inexorable course; rehearsal may be accelerated or terminated on the basis of environmental cues that signal the importance of remembering. As examples, surprising sample stimuli may accelerate rehearsal,[12,13] and "forget cues" that signal the cancellation of a retention test may terminate rehearsal.[8,14]

Within models of animal STM as an active and controlled process, it has been suggested that initiation and termination of memory processes be conceived of as following an all-or-none principle.[6,9] Thus, presentation of a sample stimulus will have the effect of changing an appropriate response instruction or code held in long-term memory from an inactive state to an active state. The longer the sample stimulus is presented, the higher becomes the probability that this state change will occur. The effect of sample presentation time on delayed matching accuracy then is explained in terms of probability of code activation, instead of in terms of growth of trace strength. Once activated, a response code may be rehearsed and remain active until presentation of comparison stimuli. However, the probability that the activated code will return to an inactive state increases as time in the absence of the sample stimulus passes. Loss of memory with an increasing retention interval and the drop in performance as repetitions are spaced farther apart are now accounted for in terms of increased probability of the response code returning to an inactive state and not in terms of a fading trace.

Within both trace strength and all-or-none models of STM, the effects of the three temporal variables I have discussed hang together in the same general way. That is, both conceptions assume that there will be an increase in memory strength or probability of memory activation the longer a sample stimulus is presented and that strength or probability of memory activation will decline the longer an animal must wait in the absence of the sample stimulus. The effects of sample stimulus presentation time, delay, and ISI are similarly understood at this general level of explanation, whether we conceive of memory as a trace continuously changing in strength or as a response code varying in its probability of activation or deactivation.

Given this background information, the remainder of this paper will deal with a fourth temporal variable studied in delayed matching to sample, the intertrial interval (ITI). An initial examination of the effects of ITI on delayed matching suggests that its effects agree quite nicely with theoretical assumptions about the effects of delay and

ISI on memory. It is now a well-established fact that the accuracy of delayed matching improves as a direct function of the ITI. This effect has been demonstrated in pigeons,[15-17] monkeys,[18] and a dolphin.[19] Most commonly, this phenomenon has been attributed to release from proactive interference (PI). Proactive interference may be defined as a process that produces forgetting by the carryover of memories between trials. Since delayed matching studies typically use only a small number of stimuli (often just two), the stimulus that served as the sample and matching comparison stimulus on the preceding trial(s) often will become the incorrect comparison stimulus on the current trial. If memories of the sample stimulus and comparison stimuli from Trial $n - 1$ and other preceding trials are retained on Trial n, these memories may interfere with choices of the correct comparison stimulus on Trial n. Two theories have been advanced as to how this might happen. One possibility suggested by Grant and Roberts[5,15,20] is that memories from preceding trials directly compete with memory of the current trial sample stimulus. If the sampe stimulus on Trial $n - 1$ becomes the incorrect comparison stimulus on Trial n, memory of the sample from Trial $n - 1$ often may yield choice of the incorrect comparison stimulus on Trial n. The other possibility is that PI arises from a failure to discriminate the temporal recency of the sample stimulus presented on Trial n from that of samples presented in Trial $n - 1$ and other preceding trials.[21,22] If an animal has learned to always choose the comparison stimulus that matches the most recently presented sample stimulus, but cannot determine whether the most recently presented sample was the one presented on Trial n or Trial $n - 1$, frequent errors in retention will occur.

Both the competition and temporal discrimination models hold that interference effects should weaken as the ITI is made longer. In the case of the direct competition model, spacing trials apart allows more time for memories from preceding trials to be forgotten. If memory is thought of as varying in trace strength, a long ITI allows traces that persist beyond the end of a trial more time to decay. From the point of view of the all-or-none conception, response codes may remain active after completion of a trial, but will have more opportunity to return to an inactive state the longer the ITI. The notion that memory is lost over a long ITI then explains the effect of spacing trials and fits well with our account of the effects of other temporal variables. The temporal discrimination model accounts for the effects of ITI somewhat differently. It is assumed that animals keep track of the age or recency of memories of sample stimuli from preceding trials. When trials are spaced far apart, the ages of memories of sample stimuli from Trials n and $n - 1$ are more discriminable than they are when trials are massed together. Lengthening the ITI then improves retention by making temporal discrimination easier.[21,22]

Attributing the effect of ITI on delayed matching to release from PI then lets us provide an integrated account of four temporal variables in terms of simple notions of memory gain and loss with the passage of time. Unfortunately for this rather tidy position, problems have begun to crop up recently with an account of the trial-spacing effect in terms of release from PI. Such an account suggests that a trial-by-trial analysis of performance should show an interaction between different types of trial sequences and the length of the ITI. Within a sequence of delayed matching trials in which only two stimuli are used, there will be some pairs of successive trials in which the same stimulus served as the sample and some pairs in which different stimuli served as the sample. Interference should be more marked with different sample stimuli than with same sample stimuli, and we would expect this interference to be most noticeable when trials are massed and least noticeable when trials are spaced. Within-sessions analyses show that this pattern is not found. Pigeons show poorer retention when the samples are reversed from Trial $n - 1$ to Trial n than they do when the sample stimuli stay the same, but this effect is approximately constant at both short and long ITIs. A further surprising aspect of this "fine-grain" analysis of trial sequences is the observation that birds carry over only a tendency to choose the comparison stimulus chosen on the preceding trial. This effect does not interact with reinforcement; that is, it does not matter whether the comparison stimulus chosen on Trial $n - 1$ was correct and reinforced or incorrect and nonreinforced.[23,24]

A further difficulty with the release from PI account of the trial-spacing effect is found in some results reported by Roberts.[23] Pigeons were tested on delayed matching trials, with ITI values of 1 or 20 sec used within different sessions. Two types of sequences of delayed matching trials were used between sessions, either a random sequence or a homogeneous sequence. The random sequence was the same as the traditional procedure in which two stimuli alternate randomly as sample and incorrect comparison stimulus. In homogeneous sequences, the same sample stimulus was used on every trial of the session. The random condition should generate PI, and the idea that a long ITI attenuates PI leads us to expect better performance at the 20-sec ITI

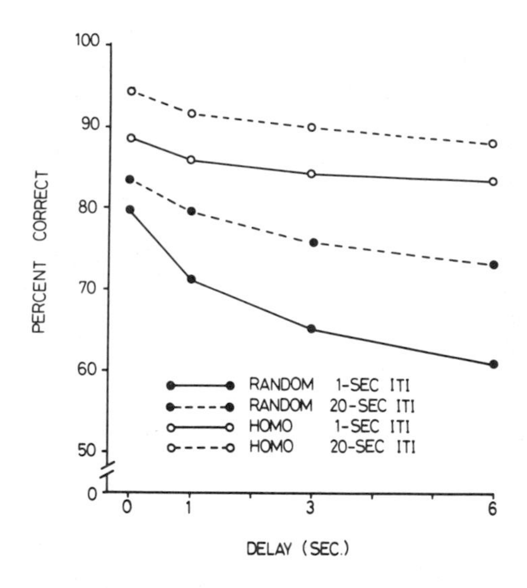

FIGURE 4. Retention curves for sessions containing random or homogeneous sequences of delayed matching trials. Within each type of sequence, performance was tested at ITIs of 1 and 20 sec.

than at the 1-sec ITI. Quite a different prediction may be made for the homogeneous sequences. In this case, memories carried over from Trial $n - 1$ and other preceding trials to Trial n should only support choice of the matching stimulus. Massing trials should facilitate performance, and we can expect that accuracy will be higher with a 1-sec ITI than with a 20-sec ITI. In other words, an interaction between ITI and type of trial sequence is predicted.

Retention curves for this experiment are shown in FIGURE 4. For the random sequences, the results are as expected, with higher retention at the 20-sec than at the 1-sec ITI. The more surprising results are seen in the homogeneous condition, in which the 20-sec ITI also led to significantly better performance than the 1-sec ITI. The finding that spacing trials still leads to higher accuracy than massing trials, even when every trial contains the same information, suggests strongly that something other than release from PI is responsible for the effect of ITI length on delayed matching.

One further problem with the PI account of the spaced trials effect will be described. Roberts and Kraemer[25] further examined the effects of ITI on delayed matching in pigeons, and used ITI values of 4, 8, 16 and 32 sec. When these ITI lengths were varied between sessions, but kept constant within a session, accuracy improved continuously as ITI increased. In other experiments, ITI was manipulated as a within-sessions variable. Within a session, ITIs of 4, 8, 16 and 32 sec were randomly placed between different pairs of successive trials. When performance was examined on trials that followed each ITI, it was found that percentage of correct responses varied little as a function of ITI length. There was improvement in accuracy from the 4-sec ITI to the 8-sec ITI but no improvement at all at ITIs of 16 and 32 sec. Quite in contrast to the effect of varying ITI between sessions, variation within sessions had little effect. The continuous increase in accuracy seen with ITI varied between sessions follows from the notion that there should be a continuous release from PI as the ITI gets longer. However, a similar effect also is predicted when ITI is changed within sessions. Enlarging the interval between Trial n and Trial $n - 1$ should provide increasing opportunity for memories from Trial $n - 1$ to be forgotten, regardless of whether the interval is varied between or within sessions. The effect of within-sessions variation in ITI appears to be beyond the scope of a release from PI mechanism.

In light of these difficulties with release from PI as an explanation of the effect of trial spacing on performance in delayed matching, we have explored alternative accounts of this effect. One possibility we have entertained is the idea that spacing trials apart may improve matching through a process of pattern perception. If we think of a sequence of trials as being a temporal pattern, performance may be enhanced the better an animal can perceive sample and comparison stimuli on each trial as integrated units separate from other trials. According to the gestalt laws of organization, spacing trials apart should lead animals to group trial events together by proximity; matching behavior may be enhanced then by a clear perception of the relationship between sample and comparison stimuli within a trial. On the other hand, when trials are massed together at short ITIs, temporal organization may break down, and animals often may fail to perceive sample and comparison stimuli on a given trial as events to be processed separately from other events in the temporal sequence. The observation that ITI variation has a much larger effect on matching accuracy when manipulated between sessions than within sessions may be understood by this hypothesis. With a constant ITI, particularly a long one, a pattern may be easily discerned, since the pattern is cyclical or repeating. When ITI is altered within sessions, the pattern is continually changing, and it may be difficult for animals to identify a clear organization among events. If this is the case, we would not expect local variations in ITI to have differential effects on matching accuracy.

An experiment was carried out to test some predictions from the pattern-perception hypothesis.[25] Pigeons were tested on delayed matching sessions within which ITI varied between trials; the mean length of the ITI varied between sessions, at lengths of 6, 12, and 24 sec. Of critical importance, the degree to which ITI varied within sessions was manipulated over three levels, high, intermediate, and none. The ITI could vary from the mean by as much as $\pm 67\%$ in the high variation condition and by as much as $\pm 33\%$ in the intermediate variation condition. In the condition with no ITI variation, the same ITI was used between all trials in a session. The pattern-perception hypothesis suggests that performance should be best with no variation in the ITI and should decline as the degree of ITI variation is increased. A further implication of the hypothesis is that the effect of between-sessions differences in the mean ITI should be more marked with no variation in ITI than with intermediate and high variation in ITI. Low variation in ITI should allow clear perception of temporal patterns as mean ITI is lengthened, whereas high variability should obscure the emergence of pattern perception. Therefore, an interaction between mean ITI length and degree of ITI variation is predicted.

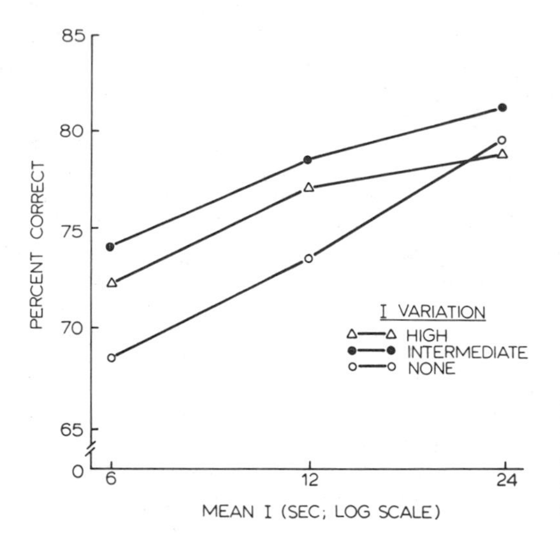

FIGURE 5. Percentage of correct responses on delayed matching plotted as a function of mean I length, with separate curves for each degree of I variation.

The findings from this experiment are depicted in FIGURE 5, in which percentage of correct choices is plotted against the mean ITI, and separate curves are shown for "high," "intermediate," and "none" conditions of ITI variation. These data directly contradict the predictions made from the pattern perception hypothesis. Instead of performance being best in the condition with no variation in ITI, this condition yields lower overall accuracy than the intermediate and high ITI variation conditions. Further, there is no suggestion of the predicted interaction between degree of variation and mean ITI length. The curves for different levels of ITI variation improve over log mean ITI lengths, generally as a set of linear and parallel functions.

In summary, the effect of ITI length on delayed matching appears to be refractory to explanation in terms of either release from PI or pattern perception. Still another type of theory is one that suggests that ITI directly affects processing of the sample stimulus and its memory. For example, Wagner has introduced a model of memory in which priming a representation of a stimulus into STM prior to presenting the stimulus will attenuate subsequent rehearsal of the stimulus event.[26–28] It may be argued that the comparison stimuli presented at the end of each delayed matching trial prime STM with a representation of the sample stimulus to be presented on the next trial.[23] If the next trial occurs immediately, the primed representation will detract from the surprise value of the sample stimulus, and little subsequent rehearsal of the sample representation then will take place over the retention interval. Retention would then be low with a short ITI. With longer ITIs, however, the probability increases that the primed representation will no longer reside in STM when the sample stimulus is presented. In this case, more vigorous rehearsal of the sample stimulus memory should occur during the retention interval, and retention should be higher than with the primed sample stimulus. Unfortunately, this hypothesis stumbles also on one of the findings already described, the observation that within-sessions variation in ITI has little effect on performance. The probability of a sample stimulus's memory being strongly rehearsed should increase as a direct function of ITI length, both with between- and within-sessions variation in the ITI. Our results suggest, to the contrary, that it may be the overall ITI length within a session that determines performance and not local or trial-to-trial variations in ITI.

Given this history of frustrated attempts to account for the ITI effect, we have approached the phenomenon recently from a somewhat different point of view.[25] We have noticed that there is a good deal of similarity between the effects of ITI on performance in delayed matching and autoshaping experiments. Autoshaping refers to Pavlovian or classical conditioning of key-pecking. The conditioned stimulus (CS) is presentation of a lit key for a period of time (T), and the unconditioned stimulus (US) is the delivery of food at the end of T. With successive pairings of the CS and US, pigeons begin to peck the lit key, even though pecking has no control over the delivery of the US. Both the rate of acquisition of autoshaping and the rate of key-pecking in trained birds increases with the ITI.[29,30] Further, rate of autoshaping has been found to be insensitive to local variations in ITI length and to improve with mean ITI length in sequences of trials using both fixed and variable ITI lengths.[29,31,32]

When ITI or interval (I) length has been manipulated systematically with T length, it has been found that rate of autoshaping increases as a power function of the I/T ratio. In addition, approximately equal rates of learning have been found with equivalent I/T ratios formed from different values of I and T. The length of T may be seen as a delay between initiation of a signal for food and its delivery. The delay used in delayed matching experiments may be seen to have a similar role. We wondered how performance in delayed matching might improve as a function of the ratio of I to delay (D) and if equivalent levels of accuracy might be found at equivalent I/D ratios made up of different values of I and D. In the experiment already mentioned, in which I was set at values of 4, 8, 16, and 32 sec between sessions, D was varied factorially with I and set at values of 0.5, 1, 2 and 4 sec. The percentage of correct responses has been plotted as a function of the I/D ratio in FIGURE 6. The I/D ratios range from 1 to 64, and there are several instances in which the same ratio is formed from different absolute lengths of I and D. Accuracy improves as a linear function of log I/D, and performance appears to be approximately constant at constant I/D ratios. These data then suggest further commonality between the effects of temporal variables upon delayed matching and autoshaping.

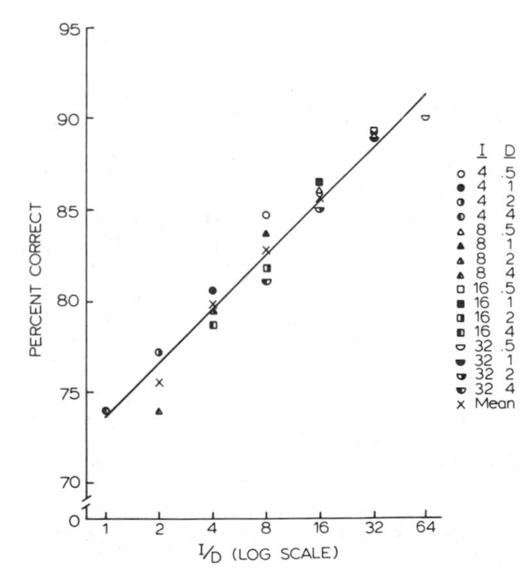

FIGURE 6. Percentage of correct responses plotted as a function of the log I/D ratio. The line represents the best fit to the means at each I/D ratio.

Gibbon and Balsam[31] and Jenkins *et al.*[32] have offered similar accounts of the effects of I and T on autoshaping. Gibbon and Balsam suggest that the periodic delivery of food establishes an overall or background expectancy of reinforcement, and that this expectancy is weaker the higher the average length of the ITI. Within trials, a trial expectancy of food is established and is spread over T, so that the trial expectancy is weaker the longer T is. A ratio-comparator mechanism is assumed to compare the level of expectancy during T with the background level of expectancy. As I/T gets larger, the ratio of expectancies will exceed a threshold value, and pecking will occur; performance should improve continuously as I/T gets larger. Jenkins *et al.* have referred to this relationship as one of relative waiting times. As I gets longer and T gets shorter, waiting time within a trial becomes short relative to overall waiting time between reinforcements, and readiness to peck increases.

The question we pose is whether similar mechanisms could be operating in delayed matching experiments. Delayed matching to sample is not Pavlovian conditioning, since reward is delivered only after choice of the matching comparison stimulus. We continue to think of delayed matching as a STM task, in which memory is increased with exposure to the sample stimulus and lost progressively over a delay or ISI. However, it may be possible that the overall temporal context in which a session takes place may exert a further influence on performance. The variables I and D in delayed matching might give rise to a comparison process similar to that postulated between background and trial expectancies in autoshaping. If the expectancy of reward within trials increases as I becomes longer, it is possible that increased expectancy would lead to improved accuracy of delayed matching. That expectancy of trial outcome may play an important role in delayed matching has been shown by recent experiments carried out by Peterson and his colleagues.[33-35] When differential trial outcomes are associated with sample stimuli, both rate of acquisition and steady-state accuracy at varying delays is considerably improved over control conditions in which differential outcomes are not present. Although the exact mechanisms by which expectancies may improve delayed matching are not yet obvious, we suggest that this theoretical avenue holds promise as an account of trial-spacing effects.

REFERENCES

1. ROBERTS, W. A. 1972, J. Exp. Psychol. **94:** 73–83.
2. BJORK, R. A. 1970. *In* Models of Human Memory. D. A. Norman, Ed. Academic Press. New York, NY.
3. MELTON, A. W. 1970. J. Verb. Learn. Verb. Behav. **9:** 596–606.
4. ROBERTS, W. A. & D. S. GRANT. 1974. Learn. Motiv. **5:** 393–408.
5. ROBERTS, W. A. & D. S. GRANT. 1976. *In* Processes of Animal Memory. D. L. Medin, W. A. Roberts & R. T. Davis, Eds. Erlbaum. Hillsdale, NJ.
6. HONIG, W. K. 1978. *In* Cognitive Processes in Animal Behavior. S. H. Hulse, H. Fowler, & W. K. Honig, Eds. Erlbaum. Hillsdale, NJ.
7. HONIG, W. K. 1981. *In* Information Processing in Animals: Memory Mechanisms. N. E. Spear & R. R. Miller, Eds. Erlbaum. Hillsdale, NJ.
8. MAKI, W. S. 1981. *In* Information Processing in Animals: Memory Mechanisms. N. E. Spear & R. R. Miller, Eds. Erlbaum. Hillsdale, NJ.
9. GRANT, D. S. 1981. *In* Information Processing in Animals: Memory Mechanisms. N. E. Spear & R. R. Miller, Eds. Erlbaum. Hillsdale, NJ.
10. ROITBLAT, H. L. 1980. Anim. Learn. Behav. **8:** 341–351.
11. HONIG, W. K. & R. K. R. THOMPSON. 1982. *In* The Psychology of Learning and Motivation. G. H. Bower, Ed. Academic Press. New York, NY.
12. MAKI, W. S. 1979. Anim. Learn. Behav. **7:** 31–37.

13. GRANT, D. S., R. G. BREWSTER & K. A. STIERHOFF. 1983. J. Exp. Psychol. Anim. Behav. Processes. **9:** 63–79.
14. GRANT, D. S. 1981. Learn. Motiv. **12:** 19–39.
15. GRANT, D. S. 1975. J. Exp. Psychol. Anim. Behav. Processes. **1:**207–220.
16. MAKI, W. S., J. C. MOE & C. M. BIERLEY. 1977. J. Exp. Psychol. Anim. Behav. Processes **3:** 156–177.
17. NELSON, K. R. & E. A. WASSERMAN. 1978. J. Exp. Anal. Behav. **30:** 153–162.
18. JARRARD, L. E. & S. L. MOISE. 1971. *In* Cognitive Processes of Nonhuman Primates. L. E. Jarrard, Ed. Academic Press. New York, NY.
19. HERMAN, L. M. 1975. Anim. Learn. Behav. **3:** 43–48.
20. GRANT, D. S. & W. A. ROBERTS, 1973. J. Exp. Psychol. **101:** 21–29.
21. D'AMATO, M. R. 1973. *In* The Psychology of Learning and Motivation. G. H. Bower, Ed. Vol. **7.** Academic Press. New York, NY.
22. WORSHAM, R. W. 1975. Anim. Learn. Behav. **3:** 93–97.
23. ROBERTS, W. A. 1980. J. Exp. Psychol. Anim. Behav. Processes **6:** 217–237.
24. ROITBLAT, H. L. & R. A. SCOPATZ. 1983. J. Exp. Psychol. Anim. Behav. Processes. **9:** 202–221.
25. ROBERTS, W. A. & P. J. KRAEMER. 1982. J. Exp. Psychol. Anim. Behav. Processes **8:** 342–353.
26. WAGNER, A. R. 1976. *In* Habituation: Perspectives from Child Development, Animal Behavior, and Neurophysiology. T. J. Tighe & R. N. Leaton, Eds. Erlbaum. Hillsdale, NJ.
27. WAGNER, A. R. 1978. *In* Cognitive Processes in Animal Behavior. S. H. Hulse, H. Fowler & W. K. Honig, Eds. Erlbaum. Hillsdale, NJ.
28. WAGNER, A. R. 1981. *In* Information Processing in Animals: Memory Mechanisms. N. E. Spear & R. R. Miller, Eds. Erlbaum. Hillsdale, NJ.
29. GIBBON, J., M. D. BALDOCK, C. M. LOCURTO, L. GOLD & H. S. TERRACE. 1977. J. Exp. Psychol. Anim. Behav. Processes **3:** 264–284.
30. PERKINS, C. C., W. O. BEAVERS, R. A. HANCOCK, P. C. HEMMENDINGER, D. HEMMENDINGER & J. A. RICCI. 1975. J. Exp. Anal. Behav. **24:** 59–72.
31. GIBBON, J. & P. BALSAM. 1981. *In* Autoshaping and Conditioning Theory. C. M. Locurto, H. S. Terrace & J. Gibbon, Eds. Academic Press. New York, NY.
32. JENKINS, H. M., R. A. BARNES & F. J. BARRERA. 1981. *In* Autoshaping and Conditioning Theory. C. M. Locurto, H. S. Terrace & J. Gibbon, Eds. Academic Press. New York, NY.
33. PETERSON, G. B., & M. A. TRAPOLD. 1980. Learn. Motiv. **11:** 267–288.
34. PETERSON, G. B., R. L. WHEELER & G. D. ARMSTRONG. 1978. Anim. Learn. Behav. **6:** 279–285.
35. PETERSON, G. B., R. L. WHEELER & M. A. TRAPOLD. 1980. Anim. Learn. Behav. **8:** 22–30.

Timing, Learning, and Forgetting

CHARLES P. SHIMP
Department of Psychology
University of Utah
Salt Lake City, Utah 84112

Within the experimental analysis of behavior, a truism is that behavior takes place in time. There is not much else about which everyone agrees regarding the role of time in operant conditioning. Some theories let long-term averages of rates of reinforcers enter directly, without any intervening psychological mechanisms, into causal explanations of behavior.[3-6] This position, which is labeled a "molar" position, stands in sharp contrast to other positions that focus on psychological events taking place over shorter periods of time, and which seek to derive "molar" phenomena from more "molecular" ones. A classic example of these different approaches is their treatment of the phenomenon of matching in concurrent interval schedules, where the percentage of responses to one alternative, averaged over several hours, roughly equals the percentage of reinforcers earned by responses to that alternative.[7] This "matching law" has been said to reflect a basic causal relation between behavior and reinforcement.[3,5,8] But local reinforcement probabilities constantly change in this context as a function of an organism's behavior, and a more molecular position holds that matching over long time periods does not reflect a fundamental law, but only reflects averaging of psychological processes that occur over shorter periods of time. Such a molecular position is especially likely to focus on the way behavior might adapt to the locally changing reinforcement probabilities. One extreme version of such local adaptation maintains that each response in a concurrent schedule situation is allocated to the alternative that momentarily has the greater probability of reinforcement. While this idea seems to describe a reasonable amount of data,[9-12] it still leaves a good deal to be desired from the viewpoint of psychological theory. This "momentary maximizing" does not, for example, explain how an organism knows which alternative is locally the more attractive, or why actual behavior does not in fact altogether accurately conform to this kind of optimal strategy. It is, in short, simply a descriptive rule borrowed from similar ideas in philosophy, economics, and evolutionary theory, rather than a psychological theory, and as such does not "interface" with the literature on associative learning or the literature on animal cognition.

These problems with the momentary maximizing idea led me some years ago to develop a model (here called AL) based on psychological processes, especially associative learning and short-term forgetting. The model had as its objective something like that of statistical learning theory[13]: a description of the psychological events taking place over relatively short periods of time which could in the aggregate explain molar phenomena. The advantages of this approach over simpler, algebraic, deterministic models of behavior were made clear during the evolution in the 1950s and 1960s of mathematical models of learning.[14,15] Some of these advantages are that a necessary distinction between learning and performance tends to be more sharply articulated[16]; variability in performance during steady-state conditions is an automatic consequence of the approach and needs neither auxiliary assumptions to be assimilated nor to be ignored as an embarrassment; acquisition of knowledge and steady-state performances are handled by the same processes and do not require different assumptions; and the theoretical assumptions deal with general issues about the nature of knowledge and its acquisition, not about performance in a particular setting, so that the basic structure of a model is widely applicable to many settings.

PREVIOUS WORK ON THE ASSOCIATIVE LEARNER (AL) MODEL

The associative learner's behavior seems most easily studied with computer-simulation techniques.[15] The widespread adoption of computers for on-line control and analysis of operant data and the concomitant acquisition of computer-related skills has made simulation a more accessible technique than it previously was, and the flexibility with which the approach permits exploration of new theoretical ideas is a strong recommendation in its favor.

Issues to which the AL has been applied include the psychophysics of timing and both molar and molecular effects of reinforcement. More particularly, the domain of the previous version included the following results: that the subjective midpoint in temporal bisection experiments lies approximately at the geometric mean[17,18]; that the Weber fraction in differential-reinforcement-of-low-rate (DRL) schedules relating average interresponse time (IRT) to a measure of the variability in IRTs is roughly constant over conditions where the reinforced IRT is varied and that, also in DRL schedules, the modal IRT is somewhat shorter than the reinforced IRT and that this shift is greater for longer than for shorter reinforced IRTs.[17] Finally, AL predicts that mean response rate in variable-interval (VI) schedules is an increasing function of overall reinforcement rate.[19] While the AL's range of applicability includes timing and both molar and molecular phenomena, some of the most notable experimental results still have not been assimilated. Both the molar matching outcome described above and various molecular IRT phenomena have not been described adequately by the model. It is to these data that AL is applied in the present chapter.

A NEW VERSION OF THE ASSOCIATIVE LEARNER MODEL

The new version of the AL is largely a simplification of the former and can be described fairly efficiently in terms of its theoretical assumptions and in terms of the computer-simulation subroutines that correspond to these assumptions.

An environmental event such as a response is assumed to activate a corresponding memory representation having numerous features. Relatively little of a detailed sort is known about these features,[20] so it is assumed merely that there are several of them and that each is at any moment either activated or deactivated. Once activated by the corresponding environmental event, each feature probabilistically becomes deactivated in each subsequent short period of time. Deactivation of one feature is independent of deactivation of any others. Thus, immediately after a stimulus occurs, all its corresponding features are assumed to be activated and then as time elapses, the number of features remaining activated decreases at a constant rate, that is, on the average, geometrically. Note, however, that on some occasions the number still activated at some particular time after a stimulus will by chance be larger than on other similar occasions. The number of features in the simulations of concurrent IRT–IRT schedules was 63, and in the simulations of concurrent VI–VI performance, it was 31. The model assumes that the number of features still activated can be a discriminative stimulus. It is as though an animal is assumed to be able to judge the clarity of the short-term memory trace of a recent event. A subject is assumed by the model to tell time by virtue of this assumption: if a trace is sharp (many features are still activated), little time is judged to have elapsed since the event; if a trace is fuzzy (many features have become deactivated), a lot of time is judged to have elapsed. These assumptions about activation and deactivation of memory traces are essentially identical to those in earlier versions of the AL.[17-19] The general way AL interrelates timing and short-time memory is similar to that proposed by D'Amato.[21]

These theoretical assumptions were embedded in two subroutines. Subroutine ACTIVATE was executed every time a response occurred: every response activated all the features in the memory representation of that response. FORGET was executed once every second after a representation was activated. Every feature that was still activated was threatened (with probability .025) with deactivation. A feature once deactivated remained so until the corresponding response occurred: there was no spontaneous remembering of forgotten features. Finally, FORGET calculated the number of features still activated in a trace.

The assumptions described so far affect remembering and forgetting, that is, the activation and deactivation of features in a memory representation. So far, the only theoretical parameter encountered is just the probability of forgetting a feature. It was equal to .025 per sec in all the simulations reported here.

Next consider the assumption regarding associative learning. Subroutine LEARN was executed every time a response occurred. LEARN took the number of currently active features, call it N, and associated N either with reinforcement with probability 1.0, if the response was reinforced, or with nonreinforcement with probability .025, if the response was not reinforced. Associative learning was all-or-nothing in the sense that associative values of 0 (an association with nonreinforcement) and 1 (an association with reinforcement) were the only permissible values. Notice that a value of N that had signaled nonreinforcement with probability 1.0 signaled reinforcement after a response was reinforced in its presence, but that a value of N signaling reinforcement changed to signal nonreinforcement, after an unreinforced response occurred in its presence, with probability of only .025.

Finally, subroutine EMIT was executed once every second and simply asked whether the current value of N was a discriminative stimulus for reinforcement or nonreinforcement. If the current N signaled reinforcement, EMIT produced a response with probability 1.0. On the other hand, if the current value of N involved an association with nonreinforcement, a response was made with probability .025. Since EMIT was executed only every second, the highest rate at which a response could occur was one per second. The simulated absolute response rates therefore sometimes were too low, but they were not the focus of attention throughout the following simulations and this feature did not seem to handicap the present work. In any case, higher frequencies of executing EMIT presumably would remove this problem.

GENERAL FEATURES AND SUMMARY OF SIMULATIONS

The simulations were executed in assembly language on an elderly PDP-12/30. Three stat-birds were simulated concurrently and independently for each schedule contingency. Ten sessions were conducted for each set of schedule parameters for concurrent IRT–IRT schedules and 15 were conducted for concurrent VI–VI schedules. Simulated session duration was approximately 1 hour.

The model required the estimation of three theoretical probabilities: the probability per second that an activated feature becomes deactivated; the probability that a given trace that signals reinforcement changes to signal nonreinforcement when a response occurs in its presence and is unreinforced; and, the probability that a response occurs in the presence of a trace that signals nonreinforcement.

Numerical values for the three theoretical parameters were selected in a crude yet apparently satisfactory manner. The investigator simply tried several combinations of values for several of the concurrent IRT–IRT contingencies until he found a combination that seemed to capture the chief qualitative features of real data. The estimated parameters are therefore in absolutely no sense the "best fitting" ones. It is presumably a coincidence that all three parameters were assigned the same numerical value of .025. Once the theoretical parameters were assigned values, all of the simulations reported here were conducted without making any changes in them.

In each simulated schedule contingency, the initialization of associative values was either such that all were associated with nonreinforcement, all were associated with reinforcement, or 1 out of 10 was associated with reinforcement. The subsequent text describes which initializations applied in each simulation.

CONCURRENT IRT–IRT SCHEDULES

One of the most obvious molecular variables, and the one that has received perhaps the most attention, is the time between two successive responses, such as key-pecks, lever-presses, or treadle-presses. These IRTs define in many situations the local temporal structure of behavior that in turn generally defines the focus of attention of a molecular analysis: Molar analyses historically look for causal relations among variables that ignore this local temporal patterning, while molecular analyses have searched for causal relations reflected in the local patterning. In practice, molar analyses often look at rate of key-pecking averaged over local patterns and molecular analyses look at local patterns, often IRTs. In passing, note the irony that molecular analyses therefore tend to involve larger units of analysis than do molar analyses.

A body of molecular data has been built up over the past 10 to 20 years showing how reinforcement affects various sorts of local patterns of behavior. Perhaps the most comprehensive molecular picture deals with IRTs. A sizable body of data shows how making reinforcement contingent on various properties of IRTs affects the way an organism temporally patterns its behavioral output. Within the larger body of IRT data, there is a subset dealing with a situation in which two classes of IRTs, a shorter and a longer class, are reinforced according to a concurrent VI–VI schedule. Thus, the schedule is an ordinary concurrent VI–VI schedule, except that reinforcement is arranged not as is usual for just a single peck on a left or right key, but for pecks on a single key that terminate IRTs falling in a shorter or a longer class. With such an arrangement, one can experimentally manipulate schedule parameters such as the absolute and relative durations of the reinforced IRTs and the absolute and relative frequencies of reinforcement for the two classes of reinforced IRTs. AL is next applied to the problem of describing the performance of animals when these schedule parameters are manipulated.

For all of the simulations reported in this section, there were 10 sessions per condition. The parameter values were as described previously. The initial associations at the beginning of each condition were all to nonreinforcement except in the conditions where the relative frequency of reinforcement was manipulated, in which case they were all to reinforcement.

RELATIVE DURATION OF REINFORCED PATTERNS

Under certain conditions the relative frequency of occurrence of an IRT approximately equals the relative reciprocal of its duration: if 2-sec and 4-sec IRTs are concurrently reinforced on the average equally often, then the relative frequency of 2-sec IRTs will be

$$\frac{1/2}{1/2 + 1/4} = .67$$

This phenomenon occurs provided that the shorter IRT is neither too short, say less than a second or so, nor too long, say more than 4 seconds or so.[12] The shorter reinforced IRT was 3–4 sec throughout the present simulations, while the longer was varied over conditions and was 5–6 sec, 6–8 sec, 12–14 sec, or 18–20 sec. The VI schedule was made richer as the longer reinforced IRT was lengthened to preserve roughly the same overall reinforcement density. Reinforcements were assigned on the average equally often to shorter and longer reinforced IRTs. FIGURE 1 shows AL's performance on such a concurrent IRT–IRT schedule. (Throughout this section on concurrent IRT–IRT schedules, a shorter or a longer IRT is considered to be one falling in the corresponding class of reinforced IRTs.) The results from the three stat-birds displayed in FIGURE 1 are extremely similar to, indeed, are perhaps not discriminably different from, those of three real birds reported by Shimp,[22] where there was considerable variability across subjects and the best-fitting straight line for the average data was $y = -.075 + 1.068x$.

ABSOLUTE DURATION OF REINFORCED PATTERNS

The kind of result shown in FIGURE 1 obtains, as noted above, provided that the shorter reinforced IRT is neither too short nor too long. As the absolute durations of both reinforced IRTs are lengthened, while their relative durations are kept equal, preference for the shorter increases from a level approaching indifference for very short reinforced IRTs, through the relative reciprocal values portrayed in FIGURE 1, to a value exceeding the relative reciprocal.[23] The point at which preference appears to

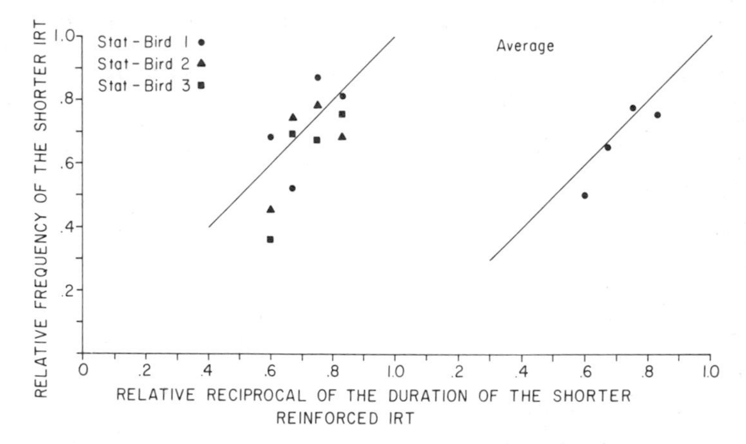

FIGURE 1. The relative frequency of the shorter of two reinforced IRTs as a function of the relative reciprocal of its duration. Each point is for the last one of 10 sessions in a condition. These points are theoretical predictions from a computer-simulation model (AL) and are similar to corresponding data from real pigeons reported by Shimp.[22]

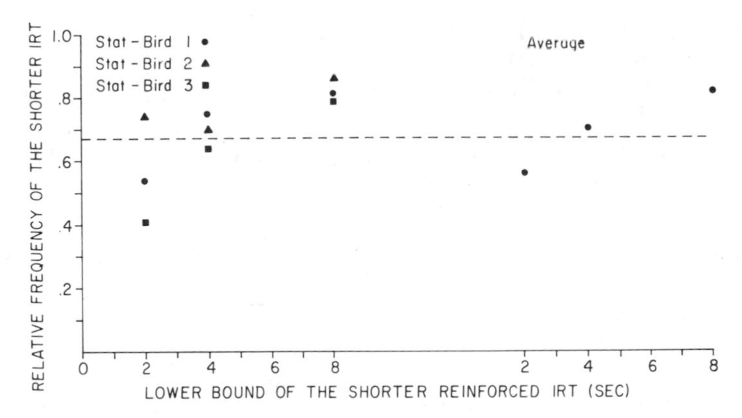

FIGURE 2. The relative frequency of the shorter of two reinforced IRTs as a function of their absolute durations, as measured by the duration of the shorter of the two. The longer was twice as long as the shorter. Each point is for the last one of 10 sessions in a condition. The horizontal line corresponds to the relative reciprocal of the shorter IRT. These points are theoretical predictions from a computer-simulation model (AL) and are similar to corresponding data from real pigeons reported by Hawkes and Shimp.[23]

exceed the relative reciprocal value is about 4 sec.[23] Thus, AL was required to perform here on a concurrent IRT–IRT schedule in these conditions: shorter and longer reinforced IRTs were 2–3 and 4–5 sec, 4–6 and 8–10 sec, and 8–11 and 16–19 sec. The relative reciprocal of the shorter reinforced IRT therefore was always .67. Shorter and longer IRTs were reinforced equally often. FIGURE 2 shows the corresponding results. As is the case with FIGURE 1, when the simulated results are compared to the behavior of real birds, it is doubtful one could tell the difference: the predicted curve rises as it should with longer absolute durations, and crosses the relative reciprocal value at approximately the correct absolute duration.

ABSOLUTE RATE OF REINFORCEMENT

The relative frequency of the shorter of two reinforced IRTs seems to depend on the total reinforcement rate.[24] When the relative reinforcements per hour are equal for each reinforced pattern, and the total reinforcements per hour are reduced, preference seems to move toward indifference, that is, toward a relative frequency of the shorter IRT of approximately .5, with some subjects showing more of an effect than others. TABLE 1 shows AL's behavior in two such cases. In each case, there is a trend for preference to move toward indifference, and as is the case across real birds, different stat-birds show the effect to different degrees. Thus, AL seems generally to predict the correct effect of total reinforcement rate on preference.

RELATIVE FREQUENCY OF REINFORCEMENT

Preference for the shorter of two reinforced IRTs increases as the relative frequency of reinforcement for that IRT increases.[25,26] No simple quantitative rule to summarize this effect has ever been articulated, but some salient features of the data can be listed.

TABLE 1. Effects of Overall Reinforcement Frequency for Concurrently Reinforced IRTs[a]

Reinforced IRTs	Total Reinforcements per Hour				Relative Frequency of Shorter IRT			
	Stat-Bird				Stat-Bird			
	1	2	3	Average	1	2	3	Average
1–2 sec and	7	14	7	9	.78	.45	.57	.60
3–4 sec	52	66	59	59	.76	.73	.67	.72
6–10 sec and	0	0	0	0	.50	.77	.60	.62
18–22 sec	21	37	26	28	.75	.79	.74	.76

[a]Data are for day 10 of each condition.

At a value of the relative reinforcement frequency of .5, preference approximates the relative reciprocal (see above) and moves toward exclusive preference for the shorter IRT when that IRT receives progressively more than half of the reinforcers. However, the curve is not symmetric around .5: when the shorter IRT receives less than half of the reinforcers, preference for that IRT remains at a fairly high level: if the individual data reported by Shimp[25] are extrapolated to estimate a y-intercept, the three values range from roughly, .35 to .56. FIGURE 3 shows that AL predicts all these effects. As with FIGURES 1 and 2 and TABLE 1, it is difficult to discriminate AL's performance from real data. Thus, AL predicts the effects on preference of both relative and overall reinforcement rate.

DEVELOPMENT OF TEMPORAL STRUCTURE

The previous results deal only with the last days of various conditions. But AL also predicts the development of temporal patterning. Unfortunately, almost nothing is

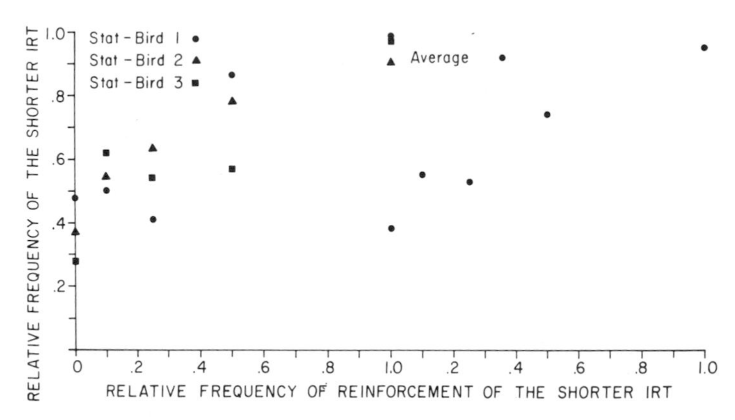

FIGURE 3. The relative frequency of the shorter of two reinforced IRTs as a function of the relative frequency of reinforcement for that IRT. The shorter and longer reinforced IRTs were 7 to 9 sec and 21 to 23 sec, respectively. Each point is for the last one of 10 sessions in a condition. These points are theoretical predictions from a computer-simulation model (AL) and are similar to corresponding data from real pigeons reported by Shimp[25] and Staddon.[26]

known about this development, so the detailed and quantitative evaluation of whether AL correctly handles acquisition is impossible. Qualitatively, AL seems basically correct: Initial IRT distributions depend on the initial associative values and may assume a variety of appearances, but these change over sessions into those summarized above. AL seems to assimilate adaptive *changes* in the local temporal patterning of behavior as well as terminal states.

CONCURRENT VI–VI SCHEDULES

These schedules have played a fascinating role in the evolution of the experimental analysis of behavior. From a field almost exclusively oriented toward an inductivist approach, it has changed in the last 10 or 15 years to one where theoretical issues are central. Herrnstein's 1970 paper contributed greatly to this change, and concurrent VI–VI schedules played a central role in that paper.[8] The critical historical event was the discovery of a quantitative result beautiful in its simplicity and perfect as an object of theoretical speculation. This discovery was that the percentage of responses to one of two keys approximately equaled the percentage of all reinforcers that were delivered for responses to that key.[7] Some investigators assigned to this relation an unparalleled theoretical significance,[5,8] while others viewed it more as a statistical artifact of causal relations involving variables other than the overall percentages of responses and reinforcers.[27] In particular, it was suggested that behavioral adaptation to the locally fluctuating reinforcement probabilities would produce local choice probabilities, the average of which was molar matching. The most extreme form of such local adaptation is called momentary maximizing and holds that each choice is to the alternative momentarily having the greater probability of reinforcement.[12,27] Retrospectively, it seems as though this view was given rather short shrift. Nevin[1] found that the probability of changing over from one key to the other did not seem to track the local changes in reinforcement (the changeover function was essentially flat), and this evidence seems to have been sufficient for all but a few theorists (Mackintosh,[28] for example) to reject the idea of local adaptation. Instead, there was a proliferation of molar accounts that assigned causal relations to molar variables and ignored the apparently unnecessary complications of local contingencies.[3,4,6,29]

An experiment by Silberberg *et al.*[11] called attention to the possible advantages of descriptions of concurrent performances in terms of local strategies. That paper, and more recent ones by Hinson and Staddon,[9,10] suggest that the early rejection by the operant community of local adaptation was premature. (Indeed this may be a useful example to be pondered by sociologists of science who deal with how unpopular ideas are perceptually ignored: a serious evaluation of local adaptation, response strategies, and the general molecular approach was functionally suppressed by the dominant molar conceptual framework in terms of which molecular issues made little sense.)

The evolutionary development of theories for concurrent VI–VI schedules seems now to have reached a point where an approach altogether different from the earlier ones is appropriate. While molecular phenomena can no longer be ignored, at the same time severe shortcomings to the idea of perfect local adaptation, as embodied in momentary maximizing, are obvious to all. (However, Dr. John Wearden in very recent and unpublished work, has developed what to my knowledge is by far the most comprehensive and general version of momentary maximizing. Such an approach where momentary maximizing is used as a descriptive device may yet serve as an important diagnostic tool in the development and evaluation of theories of concurrent performances.)

How can a molecular theory assimilate molar matching without imposing unreasonably restrictive demands on sequential performance, such as those of perfect local maximizing? How can, in other words, a molecular theory handle both molar matching and local adaptation, without requiring an unrealistically precise local adaptation? Further, how can molar matching be conceptually interrelated with the molecular phenomena such as the IRT data described previously? Preliminary answers to these questions are provided by the simulations of AL described next.

In simulating concurrent VI–VI preformance, AL kept track of how long it had been since the last response on the left key (as defined by the number of left key-peck features still active) and how long it had been since the last response on the right key (as measured by the corresponding number of still-active right-key features). These two numbers defined an ordered pair of numbers characterizing the momentary states of two memory representations, and that pair signaled reinforcement or nonreinforcement on left and right keys in basically the same manner as described previously for concurrent IRT–IRT performances. As time elapsed since a response, each of these two numbers of still-active features decreased randomly and independently in the same way N decreased in the application of AL to concurrent IRT–IRT schedules described previously. After each occasion when the traces were subjected to decay, that is once every second, AL looked up the new ordered pair in its associative memory. For expository purposes, take the example (10, 5) that means 10 features of a left-key peck are still active and 5 features of a right-key peck are still active. Note that this trace suggests that less time probably has elapsed since a left-key peck than since a right-key peck. The pair (10, 5) of active features of memory representations sometimes might have been present when either a left key-peck or a right key-peck occurred and was reinforced. Accordingly, AL would look up in its associative memory which of the two responses, left or right, was associated with reinforcement in the presence of the trace

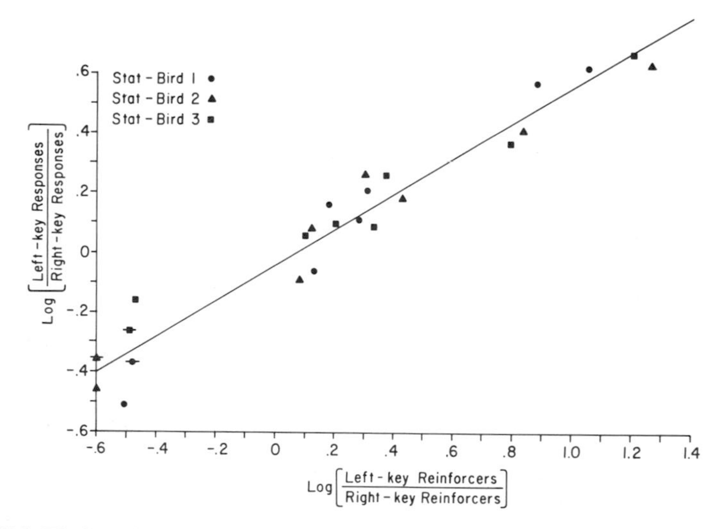

FIGURE 4. The logarithm of the ratio of left-key responses to right-key responses as a function of the logarithm of the ratio of left-key reinforcers to right-key reinforcers. Each point is for the last one of 15 sessions in a condition, except for those points marked by short horizontal lines, which are for day 14. These points are theoretical predictions from a computer-simulation model (AL) and resemble data from real pigeons as summarized by Wearden and Burgess.[31]

TABLE 2. Linear Regression Coefficients for Session 15 for Three Identical Stat-Birds

Subject	Slope	Intercept	Correlation (r)
1	.73	−0.09	.98
2	.58	−0.06	.98
3	.32	0.13	.83
Average	.60	−0.04	.97

(10, 5). If only the left response was associated with reinforcement, a left response occurred with probability 1.0. Similarly, if only the right response was associated with reinforcement, then AL made a right response. If neither response was associated with reinforcement in the presence of these traces, a response was made with probability .025 and the choice between left and right was made randomly. If both responses were associated with reinforcement, then a response was made with probability 1.0 and the choice was made randomly. The initial associative values were as follows: 90% of the traces were associated with nonreinforcement and of the remaining 10%, half were to the left and half were to the right. In all other ways, including the numerical values of the estimated parameters, the simulations were the same as those described previously.

Note that the model should adapt to local contingencies to the extent to which the times since the last left and right responses signal local reinforcement probability. Of course, in concurrent VI–VI schedules these two times are the times with which reinforcement probability is correlated (see for example, the papers by Hinson and Staddon[9,10]) so that AL should adapt nicely, but certainly not perfectly, to the local contingencies.

Can AL predict correct molar performance? FIGURE 4 shows the log of the ratio of left-key over right-key responses, plotted against the log of the ratio of left-key reinforcers divided by right-key reinforcers. The usual result is undermatching, represented by a straight line with slope less than unity.[30,31] FIGURE 4 shows just such a result. TABLE 2 summarizes these results. First, as the literature suggests it should, a straight line nicely descibes these results ($r^2 = .94$). A problem with this traditional log–log plot is its inability to handle the two important special cases where all the reinforcers are to one or the other key. While these cases do not appear in FIGURE 4, AL performed correctly: when either key had extinction programmed on it, AL responded exclusively to the other key. Second, the line suggests that there was on the average no important bias to either key. Third, the degree of undermatching varied across stat-birds even though all three had identical theoretical parameters. The slope varied considerably across the three stat-birds, with .32, the lowest, being quite low in comparison to those of real birds, but with the other two slopes, .58 and .73 being well within the usual range, as is the average slope of .60 (Ref. 31). Note also that AL produces a reasonable degree of variability over sessions: day 14, represented in FIGURE 4 by points intersected by short horizontal lines, is noticeably closer to the regression line than is day 15. In general, AL behaves in a way that at a molar level results in an unbiased undermatching that seems on average indiscriminably different from that of real birds.

The predicted undermatching is put into perspective by the work of Wearden,[32] who has shown how very small amounts of random responding added to matching can produce undermatching in concurrent performances. AL's tie-breaking procedures described previously are precisely the kinds of random mechanisms that produce undermatching, so that the slopes of the predicted curves presumably depend heavily on the form these tie-breaking rules assume.

Lastly, in connection with AL's molar performance, it is interesting to note the variability over subjects in the degree of undermatching. This predicted variability helps to interpret the variability in real data. For example, in an experiment by Rodewald,[33] as summarized by Wearden and Burgess,[31] slopes for individual birds ranged from .52 to 1.24. This variability might be interpreted primarily as evidence for individual differences in such things as deprivation level, discriminability of the alternatives, and so on.[34] However, AL's performance suggests that variability in slopes should be interpreted cautiously, since three *identical* stat-birds produced slopes ranging from .32 to .73. In the absence of a theory such as AL that predicts variability in slopes, it is exceedingly difficult to know how to interpret such variability.

Observe that the results in FIGURE 4 and TABLE 2 presumably could be somewhat improved by estimating the model's theoretical parameters from these data, but the estimates used here were the same as those used above for molecular phenomena in concurrent IRT–IRT schedules. Thus, we have unbiased undermatching at the molar level in concurrent VI–VI schedules predicted by a model with parameters estimated from molecular IRT data in a different context.

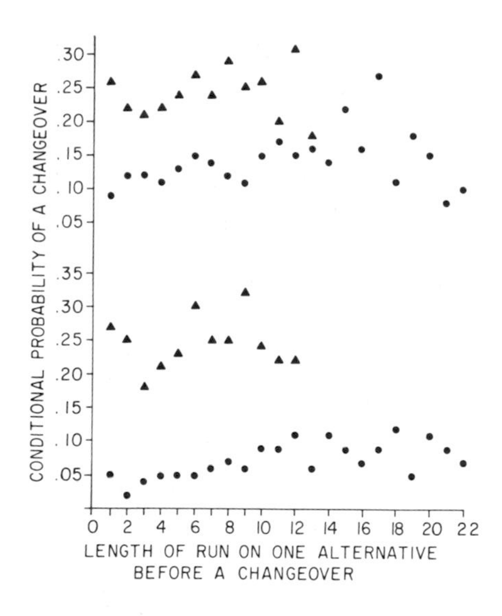

FIGURE 5. Conditional probability of a changeover to one key as a function of the length of the preceding run of responses to the other key. *Triangles* and *circles* refer to leaner and richer component schedules, respectively. The data are averages over three stat-birds for session 15 of two different conditions: concurrent VI (10 sec)–VI (25 sec) (**top**) and concurrent VI (10 sec)–VI (100 sec) (**bottom**). These theoretical points closely resemble those of real birds.[1,2]

Before leaving the results at the molar level, it is important to observe also that AL describes acquisition as well as steady-state performances. Thus, all birds showed preference at approximately 50–50 levels in the first session of each 15-session condition. Preference gradually changed over sessions to that reflected in FIGURE 4 and TABLE 2. The "learning rate" depended on the relative frequency of reinforcement: if the latter was close to .50, then performance on day 1 was not discriminably different from .50, but if the relative frequency of reinforcement was very different from .50, then even by the end of day 1, preference was already different from .50.

Perhaps the chief empirical anomaly facing a molecular view of concurrent VI–VI performance has been the flat changeover function. This curve seemed to suggest that the local temporal organization of behavior did not reveal adaptation to local reinforcement probabilities.[1,2,35,36] FIGURE 5 shows that AL produces the essentially flat changeover curves that are produced also by real pigeons. The points that appear in the curves are conditional relative frequencies for session 15 for two different conditions: the value for a run of length n is the number of times a changeover occurred after n pecks to the same key divided by the total number of runs of length n or greater. (Variability necessarily therefore increases for larger values of n, since the corresponding frequencies of occurrence are smaller.) It is clear that for all practical purposes, a straight line drawn through the curves, both those for the richer and leaner schedules, would be flat. Despite AL's molecular basis in the way reinforcement history affects which memory traces of the most-recent left and right responses are associated with reinforcement, and despite its corresponding adaptiveness to local reinforcement contingencies, AL behaves in a way that has been viewed as the most inconsistent with a molecular analysis. Hinson and Staddon[10,11] have found similar effects in the performance of real birds: Clear signs of adaptation to local reinforcement contingencies coexist with flat changeover curves. Thus, AL seems to capture both the molar matching phenomenon and the molecular temporal patterning of choices.

Consider finally a way in which AL describes the acquisition of molecular performance within an experimental condition. By virtue of the initial structure of AL's associative memory, on the first session of a condition the changeover curves are the same for both left and right keys. AL successfully describes the transition from this state to that pictured in FIGURE 5.

AL seems to have important virtues as a picture of concurrent VI–VI performances. It seems likely that AL's molar and molecular performances could not be discriminated from the real thing. So far as I am aware, no other model captures molar undermatching, flat changeover curves, and the development of those molar and molecular phenomena over sessions. AL's performance seems particularly impressive since its parameters were estimated from a molecular context many investigators seem to view as virtually unrelated to ordinary concurrent schedules.

CONCLUSIONS AND FUTURE DIRECTIONS

A model has been presented and evaluated that is designed to describe the underlying psychological mechanisms responsible for an organism's behavioral adaptation to reinforcement contingencies. AL's assumptions about forgetting and associative learning are about as simple as they could be: forgetting of a feature of a memory representation is constant over time and independent of forgetting of other features, and associative learning is all-or-nothing. Thus the theoretical assumptions are quite restrictive. Perhaps even more restrictive is the practical fact that in the present simulations, the same estimated values of the theoretical parameters were used throughout: in no case were parameters reestimated to improve the fit between predictions and data. Considering these restrictions, it seems rather striking that AL can assimilate in quantitative detail a range of empirical phenomena as diverse as that reviewed above: AL describes molar and molecular data, acquisition and steady-state data, and a previous version has assimilated results on temporal psychophysics. The model thus conceptually interrelates molar and molecular effects in operant conditioning, memory processes, timing, and the development and terminal nature of behavior.

The real-life nature of AL's performance is strikingly different from predictions from molar theories of operant behavior. AL generates an actual response protocol such as a real subject produces, and the protocol accordingly can be scrutinized in any way one wishes. A molar theory, however, often predicts only a single number, such as mean response rate or relative response rate. AL clearly shows the advantage of a molecular approach: A molecular model can handle both molecular and molar effects, while a molar model is silent about molecular effects.

One of the blessings of behavioral research is an unending array of unsolved problems. In the present case, one immediately wonders if AL can be developed to handle such things as large-scale temporal patterning on fixed-ratio and fixed-interval schedules, changeover-delay effects in concurrent schedules, short-term memory phenomena such as exposure duration, relative recency, and so on. Suitably enough, only time will tell.

SUMMARY

A computer-simulation model (AL) was developed for molecular and molar operant conditioning data. AL assumes a simple forgetting rule and all-or-nothing associations between memory representations of responses and reinforcers. AL describes acquisition and steady-state behavior in both concurrent interresponse time (IRT) schedules and in ordinary concurrent schedules. It describes functions relating the relative frequency of an IRT to IRT duration and to reinforcement frequency. AL accurately predicts for ordinary concurrent variable-interval schedules that the log ratio of responses to the two alternatives is a linear function of the log ratio of the reinforcers they deliver, and that the slope of this straight line approximates the degree of undermatching that characterizes the behavior of real animals. In addition, AL successfully describes the fact that in real data of Nevin,[1] Heyman,[2] and others, the probability of a changeover from one response to the other is roughly independent of the number of consecutive responses to the same alternative preceding the changeover. A previous version of AL assimilated data on temporal psychophysics, such as that animals bisect two temporal intervals at the geometric mean, and that the Weber fraction in DRL schedules is approximately constant. AL therefore describes transient and steady-state molecular and molar performances as well as data on temporal psychophysics.

ACKNOWLEDGMENTS

I would like to thank Dr. John Wearden for stimulating and enjoyable conversations on the development of models for operant behavior and for his comments on an earlier version of this manuscript. I am grateful to Drs. Veronica Dark and William Johnston for helpful comments on the computer-simulation model.

REFERENCES

1. Nevin, J. A. 1969. Interval reinforcement of choice behavior in discrete trials. J. Exp. Anal. Behav. **12:** 875–885.
2. Heyman, G. H. 1979. A Markov model description of changeover probabilities on concurrent variable-interval schedules. J. Exp. Anal. Behav. **31:** 41–51.
3. Baum, W. M. 1973. The correlation-based law of effect. J. Exp. Anal. Behav. **20:** 137–153.
4. Baum, W. M. 1981. Optimization and the matching law as accounts of instrumental behavior. J. Exp. Anal. Behav. **36:** 387–403.

5. RACHLIN, H. 1971. On the tautology of the matching-law. J. Exp. Anal. Behav. **15:** 249–251.
6. RACHLIN, H. 1978. A molar theory of reinforcement schedules. J. Exp. Anal. Behav. **30:** 345–360.
7. HERRNSTEIN, R. J. 1961. Relative and absolute strength of response as a function of frequency of reinforcement. J. Exp. Anal. Behav. **4:** 267–272.
8. HERRNSTEIN, R. J. 1970. On the law of effect. J. Exp. Anal. Behav. **13:** 243–266.
9. HINSON, J. M. & J. E. R. STADDON. 1981. Maximizing on interval schedules. *In* Recent Developments in the Quantification of Steady-State Operant Behavior. C. M. Bradshaw, Ed.: 35–47. Elsevier/North-Holland. Amsterdam.
10. HINSON, J. M. & J. E. R. STADDON. 1983. Hill-climbing by pigeons. J. Exp. Anal. Behav. **39:** 25–47.
11. SILBERBERG, A., B. HAMILTON, J. M. ZIRIAX & J. CASEY. 1978. The structure of choice. J. Exp. Psychol. Anim. Behav. Processes **4:** 368–398.
12. SHIMP, C. P. 1969. Optimal behavior in free-operant experiments. Psychol. Rev. **76:** 97–112.
13. ESTES, W. K. 1959. The statistical approach to learning theory. *In* Psychology: A Study of a Science. S. Koch, Ed. Vol. **2:** 380–491. McGraw-Hill. New York, NY.
14. BUSH, R. R. & W. K. ESTES, Eds. 1959. Studies in Mathematical Learning Theory. Stanford University Press. Stanford, CA.
15. BUSH, R. R. & F. MOSTELLER, Eds. 1955. Stochastic models for learning. Wiley. New York, NY.
16. ESTES, W. K. 1969. New perspectives on some old issues in association theory. *In* Fundamental Issues in Associative Learning. N.J. Mackintosh & W. K. Honig, Eds.: 162–189. Dalhousie University Press. Halifax, Nova Scotia.
17. SHIMP, C. P. 1978. Memory, temporal discrimination, and learned structure in behavior. *In* The Psychology of Learning and Motivation. G. H. Bower, Ed. Vol. **12:** 39–76. Academic Press. New York, NY.
18. SHIMP, C. P. 1981. Local structure of steady-state operant behavior. *In* Quantification of Steady-State Operant Behavior. C. M. Bradshaw, E. Szabadi & C. F. Lowe, Eds.: 262–298. Elsevier/North-Holland. Amsterdam.
19. SHIMP, C. P. 1979. The local organization of behavior: Method and theory. *In* Advances in Analysis of Behavior. Vol. 1: Reinforcement and the Organization of Behavior. M. D. Zeiler & P. Harzem, Eds.: 261–298. Wiley. Chichester, England.
20. ROITBLAT, H. L. 1982. The meaning of representations in animal memory. Behav. Brain Sci. **5:** 353–372.
21. D'AMATO, M. R. 1973. Delayed matching and short-term memory in monkeys. *In* The Psychology of Learning and Motivation. G. H. Bower, Ed. Vol. **7:** 227–269. Academic Press. New York, NY.
22. SHIMP, C. P. 1969. Concurrent reinforcement of two interresponse times: The relative frequency of an interresponse time equals its relative harmonic length. J. Exp. Anal. Behav. **12:** 403–411.
23. HAWKES, L. & C. P. SHIMP. 1974. Choice between response rates. J. Exp. Anal. Behav. **21:** 109–115.
24. SHIMP, C. P. 1970. Concurrent reinforcement of two interresponse times: Absolute rate of reinforcement. J. Exp. Anal. Behav. **13:** 1–8.
25. SHIMP, C. P. 1968. Magnitude and frequency of reinforcement and frequencies of interresponse times. J. Exp. Anal. Behav. **11:** 525–535.
26. STADDON, J. E. R. 1968. Spaced responding and choice: A preliminary analysis. J. Exp. Anal. Behav. **11:** 669–682.
27. SHIMP, C. P. 1966. Probabilistically reinforced choice behavior in pigeons. J. Exp. Anal. Behav. **9:** 443–455.
28. MACKINTOSH, N. J., 1974. The Psychology of Animal Learning. Academic Press. New York, NY.
29. CATANIA, A. C. 1973. Self-inhibiting effects of reinforcement. J. Exp. Anal. Behav. **19:** 517–526.

30. Baum, W. M. 1979. Matching, undermatching, and overmatching in studies of choice. J. Exp. Anal. Behav. **32:** 269–281.
31. Wearden, J. H. & I. S. Burgess. 1982. Matching since Baum (1979). J. Exp. Anal. Behav. **38:** 339–348.
32. Wearden, J. H. Undermatching and overmatching as deviations from the matching law. J. Exp. Anal. Behav. In press.
33. Rodewald, H. K. 1978. Concurrent random-interval schedules and the matching law. J. Exp. Anal. Behav. **30:** 301–306.
34. Baum, W. M. 1974. On two types of deviation from the matching law: Bias and undermatching. J. Exp. Anal. Behav. **22:** 231–242.
35. De Villiers, P. 1977. Choice in concurrent schedules and a quantitative formulation of the law of effect. *In* Handbook of Operant Behavior. W. K. Honig & J. E. R. Staddon, Eds.: 233–287. Prentice-Hall. Englewood Cliffs, NJ.
36. Nevin, J. A. 1979. Overall matching versus momentary maximizing: Nevin (1969) revised. J. Exp. Psychol. Anim. Behav. Processes **5:** 300–305.

A Model for Temporal Generalization and Discrimination[a]

ERIC G. HEINEMANN

Department of Psychology
Brooklyn College of the City University of New York
Brooklyn, New York 11210

This paper describes a model for memory and decision processes that has been applied to the acquisition and performance of discriminations by pigeons in a variety of situations. The phenomena to which the model has been applied include probability learning, the acquisition of easy and difficult discriminations, discrimination reversal, probabilistic discrimination learning, stimulus categorization,[1] and absolute identification of stimuli.[2]

The situations considered in the past have involved stimuli that differ in intensity. In the present paper I consider the application of the model to stimuli that differ in duration, and shall focus on variables governing performance at asymptotic levels rather than on acquisition processes.

The prototypical situation considered in the original development of the model involves the presentation, on each of a series of discrete trials, of one of two stimuli, *S*1 or *S*2, that differ in a single attribute. The subject is rewarded for making one response, *R*1, when presented with *S*1, and an alternative response, *R*2, when presented with *S*2. It is assumed that the acquisition of the discrimination between *S*1 and *S*2 involves two stages. The first stage is the presolution period, represented by a series of trials, at the beginning of training, during which there is no evidence of a developing discrimination. The animal is regarded as functioning, during this stage, as a detector of the statistical association the experimenter has arranged among the environmental stimuli, *S*1 and *S*2, selected behaviors (responses), *R*1 and *R*2, and the consequences of these behaviors (reward and nonreward, for example). During the presolution period the subject learns to attend to the stimulus dimension that is correlated with the differential consequences of its behavior. A formal treatment of this matter, based on the theory of sequential analysis developed by A. Wald[3] and his associates, has been presented by Heinemann.[4]

The presolution period ends when the subject discovers that variation in the dimension along which *S*1 and *S*2 differ is related to the outcome of its choice behavior. The theory assumes that the processes occurring after the end of the presolution period involve a memory that has a limited storage capacity (to be referred to as the limited capacity memory or LCM). The principal assumptions of the second-stage model are as follows:

1. Storage and the representation of events in memory. Information gathered on each trial of an experiment is assumed to be placed in a memory that has a limited number of storage locations. Each record placed in this memory during the course of discrimination training contains information concerning the *discriminative stimuli* that analysis done during the presolution period has shown to be predictive of the

[a]This work was supported by Grant MH18246 from the National Institute of Mental Health and by Grant 14002 from the Professional Staff Congress—City University of New York Research Award Program.

outcomes of behavior, the *response* made, and the *reward* received. One such record is entered into the LCM on every trial and is said to occupy a "storage location." The location to which each record is sent is selected randomly, and any record occupying a storage location will be destroyed ("overwritten") when a new record is entered at that location.

The experiments considered in this paper involve the presentation of sets of stimuli that differ in duration. It is assumed that the duration of each stimulus is measured on an internal clock. The durations read from this clock constitute stimuli to a sensory system that has an output proportional to the logarithm of the stimulus. This is a form of Fechner's law,

$$S = c \log t,$$

where S is subjective time, t is the duration read from the clock, the constant $c = 1/\log(1 + W)$, and W is the Weber fraction.

The clock is assumed to be perfectly accurate, but the response of the sensory system to a constant input from the clock varies from trial to trial. It is assumed that, over repeated presentations, each stimulus induces sensory effects, S, that are normally distributed with different means, $\mu_1, \mu_2, \ldots, \mu_n$, and a common standard deviation σ. The sensory effect experienced on each trial is stored in the LCM.

The responses made just before receipt of reward are assumed to be represented in memory in the form of sensory information associated with these responses, such as the visual characteristics of the device that was pecked or pressed, such as its color, position, and so forth. In some situations, such as those considered by Chase,[2] the manner in which response information is represented plays an important role in the analysis. In the present paper, however, the responses will be treated as though they were represented in memory simply by the labels $R1$ and $R2$. For now, outcomes are treated as though they were represented in memory simply as reward (positive) and nonreward (negative).

2. Retrieval. It is assumed that during each trial the subject draws a small sample of records from the LCM. The choice of response is based solely on the information contained in this sample of records. The samples are assumed to be independent random samples with the following restriction: The subject is assumed to draw records from the LCM one at a time until a fixed number of positive records (records showing that a reward was received) has been obtained. To say that the subject draws a sample of records is not intended to imply that the records in question are removed from the LCM. The idea is that these records are "looked at" or are "copied" for use in a working memory.

3. Response selection. The choice of response on each trial involves not only a sample of records from memory, but also the sensory effect present on the continuum that the analysis done during the presolution period has shown to be predictive of outcomes, for example, the currently experienced duration of a visual signal. This sensory effect will be called the "current input."

Before describing how the current input and the records drawn from the LCM are used in response selection, it is necessary to state one further assumption. The sensory effect represented on each record that has been drawn from the LCM is assumed to fluctuate rapidly over time, momentary values falling into a normal distribution whose mean represents the sensory effect experienced on the trial on which the record was formed.

The response the subject selects is the one most likely to be rewarded on the basis of the evidence contained in the sample of records drawn from the LCM. To find it, the subject gets the sum of the probability densities for each response, at the current input, and selects the response for which the density is greatest. The process is illustrated in FIGURE 1, which shows sensory effect distributions representing four records, three of which show that $R2$ was rewarded and one that $R1$ was rewarded. The process amounts to summing the heights of the $R2$ curves above the point representing the current input, doing the same for the $R1$ curves, and then determining which sum is larger.

If the subject were in the presolution period or in an experiment on probability learning, the records drawn from the LCM would contain no information concerning discriminative stimuli. Given a sample of this sort, the subject simply chooses the response with the greater probability of reward. The rule that accomplishes this is a simple one: if the sample contains more records labeled $R2$ than $R1$, make response $R2$; otherwise make response $R1$.

The model described here has been written as a computer program which was used to simulate results to be expected in three experimental situations dealing with discrimination and generalization of durations. The model has three parameters. One of these, the size of the LCM, is the primary determinant of the rate at which learning proceeds, but is virtually irrelevant to the analysis of asymptotic performance. The other parameters are θ, the size of the sample retrieved on each trial, and W, the

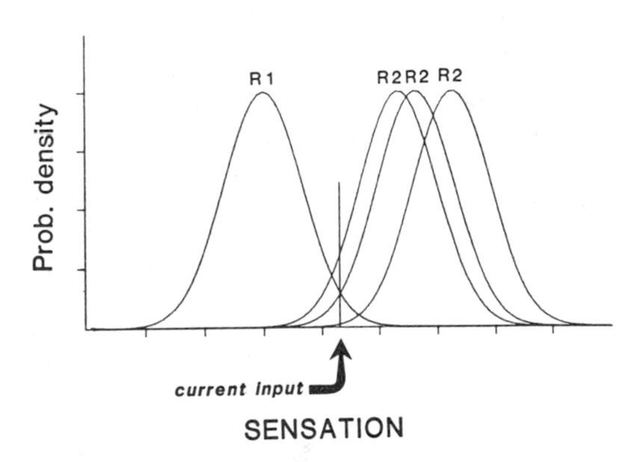

FIGURE 1. A sample of four positive records from the LCM. The choice of response is based on the probability densities at the point labelled "current input."

quantity that represents the Weber fraction in Fechner's law. In the context of the present model W functions as a sensitivity parameter. W was assigned the value of 0.25 for all simulations presented in this paper.

PSYCHOMETRIC FUNCTIONS FOR DURATION

In a well-known set of experiments by Stubbs[5] pigeons were required to classify lights differing in duration. In one experiment the birds were presented on each trial with one of 10 signal durations ranging from 1 to 30 sec. They were rewarded for pecking on one of two keys when presented with any of the durations that exceeded 5 sec, and for pecking on the other key when presented with those durations equal to or less than 5 sec.

The results Stubbs obtained for three pigeons are shown by the dashed lines and data points in FIGURE 2. The solid line represents the results of a computer simulation based on the LCM model. The psychometric function predicted by the model is a close approximation to the empirical functions obtained by Stubbs. Psychometric functions of this sort may be used, of course, to obtain indices of differential sensitivity to duration. Stubbs obtained such functions at several absolute duration ranges and found Weber's law to be true when all time values were doubled, tripled, or quadrupled. The simulation program considers only relative durations so, not surprisingly, it also yields Weber's law.

One interesting feature of the functions shown in FIGURE 2 is that the lower and upper asymptotes lie at roughly 0.05 and 0.95, respectively, instead of at 0 and 1.0, as would be expected by most theories of the psychometric function. This feature is fairly typical of psychometric functions obtained in animal experiments. It suggests that on some trials the animal's choice is not under stimulus control, that is, that the animal's choice of response is independent of the stimuli manipulated by the experimenter. One way to describe this situation is to say that the animal is not "attending" to the stimulus dimension on all trials. Heinemann, *et al.*[6] introduced the following equation which may be used to correct the data for the effects of "inattention":

$$p(R) = p(A)p(R \mid A) + (1 - p(A))p(R \mid \overline{A})$$

where $p(R)$ is the probability of one of the two choice responses, $p(A)$ the probability of attention, $p(R \mid A)$ the probability of the response given attention (given by the model for the psychometric function), and $p(R \mid \overline{A})$ the probability of the response given inattention.

The LCM model provides a deeper explanation of the phenomenon under discussion. That the asymptotes of the psychometric function lie at values other than 0 and 1.0 has to do with the fact that the sample of records the subject retrieves from memory on each trial may fail to include information about all stimulus values. Consider a trial on which a stimulus that lies at the asymptote of the psychometric function is presented, say, a very short duration for which $R1$ is the response that would earn a reward. Now assume that the sample of records retrieved from memory contains only records of trials on which $R2$ was made. In this case the subject must make response $R2$ (which is an error, of course). The smaller the size of the sample, the more frequently such instances of "missing records" will occur, and the more the asymptotes

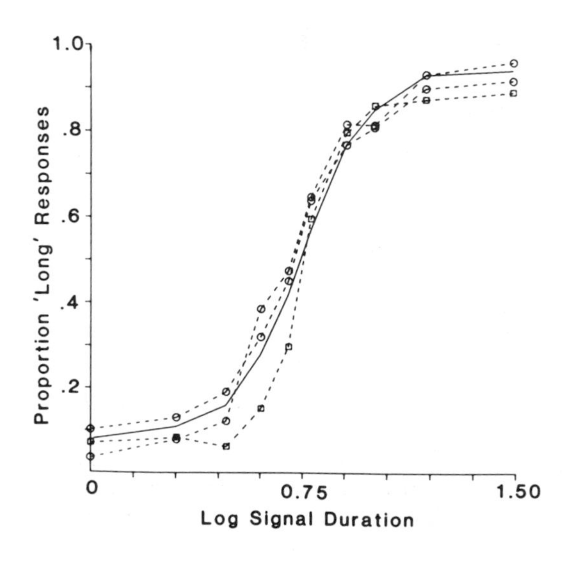

FIGURE 2. Proportion of trials on which the subject made the response designated as correct for the five longest signal durations. The *dashed lines* connecting the circles and squares represent results obtained by Stubbs[5] for three pigeons. The *solid line* represents a computer simulation with $W = 0.25$ and sample size equal to 4.

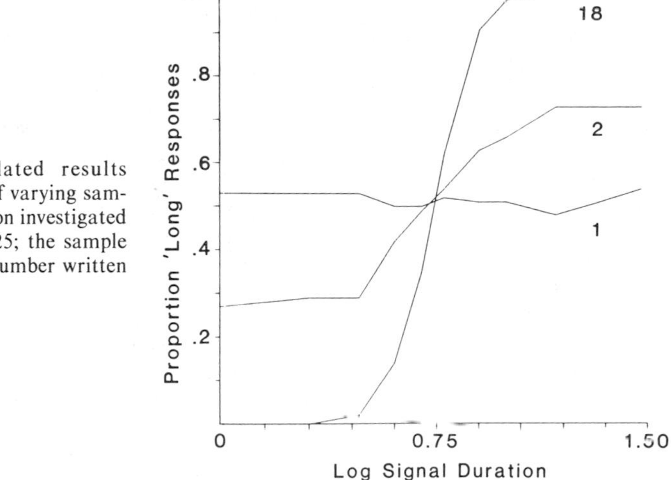

FIGURE 3. Simulated results showing the effects of varying sample size in the situation investigated by Stubbs.[5] $W = 0.25$; the sample size is given by the number written under each curve.

of the psychometric function will depart from values of 0 and 1.0.[b] These effects of varying sample size are illustrated in FIGURE 3, which shows simulated results for sample sizes equal to 1, 2, and 18.

CATEGORIZATION FOLLOWING TWO-STIMULUS TRAINING

Working with rats in a two-choice situation that was formally similar to the one used by Stubbs,[5] Church and Deluty[11] presented only two stimulus durations during training, a short one (for example, 2 sec), after which one response was rewarded, and a long one (for example, 8 sec), after which the alternative response was rewarded. Following training they presented stimuli of intermediate durations in a generalization test. The functions relating proportion of one of the choices to signal durations were very similar to those obtained by Stubbs and shown in FIGURE 2. The stimulus duration corresponding to a response proportion of 0.5, referred to as the "bisection point" corresponded quite closely to the geometric mean of the two stimulus values used in training. This was true for several different sets of training durations, as would be expected if subjective time were logarithmically related to real time. Simulations based on the LCM model also show that the bisection points fall very close to the geometric mean of the training stimuli. For example, in simulations in which the sample size was set equal to 10 and the signal durations presented in training were either 2 and 8 sec or 4 and 16 sec, the bisection points fell at 3.82 sec (geometric mean = 4 sec) and 8.13 sec (geometric mean = 8 sec), respectively.

[b]There are situations in which the asymptotes of the psychometric function fail to lie at 0 and 1.0 for quite different reasons. Among the most important of these, from a theoretical point of view, are situations in which control of behavior is shared by two or more stimulus dimensions. The principles involved have been discussed by Chase and Heinemann,[7] Heinemann and Chase,[8] and Heinemann *et al.*[9,10] Of course, if the appropriate receptors are not stimulated on every trial, then the psychometric function will also have asymptotes other than 0 and 1.0.

CATEGORIZATION OF DURATIONS IN A DISCRETE-TRIAL GO/NO-GO SITUATION

In a recent series of experiments Church and Gibbon[12] studied rats' behavior in a situation first used by Blough[13] and referred to by him as "maintained generalization." In most of the experiments the animals were presented on each trial with a visual signal having one of nine durations. A press on a single lever earned a reward if it occurred when the animal was presented with one of these durations (the $S+$), but had no consequences if it occurred following any of the other eight durations (different values of $S-$). Typically, the $S+$ was the median duration.

This situation differs from all the situations to which the LCM model has heretofore been applied in that it deals with the occurrence or nonoccurrence of a single response rather than a choice between alternative responses that are being recorded by the experimenter. The simulations of this situation assume that the animals are rewarded by the experimenter for not responding to negative stimuli. This is a

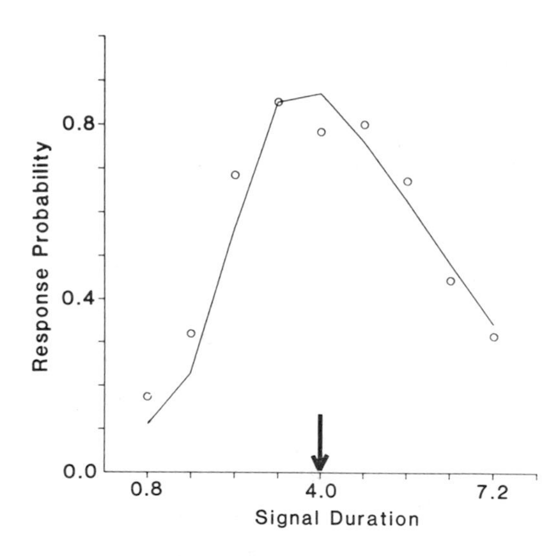

FIGURE 4. Probability of one or more bar-presses as a function of signal duration. The points represent results obtained by Church and Gibbon.[12] The *line* represents a computer simulation with $W = 0.25$ and sample size equal to 10. The *arrow* points to the positive signal duration.

simplifying assumption, probably to be replaced in the future by the assumption that, when the animal does not make the response the experimenter is recording, it is engaged in some other behavior that has some positive value for it. The simulated results that will be shown do not represent "best fits." To obtain them a single parameter, θ, was varied by trial and error until a reasonable approximation to the empirical results was obtained.

FIGURE 4 shows the outcome of one of the basic experiments done by Church and Gibbon[12] (points) together with the results of a simulation (solid line) based on an assumed sample size of 10. (As mentioned, $W = 0.25$ in all simulations shown in this paper.) FIGURE 5 shows the same results plotted on a logarithmically scaled stimulus axis. As Church and Gibbon noted, the generalization curve appears more nearly symmetric on the linear axis than on the logarithmic axis. This may seem surprising in view of the fact that the LCM model assumes Fechner's law. The greater symmetry of the curve on the linear scale reflects the linear stimulus spacing used in this

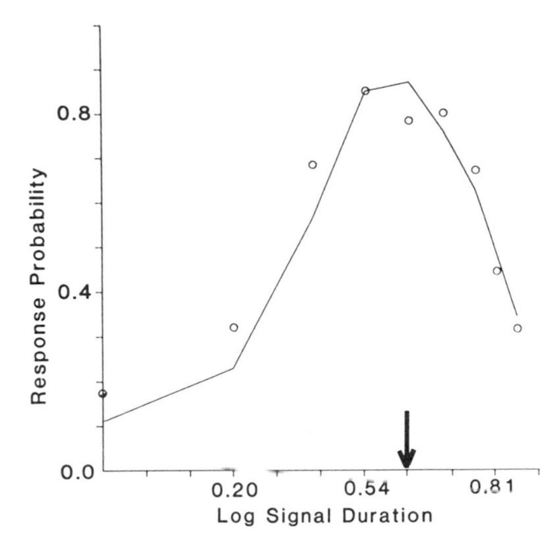

FIGURE 5. Probability of one or more bar-presses as a function of log signal duration. These data are the same as those shown in FIGURE 4.

experiment. If the nine stimuli used in the experiment are spaced at equal logarithmic distances, then the simulated generalization curve is symmetric on a logarithmic stimulus axis.

The last-mentioned theoretical result is in conflict with the empirical results of Church and Gibbon,[12] who found no substantial effect of stimulus spacing on the form of the generalization curve. This may reflect a defect in the LCM model, but it is also possible that the experimental conditions assumed in the simulation do not in fact match the conditions of the actual experiment. For example, the asymptotic levels of performance assumed in the simulation might not have been achieved in the actual experiments. This is of some importance because the same animals were trained first with one spacing and then the other.

These results, particularly as represented in FIGURE 5, suggest that the generalization curves may be leveling off at a response probability well above zero. Noting this, Church and Gibbon decided to check on the matter by extending the range of durations presented during the experiment. Their results with the extended stimulus range are shown in FIGURE 6. The squares represent the results obtained with the "normal" range of stimuli represented in FIGURE 5, the circles the results obtained with the extended range. The solid line represents the results of a simulation based on the LCM model for sample size equal to 5. Both empirical and theoretical curves clearly level off well above a response probability of zero.

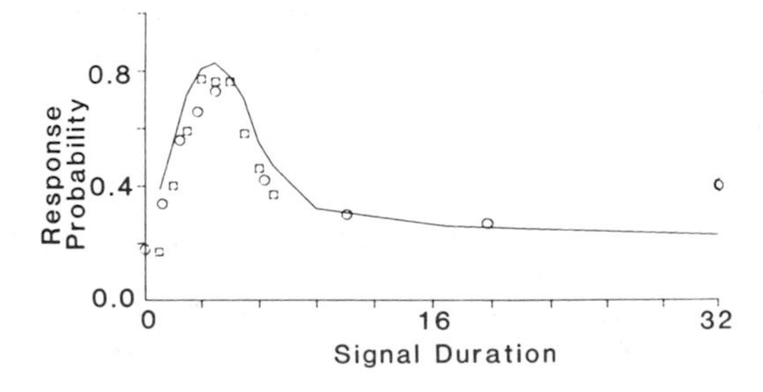

FIGURE 6. Extended stimulus range: The *circles* represent performance of rats on the extended range; the *squares* show performance on the standard range. Data are from Church and Gibbon.[12] The *solid line* represents the result of a simulation. $W = 0.25$ and the size of the sample was 5.

Heinemann and Chase[14] showed how the correction for "inattention" may be applied to go/no-go situations of the sort under discussion here, and this correction is part of the theoretical analysis proposed by Church and Gibbon.[12] According to this interpretation, when the subjects are attending to the relevant stimulus continuum (duration), they base their choice of response upon the experienced value of subjective time in accordance with a specified decision rule; when they are not attending, they base the choice of responding on factors unrelated to the value of the stimuli manipulated by the experimenter. The LCM model elucidates the notion of "inattention" and eliminates the need to estimate from the data two free parameters, the probability of attention and the probability of the response given inattention.

According to the LCM model the cause of the raised asymptote shown in FIGURE 6 is again missing records. Consider a trial on which a very long duration (say 32 sec) is presented, but the sample of records contains information concerning the response made in the presence of $S+$, but no records of responses to any durations longer than $S+$. In this case the subject will make the response appropriate to $S+$.

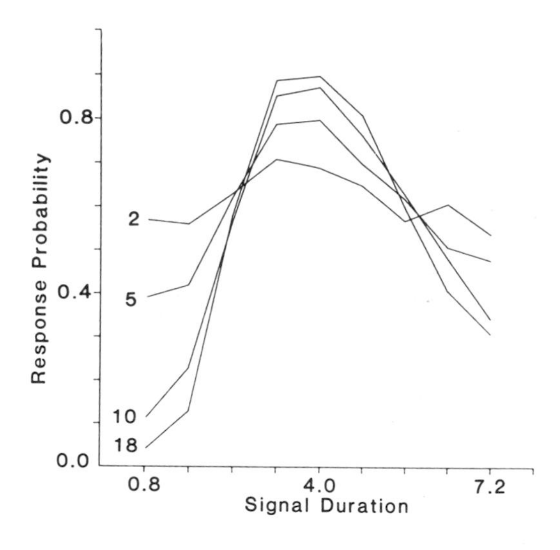

FIGURE 7. Theoretical effects of sample size on the generalization curves. Results of simulations assuming the sample sizes that are written next to each curve.

The probability that the sample will not contain information concerning responding to stimuli longer than S+ increases as the size of the sample decreases. The theoretical effect of sample size on the generalization curves may be seen in the simulated results shown in FIGURE 7. The theoretical curves become flatter and level off at progressively higher values of response probability as the size of the sample decreases. For a sample size of 1, the generalization function becomes a horizontal line at a level that matches the presentation probabilities of the positive and negative stimuli (assumed to be 0.5 in the present simulation).

Variation in the proportion of trials on which the S+ is presented has a pronounced effect on the generalization curve as shown by Church and Gibbon.[12] The results they obtained with two different presentation probabilities are shown by the points and dashed lines in FIGURE 8. The solid lines also shown in FIGURE 8 represent the results of a series of simulations assuming the presentation probabilities written next to each curve. With respect to the form of the curves and the direction of their vertical displacement as a function of presentation probability, there is good agreement

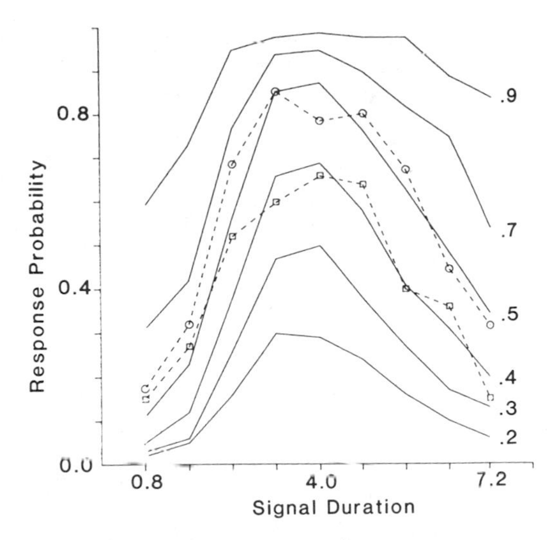

FIGURE 8. The effects of presentation probability on the generalization curves. The *solid lines* are the results of simulations assuming the sample size written next to each curve. The *dashed lines* connect data from Church and Gibbon.[12] *Squares:* presentation probability = .125; *circles:* presentation probability = .5.

between the theoretical and empirical curves. That the numerical values of the presentation proportions represented by the empirical curves and the nearest theoretical curves do not agree should not be surprising in view of the approximations made in the simulations and the fact that the experimental procedures assumed in the simulations do not quite match the actual experimental procedures.

FIGURE 9 shows the effects of varying the probability of reward for correct responses to $S+$. Again, the points represent results obtained by Church and Gibbon[12] for two values of reward probability. The solid lines represent the results of simulations for the reward probabilities written under each curve.

According to the LCM model, the effects of variations in stimulus presentation probabilities and reward probabilities have essentially the same cause. Changing the

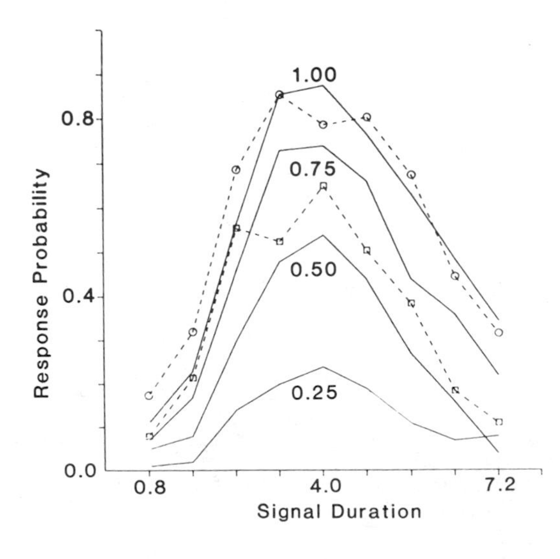

FIGURE 9. The effect of reward probability on the generalization curves. The *solid lines* represent the results of simulations assuming the reward probabilities written above each line. The *dashed lines* connect results obtained for rats by Church and Gibbon.[12] *Squares:* reward probability = .25; *circles:* reward probability = 1.0.

proportion of trials on which the $S+$ is presented is reflected directly in the number of records of successful responding to $S+$ that are contained in the memory. Since the records upon which the choice of response on any trial is based represent a random sample of the records in the LCM, any change in the proportion of one type of record in the LCM is reflected in a corresponding change in the average proportion of these records in the samples. Thus, an increase in the presentation probability of $S+$ results in an increase in the average proportion of $S+$ records in each sample and therefore an increasing probability of responding. Correspondingly, a decrease in the presentation probability of $S+$ will result in a decrease of responding.

Turning to the effect of reward probability: each unrewarded response to $S+$ results in placing a negative (uniformative) record in the memory. As a result the average sample retrieved from memory will contain fewer $S+$ records than it would if all responses to $S+$ were rewarded; and the probability of obtaining samples that contain no records of responding to $S+$ at all is correspondingly larger.

CONCLUDING REMARKS

The LCM model, which describes a large number of phenomena of intensity discrimination and generalization, also gives a reasonable account of some basic phenomena of temporal discrimination and generalization. It appears that information concerning the intensity and duration of stimuli is processed in much the same manner.

One of the virtues of the model that has been discussed is parsimony. As applied to asymptotic performance, the LCM model uses just two parameters, W and θ. It gives an account of the effects of "inattention" and it provides a rational account of the effects of "motivational" variables, such as the presentation probability of the positive and negative signals and the probability of obtaining a reward for responding to $S+$.

ACKNOWLEDGMENT

I wish to thank Sheila Chase for important advice on theoretical matters.

REFERENCES

1. HEINEMANN, E. G. 1984. A memory model for decision processes in pigeons. *In* Quantitative Analyses of Behavior, Vol. 4: Discrimination Processes. M. L. Commons, R. J. Herrnstein & A. R. Wagner, Eds. Ballinger. Cambridge, MA.
2. CHASE, S. 1984. Pigeons and the magical number seven. *In* Quantitative Analyses of Behavior, Vol. 4: Discrimination Processes. M. L. Commons, R. J. Herrnstein & A. R. Wagner, Eds. Ballinger. Cambridge, MA.
3. WALD, A. 1947. Sequential Analysis. Dover. New York, NY.
4. HEINEMANN, E. G. 1984. The presolution period and the detection of statistical associations. *In* Quantitative Analyses of Behavior, Vol. 4: Discrimination Processes. M. L. Commons, R. J. Herrnstein & A. R. Wagner, Eds. Ballinger. Cambridge, MA.
5. STUBBS, A. 1968. The discrimination of stimulus duration by pigeons. J. Exp. Anal. Behav. **11:** 223–238.
6. HEINEMANN, E. G., E. AVIN, M. A. SULLIVAN & S. CHASE. 1969. Analysis of stimulus generalization with a psychophysical method. J. Exp. Psychol. **80:** 215–224.
7. CHASE, S. & E. G. HEINEMANN. 1972. Decisions based on redundant information: An analysis of two-dimensional stimulus control. J. Exp. Psychol. **92:** 161–175.

8. HEINEMANN, E. G. & S. CHASE. 1970. Conditional stimulus control. J. Exp. Psychol. **84:** 187–197.
9. HEINEMANN, E. G., S. CHASE & C. MANDELL. 1968. Discriminative control of "attention." Science **160:** 553–554.
10. HEINEMANN, E. G., S. CHASE & C. MANDELL. 1969. Discrimination control of "attention:" Technical note. Science **164:** 198.
11. CHURCH, R. M. & M. Z. DELUTY. 1977. Bisection of temporal intervals. J. Exp. Psychol. Anim. Behav. Processes **3:** 216–228.
12. CHURCH, R. M. & J. GIBBON. 1983. Temporal generalization. J. Exp. Psychol. Anim. Behav. Processes **8:** 165–186.
13. BLOUGH, D. S. 1975. Steady state data and a quantitative model for operant generalization and discrimination. J. Exp. Psychol. Anim. Behav. Processes **1:** 3–21.
14. HEINEMANN, E. G. & S. CHASE. 1975. Stimulus generalization. *In* Handbook of Learning and Cognitive Processes, Vol. 2; Conditioning and Behavior Theory. W. K. Estes, Ed. Erlbaum. Hillsdale, NJ.

Time's Role for Information, Processing, and Normalization[a]

DOMINIC W. MASSARO

Program in Experimental Psychology
University of California, Santa Cruz
Santa Cruz, California 95064

. . . What then is time? If someone asks, I know.
If I wish to explain it to someone who asks,
I know not . . .
—ST. AUGUSTINE

As a rule sensations outlast for some little time the
objective stimulus which occasioned them.
—WILLIAM JAMES[1]

INTRODUCTION

Time serves multiple roles in our interaction with our environment. The most obvious role is in terms of information; the duration of an event provides a cue to the identity of the event. We will consider examples of duration information in the domain of speech perception. Our concern will be with how duration cues in speech are evaluated and integrated in the classification of speech patterns. Second, time is necessary for perceptual processing and restricts memory of auditory events. We will briefly review the study of perceptual processing time and the time course of preperceptual auditory memory. Finally, we will address the issue of normalizing the information available to the perceptual process. The important question will be how duration information of a speech segment is evaluated relative to the duration of the context.

DURATION INFORMATION IN SPEECH

One of the critical cues for the perception of speech is the duration of the component segments of the speech signal. At both segmental and suprasegmental levels, duration provides identifying information for a spoken utterance. At the segmental level, for example, duration can distinguish long and short vowels as in "bit" and "beet"[2]; and vowel and consonant duration are cues to voicing of a postvocalic consonant.[3–5] Duration information also serves as a supplementary cue to lexical tone in Mandarin Chinese.[6] At the suprasegmental level, word and phrase duration influence such factors as the placement of a constituent boundary of an otherwise ambiguous phrase.[7] For example, the ambiguous phrase "old men and women" is more likely to be interpreted as ("old men") ("and women") if the duration of the phrase "men and women" is long relative to the case in which it is short.[8]

[a]This work was supported in part by Grant MH-35334 from the National Institute of Mental Health.

Vowel Identification

In English, about half of the vowels are relatively short in duration and the other vowels are relatively long. It is only natural that listeners might utilize duration as a feature for vowel identification. The importance of vowel duration as an acoustic feature has been demonstrated by Ainsworth.[9] Listeners identified synthetic vowels differing in the first and second formant and duration. The number of correct identifications of the normally shorter vowels decreased as the vowel durations were increased. The converse held true for identification of the normally longer vowels.

An important question is how duration information is combined with other information about vowel identity. Duration can function as a supplemental feature to vowel identification, particularly in distinguishing between vowels with similar formant patterns. Bennett[10] independently varied the duration and formant levels of

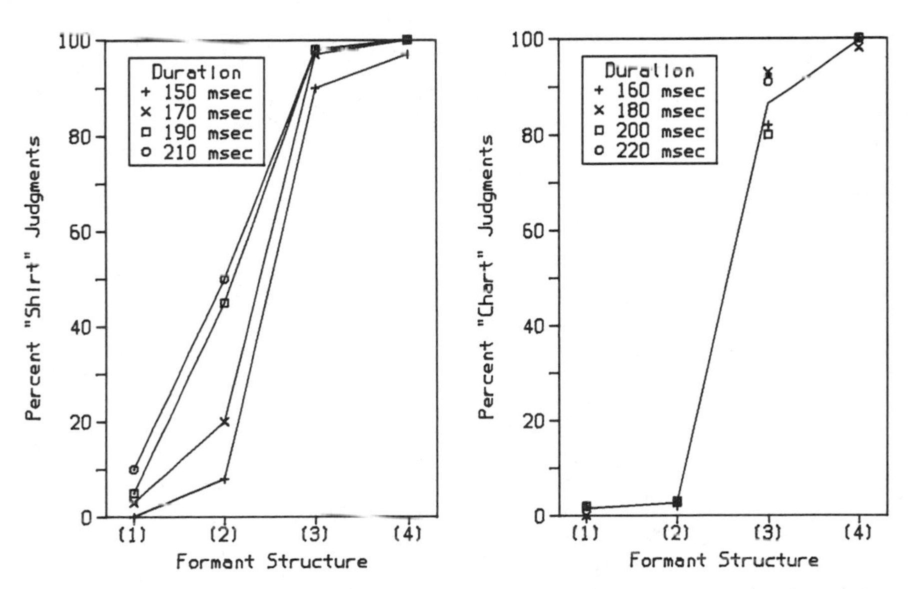

FIGURE 1. Predicted and observed percentage of vowel identification as a function of the formant values; duration of the vowel is the curve parameter. (Data from Bennett.[10])

vowels in words using synthetic speech. In one set, the formant patterns were changed in four steps from the word "shut" to the word "shirt." The formant patterns of these two vowels are fairly similar. For each of the four formant patterns, the word was synthesized with four different vowel durations. Listeners identified each of the 16 stimuli as either "shut" or "shirt." The left panel of FIGURE 1 gives the proportion of "shirt" identifications as a function of the formant structure of the vowel; duration is the curve parameter. Changes in the formant structure changed the identification from one alternative to the other. Duration had a significant, but smaller effect; its effect was most noticeable at the second level of the formant structure. We might assume that the second level of formant structure was relatively ambiguous and, therefore, the influence of duration made more apparent.

Bennett[10] replicated this experiment, using the words "chat" and "chart," respectively. The vowels in these two words differ in duration in the same way as the vowels in "shut" and "shirt," but normally have a greater difference between their formant patterns. Therefore, a four-stimulus continuum between "chat and chart" is unlikely to create an ambiguous level of formant structure. As can be seen in the right panel of FIGURE 1, vowel duration had no effect. The judgments were not ambiguous at any of the four levels of formant structure. We expect that a duration effect might have been observed if additional formant levels were created to produce more ambiguous judgments. Taken together, the results of the two experiments show that duration influences vowel identification primarily when the formant structure is ambiguous. Vowel duration has the potential of being a relevant feature in vowel recognition when it is necessary to identify vowels with relatively ambiguous patterns. The ambiguity could be either in the speech signal itself or could result from some limitation in the processing of the formant patterns.

The results of these two experiments can be described by a fuzzy logical model of perceptual recognition that will be described in the next section. The formant structure and duration are treated as independent acoustic features defining the vowel alternatives. The predictions of the results given by the model are given by the lines in FIGURE 1. The model gives a good description of the results and quantitatively describes the larger contribution of duration at the more ambiguous level of formant structure.

Integrating Multiple Durations in Speech Perception

Both vowel duration and consonant duration provide information about consonant voicing in a vowel–consonant syllable.[3-5] According to the fuzzy logical model,[28] speech perception involves feature evaluation, prototype matching, and pattern classification operations. Vowel duration and consonant duration are assumed to be independent cues at feature evaluation. The information available at feature evaluation is assumed to be continuous rather than categorical, and is represented by fuzzy-truth values between zero and one. During prototype matching, the cue values are inserted into prototypes in long-term memory and integrated (conjoined) together to arrive at a goodness-of-match of the stimulus with each prototype. For the present speech contrast, prototypes for the voiced and voiceless alternatives are defined as the conjunction (&) of vowel duration (V) and consonant duration (C) information. In addition, negation is realized by the unary complement.

voiced: long vowel and NOT (long consonant)

$$= V \,\&\, (1 - C) \tag{1}$$

voiceless: NOT (long vowel) and long consonant

$$= (1 - V) \,\&\, C \tag{2}$$

Subjects evaluate the degree to which the vowel and consonant durations are long and enter these values into the prototype definitions. The cues are treated independently and given fuzzy-truth values between zero and one. The conjunction operation is defined as multiplication.[11] To arrive at a particular judgment, the relative goodness-of-match is taken. In this case, the probability of a voiced judgment. P(Voiced), is equal to

$$P(\text{Voiced}) = \frac{V \times (1 - C)}{[V \times (1 - C)] + [(1 - V) \times C]}. \tag{3}$$

The idea of independent consonant and vowel duration cues should not be interpreted as a statistical independence (noninteraction) of the effects of these two variables.[29] Multiplying the cues together at the prototype matching operation does not violate the principle of independent cues at the feature evaluation operation. The feature value of one cue remains independent of the feature value of another cue, even though the conjunction of two or more cues is, by definition, an interactive process. Since the outcome of the conjunction depends on both values, the cues can interact statistically in their effects on performance.

In contrast to the description of the fuzzy logical model, other studies have been taken as evidence for the ratio of the consonant duration to the vowel duration as the critical cue to consonant voicing.[3] Port and Dalby[12] (experiment 2) asked subjects to identify synthetic speech stimuli as "digger" or "dicker." The initial vowel duration

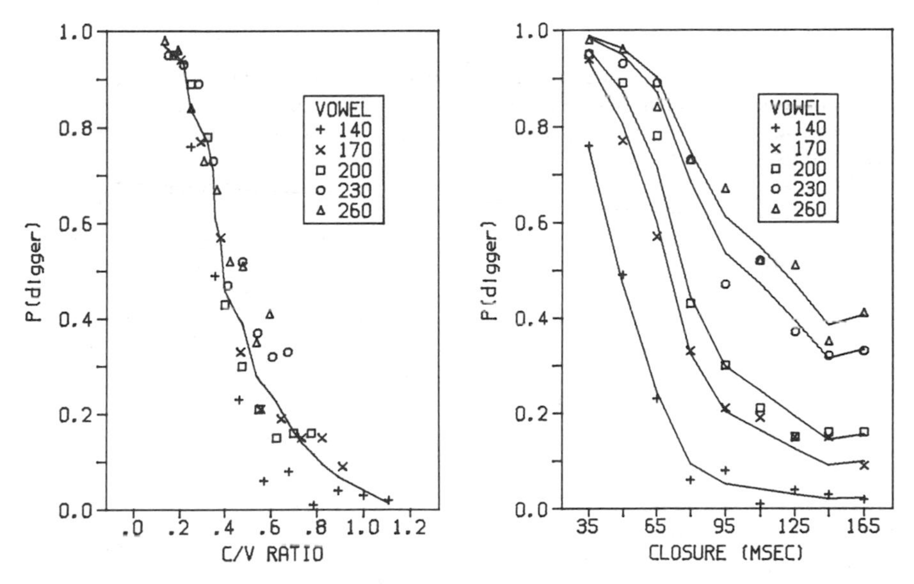

FIGURE 2. *Left panel:* Proportion of "digger" judgments as a function of the closure/vowel ratio. *Right panel:* Proportion of "digger" judgments as a function of closure duration and vowel duration. (Data from Port and Dalby.[12]).

and the medial consonant silent closure duration were orthogonally varied in a factorial design. If the consonant/vowel (C/V) ratio is the critical stimulus parameter for voicing of the medial stop, then the likelihood of a voiced response should vary *only* as a function of this ratio; the actual durations of the vowel and consonant closure should not matter. Port and Dalby[12] concluded that the judgments were determined primarily by the C/V ratio, but did not contrast this conclusion against other alternatives.

In addition to testing the fuzzy logical model, Massaro and Cohen[13] quantified the C/V ratio model and tested it against Port and Dalby's results. Both models were quantified with the same number of free parameters, which were estimated by finding those values that minimized the squared deviations between the predicted and observed results for each subject using the minimization routine STEPIT.[14] The measure of goodness-of-fit used to compare the two models was the root mean squared deviation (RMSD) between predicted and observed values. As can be seen in FIGURE 2, independent vowel duration and consonant duration cues gave a much better description of the results than did a C/V ratio (RMSD of .030 versus .078).

Massaro and Cohen[13] also demonstrated that the fuzzy logical model gave a better description than did the C/V ratio model for the individual subject identification and rating results for the /juz/ – /jus/ distinction.[4] Thus, vowel duration and consonant duration appear to provide independent cues to voicing of the consonant and the integration of the cues is adequately described by the fuzzy logical model.

Actual Duration, Perceived Duration, and Feature Values

It is reasonable to assume that perceived duration, P, is a linear function of physical duration, D.

$$P = KD \tag{4}$$

The slope, K, of the linear function might depend on a number of factors; for example, Derr and Massaro[4] found that the slope varied with the fundamental frequency pattern of the vowel. Rising or falling patterns increased the perceived duration relative to steady-state patterns (see also Lehiste[15]).

We might also expect that the slope of the function is dependent on surrounding context. Perceived duration of a given segment of sound might be longer to the extent that the surrounding segments are short. A contrast effect of this type seems to hold for both speech and nonspeech stimuli.[16,17]

Perceived duration cannot be identical to the feature values of duration cues in speech. A given perceived duration will have different consequences, depending on the segment of sound being perceived and the distinction being cued. Consider how

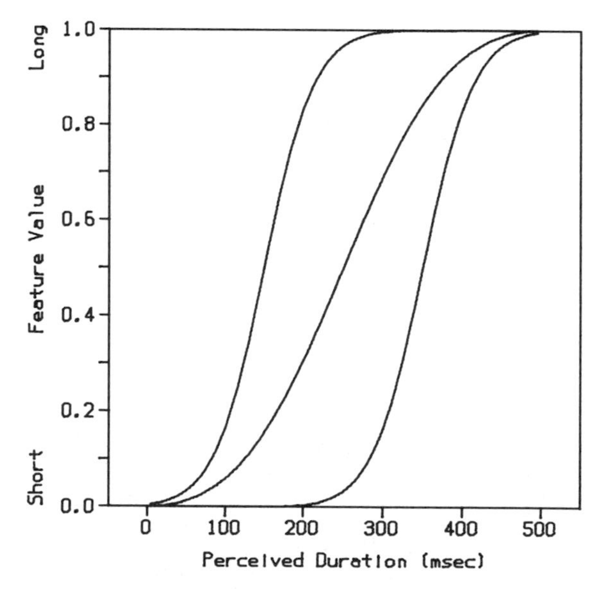

FIGURE 3. Three functions describing changes in the feature value, F, with changes in perceived duration, P.

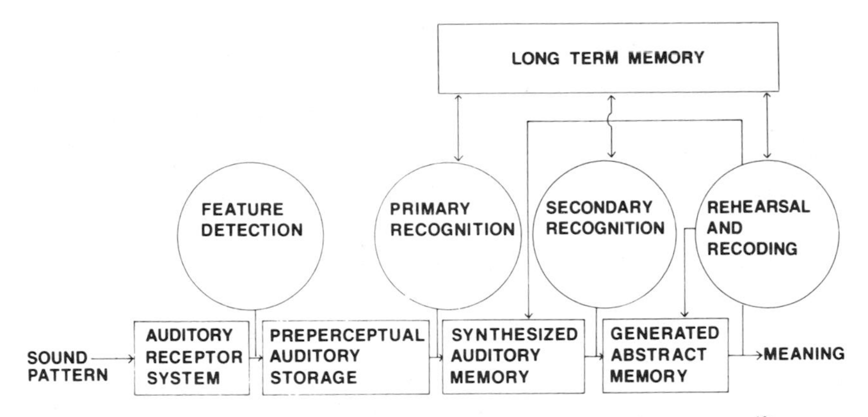

FIGURE 4. An auditory information-processing model. (After Massaro.[19])

changes in perceived vowel duration might cue the voicing of a following stop in a *VC* syllable. For very short or very long durations, a given change might not produce much of a change in the feature value. For intermediate durations, however, the same amount of change might produce a larger change in cue value. This relationship between perceived duration and feature value can be described by an ogival function relating cue value *F* and perceived duration *P* given by Equation 4

$$F = \frac{X^y}{X^y + (1 - X)^y}, \quad (5)$$

where $y > 1$. The y parameter determines the steepness of the middle section of the ogive. The values of X are assumed to be a linear function of perceived duration, P. Thus, $X = aP + b$, except that values of X less than zero are set to zero and values greater than one are set to one. Equation 5 is ogival in form when the X values fall between zero and one. The parameters a and b allow the ogive to shift horizontally along the dimension of perceived duration. FIGURE 3 plots three different ogival functions describing changes in F with changes in P. This relationship was used successfully to describe the change in the voicing feature value of a vowel–consonant syllable as a function of vowel duration and changes in fundamental frequency.[4]

PERCEPTUAL PROCESSING TIME

The model shown in FIGURE 4 has been utilized in our research in auditory information processing and speech perception.[18,19] The feature detection process transforms the sound wave pattern into acoustic features that are stored centrally in a preperceptual auditory storage. This storage accumulates the acoustic features until the sound pattern is complete and holds the features for a short time after a sound is presented. The primary recognition process resolves the featural information, which produces the phenomenological outcome of perceiving a particular sound of a particular loudness, quality, and duration, at some location in space.

The perceptual processing during the stimulus and a short period following is responsible for perceptual recognition and experience of the stimulus. A second pattern does not usually occur until the first pattern has been perceived. However, if the second pattern is presented before recognition of the first pattern is complete, the second pattern might interfere with recognition of the first. By varying the delay of the second pattern we can determine the duration of preperceptual auditory storage and the temporal course of perceptual recognition. The experimental task is referred to as a backward recognition masking paradigm. Pure tones differing in frequency, intensity, waveform, duration, and spatial location have been employed as test items in the backward masking task.[20,21]

Backward Masking

Consider a typical backward masking task. Consonant–vowel syllables /ba/, /da/, and /ga/ presented at a normal listening intensity were used as test and masking stimuli.[22] Only enough of the syllable was presented to make it sound speech-like. The 42-msec syllable had 30 msec of the consonant–vowel transition plus 12 msec of steady-state vowel. On each trial, one of the three syllables was presented followed by a variable silent interval before presentation of a second syllable chosen from the same set of three syllables. The subject's task was to identify the first syllable as one of the three alternatives. Subjects were told to ignore the second syllable, if possible.

FIGURE 5 plots the observed results in terms of discriminability (d') values for each of the three test alternatives. The d' measure provides an index of how well the subject discriminates a given test alternative from the other test alternatives in the task. Correct identification of the first speech sound increased with increases in the silent interval between the two sounds. This result shows that recognition of the consonant phoneme was not complete at the end of the consonant–vowel transition or even at the end of the short vowel segment of the sound. Syllable recognition required perceptual processing after the speech sound was terminated. These results support the idea that the speech sound is held in preperceptual auditory storage while processing takes place. A second sound interferes with additional processing of the first sound.

Perceiving Speech

The role of backward masking in the perception of speech is nicely illustrated in the identification of intervocalic stop consonants (*VCV*s). FIGURE 6 presents a stylized

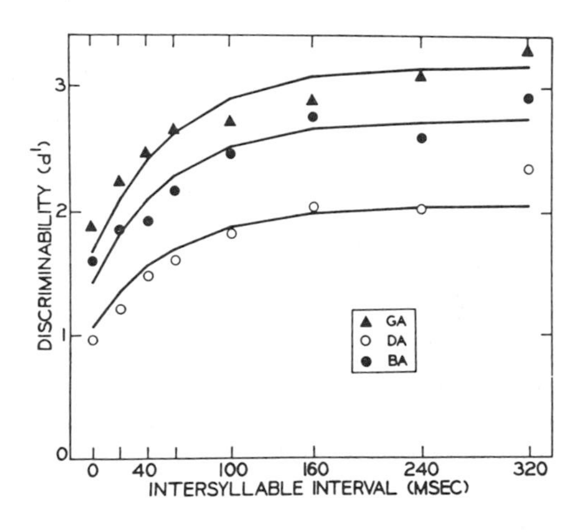

FIGURE 5. Predicted and observed discriminability of the test syllables as a function of the silent processing time available before the onset of the masking syllable. (After Massaro.[22])

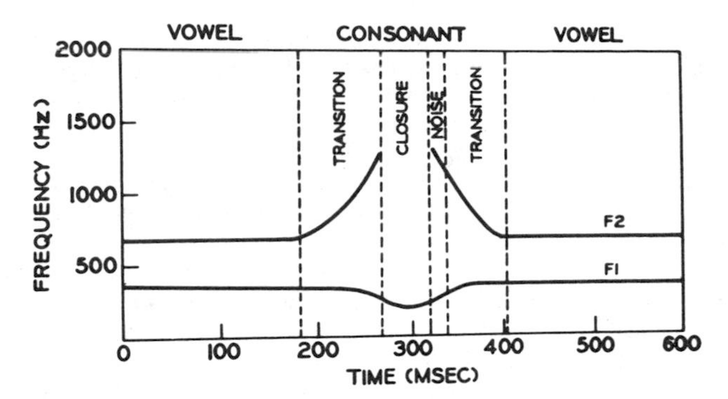

FIGURE 6. Stylized representation of a *VCV* sequence with a voiced-stop consonant containing steady-state vowel formants, transitions, closure, noise burst, and steady-state vowel formants.

spectrogram of a *VCV* with a voiced stop consonant. The steady-stated vowel is followed by a transition into the consonant, a closure period representing the closure of the vocal tract, a transition into the final vowel, and the final steady-state vowel. We refer to the segment before the closure as the *VC* and the segment after the closure as the *CV*. The *VCV* speech stimulus is an interesting one because of its redundancy and noncorrespondence with perceptual experience. We tend to hear a single consonant between the two vowels, although either the segment before closure or the segment after closure is sufficient when presented alone to produce the perception of the consonant.

Why doesn't the listener hear two consonants when processing a *VCV* speech stimulus? Following the logic of backward masking, we might expect the *CV* to "backward-mask" the processing of the *VC* to the extent the closure duration is short. Given insufficient processing time for the *VC* transition because of the quickly following *CV* transition, the *CV* transition has the largest impact on the perceptual experience.[19] More recent evidence for backward masking in perceiving speech comes from Repp.[23] Subjects had to discriminate normal *VCV* stimuli from modified *V–CV* stimuli in which the *VC* transitions were eliminated and replaced with the steady-state vowel. To the extent that the *CV* masks perceptual recognition of the segment preceding the closure, subjects should not be able to discriminate the *VCV* from the *V–CV* stimuli. The results showed that discrimination was poor at short closure durations and improved dramatically with increases in the closure interval. Repp also synthesized stimuli with contradictory *VC* and *CV* transitions; subjects had to discriminate /ab-di/ from /ad-di/ or /ab-bi/ from /ad-bi/. The results showed typical backward-masking effects. A *CV* segment can terminate processing of an earlier *VC* segment so that the *CV* segment provides the primary cues for the perception of the intervocalic consonant. Thus, the time available for processing plays an important role in perceiving speech.

Perceiving Time

The perceived duration of a stimulus is a function not only of its physical characteristics, but also of the amount of time spent processing it. A given stimulus, having constant physical characteristics, will have a variety of perceived durations, dependent upon the amount of available processing time. Massaro and Idson[24] used a backward-masking recognition paradigm, in which one of two target tones of differing durations was presented on each trial, followed after a variable silent interval by a masking tone of one of three durations. The subject identified the target tone as being long or short. The top panel of FIGURE 7 presents the average percentage of correct test-tone identifications as a function of intertone interval and masking-tone duration. Average performance improved as a negatively accelerated function of the intertone interval. This result was obtained for all three masking-tone durations. Thus, the average masking function for duration reflects similar processes to those occurring for other stimulus dimensions.

The bottom panel of FIGURE 7 illustrates that the short and long targets give different functions. Identification of the long target increases, whereas the short target gives a decreasing function with increases in the intertone interval. Massaro and Idson[24] accounted for these results within the model illustrated in FIGURE 4. The perception of duration is assumed to occur in essentially the same manner as for other

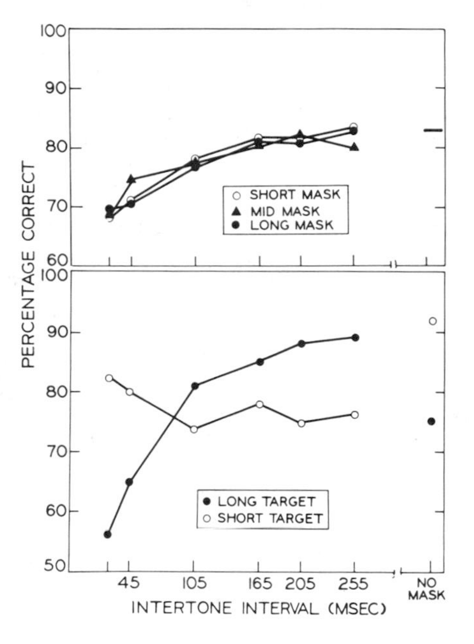

FIGURE 7. *Top panel:* Percentage of correct identifications of the target as a function of the duration of the mask and the intertone interval. *Bottom panel:* Percentage of correct identifications of each of the target tones as a function of the intertone interval. (From Massaro and Idson.[26])

attributes. However, the added assumption is made that the perceived duration of the target increases with the silent processing time after target presentation. That is, both discriminability and perceived duration are assumed to increase with additional processing time. If sufficient time is allowed for complete processing of the target, its perceived duration will be directly related to its temporal extent. If less than complete processing time is available, the target will be perceived as having a duration shorter than its asymptotic value. Therefore, with short intertone intervals the long target will be inaccurately identified as short quite often, while identification of the short target will be relatively good. With increases in the intertone interval a greater proportion of targets will come to be classified as long, simultaneously increasing accuracy on the long target and decreasing accuracy on the short target. These were exactly the results which were obtained (compare FIGURE 7).

It is also assumed that the mask adds to the perceived duration of the target. The addition is proportional to the mask duration. This assumption was supported by the strong target–mask interaction found in these studies. To the extent that the target and mask were both short, performance was better than if the target was short and the mask was long. A quantification of the model provided a good description of the perception and discrimination of tone and vowel duration.[21,24–26]

NORMALIZATION PROCESSES IN SPEECH PERCEPTION

It is now well documented that perceptual recognition of a segment of speech can be influenced by the duration of surrounding segments. Nooteboom[27] demonstrated a boundary shift between the long and short Dutch vowels /a/ and /a:/ with changes in speech rate. Longer durations were required for hearing the long vowel alternative as the speech phrase preceding the word containing the vowel was made slower. Normalization phenomena reveal the *relative* nature of acoustic cues in speech perception: The feature value of a given acoustic property must be normalized relative to other local context. In the fuzzy logical model, the context duration might lead to a normalization of acoustic feature values. This influence could result from an actual change of the psychophysical relationship between physical duration and perceived duration or from a more cognitive change of the function relating perceived duration to the feature value. Consider a typical mapping between physical duration and a feature value. For a given feature value, a shorter physical duration is necessary at a faster speech rate relative to a slower rate. As an example, it is well known that a relatively long silent closure interval cues a voiceless stop in a *VCV* context. A longer silent closure duration would be necessary for a given value of this cue to the extent that the stop is embedded in a context of a slow speaking rate.

Normalizing Vowel and Consonant Duration Cues

Port and Dalby[17] evaluated the role of speech tempo in the perception of the voicing distinction between the words "rapid" and "rabid." A trained speaker recorded the sentence "I'm trying to say 'rabid' to you," at a fast tempo and a slow tempo. The word "rabid" was removed from both sentences and cut immediately after closure of the medial /b/ and just before the release. The glottal pulsing during the closure interval was deleted and silent periods were inserted to create a continuum varying from 50 to 200 msec in steps of 10 msec. These modifications of the two versions of "rabid" were now embedded in both the fast and slow original sentences and presented to listeners for identification. Thus, listeners heard the word "rabid" taken from a fast or slow sentence and presented with one of 16 closure intervals embedded in either a fast or slow sentence. Listeners simply identified whether they heard "rabid" or "rapid" in each of the 64 possible test sentences.

FIGURE 8 gives the results of the experiment. In addition to the large effect of closure duration, both speaking rate of the original test word and the context sentence had significant influences on the identification judgments. The percentage of "rabid" judgments decreased with increases in the closure duration. There were more voiceless judgments for the test words originally spoken at a fast tempo relative to those spoken at a slow tempo. In addition, a longer closure was necessary to hear "rabid" in a fast than in a slow sentence. The predictions of the fuzzy logical model are also given in FIGURE 8. The derivation of these predictions is carried out in the next section.

A Model of Normalization

We define the two prototypes in the "rabid"–"rapid" identification task as

rabid: long vowel and NOT (long closure)

$$= V \times (1 - C) \tag{6}$$

rapid: NOT (long vowel) and long closure

$$= (1 - V) \times C, \tag{7}$$

where V and C stand for long vowel and long closure, respectively.

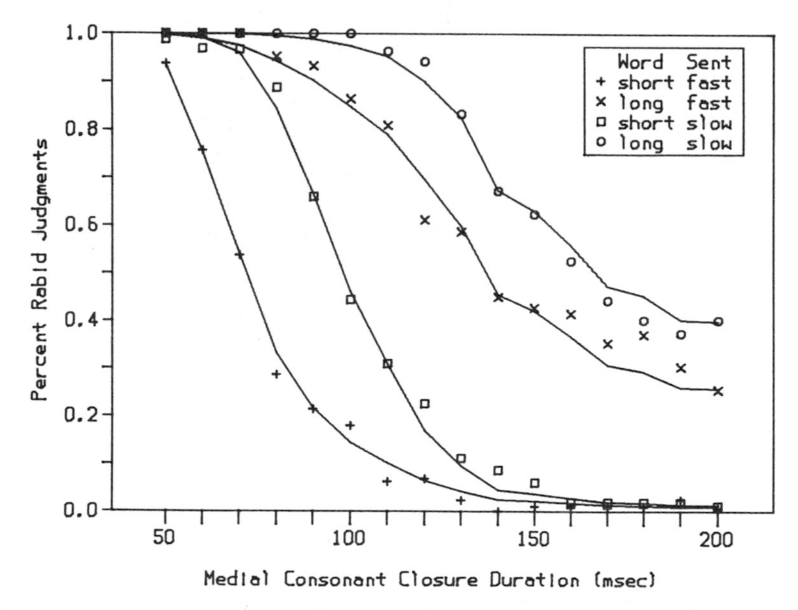

FIGURE 8. Predicted and observed identifications of word "rabid" as a function of closure duration; vowel duration and sentence duration are the curve parameters. (Data from Port and Dalby.[12])

To quantify the contribution of sentence tempo, a parsimonious assumption is that each feature value cueing duration is modified appropriately as a function of sentence tempo. A fast sentence rate will give a value more towards the long alternative for a given feature than will a slower sentence rate. The prototypes given by Equations 6 and 7 can be considered to be the representation for a normal rate; faster or slower speech rates will be represented by exponents on the feature values. In this case, the values V and C would be raised to exponents a and b, respectively.

The different exponents a and b simply allow different degrees of normalization for different features. The exponents will vary only with sentence rate. The exponents should be less than one if the sentence is faster than normal. As an example, a 0.7 value of V will give a .84 value when the exponent is .5. The slower sentence context has just the opposite influence and requires an exponent greater than one. An exponent of 2 reduces a .7 value to .49, resulting in a shift away from the long alternative.

The likelihood of a voiced response, P(rabid), would be based on the value of match to the rabid prototype relative to the sum of the values of prototype matches for rabid and rapid.

$$\text{P(rabid)} = \frac{V^{a} \times (1 - C^{b})}{[V^{a} \times (1 - C^{b})] + [(1 - V^{a}) \times C^{b}]} \quad (8)$$

The predictions of Equation 8 for Port and Dalby's results are given in FIGURE 8. The model provides an adequate description of the normalization results with an RMSD of .027.

CONCLUSION

We have discussed three roles of time in auditory information processing; the information provided by time, the time necessary for information processing, and the normalization of processing due to time. These important aspects of auditory information processing can be described within a general auditory information-processing model. The model was developed to account for dimensions of our auditory experience other than the temporal one and thus auditory time offers a strong test of the basic assumptions of the model. The critical features of the model include memory structures and functional processes, the evaluation and integration of auditory information, and the classification of sound patterns. The model accounted for a wide variety of temporal phenomena contributing to our auditory experience. Thus, auditory time provides additional evidence for both the general information-processing model and the fuzzy logical model developed primarily in nontemporal domains of auditory perception.

ACKNOWLEDGMENTS

Michael M. Cohen provided valuable assistance at all stages of the project.

REFERENCES

1. JAMES, W. 1980. The Principles of Psychology. Holt. New York, NY. (Reprinted by Dover Publications. New York, NY.)
2. PETERSON, G. E. & I. LEHISTE. 1960. Duration of syllable nuclei in English. J. Acoust. Soc. Am. **32:** 693–703.
3. DENES, P. 1955. Effect of duration on the perception of voicing. J. Acoust. Soc. Am. **27:** 761–764.
4. DERR, M. A. & D. W. MASSARO. 1980. The contribution of vowel duration, FO contour, and frication duration as cues to the /juz/–/jus/ distinction. Percept. Psychophys. **27:** 51–59.
5. RAPHAEL, L. J. 1972. Preceding vowel duration as a cue to the voicing of the voicing characteristic of word-final consonants in American English. J. Acoust. Soc. Am. **51:** 1296–1303.
6. TSENG, C., D. W. MASSARO & M. M. COHEN. Lexical tone perception: Evaluation and integration of acoustic features in Mandarin Chinese. Unpublished manuscript.

7. LEHISTE, I. 1973. Phonetic disambiguation of syntactic ambiguity. Glossa: 107–122.
8. LEHISTE, I., J. P. OLIVE & L. A. STREETER. 1976. Role of duration in disambiguating syntactically ambiguous sentences. J. Acoust. Soc. Am. **60:** 1199–1202.
9. AINSWORTH, W. A. 1972. Duration as a cue in the recognition of synthetic vowels. J. Acoust. Soc. Am. **51:** 648–651.
10. BENNETT, D. C. 1968. Spectral form and duration cues in the recognition of English and German vowels. Lang. Speech **11:** 65–85.
11. MASSARO, D. W. & M. M. COHEN. 1976. The contribution of fundamental frequency and voice onset time to the /zi/ – /si/ distinction. J. Acoust. Soc. Am. **60:** 704–717.
12. PORT, R. F. & J. DALBY. 1982. Consonant/vowel ratio as a cue for voicing in English. Percept. Psychophys. **32:** 141–152.
13. MASSARO, D. W. & M. M. COHEN. 1983. Consonant/vowel ratio: An improbable cue in speech. Percept. Psychophys. **33:** 501–505.
14. CHANDLER, J. P. 1969. Subroutine STEPIT finds local minima of a smooth function of several parameters. Behav. Sci. **14:** 81–82.
15. LEHISTE, I. 1976. Influence of fundamental frequency patterns on the perception of duration. J. Phonet. **4:** 113–117.
16. MILLER, J. L. & A. M. LIBERMAN. 1979. Some effects of later-occuring information on the perception of stop consonant and semivowel. Percept. Psychophys. **25:** 457–465.
17. CARRELL, T. D., D. B. PISONI & S. J. GANS. 1980. Perception of the Duration of Rapid Spectral Changes: Evidence for Context Effects with Speech and Nonspeech Signals. Research on Speech Perception, Progress Report 6. Department of Psychology, Indiana University, Bloomington, Indiana.
18. MASSARO, D. W. 1975. Experimental Psychology and Information Processing. Rand-McNally. Chicago, IL.
19. MASSARO, D. W. 1975. Understanding Language: An Information Processing Analysis of Speech Perception, Reading and Psycholinguistics. Academic Press. New York, NY.
20. KALLMAN, H. J. & D. W. MASSARO. 1983. Backward masking, the suffix effect, and preperceptual auditory storage. J. Exp. Psychol. Learn. Mem. Cog. **9:** 312–327.
21. IDSON, W. L. & D. W. MASSARO. 1977. Perceptual processing and experience of auditory duration. Sensory Processes **1:** 316–337.
22. MASSARO, D. W. 1974. Perceptual units in speech recognition. J. Exp. Psychol. **102:** 199–208.
23. REPP, B. H. 1978. Perceptual integration and differentiation of spectral cues for intervocalic stop consonants. Percept. Psychophys. **24:** 471–485.
24. MASSARO, D. W. & W. L. IDSON. 1976. Temporal course of perceived auditory duration. Percept. Psychophys. **20:** 331–352.
25. IDSON, W. L. & D. W. MASSARO. 1980. The role of perceived duration in the identification of vowels. J. Phonetics **8:** 407–425.
26. MASSARO, D. W. & W. L. IDSON. 1978. Target-mask similarity in backward recognition masking. Percept. Psychophys. **24:** 225–236.
27. NOOTEBOOM, S. G. 1982. Speech rate and segmental perception or the role of words in phoneme identification. *In* The Cognitive Representation of Speech. T. Myers, J. Laver, & J. Anderson, Eds. North-Holland. Amsterdam.
28. ODEN, G. C. & D. W. MASSARO. 1971. Integration of featural information in speech perception. Psychol. Rev. **85:** 172–191.
29. ANDERSON, N. H. 1974. Information integration theory: A brief survey. *In* Contemporary Developments in Mathematical Psychology, Vol. 2. D. H. Krantz, R. C. Atkinson & P. Suppes, Eds. Freeman. San Francisco, CA.

Time in Cognitive Processing and Memory: Discussion Paper

DOUGLAS L. MEDIN

Department of Psychology
University of Illinois
Champaign, Illinois 61820

INTRODUCTION

These excellent papers provide interrelated, but also wide-ranging views of time in cognitive processing and memory. My comments are organized as follows: First, I consider some of the different ways to think about time in cognitive processing and memory as reflected in these papers. Second, I provide a few detailed remarks on the specific papers and then suggest some themes that either are or might be running through this work. Finally, I suggest a "moral" that seems to follow from these themes.

ROLES FOR TIME IN COGNITIVE PROCESSING AND MEMORY

Processes in Time

The traditional approach to time is simply to focus on processes occurring in time. Time enters into performance only indirectly. An example of this approach is the strength theory discussed by Roberts and Kraemer. Strength is assumed to increase with stimulus presentation time and to decrease during retention intervals. Correct delayed matching to sample performance occurs not because pigeons choose the more *recent* stimulus, but rather because they choose the stimulus with greater *strength* (which will tend to be the more recent stimulus because there will have been less time for its strength to decrease).

Processes in time are also of interest in studies of time-limited processing, as reflected in the papers by Massaro and Michon and Jackson. The distinction between controlled or deliberate processing and automatic processing, as considered by Michon and Jackson, interacts with the extent to which processing may be limited by time (automatic processing is faster).

Time as an Attribute Entering into Processes

Another approach to time is to consider it as an attribute which may influence performance. For example, the temporal discrimination theory of D'Amato[1] assumes that delayed matching to sample performance is based directly on encoded relative recency, not strength. Almost all the papers in this section consider time as an attribute. Heinemann and Staddon each assume that time is coded fairly directly, whereas Shimp's model indirectly allows timing to result from patterns of activation and deactivation.

Time Per Se and What Influences It

A third approach is to focus on time itself and processes that might influence it. For example, Michon and Jackson consider the role of temporal and topical information in temporal-order memory, and Massaro describes factors affecting apparent duration. In addition, Heinemann is concerned with memory and decision processes in temporal behaviors.

These roles for time constitute a useful background for the remainder of this paper. The next section of this paper provides some brief comments on the papers in this section of the book.

COMMENTS ON INDIVIDUAL PAPERS

Space does not permit extensive remarks on individual papers. In addition, despite my best efforts to be critical, I have only a few reservations and suggestions. Therefore, this section will be brief.

Roberts and Kraemer provide data that represent a challenge for theories of memory in pigeons. I admire their flexibility in trying approaches ranging from strength theory to Gestalt organizational principles. Although it may be a good idea to integrate autoshaping phenomena with delayed matching performance, the I/D ratio does not seem a likely candidate for a full account of performance. In the experiment reported by Roberts and Kraemer the I/D ratio shows systematic discrepancies with performance being lower on longer delays than would be expected on the basis of the ratio.

Staddon brings out parallels between Weber fractions and Jost's law and uses this analysis to relate temporal control, delayed matching, and radial-arm-maze performance. As an interesting sidelight, Glenberg[2,3] has shown that spaced repetitions facilitate human memory whenever the retention interval is longer than the longest spacing interval. Apparently, relative time enters into basic observation on massed versus spaced repetitions. My one reservation about Staddon's paper may be a matter of taste—he does not seem to think that it is important to distinguish between alternative forms of representations of the age of an event. The papers by Shimp and Heinemann suggest that it may be important to evaluate alternative representations.

Heinemann draws a clear distinction between temporal information and processes operating on it to influence performance. His model seems to work quite well and some of the fits to data are impressive. However, two of the parameters in his model, the Weber fraction (W) and the sample size (θ), appear to be doing much the same work. A level of performance obtained with a small W and large θ could also be achieved with a larger W and a smaller θ. It is not unusual to have correlated parameters in models of this type, but usually the domain is one that can specify both parameters by mapping out something like a speed/accuracy tradeoff function. It would be nice if some corresponding function were available for evaluating temporal performance in animal experiments.

Shimp describes a model that handles both molar and molecular patterns of timing performance in impressive detail. This relatively simple model appears to represent a major advance in this area. However, there is nothing in the model that tells how the bird knows it is being presented one of the paradigms in which timing processes might apply. What happens when external cues are present as in a simple discrimination task? This comment is intended not as a criticism but rather as encouragement for Shimp to extend the model to still further situations.

Massaro gives a clear description of different roles for time in cognitive processing in describing his interesting research on speech perception. The fuzzy logical model represents a fruitful way to integrate time with other sources of information. Although this model fares quite well, it may be more accurate to say that it accounts for or reflects normalization effects rather than providing details on the underlying mechanisms associated with speech normalization.

Michon and Jackson give a wide-ranging review of their work on temporal-order information. Their observations on cognitive streaming are fascinating. They build a strong case for their argument that temporal-order information is not abstracted automatically. Since temporal-order information may derive from multiple sources, it may be that some sources are automatically encoded, but Michon and Jackson have shifted the burden of proof to the opposite camp. It would have been nice to see more attention directed at the basis of temporal-order performance. For example, Reed Hunt and his associates[4,5] have done work on the distinction between item and relational coding and the relational coding aspect of their theory represents a candidate model for temporal-order information.

The papers in this section show two encouraging trends: First, many of the papers employ formal models. Although this is not an absolute virtue, the papers provide good case histories illustrating how models can provide clarity on issues that are difficult to disentangle. The shift to simulation models does carry with it the extra burden of analyzing just what the model implies, what aspects of the model are doing the real work, and what the effects of alternative assumptions might be. Fortunately, it appears that there are several competing models and the additional problem of evaluating a model in isolation is avoided.

The other impressive trend is that the models and theories are being extended across a variety of paradigms. Each extension makes it increasingly likely that the models are capturing something important about the role of time in cognitive processing and memory.

THREE THEMES

Time is Not So Special

Implicit in the notion of treating time as an attribute is the idea that time may provide an additional source of information potentially influencing performance. Time then would have much the same status as any other attribute. Historically, time has had a more special status although not necessarily a more important one. For example, focusing on processes occurring in time demotes time to an almost epiphenomenal position.

It can be argued that treating time as special has slowed down progress in analyzing cognitive processes and memory. Although it is doubtless important to study temporal variables in classical conditioning, for example, there is also good evidence that *CS–US* similarity and spatial contiguity influence performance.[6–8] Retroactive and proactive interference depend not only on temporal relations, but also on similarity relationships.[9,10] Even retrograde amnesia is not strictly temporally bound.[11] For all of these reasons the idea that time may not be so special, implicit in the papers in this section, may help to elucidate the role of time in cognitive processes and memory.

Time Does Not Influence Performance Independently

Once one views time as an attribute, one is led to ask whether time makes an independent contribution to performance or whether time interacts with other attributes. The consensus of papers in this session is that time interacts with other attributes. In Massaro's fuzzy logical model, contributions from component dimensions are multiplied rather than added (see also Reynolds and Medin[10] for an analogous treatment of time and other attributes). Staddon has shown that overshadowing in temporal control experiments may depend on similarity relationships of various cues. Finally, Heinemann's model has an initial stage that involves learning to sort out temporal information from other attributes that potentially are relevant.

Time Can Enter into and Be Influenced by Other Processes

There is a large body of data on retrieval of information from memory.[12] One might well imagine that the temporal information associated with some episode might be altered when it is retrieved, if only because it might be difficult to distinguish the episode (for example, the last time I ate pizza) from subsequent retrievals (thinking about the last time I ate pizza) of it.[13] Perhaps the most direct demonstration of the interaction of other variables with time is seen in the studies of Michon and Jackson of cognitive streaming which indicate that semantic information interacts with (and may actually abolish) temporal-order information.

A MORAL

The themes and empirical observations running through these excellent papers suggest a moral: Whatever the advantages in the short run, over the long run it may not be a good idea to study time and temporal processing independent of other attributes and other processes.

REFERENCES

1. D'Amato, M. R. 1973. Delayed matching and short-term memory in monkeys. *In* The Psychology of Learning and Motivation. G.H. Bower, Ed. Vol. 7. Academic Press. New York, NY.
2. Glenberg, A. M. 1976. Monotonic and nonmonotonic lag effects in paired-associate and recognition memory paradigms. J. Verb. Learn. Verb. Behav. **15:** 1–16.
3. Glenberg, A. M. 1979. Component-levels theory of the effects of spacing of repetitions on recall and recognition. Mem. Cog. **7:** 95–112.
4. Einstein, G. O. & R. R. Hunt. 1980. Levels of processing and organization: Additive effects of individual-item and relational processing. J. Exp. Psychol. Hum. Learn. Mem. **6:** 588–598.
5. Hunt, R. R. & J. B. Mitchell. 1982. Independent effects of semantic and nonsemantic distinctiveness. J. Exp. Psychol. Learn. Mem. Cog. **8:** 81–87.
6. Testa, T. J. 1974. Causal relationships and the acquisition of avoidance responses. Psychol. Rev. **81:** 491–505.
7. Rescorla, R. A. 1980. Pavlovian Second-Order Conditioning: Studies in Associative Learning. Erlbaum. Hillsdale, NJ.
8. Rescorla, R. A. & P. C. Holland. 1982. Behavioral studies of associative learning in animals. Annu. Rev. Psychol. **33:** 265–308.
9. Medin, D. L., T. J. Reynolds & J. K. Parkinson. 1980. Stimulus similarity and retroactive interference and facilitation in monkey short-term memory. J. Exp. Psychol. Anim. Behav. Processes **6:** 112–115.
10. Reynolds, T. J. & D. L. Medin. 1981. Stimulus interaction and between-trials proactive interference in monkeys. J. Exp. Psychol. Anim. Behav. Processes **7:** 334–347.
11. Meyer, D. R. 1972. Access to engrams. Am. Psychol. **27:** 124–133.
12. Spear, N. E. 1978. The Processing of Memories: Forgetting and Retention. Erlbaum. Hillsdale, NJ.
13. Johnson, M. K. & C. L. Raye. 1981. Reality monitoring. Psychol. Rev. **88:** 67–85.

Introduction

JOHN R. PLATT

Department of Psychology
McMaster University
Hamilton, Ontario, Canada L8S 4K1

Psychologists have frequently found music and musicians to be fruitful subjects of study. The concerns of such studies have ranged from psychophysics to experimental esthetics. In recent years there has been a growing interest in music as a prototypical cognitive skill, and one aspect of this trend has been the development of information-processing models of timing in musical performance.

Three of the four papers in this section examine various aspects of timing and time perception in musicians. The paper by Vorberg and Hambuch deals with timing in the tapping of two-handed rhythms by amateur and professional musicians. Sternberg, Knoll, and Zukofsky examine judgment, production and imitation of fractions of a beat interval by a highly skilled musician. Shaffer analyzes the timing of actual performances by concert pianists. Each of these investigators analyzes the performance of his subjects in terms of information-processing models which share much in common with each other. The level of analytic precision and ingenuity represented by these studies is truly impressive. At the same time, some will be struck by the relatively molecular nature of these analyses and may justifiably feel a need for more holistic treatments of perception and production of rhythm in music as a complement to the present approaches.

The remaining paper in this session is very different from the other three; Hulse, Humpal, and Cynx are concerned with the ability of birds to discriminate auditory sequences on the basis of rhythmic structure, and to generalize such discriminations across variations in tempo. The goal here is not so much to illuminate psychological phenomena of music as to understand the capacity of birds to process rhythmic information. Such an understanding could contribute significantly to the study of bird song, as well as to more general issues in comparative psychology.

Timing of Two-Handed Rhythmic Performance[a,b]

DIRK VORBERG

Fachbereich Psychologie
Universität Marburg
Marburg, Federal Republic of Germany

ROLF HAMBUCH

Fachbereich Psychologie
Universität Giessen
Giessen, Federal Republic of Germany

INTRODUCTION

In this paper we present a simple stochastic model for the study of timing of skilled performance. The model is based on the notion that temporal variability and serial dependence observed in behavior sequences depend on the precision of a hypothetical central timing system as well as on the temporal jitter generated by the executing motor system. Within the theoretical framework of such a model it is possible to determine, for example, how much of the temporal variability in skilled performance is generated centrally, and how much is due to the precision lost during the execution of the motor program by the motor system. This general approach has been used first by Wing and Kristofferson[1] in their study of timing in simple tapping situations. We have proposed extensions of the basic model to repetitive rhythmic tapping; our goal was to determine whether rhythmic performance is controlled by hierarchically organized timing mechanisms, as contrasted to simple serial mechanisms.[2] Analyses of actual musical performance in terms of such a model have been reported by Shaffer.[3]

A major problem with these models is that they rest on rather strong assumptions which are not easily accessible to empirical test. Of course, the conclusions reached from applications of the models remain dubious if the validity of the underlying assumptions can be questioned. In the following, we show how to overcome some of these problems by extending the model to synchronous two-handed performance. The model we propose rests on weaker and testable assumptions; moreover, it permits rather strong inferences about the structure of the postulated timing system given that the model's validity can be established. It should be noted that several independent investigators have discovered recently how powerful the analysis of two-handed performance is in the study of timing and motor control.[3,4–6]

In this paper we focus on the properties of the central timing structures underlying synchronous rhythm production; in a companion paper[7] we investigate the statistical aspects of the motor system in more detail. The outline of the paper is as follows. First, we present the model and derive a testable prediction which is checked on data from an experiment involving two-handed rhythmic tapping. The model is then used as a

[a]The work reported here has been supported by Grant 306.82 from the Universität Konstanz.
[b]Address for correspondence: Dirk Vorberg, Fachbereich Psychologie, Gutenbergstrasse 18, D-3550 Marburg, Federal Republic of Germany.

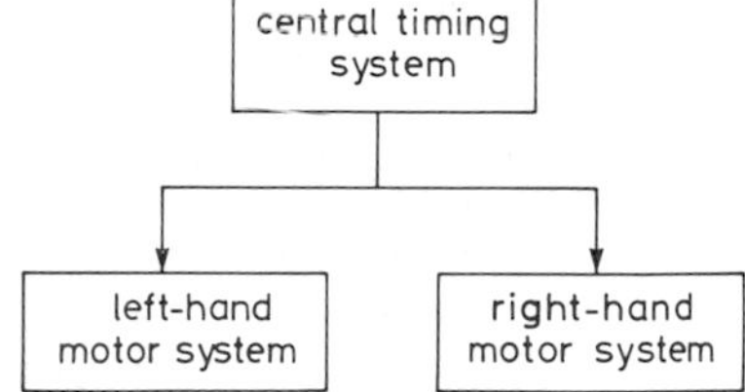

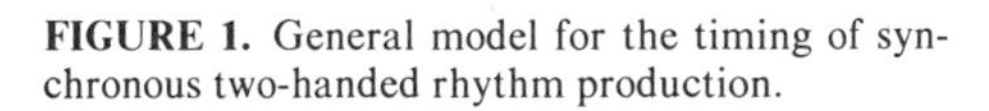

FIGURE 1. General model for the timing of synchronous two-handed rhythm production.

theoretical framework within which we try to distinguish between serial and hierarchical timing structures.

THEORETICAL FRAMEWORK

Our theoretical framework for the study of timing of synchronous two-handed rhythmic performance is shown in FIGURE 1. We assume a single central timing system generating motor commands which are sent to both the left-hand and the right-hand motor subsystem simultaneously. FIGURE 2 shows how these three systems operate in time when a repeating rhythm is beat bimanually. The tick marks on the time axis in the middle row indicate the (nonobservable) time points when the successive central commands occur. Each command triggers a response in both the left-hand and the right-hand system; however, observable responses occur only with some delay after their corresponding central commands. We assume that these *motor delays* are random variables; we denote them by L_n and R_n, where L_n (R_n) is the time difference between the occurrence of the *n*th left-hand (right-hand) response and the command that triggered it. Similarly, we assume the *timer intervals* between successive central commands as random variables denoted by T_n, $n = 1, 2, \ldots$, where T_n is the interval

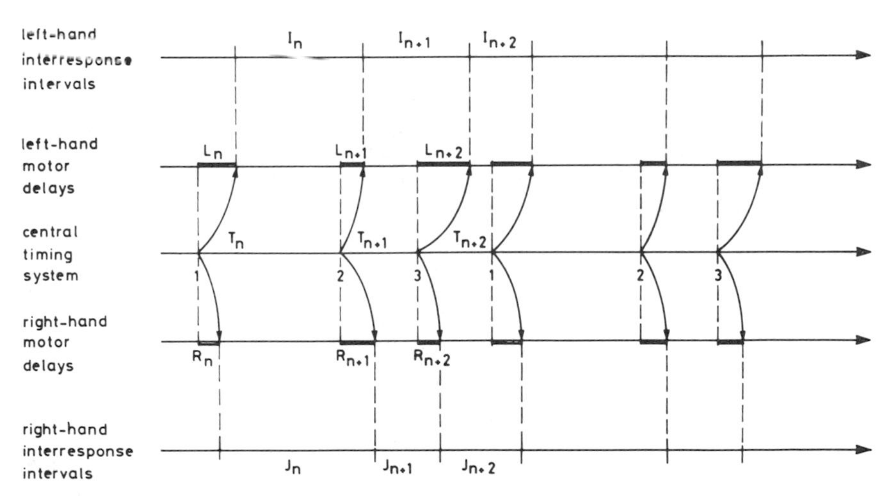

FIGURE 2. Hypothetical timer intervals (T_n) and motor delays (L_n, R_n) and their relation to the observable inter-response intervals of the left-hand (I_n) and right-hand (J_n) response sequences. A rhythm consisting of three notes (1, 2, 3) is assumed to be produced repeatedly.

bounded by the nth and the $(n + 1)$th command. The following assumptions are made:

A1: Common timing system. The same central timing commands control both motor subsystems.
A2: Independence between systems. The timer intervals, T_n, the left-hand and the right-hand motor delays, L_n and R_n, respectively, are independent of each other; for all $i, j, k > 0$, $\text{cov}(T_i, L_j) = \text{cov}(T_i, R_k) = \text{cov}(L_j, R_k) = 0$.
A3: Order preservation. The order of responses set up by the central timing system is not altered by the motor systems, that is, $P(T_n + L_{n+1} - L_n > 0) = 1$ and $P(T_n + R_{n+1} - R_n > 0) = 1$ for all n.

Some comment on the assumptions is in order.

The assumption of a common timing system which controls both hands (*A1*), though plausible, might be violated when a performer notes that one hand consistently lags behind the other and tries to compensate for this by triggering the early hand only after some delay, that is, by using separate commands for the two hands. We try to minimize the likelihood that this happens in our experiments by providing auditory feedback only about the response that occurs first.

Independence between the timer and the motor systems (assumption *A2*) might break down at fast tapping rates. If successive motor delays within a hand are correlated and if the degree of correlation depends on how closely in time two responses follow each other, then varibility of the timer intervals, T_n, will affect the dependence structure of the motor delays and therefore violate the independence assumption. For tapping at medium or slow tempos, however, this problem should be negligible.

The order preservation assumption (*A3*) holds if the variability of the difference between successive motor delays, for example, $\text{var}(L_{n+1} - L_n)$, is small relative to the corresponding mean interresponse intervals, $E(T_n + L_{n+1} - L_n)$. From previous results,[1,2] *A3* is almost certainly true.

It should be noted that independence is assumed only *between* systems, whereas the random variables *within* any system may be arbitrarily correlated. This stands in marked contrast to previous work on timing by us and others. With the exception of Wing,[8] successive motor delays were assumed to be independent.[1-3,6,9] Similarly, the intervals generated by the timing system were either assumed independent[1,6,8,9] or decomposable into independent random variables.[2,3] The main advantage of studying two-handed rhythmic performance is that it permits testing of the independence between system as well as determination of the actual amount of dependence existing within the timer intervals and the motor delays rather than taking independence between and within systems for granted.

If assumptions *A1–A3* hold, the postulated timer and motor-delay random variables can be related to the observable interresponse intervals (IRIs). Let I_n denote the IRI bounded by the nth and the $(n + 1)$th left-hand responses; correspondingly, J_n denotes the nth right-hand IRI. FIGURE 2 illustrates these definitions.

The left-hand and right-hand IRIs are tied to the timer intervals, T_n, and the motor delays, L_n and R_n, by the basic equations

$$I_n = T_n + L_{n+1} - L_n, \tag{1a}$$

$$J_n = T_n + R_{n+1} - R_n, n > 0. \tag{1b}$$

Equation 1 can be verified from FIGURE 2; its validity follows from assumptions *A1* and *A3*.

Equation 1 provides a simple tool for examining the timing system, and for separating its contributions to the data from those of the two motor systems. By analyzing the covariation between left-hand and right-hand IRIs, it is possible to infer the statistical properties of the sequence of intervals produced by the timing system, $\{T_n\}$, and thus, potentially, its structure. The intuitive idea behind our analysis is as follows:

Our primary interest is on the stochastic process, $\{T_n\}$, generated by the timing system which is accessible to observation only up to some "temporal noise" that is added by the motor systems. However, the left-hand and right-hand IRI sequences, $\{I_n\}$ and $\{J_n\}$, respectively, provide an opportunity to observe the same process $\{T_n\}$ twice, once after the left-hand and once after the right-hand motor system has operated on it, adding the independent noise sources $\{L_{n+1} - L_n\}$ and $\{R_{n+1} - R_n\}$, respectively. By comparing the left-hand with the right-hand IRI sequences, we can recover what they have in common, namely, the intervals $\{T_n\}$ generated by the timing system (see Equation 1).

The statistical analysis of the model turns out to be particularly simple if the dependence relations between the IRIs are described in terms of covariances. For random variables X and Y, the covariance is defined as $\text{cov}(X, Y) = E[(X - \mu_x)(Y - \mu_y)]$, where μ_x and μ_y are the expectations of X and Y, respectively.[c] The amount of dependence that exists between any pair of IRIs from the left-hand and the right-hand sequences is found as

$$\text{cov}(I_m, J_n) = \text{cov}(T_m + L_{m+1} - L_m, T_n + R_{n+1} - R_n)$$

(by Equation 1)

$$\begin{aligned} &= \text{cov}(T_m, T_n) \\ &\quad + \text{cov}(T_m, R_{n+1} - R_n) \\ &\quad + \text{cov}(L_{m+1} - L_m, T_n) \\ &\quad + \text{cov}(L_{m+1} - L_m, R_{n+1} - R_n) \end{aligned}$$

(by the distributivity of covariances; for example, Vorberg[10])

$$= \text{cov}(T_m, T_n)$$

(by the independence assumption $A2$).

Thus, if the assumptions of our model hold, the covariance between any pair of left-hand and right-hand IRIs equals that between the corresponding timer intervals, but is independent of the properties of the motor delays.

Two immediate consequences of this result should be noted: First, since $\text{var}(T_m) = \text{cov}(T_m, T_m)$, the variance of an interval produced by the timing system, T_m, can be determined from the covariance between the corresponding left-hand and right-hand IRIs, that is, $\text{var}(T_m) = \text{cov}(I_m, J_m)$. Second, since covariances are symmetric in the variables (for example, $\text{cov}(T_m, T_n) = \text{cov}(T_n, T_m)$), the dependence between any two IRIs is the same when the role of the hands is interchanged, that is, $\text{cov}(I_m, J_n) = \text{cov}(T_m, T_n) = \text{cov}(T_n, T_m) = \text{cov}(I_n, J_m)$.

We summarize these results:

$$\text{cov}(T_m, T_n) = \text{cov}(I_m, J_n) = \text{cov}(I_n, J_m), \qquad \textbf{(2a)}$$

$$\text{var}(T_m) = \text{cov}(I_m, J_m). \qquad \textbf{(2b)}$$

[c]The covariance equals the nonnormed product–moment correlation, that is, $\text{cov}(X, Y) = \rho_{xy}\sigma_x\sigma_y$.

Equation 2 can be used as a way to test the model empirically as well as a means of inferring the stochastic structure of the timing system from the IRIs, assuming the model to be valid. In the subsequent section, TEST OF THE MODEL, we examine whether the symmetry prediction (Equation 2a) holds for the left-hand and right-hand IRI sequences observed in two-handed rhythm production. Later sections then focus on the temporal variability and the dependence structure of the intervals produced by the timing system as revealed by the covariances between left-hand and right-hand IRIs.

The model as sketched so far is testable on two-handed tapping data without additional assumptions. Since we apply it to the production of repeating rhythms it seems natural, however, to add the assumption that the sequences of timer intervals, $\{T_n\}$, and motor delays, $\{L_n\}$ and $\{R_n\}$, are periodic in the means, variances and covariances, where the period depends on the number of notes per rhythmic cycle.

> *A4: Periodicity*. For repeated rhythmic cycles consisting of c notes, the sequences $\{T_n\}$, $\{L_n\}$, and $\{R_n\}$ are periodic in the means and covariances with period c. That is, for all $m, n \geq 1$, $E(T_n) = E(T_{n+c})$, and $\text{cov}(T_m, T_n) = \text{cov}(T_{m+c}, T_{n+c})$, and analogously for $\{L_n\}$ and $\{R_n\}$.

Since $\text{var}(T_m) = \text{cov}(T_m, T_m)$, *A4* implies periodicity in the variances as well. Note that the assumption allows the means, variances and covariances of the timer intervals and the motor delays to differ for the different responses within a cycle, thus relaxing the simplifying assumption of Vorberg and Hambuch[2] of constant motor-delay variances. Obviously, *A4* implies periodicity of the IRI sequences, $\{I_n\}$ and $\{J_n\}$.

Periodicity is not an essential assumption for the tests of the general model reported in the following section; there, it is used for economical reasons only, justifying the estimation of variances and covariances from the corresponding serial statistics computed across IRI sequences of repeating rhythmic cycles. Periodicity will become important, however, when we examine the statistical structure of the intervals $\{T_n\}$ generated by the timing system.

EXPERIMENTAL METHOD[d]

Our experimental paradigm is patterned after the continuation procedure introduced by Wing and Kristofferson[1] (see Figure 3). Each trial starts with a computer-generated rhythm auditorily presented to the subject via headphones. The rhythm keeps repeating at a fixed rate until the subject starts tapping the rhythm on Morse keys; the subject is to respond in synchrony with the computer-generated tones. This synchronization phase lasts for four repetitions of the rhythm, called *cycles* hereafter. Then the tones stop, and the subject is to continue tapping the rhythm at the given rate for 16 unaccompanied cycles. The end of the continuation phase is signaled to the subject by a tone.

Four different rhythms were employed (FIG. 4) abbreviated by 2–4–2, 2–1–3–2, 3–1–2–2, and 2–2–2–2, where the numbers given indicate the ratios of the successive time intervals between the notes of the rhythm. The tempo was kept fixed at 175 msec per eighth note. Therefore, the actual time intervals of the rhythms are obtained by multiplying the number code above by 350 msec; for example, the successive notes in rhythm 3–1–2–2 were separated by 525, 175, 350 and 350 msec.

[d] A more detailed description of the method is given in a companion paper.[7] The experiment contained one-handed as well as two-handed conditions; data from the latter only are reported here.

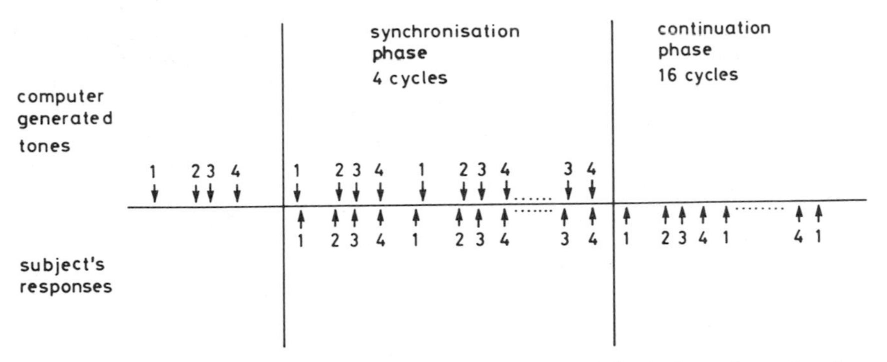

FIGURE 3. Schematic of an experimental trial; the four notes of the rhythm to be produced are denoted by 1, 2, 3, 4, 1, ...

Subjects were to beat the rhythms with both hands on two separate Morse keys on a table in front of them. They were instructed to produce the rhythm as precisely as possible, trying to keep the indicated tempo as well as close synchrony between the hands.

The tone sequences generated by the computer consisted of 50-msec sinusoids at a comfortable listening level; they were of equal pitch (400 Hz) except for the first tone of each cycle, which was two semitones up (449 Hz) to mark the cycle onset. Subjects were provided auditory feedback about their responses; immediately after being registered, taps elicited 50-msec tones an octave above the corresponding computer-produced tones. In order to prevent subjects from knowing how precisely they achieved synchronization of the hands, feedback tones were elicited only by the hand registered first.

The data from one professional musician and four amateur musicians with at least 6 years of instrumental practice will be reported. Subjects were screened from a preliminary sample of 26 persons. After two to five 1-hr training sessions, they served for 10 sessions in the main experiment. Precision of performance was constantly

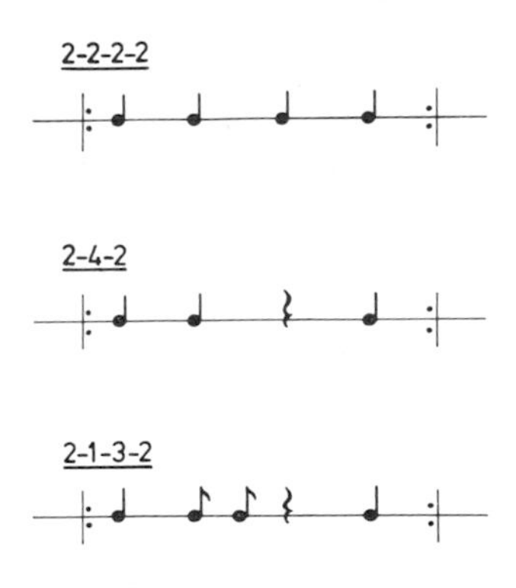

FIGURE 4. The four different rhythms employed in the experiment.

monitored during the experiment; any response sequence that deviated too much from the prescribed rhythms in terms of average tempo or the rhythmic pattern produced was to be redone immediately. During training, these criteria were gradually increased; the main experiment started when the final criteria were violated on no more than 5% of the trials. In order to keep subjects motivated, they were given visual feedback about their performance on a CRT that graphically indicated both the duration of the individual rhythmic cycles as well as the mean interresponse times within the rhythm averaged across the different repetitions. Information on whether a sequence was to be repeated was also presented via the CRT.

Each session consisted of two blocks with each of the conditions occurring twice per block. The different conditions were presented in random order except for the constraint that the two replications per condition followed each other immediately.

The data analysis of the IRI sequences $\{I_n\}$ and $\{J_n\}$ was performed separately for each subject. First, individual sequences were checked for "stationarity" by analyzing the durations of successive cycles (that is, $I_1 + \ldots I_c, I_{c+1} + \ldots + I_{2c}, I_{2c+1} + \ldots I_{3c}$, and so on) for linear trend. Any sequence was eliminated for which the corresponding t statistic exceeded the critical value at $\alpha = .25$. Overall, 13.8% of the sequences were discarded due to nonstationarity.

Variances and covariances were estimated from the corresponding serial statistics for each pair of sequences, $\{I_n\}$, $\{J_n\}$, which were then averaged across all replications produced by a subject under the given experimental condition.[e] Standard errors of the statistics were estimated from their standard deviations across replications. All statistics are based on between 26 and 38 pairs of 16-cycle IRI sequences (average 33.4) per subject; this amounts to a total of about 2670 pairs of IRI per condition, on which each covariance estimate is based.

TEST OF THE MODEL

Elsewhere[7] we have reported tests of the assumed independence between the timing and motor systems and with the predictions of the variances of the asynchronies between the two hands. In these tests, the model succeeded quite well, giving support for the general theoretical framework. Here, we examine the validity of Equation 2a as a prerequisite of our more detailed analysis of the timing system.

Consider the covariances between interresponse times of the left-hand, I_m, and the right-hand, J_n. For any $m, n > 0$, the covariance between them depends only on the statistical structure of the central timer and not on those of the motor systems, since $\text{cov}(I_m, J_n) = \text{cov}(T_m, T_n)$ by independence (Equation 2).

As was seen above, an important feature of the model is the symmetry in the dependence of the left-hand and the right-hand IRI sequences. For any two IRI pairs, (I_m, J_m) and (I_n, J_n), the model requires that $\text{cov}(I_m, J_n) = \text{cov}(I_n, J_m)$. For example, consider the fourth and the sixth IRI-pair of $\{I_n\}$ and $\{J_n\}$, that is, (I_4, J_4) and (I_6, J_6). If the model holds, the amount of covariation between I_4 and J_6 will be the same as that observed when we reserve the role of the hands, that is, between J_4 and I_6. Of course, the symmetry prediction must hold for all $m, n > 0$, that is, for the entire

[e]The estimators used are given by:

$$\widehat{\text{cov}}(I_m, J_n) = \frac{1}{r-1}\left[\sum_{k=0}^{r-1} I_{m+kc} J_{n+kc} - \frac{1}{r}\sum_{k=0}^{r-1} I_{m+kc}\sum_{k=0}^{r-1} J_{n+kc}\right],$$

where r = number of cycles per sequence, and $0 < m, n \leq c$.

cross-covariance function describing the dependence between the sequences $\{I_n\}$ and $\{J_n\}$. Note that, in general, cross-covariance functions are not symmetric; the predicted symmetry is a consequence of the mutual independence of the three subsystems of the model. Testing for symmetry is thus a way of testing the independence assumption (*A2*).[f]

The test was carried out by computing the empirical cross-covariance function, cov (I_m, J_n), that relates the IRI sequences $\{I_m\}$ and $\{J_m\}$, making use of the periodicity assumption (*A4*). For each experimental condition, and $1 \leqq m, n \leqq c + 1$, $1 \leqq |m - n| \leqq c$, $\widehat{\text{cov}}\,(I_m, J_n)$ was compared with $\widehat{\text{cov}}\,(I_n, J_m)$; c is the number of notes per rhythm, that is, $c = 3$ for condition 2–4–2 and $c = 4$ for the remaining conditions. FIGURE 5 shows the results combined across conditions. For each subject, each rhythm, and each pair (m, n), $\widehat{\text{cov}}\,(I_m, J_n)$ is plotted against $\widehat{\text{cov}}\,(I_n, J_m)$.

If symmetry holds, all points lie on the diagonal except for statistical fluctuation. Evidently, the prediction is borne out rather well since no systematic deviation from the diagonal is discernible.

This conclusion is corroborated by statistical tests; analyses of the sign of $\widehat{\text{cov}}\,(I_m, J_n) - \widehat{\text{cov}}\,(I_n, J_m)$ generally showed deviations in both directions to be about equally likely. Combined across conditions, the χ^2 statistics (d.f. = 1) obtained for the five subjects were 2.08, .42, 2.95, .86, and .24, respectively, which are all nonsignificant as is the combined statistic, $\chi^2 = 6.55$ (d.f. = 5). Similarly, combining across subjects, nonsignificant χ^2 statistics are obtained for the different rhythm conditions (1.69, .07, 1.10, and .30, respectively; d.f. = 1) and their sum (3.16; d.f. = 4). However, when the covariances, collapsed across conditions, are analyzed as a function of lag $|m - n|$, some discrepancies are found. For the five subjects, the following χ^2 statistics (d.f. = 1) resulted: for lag 1: .60, .60, 5.4 (p .025), 1.67, and 4.57 (p .05); for lag 2: .08, .08, 1.14, .60, 7.14 (p .01); and for lag 3: 2.27, .00, .33, .33, 2.78.

The evidence in favor of the symmetry prediction is thus not unequivocal. However, since neither the three discrepant results nor the nonsignificant ones deviate in a consistent direction from the expectations, we tentatively keep the model as a framework within which to examine the stochastic properties of the postulated timing system.

HIERARCHICAL VERSUS SERIAL TIMING STRUCTURES

Unequal Cycle Variances

If our general model of two-handed tapping is valid, the covariances between left-hand and right-hand IRIs reveal the statistical properties of the central timing system, since cov (I_m, J_n) = cov (I_n, J_m) = cov (T_m, T_n). By examining the stochastic structure of $\{T_n\}$ in detail, we obtain information about the timing mechanisms that underly rhythmic performance. As in Vorberg and Hambuch,[2] our main concern is whether timing is achieved by a *serial* mechanism that simply concatenates the individual intervals within a rhythm, or whether there is evidence for a *hierarchical* organization of the timing system in the sense that the duration of larger units (for example, half-cycles, full cycles) is under direct timer-control rather than controlled indirectly, as in serial timing systems, which time the intervals between successive commands only. In order to detect hierarchically organized timing mechanisms, we

[f]This is an example of the theoretical advantage of describing dependence between the intervals in terms of covariances rather than correlations: in general, symmetry will not hold for the cross-correlation function, $\rho(I_m, J_n) = \text{cov}\,(I_m, J_n)/[\text{var}\,(I_m) \cdot \text{var}\,(J_n)]^{1/2}$.

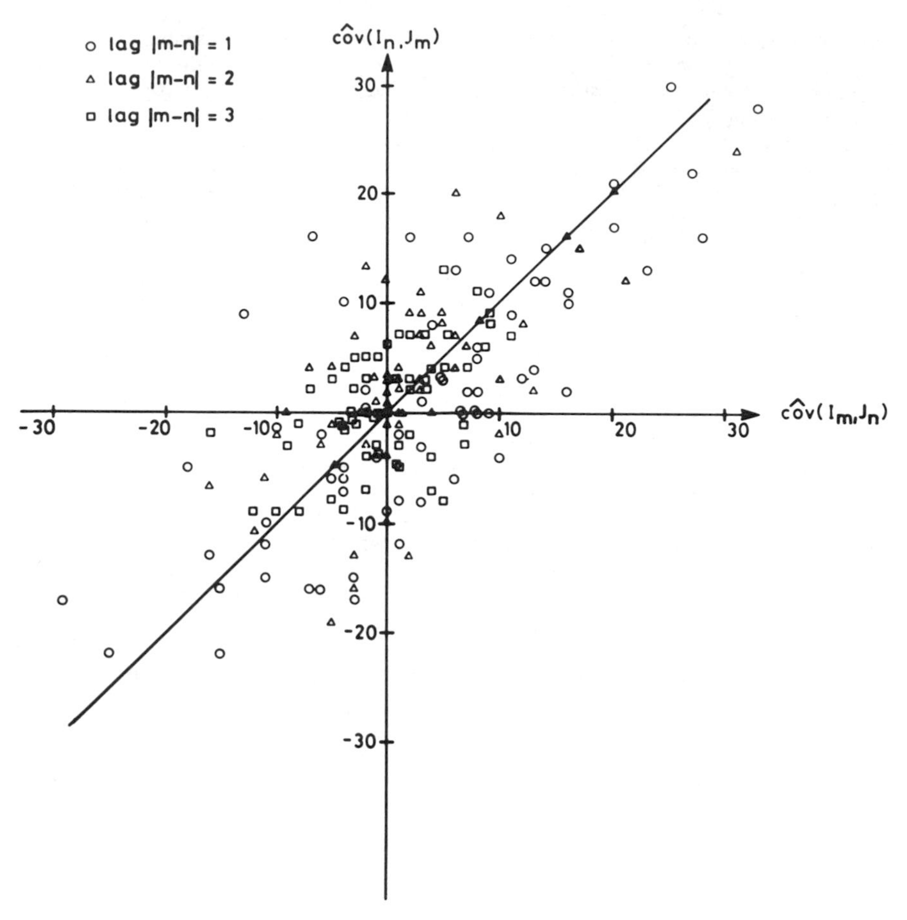

FIGURE 5. Scatter diagram of $\widehat{\text{cov}}(I_m, J_n)$ vs. $\widehat{\text{cov}}(I_n, J_m)$, $0 < m, n \leq c, 0 < |m - n| < c$. Each dot represents a different subject times rhythm combination.

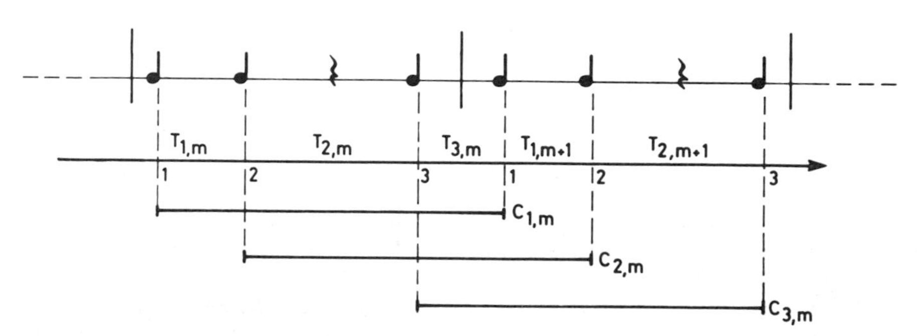

FIGURE 6. Definition of cycle durations, $C_{i,m}$, for condition 2-4-2.

examine the precision with which the system controls the duration of cycles. In a cycle consisting of c notes, there are c different ways of forming intervals that span full cycles; to determine a cycle's duration, we can either measure the interval bounded by the first notes in successive cycles, or that bounded by the second notes, and so forth. We call these overlapping intervals the alternative *cycle* durations. FIGURE 6 illustrates their definition for rhythm 2–4–2.

For convenience, we change the notation so that it indicates the periodicity of the timer intervals more clearly. Let $T_{i,m}$ denote the ith intercommand interval within the mth cycle, that is, the interval between the ith command and the following one within the cycle.[g] We then define the cycle durations $C_{i,m}$, $i = 1, c$, by

$$C_{i,m} = \begin{cases} \sum_{j=1}^{c} T_{j,m}, & i = 1 \\ \sum_{j=i}^{c} T_{j,m} + \sum_{j=1}^{i-1} T_{j,m+1}, & i > 1 \end{cases}$$

(see FIGURE 6).

Since all the $C_{i,m}$ span one full cycle, they have the same expectation, as is easily seen from *A4*. However, since they differ with respect to the note in the sequence from where they are taken, their variances need not be equal. In fact, if there exist mechanisms that directly time the duration of intervals larger than those between successive commands, the variance of cycle durations that contain those higher-order timing units will in general be smaller than of those that start and end in different higher-order units.

These intuitive ideas can be made more precise. Let us analyze the variances of the cycle durations, called *cycle variances* in the following, by going back to the definition of the $C_{i,m}$. As an example, consider cycles with $c = 3$. For $C_{1,m}$ and $C_{2,m}$, we obtain

$$\begin{aligned} \text{var}\,(C_{1,m}) &= \text{var}\,(T_{1,m} + T_{2,m} + T_{3,m}) \\ &= \text{var}\,(T_{1,m} + (T_{2,m} + T_{3,m})) \\ &= \text{var}\,(T_{1,m}) + \text{var}\,(T_{2,m} + T_{3,m}) + 2\,\text{cov}\,[T_{1,m}, (T_{2,m} + T_{3,m})]; \\ \text{var}\,(C_{2,m}) &= \text{var}\,(T_{2,m} + T_{3,m} + T_{1,m+1}) \\ &= \text{var}\,[(T_{2,m} + T_{3,m}) + T_{1,m+n}] \\ &= \text{var}\,(T_{2,m} + T_{3,m}) + \text{var}\,(T_{1,m+1}) + 2\,\text{cov}\,[(T_{2,m} + T_{3,m}), T_{1,m+1}]. \end{aligned}$$

Since $\text{var}\,(T_{1,m+1}) = \text{var}\,(T_{1,m})$ due to the periodicity of the timer intervals, we see that $\text{var}\,(C_{1,m}) \leqq \text{var}\,(C_{2,m})$ if and only if $\text{cov}\,[T_{1,m},(T_{2,m} + T_{3,m})] \leqq \text{cov}\,[(T_{2,m} + T_{3,m}), T_{1,m+1}]$. In a similar way we obtain analogous inequalities for all pairs of cycle variances. The results are summarized by

$$\text{var}(C_{1,m}) \leqq \text{var}(C_{2,m}) \tag{3a}$$

if and only if

$$\text{cov}\,[T_{1,m}, (T_{2,m} + T_{3,m})] \leqq \text{cov}\,[(T_{2,m} + T_{3,m}), T_{1,m+1}],$$

$$\text{var}(C_{1,m}) \leqq \text{var}(C_{3,m}) \tag{3b}$$

[g]The different notations are related by $T_n = T_{i,m}$ if $n = c\,(m - 1) + i$.

if and only if

$$\text{cov}\,[(T_{1,m} + T_{2,m}), T_{3,m}] \leqq \text{cov}\,[T_{3,m}, (T_{1,m+1} + T_{2,m+1})],$$

$$\text{var}\,(C_{2,m}) \leqq \text{var}\,(C_{3,m})$$

if and only if

$$\text{cov}\,[T_{2,M}, (T_{3,m} + T_{1,m+1})] \leqq \text{cov}\,[(T_{3,m} + T_{1,m+1}), T_{2,m+1}]. \tag{3c}$$

To see the implications of these inequalities, consider Equation 3a. If rhythmic performance is timed by a serial mechanism that directly controls the duration of the lowest-order intervals only, there is no reason why the sum of the second and third intervals of a cycle should covary more with the first interval of the same cycle than with that of the following one. If, on the other hand, there is a higher-order mechanism within a hierarchical timing system controlling the duration of the whole cycle,[h] that is, $C_{1,m} = T_{1,m} + T_{2,m} + T_{3,m}$, we should expect a lower covariance between the segments within a cycle, that is, $T_{1,m}$ and $T_{2,m} + T_{3,m}$, than across two cycles, that is, $T_{2,m} + T_{3,m}$ and $T_{1,m+1}$. This is because time intervals from different cycles will in general be either independent or covary slightly positively with each other due to tempo fluctuations, whereas intervals within cycles ought to covary less; if the first interval, $T_{1,m}$, happens to be smaller (larger) than average, the sum of the following two intervals, $T_{2,m} + T_{3,m}$, will be likely to be larger (smaller) than average by virtue of the higher-order timer controlling the total cycle duration, $C_{1,m}$. Similar reasoning can be applied to the remaining inequalities as well, leading to the conclusion that the variances of $C_{1,m}$, $C_{2,m}$ and $C_{3,m}$ should be about the same if the underlying timer system is serial, whereas systematic differences in the cycle variances are to be expected if rhythmic performance is controlled by a hierarchical timing system with higher-order units that time the duration of intervals between nonsuccessive commands. Of course, the analogous prediction holds for the cycle variances of rhythms with any number of notes within a cycle, c.

Particular Hierarchical Models

We can sharpen our notion of hierarchical timing structures and gain more insight into their behavior by modeling them. Previously, we[2] have considered a class of timing models defined by the following properties: (*i*) A timing structure controlling a rhythmic cycle with c notes consists of c timers that may be located on different levels of a hierarchy. (*ii*) When started, a timer generates an interval; when it expires, the timer may start other timers *on the same or on lower levels* of the hierarchy. (*iii*) A timer is started by exactly one other timer (which may be itself it if is on the top level). (*iv*) Low-level timers generate intervals that are overlapped completely by those generated on higher levels.

As an example, we examine in some detail the two-level timing structure shown in FIGURE 7. This level-2 timer controls the duration of the full cycle. The successive intervals it generates, indicated by the arrows, are random variables denoted by F_m, F_{m+1}, ... The timer restarts itself when the last time interval has elapsed. Simultaneously, a level-1 timer is started that controls the interval between the first

[h] For simplicity, we are assuming here that the top-level timer starts on the cycle's first note. The argument remains valid if timing starts from a note different from the nominal beginning of a cycle.

two notes in the cycles, i.e., $T_{1,m}$. When it expires, a second level-1 timer is triggered which determines the interval between the second and the third note, that is, $T_{2,m}$. The intervals generated by these level-1 timers are denoted by D_m and E_m, respectively.

The gist of the notion of a hierarchical timing structure is that timing control may pass from higher to lower levels, but not vice versa. Therefore, the interval between the last note in the mth cycle and the first one of the $(m + 1)$th cycle, $T_{3,m}$, is not controlled directly, but is determined by the joint operation of the level-2 and the level-1 timers. FIGURE 7 shows that $T_{3,m} = F_m - D_m - E_m$.

Hierarchical timing structures may be described in terms of the relations between the time intervals bounded by successive commands, the $T_{i,m}$, and the intervals generated by the separate timing devices. For the particular structure, these are given by

$$T_{1,m} = D_m, \tag{4a}$$

$$T_{2,m} = E_m, \tag{4b}$$

$$T_{3,m} - F_m - D_m - E_m. \tag{4c}$$

In principle, the $\{D_m\}$, $\{E_m\}$ and $\{F_m\}$ may be arbitrary sequences of non-negative

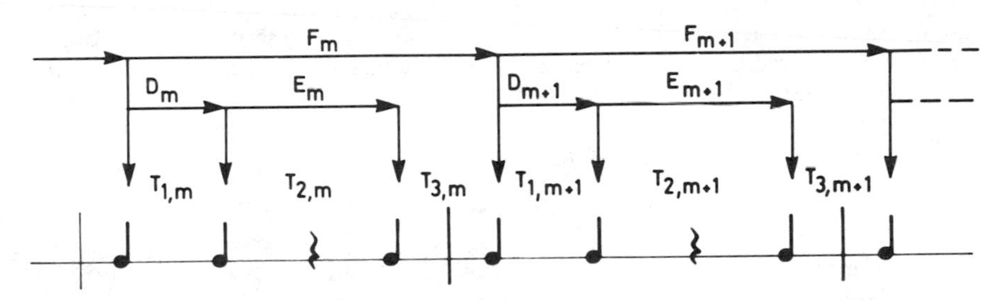

FIGURE 7. A hierarchical timing structure for cycles with $c = 3$.

random variables, except for the constraint that $P(F_m \geqq D_m + E_m) = P(T_{3,m} \geqq 0) = 1$, since otherwise the proper order of the notes within cycles would not necessarily be maintained. Of course, this is the analogue of the order preservation assumption ($A3$) that is to hold for the joint operation of the timer and the motor systems.

Consider the ordering of the cycle variances predicted by this particular model. For example, from Equation 4a, inequality **3a** can be rewritten as

$$\text{var}(C_{1,m}) \leqq \text{var}(C_{2,m}) \leftrightarrow \text{cov}(D_m, F_m - D_m) \leqq \text{cov}(F_m - D_m, D_{m+1}).$$

In Vorberg and Hambuch,[2] we assumed all the timer random variables to be mutually independent. This leads to the expected variance ordering, that is, $\text{var}(C_{1,m}) \leqq \text{var}(C_{2,m})$, since $\text{cov}(D_m, F_m - D_m) = -\text{var}(D_m)$, and $\text{cov}(F_m - D_m, D_{m+1}) = 0$ for independent D_m, D_{m+1} and F_m. It is important to note, however, that our technique of inferring the existence of hierarchical timing mechanisms from unequal cycle variances is more general and may remain valid even if independence breaks down.

For example, assume that independence only holds between intervals generated by different timers, but that successive intervals produced by the same timer covary. This still implies $\text{var}(C_{1,m}) \leqq \text{var}(C_{2,m})$ since $\text{cov}(D_m, F_m - D_m) = -\text{var}(D_m) \leqq -\text{cov}(D_m, D_{m+1}) = \text{cov}(F_m - D_m, D_{m+1})$, given that $\text{cov}(D_m, F_m) = \text{cov}(D_{m+1}, F_m) = 0$.

Another, more reasonable, assumption might be that the intervals produced within any given cycle are dependent, which would be the case if the hierarchy passed fallible information down its levels for setting the timer parameters. This is likely to cause positive dependence between the intervals within cycles, that is, cov $(D_m, F_m) \geqq 0$ and cov $(D_m, E_m) \geqq 0$ in our example, since the information used to set the level-2 timer is also fed to the level-1 timers. If independence is assumed between intervals within different cycles only, the variance of the cycle duration taken from the second note, var $(C_{2,m})$, will, under quite general conditions, still exceed that of the "natural" cycle duration, var $(C_{1,m})$. If cov (F_m, D_{m+1}) = cov $(D_m, D_{m+1}) = 0$, the result is var $(C_{1,m}) \leqq$ var $(C_{2,m}) \leftrightarrow$ cov $(D_m, F_m) \leqq$ var (D_m). By some algebra, this can be shown to be equivalent to the condition $\rho_{DF} \leqq \sigma_D/\sigma_F$. Thus, for negative as well as for some amount of positive dependence between the timer-produced intervals, the predicted variance ordering holds.

Additional inequalities relating the variances of the alternative cycle durations, $C_{1,m}$, $C_{2,m}$, and $C_{3,m}$, can be derived from (**3**) and (**4**) for this particular model, and, of course, analogously for models that embody alternative timing structures as well as for arbitrary numbers of notes per cycle.

There exist 10 different distinguishable models for rhythms with $c = 3$, and 31 for rhythms with $c = 4$ satisfying conditions (*i*) to (*iv*) above. Rather than investigating them all, we make use of the cycle variance inequalities as a diagnostic means for determining some kind of hierarchical timing structure and as a test against serially organized timing mechanisms. For the whole class of models described above it can be shown quite generally that some unequal cycle variances are to be expected unless the timing structure is serial; this prediction holds for a wide range of dependence between the different timer random variables.

Results

In order to estimate the cycle variances, var $(C_{i,m})$, covariances were computed for appropriately concatenated left-hand and right-hand IRIs; for example, for $c = 3$ the estimate of var $(C_{1,m})$ is given by $\widehat{\text{cov}}$ $(I_{1,m} + I_{2,m} + I_{3,m}, J_{1,m} + J_{2,m} + J_{3,m})$, since $C_{1,m} = T_{1,m} + T_{2,m} + T_{3,m}$.

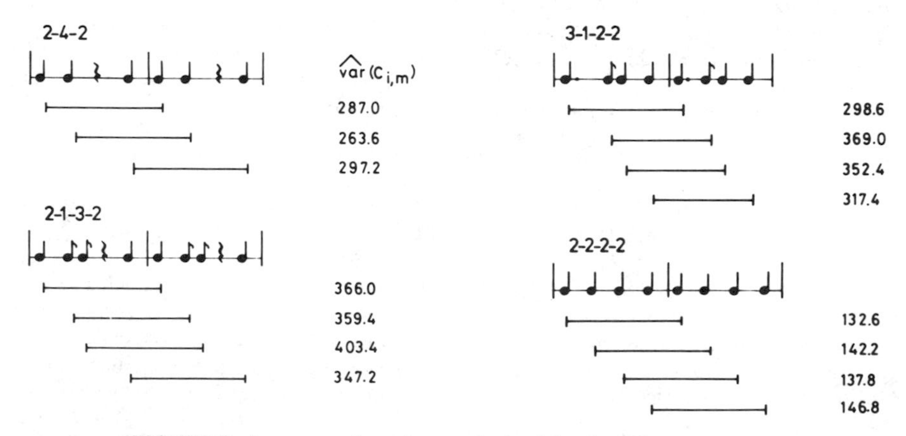

FIGURE 8. Average cycle variances obtained for the different rhythms.

TABLE 1. Central Timer Autocovariances Estimated from the Cross-Covariances Between Left-Hand and Right-Hand IRI Sequences

			Condition 2–4–2			
	$T_{2,m}$	$T_{3,m}$	$T_{1,m+1}$	$T_{2,m+1}$		
$T_{1,m}$	3.5	5.5**				
$T_{2,m}$		−8.5**	2.5			
$T_{3,m}$			−5.5**	8.0**		
			Condition 2–1–3–2			
	$T_{2,m}$	$T_{3,m}$	$T_{4,m}$	$T_{1,m+1}$	$T_{2,m+1}$	$T_{3,m+1}$
$T_{1,m}$	4.0*	2.0	5.0**			
$T_{2,m}$		−11.5***	−1.0	1.5		
$T_{3,m}$			15.0***	1.0	−2.0	
$T_{4,m}$				5.0	9.5***	−4.5
			Condition 3–1–2–2			
	$T_{2,m}$	$T_{3,m}$	$T_{4,m}$	$T_{1,m+1}$	$T_{2,m+1}$	$T_{3,m+1}$
$T_{1,m}$	−4.5	−7.5*	4.0			
$T_{2,m}$		2.5	4.0**	.5		
$T_{3,m}$			1.5	11.0**	.0	
$T_{4,m}$				16.0***	2.5	−1.0
			Condition 2–2–2–2			
	$T_{2,m}$	$T_{3,m}$	$T_{4,m}$	$T_{1,m+1}$	$T_{2,m+1}$	$T_{3,m+1}$
$T_{1,m}$	− .5	2.5	−2.5			
$T_{2,m}$		1.5	3.5*	−.5		
$T_{3,m}$			3.5***	−1.5	.0	
$T_{4,m}$				6.0***	4.0*	.5

NOTE: Asterisks indicate covariances that differ significantly from zero (*: $p < .05$; **: $p < .01$; ***: $p < .001$).

The estimates obtained for the different cycle variances, averaged over subjects, are shown in FIGURE 8. Analyses of variance (repeated measures design), performed on the square roots of these estimates, showed the differences to be significant except under the isochronous condition (2–4–2: $F(2,8) = 96.38$, $p < .001$; 2–1–3–2: $F(3,12) = 3.80$, $p < .05$; 3–1–2–2: $F(3,12) = 14.06$, $p < .001$; 2–2–2–2: $F(3,12) = 1.47$, $p > .25$). Thus, there is reliable evidence for the existence of higher-order timing mechanisms controlling the production of rhythms with unequal time intervals, whereas isochronous rhythms seem to be controlled by a simple serial timing structure. This last finding is consistent with those reported by Vorberg and Hambuch.[2] Analyzing response sequences from experimental conditions where subjects were required to tap evenly with one hand while grouping their responses by two, three, or four, we found no evidence supporting hierarchical timing structures, whereas the data were in agreement with the predictions of a serial timing structure.

The results of the cycle variance test rule out serial timing mechanisms for conditions 2–4–2, 2–1–3–2, and 3–1–2–2; however, they do not tell us much about the particular structure of the inferred timing mechanisms. Unfortunately, none of the strong hierarchical models that we have proposed[2] with timing structures composed of independent random variables is consistent with the data, as a cursory examination of the timer covariances reveals. TABLE 1 shows the $\widehat{\text{cov}}(T_{i,m}, T_{j,m+k})$, estimated from the averaged cross-covariances, $[\widehat{\text{cov}}(I_{i,m}, J_{j \cdot m+k}) + \widehat{\text{cov}}(I_{j,m+k}, J_{i,m})]/2$. For each condition, the estimated timer covariances are given for intervals no more than $c - 1$ steps apart.

Hierarchical models with independent timer random variables have problems with these data since they cannot account for positive covariances between any pair of intervals, $T_n = T_{i,m}$ and $T_q = T_{j,m+k}$, unless $n = q$. It is easy to show that, in general, cov $(T_n, T_q) \leq 0$, $n \neq q$, for all models of the class sketched above. However, as TABLE 1 shows, there are sizable positive covariances under each condition in addition to the negative ones that are expected for hierarchical models. The reliability of these findings was assessed by testing the average estimates against the corresponding standard errors. The results clearly show timing structures that are more complex than our present models can account for.

DISCUSSION

The proposed general framework for two-handed tapping has performed rather well in the test of the symmetry prediction as well as in several others reported in Vorberg and Hambuch.[7] We take this as support for the assumptions of a common timing system and of independence between the three subsystems. This conclusion is in agreement with several recent findings bearing on the assumptions.

Wing[6] has reported the application of a related model to two-handed even tapping; his model admits positive covariation between the left-hand and the right-hand motor delays by positing that they share a common component, C'_m, such that $L_m = C'_m + L'_m$, and $R_m = C'_m + R'_m$, where C'_m, L'_m, and R'_m are assumed to be independent. The model implies that cov $(L_m, R_m) = \text{var}(C'_m)$. However, Wing's analysis showed that the contribution of the postulated common component was negligible, that is, $\text{var}(C'_m) \approx 0$, implying independence between the left-hand and the right-hand motor delays.

Shaffer[3] has analyzed actual piano performance in terms of a model that is identical to ours except that the random variables are assumed to be independent within systems as well. Several of the model's predictions could not be supported; however, Shaffer concluded that the model fits the data well if the motor delays within hands are allowed to covary. Both this as well as Wing's result can be regarded as additional support for the weaker general model which requires independence between systems only.

The interpretation is valid if the two hands are controlled by the same timing system for which we do not have direct evidence. However, in view of the findings of Kelso *et al.*[4] that there is a strong tendency for the two hands to start and stop movements in synchrony even when they have to perform different movements, the common timing system assumption seems reasonable under the conditions of our experiments. This does not mean that separate timing systems for the two hands might not be involved in more complex rhythmic performance as in polymetric and polyrhythmic music, or in rubato passages in piano playing where one hand moves systematically in and out of the meter provided by the other hand.[3,11] The large amounts of practice most musicians require to achieve such levels of proficiency, however, can be taken as evidence for a single timing system under conditions where this is sufficient.

The model provides an easy way to assess the properties of the timing system directly via the cross-covariances of the IRIs between hands. This is the major advantage of the approach we advocate. In contrast to previous work on one-handed performance, our conclusions about the timing system's structure do not depend on particular assumptions about the motor delays (for example, equal variances, independence within hands). Nevertheless, our findings agree well with our earlier work[2]; both studies found no evidence in support of hierarchical structures for isochronous rhythms.

Earlier,[2] we speculated about the advantages of serial as compared to hierarchical timing structures. We pointed out that hierarchical timing necessitates control mechanisms that prevent the order of responses set up by the motor program from scrambling during its execution, whereas serial structures, directly timing the intervals between successive commands, do not face this problem. A related point is that hierarchical structures with concurrent timers will have to control longer intervals than a serial structure appropriate for the rhythm at hand. This has implications for the temporal precision with which a serial structure can control the duration of a cycle as compared to that of hierarchical ones: We might expect serial timing to be able to keep the variance of any of the cycle durations smaller than hierarchically organized timers can. This expectation is borne out in the data. The cycle variances under the isochronous condition are only about half as large as those of the remaining conditions which were concluded to be controlled hierarchically. Thus, the qualitative differences that we observed between isochronous rhythms as compared to rhythms composed of unequal intervals are paralleled by a quantitative difference in timing precision, corroborating our conclusion that rhythmic performance seems to be timed by some kind of hierarchical structure unless its notes are evenly spaced in time.

The most interesting question remains: What are the particular structures that underlie performance under conditions 2–4–2, 2–1–3–2, and 3–1–2–2? The covariance estimates (TABLE 1) seem to provide some clues. However, we must admit that we have been unable so far to extract meaningful patterns from these data. What is needed are models of timing structures that can account for positive dependencies in the timer intervals as well as for negative ones. At present, we are working on models that generalize the class of hierarchical models sketched above. We assume that each timer in a hierarchy uses information about the size of the interval to be generated which is provided by the timer above it in the hierarchy, and passes it along to those timers below that it dominates. In this way, positive dependence is introduced between the different timer random variables; its extent depends on the variability of the information used to set the timer parameters. Whether such models can give a satisfactory account of our findings remains to be seen. Before fitting them to data, it seems necessary to establish the generality of our present results by extending our analyses to situations which involve different tempos as well as more complex movement sequences. Research along these lines is currently under way.

ACKNOWLEDGMENTS

We would like to thank Barbara Hansen for running the experiments and Uwe Mortensen for a critical reading of a first draft of the paper.

REFERENCES

1. WING, A. M. & A. B. KRISTOFFERSON. 1973. Response delays and the timing of discrete motor responses. Percept. Psychophys. **14:** 5–12.
2. VORBERG, D. & R. HAMBUCH. 1978. On the temporal control of rhythmic performance. *In* Attention and Performance: VII. J. Requin, Ed. Erlbaum. Hillsdale, NJ
3. SHAFFER, L. H. 1981. Performances of Chopin, Bach, and Bartok: Studies in motor programming. Cognit. Psychol. **13:** 326–376.
4. KELSO, J. A. S., D. L. SOUTHARD & D. GOODMAN. 1979. On the coordination of two-handed movements. J. Exp. Psychol. Hum. Percept. Perform. **5:** 229–238.

5. KLAPP, S. T. 1979. Doing two things at once: The role of temporal compatibility. Mem. Cognit. **7:** 375–381.
6. WING, A. M. 1982. Timing and co-ordination of repetitive bimanual movements. Q. J. Exp. Psychol. **34A:** 339–348.
7. VORBERG, D. & R. HAMBUCH. Test of a model for two-handed rhythmic performance. Paper presented at the Symposium "Psychology of Motor Behavior," Dortmund, Germany, 1983.
8. WING, A. M. 1977. Effects of type of movement on the temporal precision of response sequences. Br. J. Math. Stat. Psychol.: **30:** 60–72.
9. WING, A. M. 1980. The long and the short of timing in response sequences. *In* Tutorials in Motor Behavior. G. E. Stelmach & J. Requin, Eds. North-Holland, Amsterdam.
10. VORBERG, D. 1978. Problems in the study of response timing. Comments on Reece's [1976] "A model of temporal tracking." Acta Psychol. **42:** 67–77.
11. SHAFFER, L. H. 1984. Timing in musical performance. This volume.

Processing of Rhythmic Sound Structures by Birds[a]

STEWART H. HULSE,[b] JOHN HUMPAL, AND JEFFREY CYNX

Department of Psychology
The Johns Hopkins University
Baltimore, Maryland 21218

INTRODUCTION

This article reports some observations on a mechanism fundamental for the adaptive behavior of both humans and animals, namely, the perception and processing of temporally organized acoustic information. For people this capacity assures that speech will be produced and understood, and that the rhythms of music will be heard and appreciated. For animals, vocalization and other forms of time-organized sound production and reception play an important role in social interaction and communication. This is true for the temporally structured sound patterns produced by crickets, and it is equally true for the complex, meaningful acoustic signals represented by the calls of monkeys and rats, and the songs of birds and whales.

Goals

Two goals guided the work, and we outline them here.

1. Steps toward a functional unit of analysis. There is, at present, no *a priori,* well-developed, psychophysically oriented technique for a meaningful analysis of naturally occurring acoustic signals. We sought an approach that, if successful, promised some useful tools with which to undertake such an analysis in the future. In particular, we designed certain acoustic stimulus patterns based on principles of rhythm and pitch perception and, with operant techniques, tested birds' capacity to discriminate between the patterns. We then distorted the patterns with certain preplanned transformations and tested for the birds' capacity to maintain the discrimination. Transformations that disrupted discrimination were assumed to tap perceptual processes that were significant for discrimination and acoustic information processing. Those that failed to disrupt discrimination, or did so to a lesser degree, were assumed to be of relatively less functional significance for the discrimination task. While the stimulus structures were utterly arbitrary with respect to their occurrence in nature, they had the virtue of a clearly specified formal organization based on well-known, physically defined acoustic dimensions. The presumption was, and remains, that there should be some generalization from principles discovered from discrimination tasks based on such complex acoustic patterns to principles governing the organization of sound patterns used for communication in nature.

[a]This research was supported by Research Grant BNS-801437 from the National Science Foundation.

[b]To whom correspondence should be addressed.

While there has been much research on the structural properties of acoustic communication in animals, as any student of birdsong, for example, knows quite well, research has generated relatively little progress in detailing the functionally significant units of the acoustic signal for the species in question. Thus, we have developed a unit of acoustic analysis, the phoneme, for human speech, but we have yet to uncover a comparable set of units for the myriad forms of animal communication for which we know functional significance exists. In the case of birdsong, for example, the sonogram provides a representation of the distribution of sound energy according to frequency and time, and there are schemes for dividing such representations into "syllables," "phrases," and other arbitrary "units" (see, for example, Greenewalt,[5] Heinz *et al.*[6] and Thorpe[25]). But, thus far, such schemes have been most productively used to describe rather gross characteristics of acoustic signals, such as their complexity (as measured, say, by number of phrases and syllables). There has been relatively less success in ascribing functional significance—semantics, if you will—to such putative units of analysis.

This is not to say that nothing at all is known about the functional significance for animals of the components of acoustic signals. That would be far from the case—for birdsong at the very least. As a case in point, Peters *et al.*[17] have studied the reaction of birds to various arbitrarily distorted forms of natural song and identified certain distortions which do or do not appear to be of functional behavioral significance. Similarly, West *et al.*[26] have studied the functionally significant components of natural birdsong that are relevant for mating behavior. Two things handicap such work, however. First, the alterations in the natural signal are arbitrary (because there is, to date, no well-defined basis for breaking the signal into meaningful pieces); and, second, functional significance is assessed primarily by searching for gross changes in some broad, naturally occurring behavioral repertoire as a function of distortions in the normally adequate stimulus (for example, changes in female sexual receptivity correlated with distortions in male song). With some important exceptions,[6,23] the powerful psychophysical methods that are available for analyzing the perceptual features of complex acoustic discrimination have not been brought to bear on these issues.

2. Information processing and cognition. Because the stimulus patterns were complex, formally organized, and based on principles of pitch and rhythm perception, our second goal was to learn something about birds' cognitive capacity to perceive, learn, and remember (that is, to process) such complex information. In particular, our goal was to acquire new information about birds' ability to process acoustic information organized in *serial patterns.* The capacity for serial organization of stimuli and responses is fundamental to the behavior of all species.[18] Recently, there has been a surge of interest in studying the cognitive capacity of animals,[10,20] and our data should add to the storehouse of relevant information.

Rhythm Structures

The design of stimulus patterns used in the research was based, first of all, on the remarks of Lashley[18] that emphasized the hierarchical, rule-based properties associated with serially arranged patterns of stimuli, such as those in human language. Lashley's emphasis stood in contrast to the simple, linear stimulus–stimulus or stimulus–response associative chains that were serving as the primary model for serially organized behavior at that time. His work stimulated the development of general, rule-based theories of human serial pattern learning[16,19,22]—theories that have been applied to the analysis not only of the human serial pattern learning per se, but also to human music perception and other forms of serial information processing.

Recently, too, there have been successful applications of such models of serial information processing to animal learning and behavior.[8,9,11]

The idea of a serial pattern grows from the concept of a stimulus *alphabet*.[12,16] An alphabet is a set of stimuli in which the elements of the set are (1) discriminable from one another, and (2) ordered according to the properties of, at a minimum, an ordinal scale. That is, operations of "less than," "equals," and "greater than" can be meaningfully applied to the elements of the set. Other alphabets may include the properties of equal-interval or ratio scales, in which case operations of addition, subtraction, multiplication, and division are meaningfully applied. A *serial pattern* of stimuli is developed by initializing the pattern at some point in the alphabet, and then applying *rules* to the alphabet to generate successive pattern elements. Using the Roman alphabet (which becomes an alphabet in the present sense by the time one has memorized the arbitrary order of the letters in early education), one can generate a pattern such as ABCDBCDECDEF by initializing at the letter A, using a "next" or "+1" rule to generate successively the next three letters, then applying a "next" rule to the entire initial *subset* of elements to generate the next subset, BCDE, and so on. This procedure produces a pattern with a linear surface structure of letters based, however, on an underlying hierarchical rule structure.

The same principles can be used in the composition of pitch and rhythm structures. For example, imagine an alphabet of temporal durations defined, in musical notation, by a 32nd note, a 16th note, a quarter note, a half note, and a whole note. Given the duration of the 32nd note to be 100 msec, the temporal duration of each successive note in the alphabet doubles that of its predecessor. A serial pattern, a rhythmic structure in this case, can be built from such an ordered series of temporal durations by initializing at, say, the 32nd note, using a "repeat" rule to generate another 32nd note, a "next" rule to generate a 16th note, a "next" rule to generate a quarter note, and so on.

The sound patterns used in the research reported in this article were derived according to the foregoing pattern-generating procedures. A large body of research has been accomplished using rule-based pitch pattern discrimination, and that is described elsewhere.[13,14] This article addresses research involving discriminations and transformations of rule-based rhythmic sound structures.

Technical Strategies for a Comparative Analysis

In human perception, rhythmic structures are not only discriminable, of course, but they remain perceptually constant when their tempo changes. That is, within very broad limits, the rhythmic structure associated with a familiar tune remains perceptually invariant as the tune's speed changes—the rhythm of "Yankee Doodle" sounds the same whether we hear the tune played fast or slow. This implies that people respond to *relational* invariances in rhythmic perception. That is so because the only feature of a rhythmic structure that remains invariant across tempo changes is the duration *ratio* of neighboring temporal intervals in the structure.[3,24]

The initial strategy adopted for the research, then, was the assumption that birds' perception of rhythmic structures under tempo changes involving constant temporal ratios would parallel human perception. That is, like people, birds would maintain a discrimination between two rhythmic structures that changed in tempo as long as the relational properties of their component temporal units remained constant. This is a strong working hypothesis, especially for animals outside the primates. It assumes that birds can (1) discriminate among rhythmic structures (that is, that rhythm is, in fact, a stimulus attribute that birds can process) and (2) respond to and generalize *relationally* across features of temporal structure.

A second strategy underlying the research guided the construction of rhythmic stimulus patterns. One could, in principle, begin research in this domain with "simple" patterns constructed, say, from one or two temporal durations presented in sequence just once, and ask animals to base a choice response on a discrimination between two such simple patterns. By "simplifying" discriminative stimuli in this fashion, however, one may, in fact, make them difficult to discriminate because they are grossly impoverished as a structured *pattern* of temporal intervals. Accordingly, our strategy was to *enrich* rhythmic stimuli by formally configuring and repeating them so as to highlight the temporal differences distinguishing them.

A third strategy stemmed from the fact that almost all research on the functional significance of birdsong has been based on studies of sound *production.* With some important exceptions identified earlier,[21,26] relatively little has been done testing directly how birds process the information they *hear* or, as we have seen, searching for the discrete features of natural acoustic stimuli to which they attend. Accordingly, we adopted the *reception* technique alluded to earlier in which birds perceived and responded to arbitrary, experimenter-generated stimulus patterns. The technique afforded great control over the stimulus patterns and their attributes, and the modes through which the birds told us what they heard.

Finally, our choice of experimental subject—the European starling (*Sturnus vulgaris*)—was guided by the need for *a priori* evidence of an ability to process arbitrary sound patterns. A mimicking species of bird, of which the starling is an excellent example, met this need very nicely, on the reasonable assumption that mimicking of both natural and nonnatural sounds provides evidence that such sounds are heard, learned, and remembered.

Problems for Analysis

The initial problem selected for analysis was to determine whether birds could discriminate a formally structured rhythmic pattern from a random, unstructured, arrhythmic pattern of temporal intervals. Given that the discrimination could be learned, the next step was to see how it would fare when the patterns were speeded or slowed in tempo. On the basis of human perception, the discrimination should occur. Furthermore, it should be maintained with tempo changes as long as a constant ratio holds between the duration of successive temporal units in the patterns.

Given that the starlings' performance gave sensible solutions to the first problem, the second was to test for perception and discrimination between two stimulus patterns that were both formally structured rhythms. In this case again, the discrimination should be maintained with temporal transformations that maintain a constant ratio between successive temporal units in the patterns.

RHYTHMIC-ARRHYTHMIC DISCRIMINATION

Stimuli

Three stimulus patterns were used to analyze the starlings' ability to process rhythmic and arrhythmic sound structures. All patterns were constructed from 2000-Hz sinusoids at a sound pressure level of approximately 70 dB measured near the bird's head in the apparatus. For two rhythmic patterns, a temporal interval of 100 msec served as a basic structural unit. For the *linear* rhythmic pattern, 100-msec units of tone alternated with 100-msec units of tone-off to produce a pattern of tones and intertone intervals generated by a simple "alternate" rule. On any trial, as described below, the pattern was presented for at least 4 sec; its total duration was determined by other programmed contingencies described below.

For the second rhythmic pattern a hierarchical rule structure was adopted. Here, a 100-msec tone alternated with a 100-msec intertone interval four times to form one subpattern. Successive four-tone subpatterns were separated by long, 800-msec interpattern intervals. The formal structure of the overall pattern thus consisted of four-tone subpatterns and 800-msec interpattern intervals joined by a simple "alternate" rule. Again, on any trial, the hierarchical pattern was presented for a minimum of 4 sec with a total duration determined by other contingencies.

The random, arrhythmic pattern consisted of a continuing series of alternating tone and intertone intervals whose durations were drawn at random by the computer from a list of durations ranging from 30 to 300 msec.

Experimental Techniques

Four food-deprived starlings were trained to discriminate the hierarchical rhythmic pattern from the arrhythmic pattern, while two were trained to discriminate the linear from the arrhythmic pattern. All birds were trained in a test cage located in a sound-proofed chamber. Three translucent keys that the birds could peck were located in a horizontal row on one wall of the cage, and a food hopper was located just below the center key. A loudspeaker through which sound patterns were delivered was located directly over and facing down into the test cage. Two small lights in the ceiling of the test cage served as "houselights" to illuminate the test cage, while ambient light was provided by a lamp located on one wall of the sound-proofed chamber. All stimuli were generated by a PDP8E computer driving a programmable oscillator. The same computer controlled experimental contingencies and recorded data for later analysis.

The birds first learned to peck the keys for food reward. After this initial training, discrimination training began. The computer began a discrimination trial by lighting the center key. A peck on this key turned off the center key light, and turned on either the rhythmic or the arrhythmic sound pattern (with $p = 0.5$) for a 4-sec "listening" period. During this period, no peck on any key had programmed consequences. Following the listening period, the two side keys lighted while the sound pattern continued. At this time, a "correct" peck on the right key if, say, the rhythmic pattern was present, turned off the sound pattern and generated 1.5-sec access to the food hopper. An "incorrect" peck on the left key, however, turned off the sound pattern and darkened the overhead houselights for a 10-sec "time-out" period. This served to delay the beginning of the 5-sec intertrial interval and the start of the next trial. The reverse contingencies held for the arrhythmic pattern, this is, a peck on the left key was correct, while a peck on the right key was incorrect. A daily session of discrimination training lasted for 1 hour during which the birds generally produced 100 to 150 trials.

Baseline Discrimination Performance

Discrimination training lasted for different birds for 117 to 119 days at which time the birds responded stably at 81 to 96% correct. There was no indication of any difference in rhythmic–arrhythmic discrimination performance for the linear as compared with the hierarchical rhythmic patterns. At this time, testing with tempo transpositions began.

Ratio Transpositions

For *ratio* tempo transpositions, the computer multiplied the tone and intertone durations used in baseline training by one of six constants designed to span a range doubling the tempo (halving tone and intertone durations) on the one hand, and halving the tempo (doubling tone and intertone durations) on the other. Intermediate multipliers were chosen to divide the tempo span into equal intervals on a log scale. Thus, the multipliers used in ratio tempo transpositions were 0.5, 0.6, 0.8, 1.0, 1.3, 1.6, and 2.0, with 1.0 representing the "multiplier" for the baseline discrimination.

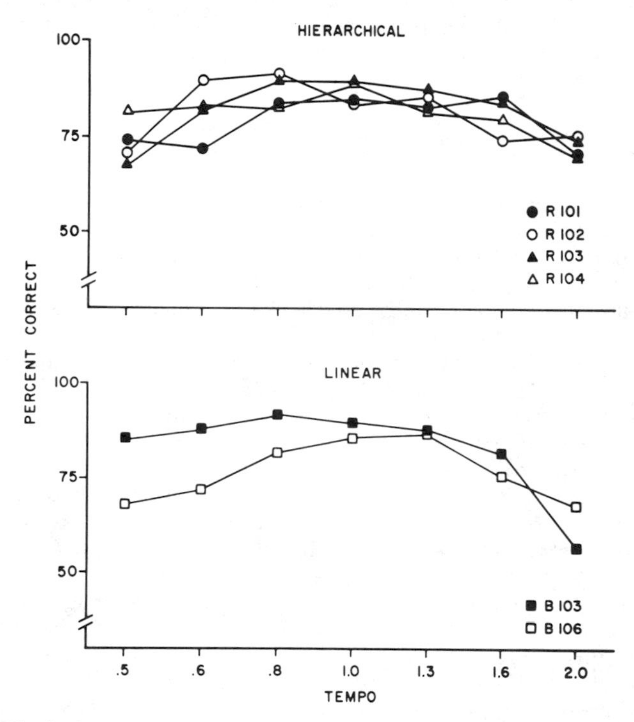

FIGURE 1. Rhythmic–arrhythmic discrimination performance for the hierarchical (*top panel*) and linear (*bottom panel*) rhythmic patterns when the tempo of the patterns was changed by a ratio transformation of tone and intertone intervals. The baseline tempo was 1.0. The 2.0 tempo halved, while the 0.5 tempo doubled the baseline tempo. Intervening tempos are arranged at equal intervals on a log scale.

Testing procedures on any given day during the tempo transfers were identical to those prevailing during original discrimination training, except, of course, for the changes in the stimulus patterns. A given test tempo was in effect for an entire daily session, and different orders of testing were used for each bird. Each test session was followed by a session on baseline conditions (a multiplier of 1.0); test and baseline sessions alternated until two observations had been made at each test tempo.

The results of the ratio transpositions are shown in FIGURE 1. Quite apparently, there was very little loss by any bird in discrimination performance relative to baseline except at the slowest tempos (that is, multipliers of 1.6 and 2.0). This was true for both the hierarchical and linear patterns, for which no reliable differences appeared in the data. These flat, horizontal generalization gradients for the tempo dimension are to be contrasted with generalization gradients for virtually any other stimulus dimension (for example, sound frequency and loudness, light wavelength and brightness) on which animals have been tested.[7] Typically, performance falls steeply on each side of the stimulus value used in baseline training.

The birds' relative loss of discrimination performance at slow tempos also finds parallels in human perception,[3,4] although the loss may have occurred because at slow tempos the birds received less information about pattern structure during the 4-sec listening period on each trial.

The birds performance on the transfers indicates a remarkable stability of the discrimination under a ratio transformation of temporal structure. The most straightforward conclusion from these data is that, like humans, starlings process temporal acoustic information such that perceptual constancy holds for changes in tempo correlated with changes in tone–intertone ratios.

Additive Transfers

Converging evidence for the foregoing conclusion would be provided by changes in tempo produced by other, *nonratio* forms of temporal transposition of the rhythmic and arrhythmic stimulus patterns. In particular, perceptual constancy is presumed to hold for humans only for ratio transfers, so an *additive* transformation in which, say, tone duration changed from session to session while intertone duration remained constant ought to lead to a deterioration in discrimination performance.

To test this proposition, three birds trained on the rhythmic–arrhythmic discrimination with the linear rhythmic pattern were transferred to transformed patterns in which either tone duration remained constant at 100 msec while baseline intertone durations varied from session to session by multiplying the baseline, 100-msec value by constants of 0.5, 0.6, 1.6, or 2.0., or intertone duration remained constant at 100 msec while the baseline 100-msec tone duration varied after the same fashion. Test sessions alternated with sessions on baseline conditions. Test conditions were counterbalanced across birds, and were presented to any given bird in a haphazard order.

Performance on the additive transfers was remarkably good and, for the most part, well above chance. FIGURE 2 shows for a representative bird that performance remained quite stable at roughly 80% correct when tone duration was held constant while intertone interval varied. Performance was also well above chance when tone duration varied while intertone interval was held constant, although performance was poor at the extreme slow tempo (multiplier of 2.0). For that condition, performance was at chance levels and significantly poorer ($p < .01$) than performance when intertone interval varied.

FIGURE 2 also gives representative data obtained when the birds were returned to a ratio transformation just after the data for the additive transformations had been obtained. The figure indicates that, at this time, performance was (1) quite similar to that obtained for original ratio transformations, and (2) much more characteristic of performance obtained in the additive transfers when tone duration as compared with intertone interval was varied.

Compared with human perception, the excellent overall performance of the birds under the additive transformations is anomalous—although, to be sure, there are no human data of which we are aware that have been obtained under precisely the same stimulus conditions. Nevertheless, the fact that the birds maintained the discrimination so well under additive conditions—where one would not expect them to do so—suggests that the excellent performance under ratio transformations may not be due to a putative perceptual constancy unique to that condition. Instead, some other characteristic of the rhythmic–arrhythmic stimulus comparison may have assured good discrimination performance under *both* ratio and additive transfers.

One possibility is that the birds in the original baseline discrimination and in both the ratio and additive transfers were responding to a *qualitative* difference between an arrhythmic pattern and a general *class* of rhythmic patterns. While sensitivity to temporal relations in the patterns must have been present for the discrimination to be solved and maintained under transformations at all, perhaps the birds were responding to some general "rhythmicity" attribute of the rhythmic patterns as they learned and transferred the discrimination. *A priori,* rhythmicity might be independent of the relative duration of tone–intertone intervals. Instead, rhythmicity might depend on simple periodic recurrence of pattern components, constrained only by the requirement

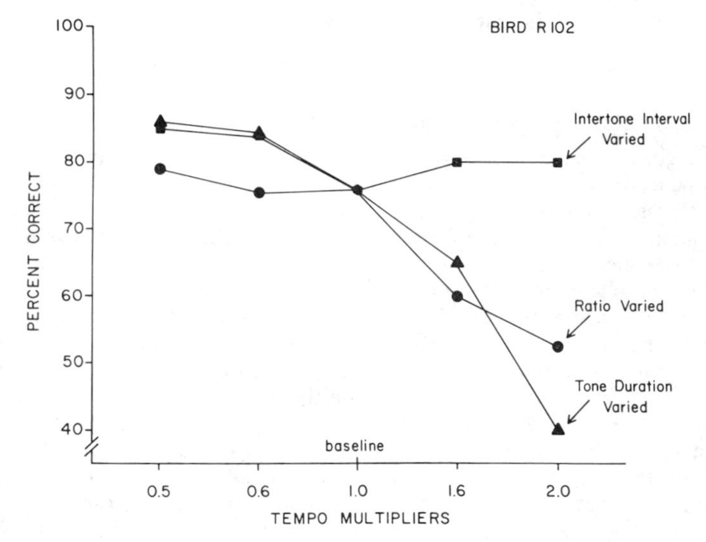

FIGURE 2. Discrimination performance for one bird under additive and ratio transformations of the baseline rhythmic–arrhythmic discrimination. Tone duration was held constant while intertone interval varied (in different sessions) according to the multipliers listed on the abscissa, or intertone interval was held constant while tone duration varied. Data also appear for tempo changes based on ratio transformations of tone duration and intertone interval.

that they be of a fixed temporal duration within any given transfer. Tests in which, say, pattern components such as intertone intervals varied randomly while tone durations remained constant would move rhythmic patterns qualitatively toward arrhythmic patterns. The result ought to be a decrease in discriminability between the class of rhythmic and the class of arrhythmic patterns. Such a result has been obtained and is reported elsewhere in this volume.[15]

RHYTHMIC DISCRIMINATION

If simple classification of rhythmic and arrhythmic patterns on a qualitative basis accounts for the rhythmic–arrhythmic discrimination just described, that would be less likely the case for a discrimination between two temporal patterns that both possessed a formally defined temporal structure. As we have just defined rhythmicity, two rhythmic structures would be perceptually equivalent in terms of their simple rhythmicity. If, therefore, a discrimination between two rhythmic patterns were to occur and maintain itself across tempo transformations, there would be definitive evidence that starlings could solve temporal pattern discriminations not only on the basis of rhythmicity, but also, in fact, on the basis of differences in formal tone–intertone temporal structure. Accordingly, a second experiment tested for starlings' ability to discriminate between two rhythmic patterns and to maintain the discrimination under a ratio transformation of tone and intertone intervals.

Stimuli

Two hierarchically organized rhythmic patterns were constructed from 2000-Hz tones. Each contained a sequence of 4 tones (subpattern 1) separated by a long intersequence interval of 700 msec (subpattern 2). The patterns differed in the structure of subpattern 1. For one pattern, subpattern 1 consisted of four tones whose durations were, respectively, 50–50–300–300 msec. For subpattern 2, the durations were 50–300–300–50 msec. The intertone durations within subpattern 1 in each case were 50 msec. Through a programming error, a third pattern was produced in which subpattern 1 consisted of tones whose durations were 50–50–300–300 msec, but where the intertone duration between the first two 50-msec tones was 300 msec instead of 50 msec. The birds began the experiment on this pattern, and rather than change conditions in midstream when the error was discovered, the experiment was completed with the unplanned stimulus structure. A subsequent replication of the experiment with the intended stimulus pattern gave virtually identical results.

Experimental Techniques

Two starlings were trained with the identical apparatus and procedure used in the rhythmic–arrhythmic discrimination, except that daily sessions were ½ hour instead of 1 hour in duration. Also, the birds have been tested only on the ratio transfer at this time, so we report data only for that condition.

Baseline Discrimination Performance

Original baseline discrimination training lasted for 60 and 71 days, respectively, at which time, the two birds were averaging 74% and 88% correct responses over a successive 3-day period. Both figures are well above chance ($p < .01$). Thus, the birds learned the baseline rhythmic discrimination very well.

Ratio Transposition

At this time, a ratio transposition was undertaken that was like that for the rhythmic–arrhythmic discriminations in all respects, save one. The computer could not reliably generate intervals sufficiently short to test for transposition at a tempo that halved the baseline tempo (that is, where duration of the shortest components of the patterns was 15 msec). Therefore, the smallest multiplier used in the transfers was 0.6

instead of 0.5. As before, sessions on the baseline tempo were interposed between sessions on one of the transposed tempos. The latter were sampled haphazardly from day to day.

FIGURE 3 shows performance on the first and last 10 trials of the first day of transfer at each tempo. For the first 10 trials, one bird, G221, maintained the discrimination at the baseline tempo and at the 1.6 tempo, but performed at chance for the remaining tempos. The other bird, R104, performed well above chance at all tempos. Both birds showed above-chance performance over the last 10 trials of the first transposition test. In all cases, there was an anomalous fall in performance at the 1.3 tempo as compared with both the baseline, 1.0 tempo, and the slower, 1.6 tempo. There was no identifiable reason for this loss in performance.

The birds learned the rhythmic discrimination, and they maintained discrimination performance under a ratio transposition of the two rhythmic patterns. The generaliza-

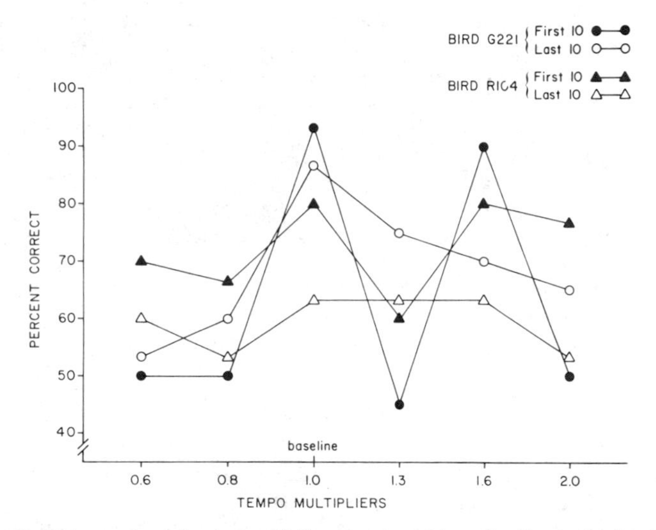

FIGURE 3. Ratio tempo transformations following acquisition of a discrimination between two rhythmic patterns. Data are percent correct for the first and last ten trials of the first test session at each tempo. Bird R104 is consistently above chance; bird G221 is above chance by the end of the first test session. The anomalous drop in performance at the 1.3 multiplier is nevertheless consistent across birds and trials.

tion gradients were not as flat for the rhythmic discrimination as they were for the rhythmic–arrhythmic discrimination, that is, discrimination performance was not maintained as well, but performance was, in general, well above chance over a range of tempos that almost doubled and halved the original baseline tempo.

GENERAL DISCUSSION

Several major conclusions emerge from the data. First and foremost, the research shows in detail for the first time that at least one avian species, the European starling, can distinguish and otherwise process formally organized temporal sound structures. This conclusion stems from several sources of converging evidence.

At the simplest level, the birds were able to learn a discrimination based on two forms of temporally organized stimuli: a rhythmic as compared with an arrhythmic sound pattern, and a rhythmic as compared with another rhythmic sound pattern. Besides the examples provided in this book, there has been a good deal of research on animals' ability to process temporal information, such as temporal duration per se.[1,2] However, this is apparently the first time for animals that a psychophysical analysis has been provided for the functionally significant components in a discrimination between structured *patterns* of temporal stimuli.

At a more complex level, the birds could maintain the discrimination, that is, show some form of perceptual constancy, when the patterns of the original discrimination were changed through ratio and additive transformations. Taken together, the data for the rhythmic–arrhythmic discrimination suggest that the perceptual constancy was nominal; that is, the birds were simply classifying the stimuli as rhythmic or not, and the ability to classify was affected primarily by transformations that modulated the underlying qualitative property of "rhythmicity." While formal definitions of "rhythmicity" may be premature, the data suggest that a workable definition would have to require that pattern components (a) recur periodically and (b) be of fixed duration within any given set of temporal conditions (for example, for any given tempo). Beyond those conditions, absolute values of tone durations and intertone intervals can apparently vary considerably from one tempo to another and still produce good discrimination. They cannot vary within a given tempo, however, because, as Humpal and Cynx[15] have shown, the introduction of tone durations that vary randomly while intertone intervals remain constant for a given tempo multiplier (or vice versa) reduces discrimination accuracy markedly.

While the principle of rhythmicity may account for the rhythmic–arrhythmic discrimination, the principle is insufficient to account for the discrimination incorporating two rhythmic structures. In the latter case, the original discrimination depended necessarily on a discrimination between two sets of specific, fixed-tone durations and intertone intervals. Furthermore, while the specific distinguishing features of the two patterns to which the birds attended remain to be determined, the critical features must have been *relational* in at least some form. That is, they must have depended upon the *relative* durations of some set of successive tone and intertone intervals. This is so because under ratio transformations relative durations remained constant—on an ordinal scale at least—while all the *absolute* temporal values changed. Further work with additive or other transformations will no doubt place boundaries on the stimulus alterations yielding relative perceptual constancy.

The fact that birds, starlings at least, are able to process rhythmic information suggests not only an interesting comparison with human rhythmic information processing, but also, as we have noted earlier, some potential applications for an understanding of natural birdsong. There is little question that the temporal structure of birdsong is functionally important in nature. There is some question, however, about the specific aspects of temporal structure that are functionally significant.[23,27] Our work suggests that an analysis of rhythmic structures and their transformations could provide useful information in this domain for an understanding not only of birdsong, but also of other forms of acoustic communication among animals.

ACKNOWLEDGMENTS

Many people contributed to the research, among them E. Burroughs, C. Crowder, R. Feuerstein, S. Fisher, J. Freitag, P. Garlinghouse, S. Gucker, C. Kwon, J. Lakind, L.

Rowe, S. V. Hulse, R. Morrison, and T. Park. We thank them all. We also thank W. T. Green, R. Wurster, and J. Yingling for important technical assistance. Special appreciation goes to Dr. C. Grue of the Patuxent Wildlife Research Center for providing experimental subjects.

REFERENCES

1. CHURCH, R. M. 1978. The internal clock. *In* Cognitive Processes in Animal Behavior. S. H. Hulse, H. Fowler & W. K. Honig, Eds.: 277–310. Erlbaum. Hillsdale, NJ.
2. CHURCH, R. M. & J. GIBBON. 1982. Temporal generalization. J. Exp. Psychol. **8:** 165–186.
3. FRAISSE, P. 1982. Rhythm and tempo. *In* The Psychology of Music. D. Deutsch, Ed. Academic Press. New York, NY.
4. GARNER, W. R. & R. L. GOTTWALD. 1968. The perception and learning of temporal patterns. Q. J. Exp. Psychol. **20:** 97–109.
5. GREENEWALT, C. H. 1968. Bird Song: Acoustics and Physiology. Smithsonian Institution Press. Washington, DC.
6. HEINZ, R., J. SINNOTT & M. SACHS. 1977. Auditory sensitivity of the redwing blackbird and the brownheaded cowbird. J. Comp. Physiol. Psychol. **94:** 993–1002.
7. HONIG, W. K. & P. URCUIOLI. 1981. The legacy of Guttman and Kalish (1956): Twenty-five years of research on stimulus generalization. J. Exp. Anal. Behav. **36:** 405–445.
8. HULSE, S. H. 1978. Cognitive structure and serial pattern learning by animals. *In* Cognitive Processes in Animal Behavior. S. H. Hulse, H. Fowler & W. K. Honig, Eds.: 311–340. Erlbaum. Hillsdale, NJ.
9. HULSE, S. H. & N. P. DORSKY. 1977. Structural complexity as a determinant of serial pattern learning. Learn. Motiv. **8:** 488-506.
10. HULSE. S. H., H. FOWLER & W. K. HONIG. 1978. Cognitive Processes in Animal Behavior. Erlbaum. Hillsdale, NJ.
11. HULSE, S. H. & N. P. DORSKY. 1979. Serial pattern learning by rats: Transfer of a formally defined stimulus relationship and the significance of nonreinforcement. Anim. Learn. Behav. **7:** 211–220.
12. HULSE, S. H. & D. O'LEARY. 1982. Serial pattern learning: Teaching an alphabet to rats. J. Exp. Psychol. Anim. Behav. Processes **8:** 260–273.
13. HULSE, S. H., J. CYNX & J. HUMPAL. J. Exp. Psychol. Gen. In press.
14. HULSE, S. H., J. CYNX, J. HUMPAL. *In* Animal Cognition. H. Roitblat, T. Bever & H. S. Terrace, Eds. Erlbaum. Hillsdale, NJ. In press.
15. HUMPAL, J. & J. CYNX. 1984. Discrimination of temporal components of acoustic patterns by birds. This volume.
16. JONES, M. R. 1978. Auditory patterns: The perceiving organism. *In* Handbook of Perception. E. C. Carterette & M. P. Friedman, Eds. Vol. **8:** 255–288. Academic Press. New York, NY.
17. PETERS, S. S., W. A. SEARCY & P. MARLER. 1980. Species song discrimination in choice experiments with territorial male swamp and song sparrows. Anim. Behav. **28:** 393–404.
18. LASHLEY, K. 1951. The problem of serial order in behavior. *In* Cerebral Mechanisms in Behavior. L. A. Jeffress, Ed. Wiley. New York, NY.
19. RESTLE, F. 1970. Theory of serial pattern learning: Structural trees. Psychol. Rev. **77:** 481–495.
20. ROITBLAT, H., T. BEVER, & H. S. TERRACE. Animal Cognition. Erlbaum. Hillsdale, NJ. In press.
21. SEARCY, W. A., P. MARLER & S. S. PETERS. 1981. Species song discrimination in adult female song and swamp sparrows. Anim. Behav. **29:** 997–1003.
22. SIMON, H. A., & K. KOTOVSKY. 1963. Human acquisition of concepts for sequential patterns. Psychol. Rev. **70:** 534–546.
23. SINNOTT, J., M. SACHS & R. HEINZ. 1980. Aspects of frequency discrimination in passerine birds and pigeons. J. Comp. Physiol. Psychol. **84:** 401–415.

24. STERNBERG, S., R. L. KNOLL & P. ZUKOFSKY. 1982. Timing by skilled musicians. *In* The Psychology of Music. D. Deutsch, Ed. Academic Press. New York, NY.
25. THORPE, W. 1961. Birdsong. Cambridge University Press. Cambridge, England.
26. WEST, M. J., A. P. KING, D. H. EASTZER & J. E. R. STADDON. 1979. A bioassay of isolate cowbird song. J. Comp. Physiol. Psychol. **93:** 124–133.
27. WOLFFGRAMM, J. & D. TODT. 1982. Pattern and time specificity in vocal responses of blackbirds *Turdus merula* L. Behavior **81**(Parts 2–4): 264–285.

Timing in Musical Performance[a]

HENRY SHAFFER

Department of Psychology
University of Exeter
Devon EX4 4QJ, England

Music exacts a strong discipline of timing from the performer, and it does this in a rather subtle way. It is constructed on an abstract periodic element, the beat, which has a meter that organizes the beats into recurrent groups, called bars or measures, and assigns accents to certain beats in each group. The durations of notes and rests in the music are given as multiples or divisions of the beat interval. Thus, the rhythms in a piece of music are temporal patterns of sound and silence, with emphasis given to notes falling on accented beats. A recommended speed of playing the music is given in its score by a tempo marking, a beat rate or a relatively vague Italian term, such as andante or allegro. The subtle point is that this formal timing information is given on the unwritten understanding that the player should be willing to distort it to give expressive life to the performance. However, such distortion should not be arbitrary, but rather should aim at revealing the underlying structure and meaning of the music. Different players may give different expressive forms to the same piece of music, and so the requirement is an ambiguous one; nevertheless, one can discriminate appropriate forms from those that are inappropriate or merely rhetorical. A musical training consists largely of learning to express the music of different composers and periods. Povel[1] gives an illustration of the expressive variety among leading exponents of the harpischord in their recordings of the same Bach prelude.

A technically interesting though unmusical way to think of a piece of music in the time domain is that it provides a schedule for all the notes to be played in a performance. Once the player begins to play, he or she is locked into this schedule until a major pause in the music is reached. There is an elasticity in the schedule to allow for expression, but in a public performance the rhythm should be preserved even if wrong notes are played. An extra timing constraint is introduced when musicians play in an ensemble. The idea of music as a schedule can be strongly conveyed in a computer printout of the performance: if it contains a time track and played notes are recorded against this, then comparing these with the score there is an immediate sense of the notes arriving in sequence at their appointed times.

In musical performance, then, we have an opportunity to study some of the most complex forms of timing. The purpose here is to describe some performances and to model the timing mechanisms.

Data on timing in a musical performance can be obtained by carrying out filtering operations on a sound recording so that the attack transients of note onsets can be determined.[1-4] With percussive instruments one can use optical methods to detect the moments at which the striking elements produce sounds. This has been used to study drumming[21] and piano playing.[5,6] The work at Iowa reported by Seashore obtained a record of hammer movements by filming onto a moving strip of calibrated film. We have used small photocells, placed in pairs opposite each hammer to detect its movements. Coded signals from the cells are fed into a computer, where they are

[a]This research was supported by Grant HR 7385/1 from the Social Science Research Council, London, England.

allocated clock times and then stored. Such recording methods do not affect the action or sound of the instrument. Using photocells in pairs we can detect the moments of note onset and offset, and from the hammer transit time between cells estimate the intensity of striking the key and hence of the sound produced.

The early studies of piano playing, reported by Seashore,[5] obtained some basically important results from the recordings of two concert pianists playing Beethoven and Chopin. They demonstrated a free use of rubato, or tempo variation, that was observed at all levels of musical units ranging from notes to phrases. A two-fold variation in bar duration was observed within a short passage of music. More interesting is that in repeat performances of the same piece of music the pattern of rubato was reproduced with remarkable fidelity, agreeing closely in absolute as well as relative timing. If the rubato was expressive, it was used in a principled way and the timekeeping achieved a high degree of precision. When asked to play the piece again metronomically, the pianist produced a more regular timing that contained the earlier pattern of rubato on a much reduced scale. This would suggest that expression rather than factors of motor production were responsible for the rubato; but unfortunately the metronomic performance was also slower.

A clearer account of the role of motor factors is found in a study by Sloboda[7] of the ways in which pianists convey the meter of a piece of music in performance. Some of the options available to the pianist are to play accented notes more forcefully or to affect their timing by playing them early or late, or by prolonging them. Sloboda constructed one-handed melodies that remained musically plausible when the sequence of notes was shifted relative to the bar lines, having the effect of altering the placement of accents. Thus, alternative melodies were produced differing only in meter. The players, not recognizing the alternatives distributed in a set of melodies, spontaneously produced rhythmic differences that reflected phrasing and accenting changes between the two versions. The role of motor factors is excluded because the note sequences were the same.

Michon[3] took advantage of the unique structure of the piece *Vexations,* by Satie, to obtain multiple repeats of a musical performance. The piece has a brief theme and two variations with an instruction to play these 840 times. One of the few ways of maintaining musical interest in the piece is to vary the tempo in the successive repeats. Michon found that the timing of note duration within a repeat was a function of its tempo. The origins of this tempo effect were unclear, however.

Clarke[8] repeated the study, paying more attention to the timing of individual notes within a repeat, and using analysis of variance rather than factor analysis to analyze the data. It was shown that the player tended to phrase the music in groups of notes, producing an allargando, or slowing, at group boundaries. Such an effect is also found in speech.[9] At the faster speeds there was a tendency to combine the notes into fewer groups, affecting the placement of phrase boundaries, and this accounted for the result obtained by Michon. Again, there is an analogy with speech, involving pitch movement, in which the number of tone groups in a sentence tends to decrease at the faster rates of speaking.[10]

Often in a piece of music a theme stated early on is repeated, immediately or following some development. On its return it has an altered musical significance and so needs different expression. Evidence of such differences in the computer recordings will be mentioned later. Here is another way of discriminating expression from motor logistics in timing. It is not that motor factors are irrelevant to timing, since a musician knows that, for instance, the choice of fingering can affect the phrasing of a passage; rather, it is that the timing produced corresponds to the musical intention.

To go more deeply into the timing in a musical performance we need a theory of the information available to the motor system constructing movements and of the timekeeping resources this system can call upon. The studies of single movements and of cyclic activities, like gait, show that they have trajectories reproducible in space and time.[11,12] We can suppose that these movements are under parametric control of the motor system itself, and hence that the motor system can act as a timekeeper, in the sense of constructing movements having a given time span or having a temporal goal.[13] It is then possible for the motor system to construct rhythmic action as a concatenation of time spans in successive movements, in which the completion of one movement triggers, or provides the reference point for the beginning of the next. This description is too simple for the skills we are interested in and has to be qualified.

In actions involving different articulators—limbs, fingers, parts of the vocal tract—in successive movements, the movements typically overlap in time. The amount of overlap in coarticulation is variable, but can be quite extensive.[14] It can also be observed that the moments at which such movements produce their effects are patterned more definitely in time than are their moments of onset.[15] We should suppose, then, that the goal of one movement provides a reference point from which the next movement, already begun, can attain its goal after a given interval. Whether the interval is specified in terms of time or of a variable, such as accent or stress, that has temporal consequences will depend on the demands of the task. Stress, for example, may be the primary variable in speech,[13] but the primary variable is more likely to be time in musical performance.

For a fluent nonrepetitive action to unfold in an orderly temporal pattern of movement, information relevant to the space–time coordinates of the successive movements must be generated ahead of the action so that it is available to the motor system when it is required. We can call this advance preparation the programming of action. A program may contain representations of output at one or more levels of abstraction, which enables a patterning of motor output at different levels of output unit. It also allows movements to flow in a smooth succession, with resulting economies of acceleration forces in muscles. It is important that a theory of motor timing should also account for fluency.

The reproducible patterning of tempo variation observed in successive piano performances of a piece implies that there is a definite way of generating or reconstructing information relevant to expression in a motor program. Reproducibility has been observed in performances that were not extensively practiced,[6,16] and so it is likely that a skilled performer has a principled basis for generating this information, although he or she may sometimes store details within a plan of a concert performance. Thus, the timing information in a motor program contains a formal schedule, obtained from the written music, together with its expressive modulations.

Given the precision observed in reproducing timing patterns, the question arises whether this can be obtained with the timekeeping resources of the motor system alone, or requires additionally a superordinate timekeeper, or clock, not directly involved in movement production. More particularly in music, is there a separate timekeeper involved in marking the timing of beats or bars? The need to construct a timed pulse for the music may arise from external constraints on timing, perhaps more severe in ensemble than in solo playing. A random error associated with the timing of each movement will, over a sequence, become cumulative: thus, in playing music in time with a metronome, the asynchrony between metronome and player can become arbitrarily large (on an ordinal count) unless it is corrected. A superordinate timekeeper whose random error is much smaller than that of the motor system can serve to regulate its temporal output. It too will drift away from the metronome over time, but the rate of drift will be of a lower order and so need less correction. With this reduced need for correction the motor system can produce a more rhythmic and fluent output.

The idea of a clock controlling a stochastic time series was used by McGill[17] to account for the timing in the train of pulses when a nerve fiber fires. He assumed that the clock was determinate, but the pulses it triggered occurred with random delays. This led to prediction of the autocovariance function of the time series of pulse intervals, one of which was that the correlation between consecutive pulse intervals should be $-\frac{1}{2}$. In their study of serial tapping, Wing and Kristofferson[18] made the weaker assumption that the clock could have a random element. The lag-one autocorrelation should then lie between $-\frac{1}{2}$ and 0, being closer to the lower bound the smaller the variance of the clock interval.

The autocovariance function is informative if a time series has a stationary rate, but becomes less so if the rate is modulated. If the modulation pattern is cyclic or is reproduced in repeated series, one can obtain covariance functions for corresponding intervals across the series and make predictions as before about the presence of clocks.[19] However, in studying musical performance, it is seldom feasible to obtain more than a few performances. To overcome this we have made use of hierarchic analyses of variance to obtain lumped estimates of covariance among interval durations. If there is a clock controlling the duration of a particular level of musical unit—the beat, bar, phrase, and so on—then the variance of this duration should be smaller than the sum of variances of durations of nested units timed independently. The interaction between performance and duration of this unit should then produce a value in the left tail of the F distribution ($F \ll 1$).[20] One cannot attach levels of statistical significance to such values because if F falls in the left tail, the data violate the independence assumption and this invalidates the test. Leaving aside this "peccadillo," the F ratio can provide a sensitive index of timing constraint, implying clock control.

This method of analysis is illustrated by a study with Sloboda as subject, in which he gave multiple performances of some of his melodies for one hand.[20] He played in succession melody A five times, then melody B five times, and then melody A another five times. In an analysis of variance of the ten playings of melody A, the variables were performance and the durations of bars and half-bars. The interaction of bar duration with performance had a value of $F = 0.45$ (d.f. = 18,27), which would have a left tail probability of $p < 0.05$. (This F value was incorrectly calculated as 0.67 in the original paper). In other words, bar duration was less variable across playings than one should expect from the variance of half-bar duration.

In a similar analysis of the playings of melody B, the variables were performance and the durations of phrase, bar, beat and note. The interactions of performance with duration produced small F ratios at the levels of the bar ($F = 0.45$, d.f. = 16, 48) and the beat ($F = 0.63$, d.f. = 48, 72).

Thus, both analyses provide evidence of a clock controlling duration at the level of the bar. The second analysis also suggests some timing control at the level of the beat. In melody B there were three beats in a bar and two notes in each beat. The two notes dividing a beat interval could be played as linked pairs, timed in parallel by the motor system with the beat as a common reference point. This would induce a negative covariance between their durations, consistent with the latter result.

A further feature of Sloboda's performances is that they contained rubato. Hence the results imply that the patterns of rubato were reproduced in the successive repeats of a melody. These results have been replicated in studies of different pianists playing music as diverse in style as a Bach fugue, a Schubert sonata, a Chopin study, and a Beethoven duet. The amount of rubato can vary with the style of the music, but its patterning is reproduced with considerable fidelity. Also, there is always a musical unit whose duration produces a small value of F in an interaction. Sometimes a unit nested within this also produces a small F ratio, and it is usually associated with groups of short notes nested within a beat.

Combining the two main results we are led to infer the existence of a clock that is programmed to construct patterns of tempo variation. The clock underlying musical performance does not behave like a regular metronome, but is capable of controlling motor timing with the same degree of precision. It may represent time by providing a sequence of discrete time markers, corresponding to beats or bars, which the motor system can use as reference points; or it may provide a more continuous representation of a time interval to mark the passage of time.

The reason for considering continuous representations of time is that musicians appear to be able to structure rhythms within recursively embedded intervals. A player can produce a rubato pattern at one level of unit and a similar pattern in a nested level of unit, both being reproducible across performances. We are currently analyzing performances by a concert pianist of a Haydn sonata. In this music the melody line moves in a hierarchy of note durations, with ornamental runs of short notes at its lowest level. The pianist made frequent use of a pattern of speeding and then slowing over runs of short notes within a beat interval, while creating a similar rubato pattern over the sequence of beat intervals within a phrase. The same pianist also played one of the *Gnossiennes* by Satie, in which the music changes from one beat to the next the number of notes in aliquot subdivisions of the beat interval, including divisions into 3, 4, 6, 7 and 8 notes. Again, these notes are played with a precise rubato pattern within a given interval. If the beat interval is under the control of a clock, it appears that the motor system knows where it is in relation to the beats and can compute a pattern of ballistic movements within the interval. We can regard all the notes played in this way as being timed in parallel from the same reference point, the beginning of the interval. The next beat, which terminates the interval, is also timed from this point and so the computation in effect scales a temporal pattern to fit the interval.

The distinction between discrete and continuous models of clock timing is in principle testable. According to discrete models the duration of the last of a sequence of notes subdividing an interval should have the largest variance, and the cross-covariance of this duration with that of the preceding note should be negative[19] because if the overall interval and most of its subintervals are independently timed, the last subinterval is determined by subtraction. According to continuous models, however, the variance and covariance for all notes in the subdivision have the same expected values.

The analyses of piano playing made in our laboratory included a performance of a Chopin study.[6] This study is an exercise in playing a polyrhythm in which the right hand plays three notes of equal duration in each beat interval while the left hand plays four. Thus the motor system has to compute different aliquot subdivisions of a beat interval for each hand. In the performance the division of a beat interval was properly achieved in each hand, but was modulated by two kinds of expressive variation of timing, one affecting tempo and the other the synchrony between the two hands. This would entail the control of three aspects of timing, the divisions of an interval, the modulation of the interval duration, and a further modulation affecting one hand relative to the other. The attempt to infer the timing mechanism of this performance was not wholly satisfactory, because it had to be based on internal consistencies of timing within the one performance.

The pianist returned a year later and gave two more performances of the same piece, making it possible to compare these with each other and with the original performance. We, again, did an analysis of variance on the timing data for the three performances to discern whether there was a clock controlling the timing. The odds were heavily against a positive result because (1) the large amount of rubato in these performances suggested spontaneous expressions of feeling rather than concern with musical architecture; (2) it seemed unlikely that the same pattern of expression would be generated on occasions so far part in time, and if the patterns differed the presence of a clock would not be detected by the analysis; and (3) in music that allows such freedom of timing there may be no need of a clock (actually, it was not Chopin but his later interpreters who encouraged such freedom).

In fact the graphs plotting the durations of successive beat intervals showed a high degree of consistency between all three performances.[16] Rubato was reproduced in the large patterns and the smaller details. A hierarchic analysis of variance on durations, obtained from the left hand, in the three performances took the durations of bars, beats (there were two beats in a bar), and notes (there were four notes in a beat interval) as its variables. The result of main interest is that the interaction of performance with beat duration had an F ratio of 0.26 (d.f. = 106, 636). Thus, contrary to immediate impression, these were highly crafted performances based upon the architecture of the music, so that it was possible to regenerate precise instructions to a clock constructing beat timing.

More surprising was the result obtained by plotting the asynchrony between hands on successive beats. Moving one hand ahead of or behind the timing of the other is analogous to bel canto or jazz singing ahead of or behind an accompaniment, which again seems spontaneous. In fact the graphs of asynchrony were in considerable agreement over the three performances: there was a definite patterning of asynchrony in which one hand would sometimes sustain a lag or lead relative to the other; also there was a tendency for one hand to be in greater lag on the first beat of the bar than on the second. We have not yet explored the musical significance of these shifting relationships between the hands in time, but there can be little doubt that this relative timing was controlled and that there was a principled basis for control.

To understand these results we need to think of two motor subsystems, one for each hand, able to construct a separate pattern of movements in space and time, and perhaps able to obtain cross-referencing information from each other. Given cross-reference, one subsystem can compute timing relative to the other and so allow the one hand to make timing excursions from the other. Alternatively, both hands may make separate reference to the clock. If one hand times its departures from the other, it should show larger fluctuations in tempo. It is worth noting that for a long period in the history of keyboard music it was accepted that the left hand should keep to metrical time and only the right hand be allowed to use rubato.

In effect the piano allows a solo musician to play a duet between the two hands. Going further, solo piano music often contains three or four distinct voices, requiring the pianist to play two voices with the different fingers of one hand. The success in doing this depends on being able to create functionally distinct computational subsystems within motor control. Coordination of timing depends on establishing hierarchic or heterarchic timing reference within the overall system.

Turning to the problem of how musicians playing together coordinate their timing, obviously the channels of timing reference are less direct between players than between the hands of a single player. Cross-referencing must now depend on listening to each other and watching for physical gestures that cue timing. This aspect of coordination can be simplified by allowing one player at any time to act as leader for the others, so that all timing relationships become dyadic. A convenient consequence of this is that we can study a duet as a paradigm of ensemble playing.

I said earlier that a condition for fluent action is that the information relevant to its sequences of movements should be programmed in advance of motor output. If this is true, then information from other players in an ensemble can take effect only after a delay. It is therefore unlikely that the players rely on continuous feedback from each other, or from a leader, to maintain coordination. Such information is more likely to be of value at points of entry or to effect a major change of tempo. A more stable basis for coordination is to have fairly precise clocks controlling their individual playings and a common agreement on the modulations of these clocks over the performance. With this arrangement the temporal drift between players can remain small enough to need only occasional correction.

Rasch[4] has analyzed recordings made of different wind and string trios and examined the degree of synchrony between players on all the notes that were played at the same time. He found that the standard deviation of asynchrony tended to be smaller in wind trios than in string trios: it ranged between 23 and 40 msec in wind trios and between 35 and 50 msec in string trios. Rasch attributed this difference to differences in rise times of the sounds made by wind and string instruments. I would suggest also that string instruments can be more expressive and that asynchrony may have been used for expressive effect in these trios.

Rasch assumed that musicians playing together try to synchronize as well as possible, but the concept of playing in time may be richer than that of maintaining synchrony and may include the possibility of a rhythmic interplay that allows players to briefly speed up or slow down relative to each other, moving them ahead of or behind the others in time. We have already shown this to occur in solo performances of Chopin, in the relative timing between the hands. We can use the same kind of analysis on ensemble playing: if the relative timing between players is controlled, it should show as a patterning of asynchrony over the time series of performance or in a comparison of asynchrony across repeat performances.

Let us look, then, at the performances obtained from two pianists playing the fairly fast rondo from a Beethoven duet on one piano. A large passage of the rondo was played twice and we are interested in comparing the two playings.[16]

The players had only a limited opportunity for joint practice, yet in the recordings they were not at all inhibited from using rubato by the constraint of playing together. The rubato was of the same order as we have found in solo performances. Furthermore, the same patterning of rubato was produced in each of the two performances, and this was true whether we compared the timing in individual performances or in combined performances, taking the first entry at a beat by either player.

There were places in the music at which the melodic rhythms were the same for both players, making it possible to compare their modulations of rhythm in detail at different levels of musical unit. There were greater differences of modulation pattern in the graphs comparing different players than in graphs comparing the repeat performances by the same player, and these differences were shown also by fitting orthogonal polynomials to the curves. The results suggest that these players engaged in some form of rhythmic interaction. Hierarchic analyses of variance were carried out on the durations of bars, half-bars, beats and half-beats (there were four beats in a bar) in these phrases, separately for each player. It was found that the interactions of bar duration with performance were of the order of $F = 0.1$ and some were as low as 0.01. Thus, although bar duration could vary freely within a performance, it was reproduced almost exactly in the repeat performance. The two results obtained from these sections of the performances may seem to contain a contradiction: The players appear to have been constrained to a high degree of timing precision by the duets, yet their performances were rhythmically different. I shall argue that the results are in fact complementary and that the second phenomenon is conditional on the first.

The graphs showing rhythmic differences between the players imply that there was a changing asynchrony between them. Graphs of these asynchronies showed that their patterning was similar in each performance. A more comprehensive picture of asynchrony was obtained from the asynchronies at the first and third beats in each bar, where possible, for all pairs of voices in the music. Each pianist played two voices and so there were six comparisons across performances.

The standard deviations of asynchrony between the hands of a player ranged from 13.8 to 22.8 msec, and between players ranged from 26.4 to 38 msec. The latter figures are close to those obtained by Rasch between players in wind trios. Graphs of asynchrony in each of the six pairs of voices all showed that the patterning was reproduced across performances. Asking the question whether reproducibility was higher within than between players, we calculated the product–moment correlations of positional asynchrony between performances to obtain measures of the agreement between corresponding graphs. All of these correlations were highly significant: The lowest ($r = 0.5$) was between the extreme voices, bass and soprano, and the highest ($r = 0.63$) was between the inner voices, tenor and alto, both involving asynchronies between players.

In the music, the theme stated at the beginning of the rondo is returned to later in the movement. The notes are the same in all four voices, but their significance, in terms of developing the musical argument, has now changed. Accordingly, the players produced different patterns of timing between these sections. Hence we can discard a hypothesis that the patterning of tempo was determined merely by motor factors of production.

To conclude, I have argued that the major role of an internal clock is to control the progress of a performance in real time. It becomes important in the context of playing in a musical ensemble. We have obtained evidence of a rather complex use of clocks controlling the individual performances in a duet. Firstly, the clock rate was modulated continually over a performance. Presumably there was an arranged agreement on modulation between the players, and their programs generated information precise enough to enable the individual clocks to remain together. This precision is reflected in the coordination between players and in the reproducibility of an individual's performance. Visual and auditory cues between players may have assisted the coordination, particularly at moments of entry from rest or preceding major changes in tempo, but given the need to program fluent performance ahead of output they could not be the sole basis of coordination. Secondly, having a secure temporal reference provided by their clocks, the players were able to produce patterns of note-timing departing from this reference to create rhythmic interplays between hands and between performers. By hypothesis, these departures were computed by the motor subsystems, constructing the sequence of movements for each hand.

REFERENCES

1. POVEL, D.-J. 1977. Temporal structure of performed music. Acta Psychol. **41:** 309–320.
2. GABRIELSSON, A. 1974. Performance of rhythm patterns. Scand. J. Psychol. **15:** 63–72.
3. MICHON, J. A. 1974. Programs and "programs" for sequential patterns in motor behaviour. Brain Res. **71:** 413–424.
4. RASCH, R. A. 1979. Synchronization in performed ensemble music. Acustica **43:** 121–131.
5. SEASHORE, C. E. 1938. Psychology of Music. McGraw-Hill. New York, NY.
6. SHAFFER, L. H. 1981. Performances of Chopin, Bach and Bartok studies in motor programming. Cog. Psychol. **13:** 327–376.
7. SLOBODA, J. A. 1983. The communication of musical metre in piano performance. Q. J. Exp. Psychol. **35A:** 377–396.
8. CLARKE, E. F. 1982. Timing in the performance of Erik Satie's "Vexations." Acta Psychol. **50:** 1–19.
9. KLATT, D. H. 1976. Linguistic uses of segmental duration in English: acoustic and perceptual evidence. J. Acoust. Soc. Am. **59:** 1208–1221.
10. LADEFOGED, P. 1975. A Course in Phonetics. Harcourt Brace Jovanovich. New York, NY.

11. COOKE, J. D. 1980. The organization of simple skilled movements. *In* Tutorials in Motor Behavior. G. Stelmach & J. Requin, Eds. North-Holland. Amsterdam.
12. SCHMIDT, R. A. 1980. On the theoretical status of time in motor-program representations. *In* Tutorials in Motor Behavior. G. Stelmach & J. Requin, Eds. North-Holland. Amsterdam.
13. SHAFFER, L. H. 1982. Rhythm and timing in skill. Psychol. Rev. **89:** 109–122.
14. KENT, R. D. & F. D. MINIFIE. 1977. Coarticulation in recent speech production models. J. Phonet. **5:** 115–133.
15. GENTNER, D. R., J. GRUDIN & E. CONWAY. 1980. Finger movements in transcription typing. Occasional Paper No. 4. Center for Human Information Processing. La Jolla, CA.
16. SHAFFER, L. H. Timing in musical solo and duet performances. In preparation.
17. MCGILL, W. J. 1962. Random fluctuations of response rate. Psychometrika **27:** 3–17.
18. WING, A. M. & A. B. KRISTOFFERSON. 1973. Response delays and the timing of discrete motor responses. Percept. Psychophys. **14:** 5–12.
19. VORBERG, D. & R. HAMBUCH. 1978. On the temporal control of rhythmic performance. *In* Attention and Peformance: 7. J. Requin, Ed. Erlbaum. Hillsdale, NJ.
20. SHAFFER, L. H. 1980. Analysing piano performance. *In* Tutorials in Motor Behavior. G. Stelmach & J. Requin, Eds. North-Holland. Amsterdam.
21. CLAYTON, A. Personal communication.

Perception, Production, and Imitation of Time Ratios by Skilled Musicians

SAUL STERNBERG AND RONALD L. KNOLL

Bell Laboratories
Murray Hill, New Jersey 07974

INTRODUCTION

Among the human skills in which timing and time perception are critical, musical skill is distinctive: for most players a notation specifies temporal pattern explicitly and provides a criterion to which performance can be compared. Because of their years of practice in the use of this notation, the behavior of professional musicians can plausibly reveal some of the ultimate capacities and constraints of human timing mechanisms. We have examined the performance of skilled musicians in three laboratory tasks designed to capture temporal aspects of music. We focused on the short time intervals—fractions of a second—that are among the shortest durations specified by musical notation. As in Western music, these intervals occurred in the context of a train of periodic beats and were defined as fractions of the beat interval. Our three tasks—perception, production, and imitation—all appear to be required of musicians during ensemble rehearsal and performance, for example. It is plausible that because players try to "keep together," ensemble experience would cause performance in the three tasks to become at least consistent and probably correct as well (that is, consistent with the notation). Neither of these expectations was borne out by our experiments; instead, we observed surprisingly large systematic errors and inconsistencies.

The principal subjects were three professional musicians: a flutist, a cellist, and Paul Zukofsky (PZ), violinist and conductor. We also obtained a small amount of corroborative data from Pierre Boulez, composer and conductor. A detailed report of our results is available, based on group data.[1] In the present paper we describe only our more interesting findings, illustrated with data from PZ, who is the most musically experienced of our principal subjects, whose performance we examined in a wider variety of procedures than the other subjects, and whose data are more consistent than theirs, both within and across experiments. The picture generated by the group data is somewhat less clear, but leads to the same conclusions.

THREE TASKS OF TEMPORAL PSYCHOPHYSICS

We used two kinds of stimuli, shown on the left of FIGURE 1. A *time pattern stimulus* contained two or more *beat clicks* separated by a *beat interval.* The beat interval was usually 1 sec. One or more of the beat clicks was followed by a *marker click* after a *fractional interval, f,* that defined a fraction of the beat. A *fraction-name stimulus, n,* was presented both as a numerical fraction and in musical notation, where a quarter note was defined as one beat. On the right of the figure are shown the two kinds of response. In making a *fraction-name response, N,* the subject would select a category such as "less than 1/8 beat" or "between 1/8 and 1/7 beat." In making a

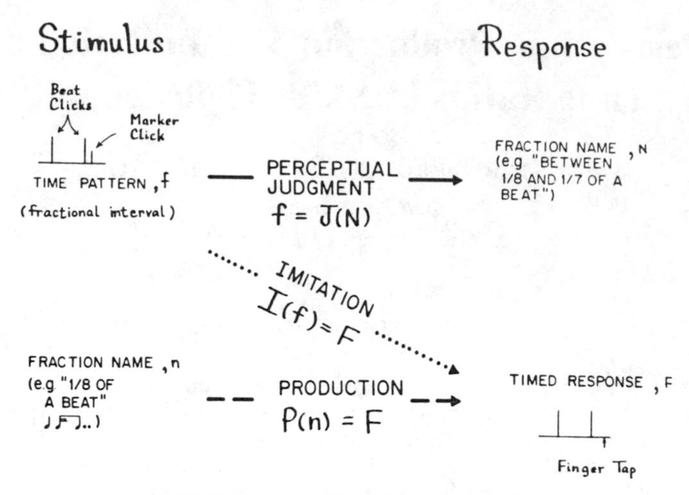

FIGURE 1. Stimuli, tasks, and responses.

timed response, the subject tapped his finger after a beat click, thereby producing a fractional interval, F, between beat click and tap.

Stimuli and responses were linked by three different tasks, as shown in the center of the figure. In *perceptual judgment* the subject assigned fraction names to time patterns. He thereby generated a *judgment function,* $f = \mathbf{J}(N)$, that maps fraction names onto their subjectively equivalent fractional intervals. In *production* the subject made a timed response to produce a fractional interval associated with a specified fraction name. He thereby generated a production function, $\mathbf{P}(n) = F$. In *imitation* (sometimes called the "method of reproduction") a time-pattern stimulus elicited a subjectively equivalent timed response. We thereby obtained an imitation function,

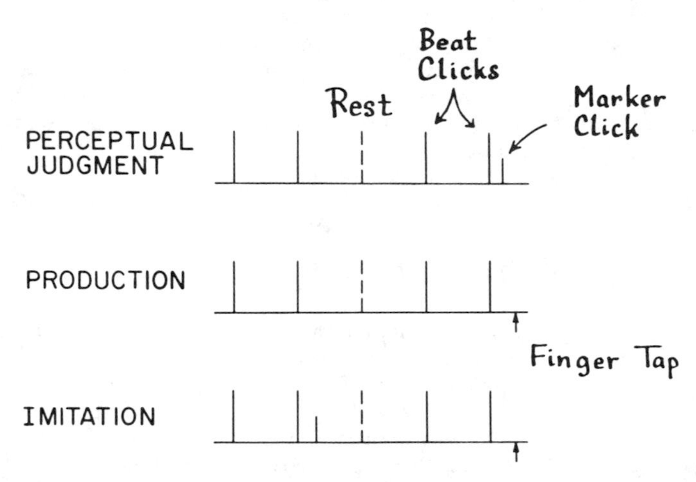

FIGURE 2. Sequences of beat clicks, marker clicks, and finger taps in Experiments $\mathbf{J}_1$, $\mathbf{P}_3$, and $\mathbf{I}_5$.

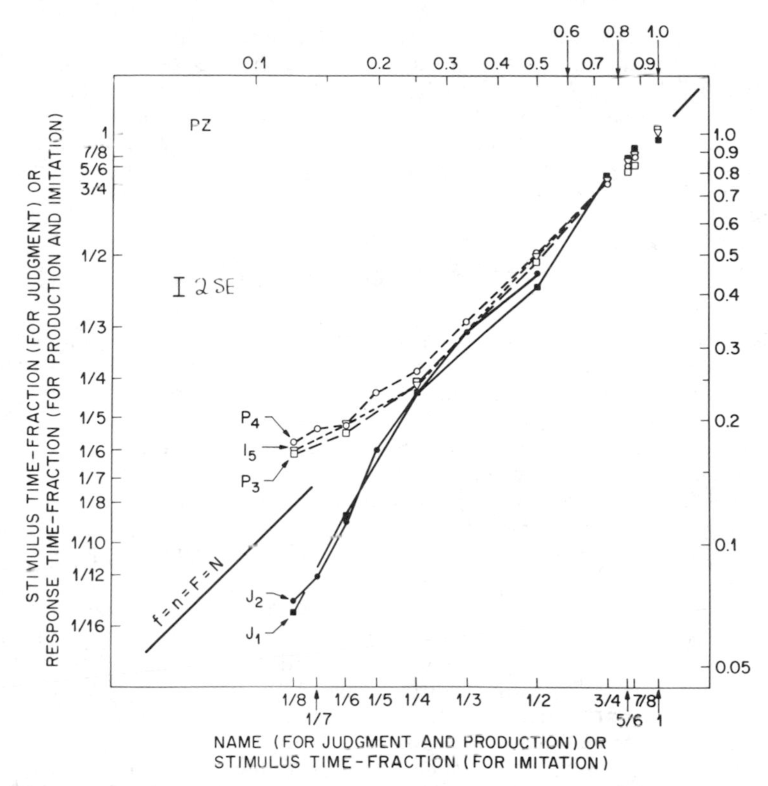

FIGURE 3. Results for subject PZ from five experiments: judgment (**J**), production (**P**), and imitation (**I**). Ordinate *(abscissa)* values are given as fractions on a logarithmic scale on the left *(bottom)* and as decimal values on the right *(top)*. Error bar indicates approximately 2 S.E. based on between-session variability. If the ordinate value is regarded as a function of the abscissa value, the functions shown are estimates of $f = \mathbf{J}^{-1}(N)$, $F = \mathbf{P}(n)$, and $F = \mathbf{I}(f)$.

$\mathbf{I}(f) = F$. To permit performance to stabilize, conditions in all experiments remained constant for at least 25 trials.

PERCEPTUAL JUDGMENT OF BEAT FRACTIONS

The stimulus pattern in our first perception experiment is shown in FIGURE 2. Two pairs of beat clicks separated by two beat intervals (a *rest*) were followed by a single marker click. For each of a set of fraction names we used an adaptive psychophysical procedure to determine the fractional interval that is subjectively equivalent to it. The resulting judgment function is labeled $\mathbf{J}_1$ in FIGURE 3. The fractional interval, *f*, is plotted as a function of fraction name, *N*. Both scales are logarithmic.[a] If stimulus and response agreed, data would fall on the straight line with unit slope. Instead, our subject radically overestimated fractions less than 1/4 beat (or 250 msec). Consider

[a] Power functions appear as straight lines on such a plot.

the case of 1/8 beat, for example. Accurate performance would assign this fraction name to a fractional interval of 125 msec; instead, it is assigned to an interval of about 67 msec, or 1/16 beat. Thus, for small fractions, the assigned fraction name is larger than the fractional interval.[b] The *N*-value grows more slowly than the *f*-value, however, coming into approximate agreement at about 1/4 beat. Despite the poor *accuracy* for small fractions, judgment *precision* is high: the difference threshold at 1/8 beat was only 4.3 msec (about 6%), for example.

PRODUCTION OF BEAT FRACTIONS

The large errors in the *perception* of small fractions make it particularly interesting to examine musicians' accuracy in *producing* such fractions. One appealing possibility is that production is mediated by a simple feedback process in which the subject judges the fraction he produces with respect to the fraction name he is trying to produce, and adjusts later productions accordingly. Suppose also that produced fractions are judged by means of the same perceptual mechanism used in the judgment task. The production function should then approximate the judgment function: $\mathbf{P}(n) = \mathbf{J}(N)$ when $n = N$. Thus, overestimation of small fractions (assigning fraction names that are too large) would lead to underproduction (producing fractional intervals that are too small).

The events on each trial in our first production experiment are shown in FIGURE 2. They are the same as in the perception experiment, except for replacement of the marker-click stimulus by a finger-tap response.[c] With each tap the subject heard the thump of his finger striking a hard surface. The production function we obtained is labeled $\mathbf{P}_3$ in FIGURE 3. The expectation from a feedback mechanism is violated dramatically. For small fractions the intervals produced are too large rather than too small. That is, we have overproduction rather than underproduction. Whereas the subject *judged* 67 msec to be 1/8 of a 1-sec beat, for example, he *produced* a mean fractional interval of 164 msec for the same fraction name. Furthermore, he seemed satisfied with his responses, and did not experience his taps as being late.[d] For small fractions in the perception task we have seen that the fraction stimulus, *f*, that is subjectively equivalent to a fraction name, *N*, changes more rapidly than the *N*-value. In contrast, the fraction, *F*, produced changes more slowly than the *n*-value.

Several analyses and experimental variants were directed at understanding the large systematic errors we had found in judgment and production, as well as their inconsistency. In the present report we consider nine of these. Others, together with these, are considered in greater detail in the full report.[1]

FURTHER ANALYSIS OF PERCEPTUAL JUDGMENT

Psychophysical Procedure

One concern is whether aspects of the judgment data may depend on special features of our psychophysical procedure. In any block of trials in the adaptive

[b] Figure 7 provides an outline of our principal findings.

[c] Note that although infrequent in earlier music, the playing of a note after the beat without playing a note on the beat is not unusual in the music of the past 60 years.

[d] This is an informal impression, not based on analysis of systematically collected data.

procedure, subjects judged time fractions with respect to just one fraction name, and the fractional intervals were concentrated in the range where judgments were the most difficult. In a second experiment we used a method more akin to traditional psychophysical scaling in which both these features were altered. The subject categorized a wide range of time-pattern fractions into one of eight categories of fraction names, ranging from "less than 1/8 beat" to "greater than 1/2 beat," in each trial block. Also in contrast to the first experiment, two beat clicks rather than only one were followed by marker clicks, providing two observations of the fractional interval on each trial. Results are labeled $\mathbf{J}_2$ in FIGURE 3. The two judgment functions are almost identical, despite the differences between procedures.[e]

Fractional Interval Defined by Subjective Onset Versus Offset of Marker

A second attempt to discover a source of the judgment errors was based on the possibility that the internal representation of a brief click may have a longer duration than the click itself. Suppose that when beat and marker clicks are close together, the subjective duration of the fractional interval is defined by the *onset* of the internal representation of the beat and the *offset* of the internal representation of the marker. Such a mechanism could then produce overestimation of the kind we observed. To test this possibility we compared judgments of our normal stimuli, in which clicks were 5-msec tone bursts, with stimuli in which the marker duration was prolonged by about 60 msec. Contrary to the idea that offset time is important, this variation produced no change in the judgments, for either small or large fractions.[f]

FURTHER ANALYSIS OF PRODUCTION

We next turn to five of our efforts to understand the systematic errors in the production task and the inconsistency between production and judgment performances for small fractions.

Opportunity for Adjustment to Feedback

As mentioned above, our subject did not report experiencing his finger-tap responses as being late for either small fractions or large. Nonetheless there may have been discrepant feedback from the perception mechanism, but too little opportunity to adjust to it, given only one production per trial. On each trial in a second production experiment the subject produced timed responses after each of ten successive beat clicks. We found no tendency for the error to be reduced over the ten successive responses. Mean produced fractions based on all the responses are labeled $\mathbf{P}_4$ in FIGURE 3. Performance agrees closely with the first production experiment; if anything, the tendency toward overproduction is slightly greater.

[e] The procedural differences did influence judgment precision, however; difference thresholds were about twice as large in the second (multiple-fraction) experiment.

[f] The measured mean change in $f = \mathbf{J}(N)$, averaged over six fraction names, was 0.6 ± 1.9 msec. There was neither a mean effect of marker duration nor an interaction of marker duration with fraction size. This finding may reflect a general property of the perception of timing and rhythm in music: the dominance of the sequence of time intervals between successive *attacks,* and the relative unimportance of *release* times.

Subjective Delay of Tap versus Click

A second potential source of inaccuracies in production might be a difference between two critical subjective delays. One is the interval between the occurrence of a click and its perceptual registration. The second is the interval (possibly negative) between our measurement of a tap and when the subject perceives it to have occurred. If these two delays differed, then direct comparison of the intervals between beat click and marker click in perception and between beat click and tap in production would be inappropriate. To estimate the difference between the two subjective delays we asked subjects to tap in synchrony with one or more beat clicks. The difference can be estimated by the mean asynchrony between tap and click; the asynchrony was small—about 10 or 20 msec, depending on the procedure—clearly too small to explain the observed effects. On the assumption that the difference between subjective delays does not depend on the beat fraction, the production data in FIGURE 3 have been corrected by these small amounts.

Improved Feedback from Finger Tap

The disparity between performances in perception and production led us to question a feedback model of production, which suggested in turn that we scrutinize the feedback itself. The feedback from tapping the finger included tactile, proprioceptive, and auditory cues, but not the marker click used in the perception task. In additional production experiments with both single- and multiple-tap procedures, but limited to the fraction 1/8 beat, each finger-tap generated a marker click. The sequence of clicks in judgment and production thereby became identical. The production performance was virtually unchanged, however; we found only a 10 msec change in the mean interval produced.

Musical Instrument Response

Another potential source of the production error might have been our choice of finger-tapping as a response. (The subject was a skilled violinist but not a skilled finger-tapper, at least at the start of these experiments.) We ran the single-response production experiment again, but now the response was to play a single note on the violin after the final beat click. We measured the onset time of the note as the subject attempted to produce the fractions 1/8, 1/2, and 1 beat. The amount of overproduction of the small fraction did not decrease. (In fact, it was nonsignificantly greater by 12 ± 11 msec for PZ.)

Minimum Reaction Time

The potential sources discussed above of the production error and the production–perception disparity had to be considered. However, we had little *a priori* reason to expect that even if they had been important, their effects would have depended on fraction size. One constraint that might have such differential effects is the existence of a minimum reaction time (RT). The minimum RT to auditory stimuli is between 100 and 150 msec. Furthermore, there are delays between excitation of a musical instrument and its acoustic response. The combination of these two effects makes it virtually impossible to produce a note 125 msec after a signal to respond (such as a beat click) when the signal is the event that initiates the response-timing process. If we assume that the timing of an offbeat response starts with the immediately preceding beat, it follows that the notation often calls for production of discriminably different response delays, some of which are less than the minimum RT. One solution would be to bias productions just as we have observed, so that for different small fractions the mean intervals produced are greater than the minimum RT, but still distinct.

A test of this explanation is provided by variances of the finger-tap delays, *Var(F)*, together with an argument suggested by Snodgrass, Luce, and Galanter.[2] We assume that as its mean increases, the variance of a response delay also increases, where the delay is measured from the event that initiates the timing process. If the responses for all fractions were timed from the final beat click, we would therefore expect *Var(F)* to increase with fraction size. Instead, we found it to vary as a *U*-shaped function of fraction size, with a minimum between $n = 1/4$ and $n = 1/2$.[g] This pattern of variability would be expected if small fractions, but not large ones, were timed from the penultimate beat; if so, overproduction cannot be explained as compensation for a limited speed of response.

PRODUCTION OF MULTIPLE SUBDIVISIONS OF THE BEAT

How can the existence of large production errors for small fractions be reconciled with our belief that musicians are able to fill a beat interval accurately with a sequence of equally spaced actions? Could the production error depend on our use of a single, isolated response?[h] To address this question we studied the three conditions shown in FIGURE 4 in a new production experiment. The beat interval here was ½ sec instead of 1 sec, incidentally testing the generality of our effects. One of the conditions required an isolated offbeat response, with a target fraction of ¼ beat, equivalent to a fractional interval of 125 msec. There were two multiple-response conditions. In one the subject started with a tap on the beat, and alternated between index fingers to fill the beat interval with quarter-beat taps. An unusual interval between the first two taps here would reveal any general distortion of subjective time near the beat. In another condition the initial on-beat tap was withheld. If overproduction depends on response isolation, then the presence of later taps within the same beat interval should eliminate it, especially since the final tap was supposed to be made on the next beat.[i]

Consider first the results from the 5-tap condition. All tap delays, and in particular the delay of the second tap, fall close to the fitted line. There appears to be no general distortion of subjective time near the beat. Furthermore, the slope of the linear

[g] Because imitation performance is virtually identical to production, in variability as well as mean, a good estimate of the effect of fraction size on the variability of tap delay is provided by *Var(F)* averaged over the two production experiments described above and the imitation experiment to be described in Section 9. Fractions that were examined in all three experiments include 1/8, 1/6, 1/4, 1/2, 3/4, 5/6, and 7/8; for PZ the corresponding RMS values of the S.D. are 21.3, 20.3, 16.7, 25.2, 32.4, 42.2, and 42.4 msec, respectively.

[h] Musicians could perhaps learn to fill an interval evenly, without accurate perception or production of isolated beat fractions, by employing judgments of evenness and synchrony, together with the ability to count actions.

[i] The first tap in the 5-tap condition provided the synchronization correction that we used to adjust all taps in the three conditions. Application of the correction in this way requires us to assume that the location of the initial subjective beat as well as the subjective beat rate depend only on the beat clicks, and are influenced neither by whether there is a tap on the beat nor by the number of taps that follow.

function is less than unity, giving a value of 435 rather than 500 msec for position 5. (Since the subject felt satisfied with his productions, we must assume that the subjective beat interval was shorter than the actual beat interval by about 13%.) In other words, with multiple subdivisions starting on the beat there is no evidence for overproduction of the small fraction. Consider next the 1-tap condition. Just as in our other experiments the response is delayed relative to the correct response time of 125 msec. Because of the shortening of the subjective beat interval, the amount of overproduction is even greater when referred to the second tap in the 5-tap condition. Consider finally the 4-tap condition. The first response—the tap in position 2—is delayed here as much as in the 1-tap condition. The overproduction effect therefore does not depend on the offbeat response being isolated. Instead, it appears to result from an onbeat response being withheld.

The displaced parallel lines provide a good description of performance in the two multiple-response conditions, indicating that every tap in the 4-tap condition is delayed

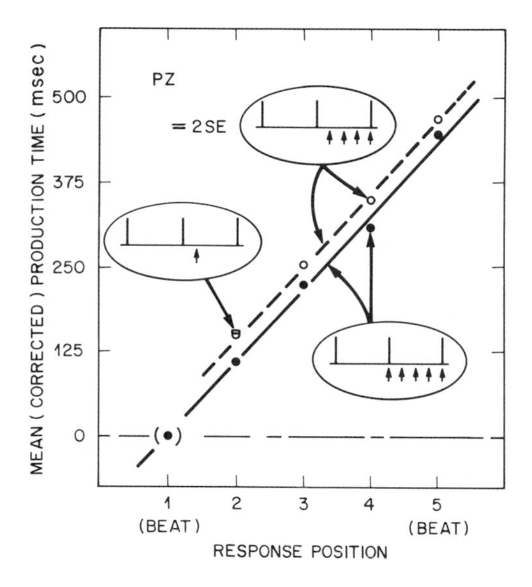

FIGURE 4. Stimulus and response patterns in three conditions in an experiment on multiple subdivisions of the beat, together with results. Mean response delays have all been corrected by the mean delay (−34.7 msec) between the second beat click and the first tap in the 5-response condition. Data from the two multiple-response conditions have been fitted by parallel lines. The displayed ± S.E. bar is appropriate for assessing adequacy of the fitted lines.

relative to the 5-tap condition, and by about the same amount. This phenomenon seems best described as *displacement of the subjective beat.* As in some circadian phenomena, the phase of a periodic process has been changed with no alteration in its period. We find it especially remarkable that the delay of the first tap is propagated all the way through the last tap, despite the presence of a final beat click.

CONJECTURES ABOUT FAILURE OF THE FEEDBACK MODEL OF PRODUCTION

The beat-displacement effect suggests one possible source of failure of the feedback model. It is plausible that subjects judge their response delays relative to the *subjective* train of beats. If so, the beat displacement associated with the delayed offbeat response would reduce the apparent delay of this response. Such an effect might explain a subject's inability to recognize and correct his overproduction. It would be too small an effect, however, to explain the major part of the discrepancy between overproduction and the underestimation we observed in the judgment task, which would require that the magnitude of beat displacement *exceed* the delay of the offbeat response, rather than merely equaling it.

The expectations from the feedback model of production that were violated depend on the assumption that the time-perception processes that accompany production are the same as those used in the judgment task. The beat-displacement effect leads us to question this assumption. The existence of the effect reminds us that in production the fractional interval terminated by the finger tap must not only be timed, but must also be placed in proper phase relation to the train of beats. Performance of small fractions in the production task therefore requires the timing of both a beat interval and a beat fraction within the same sequence. In contrast, there is no reason to believe that *judgment* of a beat fraction depends on concurrent judgment of a full beat interval.[j] It is possible that this difference between tasks contributes to the failure of the feedback model.

Results from a final judgment experiment provide weak evidence that favors this possibility, indicating that if successive long and short intervals must both be judged, the perception of at least one of them may be dramatically altered. We used click patterns like those shown in FIGURE 5, with a 1-sec beat interval, and asked whether judgment of the large interval between the final beat click and the pair of marker clicks would be influenced by the requirement also to judge the small interval between markers. On each trial the subject had to judge whether the interval between markers was large or small relative to ⅛ beat, and then also to judge whether the interval between the final beat click and the markers was large or small relative to a full beat. Judgments of the small beat fraction were very similar to performance in a single-judgment control condition. Judgments of the larger interval were enormously more variable than in its control condition, however: The difference threshold was increased by a factor of ten, from about 4% of the beat interval to about 40%.[k]

IMITATION OF BEAT FRACTIONS

Imitation of beat fractions is of interest partly because it provides a further opportunity to determine the sources of error in perception and production. In judgment and production tasks, subjects must associate beat fractions with fraction names. In imitation (see FIGURE 1), the stimulus of the judgment task is mapped onto the response of the production task; fraction names are not explicitly involved. If the errors in judgment and production are due to the input or output of fraction names, it follows that imitation performance should be accurate.[l]

Such accurate imitation of time intervals is one possibility that has been considered in previous research.[3] A second possibility that has been advanced is that imitation is

[j] Judgment of a fractional interval could use a stored representation of the beat interval. Alternatively, the beat interval might not be directly represented at all, but would determine the calibration of a mechanism that assessed beat fractions.

[k] In further work along these lines it will be important to force high accuracy in judgments of the large interval and to search for effects on both mean and variability of judgments of small fractions.

[l] The converse does not follow: if imitation were accurate, we would know only that the function that relates stimulus fractions to their internal representations must be the inverse of the function that relates internal representations to produced fractions.

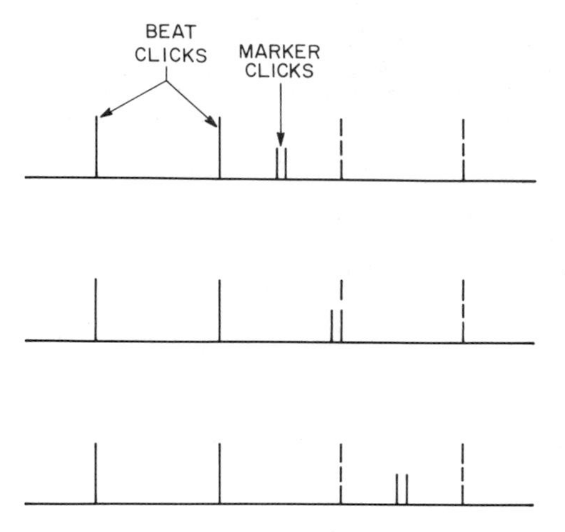

FIGURE 5. Examples of sequences of beat clicks and marker clicks in a dual-judgment experiment. *Broken lines* represent beats for which no beat click was presented.

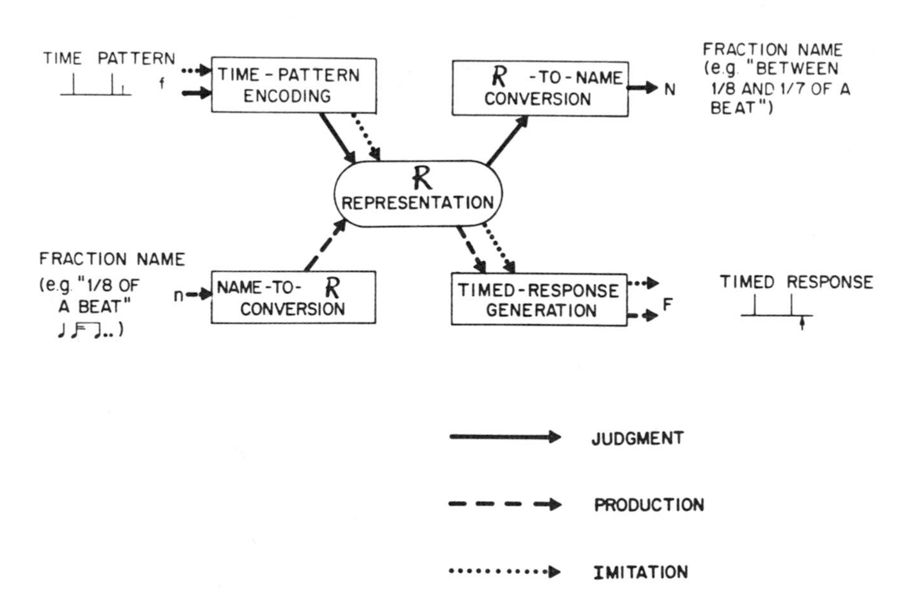

FIGURE 6. An information-flow model of perception, production, and imitation of beat fractions. The model incorporates four processes that convert time-pattern (f) or fraction-name (n) stimuli into time-pattern (F) or fraction-name (N) responses, and that make use of a common internal fraction-representation. Paths of information flow for the three tasks are represented by *arrows*.

accomplished by concatenating two processes: a judgment process that covertly assigns a name to the time pattern, and a production process that converts this name into a timed response.[4] Given our findings of overestimation as well as overproduction of small fractions, this *full-concatenation model* implies that the response fraction in imitation will be too large by the *sum* of the errors in the other two tasks; we call such an outcome *strong overimitation.*

The imitation data are labeled $\mathbf{I}_5$ in FIGURE 3. Results conform to *neither* possibility. Instead, responses to small fractional intervals in imitation were virtually identical to responses to the *names* of these fractions in production. The significance of this outcome is best explained in the context of the following model of performance in our three tasks.

AN INFORMATION-FLOW MODEL OF THE PERCEPTION, PRODUCTION, AND IMITATION OF BEAT FRACTIONS

In the skeleton model diagrammed in FIGURE 6 we limit ourselves to accounting for the relations among performances in our three tasks. Each task involves processes that perform input, output, and possibly translation functions. Parsimony leads us to postulate the minimum number of processes consistent with our data, and hence the maximum number of processes shared between tasks.

The processes underlying perceptual judgment are shown by the two upper boxes. A time-pattern stimulus generates an internal fraction representation. This representation is converted into a fraction name to generate the required response. The processes underlying production are shown by the two lower boxes. A fraction-name is converted into an internal representation of the same kind as in the judgment task. This representation is then used to produce the required timed response. A full-concatenation model of imitation would most naturally be represented by a mechanism in which the upper and lower pairs of processes made use of distinct internal representations. In such a mechanism information could not flow directly from time-pattern encoding to timed-response generation. Instead, to connect these two operations the covert output of the pair of processes used in judgment (a fraction name) would become the input for the pair of processes used in production. Because such a model can be rejected, we adopt a *partial-concatenation model* in which imitation makes use of a common internal representation, and shares only the encoding process with the judgment task and only the response-generation process with the production task.

Each of the four processes in the model can be thought of as a function or transformation that maps its input onto its output. Qualitative aspects of our data restrict the relations among these transformations. The most important of these relations is based on the close agreement of the data from production and imitation: When time-pattern and fraction-name correspond, they lead to the same timed response, so that $\mathbf{P} \simeq \mathbf{I}$. This in turn means that they must have produced the same internal representation; thus the two input transformations are the same. The internal representation therefore creates a veridical mapping between the two kinds of stimuli.[m] By a complementary argument from the finding that responses to the same fractional interval in imitation and judgment are *not* the same ($\mathbf{J}^{-1} \neq \mathbf{I}$) we conclude that the two output transformations are distinct.

Both the production and judgment tasks generate psychophysical scales that

[m] By a mapping or psychophysical scale that is veridical we mean one that associates the beat fraction $1/n$ with the fractional interval b/n, where b is the beat interval.

describe the subject's mapping of musical notation onto time fractions. As we have already seen, neither of these scales is veridical, and, moreover, they disagree with each other. Both scales are *explicit,* in that the subject's response is identified directly with one of the terms in the psychophysical function. By combining results from imitation, $F = \mathbf{I}(f)$, and production, $F = \mathbf{P}(n)$, which involve the same response, we generate an *implicit* psychophysical scale that relates musical notation, n, to time fractions, f, when both are presented as stimuli: $n = \mathbf{P}^{-1}\mathbf{I}(f)$. Our data indicate that this implicit scale is veridical, despite the systematic errors in each of the three explicit relations, **J**, **P**, and **I**.[n]

FOR SMALL FRACTIONS (< $\frac{1}{4}$ Beat)

- **PERCEPTUAL JUDGMENT**

  Results: Overestimation $f = \mathbf{J}(N) < N$

- **PRODUCTION**

  Expectation:

  Underproduction (Feedback Model) $F = \mathbf{P}(n) < n$

  Results: Overproduction $F = \mathbf{P}(n) > n$

- **IMITATION**

  Alternative Expectations:

  a) Veridical $F = \mathbf{I}(f) = f$

  b) Strong Overimitation (Full-Concatenation Model) $F = \mathbf{I}(f) >> f$

  Results: Overimitation $F = \mathbf{I}(f) > f$,

  and, for $f = n$, $\mathbf{I}(f) = \mathbf{P}(n)$

FIGURE 7. Summary of principal expectations and findings.

SUMMARY

We have described our exploration of the judgment, production, and imitation of fractions of a beat by skilled musicians, illustrating our findings with data from violinist and conductor Paul Zukofsky. For small fractions we found systematic and

[n]Given our model it is tempting to inquire whether "error" or "distortion" in any single constituent in the model can account for the performance errors in all three tasks, and their relations. Such an inquiry succeeds qualitatively: The only such single constituent that could be responsible is the internal representation, R, since this is the only constituent common to the three tasks. Suppose that for small fractions, R is "expanded" so as to correspond to a larger fraction. Examination of FIGURE 6 reveals that such an expansion, alone, would produce the three effects we observed: overestimation, overproduction, and overimitation. Quantitatively, however, this explanation fails, because it requires that $\mathbf{J}^{-1} = \mathbf{I}$, a relation we can reject reliably if we use the full range of data. The explanation might succeed, however, if we limited it to small fractions.

substantial errors. In the judgment task small stimulus fractions are associated with names that are too large (overestimation). In both production and imitation tasks the fractions produced were too large (overproduction, overimitation). A summary of our findings and of the expectations they violate is provided in FIGURE 7.

The temporal patterns we used are perhaps the simplest that qualify as rhythms, incorporating just a beat interval and a fraction. The phenomena we discovered in relation to these simple patterns, and their implications for underlying mechanisms, must be considered in attempts to understand the perception and production of more complex rhythms, as in actual music.

We explored and rejected several plausible explanations for the overestimation and overproduction of small fractions. Although we have as yet no satisfactory explanations of the errors themselves, relations among the errors have powerful implications for human timing mechanisms. The relation between the errors in judgment and production requires us to reject a feedback model of production, in which a subject uses the same processes as in the judgment task to evaluate and adjust his performance in the production task. An explanation of the inconsistency between judgment and production seems most likely to lie in a change in time perception induced by the production task. Together with the existence of systematic errors in judgment, the equality of the errors in production and imitation argues that imitation is not accomplished by concatenating all the processes used in judgment and production. Our results are instead consistent with a model containing four internal transformation processes, in which judgment and production share no process, but do involve the same internal-fraction representation, and in which imitation shares one process with judgment and another with production.

ACKNOWLEDGMENTS

We thank Paul Zukofsky, coauthor of the full report,[1] for significant contributions both as subject and collaborator. We are also grateful to Marilyn L. Shaw for helpful comments on the manuscript.

REFERENCES

1. STERNBERG, S., R. L. KNOLL & P. ZUKOFSKY. 1982. Timing by skilled musicians. *In* The Psychology of Music. D. Deutsch, Ed.: 181–239. Academic Press. New York, NY.
2. SNODGRASS, J., R. D. LUCE & E. GALANTER. 1967. Some experiments on simple and choice reaction time. J. Exp. Psychol. **75:** 1–17.
3. CARLSON, V. R. & I. FEINBERG. 1968. Individual variations in time judgments and the concept of an internal clock. J. Exp. Psychol. **77:** 631–640.
4. THOMAS, E. & I. BROWN, JR. 1974. Time perception and the filled-duration illusion. Percep. Psychophys. **16:** 449–458.

DISTINGUISHED ADDRESS

Circadian Timing[a]

JÜRGEN ASCHOFF

Max-Planck Institut für Verhaltensphysiologie
D-8138 Andechs, Federal Republic of Germany, and
Institut für Medizinische Psychologie
University of Munich
Munich, Federal Republic of Germany

INTRODUCTION

For any living system, two aspects of timing can be distinguished: Within the organism, a multiplicity of interacting processes has to be coordinated in time (as in space), and externally temporal adjustment of the organism's behavior to that of other organisms and to changing conditions in the environment is also needed. In many ways, internal as well as external timing implies the measurement of time, for example, by the decoding of impulse frequencies within the central nervous system. To serve such purposes, various mechanisms can be imagined that do not necessarily represent permanently running clocks. There are, however, four prototypes of such clocks that have evolved in adaptation to the four main periodicities in the environment: the tides, day and night, the lunar cycle, and the seasons. Each of these environmental cycles represents a temporal program that repeats itself in regular intervals and is hence predictable. By incorporating into its organization a copy of such a program, an organism is enabled to adjust its activities to the periodically changing conditions, and to be prepared in advance of ensuing tasks, that is, to measure external time by means of a biological clock.[1]

The characteristics of the four biological clocks resemble those of self-sustaining oscillations.[2] They persist in constant conditions with a frequency that slightly deviates from those of the environmental cycles that they mimic. This is indicated by the prefix "circa" used to designate circatidal, circadian, circalunar, and circannual clocks. To be useful as a time-measuring device, a circa-clock must run in synchrony with its environmental cycle. This is achieved through entraining signals from periodic factors in the environment, the zeitgebers.[3]

The most ubiquitous and of prime importance are the circadian clocks, which have been documented in all eukaryotic species from unicellular organisms up to man.[4–6] Circadian clocks not only enable the organism to occupy a "temporal niche" in the environment, but also provide internal temporal order by imposing rhythmic variations on all functions including psychomotor performance and mental activities such as time perception. A discussion of these interactions requires a brief introduction into the basic principles of circadian rhythmicity.

THE CIRCADIAN SYSTEM

The major characteristics of circadian rhythms and their entrainment by zeitgebers can readily be described on the basis of experiments performed with human subjects

[a]This paper is dedicated to Professor Colin S. Pittendrigh on the occasion of his 65th birthday.

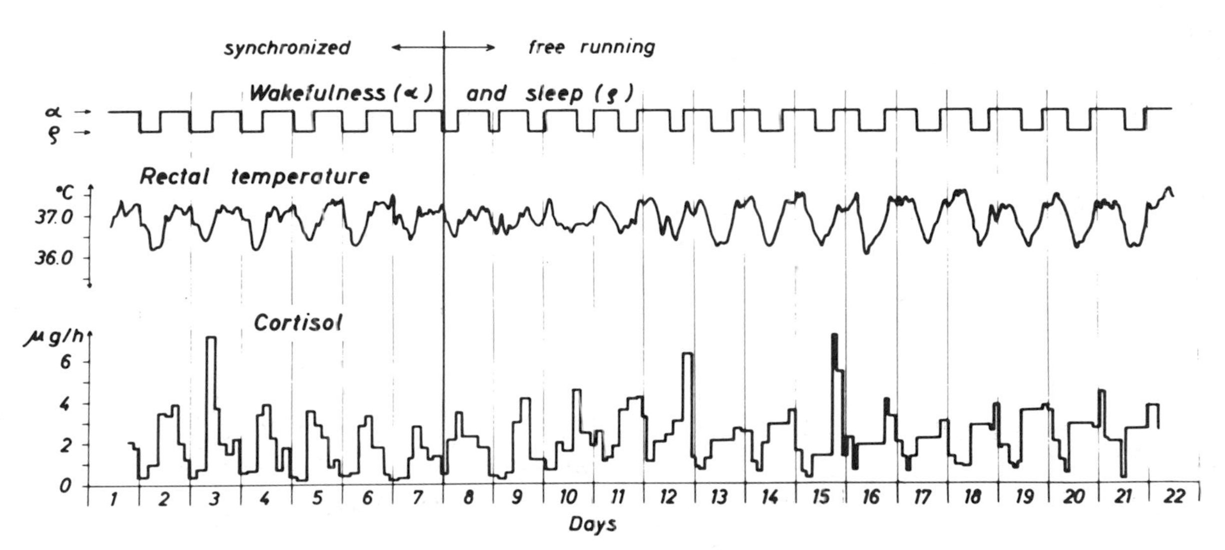

FIGURE 1. Circadian rhythms of wakefulness (α) and sleep (ρ), rectal temperature, and urinary cortisol excretion, recorded in a subject living alone in an isolation unit; for the first 7 days the subject was in contact with the experimenter (door open), but thereafter was without contact (door closed). (From Lund.[8] Reprinted by permission.)

under conditions in which external time cues are excluded or controlled by the experimenter. An underground isolation chamber has served such purpose in our institute since 1961.[7] It consists of a reasonably sized living room with bed, a small kitchen and a toilet with a shower. Subjects living in this unit for several weeks had to prepare their own meals and were usually asked to collect urine at short intervals as well as to perform a series of tests during the time they were awake. Their rectal temperature was recorded continuously, and they also indicated the times of waking up and of retiring by pressing buttons.

The results of a typical experiment are shown in FIGURE 1. For the first 7 days, the

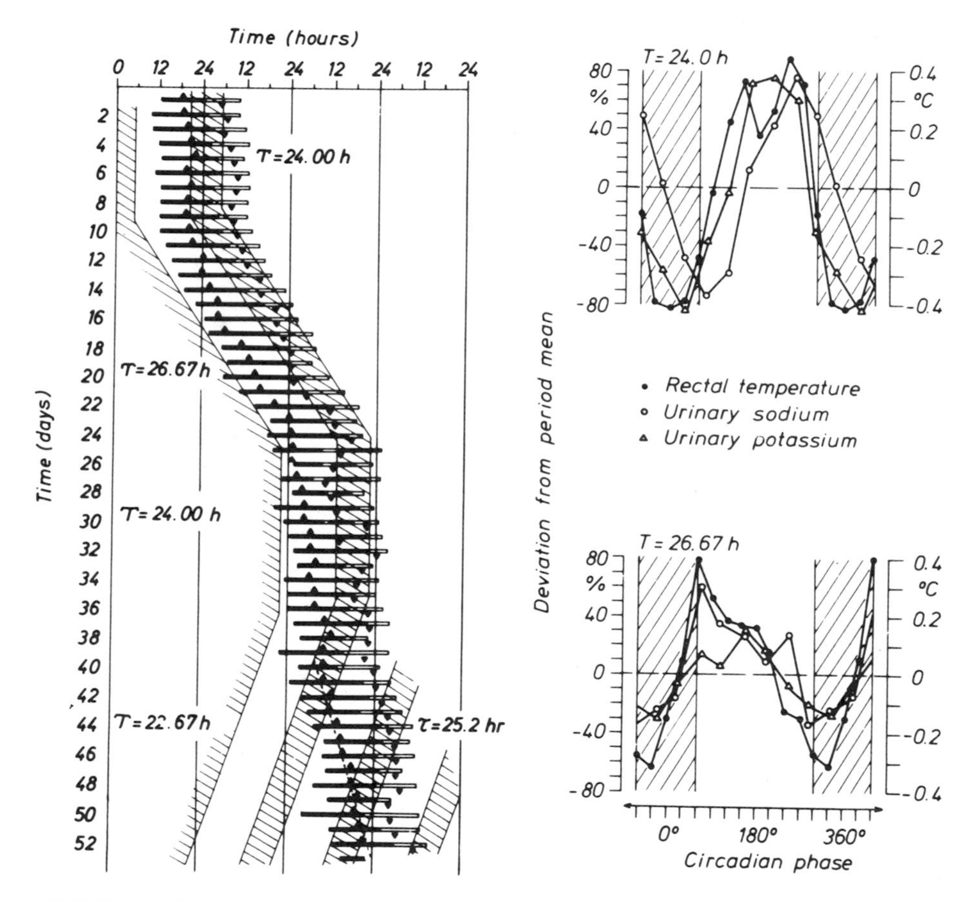

FIGURE 2. Circadian rhythms of a subject exposed in the isolation unit to an artificial zeitgeber with reading-lamp available. (*Left*) Black and white bars represent wakefulness and sleep, respectively; triangles = maxima (above the bars) and minima (below the bars) of rectal temperature; shaded area = darkness; T = zeitgeber period; τ = mean circadian period. (*Right*) Patterns of rhythms, averaged over several periods under T = 24 and T = 26.67 hr; shaded area = sleep. The abscissa is given in degree of a full circadian cycle, 360° representing 24 hr in the *upper* and 26.67 hr in the *lower* diagram. (After Aschoff *et al.*[10]; from Aschoff.[11] Reprinted by permission.)

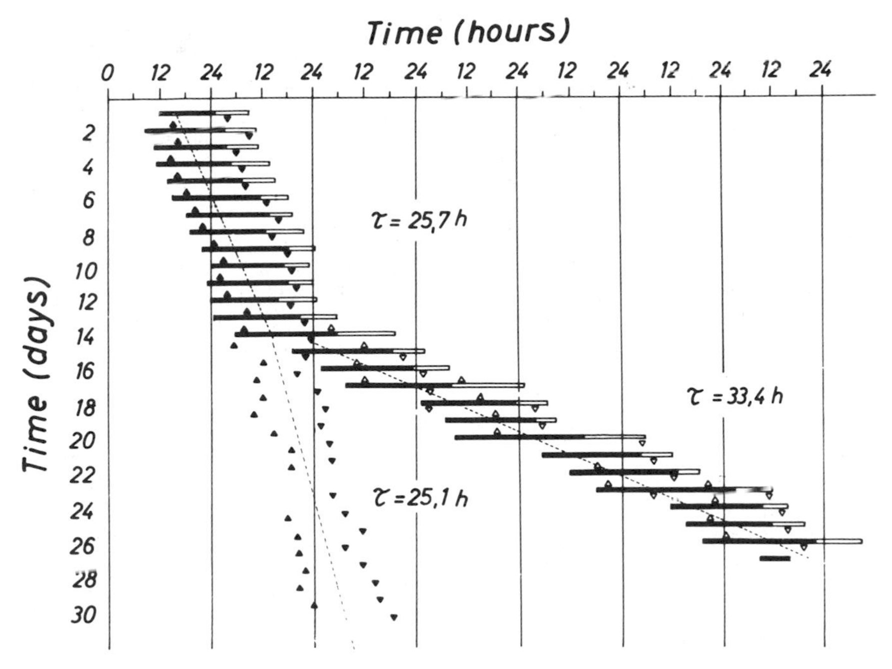

FIGURE 3. Circadian rhythms of wakefulness and sleep (black and white bars) and of rectal temperature (triangles above bars for maxima, below bars for minima) in a subject living alone in an isolation unit without time cues. Spontaneous internal desynchronization at day 16. τ = mean circadian period. (From Wever.[15] Reprinted by permission.)

door of the isolation unit was open, and the subject knew the time of day. Hence, the circadian clock was entrained to 24 hr, as shown by the rhythms in wakefulness (α) and sleep (ρ), in rectal temperature, and in the urinary excretion of cortisol. The subject was then isolated on the evening of day 7, and his rhythms were expressed as well as before, after a temporary initial disturbance. Closer inspection, however, reveals that the mean circadian period, measured for example, between successive awakenings or minima of rectal temperature, was no longer 24 hr, but lengthened to 26.1 hr, that is, the rhythm was free running. In a sample of 147 subjects, the mean period (±S.D.) of free-running rhythms was found to be 25.0 ± 0.5 hr.[9]

In the isolation unit, human circadian rhythms can be entrained by light–dark cycles complemented by gong signals at regular intervals. Using this artificial zeitgeber, we have been able to synchronize subjects to periods other than 24 hr within certain narrow limits.[10] Noteworthy in these experiments is that the subjects had the choice of whether to follow the zeitgeber or not by making use of a small reading lamp. The example presented in FIGURE 2 shows an initial entrainment to a zeitgeber period T = 24 hr, thereafter to T = 26.67 hr, and finally failure of entrainment to T = 22.67 hr. In the 24-hr day, the subject was a "late riser," waking up several hours after the light was on (see the black and white bars for representations of wakefulness and sleep, respectively). In the longer day, he became, to his own surprise, an "early riser," eventually waking up even before the light was turned on. In the final part of the experiment, the subject did not follow the short period of the zeitgeber; his rhythm started to "free-run," with a mean period of 25.2 hr. When the subject changed from a late to an early riser, he also changed his internal temporal order. In the long day, the maxima and minima of rectal temperature (see the triangles in the left diagram of FIGURE 2) were advanced relative to the sleep-wake cycle. This difference in the internal phase relationship is more obvious in the two diagrams on the right of FIGURE 2, which shows the patterns of rectal temperature and of urinary excretion during the two conditions of entrainment: In the 24-hr day, the maxima of the rhythms occurred in late afternoon, in the long day shortly after waking. Similarly, the minima were shifted, and the wave form was drastically altered in its skewness.

The dependence of the external phase relationship between rhythms and zeitgeber, and of the internal phase relationship between rhythms and the sleep-wake cycle, on the period of the entraining zeitgeber is typical for any oscillation that is driven by another oscillation.[12,13] The relevance of this rule will be discussed again in the section on anticipatory activity.

In the experiments mentioned so far, all the rhythms recorded in one subject were free-running with the same frequency, that is, they were internally synchronized with each other (compare FIGURE 1, and the last part of the experiment shown in FIGURE 2). In contrast, free-running rhythms can occasionally split into two components which run with different frequencies.[14] An example of such "internal desynchronization" is given in FIGURE 3.[15] For the first 14 days, the rhythms of this subject were free-running with a common period of 25.7 hr. At day 15, the sleep-wake cycles spontaneously lengthened to a period of 33.4 hr, while the rhythm of rectal temperature continued to free-run with a period close to 25.0 hr. Spontaneous internal desynchronization has been observed in about 30% of all our subjects studied in constant conditions. These findings suggest that the circadian system consists of a multiplicity of oscillators, differentially controlling the sleep-wake cycle on the one hand and autonomic rhythms on the other. These oscillators are normally synchronized with each other, or kept in synchrony by the zeitgebers, but they can become uncoupled under certain, not very well understood, conditions. The phenomenon of internal desynchronization is of special interest with regard to time perception in isolation (see the subsequent section TIME PERCEPTION IN THE CIRCADIAN DOMAIN).

In the past 20 years, much has been learned about circadian clocks, their cellular and molecular mechanisms,[65] oscillatory characteristics,[66] and physiology.[6] There is ample evidence that the clock represents a multioscillatory system which comprises driving (self-sustaining) and driven (self-sustaining as well as damped) units. It is also clear that central pacemakers control the whole system. They are located in the hypothalamus, for example, in the nuclei suprachiasmatici (mammals), and in the pineal body (birds and lower vertebrates), but recent findings indicate that pacemaker-like structures may also exist elsewhere in the central nervous system (for discussion see Aschoff *et al.*[16]).

CIRCADIAN CONTROL OF PERFORMANCE

It has been known for a long time that the efficiency of performance varies with time of day. Prominent examples are provided by rhythms in reaction time, as demonstrated in FIGURE 4 for optical signals (four independent studies) and for acoustical signals (two studies). It has often been emphasized that these rhythms are more or less mirror images of the rhythm in rectal temperature, and a causal relationship between temperature and reaction time has been postulated.[23] Such conclusions are dangerous because they are based on correlations between processes that are programmed in time and may well be driven independently by coupled circadian pacemakers. Several sets of data contradict the hypothesis that reaction time is dependent on body temperature.[64] In one study, multiple-choice visual reaction time was measured in seven members of a crew during a sea voyage.[24] Tests were made at 4-hr intervals during 4 to 6 consecutive days in four different months. As shown in FIGURE 5, there was a strong negative correlation (r = −0.90 to −0.97) between reaction time and body temperature. There were also considerable changes in the mean levels of reaction times and body

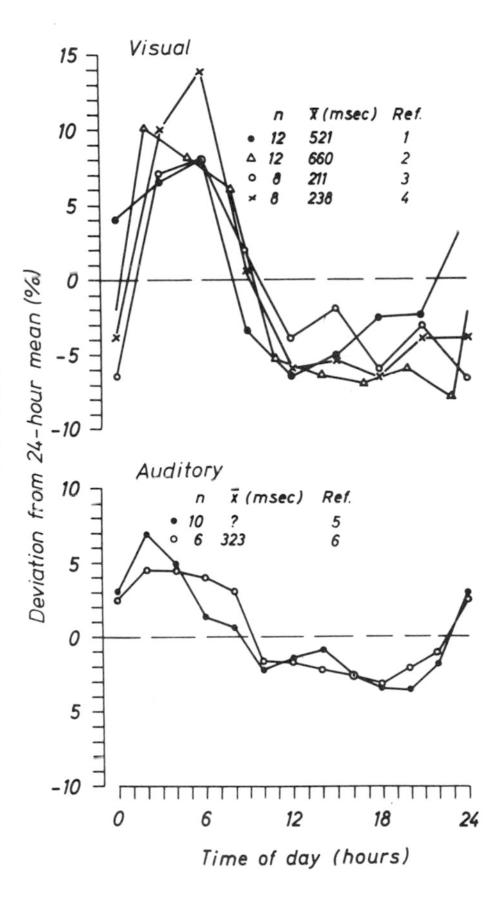

FIGURE 4. 24-hr variations in visual and auditory reaction time obtained in studies at seven different laboratories. n = number of subjects; x̄ = 24-hr mean. Sources in *upper diagram*: ● = Ref. 17; △ = Ref. 18; ○ = Ref. 19; x = Ref. 20; ▲ = Ref. 21. Sources in *lower diagram*: ● = Ref. 22; ○ Ref. 8 ("blind" subjects).

temperatures calculated on a 24-hr basis (see the dashed horizontal lines in FIGURE 5). These values were positively correlated with each other (r = 0.93). The changes in level made it possible to correlate measurements made in one subject in different months either at different times of day or at the same time of day, the latter approach excluding the time-of-day effects. As can be seen from FIGURE 6, the coefficients of correlation are all negative when data taken at different times of day are correlated (upper histogram), but become randomly distributed around zero when time-of-day effects are excluded (lower histogram). In the meantime, we have learned that it is the

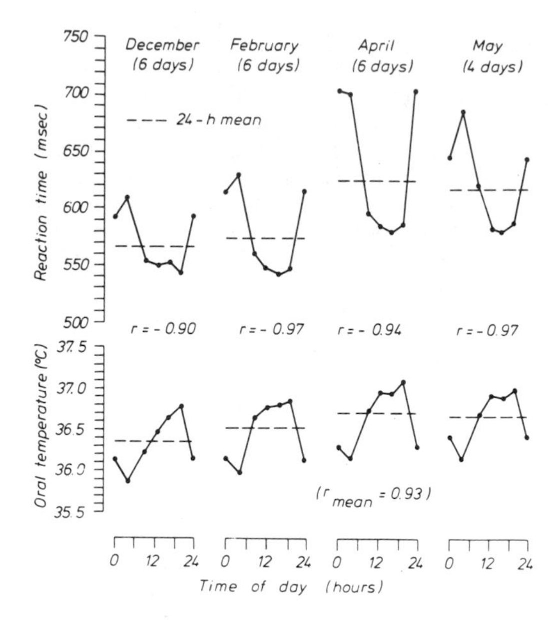

FIGURE 5. 24-hr variations in visual reaction time and oral temperature. Means of seven subjects, tested at four different months during a sea voyage. Each curve represents an average from measurements made during at least four consecutive days. (After Mann *et al.*[24])

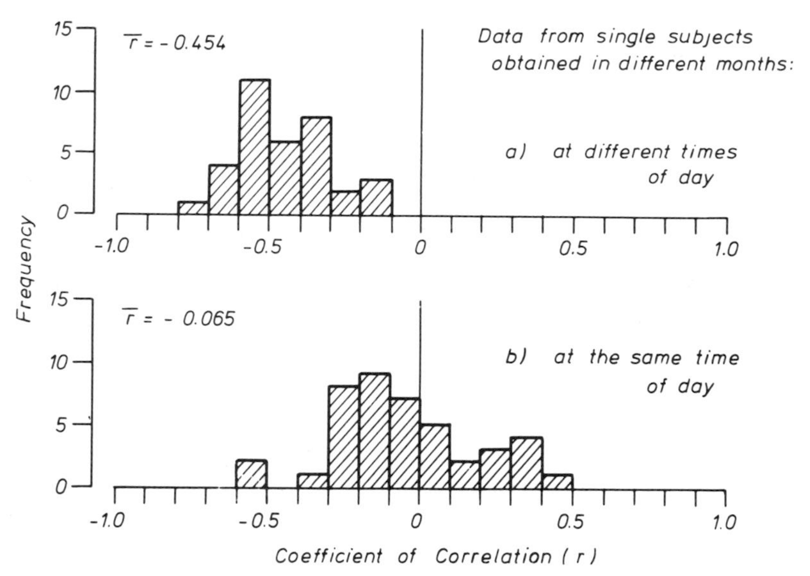

FIGURE 6. Coefficients of correlation between visual reaction time and oral temperature. Tests were made in 4-hr intervals during at least four consecutive days at four months of a sea voyage. (Data base as in FIGURE 5; after Mann *et al.*[24])

kind of task, and especially its memory load, which determines the phase relationship between the rhythms of performance and of body temperature.[25] Using a letter cancellation test, Folkhard and coworkers[26] have shown that the performance rhythm runs in phase with that of rectal temperature in a two-letter target, but counterphase in a six-letter target (FIGURE 7). These and other findings suggest that performance rhythms may be controlled by a variety of circadian oscillators.[27]

Even very simple motor performances can show strong circadian variations. The speed of tapping has been found to depend on circadian phase, both in tapping at the highest possible rate and in comfort tapping (that is, at a rate preferred by the subject).

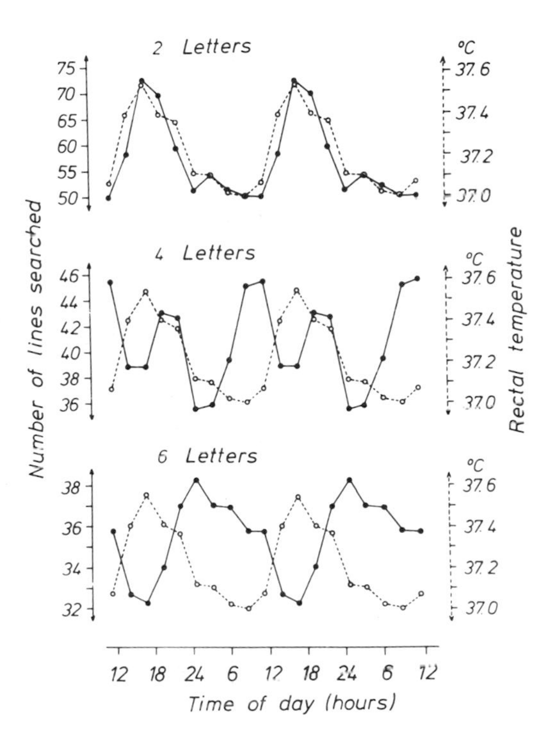

FIGURE 7. 24-hr variations in rectal temperature and in letter cancellation tests of different memory load (2-, 4-, or 6-letter target). Means from two subjects working for 18 days on a fast rotating shift schedule. (From Folkard.[25] Reprinted by permission.)

Interestingly, the data presented in FIGURE 8 indicate that females have a lower rate of fast tapping than do males, but prefer a higher speed in comfort tapping. The estimation of short time intervals also varies with time of day, as illustrated in FIGURE 9 for the production of 10-sec intervals and the reproduction of 6-sec intervals. Again, a correlation between the rhythms of time estimation and of body temperature should not be taken as evidence of a causal relationship, a hypothesis that has been considered to be supported by some findings,[30] but is contradicted by others.[31]

Most of these rhythms do not depend on the presence of a light–dark cycle. We have measured the estimation of short time intervals (production of 10 sec), together with various other performances and physiological functions, in six subjects who lived first for 4 days in a light–dark cycle (LD), and thereafter for 4 days in continuous darkness (DD), with a controlled sleep time from 23:15 to 7:30. No differences were found in either the amplitude or the phase of all rhythms between the two conditions.[32] The results of the 10-sec estimation task are reproduced in FIGURE 10. This time estimation was the only rhythm in which a slight difference was discovered: the 24-hr mean (±S.D.) was 10.84 ± 0.51 in LD, and 10.17 ± 0.46 in DD (difference significant to $p < 0.001$). I am not aware of any other study in which an effect of light or darkness on short time estimates has been demonstrated.

Furthermore, many performance rhythms persist during prolonged sleep depriva-

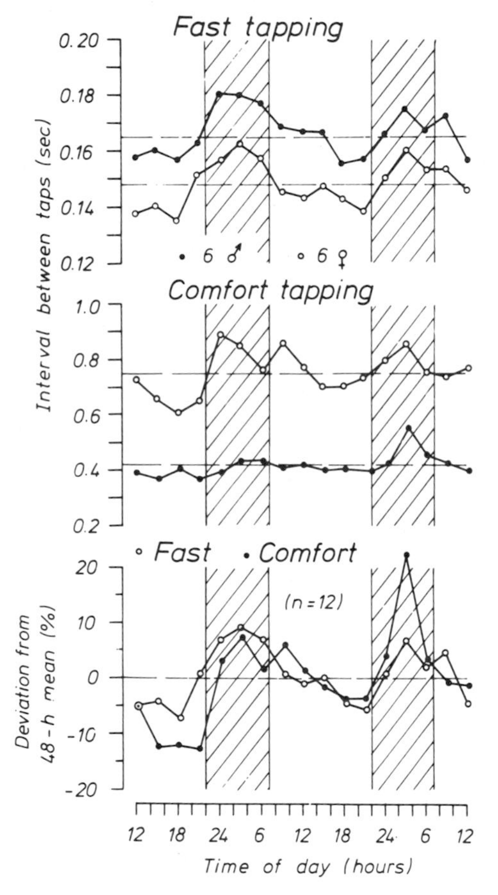

FIGURE 8. 24-hr variations in the speed of fast tapping (at the highest possible rate) and comfort tapping (at the preferred rate) measured in 3-hr intervals in six female and six male subjects during 48 hr with controlled sleep (*shaded area*). (From Winnewisser.[28] Reprinted by permission.)

tion.[17,33-37] Under those conditions, the range of oscillation (the "amplitude") may either increase, mainly due to a lowering of night values,[37] or decrease, depending on the kind of task.[36,38] An exception seems to be short time estimation, the rhythm of which usually disappears in subjects deprived of sleep. This is illustrated in the upper two diagrams of FIGURE 11; the lower two diagrams show the persistence of the tapping rhythm during sleep deprivation.

Finally, circadian rhythms in performance have been demonstrated to free-run in

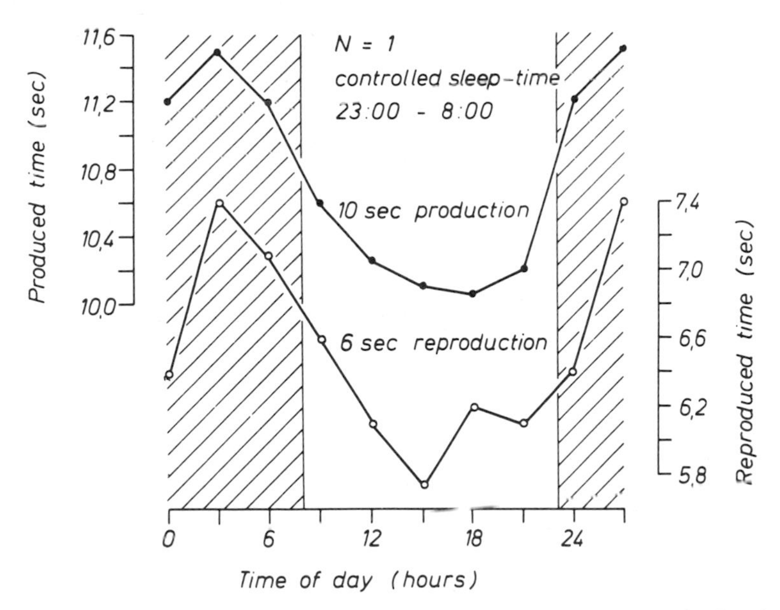

FIGURE 9. 24-hr variations in the production of 10-sec estimates, and the reproduction of 6-sec, measured in one subject during six days with a strict schedule of wakefulness and sleep (sleep time *shaded*). (From Pöppel and Giedke.[29] Reprinted by permission.)

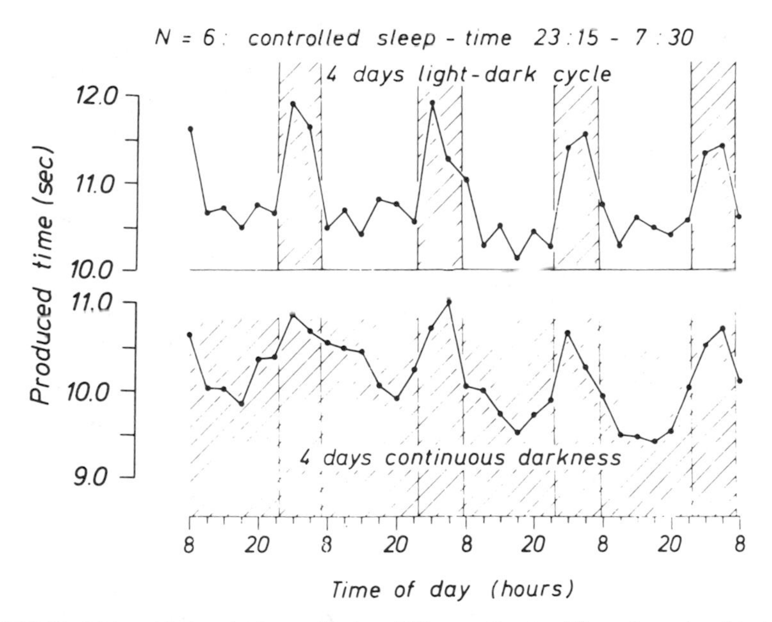

FIGURE 10. 24-hr variations in the production of 10-sec estimates. Means from six subjects who lived, first, for 4 days in a light–dark cycle, and thereafter for 4 days in continuous darkness. Sleep time under both conditions was from 23:15 to 7:30. Shaded area = darkness. (From Pöppel and Giedke.[29] Reprinted by permission.)

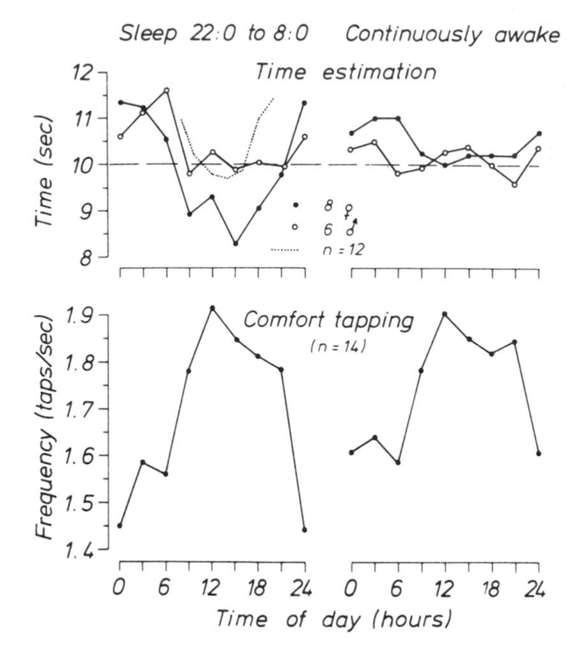

FIGURE 11. 24-hr variations in the production of 10-sec estimates and in the speed of comfort tapping, measured during a schedule with either sleep from 22:00 to 8:00 (*left*) or during sleep deprivation (*right*). Means of n subjects.

subjects living in isolation without time cues. A few examples are provided in FIGURE 12.[17] It should be pointed out, however, that the rhythm in time estimation shown in FIGURE 12 barely reached a level of statistical significance; in other isolation studies, no rhythm in the production of 10- or 20-sec estimates could be found (see the section TIME PERCEPTION IN THE CIRCADIAN DOMAIN). In cases of internal desynchronization, a performance rhythm may either follow the rhythm of rectal temperature or the sleep–wake cycle. This has been shown in subjects who were exposed in the isolation unit to a "strong" zeitgeber, that is, a light–dark cycle (and gong signals) without the availability of a reading lamp.[27] Under those conditions, the subjects were forced to adhere to the zeitgeber with their sleep–wake cycles. The results of a typical experiment are reproduced in FIGURE 13. When the period of the zeitgeber was lengthened to 28 hr, the sleep–wake cycle remained entrained, but the rhythms of rectal temperature and of performance (computation task according to Pauli) started to free-run with a mean period of 24.8 hr ("forced internal desynchronization"). After a further lengthening of the zeitgeber period to 32 hr, rectal temperature still had a period of 24.8 hr, but the computation rhythm was now in synchrony with the sleep–wake cycle.

INTERACTIONS BETWEEN ULTRADIAN AND CIRCADIAN RHYTHMS

Many animals, especially small mammals, show a rhythmic alternation between activity and rest in short intervals of 1 to 3 hr in duration.[39] These ultradian rhythms are sometimes very precise, suggesting the existence of an ultradian pacemaker. The activity of a field vole (FIG. 14), recorded under natural photoperiodic conditions between March and June at a latitude of 47°N, illustrates the regularity at which bursts of activity occur, especially in early spring, when the animal is mainly day–active. Beginning in May, nocturnal activity predominates, a shift in phase that has been observed in a variety of rodent and fish species.[41,42] In the data of March and April, a few peculiarities can be noticed: one burst of activity is strongly coupled to sunrise; the sequence of bursts becomes progressively less regular during the day; at night, the inter-burst interval (the ultradian period) is apparently longer than during the daytime.

In constant conditions, ultradian rhythms may persist with unchanged or even improved regularity. This is illustrated in FIGURE 15 by the actograms of three common voles *Microtus arvalis,* kept either in light–dark cycles (LD) or in continuous darkness (DD). In addition to the ultradian rhythms, all three animals show a circadian component (least expressed in animal 202). Several observations are noteworthy: (1) The ultradian rhythm is sometimes less pronounced in LD than in DD. (2) When free-running in DD, both components seem to remain more or less in synchrony with each other (animals 202 and 201), and a certain phase relationship is often kept between one of the ultradian bursts and the onset of the circadian activity component. (3) In a free-running rhythm, the onset of the circadian component can "jump" from one ultradian burst to a neighboring one (animal 205).

Actograms like those presented in FIGURE 14 and 15 pose several questions: (1) Is there a hierarchical order between the two components, that is, are the frequencies related to each other in integer units? (2) If the answer to (1) is negative, what effects

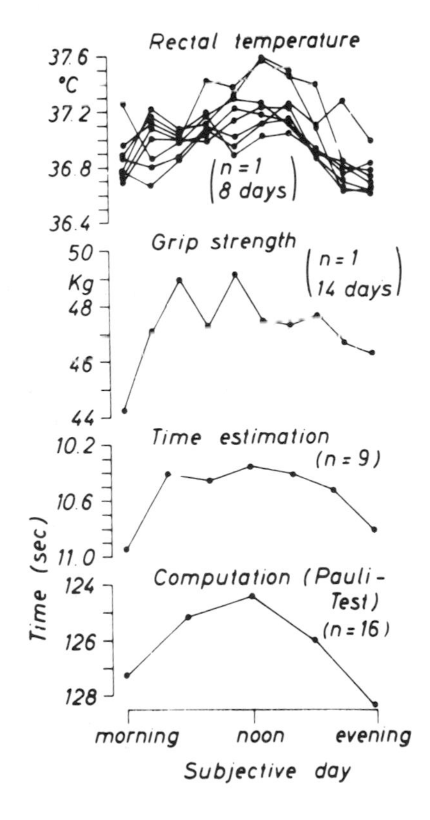

FIGURE 12. Circadian rhythms in rectal temperature, grip strength, production of 10-sec estimates and computation speed (Pauli test). Means of different subjects who lived for various time spans in an isolation unit without time cues. (From Aschoff *et al.*[17] Reprinted by permission.)

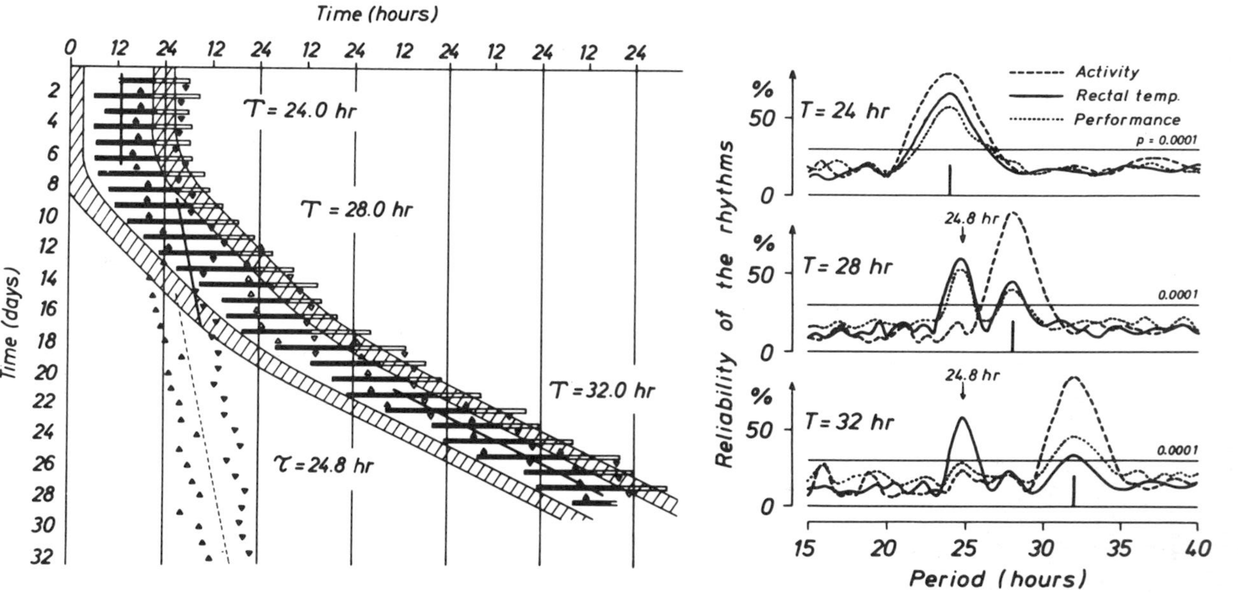

FIGURE 13. Circadian rhythms of a subject living in the isolation unit under the influence of an artificial strong zeitgeber (light–dark cycle with gong signals, no reading-lamp available). (*Left*) The rhythms of wakefulness and sleep (black and white bars), of rectal temperature (triangles for maxima and minima), and of computation speed (solid lines connect the daily phases of maximal performance speed). Shaded area = darkness. T = zeitgeber period. τ = mean circadian period. (*Right*) Period analyses of the time series computed separately for the three sections of the experiment with different zeitgeber periods. (From Wever.[27] Reprinted by permission.)

can be seen of the circadian on the ultradian rhythm? (3) Are there effects of the ultradian on the circadian rhythm?

In answer to question (1): From the analyses of many actograms, it must be concluded that an integer relationship between the two frequencies is a rare exception. In nearly all instances, ultradian periods are not a submultiple of the circadian period, and an interspecific comparison reveals that the ultradian period is positively correlated with body size,[43] while the circadian period is size-independent.

In answer to question (2): In most actograms one can see a burst of activity which is neatly coupled to the onset of the main activity component (the activity time, α). In a

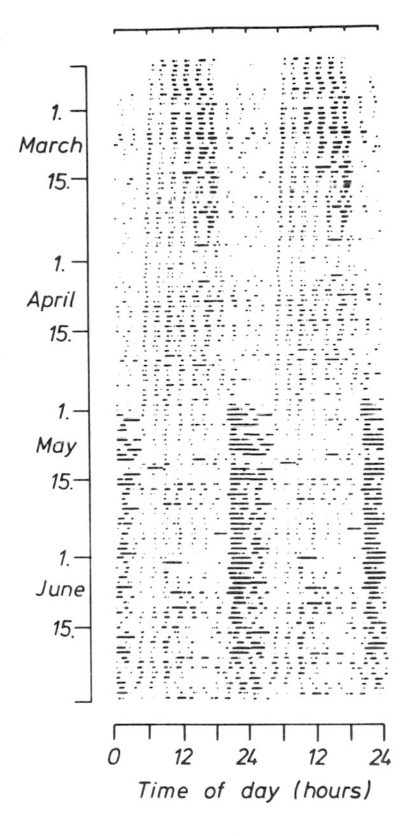

FIGURE 14. Activity pattern of a field vole, *Microtus agrestis*, recorded indoors under natural photoperiodic conditions in southern Germany. Original record plotted twice along the abscissa. (From Erkinaro.[40] Reprinted by permission.)

light–dark cycle this means that one of the bursts usually occurs either around dawn or around dusk, depending on the activity mode of the species. Furthermore, the ultradian frequency often varies with circadian time. We have analyzed this effect in more detail in the actogram of mice, *Mus musculus*, by measuring the duration of the ultradian periods, separately for the activity time α and for the rest time ρ. Data were taken from 15 animals kept in LD, and 18 animals kept in DD. As shown by the histograms in FIGURE 16, the ultradian period was consistently shorter in α than in ρ under both conditions.

In answer to question (3): The "jumping" of the main activity onset from one ultradian burst to an adjacent burst, either to the preceding, as in FIGURE 15 (animal

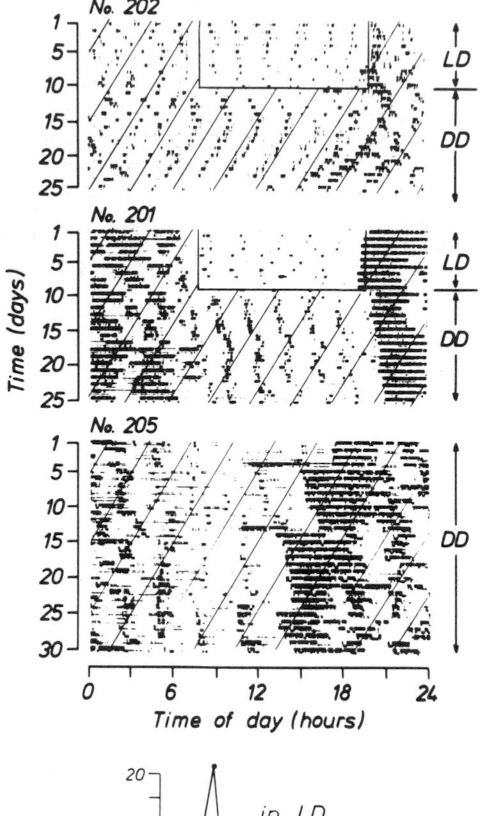

FIGURE 15. Activity patterns of three common voles, *Microtus arvalis*, recorded in the laboratory in either a light–dark cycle (LD) or in continuous darkness (DD). Courtesy of Menno Gerkema (unpublished data).

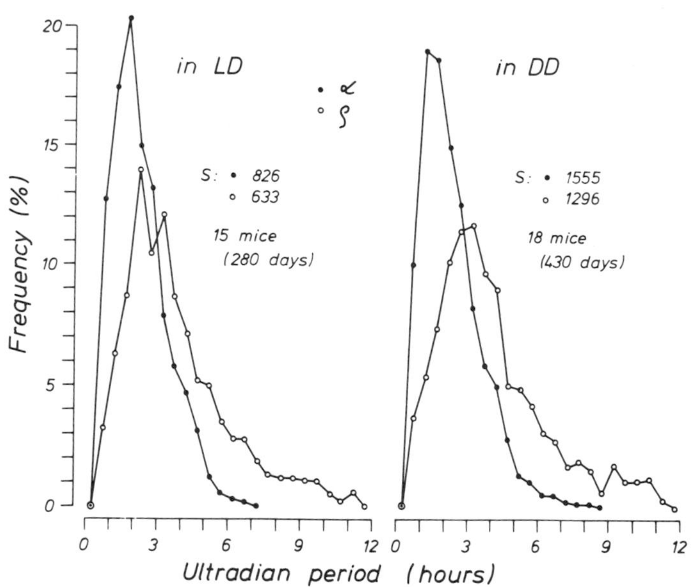

FIGURE 16. Frequency histograms of the duration of ultradian cycles in the activity of mice (*Mus musculus*) recorded in the laboratory either in a light–dark cycle (LD) or in continuous darkness (DD). The data were separately analyzed for the activity time α and the rest time ρ of the circadian rhythm. (Data from Aschoff and Meyer-Lohmann.[44])

205), or to the succeeding one, has been seen in only a few of the actograms recorded so far in constant conditions. The number of observations, however, justifies the hypothesis that there are effects of the ultradian on the circadian system.

In summary, the following conclusions can be drawn: Ultradian and circadian rhythms are usually not coupled in a strict hierarchical order; the circadian system can have a "phase-setting" effect on the ultradian rhythm and modulates its frequency; effects of the ultradian on the circadian system are suggested by the phenomenon of "jumping." It should be mentioned that all these interactions in one way or another also apply to the human circadian system and the REM/NREM cycle observed during sleep. The period of the REM/NREM cycle varies with circadian phase[45]; sleep-onset has a phase-setting effect on the first REM episode; the REM/NREM cycle is not a submultiple of the circadian period[46]; waking-up predominately occurs during

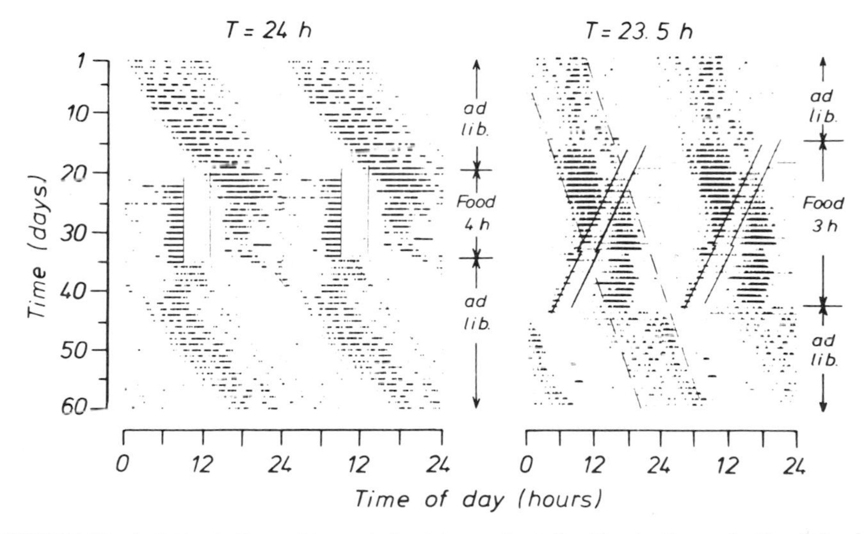

FIGURE 17. Activity rhythms of two rats kept in constant dim illumination and either fed *ad libitum* or for a few hours per day only. Meal times are indicated by two parallel *solid lines*. *Dashed lines* through onset and end of activity were drawn "by eye." T = interfeeding interval. Original records plotted twice along the abscissa. (From Aschoff *et al.*[55] Reprinted by permission.)

REM[47,48], and in a similar way sleep onset or "sleepability" may be triggered by a certain phase of an ongoing ultradian rhythm, if there is one.[49]

THE PROBLEM OF ANTICIPATORY ACTIVITY

A light–dark cycle represents the most powerful zeitgeber for the circadian rhythms of most organisms, but other environmental factors may also be effective, such as a cycle of high and low temperature[50] or of social signals.[51,52] It also has been postulated that circadian rhythms become entrained when food is offered periodically.[53,54] However, recent extensive studies have demonstrated that the activity rhythms of rats kept in otherwise constant conditions continue to free-run when *ad libitum* feeding is replaced by a schedule of offering food for a few hours per day only. The example provided in the left diagram of FIGURE 17 indicates that neither the phase nor the period of the free-running rhythm is altered by restricted feeding (RF), but it also shows a band of activity just prior to the time of feeding: the well-known anticipatory activity. Remarkably, this band of activity persists for several days after RF is again replaced by *ad libitum* feeding. From several such observations we have postulated that RF, although it does not act as a true zeitgeber, uncouples from the main circadian system a component of activity which has circadian-like characteristics.[56] If this assumption is correct, it could be expected that the duration of anticipatory activity depends on the period of the "entraining" feeding cycle (see the discussion in the previous section, THE CIRCADIAN SYSTEM, and FIGURE 2). To test this hypothesis, we have done a series of experiments in which the interfeeding interval, T, of RF was varied between 22 and 27 hr.[57] The results of an experiment with $T = 23.5$ (FIG. 17, right diagram) agree with the expectation as anticipation was shorter in $T = 23.5$ hr than in $T = 24$ hr. The summary of all results in FIGURE 18 demonstrates that the duration of anticipatory activity is positively correlated with T within the limits from 23.5 to 27 hr, and that no anticipation is left in $T = 22$ hr. Together with some other findings which cannot be discussed here in detail (see Aschoff *et al.*[57]), these findings support the notion that RF uncouples from the circadian system a subcomponent which has oscillatory capacities. The range of T values to which the subcomponent can be entrained seems to have narrow limits as is known from the circadian system.[13]

It has been reported that non-24-hr feeding cycles cannot be anticipated by rats which are entrained to 24 hr by a light–dark cycle.[58] This conclusion is based on experiments in which interfeeding intervals of 19 and 29 hr were used, that is, T values that probably were outside the range of entrainment for the subcomponent. We have repeated those experiments with T values closer to 24 hr in rats that were entrained to 24 hr by a light–dark cycle. Anticipation was seen in $T = 23.5$ as well as in $T = 25.0$ hr, and the duration of anticipatory activity depended on T in a similar way as in rats whose rhythms were free-running in constant darkness.

Anticipatory activity has often been characterized as the result of "learning," but I have not seen any explicit hypothesis on what mechanism such learning could be based. Somehow, it has to be a process of time measurement, but of what kind? The circadian system can hardly be used for this measurement, because rats anticipate feeding intervals that deviate from the period of free-running as well as of entrained circadian rhythms. The assumption of a circadian subcomponent that can be uncoupled from the main circadian system and is entrainable by restricted feeding is compatible with all observations made so far. Anticipatory time measurement cannot be based on the circadian pacemaker, as shown by the persistence of free-running rhythms during RF as well as by the observation that anticipation, with its typical dependence on T, occurs in animals that have been made arrhythmic by creating lesions in the nuclei suprachiasmatici.[59] It remains to be seen whether other pacemaker-like structures are involved.

TIME PERCEPTION IN THE CIRCADIAN DOMAIN

Subjects living in isolation units without external time cues soon become unaware of the real passage of time and even lose interest in it.[7] Experiments performed over the past 20 years with more than 250 volunteers have convinced us that only rarely can a subject judge with what circadian period he may have been living, and estimate, at the end of the experiment, how many real days he had been in the unit. Most subjects believe that they follow a more or less "normal" (that is, 24-hr) schedule of wakefulness and sleep despite the fact that the cycle usually exceeds 25 hr and can be much longer. A subject who developed a circa-bi-dian sleep–wake cycle with a period of about 50 hr estimated his stay in the unit half as long as it was in reality. In short, the passage of time is consistently underestimated.[7,60,63]

To learn more about time perception in conditions of isolation we have done several series of experiments in which the subjects were asked to press a button whenever they thought that 1 hr had passed. At each of these time points, the subjects also had to press another button for the duration of estimated 10 or 20 sec. All signals were

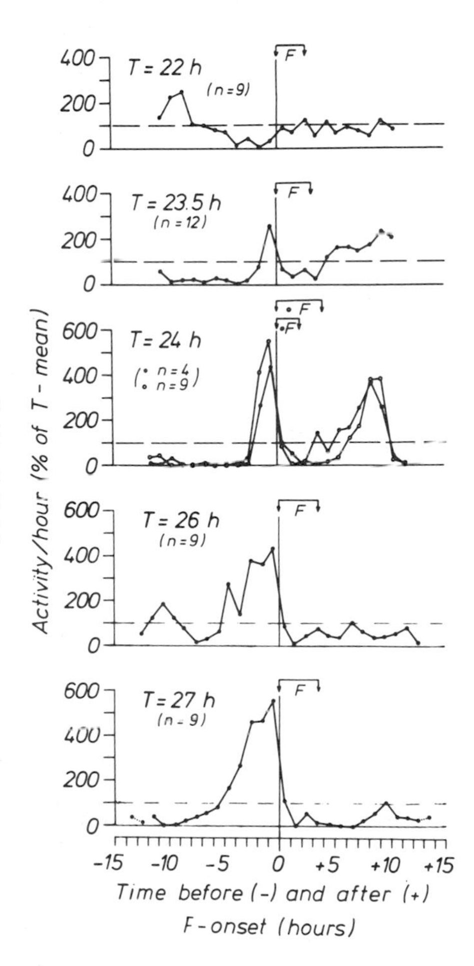

FIGURE 18. Activity patterns of rats kept in constant dim illumination and fed for a few hours per day only. F = feeding time, T = interfeeding interval. The hourly amount of activity is expressed in percent of the mean activity recorded in a full T cycle. n = number of animals. (From Aschoff *et al.*[57] Reprinted by permission.)

recorded outside the isolation unit for the full duration of the experiment, which usually lasted for 3 to 4 weeks. The data were collected many years ago, but were never published in detail. Consecutive estimates of 1-hr intervals, recorded in three subjects during 5 days of isolation, are presented in FIGURE 19. With few exceptions, the estimates were considerably longer than 1 hr. They also showed large variations within a day, but the daily means remained fairly stable. From several of the curves reproduced in FIGURE 19 one further gets the impression that there were two maxima per day, one in the morning and another one in the afternoon. On the assumption that there was an underlying bimodal distribution of estimates, we attempted to average the data from many days and several subjects. To make this possible, despite interindividual differences in period length of the free-running rhythms, we normalized the duration of wakefulness to 100%, divided it into 10 segments of equal duration (= 10% of α), and assigned each 1-hr estimate to the segment in which it ended. In doing so, we could average the data from many days irrespective of the duration of α. Finally, the 10 values were expressed in percent deviations from α-mean. The resulting curves (FIGURE 20), obtained from seven subjects in 1961–62 and seven subjects in 1965–66, support the hypothesis that 1-hr estimates have a bimodal circadian distribution. An explanation for this is offered by the concept of "filled" and "empty" time or the "level of behavior."[61] The underestimation of the passage of time is greater when subjects are

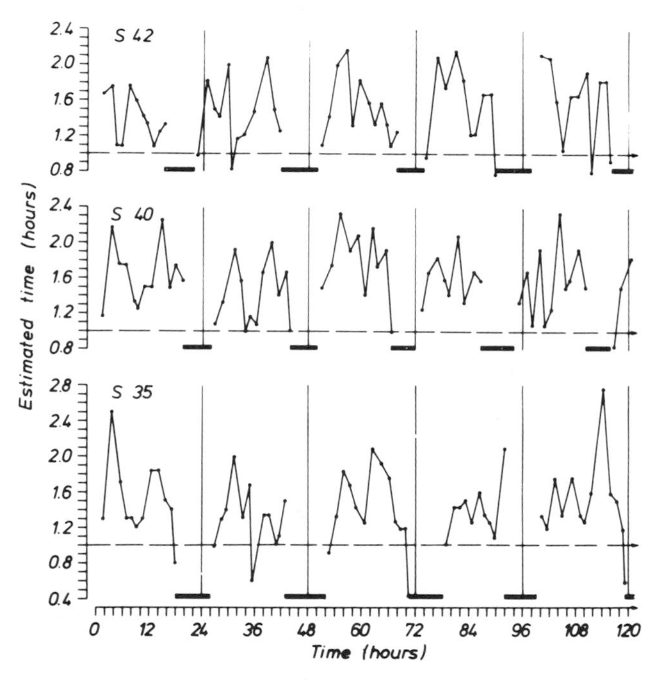

FIGURE 19. Consecutive estimates of 1 hr made by three subjects (S) who lived singly in an isolation unit without time cues. Horizontal bars = sleep.

alert and engaged in activities (late morning and afternoon), and the underestimation is smaller when subjects feel sleepy and bored, as is often the case during the "saddle" of efficiency around noon.

In a further step, we have calculated means of 1-hr estimates for every "day," and have related these values to the length of time the subject was awake during that day. As can be seen in FIGURE 21, there was a strong positive correlation between the production of 1-hr estimates and the duration of wakefulness, α, in all subjects. This is of special interest in cases of internal desynchronization where α was lengthened to 26 hr and more (subjects 5 and 7). In contrast to these very consistent findings in 1-hr estimates, the correlation between the estimation of 10 sec and α was only slightly positive in some subjects, negative in others, and also often close to zero. This is

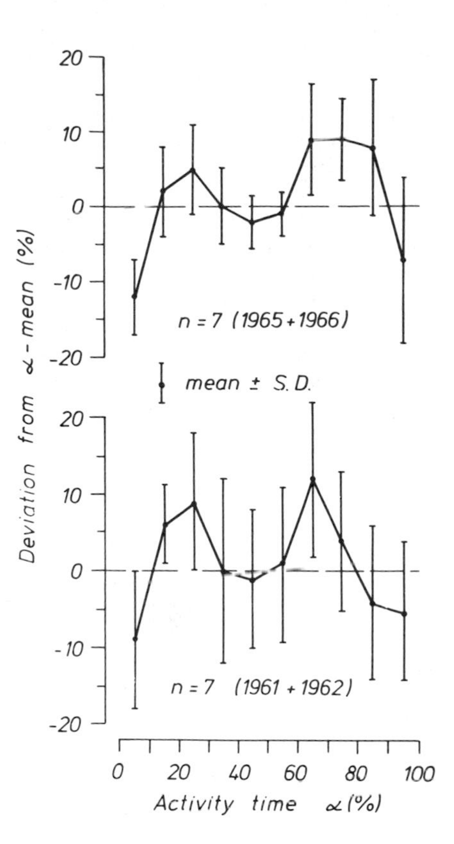

FIGURE 20. Circadian variations of 1-hr estimates. Means of two groups of seven subjects who lived singly in an isolation unit without time cues. For each single circadian period, the full time of wakefulness (activity time, α) was set as 100% and divided into ten classes of equal duration.

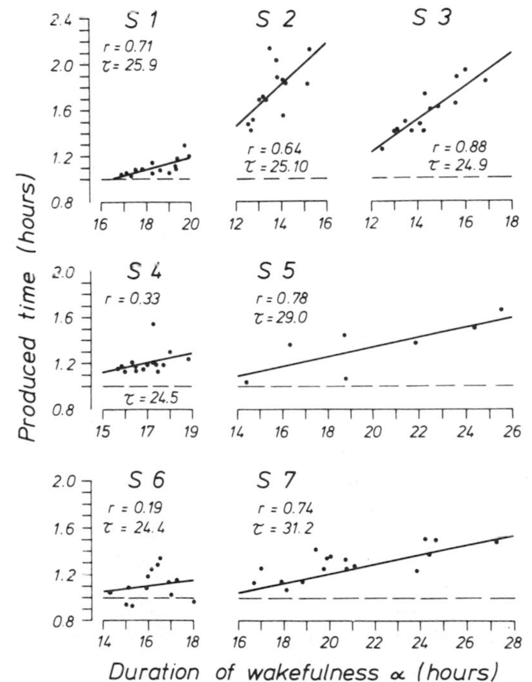

FIGURE 21. The dependence of 1-hr estimates on the duration of wakefulness, α, measured in seven subjects (S) who lived singly in an isolation unit without time cues. Each dot represents the mean of estimates made during one "day." r = coefficient of correlation; τ = mean circadian period as measured during the full experiment.

indicated in FIGURE 22 by the data from four subjects (upper row) who again showed a clear positive correlation between 1-hr estimates and α (lower row). From the same set of data, correlations were also calculated between 1-hr estimates and the circadian period τ; they were also positive, but less pronounced than those between the 1-hr estimates and α. The 10-sec estimates were not systematically correlated with τ. The frequency histograms of all coefficients of correlation, summarized in FIGURE 23, demonstrate again that, in contrast to the 1-hr estimates, the estimation of short time intervals is neither correlated with α nor with τ.

The intraindividual positive correlation between 1-hr estimates and α is also true interindividually. In the attempt to demonstrate this, we have taken into account individual differences in the mean time estimation by calculating the variation (the range) of the daily mean 1-hr estimates from the overall mean of each subject; these

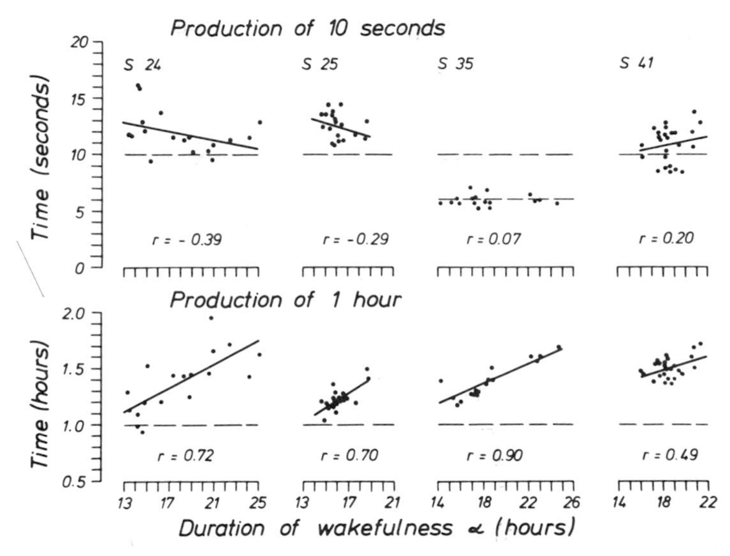

FIGURE 22. The dependence of 10-sec estimates (*above*) and 1-hr estimates (*below*) on the duration of wakefulness, α, measured simultaneously in four subjects (S) who lived singly in an isolation unit without time cues. Each dot represents the mean of estimates made during one "day." r = coefficient of correlation.

variations, expressed in percentage of the overall mean, were then related to the similarly calculated variations of α. According to FIGURE 24, the coefficient of correlation between the variation in 1-hr estimates and the variation of α, computed interindividually, was +0.575. We also calculated, separately for each subject, the extent to which the 1-hr estimates came close to being directly proportional to α, with 100% representing a 1:1-relationship. A frequency histogram of these factors of proportionality is provided in the inset of FIGURE 24.

Two conclusions can be drawn from these findings: (1) The estimation of short time intervals (10 or 20 sec) seems to be based on a process radically different from that which is used to estimate longer time intervals such as 1 hr; hence, it is not too surprising that we could not find a dependence on circadian phase in the estimation of short time intervals, although it was well expressed in the 1-hr estimates (FIG. 20). (2)

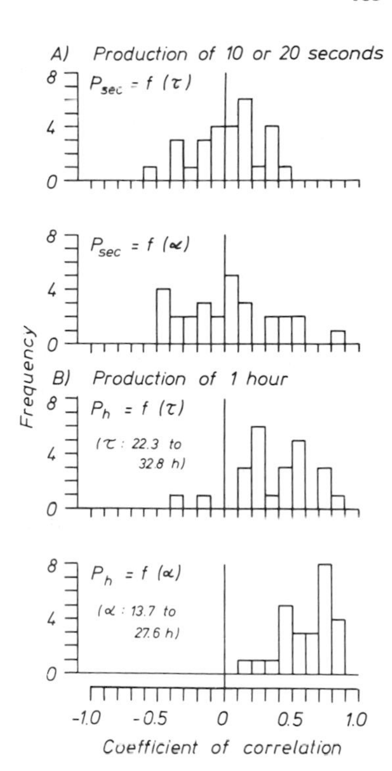

FIGURE 23. Frequency histograms of the coefficients of correlation between the production of short time estimates (P_{sec}) as well as the production of 1-hr estimates (P_h) and the duration of wakefulness (α) or the circadian period (τ), respectively. The interindividual variation of α and of ρ is given in brackets.

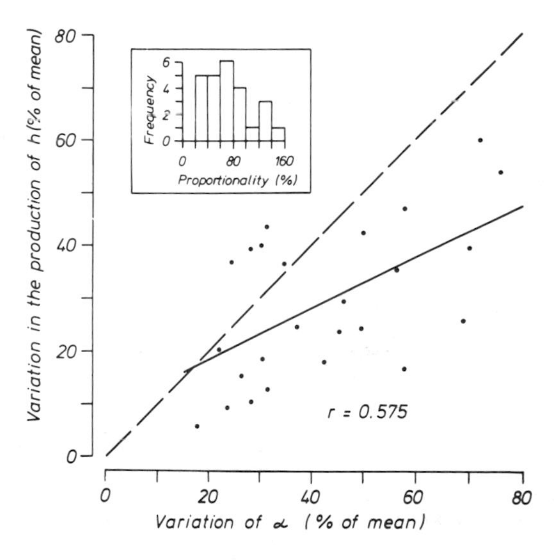

FIGURE 24. The variability of 1-hr estimates (expressed in percent of the mean of each subject) drawn as a function of the variation in wakefulness, α (expressed in percent of the mean α). Each dot represents the results from one subject who lived in the isolation unit without time cues. *Inset:* frequency histogram of factors representing the extent to which the variation in 1-hr estimates was proportional to the variation in α (100% = 1:1 proportionality).

The positive correlation between 1-hr estimates and α explains why subjects, despite their long and variable sleep–wake cycles and even during internal desynchronization, could believe that they were living on a normal 24-hr day because they produced about the same numbers of subjective "hours," irrespective of however long they were staying awake.

Intra- and interindividually, the 1-hr estimates vary less than the short time estimates. This is demonstrated in FIGURE 25 by frequency histograms which include all estimates made by six subjects (each line one subject) in the course of an experiment. The two insets of FIGURE 25 show the distribution of individual means for 10-sec estimates (18 subjects) and 1-hr estimates (25 subjects). To emphasize the difference in variability between short and long time estimation, consecutive daily means of time estimates are drawn in FIGURE 26 from 12 subjects who all made 1-hr estimates. In addition, four of these subjects were asked to produce 20-sec estimates (A), the other 8 to produce 10-sec estimates (B). The eight subjects included in (B) were arbitrarily divided into two groups of four each, according to individual differences in short time estimation. There were drastic differences in the overall means of 20-sec and 10-sec estimates (upper row of FIGURE 26). Some subjects were quite stable in their individual mean estimate, whereas others had a tendency to

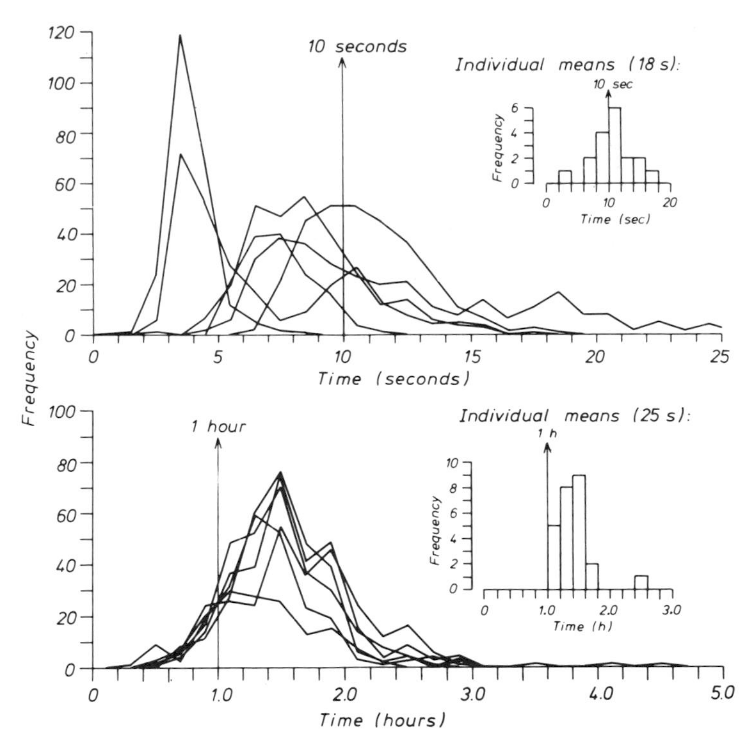

FIGURE 25. Frequency histograms of 10-sec estimates (*above*) and 1-hr estimates (*below*) measured in six subjects who lived singly in an isolation unit without time cues. Each curve includes all estimates made by a subject in the course of an experiment. *Insets:* frequency histograms of the individual means of 10-sec estimates (18 subjects) and of 1-hr estimates (25 subjects).

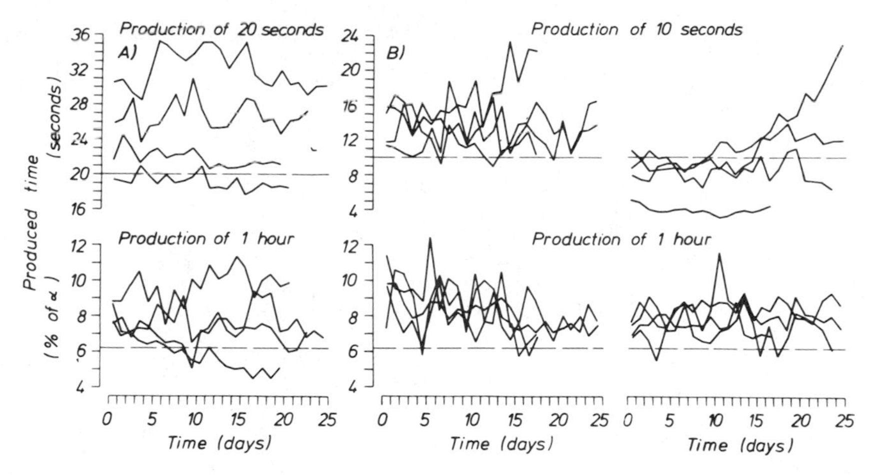

FIGURE 26. Consecutive daily means in the production of 20-sec or 10-sec estimates (*upper row*) and in the production of 1-hr estimates (*lower row*) measured simultaneously in subjects who lived singly in an isolation unit without time cues. (**A**) four subjects; (**B**) eight subjects, divided arbitrarily into two groups of four each according to individual differences in the production of 10-sec estimates.

lengthen the estimates in the course of the experiment. In the lower row of FIGURE 26, the 1-hr estimates are not given in real time but in percent of α. By this approach, the dependence of time estimation on α has been taken into account. (In a "normal" day with 16 hr of wakefulness, 1 hr amounts to 6.2% of α; see the dashed horizontal lines.) Most of the subjects kept to their individual mean throughout the experiment; in a minority an initial tendency to shorten was noticeable. All subjects except one produced means longer than 1 hr. The overall mean was about 1.4 hr, a value slightly longer than the 1.2 hr reported by Lavie and Webb[60] (mean of data from an exercising and a nonexercising group of subjects) and by Campbell[62] (data from nine subjects with 60 hr of bedrest). A more detailed analysis of time perception under conditions of prolonged temporal isolation will be published elsewhere.[68]

CONCLUDING REMARKS

There seems to be almost no process of timing and time perception that is not affected by the circadian clock in one or another way. Often, these influences result in a mere modulation of rate in, for example, the processes that underlie short-term performance. On the other hand, rhythmically timed events, such as the ultradian cycles in behavior, can be coupled to the circadian clock with unidirectional effects on the frequency of the ultradian system and bidirectional effects on the phases of both systems. Finally, of major importance, but presently least understood, are the effects of the circadian clock on the estimation of time in the range of hours. In this context, it is noteworthy that the timing of meals by subjects living in the isolation unit is correlated with the duration of wakefulness in a way similar to that observed in the 1-hr estimates. To give an example: the interval between waking up and lunch is more or less proportional to the time the subject stays awake.[67] These recent findings suggest that, during internal desynchronization, the extreme lengthening (or shortening) of the "circadian" period involves not only the sleep–wake cycle, but other physiological processes as well, possibly including metabolic rate. Under those aspects, new questions come up concerning the relationship between timing as well as time perception and the various components of the circadian system.

REFERENCES

1. ASCHOFF, J., Ed. 1981. Handbook of Behavioral Neurobiology. Vol. IV: Biological Rhythms. Plenum. New York, NY.
2. PITTENDRIGH, C. 1981. Circadian systems: General perspective. *In* Handbook of Behavioral Neurobiology. Vol. IV: Biological Rhythms. J. Aschoff, Ed.: 57–80. Plenum. New York, NY.
3. PITTENDRIGH, C. 1981. Circadian systems: Entrainment. *In* Handbook of Behavioral Neurobiology. Vol. IV: Biological Rhythms. J. Aschoff, Ed.: 95–124. Plenum. New York, NY.
4. BÜNNING, E. 1973. The Physiological Clock, 3rd ed. English Universities Press, Ltd. London.
5. MINORS, D. S. & J. M. WATERHOUSE. 1981. Circadian Rhythms and the Human. J. Wright & Sons. Bristol–London–Boston.
6. MOORE-EDE, M. C., F. M. SULZMAN & C. A. FULLER. 1982. The Clocks That Time Us. Harvard University Press. Cambridge, MA.
7. ASCHOFF, J. & R. WEVER. 1962. Naturwissenschaften **49:** 337–342.
8. LUND, R. 1974. Circadiane Periodik physiologischer und psychologischer Variablen bei 7 blinden Versuchspersonen mit und ohne Zeitgeber. Med. thesis, München.
9. WEVER, R. 1979. The Circadian System of Man. Springer Verlag. Berlin–Heidelberg–New York.
10. ASCHOFF, J., E. PÖPPEL & R. WEVER. 1969. Pflügers Arch. **306:** 58–70.
11. ASCHOFF, J. 1981. Circadian system properties. *In* Advances in Physiological Sciences. F. Obál & G. Benedek, Eds. Vol. **18:** 1–17. Akadémiai Kiadó. Budapest.
12. ASCHOFF, J. 1980. Ranges of entrainment: A comparative analysis of circadian rhythms studies. *In* Proceedings of the XIIIth Conference of the International Society of Chronobiology, Pavia. F. Halberg, L. E. Scheving & E. W. Powell, Eds.: 105–112. Il Ponte. Milan.
13. ASCHOFF, J. & H. POHL. 1978. Naturwissenschaften **65:** 80–84.
14. ASCHOFF, J., U. GERECKE & R. WEVER. 1967. Jpn. J. Physiol. **17:** 450–457.
15. WEVER, R. 1975. Int. J. Chronobiol. **3:** 19–55.
16. ASCHOFF, J., S. DAAN & G. GROOS, Eds. 1982. Vertebrate Circadian Systems: Structure and Physiology. Springer Verlag. Berlin–Heidelberg–New York.
17. ASCHOFF, J., H. GIEDKE, E. PÖPPEL & R. WEVER. 1972. The influence of sleep-interruption and of sleep-deprivation on circadian rhythms in human performance. *In* Aspects of Human Efficiency. W. B. Colquhoun, Ed.: 135–150. English Universities Press, Ltd. London.
18. RINGER, C. 1972. Circadiane Periodik psychologischer und physiologischer Parameter bei Schlafentzug. Med. thesis, München.
19. HASKE, R. 1974. Das Verhalten der Tagesrhythmik von Körpertemperatur und Leistung nach zwei Transatlantikflügen in rascher Folge. Deutsche Luft- und Raumfahrt Forschungsbericht. No. 74–55.
20. KLEIN, K. E., H. BRÜNER, *et al.* 1972. Psychological and physiological changes caused by desynchronization following transzonal air travel. *In* Aspects of Human Efficiency. W. P. Colquhoun, Ed.: 295–305. English Universities Press, Ltd. London.
21. WOJTUCZIAK-JAROSZOWA, J. 1978. Chronobiologia **4:** 363–384.
22. VOIGT, E.-D., P. ENGEL & H. KLEIN. 1968. Int. Z. Angew. Physiol. Einschl. Arbeitsphysiol. **25:** 1–12.
23. KLEITMAN, N. & D. P. JACKSON. 1951. J. Appl. Physiol. **3:** 309–328.
24. MANN, H., E. PÖPPEL & J. RUTENFRANZ. 1972. Int. Arch. Arbeitsmed. **29:** 269–284.
25. FOLKARD, S. 1981. Shiftwork and Performance. *In* Biological Rhythms, Sleep and Shift

Work. L. L. Johnson, D. I. Tepas, W. P. Colquhoun & M. J. Colligan, Eds.: 283–305. SP Medical & Scientific Books. Jamaica, New York, NY.
26. FOLKARD, S., P. KNAUTH, TH. H. MONK & J. RUTENFRANZ. 1976. Ergonomics **19:** 479–488.
27. WEVER, R. 1982. Behavioral aspects of circadian rhythmicity. *In* Rhythmic Aspects of Behavior. F. M. Brown & R. C. Graeber, Eds.: 105–171. Lawrence Erlbaum. Hillsdale, NJ.
28. WINNEWISSER, M. 1972. Die tagesperiodischen Schwankungen willkürlicher psychomotorischer Aktivität und deren Beeinflussbarkeit durch Schlafentzug. Med. thesis, München.
29. PÖPPEL, E. & H. GIEDKE. 1970. Psychol. Forschung **34:** 182–198.
30. PFAFF, D. 1968. J. Exp. Psychol. **76:** 419–422.
31. PÖPPEL, E. 1971. Stud. Gen. **24:** 85–170.
32. ASCHOFF, J., M. FATRANSKA, *et al.* 1971. Science **171:** 213–215.
33. PATRICK, G. F. W. & J. A. GILBERT. 1896. Psychol. Rev. **3:** 468–483.
34. FIORICA, V., E. A. HIGGINS, P. F. IAMPIETRO, M. T. LATEGOLA & A. W. DAVIS. 1968. J. Appl. Physiol. **24:** 167–176.
35. FRÖBERG, J., C.-G. KARLSSON, L. LEVI & L. LIDBERG. 1970. Circadian variations in performance, psychological ratings, catecholamine excretion, and urine flow during prolonged sleep deprivation. Reports from the Laboratory for Clinical Stress Research, No. 14.
36. FRÖBERG, J., C. G. KARLSSON, L. LEVI & L. LIDBERG. 1972. Int. J. Psychobiol. **2:** 23–36.
37. BUGGE, J. F. & P. K. OPSTAD. 1979. Aviation Space Environ. Med. **50:** 663–668.
38. FRÖBERG, J., C.-G. KARLSSON, L. LEVI & L. LIDBERG. 1975. Försvarsmedicin **11:** 192–201.
39. DAAN, S. & J. ASCHOFF. 1981. Short Term Rhythms in Activity. *In* Handbook of Behavioral Neurobiology. Vol. IV: Biological Rhythms. J. Aschoff, Ed.: 491–498. Plenum. New York, NY.
40. ERKINARO, E. 1969. Aquilo Ser. Zool. **8:** 1–31.
41. ERKINARO, E. 1972. Aquilo Ser. Zool. **13:** 87–91.
42. ERIKSSON, L.-O. 1978. Nocturnalism versus diurnalism—Dualism within fish individuals. *In* Rhythmic Activity of Fishes. J. E. Thorpe, Ed. Academic Press. New York, NY.
43. DAAN, S. & J. ASCHOFF. 1982. Circadian contributions to survival. *In* Vertebrate Circadian Systems: Structure and Physiology. J. Aschoff, S. Daan & G. A. Groos, Eds.: 305–321. Springer Verlag. Berlin–Heidelberg–New York.
44. ASCHOFF, J. & J. MEYER-LOHMANN. 1954. Pflügers Arch. **260:** 81–86.
45. TAUB, J. M. & R. J. BERGER. 1973. EEG Clin. Neurophysiol. **35:** 613–619.
46. SCHULZ, H., G. DIRLICH & J. ZULLEY. 1976. Arzneimittelforschung **26:** 1049–1068.
47. SCHULZ, H. & J. ZULLEY. 1980. Sleep Res. **9:** 124.
48. WEITZMAN, E. D., C. A. CZEISLER, J. C. ZIMMERMAN & J. RONDA. 1980. Sleep Res. **9:** 280.
49. LAVIE, P. & A. SCHERSON. 1981. EEG Clin. Neurophysiol. **52:** 163–174.
50. TOKURA, H. & J. ASCHOFF. 1983. Am. J. Physiol. **245:** R800–R804.
51. GWINNER, E. 1966. Experientia **22:** 765.
52. MARIMUTHU, G., R. SUBBARAJ & M. K. CHANDRASHEKARAN. 1981. Behav. Ecol. Sociobiol. **8:** 147–150.
53. SULZMAN, F. M., C. A. FULLER & M. C. MOORE-EDE. 1977. Physiol. Behav. **18:** 775–779.
54. SULZMAN, F. M., C. A. FULLER & M. C. MOORE-EDE. 1978. Am. J. Physiol. **234:** R130–135.
55. ASCHOFF, J., S. DAAN & K. HONMA. 1982. Zeitgebers, entrainment, and masking: Some unsettled questions. *In* Vertebrate Circadian Systems: Structure and Physiology. J. Aschoff, S. Daan & G. A. Groos, Eds.: 13–24. Springer Verlag. Berlin–Heidelberg–New York.
56. HONMA, K., C. V. GOETZ & J. ASCHOFF. 1983. Physiol. Behav. **30:** 905–913.
57. ASCHOFF, J., C. V. GOETZ & K. HONMA. 1983. Z. Tierpsychol. **63:** 91–111.
58. BOLLES, R. C. & J. DE LORGE. 1962. J. Comp. Physiol. **55:** 760–762.
59. STEPHAN, F. K. 1981. J. Comp. Physiol. **A143:** 401–410.
60. LAVIE, P. & W. B. WEBB. 1975. Am. J. Psychol. **88:** 177–186.

61. DEWOLFE, R. K. S. & C.-P. DUNCAN. 1959. J. Exp. Psychol. **58:** 153–158.
62. CAMPBELL, S. C. 1983. Sleep Res. 12: in press.
63. HALBERG, F., M. SIFFRE, M. ENGELE, D. HILLMAN & A. REINBERG. 1965. C. R. Acad. Sci. Paris **260:** 1259–1262.
64. WILKINSON, R. T. 1982. The relationship between body temperature and performance across circadian phase shifts. *In* Rhythmic Aspects of Behavior. F. M. Brown and R. C. Graeber, Eds.: 213–240. Lawrence Erlbaum. Hillsdale, NJ.
65. HASTINGS, J. W. & H.-G. SCHWEIGER, Eds. 1976. The Molecular Basis of Circadian Rhythms. Life Sciences Research Report 1. Dahlem Konferenzen. Berlin.
66. PITTENDRIGH, C. S. 1976. Circadian clocks: What are they? *In* The Molecular Basis of Circadian Rhythms. Dahlem Konferenzen. J. W. Hastings & H. G. Schweiger, Eds.: 12–48. Berlin.
67. ASCHOFF, J., R. WEVER & C. H. WILDGRUBER 1984. Timing of meals of subjects living without time cues. Naturwissenschaften. Submitted for publication.
68. ASCHOFF, J. 1984. Hum. Neurobiol. Submitted for publication.

PART VI. THE INTERNAL CLOCK

Introduction

RUSSELL M. CHURCH

Walter S. Hunter Laboratory of Psychology
Brown University
Providence, Rhode Island 02912

The concept of an internal clock is an important part of the explanation of the behavior of animals in many situations. Of course, empirical research in psychology consists of observation and control of environmental stimuli and observation of behavior. But to account for the influence of stimulus duration in response decisions, endogenous rhythms, and the temporal regularities of behavior that reflect temporal regularities in the environment, investigators have found it useful to postulate internal clocks. Such clocks can be assumed to have biological, psychological, or purely formal properties.

Two types of clocks have been studied: periodic and interval. Periodic clocks, like the one responsible for circadian rhythms, are continuously running, self-sustaining, and particularly sensitive near a time that repeats cyclically. They can be gradually adjusted to a new cycle, but only within narrow limits. Terman, Gibbon, Fairhurst, and Waring describe some of the properties of periodic clocks that control the circadian feeding rhythm; Silver and Bittman describe some of the properties of periodic clocks used for reproductive behavior. Interval clocks are sometimes running and sometimes stopped; they can be reset abruptly; and their temporal criteria can be adjusted within a wide range. The papers by Church, Meck, and Treisman, as well as the two previously mentioned papers, describe some of the properties of interval clocks.

The periodic and interval clocks are distinct. They have different definitions, different properties, and (as described in the articles by Terman *et al.* and Silver and Bittman) the influence of both of them can be seen simultaneously in the data. These papers describe some of the ways in which the influence of a periodic and an interval clock can combine and they may be similar to some of the ways in which the influence of two interval clocks can combine (Church).

To account for behavior, the timing mechanisms must be integrated into a more general psychological theory that may include such psychological processes as attention, arousal, learning, and response strength. The information-processing model of timing that Treisman proposed in 1963 has been developed further in the chapters by Treisman, Church, and Meck. Killeen has provided an alternative model in which performance is controlled by stimulus-specific arousal, with clock speed defined as a differential weighting function on arousal. Animal investigators are now using some of the analytical techniques of human experimenters to identify specific psychological processes. For example, Meck used unbalanced stimulus probabilities and prior entry to control attention to the duration of either a light or sound stimuli.

A major purpose of animal investigations of timing has been to obtain physiological understanding of the process. This has been a promissory note of long standing. There have been various proposals regarding the physiological basis of the pacemaker. For example, it may be related to the alpha rhythm (Treisman), arousal (Killeen), or the effective level of brain dopamine (Church). The papers in this section deal primarily with the properties of periodic and interval clocks and the interrelationship between clocks and other psychological processes intervening between stimulus and response. They may prepare the way for a physiological search for the pacemaker and other parts of an information-processing system, such as switches, accumulators, working memories, reference memories, and comparators.

Daily Meal Anticipation: Interaction of Circadian and Interval Timing[a]

MICHAEL TERMAN,[b] JOHN GIBBON,
STEPHEN FAIRHURST, AND AMY WARING

New York State Psychiatric Institute
New York, New York 10032; and
Departments of Psychiatry and Psychology
Columbia University
New York, New York

INTRODUCTION

Behavior evinces "anticipation" when it changes reliably before the occurrence of some external event. With animals, experimentation on anticipatory processes is aided by arranging events important to survival or comfort—paradigmatically, food presentations—in temporal patterns which serve as anchor points for measurable behavioral reorganization. Consider this traditional procedure: The hungry animal is given a response lever, but responses are reinforced with food only upon completion of a repeating fixed interval. An anticipatory pattern develops, with active, nonreinforced responding shortly before the interval elapses, and lower (or zero) rates beforehand. Interval duration is a factor that governs precision of the anticipatory pattern and that can evoke contrasting mechanisms of behavioral adjustment.

In one set of experiments, inspired by Skinner's[16] early investigations of the "fixed-interval scallop," the intervals are relatively short—in the seconds-to-minutes range—and the amount of food allotted per reinforcement is relatively small, to forestall satiation over repetitions of the cycle. With the development of the scalar expectancy theory[7,8] the animal's timing mechanism has been elucidated in terms of a discrimination process by which successive moments within the interval are evaluated for a reliable improvement in temporal proximity to food. Such short-interval timing shows the property of arbitrary reset: the behavior pattern is similar whether the successive onsets are fixed relative to reinforcement, or an exteroceptively signaled fixed interval is initiated upon completion of a variable intertrial interval.[13] Furthermore, response rates as well as performance-accuracy measures in temporal choice procedures[5] generally show a Weber's law-like property reflecting equivalent discriminability of time relative to the base interval, the "scalar property."[22]

In contrast to the short-interval work, another group of studies has exploited a cyclic feeding procedure with intervals in the range of whole days, and reinforcements consisting of whole meals.[20] When food availability is restricted to circadian (close to 24 hr) cycles, a major response acceleration develops in the hours before mealtime. The

[a]This research was supported in part by Grant BNS 81-19748 (to J.G.) from the National Science Foundation, Grant MH 27442 (to M.T.) and Grant MH 37528 (to J.G. and M.T.) from the National Institute of Mental Health, the New York State Psychiatric Institute Research Associate Program, and a Research Scientist Development Award (to M.T.) from the National Institute of Mental Health.

[b]Address for correspondence: New York State Psychiatric Institute, 722 West 168th Street, New York, New York 10032.

pattern is found even when daily light–dark cues are eliminated experimentally. But the circadian process is not a simple long-term analogue of the fixed-interval scalar timer. It does not show the property of arbitrary reset: The cycles must occur regularly within a few hours of the solar period, and responses to phase adjustments are gradual. Rather, it shows the self-sustaining property of an oscillator, with repetitions of the anticipatory pattern even after removal of the feeding cycle.

In this paper, we explore potential interactions between the long-and short-term timing systems as they may operate in concert, with the anticipatory patterns of both indexed concurrently by lever-pressing responses which precede a daily interval of food availability. By adding an external premeal signal, we ask whether the circadian system is relieved of the need to support protracted anticipations.

METHOD

Subjects and Apparatus

Five adult male rats of the Blue Spruce Long–Evans strain were each maintained in a ventilated, isolated chamber with constant temperature of approximately 70° F, white-noise masking, and constant illumination (LL) from a broad-spectrum Vita-Lite fluorescent lamp attenuated to approximately 5 lux by "opaqueing" with electrical tape. An enclosed testing compartment was constructed of aluminum and Plexiglas, with stainless-steel grid floor, and measured 29 × 24 × 19 cm. A response lever, food-cup, water-tube, and loudspeaker were mounted on the front wall. Feeding was arranged by delivery of 45-mg BioServ precision pellets from a magazine dispenser operated by lever presses, and water was provided by plastic tube from a wall-mounted bottle. A contact drinkometer monitored licking behavior. Under cued anticipation procedures, a soft tone from a Grayson-Stadler signal generator was presented in the premeal interval. Two of the three units included a Columbus Instruments magnetic-field general activity monitor. The chamber was opened approximately once a week for cleaning and replenishment of food and water.

Lever-pressing, water-licking, and general activity were monitored by a PDP 11/34 computer, with frequency counts stored in 5-min bins throughout the 24-hr day. In addition, individual responses were tagged in real-time over the 5 hr preceding daily mealtime. The computer controlled the time of food availability, food delivery, and auditory cue presentations, as well as handling data analyses and graphics.

Procedure

The experiments began with several weeks of continuous food availability, contingent on lever-pressing (CRF), to establish free-running patterns without food restriction. Then the animals were given a series of food-restriction conditions, in mixed orders: Under the simple feed–starve regimen, a 4-hr food-availability interval was scheduled each day at a standard time (FS 4:20). Under cued anticipation, continuous-tone presentations preceded mealtime by 4 hr, 2 hr, 6 min, or 3 min (FSQ 4, 2, 0.1, 0.05), and terminated at the start of mealtime ("nonoverlap" procedure). An FS or FSQ condition was maintained for the duration of a full free-running cycle—from one phase-coincidence with meal onset to the next phase-coincidence—so that summary curves could be derived which averaged out the varying momentary phase relation (FIG. 1B) between the free-run and mealtime. Given a typical free-running period of 24.75 hr, for example, a full-cycle would require 32 days of testing (24/0.75 = 32),

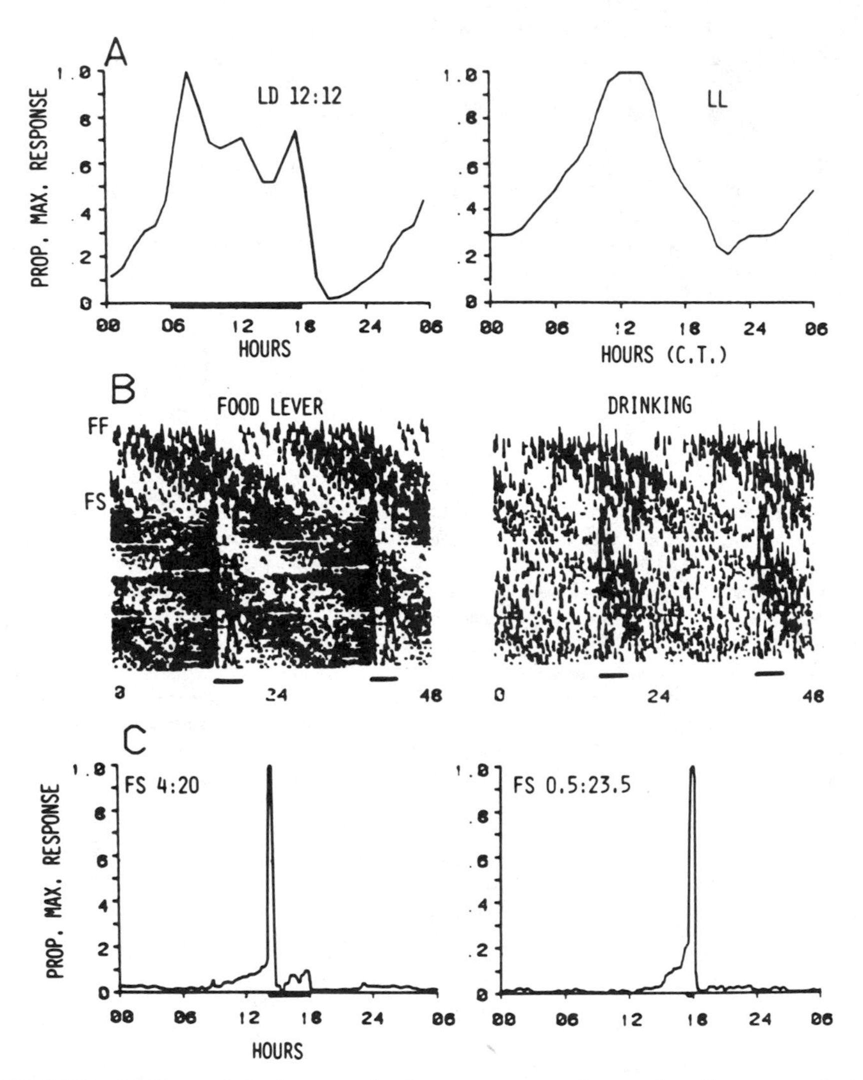

FIGURE 1. (**A**) Mean relative lever-pressing frequency, averaged across days, for a group of rats (N = 6) given free food-access (continuous reinforcement) under LD 12:12 and LL. *Abscissa* marks dark segment of LD. Curves smoothed by iterated running median, with original daily data collected in 1-hr bins (external clock time, LD; circadian time with correction for free-running periods, LL). (Adapted from Rosenwasser *et al.*[14]) (**B**) Double-plotted 3-D rasters of food-lever-pressing and water-licking under LL, for a rat first given free food-access, FF, and then transferred to a food-restriction schedule, FS 4:20. High-amplitude vertical bands under FS reflect the major daily feeding bouts. (**C**) Food-lever-pressing waveforms for a rat under FS 4:20 and FS 0.5:23.5. Data are averaged over a full free-running cycle relative to the 24-hr FS modulus. *Abscissa* marks time of food availability.

with an additional week or longer added at the start to accommodate transition effects. Occasionally, however, when the free-running activity pattern was indistinct, a condition was truncated. After FSQ tests, the uncued FS baseline pattern was re-established; the behavior pattern under FS was quite similar before and after experience with the cues, and our presentation concentrates on the post-cue subset. In a further test of cue control, one animal was tested in an "overlap" procedure under FSQ 4, in which the predictive auditory signal was extended through mealtime and conterminous with it.

Subsequently, the animals were given a meal-omission procedure, in which the scheduled daily interval of food availability under uncued FS was cancelled on occasional "probe days." This enabled measurement of response patterns throughout the meal expectancy interval, unmasked by actual ingestion. In individual cases baseline mealtime durations of 4 hr, 2 hr, and 0.5 hr were given (FS 4:20, 2:22, and 0.5:23.5).

RESULTS AND DISCUSSION

Effects of Food Restriction

An animal's ingestive pattern necessarily differs when food is in limitless supply (as with continuous reinforcement around the clock) and when feeding is restricted to certain times of day (as under the FS procedure). Even under free access to food, though, distinctive temporal patterns of feeding and drinking typify a species, and are dependent on ambient lighting conditions. Diurnal monkeys, for example, eat almost exclusively in the light, given daily light–dark (LD) alternations.[18] Nocturnal rats, in contrast, are not as strongly LD phase-dependent, as illustrated in FIGURE 1A (left). Under LD 12:12 rats show a predominant trimodal free-feeding pattern, with the first daily peak shortly after dark onset, a middle-night peak of smaller magnitude, and a final major peak shortly preceding light onset.[14] This light-entrained oscillation pattern shows its nadir early in the light, and the probability of feeding then increases gradually across the remaining daytime hours. The rat's drinking behavior is more strictly tied to the daily dark segment,[15,21] without a diurnal anticipatory acceleration. Although drinking is therefore not strictly locked to dry-food ingestion, prandial drinking bouts are common, and the nocturnal drinking pattern shows a trimodality tightly phase-locked to feeding.

These LD-entrained ingestive patterns change when the animal is exposed to continuous dim illumination (LL) without daily dark segments (a procedure that supports free-running circadian rhythmicity) with periods generally greater than 24.00 hr for the rat, but typically less than 25.00 hr. (The free-running period is set, within limits, by illumination level.[1,19]) FIGURE 1A (right) shows a group ingestion waveform obtained under LL for the same animals previously entrained under LD. The abscissa is scaled in "circadian time," for which 1 hr C.T. equals the free-running period (determined separately for each animal) divided by 24. A prominent oscillation pattern persists, but the discrete trimodality found within objective nights under LD are absent within the subjective nights (that is, high-activity phase) of LL. The animals show a relatively continuous waveform peaking in subjective night, but the drop in subjective day does not reach the near-zero levels seen under LD. The gradual rise in anticipation of subjective night is retained. As under entrainment, drinking remains low throughout the subjective day, and is tightly phased-locked to feeding within subjective night (data not shown here).

In a sense, then, animals given unlimited food access, whether in LD or LL show distinctly favored ingestion intervals, and thus some self-imposed food restriction. When the time of food availability is restricted experimentally, the daily ingestion pattern must accommodate, and a pattern of food-seeking behavior emerges that enables the animal to receive the daily ration whenever it is scheduled. FIGURE 1B shows an animal's adjustment to such a schedule, FS 4:20, after several weeks of unlimited food access under LL. A free-running period of 24.75 hr (ascertained by a χ^2 periodogram analysis[17]) is indexed in the FF raster by the diagonal trend in high-amplitude feeding bouts across days. The double-plotted format spans 48 hr on the abscissa, with each day's data presented first between 24 and 48 hr on one line, and again between 0 and 24 hr on the line below, a method that facilitates inspection of the free-run across the midnight boundary at the center of the figure. The corresponding drinking record differs in its fine grain (with high-amplitude licking bouts alternating with extended rests) but the period is also 24.75 hr, indicating that under the free-run the two ingestive variables are controlled by a common circadian timer.

When FS is introduced, and feeding is confined to the interval between 1500 and 1900 hr (yielding an exact 24 hr external cycle), the behavior pattern rapidly adjusts into rhythmically dissociated subsets: A major daily feeding bout emerges at the moment of onset of food availability; several hours before onset nonreinforced lever-pressing shows a gradual acceleration; and the underlying non-24-hr free-run persists in continuously varying phase relation to the feeding interval as a function of days.[3] Across a set of such samples the free-running period tends to shorten slightly under FS relative to the LL baseline, indicating a coupled interaction between the two concurrent rhythms. When the free-run is phase-coincident with mealtime, the animal is more likely to take multiple meals during the 4-hr food-availability interval, and thus to ingest more per day than at times of phase-displacement. Similarly, the rate of lever-pressing within the daily anticipation interval (approximately 6 hr preceding mealtime) is generally higher during phase-coincidence of the two rhythms, and during phase-displacement anticipatory responding may even be absent on an occasional day, and the meal initiation delayed. As with lever-pressing, the circadian drinking pattern persists under FS at a period close to that of the preceding free-run. The major daily drinking bouts are prandial bursts synchronized to food intake. Drinking, however, is not confined to the restricted feeding interval, and usually fails to show a 24-hr anticipation function.

Because of the dynamically changing phase relation between the free-run and food-sychronized anticipation, waveform summaries of FS behavior patterns were obtained by pooling response counts across days for the duration of a full cycle from one phase-coincidence of the dissociated rhythms to the next, thus averaging out the free-running component. FIGURE 1C shows waveforms obtained under 4-hr and 0.5-hr durations of food-availability. By necessity, reinforced lever-pressing is tightly packed into restriction intervals; 4 hr permits occasional further meals before termination of the interval, while 0.5 hr does not. Lever-pressing rates, including those seen in premeal anticipation, are relatively much lower during the daily extinction segment than at mealtime. Our subsequent quantitative treatment of the anticipation timer excludes the feeding interval.

The effect of momentary free-running phase on the form of the anticipation functions, which exclude mealtime, is illustrated in FIGURE 2A. When a 2-week subset of FS data is averaged for a segment of approximate phase-coincidence between mealtime and peak free-running phase—selected by visual inspection of the rasters—an orderly daily acceleration, starting from near-zero rates, is revealed across the final 8 hr of the deprivation interval (FIGURE 2A, left). When a contrasting subset is analyzed for a segment in which the peak free-running phase falls approximately 5 to 9

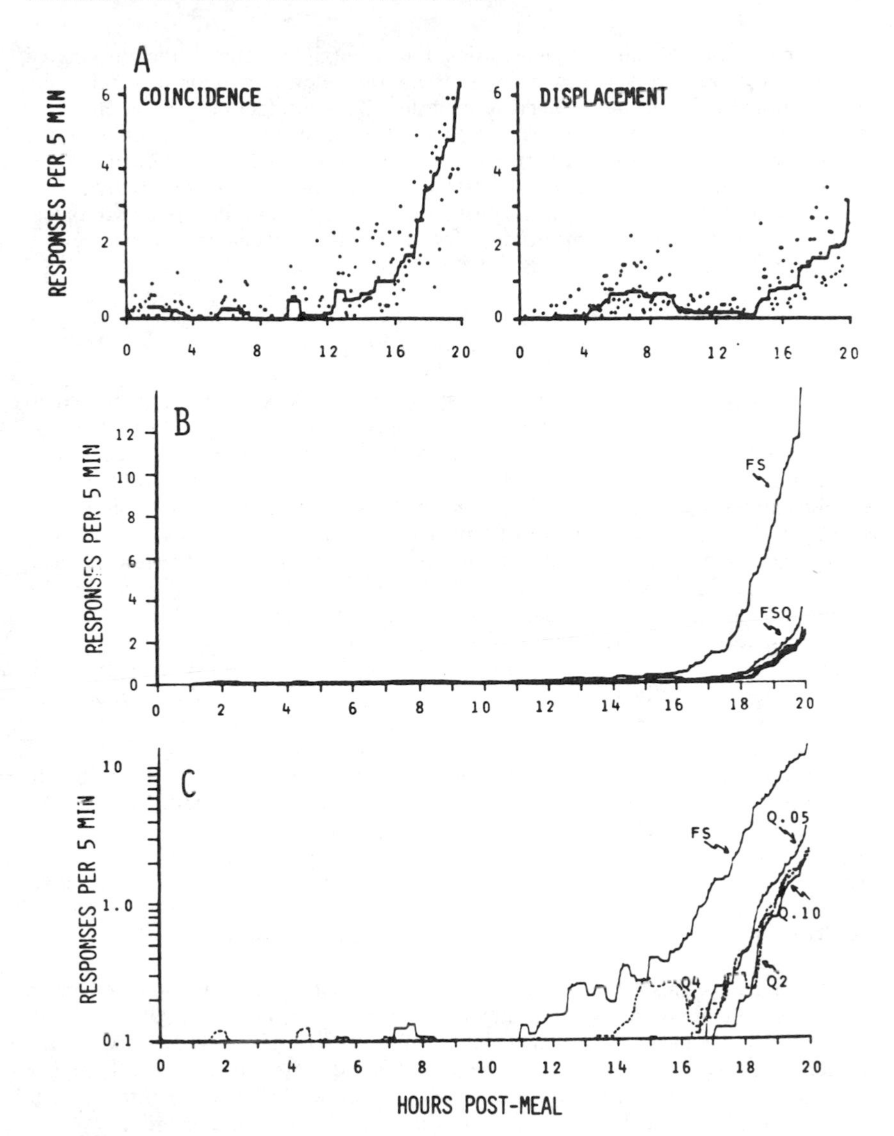

FIGURE 2. (**A**) *Left panel:* FS 4:20 meal-anticipation function for an individual animal for a 2-week segment in which the free-run is in approximate phase-coincidence with mealtime. *Ordinate* plots mean lever-pressing frequency per day within the 5-min data collection bins. The daily 4-hr feeding interval, from 20 to 24 hr, is not represented. The curve is fit by iterated running median. *Right panel:* Corresponding record for a 2-week segment of phase-displacement, in which the free-run peaks approximately 5 to 9 hr after mealtime. (**B**) Group average meal-anticipation functions under uncued FS 4:20 (second determinations only, after cue experience; N = 3), FSQ 4 (N = 5), FSQ 2 (N = 3), FSQ 0.1 (N = 4), and FSQ 0.05 (N = 4). On these coordinates, the Q 0.05 function can be distinguished from the other cue functions by slightly higher terminal rates. Within-cue data for Q 0.05 and Q 0.10, which show marked accelerations, are not included here (but see FIGURE 3). (**C**) Log scale transform of data shown in (**B**). The Q 4 function can be distinguished from the other cue functions by a pre-cue acceleration ending at 16 hr; the other cue functions begin their ascent between 16 and 17 hr.

hr past mealtime—within the deprivation interval (right panel)—the dissociated rhythmic processes are revealed by a transient rate elevation prior to a relatively attenuated and delayed anticipatory acceleration. This interaction is apparently not simply additive, and we are investigating multiplicative rules for combining the free-running and anticipatory components (seen as separate processes during phase displacement) to predict the function at phase coincidence. In further analyses, we bypass the problem of varying local phase relations by averaging over a complete free-running cycle relative to mealtime, which elucidates the generalized form of the anticipation functions.

Cued Versus Uncued Anticipation

The addition of an auditory cue predictive of mealtime, we reasoned, might serve to relieve the animal of the need for protracted nonreinforced anticipatory responding. The cue might foster engagement in other activities prior to its onset, and then initiate a short-interval timer, under which the animal's estimates of the time left to reinforcement would lead to a sharpened acceleration targeted at mealtime. The cueing paradigm might be thought to establish a discriminative stimulus, in analogy to trial-by-trial procedures previously studied in the seconds-to-minutes range.[10] The addition of an "external clock" in fixed-interval schedules of reinforcement, for example, serves to sharpen anticipatory accelerations in response rate.[6]

Alternatively, the cueing paradigm is suggestive of Pavolvian delay conditioning on a grand scale: With the cue construed as a conditioned stimulus (CS), the proposition that CS effectiveness is a function of the ratio of the background time between feedings—or "cycle time"—to the duration of the CS[9] prompted an extension to the circadian time range. At one extreme of the cue durations we tested, under FSQ 4, the cycle-to-trial ratio is 5:1 (computed by dividing the interfeeding interval, 20 hr, by the cue duration, 4 hr)—a value that supports only moderate conditioning in the short-interval range. At the other extreme, FSQ 0.05, the ratio of 400:1 indexes excellent temporal predictiveness, indeed. (Following the method of Balsam and Payne,[2] the ratio calculation excludes the feeding interval.) In this conceptual framework CS effectiveness is thought to depend upon the subject's comparison of the overall rate of food-availability (here, across days) with that associated locally with the cue.

FIGURE 2B presents a set of group-averaged anticipation functions obtained under the different cue durations, with the uncued FS curve for comparison. All responding shows a curvilinear increase in the final hours before mealtime, but the FS curve begins its steep ascent earlier (at about 16 hr post-meal) and maintains consistently higher rates than do the cues. By comparison, the FSQ curves, which begin their steep ascent at about 18 hr post-meal, are not readily differentiable on these coordinates. The slightly higher rates under Q 0.05 we think stems from sampling error. The accelerations begin hours before the short-cue presentations: The animals do not wait for cue onset to begin "probing" for daily meals, and they certainly do not wait for cue termination, even though that might serve as an unambiguous signal for food availability and is a feature common to all cue durations. (The relatively steeper FS function is obtained both before and after experience with the cues, and so cannot be ascribed to a training order effect.)

Differentiation of the anticipatory patterns becomes clearer when response rates are plotted on a log scale (FIG 2C), which magnifies trends within the lower rates. On these coordinates the final ascent of both FS and FSQ curves is approximately linear and parallel, indicative of a common exponential acceleration for which the cues serve simply to lower the base level. This makes intuitive sense if we construe the animals to register by their lever presses expectations of mealtime onset, which coincides with cue termination. The added information inherent in cue termination permits more "conservative" probing, yet the 20-hr temporal discrimination problem is unchanged.

FIGURE 2C does, however, reveal fine differences in anticipation patterns across cue duration. The Q 4 function begins to rise at approximately 14 hr post-meal, that is, 2 hr before cue onset. At 15 hr post-meal, in fact, Q 4 and FS rates momentarily converge, but Q 4 rate then levels off, and drops at cue onset. Q 2 shows a similar, but smaller, cue-onset suppression. The initial Q 2 and Q 4 accelerations may be construed as anticipations of cue onset, and the subsequent suppression as a cue-supported relief from the need to anticipate mealtime, given that the cue is timed, too, and, when close to its onset, reliably indicates that feeding is "impossible."

Circadian anticipation is thus modified complexly by the addition of premeal cues. The presence of the cues does serve to suppress responding before cue onset, indicating that in some sense the rats come to rely on temporal information inherent to the cueing procedure, and not only in direct response to cue presentation. This pattern supports the notion of a decreased "need to respond" afforded by the forthcoming cue. The most predictive cues (that is, those in the minutes range) show the largest ascents before mealtime; yet cue control is not predominant, for these ascents still begin several hours "too early."

Pre- and post-cue rate trends are analyzed further in FIGURE 3. The similarity of (logged) slope across the FS and pre-cue accelerations is likely to reflect the FS timing mechanism. The hour cues show remarkable convergence in terminal slopes, parallel to FS but nearly 1 log unit lower. Early suppression is shorter within Q 2 than within Q 4; were the two cue-suppression durations similar, it might be argued that the effect was unconditioned, and primarily dependent on the time since cue onset, not the time to subsequent feeding. Rather, the data argue for timing of the cue. The magnitude and duration of cue-onset suppression do not follow a simple scalar rule, however: the Q 4 effect is *more* than twice that of Q 2. However, if proximity to mealtime is an interacting factor, forcing the convergence of terminal rates, the Q 2 effect might be attenuated. Another alternative is akin to Pavlovian "inhibition of delay": Given extensive training, the conditioned response is withheld early in the CS interval. However, a strict parallel seems unlikely, since the Q 4 and Q 2 suppression edges appeared very rapidly (within two or three cue-meal pairings) and did not migrate forward with further training.

The notion of direct suppression by the cues, if correct, would be based on a cue-onset time anchor. But if the cue is viewed as initiating a short-interval timer, scalar expectancy theory[7] suggests cue termination, at the moment of reinforcer presentation, as a more appropriate anchor for comparisons across cue durations. Response trends within the minutes cues (FIG. 3, right) make this point more clearly than those within the hours cues, because the shorter cues occur so closely within the circadian cycle, virtually eliminating any possible interactive effect of FS phase. Given that the 6- and 3-min cues are so proximal to mealtime, they induce no cue-onset suppression. Rather, response rates show steep accelerations shortly after cue onset, leveling off midway into each interval—a typical short-interval timing pattern (allowing for the differences in final pre-cue rates under Q 0.10 and Q 0.05, approximate superposition is achieved within the interval). And even with the cue-supported response bursts just before mealtime, the terminal anticipation rates fall well below the uncued FS curve.

In contrast, the Q 4 and Q 2 curves (FIG. 3, middle) would not reach superposition in proportional time: their final ascent superimposes in real-time, so the Q 2 function would rise too early for a scalar fit. This we believe reflects an interaction with the

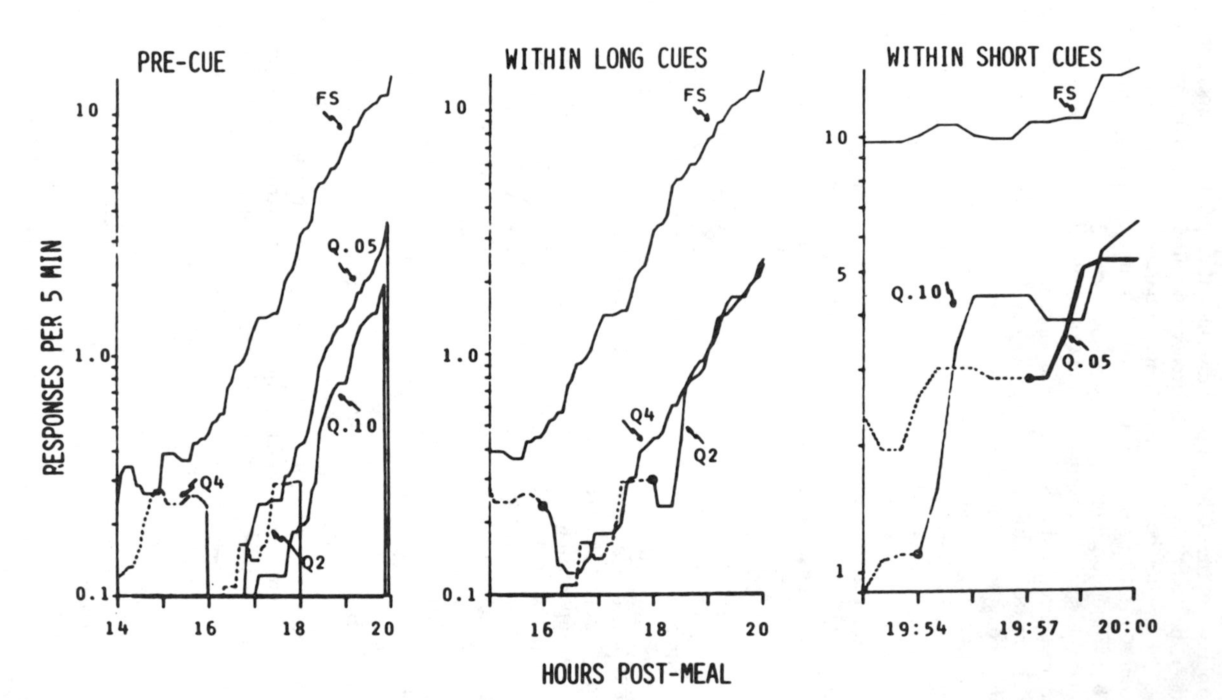

FIGURE 3. Group average anticipation functions for pre-cue and within-cue intervals (log scale; for Ns, see legend to Figure 2). Four-hour mealtime begins at 20 hr. Curves smoothed by iterated running median. **Left panel:** Pre-cue accelerations, and uncued FS comparison. Vertical lines denote time of cue onset. **Middle panel:** Accelerations within the hours cues, and uncued FS comparison. *Solid circles* denote time of cue onset, with leftward extensions (*dashed lines*) into the pre-cue interval. **Right panel:** Accelerations within the minutes cues, and uncued FS comparison. *Solid circles* denote time of cue onset, with leftward extensions (*dashed lines*) into the pre-cue interval.

concurrent FS timer, which encompasses a broader range relative to the daily cycle under Q 4 (0.16 cycle) than under Q 2 (0.08 cycle). The FS timer is in a relatively higher state—closer to mealtime—within Q 2 than within Q 4. Thus, when expressed in proportional terms, the Q 2 recovery function shows an earlier rise than the Q 4 function. In contrast, the shorter cues in the minutes range span relatively tiny fractions of the daily cycle (Q 0.05, 0.002 cycle; Q 0.10, 0.004 cycle). Because these cues make contact with the FS timer only in the terminal moments before mealtime, an FS-timer-induced differential in the scalar cue-timing functions is absent.

Not all behavior reflects an initial quieting effect of cue onset in the hours range. As illustrated for one animal in FIGURE 4, under Q 2 general motor activity accelerates very gradually between 10 and 18 hr post-meal, and then joins the lever-pressing function in its steep premeal ascent. Only the discriminated operant evinces near-to-total quiet for approximately 16 hr post-meal, as well as anticipation of cue onset, and cue-onset suppression.

The importance of cue termination in setting the level of the anticipation functions is revealed by our use of the cue-overlap procedure in one animal (FIG. 5) that had

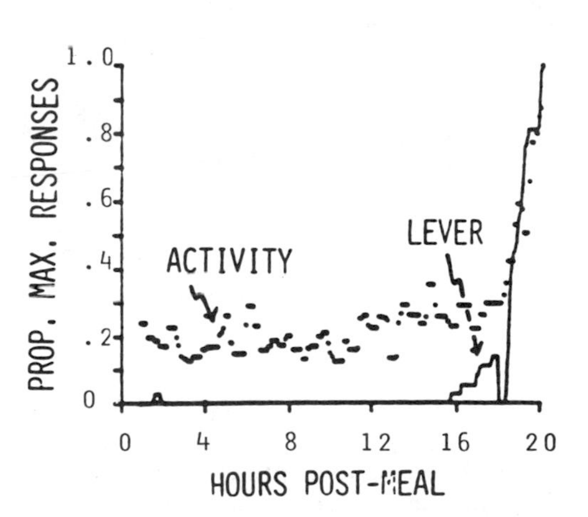

FIGURE 4. Comparison of general activity and lever-pressing for a rat under FSQ 2. Note relative quiet in lever-pressing before 16 hr, and cue-onset response suppression at 18 hr not seen in general activity.

previously received the standard Q 4 terminating at meal onset. Both cue-meal pairing procedures might be considered varieties of Pavlovian delay conditioning on the hours scale. The differences between them have not received systematic attention in short-interval Pavlovian studies, perhaps because reinforcer duration is arbitrarily brief. Here, however, the position of cue termination is critical. The overlap procedure denies the animal the reliable "announcement" of mealtime inherent in standard premeal cue termination. As a result, terminal rates under Q4 overlap closely match that of uncued FS, although pre-cue rates and the duration of suppression closely match that of regular Q 4. The overlap procedure thus maintains the standard features of cued premeal interval timing, but requires an upward level adjustment with more active probing of mealtime onset.

Cue Transitions

Further evidence for operation of a cue-supported interval timer, distinct from the circadian FS timer, is gleaned from transitional performances in which the animal,

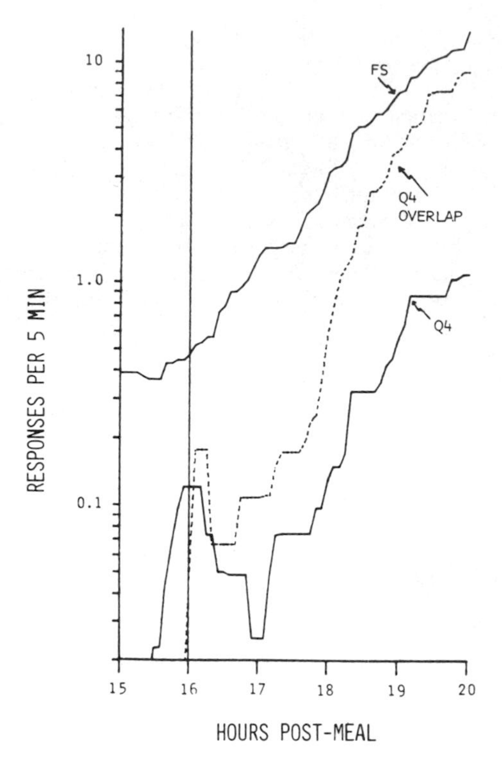

FIGURE 5. Comparison of anticipation functions for a rat under FSQ 4 in which the cue termination coincided with (a) meal onset, our standard paradigm; and (b) meal termination, an "overlap" paradigm. The terminal rate under Q 4 overlap approaches that of the uncued FS comparison curve.

given experience with a relatively short cue (such as Q 0.10), is switched to a longer one (such as Q 2). At first exposure to FSQ 2, then, the cue is presented earlier than expected on the basis of prior training, and the moment of onset is phase-advanced relative to the FS timer, which remains unchanged across FSQ 0.10 and 2. Furthermore, the new cue continues longer than expected on the basis of prior training: the Q 0.10 interval timer predicts reinforcement after only 6 min, and now reinforcement is withheld for 2 hr. This latter feature will force reorganization of the interval-timing system, which, as we have discussed above, is anchored to cue termination.

The situation bears similarity to the probe trials of Roberts'[13] "peak procedure," in which a short-interval cue (in the seconds range, and with no fixed phase relation to a circadian timer) is greatly extended and reinforcement is omitted: in that case, responding accelerates toward the expected moment of reinforcement, and then gradually decelerates beyond. The time of the obtained peak rate is, precisely, the interval timer's anchor.[22,23]

Our situation differs in that all of the cues, short and long, greatly exceed Roberts' in duration. But more importantly, the short-to-long cue transition procedure introduces a predictive discrepancy between two timers (short-cue and FS) which had previously shared a common anchor point at mealtime. After the transition, the cue is presented prematurely from the point of view of the FS timer, and a cue-induced anticipatory burst does not pay off with food. At the same time, the transition procedure extends cue duration, with the new long cue conterminous with the FS deprivation interval. How is the predictive discrepancy between short-cue and FS timers reflected in the anticipation pattern? And how does the long-cue anticipation pattern finally emerge against this complicated background?

To analyze these effects we examined anticipation functions for subsets of days following cue transitions, rather than averaging across an entire cycle of the free-run relative to FS as in the steady-state functions. FIGURE 6A shows the course of events for a rat that was transferred from FSQ 0.10 to 2. In the first week of the long cue (FIG. 6A, left), the anticipation function began its ascent in the hour preceding cue onset, and upon cue onset there was a sudden, high-magnitude acceleration toward the highest rates of the day, peaking approximately 1 hr later. Subsequently response rate declined to an intermediate level and fell even further as mealtime approached. At this early post-transition stage we conclude that cue onset 2 hr before mealtime triggered the short-interval timer previously established at 6 min before mealtime, and when food was not forthcoming within the next hour—during which the major daily meal bout would have occurred relative to the Q 0.10 timer—response rate declined. Even after this decline, however, response rates did not approach zero, because the FS timer—incrementing toward 20 hr—continued in its course. One can imagine a set of underlying timing distributions during these final hours, in which a continued decline in the short-interval function, in interaction with the FS-supported circadian accelera- tion, produced the intermediate rates between 17 and 20 hr.

The next subset of days (FIG. 6A, middle) shows the gradual acquisition of a cued anticipation pattern appropriate to the new FSQ 2 condition. Though the Q 0.10 burst persisted for many days after the transition, the terminal limb of the anticipation function again began to show increased rates. This segment reflects the development of a Q 2 interval timer conterminous with the FS deprivation interval. The presence of interval timing appropriate of two cue durations, yet triggered by the same cue, favors

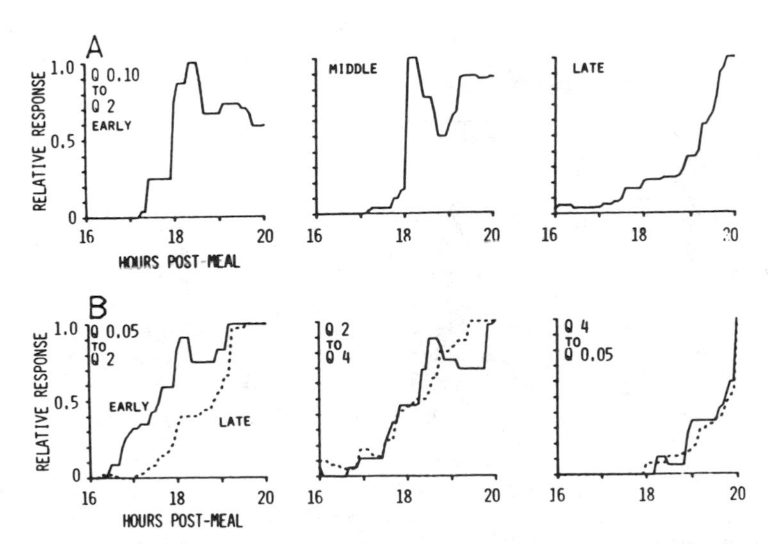

FIGURE 6. (**A**) Anticipation functions for a rat with preceding experience under FSQ 0.1, then given FSQ 2 (short-to-long cue transition). From *left to right,* panels show development of Q 2 control, with slow disappearance of transient acceleration at cue onset—an FSQ 0.1 aftereffect. Curves smoothed by iterated running median. (**B**) Anticipation functions for rats given short-to-long (Q 0.05-to-2, 2-to-4) and long-to-short (Q 4-to-0.05) cue transitions, as sampled in the first week after the transition and again several weeks later.

an interpretation in terms of "simultaneous temporal processing," as has been shown recently within the seconds range.[11] Furthermore, the persistent excitation pattern supported by the Q 0.10 timer at the onset of Q 2 must be contrasted with the cue-edge suppression normally found under steady-state FSQ 2 and 4 (FIG. 3), further evidence that the auditory signal does not have a nondifferential unconditioned quieting effect on the behavior.

In the late FSQ 2 sample (FIG. 6A, right), a more coherent anticipatory performance is finally achieved: the Q 0.10 transient burst is extinguished, and the monotonic acceleration pattern is recaptured, under joint control of the Q 2 and FS timers. Substantially similar results were obtained across all short-to-long cue transitions, as illustrated in FIGURE 6B (left and middle panels) for early and late post-transition samples for another rat within Q 0.5-to-2 and Q 2-to-4 sequences. Notice that in the latter case the transition occurs between two cues that are both within the hours range, and the transient Q 2-supported acceleration is appropriately delayed beyond the moment of cue onset (compare suppression intervals in FIGURES 3 and 4).

The contrasting case of a long-to-short cue transition (FIG. 6B, right) fails to show a transient acceleration at cue onset, and the early and late samples are quite similar. In this case, the former cue (4 hr) is withheld, so the Q 4 timer is not initiated at the expected moment within the deprivation interval. The new, shorter cue appears later, at 3 min before mealtime. While it is possible that the formerly established Q 4 timer is triggered after that delay, its expression is masked (or it experiences a premature reset) by the intervention of mealtime.

Meal-Omission Gradients

A key to the study of interval-timing mechanisms is the precise location of the subjective temporal anchor point; the animal is considered to make continual judgments of the "time left" to reinforcement.[10] In the present situation under FSQ, both interval and circadian timers are presumed to share an anchor point in the steady state, that is, the time of daily food availability. The major meal of the day occurs as soon as food is available (FIG. 1C), and both interval and circadian anticipatory processes presumably serve to increase the likelihood that the daily ration will be taken at the earliest opportunity. When the circadian free-run is phase-coincident with mealtime, and time permits (as in FS 4:20), the animal may take multiple meals, separated by pauses and drinking bouts, within the time of food availability. Over many months of testing, some of these secondary meals will be aborted by the termination of the 4 hr of food availability, and thus the animal gradually comes to experience both leading and trailing edges of mealtime, and to establish an expectation interval. Under highly restrictive mealtimes, as in FS 0.5:23.5, there is no time for secondary meals, and the animal is likely to confront the 30-min trailing edge of mealtime almost every day.

When the meal expectation interval is as broad as several hours, one might guess the temporal anchor point for the circadian anticipation process to be rather indistinct, especially as compared to the brief reinforcement intervals typical of short-interval timing paradigms (for example, a single pellet delivery or 3-sec access to the grain hopper). Or if not indistinct, the anchor point might be determined complexly by the varying moments within mealtime in which the animal experiences reinforcement.

On the other hand, it is possible that mealtime onset is the salient anchor point regardless of the duration of food availability. The intervention of the daily feeding bouts, abutting upon the terminal segment of the daily anticipatory acceleration, makes it difficult to sort out the alternatives. An earlier study showed that the FS anticipatory pattern has the property of a true circadian oscillator—and is not reset by the daily meal in the manner of an hourglass interval timer—by eliminating meals altogether for several days after extensive FS 4:20 training.[3] An anticipatory acceleration occurred each day at the appropriate circadian phase, followed by a gradual deceleration. Such an omission procedure affords a view of the animal's behavior throughout the expectation interval uncontaminated by ingestion per se; but it is not a viable long-term procedure for reasons of the animal's health, and with successive omissions the expectation interval might well be degraded. (In another set of experiments[12,24] animals were given free food–access following FS training, which might be expected to obliterate control by the phase of restricted feeding. But when food was subsequently omitted, the animals evinced remarkable memory of the former salient circadian phase in precisely timed activity bursts, which suggests that once established, the expectation interval is not easily degraded.)

Our exploration of the temporal anchor point (or interval) for circadian anticipation maintains expectation by interspersing a maximum of two meal-omission probe days per week within the FS baseline. On any day the animal cannot know if a meal will be forthcoming, but if it is, it will always occur at the standard FS phase. The procedure thus supports equivalent premeal anticipations on baseline and probe days, but markedly different patterns during and shortly after the meal interval. FIGURE 7A shows paired rasters in which the baseline and probe data are separately displayed for one subject. Normal FS data are presented in the upper panel, with blanks for 6 hr beginning at mealtime on omission probe days. Expectation interval data are presented in the lower panel, with blanks for the 4 hr of mealtime on days when food was received. Both rasters show the concurrent free-run, with a period greater than 24.00 hr, traveling across the mealtime boundaries over successive days. Baseline and probe days stand in marked contrast, with the latter distinguished by a continuation of anticipatory responding into mealtime, and beyond. One might guess that responding would be greatly diminished after the leading edge of mealtime has passed and responses have not been reinforced, at least after several exposures to the omission procedure. Alternatively, if a subjective temporal anchor point at mealtime onset were indistinct or variable, responding might overshoot the leading edge in proportion to the variance of the memory for circadian phase, but still decline before long. The protracted response accelerations obtained, however, belie either possibility.

Summary omission gradients, averaged across probe days, are shown for two animals in FIGURE 7B, one under FS 4:20 and the other under FS 2:22. It is clear that much higher and more protracted rates are sustained when food is omitted than during the anticipation segments. In each case responding does not end for several hours beyond the baseline food-availability interval. The functions show several minor peaks within expected mealtime. Under FS 4:20 response rate rises throughout the interval, and the major decline is initiated near the trailing edge of mealtime. The FS 2:22 curve is similar in form and width, but is phase–advanced relative to FS 4:20, so that the major decline also occurs near the trailing edge.

These results suggest that the true timing peak—or anchor point—occurs near the time feeding would normally stop, rather than at its onset. The ascending limb to the left of the peak suggests a broad band of time during which food is expected, while the declining limb may index the animal's subjective estimate of the time of day after which food is no longer "possible." While peak rate might be located past the leading edge of mealtime even if the animal were anticipating the moment of onset, the meal duration variable reveals an explicit appreciation of the food-availability interval. Thus, changing the position of the trailing edge of mealtime for an individual animal by reducing the food-availability interval—as under an FS 4:20-to-0.5:23.5 transi-

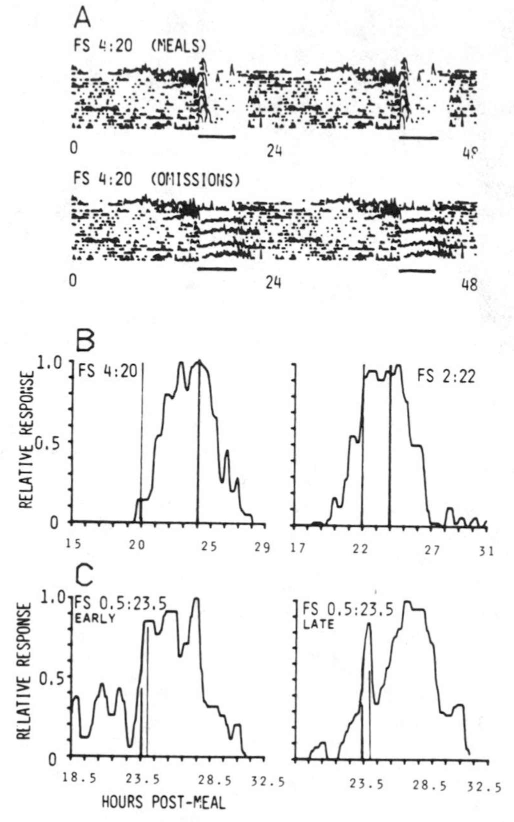

FIGURE 7. (**A**) Double-plotted 3-D rasters of food-lever-pressing by a rat under the FS 4:20 omission-probe procedure. Upper raster includes full days with meal presentations (as indicated by high-amplitude vertical band at 1500 hr) and excludes 6 hr of omission days, beginning at expected mealtime. Lower raster includes full omission days (as indicated by the anticipatory accelerations extending into expected mealtime) and excludes the 4-hr mealtime of days with feeding. (**B**) Omission gradients for two rats (FS 4:20 and 2:22) obtained by averaging data across omission-probe days. *Vertical lines* denote expected mealtime. Curves smoothed by iterated running median. (**C**) Transitional omission gradients for a rat with preceding experience under FS 4:20 (**B**, *left*), then reduced to a half-hour of food availability beginning at the same time of day, under the FS 0.5:23.5 omission-probe procedure. Late sample shows development of F 0.5 control superimposed on the formerly established F 4 pattern.

tion—has a major impact upon the omission-gradient peak, and the moment of rapid decline. FIGURE 7C charts the course of such a transition for the rat whose FS 4:20 baseline gradient is shown in FIGURE 7B (left). The gradient is thrown into disarray during the first week of the transition (FIG. 7C, left); the peak near the trailing edge loses its distinctiveness, and responding is high throughout the former 4-hr availability interval. The decline at the former trailing edge is maintained, however, and there is little evidence for differential control by the newly established 0.5-hr availability interval. Within another week a trend is clarified: There is a sharp acceleration rising throughout the newly established 0.5-hr expectation interval, and a precipitous drop in rate just after the trailing edge is passed. Control by the former 4-hr interval is not extinguished, however, and the gradient continues to rise toward its former peak, after which it declines to baseline rates as under FS 4:20. Extended training with the shorter mealtime might eventually cause the 4-hr gradient to dissipate, though it is possible that the former 4-hr anchor point might be behaviorally tagged indefinitely.[24]

AN INTERACTION HYPOTHESIS

In summary, some key features of these data are: (a) persistence of the LL-supported free-run concurrent with the FS-supported anticipation rhythm, as seen also in previous studies[4]; (b) relatively broad and high-amplitude anticipation functions under FS versus FSQ; (c) heightened anticipatory activity when the free-run and mealtime are in approximate phase-coincidence; (d) persistence of accelerations prior to cue onset, attributable to the FS timer; (e) approximate scalarity of acceleration patterns within the minutes cues; (f) retention of transient accelerations appropriate to the duration of a discontinued cue, upon transfer to a longer cue; and (h) peaking of FS meal-omission gradients toward the middle, or termination, rather than onset of expected mealtimes.

If viewed as the resultant of three classes of internal timer, the anticipation pattern falls into place: The *circadian free-running timer* varies continuously with the rhythm supported by the feeding schedule. We think the free-run influences the momentary level of anticipatory and ingestive activity rather than the accuracy of mealtime judgments. The *circadian anticipation timer* is phase-locked to the free-run when food is in unlimited supply, but dissociates when feedings are scheduled at a contrasting circadian period (not necessarily 24.00 hr as in our study; see Aschoff[25]). Brain-lesion studies point to separate physiological substrates for free-running and anticipatory systems (see Boulos and Terman[4]). The *cue-supported interval timer,* unlike the oscillation systems, has the property of arbitrary reset, and thus varies with cue, phase, and duration. For cues within the minutes range, the phase range of cue exposure relative to the 24 hr day (established by FS) is very nearly equivalent, and the scalar property is approximated for within-cue anticipations. In contrast, longer cues cover a broader phase range, and would require a correction for differential contact with the FS timer to observe superposition of within-cue anticipations. This process is capable of timing multiple intervals concurrently, as seen in our study as aftereffects of cue transitions. Further studies, using cued meal-omission probes, will show whether the interval timer is anchored to mealtime onset or termination; it is entirely possible that in standard trial-by-trial interval-timing tasks, with relatively brief and discrete reinforcements, temporal judgments are anchored to the moment of reinforcer termination.

The covariation of circadian anticipation and interval timing is clarified by grouping daily data across a full cycle of the free-run relative to the feeding period, and thus averaging out the non-24-hr trend. It seems likely to us that the remaining interaction, which forms the FSQ anticipation functions, may be described at the level of response evocation, with both timing mechanisms operating relatively independently. During the cue interval the subject responds only when both timers indicate that food is proximal. If, at a given moment, only one of the timers registers a mealtime detection, the response mechanism is not activated, but rather awaits proximity on the other timer as well. As a result, responding within the cue intervals is lower than responding under uncued FS at comparable time points. That the level is regained under the overlap procedure points to the salience of cue termination in measuring proximity to meal onset. The interaction of FS and cue processes is further revealed in comparisons of pre-cue response rates with uncued rates at corresponding phases, seen most clearly for cues in the minutes range (FIG. 3): while the acceleration patterns are quite similar at these times, the level is considerably lower in the pre-cue case. Since response rates even in the absence of the cue are a function of (impending) cue presentation, the animal's appreciation of this added source of temporal information offers further opportunity to redirect behavior when food-seeking is unlikely to pay off.

SUMMARY

Both short-interval and circadian timing systems support anticipatory response accelerations prior to food reinforcement. In the first case, the behavior pattern is determined by a scalar timing process with an arbitrary-reset property. In contrast, under daily cycles of food-availability, behavior reflects a self-sustaining oscillation. With rats as subjects, the concurrent operation of timing of both kinds was studied by addition of premeal auditory cues on the circadian baseline, in the absence of a day–night illumination cycle. Cues within both minute and hour ranges served to lower the level of premeal anticipatory responding, although exponential accelerations were similar to the uncued case. Cues within the minutes range yielded interval-timing functions that reflected approximate superposition. Cues within the hours range suppressed respondings at their outset, in proportion to cue duration. When one of the shorter cues was suddenly lengthened, short-interval accelerations appeared at inappropriate circadian phases. When a premeal cue was extended through mealtime, anticipation rates increased markedly, suggesting that cue termination at the start of mealtime is a potent anchor for premeal anticipation regardless of cue duration. By use of meal-omission probes without external cues, peak rates were located after the onset of expected mealtime, often near its termination. The results suggest interactions between the scalar interval timer and the circadian anticipation timer, as modulated by the circadian free-run timer.

REFERENCES

1. ASCHOFF, J. 1979. Influences of internal and external factors on the period measured in constant conditions. Z. Tierpsychol. **49:** 225–249.
2. BALSAM, P. D. & D. PAYNE. 1979. Intertrial interval and unconditioned stimulus durations in autoshaping. Anim. Learn. Behav. **7:** 477–482.
3. BOULOS, Z., A. M. ROSENWASSER & M. TERMAN. 1980. Feeding schedules and the circadian organization of behavior in the rat. Behav. Br. Res. **1:** 39–65.
4. BOULOS, Z. & M. TERMAN. 1980. Food availability and daily biological rhythms. Neurosci. Biobehav. Rev. **4:** 119–131.
5. CHURCH, R. M. & J. GIBBON. 1982. Temporal generalization. J. Exp. Psychol. Anim. Behav. Processes. **8:** 165–186
6. FERSTER, C. B. & B. F. SKINNER. 1957. Schedules of Reinforcement. Appleton–Century–Crofts. New York, NY.
7. GIBBON, J. 1977. Scalar expectancy theory and Weber's Law in animal timing. Psychol. Rev. **84:** 279–325.
8. GIBBON J. 1981. On the form and location of the psychometric bisection function for time. J. Math. Psychol. **24:** 58–87.
9. GIBBON J. & P. D. BALSAM. 1981. Spreading association in time. *In* Autoshaping and

Conditioning Theory. C. M. Locurto, H. S. Terrace & J. Gibbon, Eds.: 219–253. Academic Press. New York, NY.
10. GIBBON J. & R. M. CHURCH. 1981. Time left: Linear versus logarithmic subjective time. J. Exp. Psychol. Animal Behav. Processes. **7:** 87–108.
11. MECK, W. & R. M. CHURCH. 1984. Simultaneous temporal processing. J. Exp. Psychol. Animal Behav. Processes. **10:** 1–29.
12. MORI, T., N. KATSUYA & H. NAKAGAWA. 1983. Dependence of memory of meal time upon circadian biological clock in rats. Physiol. Behav. **30:** 259–265.
13. ROBERTS, S. 1981. Isolation of an internal clock. J. Exp. Psychol. Animal Behav. Processes. **7:** 242–268.
14. ROSENWASSER, A. M., Z. BOULOS & M. TERMAN. 1981. Circadian organization of food intake and meal patterns in the rat. Physiol. Behav. **27:** 33–39.
15. ROSENWASSER, A. M., Z. BOULOS & M. TERMAN. 1983. Circadian feeding and drinking rhythms in the rat under complete and skeleton photoperiods. Physiol. Behav. **30:** 353–359.
16. SKINNER, B. F. 1938. The Behavior of Organisms. Appleton–Century–Crofts. New York, NY.
17. SOKOLOVE, P. G. & W. N. BUSHNELL. 1978. The chi square periodogram: Its utility for analysis of circadian rhythms. J. Theoret. Biol. **72:** 131–160.
18. SULZMAN, F. M., C. A. FULLER & M. C. MOORE-EDE. 1978. Comparison of synchronization of primate circadian rhythms by light and food. Am. J. Physiol. Reg. Proc. **234:** 130–135.
19. TERMAN, J. & M. TERMAN. 1980. Effects of illumination level on the rat's rhythmicity of brain self-stimulation behavior. Behav. Br. Res. **1:** 507–519.
20. TERMAN, M. 1983. Behavioral analysis and circadian rhythms. *In* Advances in Analysis of Behaviour. M. D. Zeiler & P. Harzem, Eds. Vol. **3:** 103–141. Wiley. Chichester, England.
21. ZUCKER, I. 1971. Light-dark rhythms in rat eating and drinking behavior. Physiol. Behav. **6:** 115–126.
22. GIBBON, J., R. M. CHURCH & W. H. MECK. 1984. This volume.
23. ROBERTS, S. & M. HOLDER. 1984. This volume.
24. ROSENWASSER, A. 1984. This volume.
25. ASCHOFF, J. 1984. This volume.

Reproductive Mechanisms: Interaction of Circadian and Interval Timing[a]

RAE SILVER[b] AND ERIC L. BITTMAN[c]

Department of Psychology
Barnard College
Columbia University
New York, New York 10027

INTRODUCTION

In order to reproduce successfully, animals must coordinate social, internal, and environmental stimuli in time. Reproductive systems have been studied following the systematic manipulation of photoperiodic and hormonal stimuli. These mechanisms thus lend themselves to analysis of timing events on a physiological as well as on a formal level. In this paper, we are concerned with neuroendocrine systems that govern seasonal breeding, ovarian cyclicity, sexual behavior, pregnancy, and parental care. In each of these aspects of reproduction, a timing process with intrinsic daily periodicity plays a critical role. At various levels in each system, however, fundamentally different timing processes may also participate. We consider the relationships and interactions between these various timing processes.

Extensive analysis of daily rhythms provides a framework for the study of the timing of reproductive events. Such rhythms are termed circadian if they sustain a period of close to 24 hours when the animal is maintained in aperiodic conditions. Circadian rhythms tend to conserve their periods over a wide range of ambient temperatures. In the animal's natural environment endogenous rhythms are synchronized by physical oscillations, known as zeitgebers. The most important of these is the light–dark cycle. Synchronization of an endogenous rhythm by an environmental cycle is known as entrainment. Entrainment is possible only over a limited range of periods close to that of the endogenous rhythm. The normal expression of many circadian rhythms in several mammals has been shown to depend upon the suprachiasmatic nucleus of the hypothalamus. When this brain structure is destroyed, functions including locomotor activity, eating, drinking, sleep, and body temperature no longer show 24-hr periodicity under constant conditions and often cease to be normally entrained by the light–dark cycle.

Animals often need to measure and respond to time periods that lie well outside the circadian range. In the case of ultradian cycles (with periods considerably shorter than 24 hours), we may distinguish between events that do and those that do not sustain repeated cycles. In this paper we use the term *interval timer* to refer to a mechanism that times behavioral or physiological events of durations shorter than 24 hours and that do not oscillate or automatically reset. Two characteristics of interval timers are that they can be quickly stopped and restarted and that they can measure duration at

[a]This research was supported by grants from the Whitehall Foundation and by Grant 02983 (to R.S.) from the National Institute of Mental Health.

[b]To whom correspondence should be addressed.

[c]Present address: Division of Neuroendocrinology, The Rockefeller University, 1230 York Avenue, New York, New York 10021.

any time of day.[1,2] Interval timers may differ qualitatively from the mechanisms responsible for circadian timing.

In all of the instances we consider, the nervous and endocrine systems act in a coordinated manner (FIG. 1). It is well established that the brain secretes chemical messengers (neurohormones such as gonadotropin-releasing hormone [GnRH] into the specialized blood supply of the pituitary gland. The pituitary in turn secretes protein hormones which affect the gonads, among other bodily tissues. In response, the

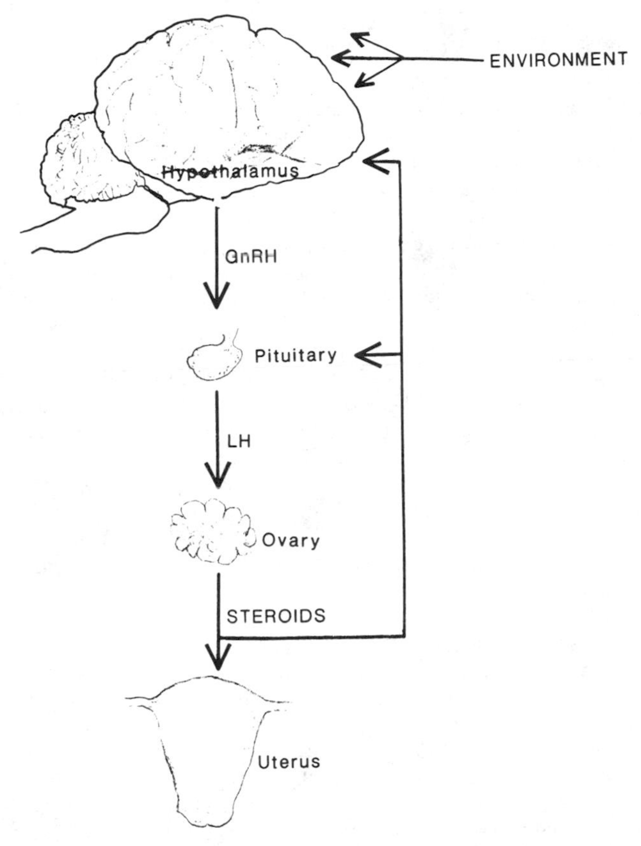

FIGURE 1. Diagrammatic representation of the hypothalamo–pituitary–gonadal axis. Some of the reproductive hormones produced at each level are illustrated and the feedback relationships discussed in the text are shown. GnRH = gonadotropin-releasing hormone; LH = luteinizing hormone.

gonads secrete steroid hormones such as testosterone, estradiol, and progesterone into the bloodstream. These steroids have multiple functions. Not only do they control target tissues such as the uterus, but they also feed back upon both the brain and the pituitary to influence the rate of their own secretion. Thus, internal messages arriving from the circulation, as well as external influences from the animal's social and physical environment, are channeled through the hypothalamic region of the brain to regulate reproductive function.

The joint participation of circadian and interval timing mechanisms, acting through the neural and endocrine systems, will be introduced by the example of a reproductive behavior, parental care in doves. The interaction of these timers in the control of mammalian fertility, in which physiological mechanisms are better understood, will then be discussed. Finally, we will explore the functional consequences of the use of both circadian and interval timers and the mechanistic relationships between these different processes.

BEHAVIOR: TIME SHARING BY PARENT DOVES (*Streptopelia risioria*)

Columbiform birds exhibit highly stereotyped parental behavior. Males and females share in all phases of caring for the eggs and young. The eggs are always kept covered; the male incubates for a block of time (6–8 hr) during the middle of the day and the female incubates for the rest of the time.[3] This behavior is expressed only in paired birds. If one or the other mate is removed, the remaining parent incubates almost continuously for several days, but eventually abandons the nest.[4] If the nest bowl is split in half by a wire mesh and each half is provided with eggs, each parent will sit on its half of the nest much of the time and will not display cooperatively timed incubation behavior.[5]

What sort of timing mechanisms produce this social division of incubation schedules? Manipulation of the light–dark cycle implicates a circadian mechanism.[5] When the photoperiod is phase-shifted during early incubation, parent birds rapidly adjust their schedule such that within 1 to 2 days the male once again sits during the middle of the light period. In constant dim illumination, incubation bouts "free-run" with the onsets of both male and female sitting generally occurring at intervals of less than 24 hours. Exposure to constant bright light disrupts the incubation pattern in a manner comparable to its effects on other avian circadian rhythms[6]: short irregular bouts of sitting occur. As in the case of other circadian rhythms, manipulation of daylength influences the phase angle of transitions between male and female incubation bouts.[7]

A series of experiments suggests that sit-bout duration is determined by an interval timing mechanism. Consider a model pair in which the male sits from 1000 hr to 1600 hr. If the male is prevented from starting his sit-bout until 1200 hr, will the end of the sit-bout occur at the usual time of day (that is, 1600 hr) or after the usual interval (that is, 1800 hr)? It was found that the female enters the nest area at the usual time of day (1600 hr) and attempts nest exchange (FIG. 2). The male, however, refuses to leave and possession of the nest is disputed for a period. In a series of such experiments, the onset of male incubation bouts was experimentally delayed by intervals of 0, 3 or 6 hr, and the duration of the ensuing bout was measured. The results indicate that the time of the male–female nest exchange is achieved by a compromise between the duration estimate of the male and the circadian phase of the female.[8] In parallel experiments in which the female was prevented from starting her sit-bout at the usual time, a similar result was obtained. The female would refuse to leave the nest until her sit-bout had reached its normal duration, while the male would try to sit at his usual circadian phase. It is proposed that a circadian system determines the time at which sit-bouts begin for both partners. Once a bout has been initiated, the sitting partner measures the sex-typical duration of an interval which begins at the onset of sitting. For several reasons this behavior suggests the operation of an interval timer. First, male–female nest exchanges can be quickly reset following phase shifts of the light–dark cycle with very few transients. Second, such shifts can be accomplished by restricting access to the nest in the absence of any change in the light–dark cycle; it is unlikely that the gate constitutes a zeitgeber because it has no lasting effect on the entrained phase of sitting[9] (Kahn, personal communication). Finally, the interval between the onset and end of incubation is largely conserved when access to the nest is delayed. This would not be expected if a circadian mechanism independently determined the beginning and end of each incubation bout. In summary, parent doves appear to measure the passage of time by using two kinds of timers.

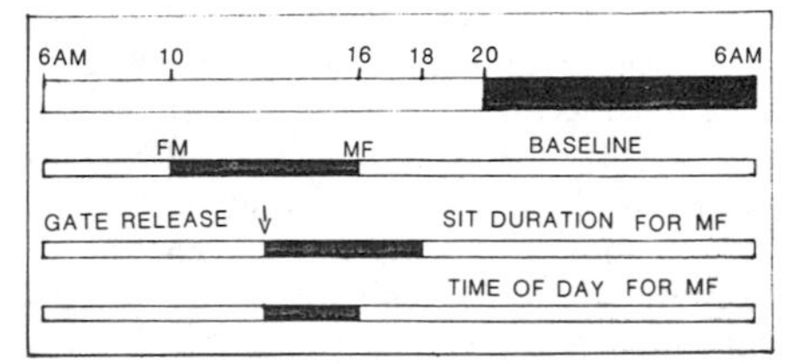

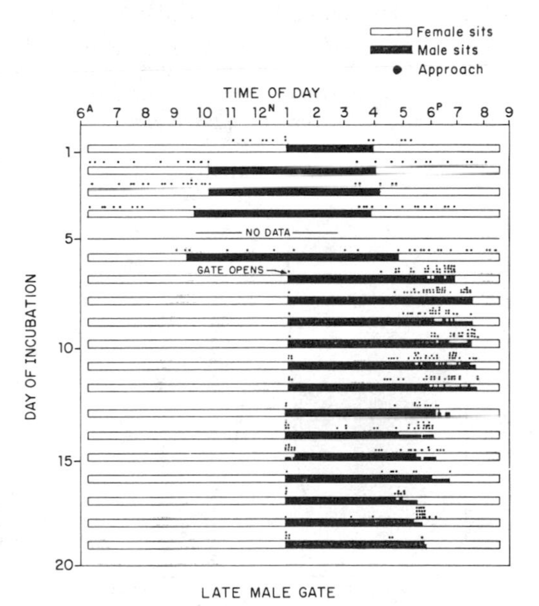

FIGURE 2. Design (*upper panel*) and results (*lower panel*) of an experiment testing determinants of timing of male-to-female transitions in nest exchange during incubation in ring doves.

(*Upper panel*): Light–dark regimen (L:D 14:10) is indicated at the top of the figure; *shaded bar* indicates darkness. Baseline condition indicates typical *ad libitum* incubation schedule; male sitting is indicated by *solid bar;* female sitting is indicated by *open bar.* FM, female–male transition; MF, male–female transition. Remainder of panel gives predicted results of experiments in which a gate separating the partners is closed during the night and opened at 1300 hr. If the male were to sit for his usual duration the male-to-female exchange would occur at 1900 hr, as shown in the middle diagram. However, if time of day controlled this exchange, it should occur as in the baseline condition at 1600 hr. See text for details.

(*Lower panel*): Timing of incubation and of approaches to the nest area by a representative pair in the delay-gate experiment diagrammed above. *Open bars* indicate female sitting; *solid bars* indicate male sitting. *Dots* indicate approaches to the nest by the nonincubating partner.

PHOTOPERIODIC TIME MEASUREMENT

Animals of the temperate zone experience strong selective pressure to ensure that young are born at the time of year when they are most likely to survive. Many of these animals measure daylength in order to restrict reproduction to the appropriate season. Circadian rhythms have been implicated in both formal and physiological analyses of the process of photoperiodic time measurement. Evidence suggests, however, that noncircadian processes involved in the timing of shorter durations might also be critical to the reproductive response to daylength.

Formal Analysis of Photoperiodic Time Measurement

How is the accurate measurement of daylength accomplished? One possibility is that the time-dependent accumulation of some light-sensitive product is used to judge whether night length is long or short. According to this model, the system works like an hourglass; when sufficient product is accumulated, it is concluded that the night is long. Each exposure to light inverts the hourglass and measurement of the ensuing dark period starts anew.[10] An alternative model proposes that an endogenous self-sustaining circadian oscillation participates in the measurement of photoperiod. One form of this hypothesis, originally devised by Bünning,[11] argues that a phase of sensitivity to light recurs at approximately 24-hr intervals in constant darkness. When the animal is entrained to a light–dark cycle, the photosensitive phase is positioned so that it is struck by light only when daylength surpasses a particular threshold. Another version of the circadian hypothesis, the "internal coincidence" model, is derived from several lines of evidence which argue that there are multiple 24-hr oscillators.[12] In this formulation,[13] the temporal relationship between classes of circadian oscillations entrained to dawn and dusk varies systematically with daylength, allowing the measurement of photoperiod. A common feature of these circadian models which critically distinguishes them from hourglass hypotheses is that they emphasize the self-sustaining nature of the oscillation underlying measurement of the interval of darkness. Data bearing on these two hypotheses have been provided in studies in which the light–dark cycle is systematically manipulated.

In "night interruption" experiments, the animal is exposed to a short period of light (often 6 or 8 hr) followed by a longer span of darkness which is punctuated by a brief

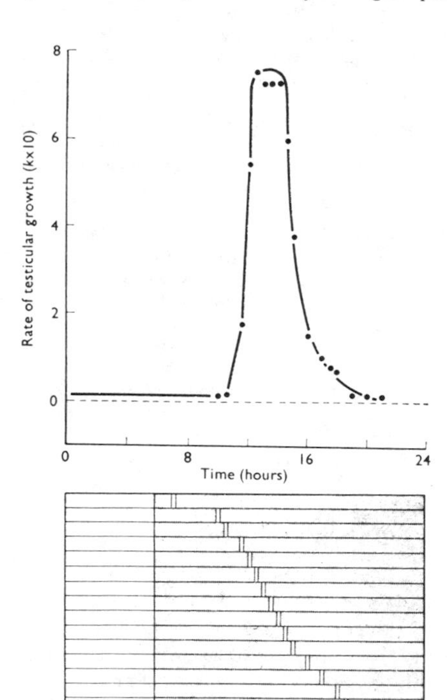

FIGURE 3. Design (*bottom*) and results (*top*) of an interrupted night experiment. Effect of 15-min light-breaks on the rate of testicular growth in Japanese quail maintained under a 6-hr photoperiod. (Data of Follett and Sharp.[113])

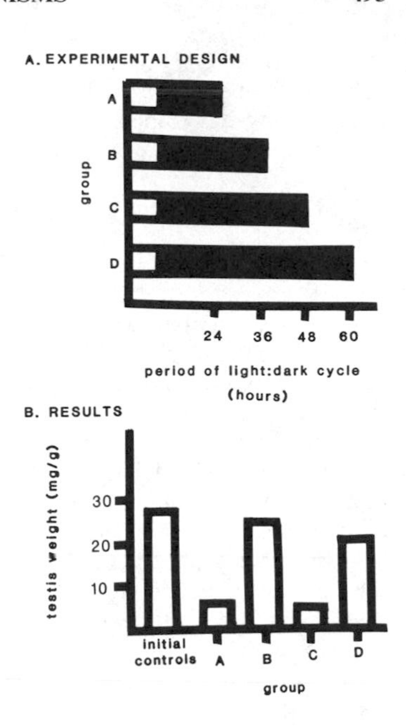

FIGURE 4. Design (*top*) and results (*bottom*) of a resonance experiment. (**A**) Open bars indicate light and shaded bars indicate darkness in experimental photoperiods to which hamsters were exposed for 89 days. For hamsters in group A, a 6-hr light period was repeated once every 24 hr (L:D 6:18); in group B, once every 36 hr (L:D 6:30); in group C, once every 48 hr (L:D 6:42); and in group D once every 60 hr (L:D 6:54). (**B**) Testis weight (mg/g body weight) of initial controls (maintained at L:D 14:10) and groups A–D at termination of experiment. Photostimulation of testicular function occurred only in the 36- and 60-hr cycles, indicating the participation of a circadian rhythm of photosensitivity in the hamster's reproductive response to photoperiod. (Data of Elliott *et al.*[17])

burst of light (FIG. 3). The timing of the second pulse of light may be varied so that the photoschedule resembles a "skeleton" of a long or short day from which several hours of light have been deleted; when the main photoperiod and the night interruption are of equal duration, the regime is termed a "symmetrical" skeleton photoperiod. In these experiments, the total amount of light to which the animals are exposed in any 24-hr period is similar to that of a winter day. Night interruptions placed well into the dark period evoke reproductive responses typical of the long daylength of summer. In short-day breeders, such as sheep,[14] such night interruptions induce a nonbreeding state, while in long-day breeders, such as hamsters,[10,15,16] such night interruptions induce reproductive activity. These results indicate that the absolute amount of light is not critical to photoperiodic time measurement; rather, the time at which light falls is important. The results of interrupted night experiments, however, are consistent with both the circadian and the hourglass interpretations.

In "resonance" experiments, short periods of light (typically 6–8 hr) are alternated with periods of darkness whose durations differ between experimental groups (FIG. 4). Thus, the total period of the photocycle (sum of the lengths of the light and dark phases) varies systematically from 24 hr. In both long- and short-day breeders, the results of this manipulation reveal a striking regularity. When the length of the cycle is a multiple of the circadian period (for example, 24 or 48 hr), the reproductive response is that typical of short days. When the period of the cycle differs by 12 hr from the circadian period, a long-day response occurs.[17–19] Since reproductive responses vary with the period of the light cycle long after the critical night length of a 24-hr day is surpassed, an hourglass mechanism is ruled out. These results are instead interpreted as evidence for the operation of a circadian clock. During sustained intervals of darkness, such as those that occur in resonance cycles of long periods, the photosensitive phase is postulated to recur at intervals of approximately 24 hours due to its self-sustaining nature as a circadian oscillation. Alternatively, internal phase relationships between dawn and dusk oscillators may resemble those resulting from exposure to 24-hr cycles. Only when the short period of light coincides with the subjective night, as when the period of the light–dark cycle deviates from some multiple of 24 hours, does the long-day response occur.

Physiological Analysis of Photoperiodic Time Measurement in Mammals

Which structures are involved in the response to daylength? Light–dark information is relayed from the retina to the suprachiasmatic nucleus (SCN) of the hypothalamus. A pathway has been traced from the SCN to sympathetic neurons of the superior cervical ganglion which innervate the pineal gland[20–23] (FIG. 5). Destruction of the SCN eliminates the reproductive responses to photoperiod.[24–26] Removal of the pineal gland, or its denervation by superior cervical ganglionectomy, also has this effect on the photoperiodic responses of long- and short-day breeders.[23,27,28]

The pineal gland has no efferent neural connections in mammals and controls reproduction by humoral means. Which pineal product is responsible for these photoperiodic responses? Attention has been focused on the indoleamine, melatonin, whose synthesis and/or secretion is entrained by the light–dark cycle and free-runs in the absence of photic cues.[29,30] The nocturnal rise in serum melatonin is eliminated after disruption of the neural pathway between the SCN and the pineal and by removal of the pineal gland.[31–34]

An important clue to the role of melatonin in photoperiodism is provided by observations in Djungarian hamsters and sheep that the duration of its synthesis or secretion is proportional to night length under entrained conditions.[29,35,36] The ability of 1-min interruptions in the night to produce long-day reproductive responses is

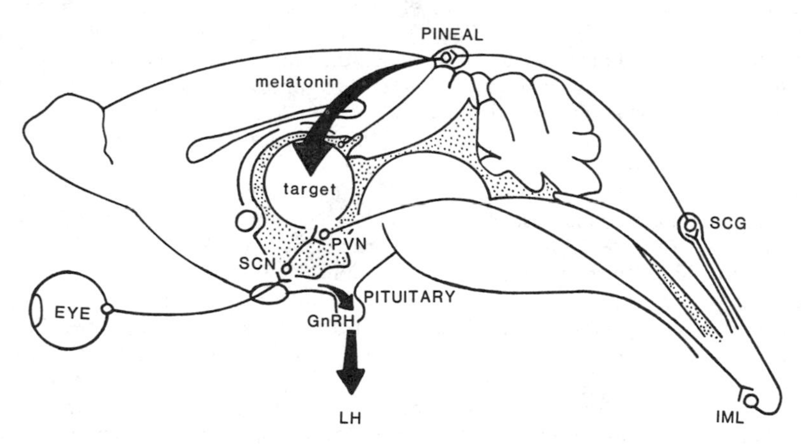

FIGURE 5. Anatomy of the pathway which transduces photoperiodic signals in the golden hamster. Light–dark information is relayed from the retina via a retinohypothalmic tract to the suprachiasmatic nucleus of the hypothalamus (SCN). A projection to the paraventricular nucleus (PVN) contacts neurons which project directly to the intermediolateral cell column of the spinal cord (IML). These cells in turn drive noradrenergic neurons in the superior cervical ganglion (SCG) which innervate the pineal gland. Melatonin is synthesized under the control of this sympathetic pathway; it is secreted into the systemic circulation during the night and is detected by a neural target which remains to be identified. This information is used to modify the pattern of secretion of the hypothalamic peptide, gonadotropin-releasing hormone (GnRH), which drives pituitary function.

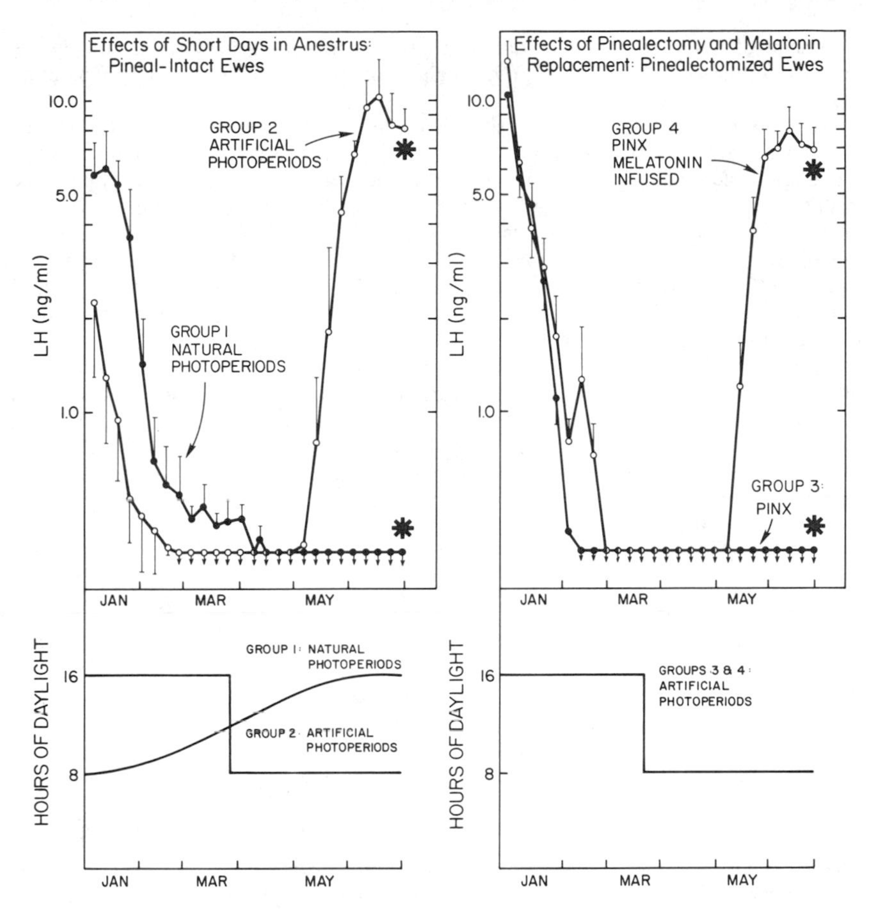

FIGURE 6. Role of pineal melatonin secretion in the photoperiodic response of sheep. Serum LH (mean ± S.E.M. of twice-weekly samples in ovariectomized, estradiol-treated ewes) reflects reproductive competence. (*Left*) Pineal-intact ewes: ● = pineal-intact controls maintained outdoors; ○ = pineal-intact controls maintained in long photoperiods (16L:8D) for the first 90 days and abruptly shifted to short days (8L:16D) for the remainder of the experiment. (*Right*) Pinealectomized ewes: ● = pinealectomized ewes not treated with melatonin; ○ = pinealectomized ewes infused nightly with physiological levels of melatonin. The photoperiods are illustrated at *bottom* of each panel. (Modified from Bittman *et al.*[36])

correlated with the shortening of the nocturnal period during which melatonin is synthesized in the pineal glands of Djungarian hamsters.[37] In order to test the hypothesis that photoperiod acts through melatonin secretion to drive seasonal rhythms of reproduction, timed melatonin replacement experiments have been performed in pinealectomized animals.[36,38] Such treatments restore effects of day-length upon reproductive function (FIG. 6). Melatonin infusions of short duration mimic the effects of short nights, while infusions of long duration mimic the effects of long nights. Significantly, the effects of melatonin are independent of the photoperiod in which pinealectomized animals are housed.[38,39] This indicates that pineal melatonin determines the direction of the reproductive response, rather than merely allowing the expression of a measurement of daylength which might be performed outside the pineal. In other words, melatonin drives the photoperiodic response rather than acting as a permissive influence.

What implications do these physiological studies have for our conception of the interdependence of circadian and interval timing processes? A model involving a circadian-based signal (melatonin) whose duration encodes night length implies the existence of a receptor system which can decode the duration of melatonin. The receptor system is proposed to be independent of the signal generator. The circadian properties of the signal generator may be adequate to explain the results of resonance experiments and reflect the entire role of circadian oscillations in the measurement of photoperiodic time. The nature of the receptor system that performs this measurement of melatonin duration is unknown; it may depend critically on a circadian component or have the properties of an hourglass or other interval mechanism. For example, an influence of melatonin on the phases of covert circadian oscillations, the existence of an entrainable melatonin-sensitive phase, or the progress of an oscillator during the time elapsed between the onset and the offset of melatonin secretion may be involved. Alternatively, an hourglass receptor mechanism may be activated at the onset of

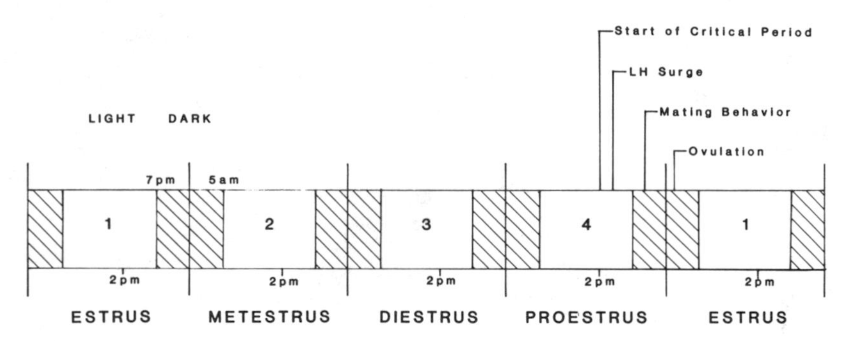

FIGURE 7. Timing of the LH surge, behavioral heat, and ovulation during the 4- and 5-day cycle of the rat. Also indicated is the beginning of the 2-hr "critical period" of proestrus during which injection of barbiturates is capable of producing a 24-hr delay in the LH surge and ovulation. (Modified from Nequin *et al.*[59])

nocturnal melatonin secretion and the time-based accumulation of some melatonin-dependent product would begin. In the course of a long night, sufficient product would build up for a threshold to be reached.

Little evidence exists to support the circadian timing of a receptor system measuring melatonin duration. Golden hamsters in which overt circadian rhythmicity has been eliminated by destruction of the SCN remain responsive to melatonin injections.[40] When the pineal gland of golden hamsters is denervated or removed, exogenous melatonin is capable of precipitating gonadal regression regardless of the time of day of its administration so long as the injections are spaced to provide a sustained daily rise of melatonin.[41] Melatonin injections even produce gonadal regression in golden hamsters housed in constant light. The significance of these observations is clouded, however, by a controversy about the role of melatonin duration in this species.[35,42] Further evidence for interval timing of melatonin duration is provided by the results of discontinuous infusion studies in Djungarian hamsters.[28] When melatonin administration is interrupted for as little as 2 hr, measurement of duration begins anew once treatments are restarted. This feature of the response is reminiscent of the reset properties of an interval timer. Finally, the ability of pinealectomized sheep and Djungarian hamsters to respond to patterns of melatonin that are inappropriate to the daylength in which they are housed[38,39] also argues against the importance of a light-entrainable circadian mechanism in the response system. In summary, photoperiodic control of mammalian reproduction is likely to involve both circadian and interval timers.

THE LH SURGE: OVULATION AND HEAT

The best-studied reproductive system in which circadian and interval timers interact is the process that results in ovulation and behavioral heat in the rat. A coordinated series of neuroendocrine events has been explored at anatomical, cellular and biochemical levels, allowing a detailed description of the "clock" mechanisms that culminate in the characteristic timing of ovulation and heat.

Ovulation

Ova are released from mature ovarian follicles at 4- or 5-day intervals in this species (FIG. 7). Ovulation is triggered by the dramatic discharge of luteinizing hormone (LH) from the pituitary gland. The basic mechanism generating this surge has long been known to involve an estrogen-sensitive neural system which operates under circadian control. In a series of classical experiments Everett and Sawyer[43] demonstrated that the injection of barbiturate between 1400 and 1600 hr of the afternoon preceding ovulation delayed this event by 24 hr in animals housed in LD14:10 with lights on at 0500 hr. Injections before or after this critical period were ineffective. More recent work demonstrates that the period of estrous cycles in some rodents (rats, hamster) is governed by a light-entrained circadian oscillator.[44-48] While the timing of the preovulatory LH surge is gated by the circadian system, the expression of the surge depends upon the secretion of steroid hormones secreted by ovarian follicles as they ripen. In effect, the brain and pituitary monitor the condition of the ovary so that the LH surge signal is generated only when a sufficient crop of follicles is mature enough to ovulate. Specifically, ovarian estradiol must be secreted in order to meet the requirements of the surge system.[49] In a sense, neuroendocrine tissues measure the concentration and duration of the estradiol signal.

The timing properties of this system have been thoroughly examined in a preparation in which the endogenous source of estradiol is removed by ovariectomy. This hormone is then replaced in physiological amounts, varying the temporal parameters of administration. The circadian basis of the LH surge system is demonstrated by the fact that preovulatory discharges are constrained to occur only at a particular time in the afternoon. The latency between the estradiol stimulus and the surge, however, is variable. In an acutely ovariectomized rat, administration of physiological doses of estradiol for periods of 7 hr or longer, beginning at 0900 hr, results in a surge of LH at about 1600 hr on the following day. When administration of the same dose of estradiol is initiated at 1400 hr or later on the day after ovariectomy, the LH surge is delayed by an additional day.[50]

Another circadian feature of this system is its oscillatory nature; when blood samples are collected from long-term ovariectomized rats on multiple days after a 29.5-hr estradiol stimulus, surges of LH continue to occur at intervals of 24 hr for at least 10 days.[51,52] Significantly, the time of day of the surge corresponds to that seen in the ovary-intact control rat. Furthermore, estradiol-induced surges can be blocked by barbiturates given during the critical period of the preceding afternoon. These data suggest that a physiological surge mechanism is activated by the exogenously administered steroid and indicate that the repeated daily surge signal underlies the circadian, light-entrained property of the rat estrous cycle.

Lesions of the SCN, which interfere with general circadian organization, eliminate the ability of estradiol to trigger daily LH surges.[53-55] The SCN is believed to regulate the generation of LH surges by another area of the brain—the preoptic area—upon which estradiol acts directly.[56] Once the system has received adequate estradiol priming, progesterone may bypass the circadian mechanism and advance the onset of the LH surge. Progesterone-induced surges appear to rely on a neural substrate distinct from the SCN.[57] Since in the normal course of events only low levels of progesterone are secreted by the ovary until well after the LH surge, this action of progesterone may be regarded as a "backup" mechanism to insure ovulation of a full complement of eggs. Progesterone released after the onset of the surge also serves the important function of restricting the LH discharge to a single day.

A description of the circadian character of the ovulation-eliciting LH surge system does not completely describe timing events critical to the estrous cycle. The properties of the system that monitors the temporal parameters of the estradiol stimulus suggest that a qualitatively different timing mechanism operates at this step. By implanting and removing a hormone-bearing capsule and by varying the interval over which the capsule is withdrawn, the effects of a "skeleton" estradiol stimulus have been compared with the effects of continuous administration over periods ranging from 9–27 hr.[58] Although estradiol will trigger an LH surge when continuously present for 7 hr or longer, two exposures of 3-hr duration can trigger surges only if separated by intervals of greater than 3 hr or less than 21 hr.[50] Although the output of the system (LH surge and consequent ovulation) shows strict circadian periodicity, there is no evidence that the system that times the duration of the critical estradiol stimulus has a circadian basis. Since this mechanism shows no oscillatory properties or 24-hr time frame and measures the duration of estradiol stimuli presented at any time of day, it fits the description of an interval timer.

Heat

The LH surge that produces ovulation also dictates the timing of behavioral receptivity, or heat, in the female. In the intact rat, heat is restricted to the fertile portion of the ovarian cycle as a result of the endocrine events that accompany ovulation. Preovulatory estradiol primes a neural mechanism over several hours or days; the subsequent release of progesterone from the ovulating follicles triggers behavioral receptivity.[59] As a result of the circadian gating of the LH surge system, periovulatory levels of progesterone become adequate to induce heat in the estradiol-primed animal on the evening of proestrus. Although heat is thus confined to a fixed time of day in the intact rat, this is probably due to the characteristics of the LH surge rather than to the estradiol priming system.

As in the case of ovulation, the temporal characteristics of the estradiol signal necessary to induce heat upon secretion of progesterone may be measured by an interval timer. These temporal parameters have been examined by administration of estradiol to ovariectomized rats. Significantly, heat can be induced by progesterone at any time of day in rats exposed to estradiol for a sufficient duration. In a series of skeleton experiments similar to those described above, estradiol was administered to ovariectomized rats in various patterns[60,61] (FIG. 8). Progesterone was injected 20 hr after the onset of estradiol treatment and heat was tested 4 hr later. When continuously administered, a minimum of 6 hr of estradiol treatment was required in order for subsequent progesterone to produce heat. Receptivity could also be induced, however, by two 1-hr presentations of estradiol priming so long as these were separated by at least 4 hr, but by no more than 13 hr. These treatments were studied in relation to

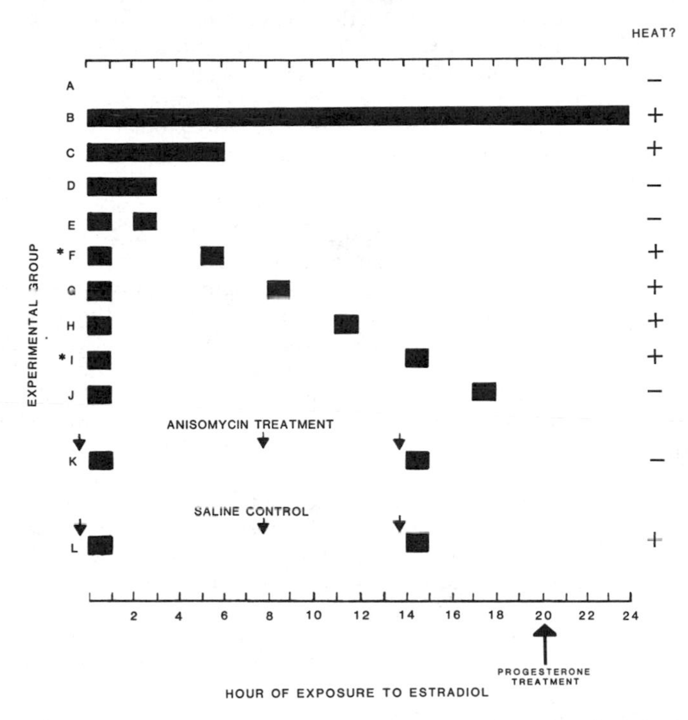

FIGURE 8. (*Top*): Effects of various patterns of estradiol (E_2) treatment of the induction of behavioral receptivity by a subsequent progesterone (P) injection in the rat. Estradiol was administered by subcutaneous implantation of a Silastic capsule which maintained proestrus serum levels; implants remained in place for the periods indicated by the horizontal *black bars*. Females were tested with an experienced male rat 24 hr after the initiation of E_2 treatment. Four hours before testing, each animal received 500 μg P. Column at right (Heat?) indicates results as determined by lordosis quotient (proportion of mounts that elicited lordosis; + indicates lordosis quotient > 15). Group A is a control (no E_2). Asterisks indicate groups in which minimum and maximum periods of estradiol withdrawal resulted in heat. (Modified from Parsons *et al.*[60,65]) (*Bottom*): Effects of blocking protein synthesis by administration of anisomycin on the induction of heat by discontinuous administration of estradiol. Administration of anisomycin at any of the times indicated by arrows blocked receptivity; inhibition of protein synthesis at other times was ineffective. (Data of Parsons *et al.*[65])

estradiol action in brain cell nuclei and were positively correlated with the induction of progesterone receptors in the hypothalamus and preoptic area. These investigators[60,61] suggest that when present for an adequate interval, estradiol induces events that culminate in the synthesis of brain progesterone receptors. When progesterone receptors reach adequate levels (18 hr after the estradiol stimulus), progesterone of ovarian origin activates sexual receptivity.[62]

The properties of the interval timer that determines whether the animal is adequately primed for heat bear close resemblance to the system whereby estradiol triggers LH surges. In both, the effective dose of estrogen is similar to proestrus levels (about 40 pg/ml), can be discontinuously presented, and can be withdrawn for similar intervals. Analysis of retention of the hormone in neuroendocrine cell nuclei shows that estradiol receptor occupancy may fall below 3.4% without interrupting the priming process. It is unclear, however, whether this minimal occupancy can continuously maintain critical biochemical processes, or whether these intracellular events continue when estradiol has functionally vanished from the cell nucleus.[63]

What estradiol-triggered events account for these temporal properties? Many of the effects of steroid hormones appear to be mediated by protein synthesis, and the time course of this intracellular process may explain some of the regularities observed in behavioral and feedback actions.[64] If the protein synthesis inhibitor, anisomycin, is given 15 min before either presentation of estradiol on the skeleton schedule or at the midpoint between the two treatments, heat is blocked[65] (FIG. 8, bottom). Estradiol-induced production of proteinaceous progesterone receptors occurs with a latency that matches that of heat induction and is activated by temporal patterns of estradiol that correspond to the skeleton stimuli adequate to prime for heat and to induce surges. Since LH surges can be generated in the absence of progesterone, it is not clear why their induction by estradiol should exhibit a similar latency. One possibility is that estradiol induces the synthesis of functionally different specific proteins along similar time courses in different brain regions. In experimental conditions under which estradiol is continuously elevated in ovariectomized rats, heat can be expressed on successive nights in the absence of ovarian progesterone. Such rhythmicity is dependent upon the SCN.[66] This suggests that a circadian mechanism has the potential to regulate the expression of heat independently of the induction of ovulation and the occupancy by progesterone of estradiol-induced receptors.

Regardless of the specific nature of the proteins involved in the induction of heat and LH surges, a model may provide insight into the timer which monitors the state of the follicles in the cycling rat. In one formulation (FIG. 9, upper part), the estradiol may initiate a sequence of events, some of which are estradiol-dependent. The interval over which estradiol can be withdrawn can be explained by the latencies with which a chain of sequentially linked processes leads to a subsequent estradiol-dependent step. In an alternative model (FIG. 9, lower part), estradiol may induce the production of substances with characteristic latencies and limited half-lives. In order to sustain adequate levels of a putative intermediate substance, estradiol has to be reintroduced to the system before the intracellular signal decays below a critical level. Either model can explain the maximal and minimal interval over which estradiol can be withdrawn, yet still produce a response in skeleton experiments. Both models involve an interval timing mechanism.

PSEUDOPREGNANCY AND PREGNANCY

In some female mammals (for example, humans, guinea pigs), each ovulation is invariably followed by a sustained period during which progesterone secreted by the postovulatory follicle, or corpus luteum, insures the capacity of the uterus to support an implanted embryo. In other mammals, including the rat, readiness for pregnancy does not follow ovulation unless the animal mates. Fertilization of the ovum is thus not adequate to ensure pregnancy. Stimulation of the uterine cervix is both necessary and sufficient to activate the ovarian events that provide a uterine environment hospitable to implantation. If no mating occurs in these animals, the corpus luteum dies, progesterone falls rapidly, and a new estrous cycle begins. Conversely, if the female mates with an infertile male, or if her cervical region is stimulated with a glass rod by a curious experimenter, she will enter a "false pregnancy" (pseudopregnancy). Despite the absence of a conceptus, the endocrine state mimics that of the first 12–13 days of pregnancy. A series of behavioral and physiological experiments demonstrate that both interval and circadian timing processes participate in the induction and maintenance of pregnancy or pseudopregnancy[67] (FIG. 10).

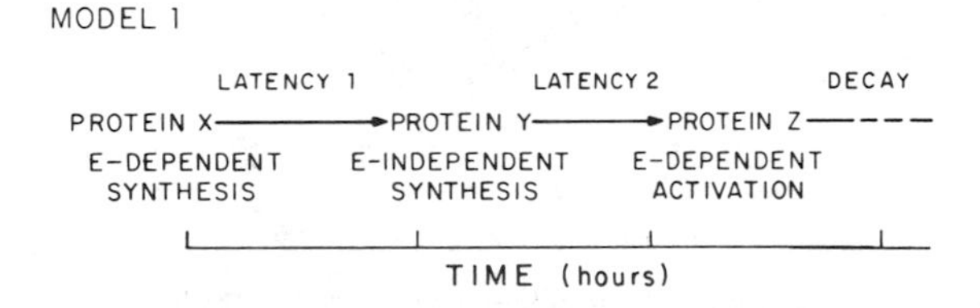

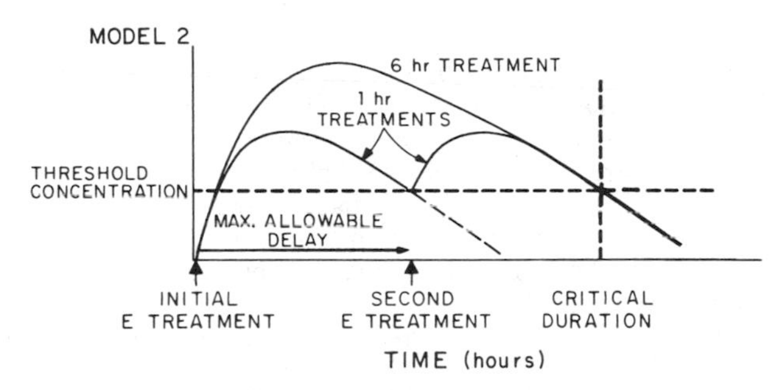

FIGURE 9. Interval timing: Two possible modes of estradiol (E) action. In model 1, estradiol initiates a chain of events by triggering the synthesis of protein X. This substance in turn triggers the synthesis of a second protein, Y, with a fixed latency and by a process that does not depend on estradiol. After a second latent period, intermediate Y leads to the production of protein Z, upon which estradiol acts to trigger events leading directly to heat and ovulation. Protein Z decays with a half-life which determines the maximal interval that can intervene between two estradiol treatments if heat and ovulation are to result. (Modified from the "domino theory" of Harris and Gorski.[105]) In model 2, only one estradiol-dependent protein is hypothesized. Levels of this protein X must remain continuously above a threshold for a critical duration in order for heat and ovulation to be induced. The amount of protein induced, and consequently the interval over which it decays to subthreshold concentrations, depends upon the strength and duration of the initiating estradiol stimulus. These parameters thus define the maximum allowable delay between two brief discontinuous presentations of estradiol.

The temporal pattern of cervical stimulation is a critical parameter of the stimuli necessary to trigger pseudopregnancy. The copulatory behavior of the male rat consists of a series of repeated mounts, intromissions, and dismounts. After eight or more such bouts, intromission is followed by ejaculation. This series is repeated several times in one mating session. The number of intromissions received by the female is important in determining the probability of pregnancy.[68,69] Ninety percent of females receiving the normal complement of intromissions before ejaculation become pregnant. Only 20% of females receiving a reduced number of intromissions before ejaculation become

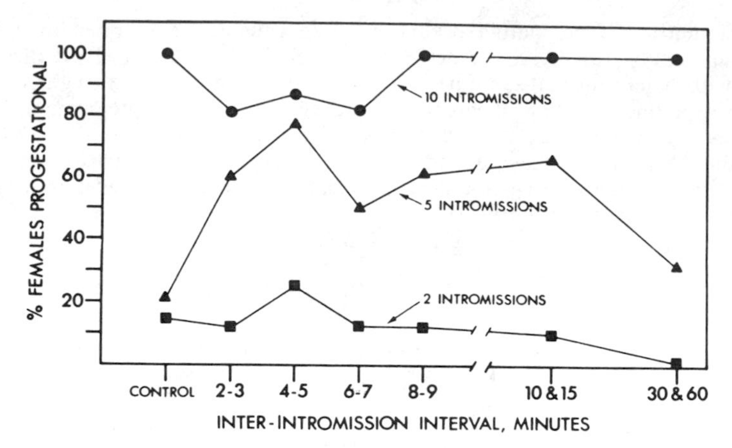

FIGURE 10. Effects of different interintromission intervals on the induction of pseudopregnancy in female rats. (After Edmonds *et al.*[70])

pregnant. Successful pregnancy or pseudopregnancy depends not only on the number of penile ineromissions but also on the inter-intromission interval. Although the mechanism and central nervous site at which afferent information provided by cervical stimulation is processed are unknown, the minimal temporal requirements for effective cervical stimulation and the outer limits of storage capacity have been delineated.[70] When an inter-intromission interval of 4–5 min is enforced, a higher portion of females becomes pseudopregnant than when 40-sec inter-intromission intervals occur. At the other extreme, the effectiveness of consecutive intromissions declines progressively at intervals of 30 min or 1 hr. The intervals over which intromissions can summate presumably reflect a decay function in the central nervous system which suggests the properties of an interval timer (FIG. 11). In support of this idea, data indicate that this timer can be started at any time. Thus, stimulation of the cervix on diestrus leads to pseudopregnancy which is delayed until after ovulation has occurred and corpora lutea have been formed.[71] The ability of cervical stimuli to induce pseudopregnancy at various circadian phases has not been directly tested.

The pseudopregnant state induced by cervical stimulation is characterized by twice-daily surges of the pituitary hormone, prolactin.[72] Prolactin thus released serves to stimulate the corpus luteum to produce progesterone, which in turn acts on the uterus to promote pregnancy or pseudopregnancy.[73,74] If optimal cervical stimulation is provided at any time of day (for example, by the male or by two 30-sec applications of electrical stimulation), the prolactin release system continues to surge for a preset interval of 10 days.[75–77] If suboptimal stimulation is provided (for example, one 30-sec application of electrical stimulation), the prolactin release system surges for fewer days. These twice-daily surges in prolactin are entrained to the light–dark cycle (FIG. 12); they free-run in blind animals housed in constant darkness and remain phase-locked to one another, suggesting that the prolactin rhythm is truly circadian. Lesions of the suprachiasmatic nucleus abolish the cervically stimulated prolactin surges, indicating further similarity to other circadian rhythms.[78] Unlike some other circadian rhythms, the prolactin rhythm is not manifested spontaneously, but occurs only after exteroceptive stimulation of the cervix. As in the case of the LH surge system, the temporal properties of a triggering event measured by an interval timer may determine whether a continuously operating circadian function is expressed. Alternatively, cervical stimulation may activate a system that then runs with circadian periodicity. In either case, the reproductive outcome relies on the combination of an interval timer, which cumulates cervical stimuli, and a circadian timer, which generates prolactin surges.

FUNCTIONAL CONSEQUENCES OF A TWO-TIMER SYSTEM

Circadian Response with Linked Interval and Circadian Timers

We have examined a number of reproductive paradigms in which both circadian and interval timers participate. Under appropriate experimental conditions, either 24-hr endogenous oscillators or shorter-interval nonoscillatory timers appear to determine the temporal parameters of the reproductive response. While interval timers may be critical at one or more levels, the final output has circadian properties in each of the systems that we have considered. The signal duration that is measured may originate inside the animal (as in the case of heat and LH surges) or in the outside world (as in the cases of pseudopregnancy, photoperiodism, and parental care). The origin of the stimulus imposes no constraints upon the participation of circadian and interval properties in the final response.

Nor is the order of linkage of circadian and interval timers invariant; we have identified examples in which either mechanism precedes the other. For instance, in seasonal breeding, we have proposed that a circadian timer relays information about the photoperiodic stimulus and is interpreted by an interval timing response system. In contrast, in the induction of pseudopregnancy an interval timer gathers information

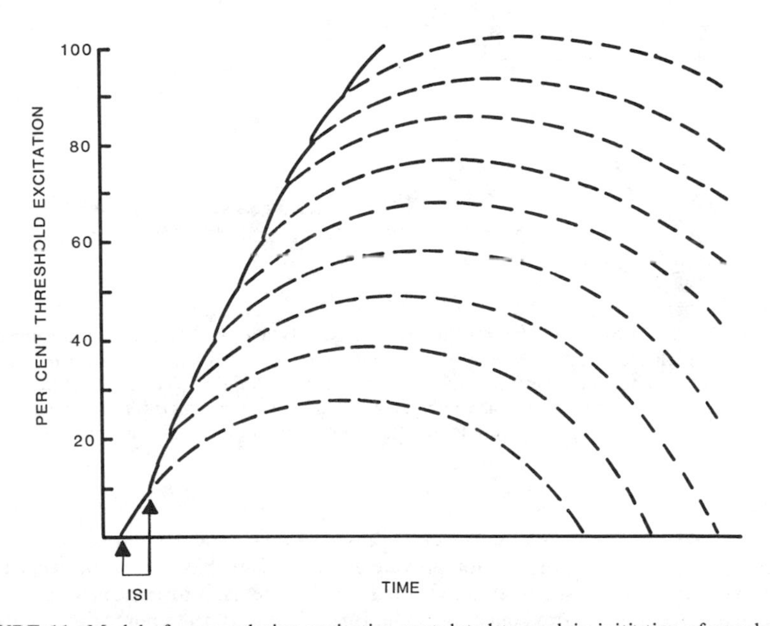

FIGURE 11. Model of accumulating excitation postulated to explain initiation of pseudopregnancy by cervical stimulation. Note that interstimulus interval determines the number of intromissions required to reach threshold for induction of pseudopregnancy. ISI, interstimulus interval. (Modified from Bermant and Davidson.[115])

about the cervical stimulus and triggers a response that is circadian in nature. The advantages and functional consequences of linking circadian and interval systems in tandem may differ between these two types of organization.

Interval–Circadian Systems

Where the interval time measurement precedes the circadian event, the interval timer establishes WHETHER permissive conditions have been met. The circadian timer

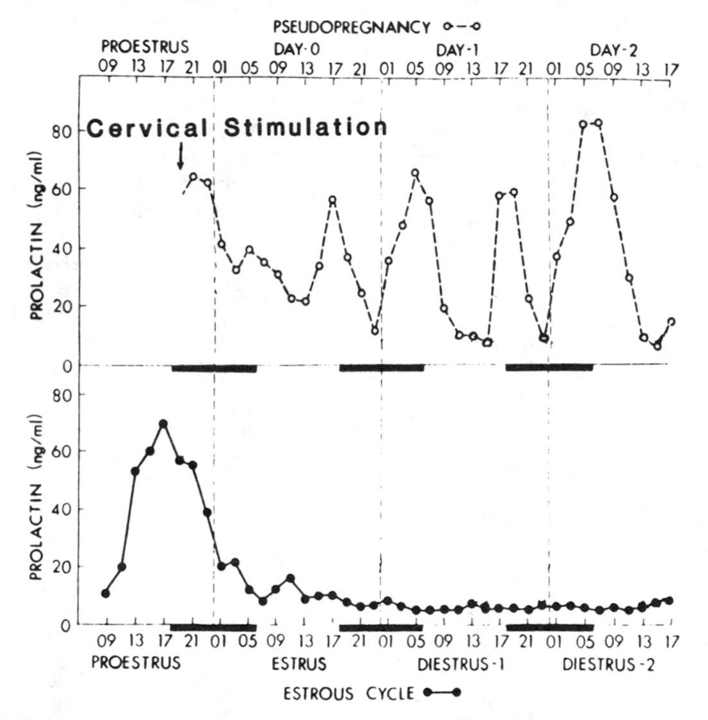

FIGURE 12. Serum prolactin levels during the estrous cycle and early pseudopregnancy. The numbers along the *top* and *bottom abscissas* represent the time of day. *Black bars* indicate darkness; the *dashed line* bisecting each bar represents midnight. Cervical stimulation was administered at 1900 hr on proestrus to animals shown in the *top panel;* females shown in the *bottom panel* received no cervical stimulation. Each point represents the mean of 5–6 decapitated animals. (Modified from Smith *et al.*[74])

determines WHEN the outcome will occur. Thus the two timing processes are functionally independent, and in conjunction provide a mechanism that allows the response to occur in the appropriate sequence and time. In such systems, information accumulated by the interval timer is stored once threshold is reached until the appropriate phase for activation of the circadian effector system. It is not known whether the storage function is part of the interval timer or represents a separate intervening mechanism.

In some instances of interval timing, an increase in the strength of the stimulus can compensate for a decrease in its duration or vice versa. Such systems may be viewed as measuring the area under a curve of stimulus intensity versus time. In the induction of pseudopregnancy, intense (electrical) stimulation of the cervix is effective at 30-sec durations, while weaker stimuli are effective only when applied over longer intervals. The interval timer (or some other mechanism) then stores information, including strength and duration of the stimulus, and the circadian clock determines the timing of prolactin output.[71]

The LH surge mechanism constitutes another hybrid system in which circadian and interval timers fulfill different functions. The properties and distinct physiological organization of these timers are clearly illustrated by a comparison between the rat and the monkey. The rat combines interval and circadian mechanisms in regulating the LH surge, as described above. The rhesus macaque does not employ a circadian timer and can ovulate at any circadian phase. In the rat, an interval system operating at the pituitary level may be sensitized by a high estradiol concentration after a short interval. An LH surge will not occur, however, until the circadian neural signal is received.[50] In the monkey, a stronger estradiol challenge acting at the pituitary will precipitate an earlier LH surge with no gating of the response by a circadian system.[79,80] Thus a two-timer system may serve to check the expression of tradeoffs between stimulus strength and duration.

A system which employs an interval timer may measure duration independent of stimulus strength by setting the effective concentration at which an intracellular event is triggered within narrow limits and allowing the stimulus effects to decay rapidly. For example, the photoperiodic system which measures night length requires a melatonin stimulus that is nearly continuous over a critical duration that corresponds to the threshold night length.[28] In contrast, in the LH surge and heat systems, estradiol may be withdrawn for periods of several hours and still be effective (see above). In this analysis, quantitative differences in the kinetics of processes triggered or activated by hormones in target tissues may explain the apparently qualitative differences in the accuracy and strength–duration characteristics of such timers.

Circadian–Interval Systems

When linked in the opposite order, the concatenation of circadian and interval timers may have different functional consequences. In such arrangements, the circadian system may act as a buffer between the stimulus to be measured and the interval timer. Here, problems of strength–duration tradeoffs in the latter mechanism may not normally arise. For example, the circadian properties of the system that transduces daylength into pineal melatonin production may restrict the characteristics of the signal with which the interval timer is presented so that errors are minimized. Evidence exists for an upper limit to the amplitude and duration of the rise in melatonin secretion. While the circadian mechanism may provide a relatively accurate coding of night length,[81] however, the estimate of duration will depend on the error in each of the concatenated timers. Measurement of night length may therefore be less accurate as a consequence of combining circadian and interval mechanisms. While the two mechanisms may correct each other in some way, it is also possible that error in the resulting time estimate will be additive.[81,82]

How are different kinds of timers used in physiological organization? Interval timers may allow conversion of circadian-based signals into alternative forms of coding which carry information for effector systems. The clearest example concerns the system by which melatonin duration determines the secretory pattern of LH in the ewe.

The reproductive axis is driven by a neural ultradian oscillator which discharges GnRH into the hypothalamohypophyseal portal system. Each GnRH pulse triggers an episode of LH release from the pituitary, which in turn elicits a discharge of estradiol from the ovary. Studies in various species indicate that quantitative changes in the frequency of the GnRH oscillation may account for qualitative changes in reproductive state culminating in puberty,[83] anorexic amenorrhea,[84] or menstrual cyclicity.[85] In sheep, photoperiod acts through the pineal to regulate the frequency of these ultradian oscillations. During anestrus, the LH pulse generator functions at a greatly reduced

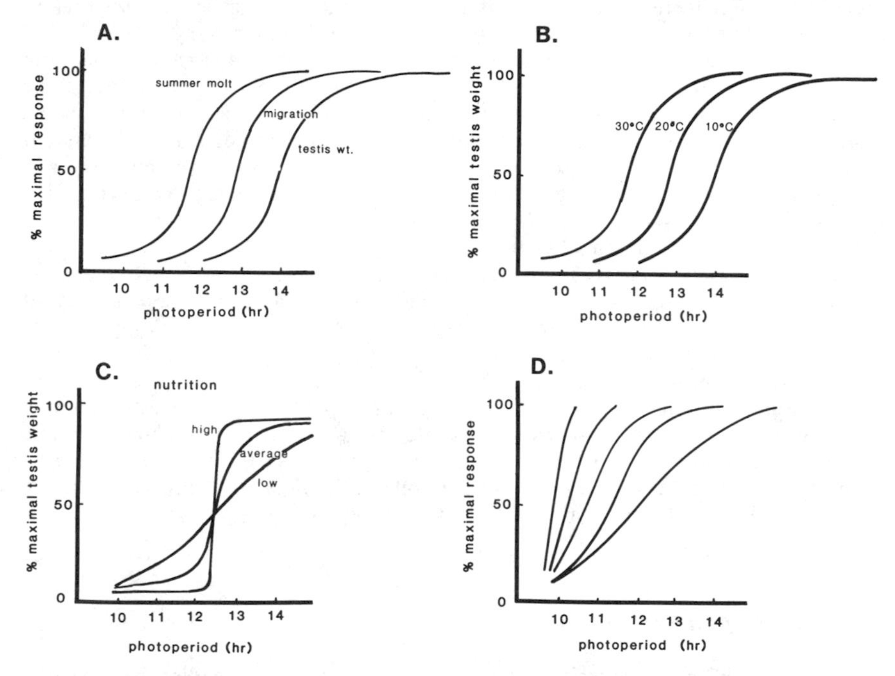

FIGURE 13. Schematic representation of the interactions of one circadian and one or more interval timers. (**A**) Multiple interval timers, each measuring the same circadian signal, control several different functions so that each occurs at its appropriate time of year. (**B**) A single interval timer allows relevant environmental cues (in this instance, temperature) to set the photoperiodic threshold. (**C**) The precision of an interval timer may be influenced by relevant environmental cues (in this case, nutrition) to determine the slope of the photoperiod response curve without any alteration in photoperiodic threshold. (**D**) The variance properties of the interval timer may change systematically as a function of the duration measured.

frequency relative to that characteristic of the breeding season. Significantly, episodic administration of exogenous hormone to raise LH pulse frequency to that of the breeding season can restore reproductive competence despite exposure to inhibitory daylengths.[23] The interval timer may measure the circadian-based melatonin signal and transduce the information carried in its duration into an ultradian LH frequency code to which the gonad can respond.

What functional capabilities are conferred by the use of two kinds of timers? We consider three models in which a single circadian oscillator linked to one or more interval timers confers flexibility on photoperiodic organisms. One possibility is that this arrangement allows a single circadian pacemaker to drive multiple outcomes independently and in correct temporal order. For example, a variety of seasonal changes, including alterations in pelage, fat stores, migratory activity, as well as in reproduction, may all be driven by changes in daylength.[86-89] It may be advantageous to execute each of several seasonal responses at a different critical photoperiod (FIG. 13A). For example, Lynch and Gendler[89] have shown that testicular growth is triggered by daylengths of greater than 13 hr of light in the white-footed mouse (*Peromyscus leucopus*), while molt to summer pelage and cessation of spontaneous daily torpor occur at much shorter photoperiods. Each response may use a common circadian signal; a variety of interval timers, each linked to a different physiological system, may provide a degree of independence of outcomes which would not be possible if the circadian signal directly drove the responses with no buffering interval timer.

A second advantage of multiple timers in circadian–interval systems may be that this arrangement allows relevant but unpredictable environmental cues to bias different seasonal responses so that they occur earlier or later in the year. Thus the use of interval timers may permit interactions between photoperiod and thermal, social, nutritional, or other cues which are known to affect the timing of reproduction,[90-92] but to which circadian rhythms may be relatively insensitive. Through their impact on the interval timer, nonphotoperiodic cues which have little access to the circadian oscillator may thus influence the reproductive response. In one model, photoperiod response curves would be shifted to the left or right (FIG. 13B) as a function of environmental influences such as temperature. Pittendrigh[93] has proposed an alternative formulation using a model involving only circadian oscillations which are differentially and incompletely temperature-compensated to explain how critical photoperiods can differ with temperature. Critical experiments must be devised to distinguish between these alternatives.

Another possible advantage of the inclusion of multiple interval timers is that they allow different responses to differ in their dependence on daylength. This results in photoperiod response curves with similar thresholds, but different slopes (FIG. 13C). A more shallow slope, achieved by the inclusion of interval mechanisms in the timing process, would permit nonphotoperiodic cues to influence the seasonal response across a wider range of intermediate daylengths.

In illustrating the circadian–interval concept, we have assumed that precision does not change as the duration of the interval measured lengthens. It is possible, however, that the precision of interval timers may be dependent on the photoperiodic threshold. Where the critical photoperiod for a given response varies with some environmental factor, analysis of its precision at, say, different temperatures may reveal scalar, Poisson, or other properties. In other words, the variance of the response may be a function of the interval being measured[2,82] (FIG. 13D). To our knowledge, none of the foregoing analyses have been performed.

Mechanistic Relationship of Interval and Circadian Timers

Is there a relationship between physiological systems that time different durations? This question has been asked about endogenous rhythms whose periods approximate tidal, lunar, and annual periodicities.[94] In each of these cases, current evidence argues against a circadian basis of lower frequency cycles.[95-97] The dependence of the 4-day rat and hamster estrous cycles on a circadian oscillation constitutes a salient exception to this generalization.

In the case of interval timers which participate in reproductive processes, little relevant data exist about mechanisms. It is not known whether any such timers are temperature-compensated, as are circadian oscillators. Furthermore, various circadian rhythms have been shown to be influenced by steroids, lithium, heavy water, neuroleptic drugs, and agents that influence adenyl cyclase, membrane potentials, and protein synthesis.[98–101] The sensitivity of interval timers to many of these substances remains to be explored (but see Meck[102]). The absence of such data restricts our understanding of the relationship between the two types of timers. It is known, however, that at least some circadian rhythms persist despite the blockade of protein synthesis by antibiotics.[100] In contrast, some interval timers (for example, the process critical to the measurement of estradiol duration discussed above) may depend heavily upon protein synthesis. Parallels between interval and circadian timers in temperature compensation, drug sensitivity, or variance properties may reveal properties central to time-measuring mechanisms and indicate whether they depend on a common pacemaker. It is possible, however, that various timers are similar only in their dependence on fundamental metabolic or respiratory processes. Such a commonality would provide little insight into mechanisms basic to the measurement of time.

At least some circadian and interval timers can be dissociated on a physiological level by brain lesions. Destruction of the SCN disrupts a number of circadian functions (see the INTRODUCTION). Among the cases we have considered, SCN lesions eliminate prolactin and LH surges, day–night differences in heat in estradiol-treated rats, and photoperiodic control of reproduction.[24,26,37,66,103,104] In each of these cases, the interval timer may be spared by this lesion. Thus, SCN-lesioned Syrian hamsters which do not sustain free-running locomotor rhythms remain reproductively sensitive to exogenous melatonin and respond differently to acute than to sustained elevations in its concentration.[40] Their accuracy in discriminating patterns of melatonin requires further examination. Similarly, LH surges and heat can be triggered by progesterone in SCN-lesioned rats exposed to estradiol for adequate intervals.[53,57] This is consistent with the observation that interval measurement of estradiol levels may take place at other neural levels as well as in nonneural target tissues such as the pituitary and uterus.[56,105,106] The effect of SCN lesions on the critical duration and concentration of estradiol priming of target tissues should be studied.

Some behavioral and endocrine ultradian cycles can survive SCN lesions in rodents.[107] Thus, the SCN may act to integrate or couple multiple oscillators, some of which may have periods shorter than 24 hr, as well as to provide a daily window during which ultradian rhythms may be expressed.[108–111] In *Drosophila,* compelling evidence supports the notion of a common genetic basis of a 60-sec song rhythm and a circadian oscillation of locomotor activity and eclosion.[112] While the reproductive interval timers we have considered typically time ultradian durations, they show no such oscillatory properties and may not bear such a relationship to the circadian system.

Conclusion

Questions about temperature compensation, susceptibility to drugs, and the tradeoff between strength and duration raise the issue of whether the interval-measuring systems we have described actually qualify as timers. Such systems may differ considerably from circadian timers in their accuracy with regard to real time. Unlike interval timers discussed elsewhere in this volume, the interval mechanisms we have described may not generate proportional or graded estimates of the passage of time and may have little flexibility in resetting critical durations. On a mechanistic level, they may reflect the time required for the satisfaction of important intermediate biochemical steps, or the achievement of critical thresholds in a complex reproductive response. In a functional sense, they fit the description of a timer in that they measure the duration of a critical interval. These interval timing mechanisms, acting in concert with circadian systems, are essential to the coordination and synchronization of reproductive events.

ACKNOWLEDGMENTS

We wish to thank the following individuals for comments on earlier drafts of this manuscript: Jürgen Aschoff, Peter Balsam, Robert Kahn, Lewis Krey, Alan Rosenwasser, Ben Rusak, George Wade, and especially John Gibbon.

REFERENCES

1. CHURCH, R. M. 1978. The internal clock. *In* Cognitive Processes in Animal Behavior. S. H. Hulse, H. Fowler & W. K. Honig, Eds. Wiley. New York, NY.
2. GIBBON, J. 1977. Scalar expectancy theory and Weber's law in animal timing. Psychol. Rev. **84:** 279–329.
3. WALLMAN, J., M. B. GRABON & R. SILVER. 1979. What determines the pattern of sharing of incubation and brooding in ring doves? J. Comp. Physiol. Psychol. **93:** 481–492.
4. SILVER, R. & J. GIBSON. 1980. Termination of incubation in doves: Influence of egg fertility and mate loss. Horm. Behav. **14:** 93–106.
5. SILVER, R. & S. NELSON. 1984. Social factors influence circadian cycles in parental behavior of doves. Physiol. Behav. In press.
6. BINKLEY, S. 1977. Constant light effects on the circadian locomotion rhythm of the house sparrow. Physiol. Zool. **108:** 170–181.
7. KAHN, R. M. 1984. Effects of photoperiods on timing of incubation in ring doves. This volume.
8. GIBBON, J., M. MORRELL & R. SILVER. 1984. Two kinds of timing in the circadian incubation rhythm of the ring dove. Submitted for publication.
9. BALL, G. F., & R. SILVER. 1983. Timing of incubation bouts by ring doves *Streptopelia risoria*. J. Comp. Psychol. **97:** 213–225.
10. HOFFMAN, R. A. & H. MELVIN. 1974. Gonadal responses of hamsters to interrupted dark periods. Biol. Reprod. **10:** 19–23.
11. BUNNING, F. 1936. Die endogene tagesrhythmic als Grundlage der photoperiodischen Reaktion. Ber. Dtsch. Bot. Ges. **54:** 590–607.
12. ASCHOFF, J., & R. WEVER. 1976. Human circadian rhythms: A multioscillator system. Fed. Proc. **35:** 2326–2332.
13. PITTENDRIGH, C. S. 1972. Circadian surfaces and the diversity of possible roles of circadian organization in the photoperiodic induction. Proc. Natl. Acad. Sci. USA **69:** 2734-2737.
14. RAVAULT, J. P. & R. ORTAVANT. 1977. Light control of prolactin secretion in sheep. Evidence for a photoinducible phase during a diurnal rhythm. Ann. Biol. Anim. Bioch. Biophys. **17:** 459–473.
15. RUDEEN, P. K. & R. J. REITER. 1980. Influence of a skeleton photoperiod on reproductive organ atrophy in the male golden hamster. J. Reprod. Fertil. **60:** 279–283.
16. EARNEST, D. J. & F. W. TUREK. 1983. Effect of one-second light pulses on testicular function and locomotor activity in the golden hamster. Biol. Reprod. **28:** 557–565.
17. ELLIOTT, J. A., M. H. STETSON & M. MENAKER. 1972. Regulation of testis function in golden hamsters: A circadian clock measures photoperiodic time. Science **178:** 771–773.
18. GROCOCK, C. A. & J. R. CLARKE. 1974. Photoperiodic control of testis activity in the vole, *Microtus agrestis*. J. Reprod. Fertil. **43:** 461–470.
19. ALMEIDA, O. F. X. & G. A. LINCOLN. 1982. Photoperiodic regulation of reproductive

activity in the ram: Evidence for the involvement of circadian rhythms in melatonin and prolactin secretion. Biol. Reprod. **27:** 1062–1075.
20. MOORE, R. Y. 1979. The anatomy of central neural mechanisms regulating endocrine rhythms. *In* Endocrine Rhythms. D. T. Krieger, Ed.: 63–87. Raven Press. New York, NY.
21. SWANSON, L. W. & W. M. COWAN. 1975. The efferent connections of the suprachiasmatic nucleus of the hypothalamus. J. Comp. Neurol. **156:** 143–164.
22. SWANSON, L. W. & P. E. SAWCHENKO. 1980. Paraventricular nucleus: A site for integration of neuroendocrine and autonomic mechanisms. Neuroendocrinology **31:** 410–417.
23. BITTMAN, E. L. 1984. Melatonin and photoperiodic time measurement: Evidence from rodents and ruminants. *In* The Pineal Gland. R. J. Reiter, Ed.: 155–192. Raven Press. New York, NY.
24. RUSAK, B. & L. MORIN. 1976. Testicular responses to photoperiod are blocked by lesions of the suprachiasmatic nuclei in golden hamsters. Biol. Reprod. **15:** 366–374.
25. STETSON, M. H. & M. WATSON-WHITMYRE. 1976. Nucleus suprachiasmaticus: The biological clock in the hamster? Science **191:** 197–199.
26. PRZEKOP, F. & E. DOMANSKI. 1980. Abnormalities in the seasonal course of oestrous cycles in ewes after lesions of the suprachiasmatic area of the hypothalamus. J. Endocrinol. **85:** 481–486.
27. LINCOLN, G. A. 1979. Photoperiodic control of seasonal breeding in the ram: Participation of the cranial sympathetic nervous system. J. Endocrinol. **82:** 135–147.
28. GOLDMAN, B. D. 1984. The physiology of melatonin in mammals. *In* Pineal Research Reviews, Vol. 1. R. J. Reiter, Ed.: 145–182. A. R. Liss, New York, NY.
29. ROLLAG, M. D. & G. D. NISWENDER. 1976. Radioimmunoassay of serum concentrations of melatonin in sheep exposed to different lighting conditions. Endocrinology **98:** 482–489.
30. TAMARKIN, L., S. M. REPPERT, D. C. KLEIN, B. PRATT & B. D. GOLDMAN. 1980. Studies on the daily pattern of pineal melatonin in the Syrian hamster. Endocrinology **107:** 1525–1529.
31. KLEIN, D. C. & R. Y. MOORE. 1979. Pineal N-acetyltransferase and hydroxy-indole-O-methyltransferase. Control by the retinohypothalamic tract and suprachiasmatic nucleus. Brain Res. **174:** 245–262.
32. LEHMAN M. N., E. L. BITTMAN & S. WINANS-NEWMAN. 1984. Role of the hypothalamic paraventricular nucleus in neuroendocrine responses to daylength in the golden hamster. Brain Res. In press.
33. LINCOLN, G. A., O. F. X. ALMEIDA & J. ARENDT. 1981. Role of melatonin and circadian rhythms in seasonal reproduction in rams. J. Reprod. Fertil. (Suppl. **30**): 23–31.
34. BITTMAN, E. L., F. J. KARSCH & J. W. HOPKINS. 1983. Role of the pineal gland in ovine photoperiodism: Regulation of seasonal breeding and negative feedback effects of estradiol upon LH secretion. Endocrinology **113:** 329–336.
35. GOLDMAN, B. D., D. S. CARTER, V. D. HALL, P. ROYCHOUDHURY & S. M. YELLON. 1982. Physiology of pineal melatonin in three hamster species. *In* Melatonin Rhythm Generating System: Developmental Aspects. Steamboat Springs Symposium. D. C. Klein, Ed.: 210–223. Karger. Basel.
36. BITTMAN, E. L., R. J. DEMPSEY & F. J. KARSCH. 1983. Pineal melatonin secretion drives the reproductive response to daylength in the ewe. Endocrinology **113:** 2276–2283.
37. HOFFMANN, K., H. ILLNEROVA & J. VANECEK. 1981. Effect of photoperiod and of one minute light at night-time on the pineal rhythm of N-acetyl transferase activity in the Djungarian hamster, *Phodopus sungorus*. Biol. Reprod. **24:** 551–556.
38. CARTER, D. S. & B. D. GOLDMAN. 1983. Antigonadal effects of timed melatonin infusion in pinealectomized male Djungarian hamsters. (*Phodopus sungorus*): Duration is the critical parameter. Endocrinology **113:** 1261–1267.
39. BITTMAN, E. L. & F. J. KARSCH. 1984. Nightly duration of pineal melatonin secretion determines the response to daylength in the ewe. Biol. Reprod. In press.
40. BITTMAN, E. L., B. D. GOLDMAN & I. ZUCKER. 1979. Testicular responses to melatonin are altered by lesions of the suprachiasmatic nuclei in golden hamsters. Biol. Reprod. **21:** 647–656.

41. GOLDMAN, B. D., V. HALL, C. HOLLISTER, P. ROYCHOUDHURY, L. TAMARKIN & W. WESTROM. 1979. Effects of melatonin on the reproductive system in intact and pinealectomized male hamsters maintained under various photoperiods. Endocrinology **104:** 82–88.
42. WATSON-WHITMYRE, M. & M. H. STETSON. 1983. Simulation of peak pineal melatonin release restores sensitivity to evening melatonin injection in pinealectomized hamsters. Endocrinology **112:** 763–765.
43. EVERETT, J. W. & C. H. SAWYER. 1950. A 24-hour periodicity in the LH-release apparatus of female rats disclosed by barbiturate sedation. Endocrinology **47:** 198–218.
44. FITZGERALD, K. M. & I. ZUCKER. 1976. Circadian organization of the estrous cycle of the golden hamster. Proc. Natl. Acad. Sci. USA **73:** 2923–2927.
45. MCCORMACK, C. E. & R. SRIDARAN. 1978. Timing of ovulation in rats during exposure to continuous light: Evidence for a circadian rhythm of luteinizing hormone. J. Endocrinol. **76:** 135–144.
46. COLOMBO, J. A., D. M. BALDWIN & C. H. SAWYER. 1974. Timing of the estrogen-induced release of LH in ovariectomized rats under an altered lighting schedule. Proc. Soc. Exp. Biol. Med. **145:** 1125–1127.
47. MOLINE M. L., H. E. ALBERS, R. B. TODD & M. C. MOORE-EDE. 1981. Light-dark entrainment of proestrous LH surges and circadian locomotor activity in female hamsters. Horm. Behav. **15:** 451–458.
48. ALLEVA, J. J., M. J. WALESKI & F. R. ALLEVA. 1971. A biological clock controlling the estrous cycle of the hamster. Endocrinology **88:** 1368–1379.
49. FERIN, M., A. TEMPONE, P. E. ZIMMERING, & L. VAN DE WIELE. 1969. Effect of antibodies to 17-beta estradiol and progesterone on the estrous cycle of the rat. Endocrinology **85:** 1070–1078.
50. MCGINNIS, M. Y., L. C. KREY, N. J. MACLUSKY & B. S. MCEWEN. 1981. Steroid receptor levels in intact and ovariectomized estrogen-treated rats: An examination of quantitative temporal and endocrine factors influencing the efficacy of an estradiol stimulus. Neuroendocrinology **33:** 158–165.
51. LEGAN, S. J., G. A. COON & F. J. KARSCH. 1975. Role of estrogen as initiator of daily LH surges in the ovariectomized rat. Endocrinology **96:** 50–56.
52. LEGAN, S. J. & F. J. KARSCH. 1975. A daily signal for the LH surge in the rat. Endocrinology **96:** 57–62.
53. GRAY, G. D., P. SODERSTEN, D. TALLENTIRE & J. M. DAVIDSON. 1978. Effects of lesions on various structures of the suprachiasmatic-preoptic region on LH regulation and sexual behavior in female rats. Neuroendocrinology **25:** 174–191.
54. RAISMAN, G. & K. BROWN-GRANT. 1977. The suprachiasmatic syndrome. Endocrine and behavioral abnormalities following lesions of the suprachiasmatic nuclei in the female rat. Proc. Roy. Soc. Proc. London **198:** 297–314.
55. KAWAKAMI, M., J. ARITA & E. YOSHIOKA. 1980. Loss of estrogen-induced daily surges of prolactin and gonadotropins by suprachiasmatic nucleus lesions in ovariectomized rats. Endocrinology **106:** 1087–1092.
56. GOODMAN, R. L. 1978. The site of the positive feedback action of estradiol in the rat. Endocrinology **102** :151–159.
57. WIEGAND, S. J., E. TERASAWA, W. E. BRIDSON & R. W. GOY. 1980. Effects of discrete lesions of preoptic and suprachiasmatic structures in the female rat. Neuroendocrinology **31:** 147–157.
58. KREY, L. C. & B. PARSONS. 1982. Characterization of estrogen stimuli sufficient to initiate cyclic luteinizing hormone release in acutely ovariectomized rats. Neuroendocrinology **34:** 315–322.
59. NEQUIN, L. G., J. ALVAREZ & N. B. SCHWARTZ. 1975. Steroid control of gonodotropin release. J. Steroid Biochem. **6:** 1007–1012.
60. PARSONS, B., B. S. MCEWEN & D. PFAFF. 1982. A discontinuous schedule of estradiol treatment is sufficient to activate progesterone-facilitated feminine sexual behavior and to increase cytosol receptors for progestins in the hypothalamus of the rat. Endocrinology **110:** 613–619.
61. SODERSTEN, P., P. ENEROTH & S. HANSEN. 1981. Induction of sexual receptivity in

ovariectomized rats by pulse administration of oestradiol-17 beta. J. Endocrinol. **89:** 55–62.

62. PARSONS, B., N. J. MACLUSKY, L. KREY, D. W. PFAFF & B. S. MCEWEN. 1980. The temporal relationship between estrogen-inducible progestin receptors in the female rat brain and the time course of estrogen activation of mating behavior. Endocrinol. **107** :774–779.
63. BLAUSTEIN, J. D., S. D. DUDLEY, J. M. GRAY, E. J. ROY & G. N. WADE. 1979. Long-term retention of estradiol by brain cell nuclei and female rat sexual behavior. Brain Res. **173:** 355–359.
64. PFAFF, D. W. & B. S. MCEWEN. 1983. Actions of estrogens and progestins on nerve cells. Science **219:** 808–814.
65. PARSONS, B., T. C. RAINBOW, D. W. PFAFF & B. S. MCEWEN. 1982. Hypothalamic protein synthesis essential for the activation of the lordosis reflex in the female rat. Endocrinology **110:** 620–624.
66. HANSEN, S., P. SODERSTEN, P. ENEROTH, B. SREBRO & K. HOLE. 1981. A sexually dimorphic rhythm in oestradiol-activated lordosis behaviour in the rat. J. Endocrinol. **89:** 63–69.
67. ADLER, N. T. 1983. The neuroethology of reproduction. *In* Advances in Vertebrate Neuroethology. N. T. Adler, R. Capranica, D. Engle and J. Ewert, Eds.: 1033–1061. Plenum Press. New York, NY.
68. ADLER, N. T. 1969. Effects of the male's copulatory behavior on successful pregnancy of the female rat. J. Comp. Physiol. Psychol. **69:** 613–622.
69. TERKEL, J. & C. H. SAWYER. 1978. Male copulatory behavior triggers nightly prolactin surges resulting in successful pregnancy in rats. Horm. Behav. **11:** 304–309.
70. EDMONDS, S., S. R. ZOLOTH & N. J. ADLER. 1972. Storage of copulatory stimulation in the female rat. Physiol. Behav. **8:** 161–164.
71. BEACH, J. E., L. TYREY & J. W. EVERETT. 1975. Serum prolactin and LH in early phases of delayed versus direct pseudopregnancy in the rat. Endocrinology **96:** 1241–1246.
72. GUNNETT, J. W. & M. E. FREEMAN. 1983. The mating-induced release of prolactin: A unique neuroendocrine response. Endocrin. Rev. **4:** 44–61.
73. FREEMAN, M. E., M. S. SMITH, S. J. NAZIAN & J. D. NEILL. 1974. Ovarian and hypothalamic control of daily surges of prolactin secretion during pseudopregnancy in the rat. Endocrinology **94:** 875–882.
74. SMITH, M. S., B. K. MCLEAN & J. D. NEILL. 1976. Prolactin: The initial luteotropic stimulus of pseudopregnancy in the rat. Endocrinology **98:** 1370–1377.
75. GOROSPE, W. C. & M. E. FREEMAN. 1981. The effects of various methods of cervical stimulation on continuation of prolactin surges in rats. Proc. Soc. Exp. Biol. Med. **167:** 78–81.
76. GOROSPE, W. C. & M. E. FREEMAN. 1982. The imprint provided by cervical stimulation for the initiation and maintenance of daily prolactin surges: Modulation by the uterus and ovaries. Endocrinology **110:** 1866–1870.
77. SMITH, M. S. & J. D. NEILL. 1976. A critical period for cervically stimulated prolactin release. Endocrinology **98:** 324–328.
78. BETHEA, C. L. & J. D. NEILL. 1980. Lesions of the SCN abolish the cervically stimulated prolactin surges in the rat. Endocrinology **107:** 1–5.
79. KARSCH, F. J., R. F. WEICK, W. R. BUTLER, D. J. DIERSCHKE, L. C. KREY, G. WEISS, J. HOTCHKISS, T. YAMAJI & E. KNOBIL. 1973. Induced LH surges in the rhesus monkey: Strength-duration characteristics of the estrogen stimulus. Endocrinology **92:** 1740–1747.
80. KNOBIL, E. 1980. The neuroendocrine control of the menstrual cycle. Recent Prog. Horm. Res. **36:** 53–88.
81. PITTENDRIGH, C. S. & S. DAAN. 1976. A functional analysis of circadian pacemakers in nocturnal rodents. I. The stability and lability of spontaneous frequency. J. Comp. Physiol. A. **106:** 223–252.
82. GIBBON, J. G. & R. M. CHURCH. 1983. Sources of variance in an information processing theory of timing. *In* Animal Cognition. H. L. Roitblat, T. G. Bever & H. S. Terrace, Eds. Lawrence Erlbaum. Hillsdale, NJ.

83. RYAN, K. D. & D. L. FOSTER. 1980. Neuroendocrine mechanisms involved in onset of puberty in the female: Concepts derived from the lamb. Fed. Proc. **39:** 2372–2377.
84. MARSHALL, J. C. & R. P. KELCH. 1979. Low dose pulsatile gonadotropin-releasing hormone in anorexia nervosia: A model of human pubertal development. J. Clin. Endorcinol. Metab. **49:** 712–718.
85. POHL, C. R., D. W. RICHARDSON, J. S. HUTCHISON, J. A. GERMAK & E. KNOBIL. 1983. Hypophysiotropic signal frequency and the functioning of the pituitary-ovarian system in the rhesus monkey. Endocrinology **112:** 2076–2080.
86. KING, J. R. & D. S. FARNER 1963. The relation of fat deposition to zugunruhe and migration. Condor **65:** 200–223.
87. HOFFMANN, K. 1973. The influence of photoperiod and melatonin on testis size, body weight, and pelage colour in the Djungarian hamster *Phodopus sungorus*. J. Comp. Physiol. **95:** 267–282.
88. LYNCH, G. L. & A. L. EPSTEIN. 1976. Melatonin-induced changes in gonads, pelage, and thermogenic characters in the white-footed mouse, *Peromyscus leucopus*. Comp. Biochem. Physiol. **53C:** 67–68.
89. LYNCH, G. R. & S. L. GANDLER. 1980. Multiple responses to a short day photoperiod in the mouse, *Peromyscus leucopus*. Oecologia **45:** 318–321.
90. LYNCH, G. R., E. WHITE, R. GRUNDEL & M. S. BERGER. 1978. Effects of photoperiod, melatonin administration, and thyroid blocker on spontaneous daily torpor and temperature regulation in the white-footed mouse, *Peromyscus leucopus*. J. Comp. Physiol. **125:** 157–163.
91. DESJARDINS, C. & M. J. LOPEZ. 1983. Environmental cues evoke differential responses in pituitary-testicular function in deer mice. Endocrinology **112:** 1398–1406.
92. WINGFIELD, J. C. 1980. Fine temporal adjustment of reproductive functions. *In* Avian Endocrinology. A. Epple & M. Stetson, Eds.: 367–389. Academic Press. New York, NY.
93. PITTENDRIGH, C. S. 1981. Circadian organization and the photoperiodic phenomena. *In* Biological Clocks in Seasonal Reproductive Cycles. B. K. Follett & D. E. Follett, Eds. 1–35. John Wright & Sons. Bristol.
94. ASCHOFF, J. 1981. A survey on biological rhythms. *In* Handbook of Behavioral Neurobiology. Volume 4: Biological Rhythms. J. Aschoff, Ed.: 3–10. Plenum. New York, NY.
95. ENRIGHT, J. T. 1972. A virtuoso isopod: Circa-lunar rhythms and their tidal fine structure. J. Comp. Physiol. **77:** 141–162.
96. NEUMANN, D. 1981. Tidal and Lunar Rhythms. *In* Handbook of Behavioral Neurobiology. Vol. 4: J. Ashoff, Ed.: 351–380. Plenum Press. New York, NY.
97. GWINNER, E. 1981. Circannual rhythms: Their dependence on the circadian system. *In* Biological Clocks in Seasonal Reproductive Cycles. B. K. Follett & D. Follett, Eds.: 153–170. John Wright & Sons. Bristol.
98. WIRZ-JUSTICE, A., G. A. GROOS & T. A. WEHR. 1982. The neuropharmacology of circadian timekeeping in mammals. *In* Vertebrate Circadian Systems: Structure and Physiology. J. Aschoff, S. Daan & G. A. Groos, Eds.: 183–193. Springer. Berlin.
99. JACKLETT, J. W. 1978. The cellular mechanisms of circadian rhythms. Trends Neurosci. **1:** 117–119.
100. HASTINGS, J. W. & H. G. SCHWEIGER, Eds. 1976. Molecular Basis of Circadian Rhythms. Dahlem Konferenzen, Berlin.
101. TAKAHASHI, J. S. & M. ZATZ. 1982. Regulation of circadian rhythmicity. Science **217:** 1104–1111.
102. MECK, W. H. 1983. Selective adjustment of the speed of internal clock and memory processes. J. Exp. Psychol. Anim. Behav. Processes **9:** 171–201.
103. BETHEA, C. L. & J. D. NEILL. 1979. Prolactin secretion after cervical stimulation of rats maintained in constant dark or constant light. Endocrinology **104:** 870-876.
104. YOGEV, L. & J. TERKEL. 1980. Effects of photoperiod, absence of photic cues, and suprachiasmatic nucleus lesions on nocturnal prolactin surges in pregnant and pseudopregnant rats. Neuroendocrinology **31:** 26–33.
105. TER HAAR, M. B. & P. C. B. MACKINNON. 1973. Changes in serum gonadotrophin levels, and in protein levels and in vivo incorporation of ^{35}S-methionine into protein of discrete

brain areas and the anterior pituitary of the rat during the oestrous cycle. J. Endocrinol. **58:** 563–576.

106. HARRIS, J. & J. GORSKI. 1978. Evidence for a discontinuous requirement for estrogen in stimulation of deoxyribonucleic acid synthesis in the immature rat uterus. Endocrinology **103:** 240–245.
107. RUSAK, B. 1977. The role of the suprachiasmatic nuclei in the generation of circadian rhythms in the golden hamster, *Mesocricetus auratus*. J. Comp. Physiol. **118:** 145–164.
108. DAAN, S. & S. SLOPSEMA. 1978. Short-term rhythms in foraging behaviour of the common vole, *Microtus arvalis*. J. Comp. Physiol. Ser. A **127:** 215–227.
109. DAVIS, F. C. & MENAKER, M. 1980. Hamsters through time's window: Temporal structure of hamster locomotor rhythmicity. Am. J. Physiol. **239:** R149–R155.
110. RASMUSSEN, D. D. & P. V. MALVEN. 1981. Relationship between rhythmic motor activity and plasma luteinizing hormone in ovariectomized sheep. Neuroendocrinology **32:** 364–369.
111. WATANABE, K. & T. HIROSHIGE. 1981. Phase relation between episodic fluctuations of spontaneous locomotor activity and plasma corticosterone in rats with suprachiasmatic nuclei lesions. Neuroendocrinology **33:** 52–59.
112. KYRIACOU, C. P. & J. C. HALL. 1980. Circadian rhythm mutations in *Drosophila melanogaster* affect short-term fluctuations in the male's courtship song. Proc. Natl. Acad. Sci. USA **77:** 6729–6733.
113. FOLLETT, B. K. & P. J. SHARP. 1969. Circadian rhythmicity in photoperiodically induced gonadotropin release and gonadal growth in the quail. Nature **223:** 968–971.
114. ELLIOTT, J. A. & B. D. GOLDMAN. 1981. Seasonal Reproduction: Photoperiodism and biological clocks. *In* Neuroendocrinology of Reproduction. N. T. Adler, Ed.: 377–423. Plenum Press. New York, NY.
115. BERMANT, G. & J. M. DAVIDSON. 1974. Biological Bases of Sexual Behavior. Harper and Row. New York, NY.

Incentive Theory III: Adaptive Clocks

PETER R. KILLEEN

Department of Psychology
Arizona State University
Tempe, Arizona 85287

A number of parametric studies have revealed a very orderly process of acquisition for autoshaped responding, and a strong dependence of the rate of acquisition on the durations of the trial (CS) and of the intertrial interval (ITI). In this article I review a few principles of reconditioning, and propose a way of combining them to account for the acquisition data.

The first principle is that incentives excite animals,[1] with the delivery of each incentive increasing the probability of a response, either instrumental (terminal), or adjunctive (interim). A simple model of the cumulation of arousal during conditioning (and its "decumlation" during extinction) is isomorphic with a model of how animals average[2,3]:

$$A_t = (1 - \beta)A_{t-1} + \beta x. \tag{1}$$

This is an exponentially weighted moving average, with A_t the average at time t, β the currency parameter, A_{t-1} the average at the previous unit of time, and x_t the "input" at time t—unity during an incentive and zero at other times.

When do animals strike the average? There are several plausible models; I assume that they are continually updating their average. These instantiations of the averaging model give us the picture in FIGURE 1, showing how the average changes as a function of the presentation and removal of an incentive, for two different values of beta. (If we take t to be measured in seconds, the values of beta actually obtained with animals are an order of magnitude smaller than these). For larger values of beta, the averager is more responsive to current input, and the system is "faster." For smaller values, the averager integrates over longer periods of time, and the system is "slower" and more stable.

The second principle is that animals cannot respond faster than some ceiling rate; any theory of how animals translate arousal into action must take that into account. One way of accommodating that principle is to assume that the proper time base for measures of response strength is not real time, but the time available for a response. If each response requires delta seconds for its emission, an animal responding at P responses per second has available only $1 - \delta P$ seconds for the $P + 1^{st}$ response in that time unit. If that additional response occurs, it reflects a more than proportionate effort to emit it. Our measure of response strength is then:

$$S = kR/(1 - \delta R), \tag{2}$$

and the predicted response rate for a given strength is:

$$R = S/(k + \delta S). \tag{3}$$

This second principle asserts that response rate is a severely nonlinear measure of response strength, and the above equations are offered as a first-order correction.

The third principle is the partial reinforcement extinction effect (PREE): Animals

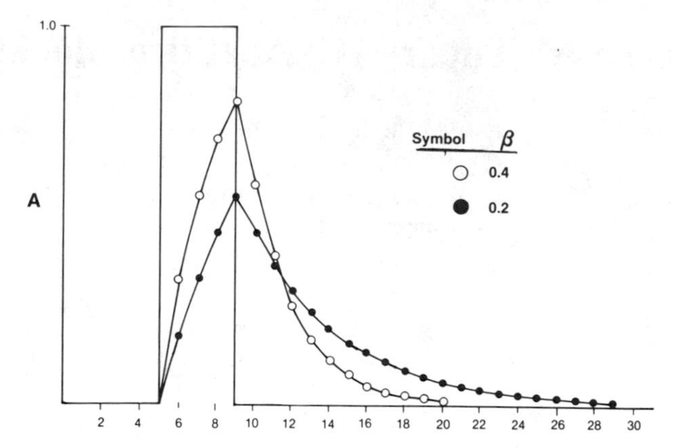

FIGURE 1. The response of an exponentially weighted moving average (EWMA) to an impulse occurring at $t = 5$, for two values of beta. (From Killeen.[2] Reprinted by permission.)

trained on a schedule of intermittent reinforcement respond through periods of extinction with greater persistance than do animals trained on a schedule of continual reinforcement; in the parlance of foraging theory, "giving up time" is greater in the former case. It is as if training with intermittent reinforcement shifted an animal to a smaller currency parameter—in FIGURE 1 from a beta of 0.4 to a beta of 0.2. The currency parameter thus cannot be a "hard-wired" characteristic of the organism, but must itself be affected by the conditions of reinforcement.

The first two principles, and models that accommodate them, are adequate to characterize the time course of acquisition and extinction of behavior, but the third principle is necessary to derive the correct parameter values for those curves, as I demonstrate below.

Referring to Equation 1, in reinforcement is discontinued at $t = 0$ and the average

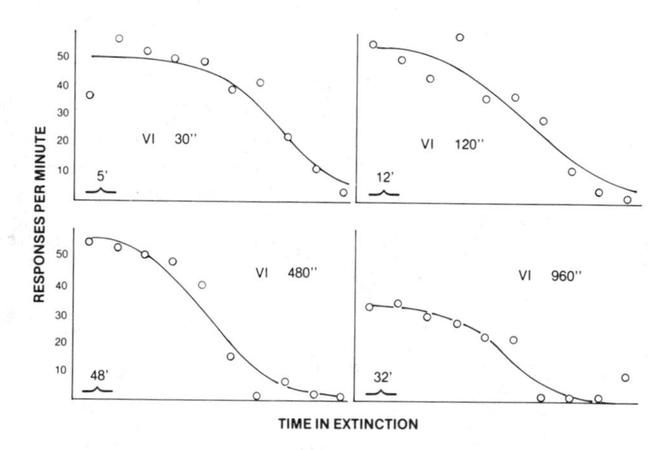

FIGURE 2. Response rates in extinction for one pigeon after exposure to various variable-interval schedules. The theoretical curves are from Equation 5. (From Killeen.[2] Reprinted by permission.)

then is A, at $t = 1$ it is $(1 - \beta)A$, at $t = 2$ it is $(1 - \beta)^2 A$, and at $t = t$ it is $(1 - \beta)^t A$. This may be written as:

$$A_t = Ae^{-\beta t}, \qquad \text{for } \beta \ll 1. \tag{4}$$

If we combine Equations 2, 3, and 4, and take A as the basis of response strength ($S = A$), we obtain predictions for the time course of extinction:

$$R = (\delta + e^{\beta t}/kA)^{-1}. \tag{5}$$

FIGURE 2 shows that this logistic equation accommodates extinction data. At the same time it illustrates the third principle, for the time scales on each of the graphs must change to accommodate the changes in persistence, reflecting the PREE. This is better seen in FIGURE 3, where the time constants (the reciprocal of the currency

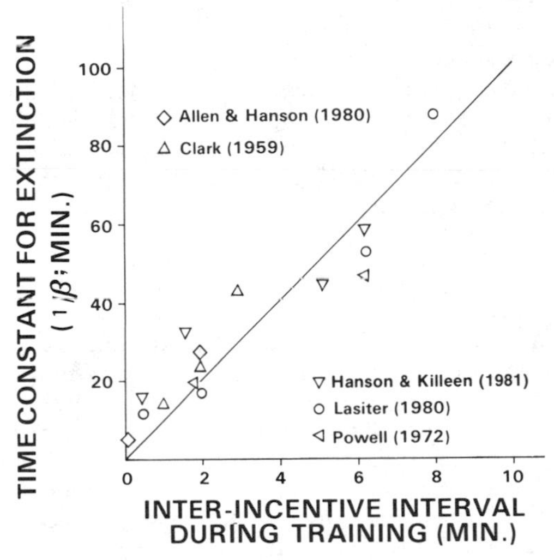

FIGURE 3. The value of the time constant ($\tau = 1/\beta$) as a function of the interval between incentives during conditioning. (After Killeen.[3])

parameters) of several extinction studies are plotted as a function of the interreinforcement interval during training.

The fourth principle is that an animal can attend to only one thing at a time, and what it is attending to at the time of reinforcement is what gets conditioned. What does an animal attend to? Signals of reinforcement. Stimuli become signals through conditioning, so that a positive feedback loop is implicated. We capture this fourth principle by letting the probability of attention to a stimulus (and therefore probability of conditioning on a trial) be proportional to the relative strengths of the signals present then.

We may now combine these principles for our theory of conditioning.

Upon the delivery of an incentive, we invoke Equation 1 to update the average arousal associated with the stimulus the animal had been attending to. Let us take the value of the reward (x) to be unity, and the probability that the animal had been attending to the CS (rather than to the background) to be equal to its relative strength:

$$p_{cs} = A_{cs}/(A_{cs} + A_{bg}). \tag{6}$$

After the delivery of the incentive, Equation 1 is iterated, and may be transformed to an exponential function as we did for Equation 4:

$$A'_{cs} = (A_{cs} + p_{cs})e^{-\beta t} \tag{7}$$

where A' is the new average, and t is the duration of the trial. This equation states that the old average is incremented by the value of the incentive (1) times the probability that the animal had been attending to the CS (p_{cs}), and this quantity is decremented by decay from the onset of the *CS* until the incentive is delivered (t).

A similar equation is written for A_{bg}, but with t set equal to the duration of the background stimuli, that is, the interreinforcement interval, and with p_{cs} being replaced by $p_{bg} = 1 - p_{cs}$. These measures of strength are compared at each trial. When the association for the CS exceeds that for the background by a threshold amount (DL), a response occurs to the CS.

One last step is necessary to make this a complete model. We have noted that the currency parameter is not constant, but approaches proportionality with the rate of reinforcement (FIGURE 3). We do not know how quickly or in what manner the animal rescales its clock, but rescale it it clearly does. Equation 1 is perhaps the simplest way of getting from one parameter value to another, so I invoke it as a model for changing the scale value of the clock:

$$\beta' = (1 - \gamma)\,\beta + \gamma\, c/T, \tag{8}$$

where c is the slope of the function relating the average interincentive interval to the asymptotic time constant (compare FIGURE 3), and gamma tells us how quickly the speed of the clock is updated. I labored with this model for some time before recognizing that it was fundamentally wrong. Equation 8 has the animal averaging rates of reinforcement, not interreinforcement intervals, and it generates an asymptotic clock speed proportional to the harmonic mean of the interreinforcement intervals. The data suggest, however, that animals must operate on and update not the currency parameter, but its reciprocal (the time constant, τ):

$$\tau' = (1 - \gamma)\,\tau + \gamma\, cT, \qquad \text{where } \tau = 1/\beta. \tag{9}$$

Note that we take T to be determined by the time between incentives, not the duration of the trial nor of the intertrial interval. The clock runs in real time. Its speed is adjusted during the receipt of reinforcement, and only then.

These three equations instantiate our theory of conditioning: on each trial Equations 6, 7 (for both CS and BG), and 9 are evaluated, and a response is predicted when the difference in strengths exceeds a threshold value. There are numerous parameters, but only two need to be modified to account for much of the autoshaping data: the threshold, DL, and the adaptive parameter, gamma. We may freeze c at a value of 16 for all experiments, and take the starting value for A_{bg} to be zero. Where there is explicit hopper training before conditioning trials, the three equations are iterated the proper number of times, with clock speed incrementing, but with only A_{bg} accruing strength. (Where the ITI during pretraining was unspecified, I assume it to be 15 sec). I start the time constant at zero. This rather arbitrary initial default has some intuitive appeal in that the animal starts the experiment with a very fast clock, one that is maximally responsive to the new environment and minimally sensitive to previous conditioning histories. Other initial values, such as the 24-hour feeding cycle of the animals during deprivation conditions, were not generally as good. The initial value for p_{cs} is zero during hopper training, but must take a nonzero value when the CS is first introduced. Fortunately the model is robust over the initial value of this parameter, so that it may generally be assigned any value between .01 and .99 without affecting the goodness-of-fit. In the following analysis I employ a starting value for A_{cs} of 0.2, which starts p_{cs} somewhere in the middle of the above range, with the exact value depending on the amount of pretraining and the value of A_{bg} established by it. Of course, if a CS is thoroughly habituated before its introduction, the probability of attending to it may fall to such a level that conditioning becomes impossible.

TABLE 1. Parameters of the Adaptive Clock Model

Study	Gamma	DL	ω^2
Terrace *et al.*[4]			
Exp. 1	0.00135	4.49	.907
Exp. 2	0.00135	4.66	.899
Gibbon *et al.*[5]			
Exp. 1	0.00135	6.20	.789
Exp. 2	0.00135	5.75	.801
Gibbon *et al.*[7]			
Partial reinforcement	0.00135	8.00	.871
Downing and Neuringer[8]			
Pretraining	0.00280	8.54	.998
Tomie[9]			
Pretraining	0.00315	9.65	.702
Balsam and Schwartz[11]			
Pretraining	0.00255	3.94	.812
Jenkins *et al.*[13]			
Exp. 1–6	0.00085	2.17	.556
Exp. 7–8	0.00085	0.00	.876

NOTE: The parameter c was always 16, and the starting value for A_{cs} was 0.2 in all experiments except that of Downing and Neuringer,[8] where it took a value of 0.0010. ω^2 is the proportion of variance accounted for by the model, using the logarithm of the acquisition scores. Medians of individual scores were used where possible. Unless otherwise noted in the text, if more than one of a group of animals failed to respond within a condition, and only a measure of central tendency was given that did not include those animals, those data were omitted from the analysis.

COMPARISON WITH DATA

Terrace, Gibbon, Farrell and Baldock[4]

Terrace and his associates kept the CS constant at 10 sec and varied the intertrial interval (ITI; the time from the end of one incentive to the beginning of the next CS) between values of 5 sec and 400 sec. In Experiment 1 the response key was dark between trials, and in Experiment 2 it was lit with a different color. Because only the ITI was varied, we may graph the number of trials before the first response as a function of the ITI. A straight line through these points misses the extremes, projecting a median of 141 trials for a 5-sec ITI (only one of four subjects responded in that condition, and that one on the 272nd trial). The adaptive clock model does better; the parameters are given in TABLE 1. The improvement is greatest at short ITIs; for example, the model predicts 300 trials for a 5-sec ITI.

Gibbon, Baldock, Locurto, Gold, and Terrace[5]

Gibbon and his associates varied both the CS and the ITI over large ranges. In the first experiment, the durations of the ITIs were variable, and in the second they were fixed. Incentive theory accounts for the distribution of data better than a power function (see TABLE 1), although not as well as one might like. (The investigators excluded 5 data points from their analysis: three with trial durations under 4 sec, and

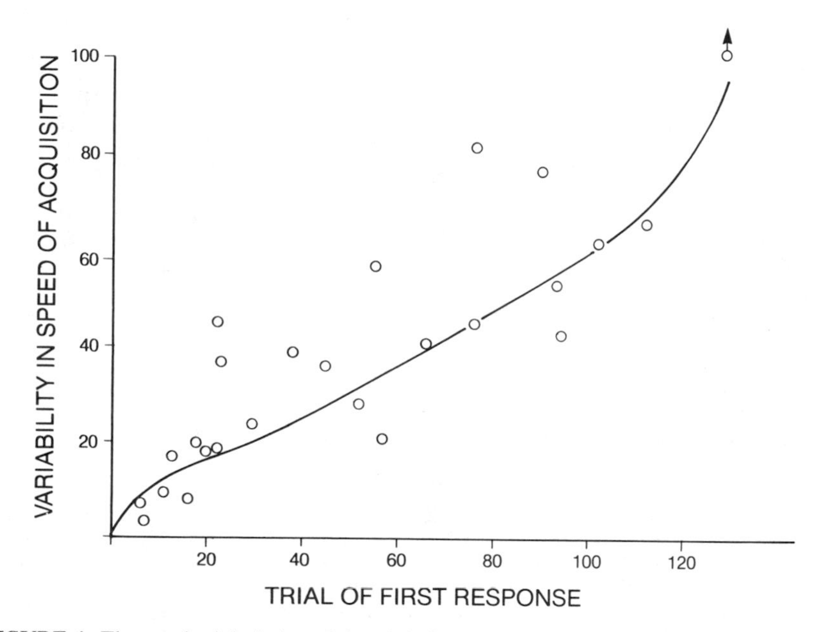

FIGURE 4. The standard deviation of the trial of the first response, plotted as a function of the predicted trial of the first response. (The data are from the first experiment reported in Gibbon *et al.*[5]; the curve is the predicted range of the trial numbers when the value of DL is varied by ± 25%.)

two with extremely large ratios of ITI to CS. At very short trial values, the pigeons move directly to the hopper—they "goal track." I also exclude those points, but not the other two). Some of the variability in these data may be caused by the variable amount of pretraining that the subjects received (20 to 200 trials): in calculating the starting values I assumed an average of 50 trials for all.

If some of the variance among subjects in the rate of acquisition is due to differences in the values of the threshold, we may select some arbitrary range for that difference and see what degree of variability the model predicts for the different experimental conditions. I evaluated the model for threshold values of ± 25% of DL, and plotted the range of predicted trial of first peck (FIGURE 4, curve) as a function of the trial predicted using a threshold of exactly DL. The data are the standard deviations of the acquisition scores for Experiment 1. The small sample size causes substantial scatter, but the predicted increase in variability with increases in the number of trials to acquisition is apparent. The reason for this increase in variability is that under extended training, the growth of strength for CS and for background becomes nearly parallel, so that any given change in threshold will be reflected in a larger range of acquisition scores. Another potential source of variability may be found in the variable schedules of intertrial intervals; several long intertrial intervals in a row might weaken background control enough to bring about an early autoshaped response.[6] However, the parameters for gamma are very conservative, so that variability in A_{bg} is highly damped; in fact, the model predicts no greater variability in this condition, and the data show no evidence of it either.

Gibbon, Farrell, Locurto, Duncan and Terrace[7]

Gibbon and his associates omitted the food on a fraction of the trials, for a range of different intertrial intervals. This experiment permits us to resolve some fundamental issues concerning the assumptions of the theory: Do pigeons adjust their time constant on trials when reinforcement is omitted? Do they adjust it to the intertrial interval or to the interreinforcement interval? Does their average arousal change on nonreinforced trials? I evaluated the fit of the model to the data, using all permutations of these assumptions. Only two sets of assumptions permitted a good fit to the data, and they were equally good, both accounting for 87% of the data variance. The first construction was that nothing happened on nonreinforced trials, except that an additional count was added to the acquisition score. The other, and more plausible, construction was that the average arousal changed on every trial (decrementing on every trial, incrementing only on reinforced trials), but that the clock speed was adjusted only on reinforced trials, and then the adjustment was based on the time since the last reinforcement, not on the ITI. As in the previous studies, a variable number of training trials were administered before the autoshaping regimen began, and I have set that value at 50 (the model is robust over a plausible range of values). The parameters are shown in TABLE 1.

Downing and Neuringer[8]

These researchers explicitly studied the effect of hopper training trials on acquisition. They found that number of trials to acquisition was a U-shaped function of the number of training trials, with 1,10,100, and 1000 trials entailing 54, 43, 19, and 145 acquisition trials. Incentive theory can accommodate these data only by assuming a very small value for A_{cs} at its first introduction (that is, the model asserts that the probability of attention to the CS on the first trial is very small). This is the first study where the starting value of A_{cs} (and thus for p_{cs}) matters, and that is because of the nonmonotonic change in acquisition scores. According to the model, increasing the number of training trials slows the clock (increases τ); this is beneficial to conditioning because it speeds the cumulation of arousal, and since the CS is gaining strength at a faster rate than the BG, speeding the clock differentially benefits the CS. But increasing the number of pretraining trials also has a deleterious effect because it permits the BG a head-start on accruing strength, and concomitantly of capturing the attention of the animals. For the parameter values used to fit these data the first factor (modification of clock speed) outweighs the second for up to 100 pretraining trials. But as pretraining continues, the starting value for A_{bg} becomes so large relative to the starting value of A_{cs} that it requires many trials to shift attention to the CS. Because this is a probabilistic process, and just one trial of attention to the CS will have a dramatic effect on the strength of the CS and on the attention subsequently given to it (with responding occurring soon thereafter), we expect a larger variance in the acquisition scores here than for any other condition. (In the other conditions, incremental strengthening dominates the processes, not probabilistic shifts in attention). This was found.

The fit of the model to the data was almost perfect, but that is not surprising, given three parameters and only four data points. Still, I know of no other models that have been elaborated enough to grapple with such data. The values for gamma and DL are relatively close to the values from other experiments. The small starting value for A_{cs} (0.0010) may be idiosyncratic to these subjects, Cornish chickens 4 days old at the start of the experiment. The range of values acceptable for all of the other studies (0.01 to 1.0)—quite large, but definitely not including the value used in this study—suggests that the upturn at 1000 pretraining trials will not be found for mature pigeons.

Tomie[9]

Tomie replicated part of Downing and Neuringer's study to see whether changing the background stimuli after pretraining would minimize its deleterious effects. Our predictions are straightforward: removing the stimuli that had accrued strength during pretraining should greatly facilitate training by minimizing competition with the CS. This effect should be greatest for the group that received extensive overtraining, and less for the group that received only basic hopper training (the "original learning control" or OLC group; both that group and the experimental group received 50 hopper training trials. I have assumed a 45-sec ITI during this condition, the value used in other studies by this author[10]). Tomie plotted the average number of trials with one or more responses in each block of ten trials during autoshaping. This way of reporting results provides many more data than the trial of the first response, and is therefore a more efficient use of the data. In fact, we might invoke Equation 5 to fit the complete acquisition curves. Here, however, I have merely interpolated for the estimated trial of the first response. The obtained (and predicted) values for the same/different background are: OLC, 20/10, (25/14); overtraining: 65/17 (87/10). I assumed a starting value of 0.2 for A_{bg}, the same as in all other studies using pigeons; because of the large number of pretraining trials in this study, however, the range of acceptable values is smaller, needing to be somewhere between 0.1 and 0.5.

The two effects of pretraining—the beneficial effect of speeding the clock and the deleterious effect of increasing A_{bg}—are thus separable. Note that we predict fewer trials to acquisition in the different condition for overtraining than for OLC. That prediction was not born out here, but it was in another study by Tomie[10] that involved four groups: an OLC group and three groups with an additional 600 pretraining trials. Two of these latter groups then underwent two sessions of extinction, one in the training context, and one in a novel context. Remembering that extinction does not affect the speed of the clock, we predict fastest conditioning for the group extinguished in the same context, next for the OLC group, and slowest for the groups extinguished in the novel context and the group not extinguished (with differences there depending on the similarity of the novel context and the standard one). His FIGURE 2 shows that this is just what he found.

Balsam and Schwartz[11]

This is another study of the effects of pretraining. Pigeons were given 150 hopper training trials in a chamber lined with cardboard and then 4, 8, 64, 128 or 256 pretraining trials in the unlined chamber, followed by autoshaping in the unlined chamber. We predict that hopper training should speed up the clock, but, since the BG was changed for the subsequent conditions, it should have no effect on A_{bg}. Pretraining should affect both A_{bg} and clockspeed (although the latter should be a substantial distance toward asymptote). The net effect of pretraining should thus be to hinder conditioning. Exact predictions are made complex by the investigators' tactic of varying hopper duration during pretraining to keep total access to food constant. The effects of hopper duration on conditioning are strongly concave, with little additional benefit derived from durations longer than 3 or 4 sec: three 3-sec hopper presentations are much more effective than one 9-sec presentation. As a first pass, we may assume that the effects of all hopper durations were equal. Optimal fits of the model under that assumption account for 75% of the data variance. Alternatively, we may invoke a model for the effects of hopper duration derived for the study of choice behavior.[12] That model takes the incentive value of D seconds of access to grain to be equal to $1 - e^{-\lambda D}$, and placed the value of lambda at about 0.7. Using that same "utility" function improves the accuracy of the predictions to 81%.

Jenkins, Barnes, and Barrera[13]

These investigators studied a range of experimental conditions that permitted them to address the question of why autoshaping depends on trial spacing. If the answer given by incentive theory and the associated model of adaptive clocks is correct, we should be able to predict, at least qualitatively, their results. However, it is difficult to make only quantitative predictions with our theory, because the complex interactions of clock speed and differential accumulation of arousal makes a number of different qualitative predictions possible for any situation. It is necessary to pin down the model by giving it explicit experimental parameters and then requiring explicit numerical predictions.

In analyzing the data of Jenkins and his associates, there are too few data points per experimental condition to find optimal parameter values in each case. Therefore I have used the same values for most of the experiments. The authors employed 16 sessions (640 trials) of "preliminary tray training" in the first study, and I assumed 5 "in other experiments [where] far fewer preliminary sessions were used."

The first study examined ITIs of 30 sec and 300 sec, and found acquisition in 55 and 10 trials, respectively. The model predicts 80 and 12 trials. (These investigators report medians of subjects that were conditioned; medians over all subjects would be larger, especially when many subjects failed to become conditioned. If we assume that all remaining subjects would have become conditioned immediately after the experiment was terminated for them, and we calculate the weighted geometric mean of their scores and the reported scores, the estimates increase to 90 and 14). In the second study the authors scheduled four extra feedings during the 300-sec ITI. They found that the location of the extra feedings within the ITI did not matter, and that on the average the animals required 22 trials before initiating a keypeck. The model is applied by iterating Equation 7 for A_{bg} four additional times within each trial, which leads to a prediction of 16 trials. The third study involved very long trials, with a feeding immediately before the CS in one condition, and no prior feeding in the other. Half the animals reached criterion by 5 trials (first condition) and by 8 trials. The model predicts 5 trials for both conditions. In the fourth study an unpaired CS was presented on three occasions during the ITI. The location of the CS did not matter, and the average number of trials to criterion was 32. In the model, we iterate the extinction of A_{cs} (Equation 4, with t = CS) an additional three times during each trial, and predict 6 trials to acquisition. This is the one prediction that is in serious error. I discuss this problem below. In the fifth experiment, the animals were trained on ITIs of 30 or 300 sec. Half of the subjects began their first trial after waiting 30 sec (W30), and half after waiting 300 sec (W300). For the 30-sec group the wait manipulation had no effect, with both groups responding on about the 32nd trial. The model predicts 25. For the 300-sec group, the median number of trials to criterion was 6 for the W30 and 2 for the W300. The model predicts 6 and 6. In Experiment 6, the investigators employed a type of variable ITI scheduling, with an average ITI of 300 sec, but with one 22-min interval and five 30-sec intervals. The model predicts autoshaping by the ninth trial, while the data show that it occurred by the eighth trial.

The next two experiments employed only a single trial per session, and demonstrated much faster acquisition than the previous studies. It is necessary to reduce the threshold parameter to a value of zero in order to capture the data. In Experiment 7, the sessions lasted for 15 min, with the chamber dark during the first 7 min, the last 7 min, both, or neither. I treat the first two conditions as the same (excluding the BO from calculation of T), and predict 3 trials to acquisition; the authors measured 5 trials and 2 trials. With no blackout, they found acquisition by the fourth trial; the model predicts 2. With two blackouts they found acquisition by the 18th trial; taking the functional duration of BG to be $T = 38$ sec, the model predicts 20 trials. (The BG remained on for an additional 22 sec after the feeding; since no other feedings were possible after the single one, it is reasonable to presume that trace inhibition from the feeding caused it to function as a different stimulus. Had the 22 sec been included, the prediction would drop to 16 trials).

In Experiment 8 Jenkins and associates paired the CS with a feeding only once in the middle of a session, and preceded and followed that trial with 15 unpaired feedings coming once every 30 sec, once every 300 sec, or with 7.5 min with no additional feedings. They found that only one of five pigeons were conditioned by the end of the experiment (30 sessions) in the first condition; the model predicts they should have been conditioned, on the average, by the 55th session. The authors found acquisition by the sixth trial in the low-density feeding condition, and by the third trial in the no-additional-feeding condition; the model predicts 6 and 2 trials.

The quality of these predictions across experimental conditions, and with few changes in theoretical parameters, is generally very good, and certainly the predictions are within the standard error of the mean of the data in almost all cases. The exception is Experiment 4, where the predictions were grossly in error. I do not know why they were. A similar experiment was conducted by Gibbon and Balsam[14] who compared groups with a 10-sec CS, a 5-sec CS, and a 5-sec CS with an additional 5-sec stimulus at the beginning or middle of the interval. The latter two groups did not differ in trials to acquisition (both required about 58). The first group required 40 trials, and the second 23. Using the same parameters as in the first study by Gibbon *et al.* (but with DL slightly higher: 6.6) generates reasonable predictions of 49, 49, and 23. The map between theory and this experimental manipulation thus remains uncertain.

DISCUSSION

There are three sources of error in the map between data and theory. The first is the large variability intrinsic to the data. According to the theory, much of this variability is due to the slow convergence of two nearly asymptotic curves, with small variations in attention or threshold having a magnified effect on responding, in much the same way that the movement of nearly parallel lines closer by a millimeter can move their point of intersection by many meters. This type of error is greatest when acquisition is slow, and the curves are more nearly parallel; note in FIGURE 4 that when acquisition occurs on the 100th trial, the standard deviation is about 65; if this estimate was based on 5 subjects, there is only a 50% chance that the population mean lies between 80 and 120. This can be improved with very large Ns (with 10 subjects, the probable error in estimating the mean reduces to ± 14 trials), or with a theory that lets one pool information across different conditions (such as a regression model; of course, the linear assumptions that involves fail in the extremes. Regression's utility as an exploratory tool does not make it a substitute for grounded theory).

There are some ways of reducing this source of error other than increasing the number of subjects. The use of an acquisition criterion other than the first response provides more reliable estimates of the point at which responding goes suprathreshold. This may be carried a step further in the following technique: Plot the cumulative number of trials on which a response has occurred against trial number, and interpolate for the trial of the first response. This assumes linearity, but over such a small range that more good will be done in attaining a reliable estimate than harm will occur from the introduction of bias. The technique assumes that there are not important sources of curvilinearity, such as a positive acceleration due to instrumental conditioning. Incentive theory does not invoke the hypothesis of instrumental conditioning, and from its viewpoint one would be surprised to see a positive acceleration.

Finally, we must admit that the trial of the first response is a very inefficient dependent variable; weeks of housing, establishment of *ad libitum* weights, reduction to 80%, pretraining, and final autoshaping bring forth a single datum, and not a very reliable one at that. The problem is in the end not an empirical one—it is a problem with the theory that must insist upon this very expensive input. The trial of the first response is merely the first point on an acquisition function; a theory that treats the function as a whole will have available a hundred-fold more data from a single additional session. Although we often hear the cry for more data, here it is more theory that is needed. I have outlined one appropriate theory in this paper, but have not yet begun to test it against the available data.

The second source of error lies in ambiguity about some of the experimental parameters. The intertrial interval during pretraining and the number of pretraining trials are often unspecified. If they are kept constant across subjects, the major effect will be on the particular values of the optimal theoretical parameters for the model. If these experimental parameters vary between subjects or conditions, however, they will be a source of variance that is not reducible, unless their values are reported and can thus be included in the model.

The third source of error lies in the nature of the theory, and its adequacy in representing the data; the theory could be wrong. Incentive theory and its model of adaptive clocks have gone through many revisions. Some were mere errors, others were plausible but incorrect instantiations (for example, Equation 8). In some cases, I had to revise provisional assumptions that subtly lost their provisional status. (For example, I kept the value of c constant at the value found in FIGURE 3, 10.0, for this provided an excellent fit to most of the data. However, it also required different values for gamma in situations where gamma should not have changed [such as between conditions in the first two studies listed in TABLE 1]. A larger value for *c* removed that irregularity. Optimization across experiments can only come after the major variables are identified and accounted for within studies: Theory construction is an iterative process.)

Some parts of the model are not essential for the majority of the parametric data. This is the case with the attention assumption, and the consequent requirement that the starting value of A_{cs} be set somewhere greater than zero. This is a reasonable modification, but it introduces a new parameter that comes into play only in a few conditions where there are many pretraining trials, and A_{bg} becomes large enough to have a chance of competing with A_{cs} for the animals' continued attention. I developed this part of the model after Herb Jenkins impressed upon me the "path dependence" of autoshaping, with conditioning being difficult or impossible if the CS is thoroughly habituated. But I believe that the attentional mechanism is an important addition, for it may help to make sense out of a number of different types of studies, such as Williams's experimental analysis of the blocking of a conditional discrimination by preexposure to one of the stimuli, and the interaction of that effect with the locus of the stimulus,[15] and Roberts and Kraemer's[16] observations of the dependence of accuracy on the ITI in matching to sample experiments. In a subsequent article I modify Equation 7 so that extinction of a stimulus occurs only while an animal is attending to it. For a CS, t is replaced by p^*_{bg} CS. For the background, t is replaced by BG-CS + p^*_{bp} CS, which assumes complete attention to the background before the onset of the CS, and partial (or probabilistic) attention during the CS. The only effect of these modifications on the analysis of autoshaping data is to increase the values of gamma by 10%.

A provocative consequence of the theory is that the speed of the animal's clock is always changing as a function of its rate of reinforcement. Although this variation will cancel out of some psychophysical measures,[17] it will not always cancel. The implications are profound: Animals drift through a relativistic world in which they have no access to absolute measures of time other than gross circadian rhythmicity. Yet, some data support this relativistic implication. Treisman noted that "facilitation of the pacemaker by the specific arousal center affects [increases] the rate at which pulses are produced."[18] Unless there are multiple clocks available to animals, incentive theory predicts that their estimates of time will vary as a function of intertrial intervals and as a function of exposure to conditioning and extinction extraneous to the events being timed. This implication will soon be tested.

SUMMARY

Incentive theory is extended to address the phenomenon of autoshaping. To do so, it is necessary to permit the speed of the animal's internal clock to vary with rates of reinforcement; clock speed is the basis for the animal's calculations of reinforcement densities. This notion of an "adaptive clock" is consistent with other effects, such as the partial-reinforcement extinction effect, and permits us to deal with the various experimental manipulations that are found in autoshaping experiments from a unified perspective.

REFERENCES

1. KILLEEN, P. R. 1979. Arousal: Its genesis, modulation and extinction. *In* Reinforcement and the Organization of Behavior. M. D. Zeiler & P. Harzem, Eds.: 31–78. Wiley. New York, NY.
2. KILLEEN, P. R. 1981. Averaging theory. *In* Quantification of Steady-State Operant Behaviour. C. M. Bradshaw, E. Szabadi & C. F. Lowe, Eds.: 21–34. Elsevier. New York, NY.
3. KILLEEN, P. R. 1982. Incentive theory. *In* Nebraska Symposium on Motivation, 1981. D. J. Bernstein, Ed.: 169–216. University of Nebraska Press. Lincoln, Nebraska.
4. TERRACE, H. S., J. GIBBON, L. FARRELL & M. D. BALDOCK. 1975. Temporal factors influencing the acquisition and maintenance of an autoshaped keypeck. Anim. Learn. Behav. **3:** 53–62.
5. GIBBON, J., M. D. BALDOCK, C. LOCURTO, L. GOLD & H. S. TERRACE. 1977. Trial and intertrial durations in autoshaping. J. Exp. Psychol. Anim. Behav. Processes **3:** 264–284.

6. LUCAS, G. A. & E. A. WASSERMAN. 1982. US duration and local trial spacing affect autoshaped responding. Anim. Learn. Behav. **10:** 490–498.
7. GIBBON, J., L. FARRELL, C. M. LOCURTO, H. J. DUNCAN & H. S. TERRACE. 1980. Partial reinforcement in autoshaping with pigeons. Anim. Learn. Behav. **8:** 45–59.
8. DOWNING K. & A. NEURINGER. 1976. Autoshaping as a function of prior food presentations. J. Exp. Anal. Behav. **26:** 463–469.
9. TOMIE, A. 1981. Effect of unpredictable food on the subsequent acquisition of autoshaping: Analysis of the context-blocking hypothesis. *In* Autoshaping and Conditioning Theory. C. M. Locurto, H. S. Terrace & J. Gibbon, Eds.: 181–215. Academic Press. New York, NY.
10. TOMIE, A. 1976. Interference with autoshaping by prior context conditioning. J. Exp. Psychol. Anim. Behav Process. **2:** 323–334.
11. BALSAM, P. D. & A. SCHWARTZ. 1981. Rapid contextual conditioning in autoshaping. J. Exp. Psychol. Anim. Behav. Processes **7:** 382–393.
12. KILLEEN, P. R. 1984. Incentive theory and amount of reinforcement. J. Exp. Anal. Behav. In press.
13. JENKINS, H. S., R. A. BARNES & F. J. BARRERA. 1981. Why autoshaping depends on trial spacing. *In* Autoshaping and Conditioning Theory. C. M. Locurto, H. S. Terrace & J. Gibbon, Eds.: 255–284. Academic Press. New York, NY.
14. GIBBON, J. & P. BALSAM. 1981. Spreading association in time. *In* Autoshaping and Conditioning Theory. C. M. Locurto, H. S. Terrace & J. Gibbon, Eds.: 219–253. Academic Press. New York, NY.
15. WILLIAMS, B. A. 1982. On the failure and facilitation of conditional discrimination. J. Exp. Anal. Behav. **28:** 265–280.
16. ROBERTS, W. A. & P. J. KRAEMER. 1982. Some observations of the effects of intertrial interval and delay on delayed matching to sample in pigeons. J. Exp. Psychol. **8:** 342–353.
17. GIBBON, J. 1981. Two kinds of ambiguity in the study of psychological time. *In* Quantitative Analysis of Behavior: Discriminative Properties of Reinforcement Schedules. M. L. Commons & J. A. Nevin, Eds.: 157–189. Ballinger. Cambridge, MA.
18. TREISMAN, M. 1963. Temporal discrimination and the indifference interval: Implications for a model of the "internal clock". Psychol. Monogr. **77**(13): 1–31.

Attentional Bias between Modalities: Effect on the Internal Clock, Memory, and Decision Stages Used in Animal Time Discrimination[a]

WARREN H. MECK

Walter S. Hunter Laboratory of Psychology
Brown University
Providence, Rhode Island 02912

When more than one signal modality is used in a temporal discrimination task the effect of an attentional bias between modalities can be analyzed in terms of internal clock, memory, and decision stages. Evidence from a number of experiments with rats indicates that the latency to start an internal clock and the selection and integration of temporal criteria and response rules for different modalities depends on the subject's attentional bias toward one stimulus modality. Attentional bias can be induced either by the method of unbalanced stimulus probabilities[1] or by the prior-entry method.[1,2] Knowledge of the subject's attentional bias between sensory channels (auditory and visual, for example) allows the prediction of the form and horizontal placement of psychophysical functions that relate the subject's response to the physical duration of a stimulus. A proposal is made for the integration of attentional bias effects and the information-processing model of animal timing behavior described by Gibbon and Church.[3]

INTRODUCTION

Gibbon and Church[3] have proposed that animals use at least three distinct information-processing stages to make temporal discriminations. The stages have been considered distinct to the extent that any one of them can be changed without changing any of the others.[4,5] During the first stage a stimulus initiates the closure of a mode switch that gates pulses from a pacemaker into an accumulator. The pacemaker–switch–accumulator system (clock stage) is used to transform physical time into psychological time (clock-reading); during the second stage (memory stage) the clock-reading may be stored in working memory; during the third stage (decision stage) the clock-reading is compared with the reference memory values of stimulus durations, after which a particular response was reinforced in the past, and a response rule appropriate for the task is applied; for example, a rat responds on the left lever if its clock-reading is less than some value and responds on the right lever if its clock-reading is greater than some value. A fourth stage where stimulus–response

[a]The research presented here was supported in part by Research Grant BNS 82-09834 from the National Science Foundation and by Research Grant MH 37049 from the National Institute of Mental Health.

sequences are classified and stored in reference memory according to trial outcome (for example, reinforcement or nonreinforcement) would also appear necessary, and this process may be incorporated into the memory stage outlined above.

In describing their information-processing model of animal time discrimination Gibbon and Church did not consider how an attentional bias to a particular sensory channel (such as the auditory or visual) affects the operation of the processing stages outlined above. Past research on animal time discrimination has focused on within-modality factors rather than between-modality factors. The purpose of the present chapter is to determine whether an attentional bias to a particular sensory channel influences the behavior of animals trained on two very different temporal discrimination procedures in a similar manner. More specifically, the purpose was to determine whether such an attentional bias affects the same information-processing stages in two different timing tasks.

The two temporal discrimination procedures described here have been used previously to evaluate within-modality processing stages. The first procedure is a psychophysical bisection task designed for use with rats. In this procedure, after an intertrial interval (ITI), a stimulus of a particular duration is presented and then two response levers are inserted into the lever box. If, for example, a 2-sec stimulus occurred, a left ("short") response is reinforced; if, for example, an 8-sec stimulus occurred, a right ("long") response is reinforced. The two reinforced signals are typically presented with a probability of .25 on each trial. On the remaining trials, signals of intermediate duration are randomly presented, each with equal probability. Neither the left nor the right response is followed by food in the case of these intermediate signal durations. In this task the proportion of "long" responses rises as a function of signal duration in a sigmoidal fashion. Previous work has indicated that (*a*) the point of subjective equality is near the geometric mean of the two reinforced signal values; (*b*) the difference limen is proportional to the durations being timed (that is, Weber's law); (*c*) the psychophysical functions are symmetrical on a logarithmic time scale; and (*d*) the functions for different signal ranges are equivalent when scaled in log units relative to the geometric mean of the two reinforced values. This is the proportionality result.[4,6,7] A quantitative model of this behavior has been developed that fits the data and ties the model parameters to psychological processes.[4,8]

Representative median data from the bisection procedure with three different signal ranges are shown in FIGURE 1. The two reinforced signal durations for the 10 rats in each group were 2 versus 6 sec, 2 versus 8 sec, or 2 versus 12 sec. During testing with five unreinforced signals of intermediate duration, the signals were spaced at equal logarithmic intervals between the two reinforced durations (2.0, 2.4, 2.9, 3.5, 4.2, 5.0, and 6.0 sec for the 2–6 range; 2.0, 2.6, 3.2, 4.0, 5.0, 6.4, and 8.0 sec for the 2–8 range; and 2.0, 2.7, 3.6, 4.9, 6.6, 9.0, and 12.0 sec for the 2–12 range). In this experiment the "short" reinforced duration was held constant at 2 sec and the "long" reinforced duration was varied (for example, by 6, 8, and 12 sec) in a manner similar to that of Platt and Davis.[9] In previous work from our laboratory both the "short" and "long" reinforced signal durations were varied so as to hold constant the ratio of the "short" to the "long" signal.[4,6,7]

The second procedure described here is a modified fixed-interval schedule called the peak procedure. In this task, after an intertrial interval, a signal occurs and, on some trials, food is primed after a fixed duration and the rat's next lever press is reinforced; on other trials (called peak trials) no food is available and the trial lasts for a relatively long time after the fixed duration. As a result, during peak trials the response rate of the rat initially increases as a function of the time since signal onset, as it does during standard fixed-interval training. After the time that food is sometimes available, however, response rate decreases in a fairly symmetrical fashion when it is

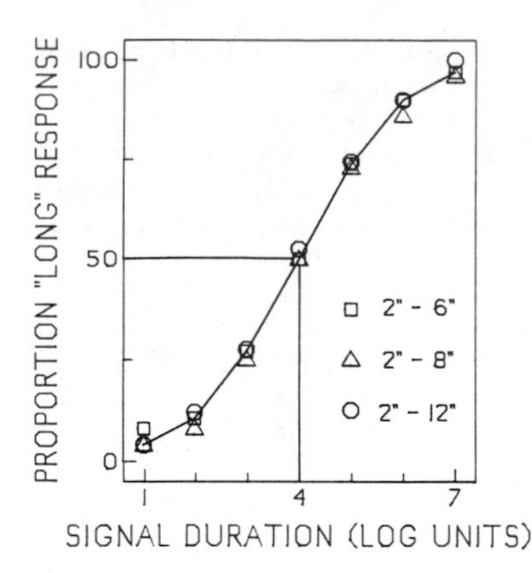

FIGURE 1. Median data from the temporal bisection procedure with three different signal ranges; 2 versus 6 sec (*squares*), 2 versus 8 sec (*triangles*), and 2 versus 12 sec (*circles*).

plotted on a linear time scale. Typically, half of the trials are peak trials. The time the response rate is maximal is called the "peak time." It occurs near the time that food is maximally expected and serves as a measure of the rat's temporal criterion, which is a value stored in reference memory. The response rate at the peak time is called the "peak rate." Roberts[5] has shown that peak time and peak rate are independent measures of performance, that is, there are treatments that affect one measure without affecting the other.

Representative median data from 10 rats trained on the peak procedure are shown in FIGURE 2. In this example, reinforcement sometimes occurred following the rat's first lever response after a white-noise signal had been present for 20 sec. On peak trials the signal was presented for 50 sec and ended without reinforcement.

Presumably, there are many ways to direct a subject's attention to a particular sense modality and thereby cause perceptual and response biases. Two methods that have been used to control and study attentional bias in human subjects were employed here with rats as subjects. The first method, adapted from the work of LaBerge and his colleagues,[1,10] uses unbalanced stimulus probabilities to direct the subject's attention to a particular sensory channel. On each trial either an auditory or a visual signal occurred and the subject classified the auditory signal as X_1 or Y_1, or classified the visual signal as X_2 or Y_2. When a block of trials all contained signals from one modality (pure modality trial blocks), the subjects knew the modality that would be used; and when a block of trials contained signals from both modalities (mixed modality trial blocks), the subjects were uncertain which modality would be used on the next trial. LaBerge was able to induce an attentional bias to one of the stimulus modalities in the mixed modality trial blocks (for example, the auditory) by increasing its probability of occurrence. When one modality occurred on 85% of the trials, subjects came to expect that modality and focused their attention on the relevant sensory channel. This attentional bias was measured by the time required for subjects to identify expected and unexpected signals. The time required to identify signals from the unexpected stimulus modality was significantly greater with unbalanced stimulus probabilities compared to the pure modality baseline trials. This effect occurred whether the unexpected stimulus occurred on the auditory channel or the visual channel.

The second method used here to induce an attentional bias is the prior-entry method. The logic of this method has a long history and was described by Titchener[11] as one of his seven laws of attention (law of prior entry). The basic idea is that through prior experience with one stimulus modality subjects may become predisposed to that stimulus. Thus, if subjects are presented with a warning cue prior to a signal presentation, attention may be directed to the modality of the cue and influence the processing of the following signal presentation. Stimuli toward which subjects are predisposed will dominate over other stimuli and will require less time to be processed than a like stimulus. In the case where the modality of the cue and the modality of the signal agree, processing of the signal may be facilitated; if the modality of the cue and the modality of the signal disagree, the processing of the signal may be inhibited or interfered with.

Sternberg *et al.*[2] have used the prior-entry method to study the effect of attentional bias on the perceived temporal order of two successive stimuli. In their study, attentional bias to a particular stimulus modality (such as the auditory or visual) was induced by a cue stimulus and measured by a concurrent reaction-time task. The prior-entry effect on attention was reflected in the lead time required to perceive that the "unfavored" signal in a sequence of two signals from different modalities occurred first rather than simultaneously or second. Attentional bias produced by the modality of the cue stimulus substantially increased this lead time for the "unfavored" signal. On the basis of a lack of significant change in the prior-entry effect as a function of stimulus intensity, Sternberg *et al.*[2] suggested that this type of attentional bias does not function at the level of the sensory channel, but at the level of a central order-decision stage. Other investigators have combined the prior-entry method with the method of unbalanced stimulus probabilities in order to study the effect of attentional bias on perception and decision processes.[1,10,12]

UNBALANCED STIMULUS PROBABILITY METHOD: BISECTION PROCEDURE

The first experiment to be described combined the temporal bisection procedure with the method of unbalanced stimulus probabilities. After being pretrained to make the

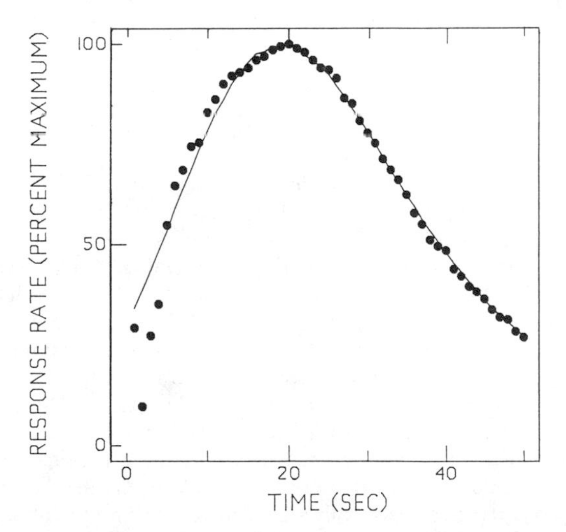

FIGURE 2. Median data from the temporal peak procedure with reinforcement sometimes available after 20 sec. Only data from no-food (peak) trials are shown.

lever response, 30 rats were trained for 15 days with pure modality trial blocks with signals separated by a fixed ITI of 40 sec. The signals to be timed were either a light stimulus (6- or 7.5-W houselight on) or a sound stimulus (80 dB re 20 $\mu N/m^2$ white-noise on). The same response rule was used for both light and sound, for example, a left lever response was reinforced if the 2-sec signal had been presented and a right lever response was reinforced if the 8-sec signal had been presented. All rats were presented with both modalites during a 3-hr session. One of the modalities, randomly selected, was used during the first 1.5 hr of the session and the other modality was presented during the second 1.5 hr of the session. (See Meck[4] for additional details of the general method.) The 30 rats were then divided into three test groups of 10 rats each. One group of rats was presented with mixed modality trial blocks in which the light signal was randomly selected to occur on 20% of the trials and the sound signal was presented on the remaining 80% of the trials (group A). Another group of 10 rats was presented with mixed modality trial blocks in which the light signal was randomly selected to occur on 50% of the trials and the sound signal was presented on the remaining 50% of the trials (group B). The final group of rats was presented with mixed modality trial blocks in which the light signal was randomly selected to occur on 80% of the trials and the sound signal was presented on the remaining 20% of the trials (group C). (See TABLE 1 for an outline of the test

TABLE 1. Unbalanced Stimulus Probabilities

Test Condition	Light Signal–Sound Signal
A	20% L–80% S
B	50% L–50% S
C	80% L–20% S

conditions.) The major question was whether unbalanced stimulus probabilities would affect the steady-state temporal bisection functions for the unexpected signal modalities.

Pure Modality Sessions

The results for the pure modality sessions are shown in the left panel of FIGURE 3. Open symbols indicate the method of unbalanced stimulus probabilities; triangles indicate light-signal trials and circles indicate sound-signal trials. As expected, the observed psychophysical functions relating the proportion of "long" responses to signal duration for pure modality trial blocks were virtually identical for both light and sound signals. The median point of subjective equality (PSE) taken across all 30 rats was 4.12 ± .05 sec for light trials and 4.06 ± .03 for sound trials, a nonsignificant difference and close to the geometric mean of the reinforced durations (4.0 sec). (Numbers are the median and plus or minus the median absolute deviation.) The smooth functions near the data points are derived from the scalar timing theory developed by Gibbon.[8] An application of this theory to temporal bisection data is provided by Meck.[4] With one exception, noted below, the parameter values used for the theory fits were identical to the values used for the saline training condition in Experiment 1 of Meck.[4] The theoretical fits in each of the figures accounted for at least 96% of the variance.

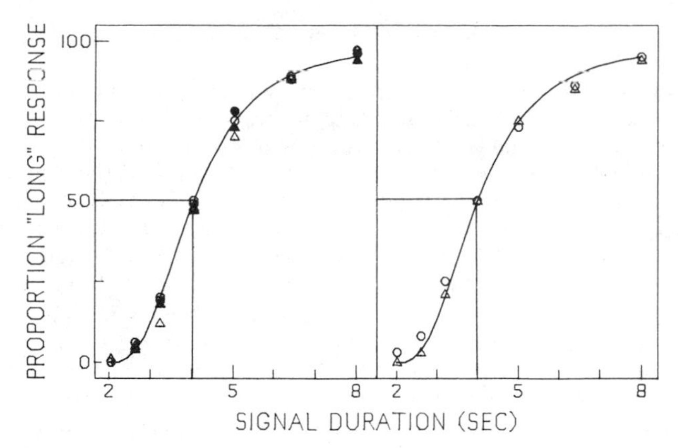

FIGURE 3. Median data from the temporal bisection procedure with the method of unbalanced stimulus probabilities (*open symbols*) and the prior-entry method (*solid symbols*). *Triangles* indicate light-signal trials and *circles* indicate sound-signal trials. The **left** panel shows the results for the pure modality sessions from the method of unbalanced stimulus probabilities and the results for the agreement of cue and signal modality from the prior-entry method. The **right** panel shows the results for the mixed modality sessions from the method of unbalanced stimulus probabilities with 50% light signals and 50% sound signals.

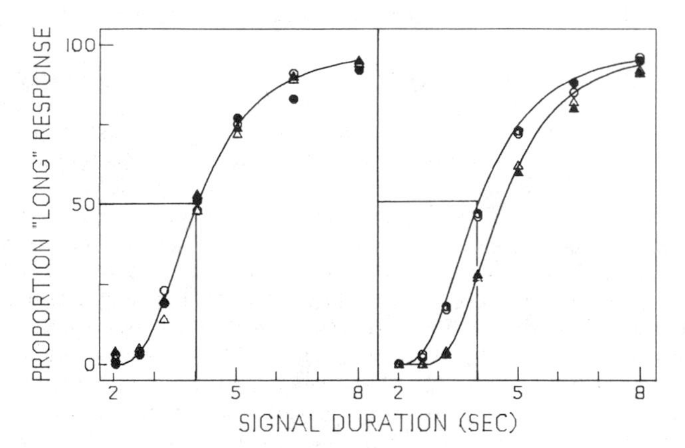

FIGURE 4. Median data from the temporal bisection procedure with the method of unbalanced stimulus probabilities (*open symbols*) and the prior-entry method (*solid symbols*). *Triangles* indicate light-signal trials and *circles* indicate sound-signal trials. The **left** panel shows the results for the mixed modality sessions from the method of unbalanced stimulus probabilities with 80% light signals and 20% sound signals. The data for the light cue test condition from the prior-entry method are also shown for both light and sound signals. The **right** panel shows the results from the mixed modality sessions from the method of unbalanced stimulus probabilities with 20% light signals and 80% sound signals. The data for the sound cue test condition from the prior-entry method are also shown for both light and sound signals.

Mixed Modality Sessions

The results from the mixed modality sessions were different from the results for the pure modality sessions. The steady-state data for group A are shown in the right panel of FIGURE 4 (symbols are the same as above). The probability of a "long" response as a function of signal duration indicated that for light signals (20%) the median PSE was 4.76 ± .05 sec, a value significantly greater than the median PSE obtained for pure modality blocks of light trials (100%) and also significantly greater than the median PSE for sound signals (80%) from the same mixed modality condition. The steady-state psychophysical function for mixed modality sound signals (80%) was indistinguishable from the sound signal function obtained from the pure modality trial blocks; the PSE for the mixed modality sound signals was 4.09 ± .04 sec.

The steady-state data for group B are shown in the right panel of FIGURE 3 (symbols are the same as above). The probability of a "long" response as a function of signal duration indicated that for light signals (50%) the median PSE was 4.09 ± .05 sec, virtually identical to the PSEs obtained for pure modality light and sound trial blocks. The same was true for sound signals (50%) from the mixed modality condition; the median PSE for mixed modality sound signals was 4.13 ± .07 sec.

The steady-state data for group C are shown in the left panel of FIGURE 4 (symbols are the same as above). The probability of a "long" response as a function of signal duration indicated that for light signals (80%) the median PSE was 4.08 ± .04 sec, virtually identical to the PSEs obtained for pure modality light and sound trials. The same was true for sound signals (20%) from the mixed modality condition; the median PSE for mixed modality sound signals was 4.05 ± .06 sec.

PRIOR-ENTRY METHOD: BISECTION PROCEDURE

The prior-entry method was interfaced with the temporal bisection procedure in the following way. After an ITI of 40 sec a cue stimulus was presented for 1 sec. Then, after a random interval (geometrically distributed) with a minimum of 2 sec and a mean of 15 sec, a signal occurred in a manner similar to the temporal bisection procedure outlined above for the unbalanced stimulus probability method. During training conditions the modalities of the cue stimulus and the signal were in agreement (for example, light cue and light signal) and each modality was selected on each trial to occur with a probability of .5. Thus, the cue served to indicate in advance which signal modality would occur on the current trial. Ten rats were trained on this procedure for 15 days. When test conditions began, a random half of the trials were identical to training trials. On the remaining trials the agreement between the modality of the cue stimulus and the modality of the signal was changed (for example, light cue and sound signal). Signal durations continued to be presented as previously described for the temporal bisection procedure[4] (see TABLE 2). The major question was whether the disagreement between the cue modality and the signal modality on test trials would affect the temporal bisection functions when plotted as a function of the "miscued" signal durations.

The results for the trials on which the modality of the cue and the signal were in agreement are shown in the left panel of FIGURE 3. Solid symbols indicate the method of prior entry; triangles indicate light-signal trials and circles indicate sound-signal trials. As expected, the observed psychophysical functions for cued signals were nearly identical for both light and sound signals when the modality of the cue and the signal were in agreement. These same functions were virtually indistinguishable from the pure modality block trials of the unbalanced stimulus probability method also shown in the left panel of FIGURE 3, that is, the PSEs were not reliably different from 4.0 sec.

The data from the light cue test condition are shown in the left panel of FIGURE 4. Triangles indicate light signals cued by light and circles indicate sound signals cued by light. The probability of a "long" response as a function of a light signal indicated that for light cueing the median PSE was 4.11 ± .05 sec and the probability of a "long" response as a function of a sound signal indicated that for light cueing the median PSE was 4.06 ± .03 sec. Neither of these measures was significantly different from each other or from previous control conditions.

The data from the sound cue test condition are shown in the right panel of FIGURE 4. Triangles indicate light signals cued by sound and circles indicate sound signals cued by sound. The test results from the "miscued" light signals were different from the results for "miscued" sound signals. The probability of a "long" response as a function of a light signal indicated that for sound cueing the median PSE was 4.71 ± .05 sec, signficantly greater than control conditions in which cue and signal modalities were in agreement. The probability of a "long" response as a function of a sound signal indicated that for light cueing the median PSE was 4.13 ± .04 sec, not significantly different from the previously described control conditions.

What are the effects of attentional bias? With both the method of unbalanced stimulus probabilities and the method of prior entry a bias toward sound signals shifted

TABLE 2. Prior-Entry Method

Test Condition	
Cue	Signal
Light	Light
	Sound
Sound	Light
	Sound

the psychophysical function for light signals rightward. This rightward shift of the temporal bisection function was fit with the theoretical function[4,8] with the assumption that the attentional bias for sound produced a latency of .6 sec for rats to begin timing light signals. Thus, the interpretation is that an attentional bias to attend to the auditory stimulus delayed the light signal from starting the internal clock. A reduction in the intensity of the sound signal (for example, 40 dB [re 20 $\mu N/m^2$]) did not reduce the magnitude of the effect. This suggests that an attentional bias does not function at the level of the sensory channel, but at a more central mechanism such as the mode switch. (See References 3 and 13 for details of the mode switch.)

The methods used to produce an attentional bias toward sound signals did not produce the same effects for light signals. A light bias, assuming that one was present, did not delay a sound signal from starting the internal clock. This effect is similar to the asymmetrical effects observed between light and sound in the cross-modal transfer of duration.[14] It is reasonable to assume that the same attentional bias effects described above will occur when rats trained exclusively with sound signals are transferred to experiments with light signals and also when rats trained exclusively with light signals are transferred to those with sound signals. Using a psychophysical choice procedure similar to the one described above, Meck and Church[14] found that cross-modal transfer from light to sound was greater than from sound to light and that rats transferred from sound-signal experiments to those using light signals, but not those rats transferred from light-signal experiments to those using sound signals made more erroneous "short" responses than erroneous "long" responses. Similar differences between light- and sound-signal durations have been observed with rats[15] and human subjects.[16]

Posner,[17] in describing human information processing, proposed that under some conditions auditory signals will have automatic access to information analyzers and visual signals will not. According to Posner's proposal, to initiate visual processing, subjects must learn to direct attention to the visual stimulus; no such learning is required for auditory processing. Thus, there would be greater transfer of training from light signals to sound signals than from sound signals to light signals. Finally, according to Posner's proposal, the time to process a visual signal is longer and more variable than the time to process an auditory signal. If this were the case in the rats trained with sound signals and transferred to light signals, the light would be slower to initiate the timing process. This would produce a shorter subjective time, which results in the observed asymmetrical error rates for the group transferred from sound to light, that is, more "short" than "long" errors. This same asymmetrical effect on the latency to initiate the timing process can account for the rightward shift of the light-signal function when the animal's attention is directed toward the auditory modality in either the unbalanced stimulus probability method or the prior-entry method. It also explains the lack of an effect on the sound-signal function when the animal's attention is directed to the visual modality.

The method of unbalanced stimulus probabilities and the prior-entry method produced essentially the same results. This suggests that unbalanced stimulus probabilities caused animals to expect that the more probable stimulus would occur on the next trial and that these expectations were functionally equivalent to the expectations produced by the first stimulus in the paired associate sequence used in the prior-entry method. The function of these expectations was to direct the animal's attention to the sensory channel appropriate to the expected stimulus, which facilitates information processing. (See the cost–benefit analysis of attention described by Posner.[17])

Previous work on the detection of the position of a visual signal has shown that when human subjects are cued on each trial about the location of a signal that is to follow the cue, they show stronger expectancy effects for the location of the signal than when a particular position is held constant for a block of trials. This result possibly reflects the active nature of expectancy for within-modality factors.[18] At present, it is uncertain whether such a result would occur for between-modality factors, although the results from the unbalanced stimulus probability method suggest that, at least under some conditions, there is an equivalent expectancy effect for cued and uncued, but expected, signals. This equivalence may be due in part to the temporal certainty of the signals in the uncued conditions as a result of the fixed ITI. Temporal uncertainty (random ITI) can lead to an increase in the processing time required for uncued signals relative to the processing time required for cued signals.[18,19]

PRIOR-ENTRY METHOD: PEAK PROCEDURE

The second experiment to be described combined the peak procedure for time discrimination with the prior-entry method. Ten rats were given extensive experience on the peak procedure with a signal of one modality (light, for example) that indicated that food might be primed 10 sec after signal onset and a signal of another modality (such as sound) that indicated that food might be primed 30 sec after signal onset. The rats were then trained with the prior-entry method. After an ITI of 130 sec a cue stimulus was presented for 1 sec. Then, after a random interval (geometrically distributed) with a minimum of 2 sec and a mean of 15 sec, a signal occurred in a manner similar to the peak procedure described above, except that one modality (such as light) signaled that food, if available, would be primed in 10 sec for a response on one lever (the left lever, for example) and another modality (such as sound) signaled that food, if available, would be primed in 30 sec for a response to the other lever (the right lever, for example). The association between the temporal criteria and the signal modalities was counterbalanced across rats. During training conditions the modalities of the cue stimulus and the signal were in agreement (the light cue and light signal, for example) and the stimulus modality was selected on each trial to occur with a probability of .5. Thus, rats could use the cue stimulus to predict which signal modality would next occur, and therefore, which temporal criterion would be appropriate for timing the upcoming signal. After a minimum of eleven days of training with cue and signal modalities in agreement, rats were transferred to test conditions (TABLE 2). During the seven test sessions a randomly selected 75% of the trials were identical to training trials. The remaining 25% of the trials were peak trials for which the agreement between the modality of the cue stimulus and the modality of the signal was

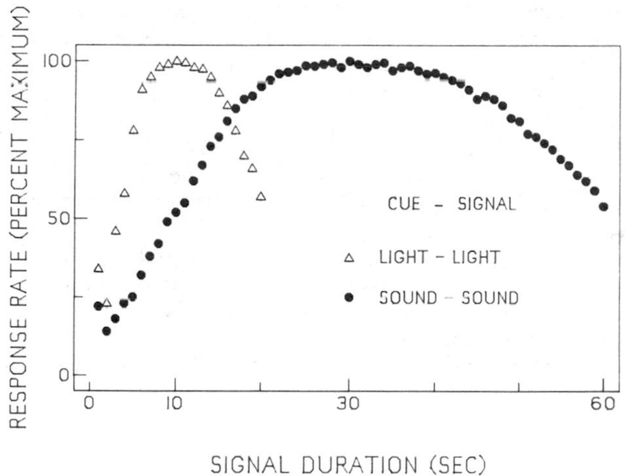

FIGURE 5. Median baseline data from the temporal peak procedure with the prior-entry method in which the modality of the cue and the modality of the signal were in agreement. *Open symbols* indicate responses to the left lever and *solid symbols* indicate responses to the right lever. *Triangles* indicate light-signal trials for which a 10-sec criterion was trained on the left lever. *Circles* indicate sound-signal trials for which a 30-sec criterion was trained on the right lever.

changed (for example, light cue and sound signal). The major question was whether the modality of the cue stimulus would influence which temporal criterion and response rule rats would use to time the signal. An attentional bias toward the modality of the cue stimulus might cause rats to use the temporal criterion and response rule appropriate for the modality of the cue and ignore the modality of the signal. On those test trials in which the cue modality and the signal modality were different, the modality of the cue stimulus would indicate a different temporal criterion and a different response rule from the signal. This difference allowed a dissociation between the behavioral control due to the cue and the behavioral control exerted by the signal.

The results for the trials on which the modality of the cue and the signal were in agreement are shown in FIGURE 5. Open symbols indicate responses to the left lever and solid symbols indicate responses to the right lever. Triangles indicate light-signal trials and circles indicate sound-signal trials. As expected, the observed peak functions for cued signals were fairly symmetrical on an arithmetic scale and peak times were not significantly different from 10 sec for the light signal and 30 sec for the sound signal.

Peak response rates were also very similar for the two reinforced values, 61 ± 5 responses/min for the 10-sec function and 56 ± 6 responses/min for the 30-sec function.

The data from the test condition where the cue modality and signal modality were in disagreement are shown in FIGURE 6. Open symbols again indicate responses to the left lever and solid symbols indicate responses to the right lever. Triangles again indicate light-signal trials and circles indicate sound-signal trials. Again, the observed peak functions were very close to 10 sec and 30 sec, but a behavioral dissociation between the cue and signal modalities was evident. Rats used a temporal criterion and response rule appropriate for the signal modality that they *expected* would follow the cue stimulus, but they timed the duration of the signal. That is, the temporal criterion and response rule appropriate to the cue modality was used to time the signal even though this information was inappropriate for the modality of the signal. From this result we can infer that the consistent relation between the cue modality and the signal modality during training led rats to bias their attention toward the modality of the cue when it occurred in order to "prepare" for the upcoming signal. This "preparation" involved the selection of a temporal criterion and a response rule to be used while timing the signal. Thus, during test conditions, the dissociation of the cue modality from the signal modality allowed the observation of the attentional bias produced by the prior-entry method. The attentional bias produced by the cue stimulus apparently prevented the rats from noticing the modality of the signal and adjusting their temporal criterion and response rule to be appropriate to the modality of the signal being timed.

The data from the prior-entry method with the peak procedure suggest that attentional bias can affect at least two of the information-processing stages used to make temporal discriminations. When a sound signal was unexpected, that is, cued by light, there was no apparent latency to initiate timing, but a temporal criterion and a response rule appropriate for light signals was used. This result strongly suggests that

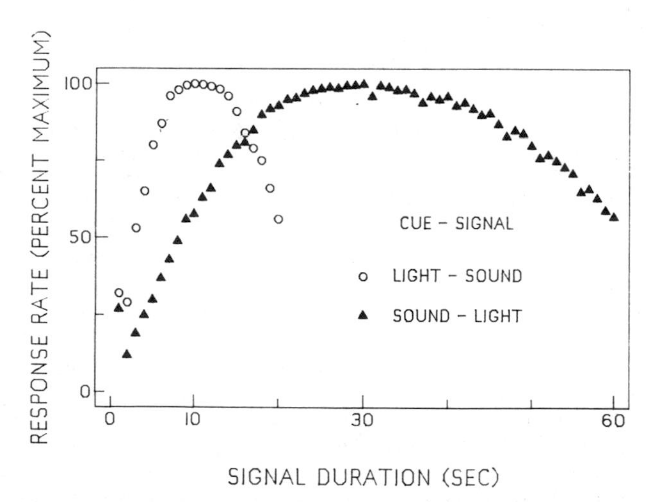

FIGURE 6. Median test data from the temporal peak procedure with the prior entry method in which the modality of the cue and the modality of the signal were in disagreement. *Open symbols* indicate responses to the left lever and *solid symbols* indicate responses to the right lever. *Triangles* indicate light-signal trials for which a 10-sec criterion was trained on the left lever. *Circles* indicate sound-signal trials for which a 30-sec criterion was trained on the right lever.

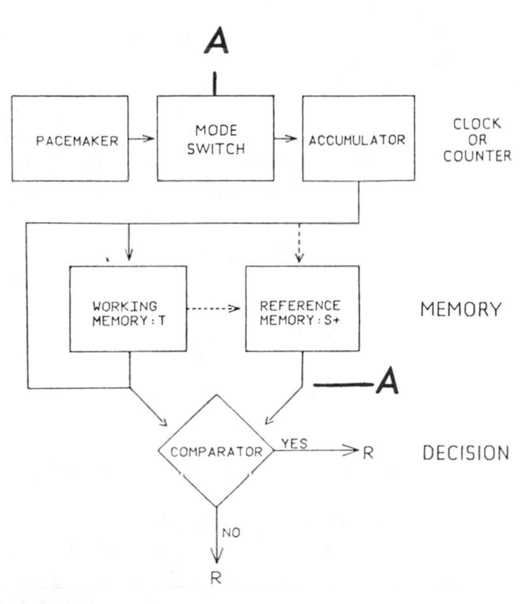

FIGURE 7. An information-processing model for temporal discrimination that illustrates the relation of attentional processes (A) to the different stages. The *top row* outlines a clock stage with a pacemaker generating pulses that are switched into an accumulator. The *middle row* outlines a memory stage composed of a working memory and a reference memory. The *bottom row* outlines a decision stage where a comparison is made between a representation of the current trial's clock-reading and sampled representations of previously acquired temporal criteria.

the attentional bias or expectation produced by the modality of the cue stimulus influenced the decision stage. On test trials it was observed that the presentation of a light cue led rats to sample a temporal criterion and a response rule from reference memory appropriate for light signals and to use this information to control their response during the following signal presentation regardless of the modality of the signal. Thus, although the onset of a specific signal modality initiated the timing process, the modality of the cue, rather than the modality of the signal, determined the temporal criterion and the response rule that animals use to process the signal. Apparently a single sample of the temporal criterion and the response rule is taken from reference memory on each trial and under the training conditions described above, a cue stimulus can initiate this process. If more than one sample of the temporal criterion and the response rule were taken for a trial, the temporal criterion and response rule would be updated and corrected during the signal presentation. This effect of attentional bias on the decision stage is diagrammed in FIGURE 7 at the level where the decision process samples from reference memory.

The second information-processing stage where the effects of attentional bias were observed was at the mode switch in the clock stage. This effect was indicated by the asymmetrical modality effect observed for the location of the peak time. When a light signal was unexpected, that is, cued by sound, there was a median increase in peak time for the 30-sec function of 1 ± .3 sec as compared to the baseline peak functions. This increase in peak time for the light-signal function suggests that there was a latency to begin timing of about 1 sec when the light signal was unexpected. The rightward shift in peak time is similar to the rightward shift observed in the PSE of the temporal bisection functions when the unbalanced stimulus probability and prior-entry methods were used. A latency to begin timing caused by attentional bias was used as an explanation for both effects. Additionally, a temporal criterion and a response rule appropriate to the sound signal was used when light signals were cued by a sound stimulus. This result suggests that the attentional bias or expectation produced by the modality of the cue stimulus influenced the decision stage in a similar manner to the effect of unexpected sound signals described above. The proposed effect of attentional bias on the clock stage is diagrammed in FIGURE 7 at the point where the mode switch gates pulses from the pacemaker into an accumulator.

SUMMARY

This paper makes two main points:

1. Both the presentation of unbalanced stimulus probabilities and the insertion of a predictive cue prior to the signal on each trial apparently induces a strong bias to use a particular stimulus modality in order to select a temporal criterion and response rule. This attentional bias toward one modality is apparently independent of the modality of the stimulus being timed and is strongly influenced by stimulus probabilities or prior warning cues. These techniques may be useful to control trial-by-trial sequential effects that influence a subject's perceptual and response biases when signals from more than one modality are used in duration discrimination tasks.
2. Cross-procedural generality of the effects of attentional bias was observed. An asymmetrical modality effect on the latency to begin timing was observed with both the temporal bisection and the peak procedure. The latency to begin timing light signals, but not the latency to begin timing sound signals, was increased when the signal modality was unexpected. This asymmetrical effect was explained with the assumption that sound signals close the mode switch automatically, but that light signals close the mode switch only if attention is directed to the light. The time required to switch attention is reflected in a reduction of the number of pulses from the pacemaker that enter the accumulator.[17,19]

One positive aspect of this work is the demonstration that procedures similar to those used to study human cognition can be used with animal subjects with similar results. Perhaps these similarities will stimulate animal research on the physiological basis of various cognitive capacities. Animal subjects would be preferred for such physiological experimentation if it were established that they possessed some of the cognitive processes described by investigators of human information processing.

One of the negative aspects of this work is that only one combination of modalities was used and variables such as stimulus intensity, stimulus probability, and range of signal durations have not been adequately investigated at present. Future work might test additional combinations of modalities and vary stimulus intensity and stimulus probability within a signal detection theory (SDT) framework to determine the effects of these variables on attentional bias.[18,20]

ACKNOWLEDGMENTS

I express my appreciation to Russell M. Church and Fariba N. Komeily-Zadeh for their advice on material relevant to this chapter.

REFERENCES

1. LaBerge, D., P. van Gelder & J. Yellott, Jr. 1970. A cueing technique in choice reaction time. Percept. Psychophys. **71:** 57–62.
2. Sternberg, S., R. L. Knoll & B. A. Gates. 1971. Prior entry reexamined: Effect of attentional bias on order perception. Paper presented at the annual meeting of the Psychonomic Society, November 1971, St. Louis, Missouri.
3. Gibbon, J. & R. M. Church. Sources of variance in an information processing model of timing. *In* Animal Cognition. H. L. Roitblat, T. G. Bever & H. S. Terrace, Eds. Erlbaum. Hillsdale, NJ. In press.
4. Meck, W. H. 1983. Selective adjustment of the speed of internal clock and memory processes. J. Exp. Psychol. Anim. Behav. Processes **8:** 171–201.
5. Roberts, S. 1981. Isolation of an internal clock. J. Exp. Psychol. Anim. Behav. Processes **7:** 242–268.
6. Church, R. M. & M. Z. Deluty. 1977. The bisection of temporal intervals. J. Exp. Psychol. Anim. Behav. Processes **7:** 242–268.
7. Maricq, A. V., S. Roberts, & R. M. Church. 1981. Methamphetamine and time estimation. J. Exp. Psychol. Anim. Behav. Processes **7:** 18–30.
8. Gibbon, J. 1981. On the form and location of the psychometric bisection function for time. J. Math. Psychol. **24:** 58–87.
9. Platt, J. R. & E. R. Davis. 1983. Bisection of temporal intervals by pigeons. J. Exp. Psychol. Anim. Behav. Processes **9:** 160–170.
10. LaBerge, D. H. 1973. Identification of two components of the time to switch attention: A test of a serial and a parallel model of attention. *In* Attention and Performance: IV. S. Kornblum, Ed. Academic Press, New York, NY.
11. Titchener, E. B. 1908. Lectures on the Elementary Psychology of Feeling and Attention. MacMillan. New York, NY.
12. Boulter, L. R. 1977. Attention and reaction time to signals of uncertain modality. J. Exp. Psychol. Hum. Percept. Perform. **3:** 379–388.
13. Meck, W. H. & R. M. Church. 1983. A mode control model of counting and timing processes. J. Exp. Psychol. Anim. Behav. Processes **9:** 320–334.
14. Meck, W. H. & R. M. Church. 1982. Abstraction of temporal attributes. J. Exp. Psychol. Anim. Behav. Processes **8:** 226–243.
15. Roberts, S. Cross-modal use of an internal clock. J. Exp. Psychol. Anim. Behav. Processes **8:** 2–22.
16. Goldstone, S. & J. L. Goldfarb. 1966. The perception of time by children. *In* Perceptual Development in Children. A. H. Kidd & J. L. Riviore, Eds. International Universities Press. New York, NY.
17. Posner, M. I. 1978. Chronometric Explorations of Mind. Erlbaum. Hillsdale, NJ.
18. Posner, M. I., C. R. R. Snyder, & B. J. Davidson. 1980. Attention and the detection of signals. J. Exp. Psychol. General. **109:** 160–174.
19. Meck, W. H. 1981. Directed attention and automatic activation of psychological pathways. Manuscript submitted in partial fulfillment of the requirements for the Ph.D. degree at Brown University, Providence, RI.
20. Green, D. M. & J. A. Swets. 1974. Signal Detection Theory and Psychophysics. Krieger. Huntington, NY.

Temporal Rhythms and Cerebral Rhythms[a]

MICHEL TREISMAN

Department of Experimental Psychology
University of Oxford
Oxford OX1 3UD, England

Hoagland[1] appears to have been the first to suggest that the ability to make temporal judgments may depend on the possession of a temporal pacemaker or internal clock, analogous to the pacemaker cells responsible for many physiological rhythms. This valuable insight has stimulated much research. To get full benefit from the hypothesis, however, two things must be borne in mind. The first is that the hypothesis of a pacemaker needs to be supplemented by further mechanisms to explain performance: I shall use the term "pacemaker" for the source of the temporal reference frequency, and the term "internal clock" for the complete set of mechanisms. The second consideration is that an adequate model should aim at explaining not just one facet of time judgment, but as many of its features as possible.

A model for the internal clock which I have previously put forward[2] attempted to account for several features of temporal judgment which seemed likely to be important. These were: (a) The Weber function for temporal discrimination. (b) The "central tendency of judgment" (if time intervals of different length are reproduced, positive constant errors are commonly obtained for short intervals and negative for long, or there is a difference in this direction). (c) The "lengthening effect"[2] (during the course of a session, productions or reproductions of a given interval, or the point of subjective equality, tend to increase, while verbal estimates of a standard interval diminish). (d) Differential lengthening (this does not proceed at the same rate for different standard intervals; the proportionate increase tends to be greater, the shorter the standard interval).

The model for the internal clock is illustrated in FIGURE 1.[2] It includes a temporal pacemaker, a counter, a memory store, and a comparator. The assumption was also made that the level of activation or arousal of the temporal pacemaker may vary. This state of the pacemaker was referred to as "specific arousal," to avoid any implication that it is necessarily related to any state of general arousal that may exist. Specific arousal of the pacemaker may be modulated by external influences. Raised arousal increases the rate of the pacemaker and reduced arousal slows it.

If the pacemaker runs more slowly during a reproduction, the latter will be lengthened. If it runs more quickly, the reproduction will be shorter. Thus, increased temporal arousal will shorten reproductions, and decreased arousal will lengthen them. The relation will be the opposite for verbal estimates of a standard interval.

This model is described more fully elsewhere.[2] With appropriate rules for operation it appears capable of generating the findings referred to above, and others.

AROUSAL AND THE INTERNAL CLOCK: THREE HYPOTHESES

In the model, the temporal pacemaker is subject to specific arousal. The object of the present study was to investigate this further. The assumption of specific arousal implies

[a]This work was supported by a grant from The Medical Research Council of Great Britain.

that the pacemaker may be affected by external sources which we assume, at least to start with, may have effects similar to those traditionally associated with the concept of arousal. Several lines of evidence support this hypothesis. For example, auditory stimuli are usually more arousing than visual; intense stimuli more arousing than weak ones; and fear of danger is arousing. In each case, the temporal pacemaker is speeded.[2-4] Such observations are consistent with what we shall call:

(a) The specific arousal hypothesis. This claims that there may be, but not necessarily will be, similarities between the effects of external influences on temporal arousal, and those they are believed to produce on general arousal. But otherwise these hypothetical processes are unrelated.

In contrast to this, perhaps the simplest possible hypothesis would unify the two

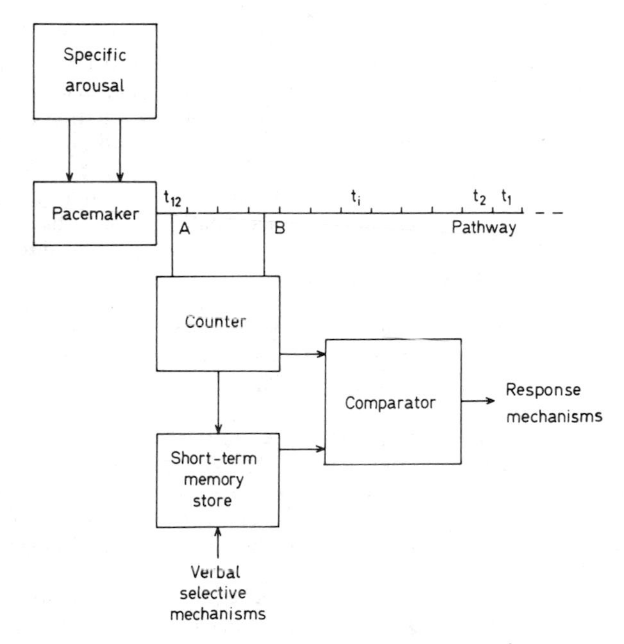

FIGURE 1. A model of the internal clock.[2]

concepts by assuming that "specific" arousal is simply the local action of "general" arousal, and that there is no need for any more complex assumption. We call this:

(b) The correlated arousal hypothesis. This assumes that factors that modulate the arousal of the subject as a whole act also on the temporal pacemaker, in a similar way, to produce corresponding states of high or low arousal in it, causing it to run correspondingly fast or slow.

To test this hypothesis we need to examine concurrent variations in temporal judgment and in an indicator of general arousal. An attractive measure for this purpose is the alpha rhythm of the electroencephalogram: it is continuously observable by methods to which the average subject adjusts, and it has two aspects, both of which are continuously variable and both of which are believed to relate to the subject's state of general arousal: its frequency and its prevalence, that is, the extent to which it can be seen in the record.

The alpha rhythm is undoubtedly the best known rhythmic activity that has been described in the nervous system. Its source is unknown. Many suggestions have been put forward, including the proposal that it is controlled by a pacemaker.[5] The extent to which it is present in the electroencephalogram has often been used as an index of arousal. It is customary to order different degrees of "general arousal," often not distinguished from the sleep–waking cycle, along a continuum, and to associate them with characteristic EEG patterns. Of the states distinguished in this way by Lindsley,[6] the three most likely to be observed in waking subjects are (*i*) "alert attentiveness," in which fast low-amplitude waves dominate the encephalogram; (*ii*) "relaxed wakefulness," in which the record becomes more synchronized and the alpha rhythm is most strongly represented; and (*iii*) "drowsiness," which is marked by a reduction in alpha activity, the record becoming flat with occasional slow waves. Thus, as general arousal decreases, the proportion of alpha in the record first rises and then falls.

There is evidence that the frequency of the alpha rhythm may also vary as a function of arousal. Monotony, immobilization, and sensory or perceptual deprivation reduce the frequency, and it is low on waking and in drowsy subjects.[7]

The correlated arousal hypothesis will predict an inverse relation between alpha frequency and time productions. But so also will a third hypothesis, which has been proposed by a number of investigators:

(*c*) *The common pacemaker hypothesis.* The alpha rhythm and the frequency of the temporal clock might be determined by a single common pacemaker. This hypothesis derives from the work of Hoagland[1] and is of sufficient interest and has been proposed sufficiently often[8–10] to warrant examination.

A number of investigators have examined this hypothesis. Werboff[8] found anomalous relations between EEG frequency and time productions. Unfortunately, he derived EEG frequency from the number of waves per second without regard for the specific frequencies. Surwillo[9] reported a correlation of 0.235 between time productions and the period of the alpha rhythm. This is a relatively low value, derived from a single session. Holubar[10] attempted to use photic driving to alter the frequency of the brain rhythm and so alter temporal performance. To measure the latter he used temporal conditioning of galvanic skin and EEG responses. He showed that in four of 15 subjects, exposure to flicker frequencies of 7 or 14–15 Hz reduced inter-response intervals by approximately half. He concluded that the alpha rhythm constitutes the temporal pacemaker.

A difficulty is that temporal conditioned reflexes are subject to disinhibition, resulting in premature production of the conditioned response. It is possible that Holubar's flicker had its effects in this way.

Adam *et al.*[11] obtained negative evidence. Low doses of anesthetics prolonged time productions, but changes in alpha were minor. However, the effects of anesthetics are not well understood. Thus, it is conceivable that they may have affected decision processes rather than the temporal pacemaker.

It appears there is some evidence in each direction, but in no case is it completely conclusive.

PREDICTIONS OF THE THREE HYPOTHESES

To examine these hypotheses, an experiment was planned in which subjects would repeatedly produce a standard time interval. The durations of their time productions (*T*), the prevalence of alpha during the time production (*P*: the proportion of time alpha is visibly present in the record), and the frequency of this alpha rhythm (*F*) would be measured. These results could then be related to the following predictions of the three hypotheses.

(a) Specific arousal hypothesis. Since different mechanisms have independent dimensions of arousal, this allows a variety of relationships. In particular, two measures may be either positively or negatively related on different occasions if they reflect specific arousal of different mechanisms which may respond in characteristic ways, sometimes in like manner, sometimes oppositely, to the given external circumstances.

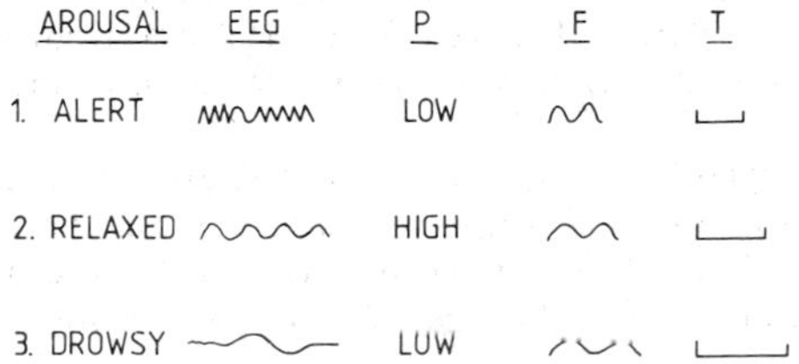

AROUSAL	EEG	P	F	T
1. ALERT		LOW		
2. RELAXED		HIGH		
3. DROWSY		LOW		

PREDICTED CORRELATIONS

HYPOTHESIS	T–F	T–P	F–P
Correlated Arousal			
1 - 2	–	+	–
2 - 3	–	–	+
Common Pacemaker			
1 - 2	– –		–
2 - 3	– –		+

FIGURE 2. (*Top*) Schematic representation of the relations envisaged by the correlated arousal and common pacemaker hypotheses. The appearance of the EEG, proportion of alpha (P), frequency of the alpha rhythm (F), and duration of temporal productions (T) are shown for the three states of highest general arousal, on the assumption that the frequency of the temporal pacemaker is either correlated with or identical to the alpha frequency. (*Bottom*) The partial correlations that may be derived from these assumptions. For the correlated arousal hypothesis, variation in arousal should produce a negative correlation between T and F, variation between states 1 and 2 (alert–relaxed) should produce a positive relation between T and P, but variation between the relaxed and drowsy states should produce a negative T–P correlation. Other correlations are derived in the same way.

(b) Correlated arousal hypothesis. The assumptions of the hypothesis are shown schematically in FIGURE 2, together with the correlations predicted by it. Three states of arousal are shown. Since P is high in the relaxed state and low in the other two, correlations involving P will be opposite in direction if variation in arousal during a session is mainly between "alert attentiveness' and "relaxed wakefulness," or between the relaxed and drowsy states. Significant variation between all three states should produce low correlations.

Since increase in the frequency of the temporal pacemaker will arise from its specific arousal, and increase in alpha frequency from general arousal, and these will vary together, the hypothesis predicts a negative correlation between *T* and *F* (or a positive correlation between *T* and *Per*, where *Per*, the alpha period, is the reciprocal of *F*), but its magnitude will not necessarily be high. The relations predicted between *T* and *P*, and between *F* and *P* are shown in the figure.

(c) Common pacemaker hypothesis. This predicts that the correlation between *T* and *Per* should be not only positive but high (or the *T–F* correlation negative and high) at all times. The hypothesis assumes that the common pacemaker is active whether or not alpha can be seen in the electroencephalogram, its intermittent absence from the latter being due to unrelated causes. Thus it makes no prediction for the correlation between *T* and *P*. The relations between *F* and *P* derive from beliefs about general arousal and are therefore the same for the correlated arousal and common pacemaker hypotheses.

Both the last two hypotheses require a positive relation between *T* and *Per*, but the common pacemaker hypothesis also requires that this should be high. A nonsignificant correlation between *T* and *P* is compatible with both hypotheses, but significant correlations, positive or negative, between *T* and *P* will favor the correlated arousal hypothesis.

OBSERVATIONS ON THE ALPHA RHYTHM AND TIME PRODUCTIONS

An experiment was run in which the alpha rhythm and prevalence of alpha were measured during time productions.

Apparatus

The subject was seated in a dark, sound-shielded cubicle. A 500-Hz tone generated by an advance oscillator could be delivered to him through Brown type K earphones. The subject could terminate a presentation of the tone by pressing a reaction key. Silver scalp electrodes were attached to the left parietal and occipital regions 4 cm from the midline, using electrode jelly and collodion, and an earth electrode was attached to the left forearm or earlobe.[12] The EEG was recorded on one channel of an AEI Mark III four-channel pen oscillograph at a paper speed of 3 cm per second, and the subject's time productions were recorded on a second channel. An electronic timer was used to vary the interstimulus interval between the end of one time production and the beginning of the next.

Procedure

About 20 minutes were usually spent in introducing the subject to the procedure and attaching the electrodes. The subject's absolute threshold for the 500-Hz tone was then found, and the intensity of the tone was set at 50 db SL. One or two examples of a 4-sec interval, presented at 4-sec durations of the tone, were then presented. Each trial began with the onset of the tone. The subject terminated it after what he estimated to be exactly four seconds, by pressing the reaction key. He was asked to keep his eyes shut and not to count. The interstimulus interval varied randomly between 3 and 8 seconds. The session proper lasted 35–55 minutes.

Subjects

Five students and two laboratory technicians served as subjects. Five subjects served once and two served twice, giving nine sessions in all.

RESULTS

The duration of each time production (*T*) in seconds was recorded, and the section of the EEG trace corresponding to the time production was scored by eye to give measures of *P*, the proportion of the trace during which alpha activity was detectable in the trace, and *F*, the alpha frequency. *Per* is given by the reciprocal of *F*. The alpha rhythm was present on 99.4% of trials. Five percent of trials could not be scored because of artefacts or equipment failure and were omitted from further analyses, no attempts being made to correct for their absence. The mean number of trials per session was 227.

The mean values of *T*, *F* and *P* are shown for each session in TABLE 1. Lengthening

TABLE 1. Mean Results for Nine Sessions

	Mean Values			Lengthening (%)		Coefficients of Variation	
Session	*T*	*P*	*F*	*T*	*F*	*T*	*F*
1a	4.55	0.84	8.65	28.1	−1.6	0.19	0.032
1b	4.57	0.84	8.64	−16.2	3.5	0.17	0.032
2a	7.43	0.67	10.67	7.4	1.1	0.13	0.030
2b	2.28	0.71	10.29	13.5	−2.4	0.15	0.051
3	3.07	0.86	9.40	51.2	−2.9	0.19	0.034
4	6.10	0.66	10.10	96.4	−6.9	0.30	0.077
5	7.07	0.62	10.22	77.6	0.7	0.21	0.026
6	6.83	0.74	11.13	5.5	0.1	0.16	0.023
7	5.15	0.65	10.26	−16.8	0.1	0.21	0.030
Mean	5.23	0.73	9.93	27.4	−1.7	0.19	0.037

was measured as the percentage increase in *T*, or *F*, from the first 50 to the last 50 time productions in each session. Coefficients of variation for *F* and *T* are also given.

The usual lengthening effect is shown: in seven sessions *T* increased by more than 5% by the end of the session; in two it decreased. Lengthening may be attributed to a fall in specific arousal during the session, and the rarer finding of decrease in length of time productions to an increase in specific arousal.

The common pacemaker hypothesis predicted a strong positive relation between *T* and *Per*. The present results tell against that hypothesis: if a common pacemaker determines both *T* and *F*, then variation in one measure should be of the same order as variation in the other. But the lengthening of *T* is unrelated to and an order of magnitude greater than the changes in *F*, and the coefficients of variation for *T* are about five times as great as those for *F*.

The partial correlations for each session are given in TABLE 2. The common pacemaker hypothesis requires high negative correlations between *T* and *F*. But although seven of the nine *T–F* correlations are negative, they are far from approach-

TABLE 2. Partial Correlations

Session	*T–F*	*T–P*	*F–P*	Duration (min)
1a	−0.10	0.05	−0.09	35
1b	0.22**	0.06	−0.01	45
2a	0.25**	−0.16*	−0.08	40
2b	−0.13*	0.06	0	45
3	−0.36**	−0.30**	−0.14*	50
4	−0.28**	0.16*	−0.18*	45
5	−0.15	−0.43**	−0.23**	50
6	−0.04	0.03	0.29**	50
7	−0.07	−0.03	0.05	55

NOTE: Significance (two-tailed): *= $p < 0.05$; **= $p < 0.01$.

ing −1.0. Four negative correlations are not significant and two highly significant partial correlations are positive. The common pacemaker hypothesis suggests that *T* and *P* should be unrelated, but in four sessions the two variables are significantly correlated. These results, too, argue against a common pacemaker.

The correlated arousal hypothesis predicted that positive or negative correlations between *T* and *P* might occur, and so they do. But this hypothesis also predicts that if general arousal rises, alpha frequency will increase and time productions get shorter. The significant positive correlations between *T* and *F* are a severe embarrassment for this hypothesis. Comparison between the correlations for different pairs of variables presents a further difficulty. As FIGURE 2 illustrates, if the subject varies mainly between two levels of arousal, and so gives significant *T–P* and *F–P* correlations, these should be of opposite sign. But in two of the three sessions in which both partial correlations are significant, they have the same sign.

Although neither of the two hypotheses linking *F* and *T* is supported, the data are not wholly negative. Of 27 partial correlations, 13 are significant, and for each pair of variables significant correlations occur in both directions, a finding that requires explanation. The absence of a simple pattern to these results suggests that there may be underlying relations which product–moment correlations are not well fitted to reveal. We now turn to different procedures to examine this possibility.

FURTHER FEATURES OF THE DATA

A significant correlation might occur in a given session simply because there is a consistent slow drift in the value of each variable. Such drifts might be quite independent, sometimes proceeding in the same, and sometimes in opposite directions, and giving rise to positive or negative correlations accordingly. To examine the relations between the variables more closely, the moving average partial correlation between each pair of variables in each session was computed. For this purpose the product–moment correlations for all sections of the record 15-trials long were calculated and plotted against the midpoints of the corresponding ranges. Thus the correlations for trials 1–15, 2–16, and so on, are plotted against trials 8, 9, and so on, as in FIGURE 3, where representative samples of such records are shown for the three pairs of variables. The results not shown resemble those in the figure.

The striking observation is the frequent appearance of more or less regular oscillations between positive and negative cross-correlation. This apparent periodicity is not in some way an artefact of the use of ranges of length 15. For session 5, the two curves shown were calculated using ranges of 10 or 20 trials: they share similar main features which occur at similar positions, although the 20-trial curve is, of course, smoother than that for the shorter range.

These graphs suggest that whether an overall correlation is positive or negative may depend simply on whether a particular session samples more of the peaks or of the troughs of slow oscillations in the relation between the two variables. If so, overall correlations cannot throw much light on the underlying mechanisms, and we must turn instead to the problem of explaining the oscillations shown in FIGURE 3.

On examining plots of each variable against trials, it seemed that spontaneous oscillations were present in all of them. FIGURE 4 illustrates the curves for session 1a, a session that gave no significant partial correlations. The curves are smoothed: each point represents the mean of seven neighboring readings. Alpha period has been plotted rather than F: on both the common pacemaker and correlated arousal hypotheses, *Per*

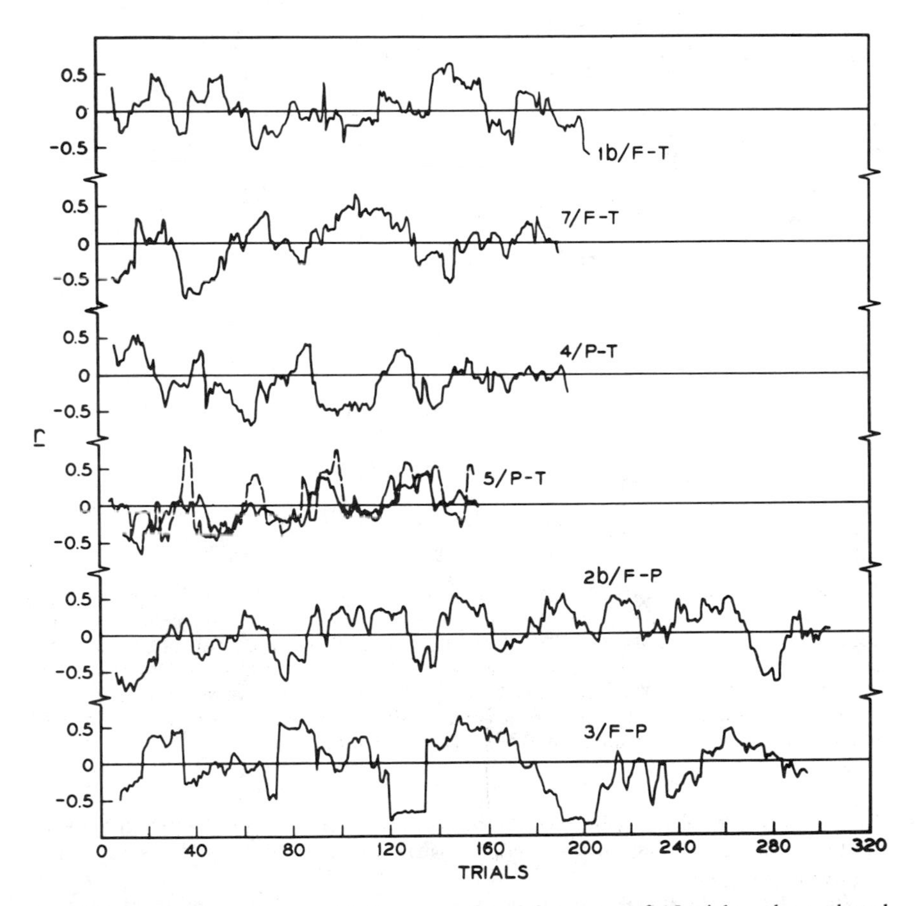

FIGURE 3. Moving average correlations were found for ranges of 15 trials and are plotted against the midpoints of these ranges. The results for two sessions are shown for each possible pair of variables. For session 5 two curves are shown, one for a range of 10 (*dashed curve*) and one for a range of 20 (*continuous curve*).

should vary in the same direction as *T*. It can be seen that the three variables show marked but rather different patterns. *T* and *Per* do not show a strong positive relation. Indeed, between appoximately trials 50 to 100, they vary in almost a mirror-image fashion.

Further comparisons between *T* and *Per* are shown in FIGURE 5. Sessions 2b and 4 gave significant negative *T–F* partial correlations, implying a positive relation between *T* and *Per*; sessions 6 and 7 gave nonsignificant correlations. The positive relation between *T* and *Per* in session 4 is probably determined by the rising trend in the first

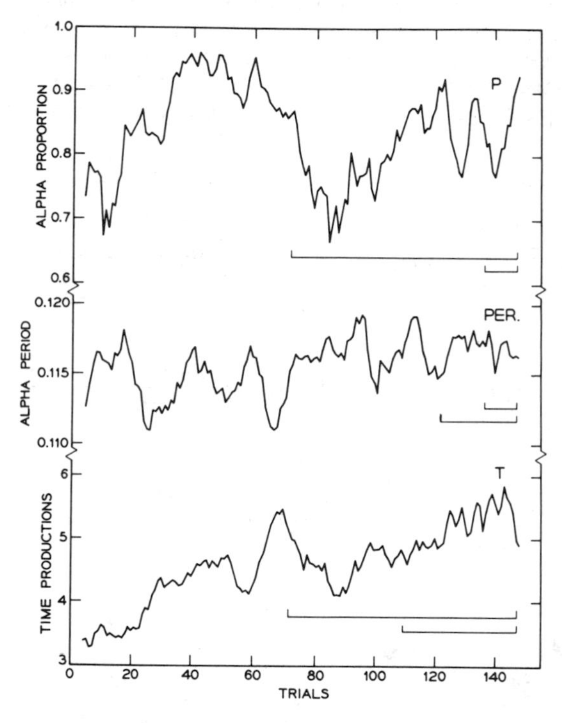

FIGURE 4. Results for session 1a. The prevalence of alpha in the record (*P*), the alpha period in seconds (*Per*), and the length of time productions in seconds (*T*) are plotted against trials. The curves were smoothed over 7 points. That is, the mean for trials 1–7 is plotted against trial 4, the mean for 2–8 against trial 5, etc. The two horizontal lines below each trace represent, from above down, the periods of the spontaneous oscillations in the range 2–15 cycles per session that were ordered first and second by the power spectrum analysis in each case. These are, *P*: 17.5 and 2.5 minutes; *Per:* 2.5 and 5.8 minutes; and *T*: 17.5 and 8.8 minutes.

third of the record; after about trial 80 the traces suggest a negative relation between the two variables. In session 6 the main positive peaks in *T* seem to correspond to positive peaks in *Per*.

FIGURE 6 shows *P* and *T* traces for two sessions which gave significant negative partial correlations. There is a rough mirror-image relation between the variables in session 5.

The data in FIGURES 3, 4, 5 and 6 suggest, first, that the three variables are subject

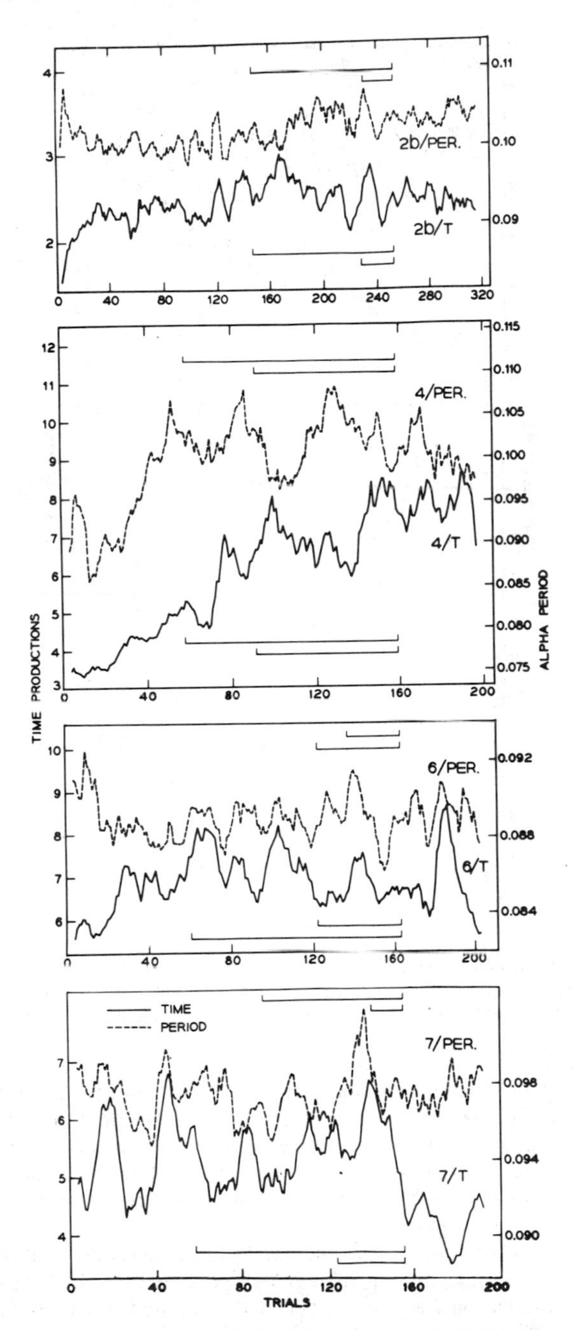

FIGURE 5. *T* and *Per* in seconds (the scale for the first on the *left* and for the second on the *right*) are plotted against trials in the same way as in FIGURE 4, for sessions 2b, 4, 6 and 7. The periods of spontaneous oscillations in the range 2–15 cycles per session which were ranked first and second are shown for each trace.

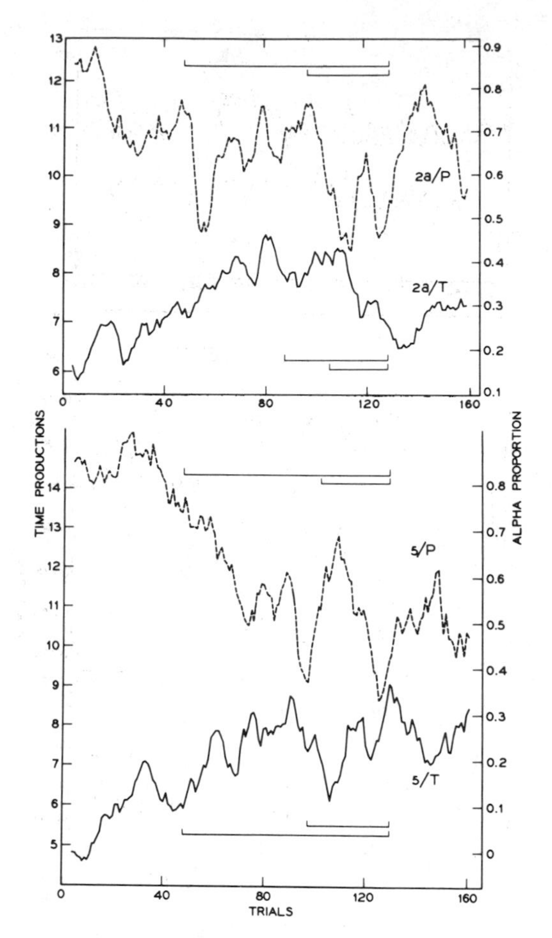

FIGURE 6. *T* in seconds (scale on *left*) and *P* (scale on *right*) are plotted against trials. As in the two previous figures, the curves are smoothed over 7 trials, and the periods of the two oscillations ranked first and second in the range 2–15 cycles per session are shown for each trace.

to more or less regular fluctuations, and second, that the changes in two measures may be related. These fluctuations will be referred to as "oscillations" rather than as "rhythms," to avoid any possible confusion with the term alpha rhythm.

To examine these oscillations, power spectrum analyses were performed on the measurements for each variable in each session, and cross-spectral analyses were calculated for pairs of variables.[13]

Power densities were examined for oscillation frequencies taken at intervals of one cycle per session, for a range of frequencies corresponding in each case to 2 to 60 cycles per hour. Lower oscillation frequencies were not examined since spontaneous oscillations might not be distinguishable from responses to major events of the experimental session, such as its start or ending. Oscillations above 60 cycles per hour were not examined since in some cases these might derive directly from the succession of trials and responses, which occurred at rates of 4–7 a minute.

Two examples of power spectrum analyses are given in FIGURE 7, which shows the power density (scaled to have a maximum of 1.0) at each oscillation frequency for the range 2–26 cycles per session. This corresponds to oscillation periods of 17.5 to 1.4 minutes for session 1a, and 22.5 to 1.7 minutes for session 4. The corresponding data are shown in FIGURES 4 and 5. The analyses suggest that there may be a tendency for greater representation of low-frequency oscillations in the *T* record than in *P*, and in *P* than in *F*. There is no suggestion of shared frequencies for session 1a, but in session 4 there is a peak at 9 cycles per session in *P* and *T* and a peak at 24 cycles per session in all three records.

The oscillation frequencies can be rank-ordered by power density, the frequency with the highest density, in the range examined, being given the rank "1." TABLE 3 gives the first and second dominant oscillation in each record. It appears that as we go from *T* to *P* to *F*, and as rank decreases, the oscillations become higher in frequency.

To look at this further, the five dominant oscillations in the range 2–60 cycles per hour were combined for the nine sessions and cumulative curves drawn giving the proportion of dominant oscillations having frequencies less than or equal to the abscissa value. The cumulative curves for *T*, *P* and *F* are shown in the upper panel of

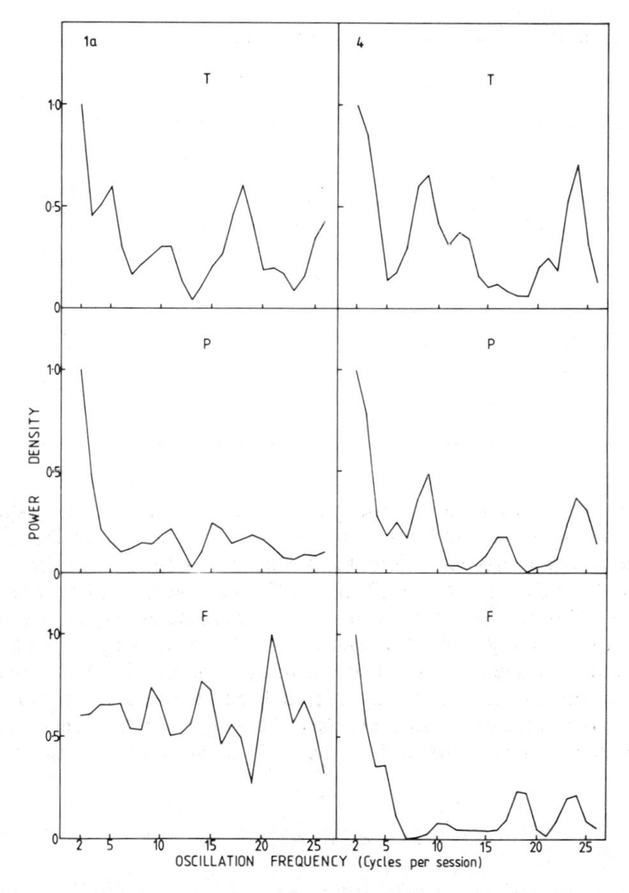

FIGURE 7. Power spectrum analyses for sessions 1a and 4. The power density (scaled to have a maximum of 1.0) is shown for oscillation frequencies in the range 2–26 cycles per session for *P*, *F* and *T*.

FIGURE 8. We see that low frequencies are more strongly represented for *T* than for *P*, and more strongly for *P* than for *F*. The medians of these frequencies are given in TABLE 4. On analysis of variance of the first four dominant oscillation periods rank order was significant ($F(3,24) = 4.78$, $p < .01$), the highest ranks having the lowest frequencies, and variable was significant ($F(2,16) = 4.43$, $p < 0.05$): *T* has the lowest and *F* the highest frequencies.

This finding presents a problem for general arousal theory. We saw (see FIGURE 2) that a cycle from the alert to the drowsy state and back again should give two cycles of variation in *P* for one in *F*. The finding that major oscillations in *P* are of lower frequency than those in *F* is discordant with this.

Do the different variables "choose" their dominant oscillations independently, or are the same oscillations dominant in two or more records more often than would be expected by chance? Inspection of TABLE 3 suggests this may be the case. To test this, for each subject and for each pair of variables, the number of frequencies ranking in the first five for each variable and common to both was found. If *x* common dominant frequencies were obtained for a given session, the probability of obtaining a number as

TABLE 3. Dominant Oscillations (Cycles per Hour)

Measure:	*T*		*P*		*F*	
Rank:	1	2	1	2	1	2
Session						
1a	51.1	3.4	3.4	5.1	35.8	37.5
1b	2.7	8.0	2.7	14.7	2.7	40.0
2a	3.0	37.5	4.5	3.0	3.0	4.5
2b	2.7	4.0	14.6	13.3	46.5	2.7
3	3.6	2.4	25.1	26.3	46.6	39.5
4	2.7	4.0	2.7	4.0	2.7	6.7
5	4.8	2.4	2.4	7.2	19.2	3.6
6	6.0	7.2	13.9	12.9	47.8	49.0
7	2.2	3.3	14.1	13.0	48.8	49.9
Median:	3.0	4.0	4.5	12.9	35.8	37.5

great as *x* or greater by random processes was calculated from the hypergeometric distribution, and these probabilities were combined for the nine sessions.[14] The mean numbers of common frequencies were, for *T* and *P*: 1.44 ($p < 0.01$); for *T* and *F*: 1.00 (not significant); and for *F* and *P*: 1.22 ($p < 0.05$). The mean number to be expected by chance is 0.56. This provides some evidence that frequencies may be shared.

These analyses establish that the fluctuations that were seen in FIGURES 3 to 6 are not simply due to random noise. The oscillation frequencies that predominate in individual records are not randomly selected but tend to come from the lower frequencies in the range examined, this effect being greater for *T* than *P*, and greater for *P* than *F*. It also appears that different variables may share common frequencies.

CROSS-SPECTRAL ANALYSES

If the same oscillation frequency dominates two variables, and this is due to coincidence only, then the phase relation between the two occurrences of this frequency

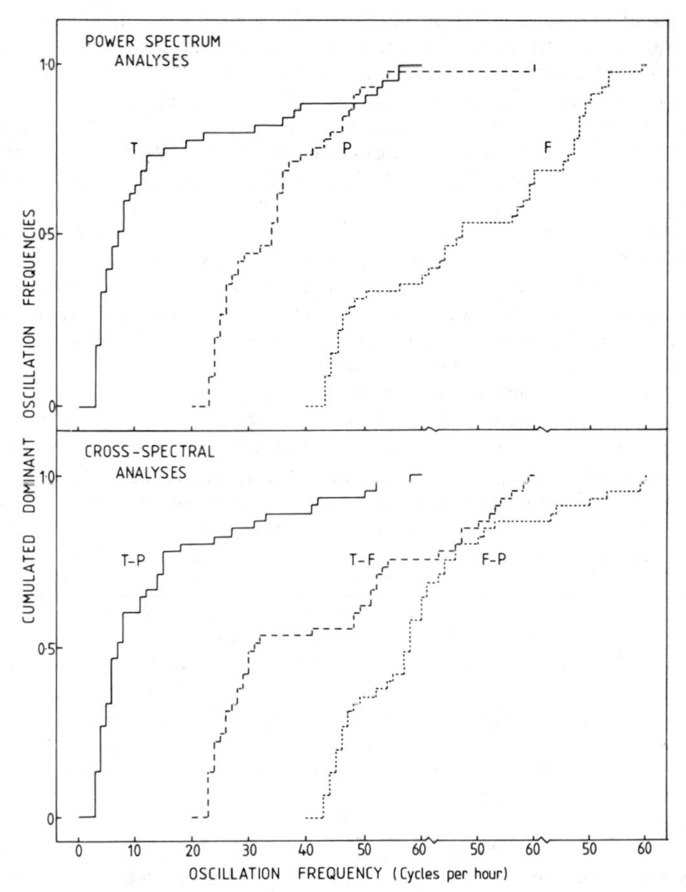

FIGURE 8. The five dominant frequencies in the power spectrum analysis for each record for each variable were expressed as cycles per hour and cumulated. The cumulative curves, which represent the proportion of the dominant oscillations lying at or below the oscillation frequencies represented on the abscissa, are plotted in the *upper panel.* A similar analysis was performed on the cross-spectral analyses for each pair of variables, and the cumulative curves for the cross-spectral analysis of variables *T* and *P* (*T–P*), for *T–F* and for *F–P* are shown in the lower panel of the figure.

TABLE 4. Median Dominant Oscillation Frequencies (Ranks 1 to 5)

Measure:	*T*	*P*	*F*
	6.59	13.09	26.07
	T–P	*T–F*	*F–P*
	6.59	10.15	16.87

should be randomly determined. But if the oscillation is produced in both records by a common cause, the phase relations that may occur may be more restricted. The parallel or mirror-image relations seen in some of the earlier figures, in which major peaks or troughs occurred at more or less the same time for different variables, suggest that common frequencies may occur at phase relations near 0° or 180°. Cross-spectral density and phase spectra were found for each pair of variables for each subject.

The five dominant frequencies in the range 2–60 cycles per hour were found for each variable pair, and cumulative curves for the nine sessions are shown in the lower panel of FIGURE 8. The median dominant shared frequencies are given in TABLE 4. The cumulative curve for *T* and *P* rises rapidly, as do the curves for *T* and for *P* in the upper panel, and it has a low median, like the *T* curve. The *T–F* curve appears to be bimodal, with an initial rapid rise followed by an intermediate plateau and a further rise, and the *F–P* curve rises at a rate intermediate between curves *F* and *P*. These results suggest that shared frequencies may come from both sources.

An analysis of variance of the four dominant oscillation periods for each variable pair showed rank significant ($F(3,24) = 10.56$, $p < 0.001$), the highest ranking oscillations having the lowest frequencies. Variable pair approached significance ($F(2,16) = 3.17$, $p = 0.07$), the mean oscillation periods decreasing in the order *T–P*, *T–F*, and *F–P*, and there was a significant interaction between rank and variable pair ($F(6,48) = 2.79$, $p < 0.025$): oscillation frequencies were lower for *T–P* than for *F–P*. For *T–F* they were more similar to *T–P* at ranks 1 and 2, and more similar to *F–P* at ranks 3 and 4.

If an oscillation frequency is shared by variables *A* and *B*, ϕ_{AB} is the phase angle by which that frequency in *B* is in advance of the same frequency in *A*. In view of the suggestions of bimodality in FIGURE 8, and the tendency for the lowest frequencies to have the greatest amplitude, the oscillations were divided into a low-frequency range (2–20 cycles per hour) and a high-frequency range (20–60 cycles per hour), and for each session and each variable pair the five dominant low-frequency and the five dominant high-frequency oscillations were determined. The phases of these oscillations, 90 for each variable pair, are plotted in FIGURE 9, in which they are grouped in ranges of 45°.

The phase plots show that relative phases are not uniformly distributed around the circle, as would be the case if the oscillations were independently set up for each variable, or randomly determined. Instead, each conjunction of variables prefers certain phases. For *T–P*, positive phases are more strongly represented than are negative phases. For *T–F*, there is a lesser tendency in the same direction. *F–P* tends to prefer negative phases. In each case the longest arm of the plot is diametrically opposite the shortest arm.

The length of each arm to the dotted line indicates the contribution of the first four dominant shared oscillations in each record. The addition beyond the dotted line is the contribution of oscillations ranked fifth. In both pairings which include *F* these frequencies are more strongly represented at 0° than at other phases; otherwise, they are roughly uniformly distributed around the circle, suggesting that a considerable proportion of oscillations of this low dominance are independently or randomly determined for each variable. Where there is a common external source for such low-ranking frequencies, we may suppose it is equally remote from the generators of both our measures. If so, the phase should be 0° or 180°, depending on the direction of the effect. The slight preponderance at 0° for *T–F* and *T–P* suggests that such remote effects increase *T* and *F* or *F* and *P* concurrently.

On analysis of variance, performed on the first four dominant frequencies, the difference in the distribution of phases between variable pairs was significant ($F(2,16) = 3.66$, $p < 0.05$). Rank was not significant, and there was no significant difference in the distribution of phases over variable pairs for the low-frequency range (2–20 cycles per hour) and the high-frequency range (20–60 cycles per hour; $F(1,8) = 0.02$).

These results have a number of implications. First, the existence of preferred phase relations indicates that some structural principle is at work relating the expression of at least some of the common frequencies. Second, consideration of the patterns shown by the phase plots indicates a restriction on the extent to which oscillations with a common origin are shared by all three of the variables studied here. If a given frequency is shared by all three variables, then the relative phase between P and F must conform to

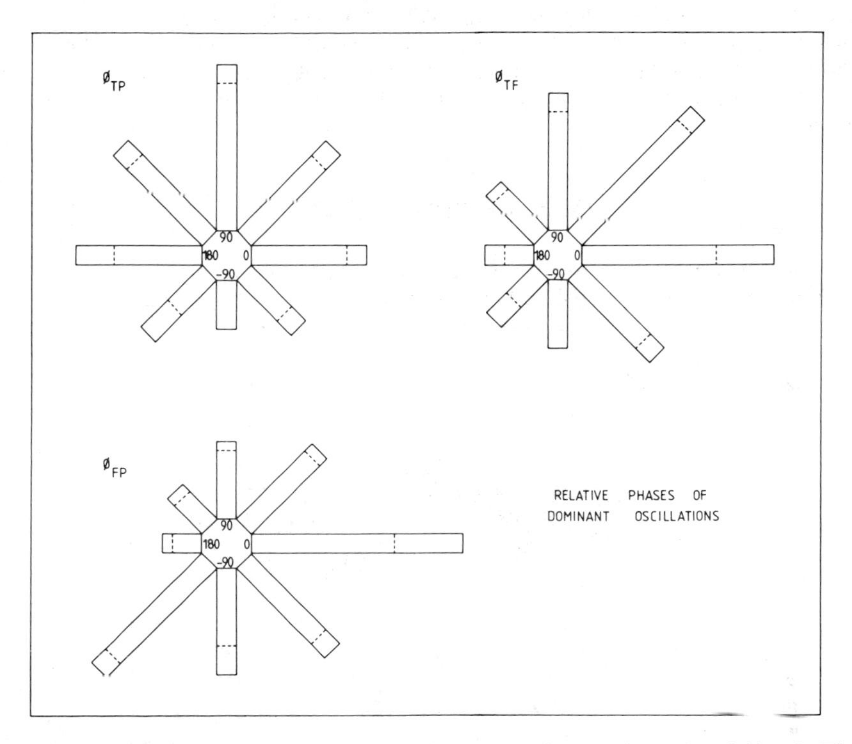

FIGURE 9. Plots of the phases between shared dominant oscillations for each variable pair. The length of each arm is proportional to the number of oscillations having relative phases lying within a range of 45°. Thus $\emptyset_{TP}$ represents the phase by which a given frequency in P is in advance of the same frequency in T. The arm at 0 represents the proportion of shared frequencies (for the nine sessions) in which $-22.5° < \emptyset_{TP} \leq 22.5°$. The total length of each arm is determined by the five dominant shared frequencies for each record. The length up to the *dotted line* is based on the first four dominant shared oscillations.

$\emptyset_{FP} = \emptyset_{TP} - \emptyset_{TF}$. It follows that if $\emptyset_{TP}$ is on average more positive than $\emptyset_{TF}$ (as is the case for the phases plotted in FIGURE 9), then $\emptyset_{FP}$ should be positive. But we see that it tends to be negative.

Third, the tendency for $\emptyset_{TF}$ to take a zero or small positive value provides further evidence against the common pacemaker hypothesis. It implies that a common oscillation causing a rise in alpha frequency will shortly thereafter be followed by lengthening of time productions, that is, a slowing of the temporal pacemaker.

Finally, the finding that the distribution of phases is almost identical for the low-frequency and high-frequency ranges of oscillations is of particular interest because it shows that we are not looking at the transmission of effects which depend simply on the lapse of time, and therefore will give larger phase differences for higher frequencies—but at effects that depend on oscillation frequency as such.

It would be unwise to attempt to derive directions of effects from these results. We see that a large proportion of frequencies shared between T and P have ϕ_{TP} positive. This could mean that a factor determining an increase in the prevalence of alpha shortly thereafter determines a slowing of the temporal pacemaker. But the reverse could be true, at the complementary lag. Also, if we consider not P but $(1 - P)$, the sign of the phase difference will reverse, and the implication may then be that a factor slowing the temporal pacemaker shortly later determines a decrease in P.

The results above lead immediately to two further questions which will now be considered.

TWO FURTHER QUESTIONS

We may ask: First, are the oscillations seen above produced only during the process of making time productions—induced in some way by the mechanism then active—or are

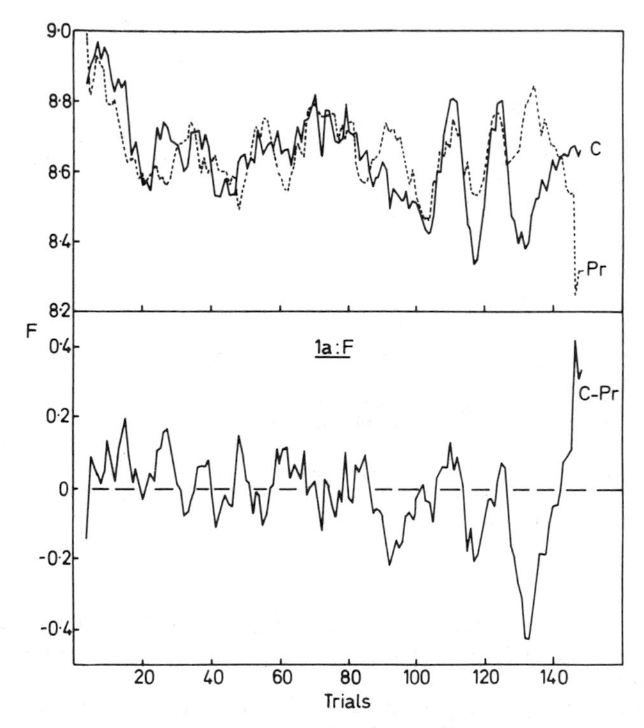

FIGURE 10. Session 1a. The *upper panel* shows values of F recorded during the 3 seconds immediately prior to a trial (*Pr, dashed lines*) or concurrent (*C, continuous lines*) with the performance of the time-production task. Values are moving averages, taken over a range of 7 trials. The *lower panel* gives the difference ($C - Pr$) between the two upper curves.

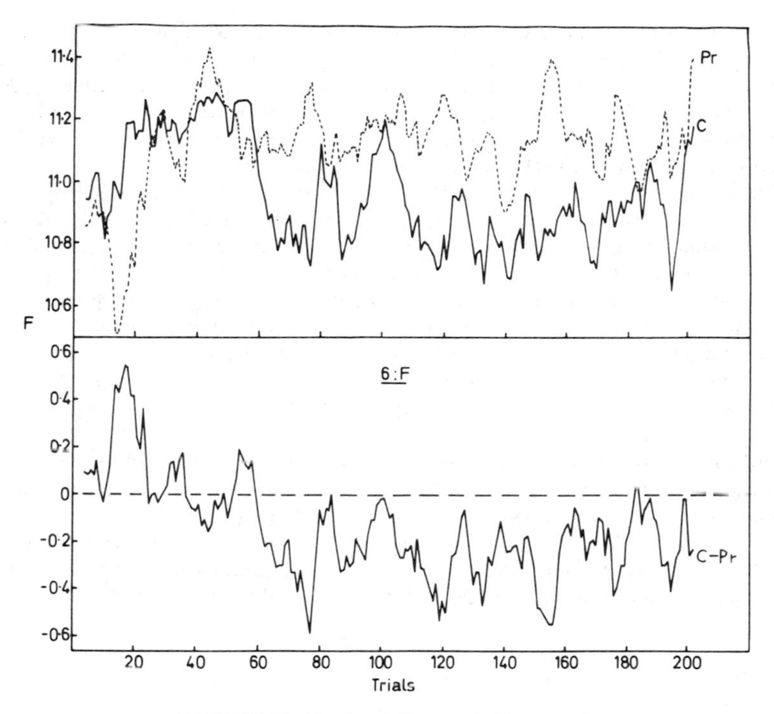

FIGURE 11. Session 6. Data as in FIGURE 10.

they continuously present, whether or not the subject is engaged in this task? Second, do oscillations affect only the three variables we have recorded, or can they be seen in other measures?

To examine the first question, values of P and F were measured for the 3 seconds immediately preceding each trial for two sessions (1a and 6). Some results are shown in FIGURES 10, 11, and 12. The traces obtained concurrent (C) with the time production task and prior (Pr) to it are plotted in the upper panel of each figure, and the differences between these values ($C - Pr$) are plotted in the lower panel.

Oscillations are shown by prior measures as well as by concurrent measures. But although the traces are similar, they are not simply parallel. The divergences between them are revealed by plotting the difference between the two traces ($C - Pr$): These plots show well-marked relatively regular oscillations. These represent frequencies that are of greater amplitude in one trace than in the other. In FIGURES 10 and 12 the C and Pr traces are generally similar in form. In FIGURE 11, $C - Pr$ shows a slow drift—initially the experimental task tends to increase F, later it reduces it.

For F in session 1a (FIGURE 10) the dominant trace (in the range 2–60 cycles per hour) for the prior record has a frequency of 11.9 cycles per hour. This frequency is weak, if present at all, in the current trace: it ranks only 29th. But a power spectrum analysis of the difference between the two records, $F(C - Pr)$, again assigns 11.9 cycles per hour the highest rank, showing that it is strong in one record and weak or absent in the other. Thus, it appears that a marked resting variation of about 12 cycles per hour in the alpha frequency was suppressed during performance of time productions in session 1a.

FIGURE 11 shows similar results for F in session 6. The clearly marked oscillation in

$F(C - Pr)$, plotted in the lower panel, is reminiscent of that seen in FIGURE 10. The power spectrum analysis for this difference curve assigned the highest rank to a frequency of 11.9 cycles per hour. This oscillation frequency was negligible for $F(C)$, ranking 37th, but for $F(Pr)$ it was 8th. Here again it seems that a variation in alpha frequency occurring about 12 times an hour is manifest in the inter-trial intervals, but is suppressed during task-performance.

Similar analyses were made for the prevalence of alpha. In session 1a, the two highest ranking oscillations in the current trace were 3.4 and 5.1 cycles per hour (corresponding to 2 and 3 cycles per session); but these ranked 12th and 30th, respectively, in the prior trace. The dominant prior oscillation was 47.7 cycles per hour; this ranked 26th in the current trace. These three oscillation frequencies were dominant for $P(C - Pr)$. This suggests that the two low frequencies may have represented adaptations to the experimental session which occurred during the performance of the task but not in the intervals between time productions. But a rapid rhythm with a period a little over a minute occurred in the resting state, but was suppressed during performance.

FIGURE 12 shows the corresponding results for session 6. The difference curve for prevalence of alpha has a strongly marked, fairly rapid oscillation. This represents a dominant frequency of 15.5 cycles per hour. This oscillation is shown both by $P(C)$ (ranking 2nd) and $P(Pr)$ (ranking 3rd). It appears more strongly marked in $P(C)$ and leads by 52° in $P(Pr)$. Thus, its prominence in the difference trace may be due to a phase lag and possibly an increase in amplitude induced by task-performance.

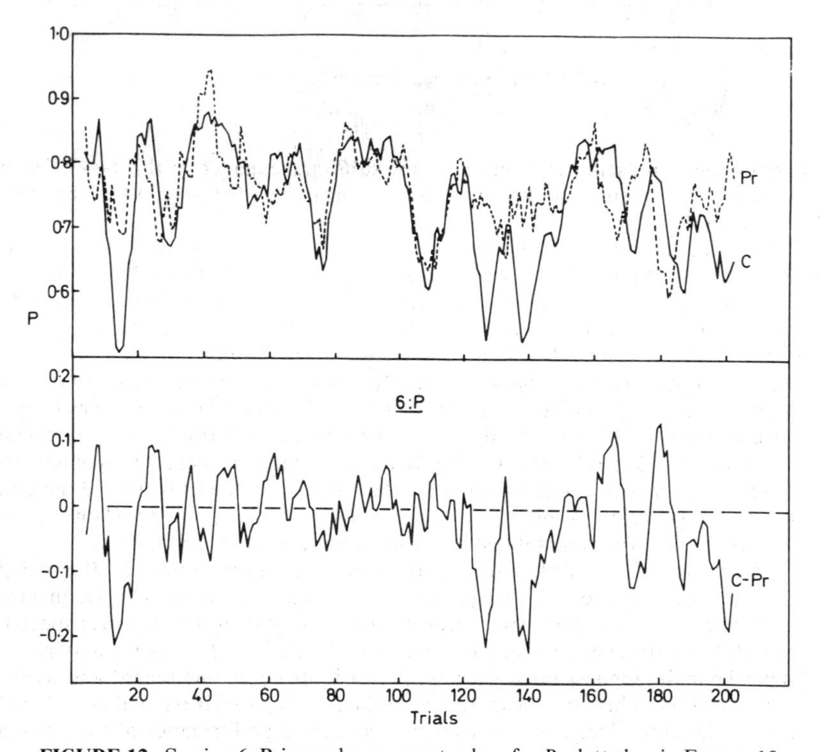

FIGURE 12. Session 6. Prior and concurrent values for P, plotted as in FIGURE 10.

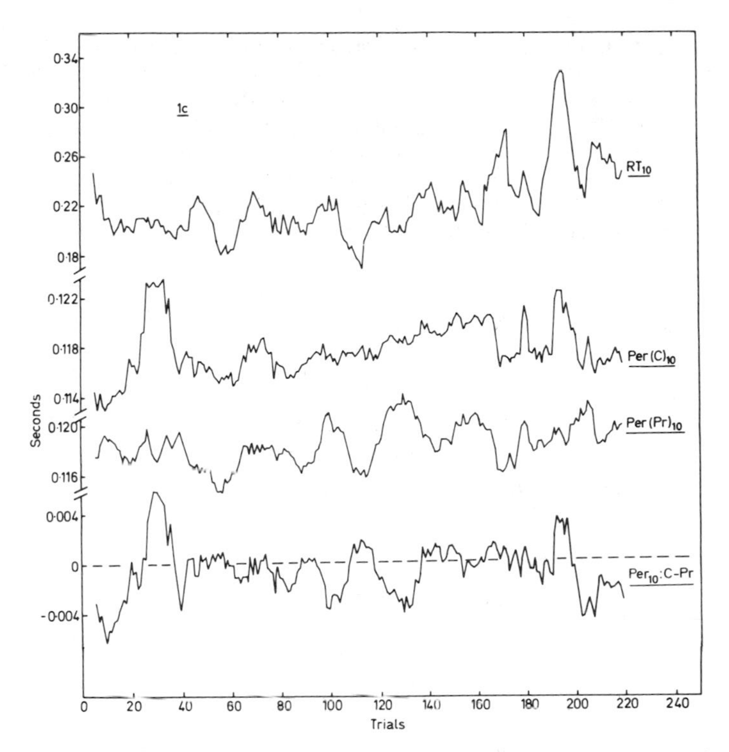

FIGURE 13. Session 1c. The simple reaction time to the onset of a tone, RT_{10}, the moving average taken over a range of 10 trials, is plotted against trials, as are *Per*(*C*), the average alpha period during 2 sec commencing at the start of the tone; *Per*(*Pr*), the alpha period for the 2 sec prior to the onset of the tone; and Per_{10}:C – *Pr*, the difference between these two measures.

These results suggest that the system producing time intervals may amplify, suppress or phase-shift oscillations. The oscillations dominating the difference curves have periods much longer than a single trial. Time production involves activity both of the temporal pacemaker and of motor response mechanisms. Oscillations in the lengths of time productions must be attributed to the temporal pacemaker: they are far greater than any concurrent variations in reaction time.[2] But it is possible that effects may also arise in the motor response mechanisms themselves. Two further sessions were run in which subjects made simple reaction times, and their prior and concurrent alpha frequency was recorded.

In each case the session followed shortly after a time production session. Session 1c followed 1b, and 5b followed 5. The procedure was the same as before, except that the subject was asked to press the reaction key as soon as the stimulus came on. In session 1c the stimulus was the onset of the 50 dB SL tone. In session 5b it was the onset of a continuous light produced by a Dawe stroboscope. Thus, any similarity in the results cannot be attributed to the nature of the stimulus. In each session the subject sat with his eyes closed, and his response terminated the stimulus. Session 1c lasted 25 minutes and session 5b 20 minutes; trials were given at an average rate of 9 trials per minute and 8.2 trials per minute, respectively. Four percent of trials were lost because of artefacts or failures. No correction was made for their loss. Reaction time (*RT*) was measured as the interval from the onset of the stimulus to the onset of the muscle contraction artefact from the subject's forearm. On each trial the mean period of the alpha rhythm was scored for a 2-sec interval immediately preceding the stimulus—*Per*(*Pr*)—and for the 2-sec period beginning at the stimulus onset, giving *Per*(*C*). Of course, production of the response occupied only a small proportion of this period. *P* was not recorded since the response artefact would have made *P*(*C*) uncertain. The results are shown in FIGURES 13 and 14. *Per* is plotted rather than *F* since it has been suggested that alpha period may determine reaction times.[15]

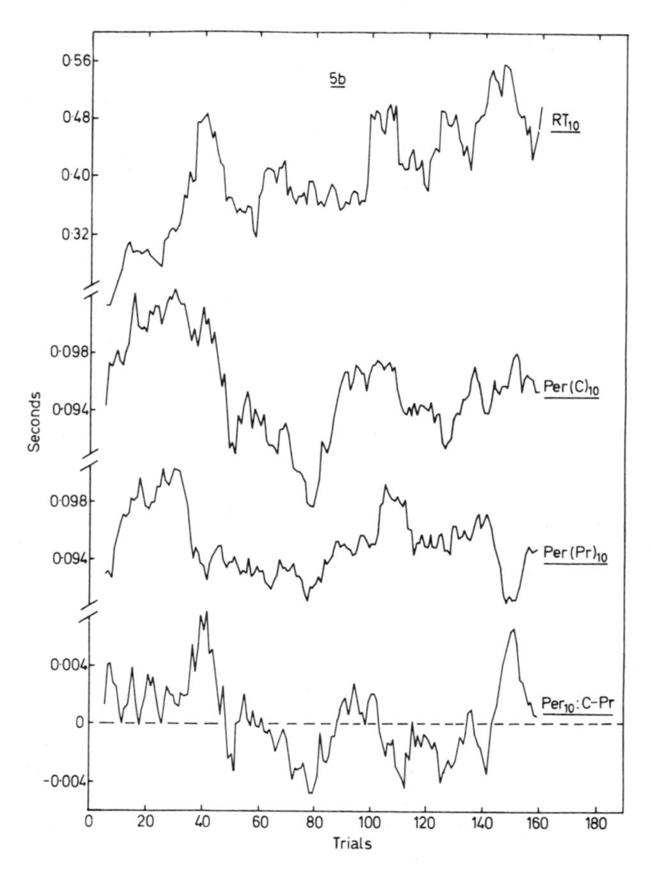

FIGURE 14. Session 5b. As FIGURE 13; reactions were made to the onset of a light.

The power spectra for session 1c give results for *F*(*C*) and *F*(*Pr*) which are similar but not identical. In both cases the dominant oscillation is 40.8 cycles per hour. For reaction time the two dominant frequencies are 28.8 and 26.4 cycles per hour. *F*(*C* – *Pr*) is dominated by 28.8, 26.4 and 40.8 cycles per hour. Thus similar frequencies can be found in all the variables. The presence of the frequencies 28.8 and 26.4 cycles per hour in the difference curve does not result from their suppression in one of the *F* curves: they are present in both *F* curves in similar strength, but 28.8 cycles per hour leads in the prior trace by 109° and 26.4 leads by 118°.

In session 5b, low frequencies (6.0 and 9.0 cycles per hour) are strongly represented for all the variables. The dominant frequency for *RT* is 42.0 cycles per hour, and for *F*(*C* – *Pr*) it is 30.0 cycles per hour. This frequency is present in the current and prior *F* traces at similar strength, but leads in the current trace by 94°.

It is evident that *RT* is not simply determined by alpha period. Spontaneous variations occur in *RT* that are not shown in *Per*(*C*) or *Per*(*Pr*), and oscillations occur in the latter that are not reflected in the former. The product–moment correlation between *RT* and *Per*(*C*) is $r = 0.15$ ($p < 0.05$) for session 1c, and $r = -0.06$ (NS) for session 5b. Surwillo[15] reported mean intraindividual correlations between reaction time and EEG period of 0.30; and Woodruff[16] reported correlations ranging from –0.31 to 0.35, with a mean of 0.02. In view of the occurrence of oscillations, closer agreement between experiments would not be expected.

We see that reaction time is also subject to oscillations, and that the reaction time task is sufficient to induce differences between the oscillations shown by the current and immediately preceding alpha rhythm.

DISCUSSION

We have found that oscillations, in the range 2–60 cycles per hour studied here, and presumably in other ranges, can be seen in alpha frequency, the prevalence of alpha in the record, in time productions, in simple reaction time, and in the relations between measures of *F* or *P* taken during or immediately prior to performance of a task. These are unfamiliar phenomena, whose implications for the understanding of behavior remain to be examined. They require explanation: unfortunately, there is no space here to discuss this. A theoretical treatment will be presented elsewhere that accounts for the occurrence of such oscillations.[17] It deals with the processes by which competing response mechanisms share out the control of behavior. Each such mechanism is assumed to vary in its specific arousal. The oscillations in *T*, *F* and *P* would reflect variations in the specific arousal of the mechanisms that determine them.

CONCLUSIONS

1. The common pacemaker hypothesis cannot be sustained. The stability of the alpha rhythm is much greater than that of temporal productions over the course of a session. The variability of the temporal pacemaker greatly exceeds that of the alpha rhythm, as shown by the coefficients of variation. The pattern of correlations between *F* and *T* is inconsistent with the hypothesis. We have also found that the variation in *T* and *F* includes unshared oscillation frequencies, which cannot be explained if they derive from a single pacemaker. As the temporal pacemaker is at present understood, the hypothesis of identity with an alpha rhythm generator must be rejected.

2. The hypothesis that specific temporal arousal is simply the local expression of general arousal is not supported. This hypothesis assumes that *F* and *P* can be taken as indices of general arousal, and therefore that the specific arousal of the temporal pacemaker can be predicted from them. But the correlations obtained are incompatible with the predictions of this theory. The relations found between *F* and *P* also present grave difficulties for the hypothesis of a single dimension of arousal, of which these are indicators.

3. Spontaneous oscillations were found to occur in measures of alpha rhythm, alpha prevalence, and temporal productions taken concurrently. Oscillations in the

range 2–60 cycles per hour were studied by power spectrum and cross-spectral analyses.

4. Dominant oscillations are of lower frequency for *T* than for *P*, and for *P* than for *F*. They also tend to be of lower frequency for high-ranking oscillations.

5. Pairs of variables may share oscillations. Dominant shared oscillations tend to be representative of both parent variables. Their frequencies increase as their rank declines. *P* tends to lead *T* in phase, *F* tends to lead *T* in phase but more weakly, and *F* tends to lead *P* in phase. This applies equally to shared oscillations of low or high frequencies.

6. Task performance affects the oscillations. Common oscillations are found in prior and concurrent traces of *F* and *P*, but there are also differences in the oscillations they show. In two sessions, a frequency of 12 cycles per hour in the prior *F* trace was in each case suppressed in the concurrent *F* trace.

7. Oscillations are found both in reaction time and in concurrent and prior alpha frequency, but there is no strong relation between these variables. In two reaction time sessions, an oscillation of 29–30 cycles per hour in *F* was phase-shifted between the prior and current traces.

8. The existence of independent fluctuations in *F* and *P* indicates that we must reject a simple model of alpha production in which the frequency and prevalence of alpha both reflect the degree of activation of a single alpha rhythm generator. To some extent at least, alpha frequency and prevalence are determined by different systems.

SUMMARY

A model for the internal clock is briefly described. It includes a temporal pacemaker whose rate determines time judgments, and whose frequency is affected by arousal specific to it. Three hypotheses relating time judgments and the alpha rhythm are considered: (*a*) They may be wholly independent, each reflecting the specific arousal of the mechanism determining it. (*b*) The alpha rhythm may be an index of a state of general arousal which also acts on the temporal pacemaker. Because of this common influence, the alpha frequency, and the proportion of alpha in the electroencephalogram, may be correlated with the speed of the temporal pacemaker. (*c*) The same pacemaker may be common to the internal clock and an alpha rhythm generator.

Concurrent observations on alpha frequency, alpha prevalence, and temporal productions show that there are no simple relations between these measures such as might support the general arousal or common pacemaker hypotheses. However, relations are found between the variables. More or less regular oscillations occur in their values, some of which are common to two or more of the variables studied. These phenomena are further investigated and described.

ACKNOWLEDGMENTS

I would like to thank T. Walsh and T. R. Watts for assistance in running the experiment, A. Faulkner for doing the power spectrum analyses, and D. A. Allport for useful comments.

REFERENCES

1. Hoagland, H. 1933. The physiological control of judgments of duration: Evidence for a chemical clock. J. Gen. Psychol. **9:** 267–287.

2. TREISMAN, M. 1963. Temporal discrimination and the indifference interval: Implications for a model of the "internal clock." Psychol. Monogr. **77:**1–31. (whole no. 576).
3. GOLDSTONE, S., W. K. BOARDMAN & W. T. LHAMON. 1959. Intersensory comparisons of temporal judgments. J. Exp. Psychol. **57:** 243–248.
4. LANGER, J., S. WAPNER & H. WERNER. 1961. The effect of danger upon the experience of time. Am. J. Psychol. **74:** 94–97.
5. BRUMLIK, J., W. B. RICHESON & J. ARBIT. 1967. The origin of certain electrical cerebral rhythms. Brain Res. **3:** 227–247.
6. LINDSLEY, D. B. 1960. Attention, consciousness, sleep and wakefulness. *In* Handbook of Physiology. Section 1: Neurophysiology. Vol. III. J. Field, Ed.: 1553–1593. American Physiological Society. Washington, DC.
7. LOOMIS, A. L., E. N. HARVEY & G. HOBART. 1936. Electrical potentials of the human brain. J. Exp. Psychol. **19:** 249–279.
8. WERBOFF, J. 1962. Time judgment as a function of electroencephalographic activity. Exp. Neurol. **6:** 152–160.
9. SURWILLO, W. W. 1966. Time perception and the "internal clock": Some observations on the role of the electroencephalogram. Brain Res. **2:** 390–392.
10. HOLUBAR, J. 1969. The Sense of Time. MIT Press. Cambridge, MA.
11. ADAM, N., B. S. ROSNER, E. C. HOSICK & D. L. CLARK. 1971. Effect of anesthetic drugs on time production and alpha rhythm. Percept. Psychophys. **10:** 133–136.
12. HILL, D. & G. PARR Eds. 1950. Electroencephalography. Macdonald. London.
13. BENDAT, J. S. & A. G. PIERSOL. 1966. Measurement and Analysis of Random Data. Wiley. New York, NY.
14. FISHER, R. A. 1954. Statistical Methods for Research Workers, 12th ed. Oliver and Boyd. Edinburgh.
15. SURWILLO, W. W. 1963. The relation of simple response time to brain-wave frequency and the effects of age. EEG Clin. Neurophys. **15:** 105–114.
16. WOODRUFF, D. 1975. Relationships among EEG alpha frequency, reaction time, and age: A biofeedback study. Psychophysiology **12:** 673–681.
17. TREISMAN, M. 1984. A theory of response selection. Psychol. Rev. In press.

Properties of the Internal Clock[a]

RUSSELL M. CHURCH
Walter S. Hunter Laboratory of Psychology
Brown University
Providence, Rhode Island 02912

In many testing procedures, the behavior of an animal is a function of the time since a stimulus began. One type of explanation of such behavior is a psychological process model that includes several modules, including an internal clock. The purpose of this paper is to describe the properties of this internal clock, and other parts of a psychological process model, on the basis of a review of some relevant experimental studies.

SOME ANIMAL TIMING PROCEDURES

Because our aim is to describe properties of functional parts of an animal, not properties of a procedure, it is desirable to examine the performance of animals in several different procedures with different characteristics. Most of the research in our laboratory has used one of three procedures: the temporal generalization, the peak, or the bisection procedure. The subjects were rats. The testing apparatus consisted of 10 standard operant boxes, each with two retractable levers. Computers were used in all phases of the experiment: to control the procedures, record the results, analyze the data, and so forth (More detail on the use of computers in these experiments is available.[1])

For the *temporal generalization procedure,* on each trial a signal is presented for some duration, and then a lever is inserted into the box. If the signal is of some particular period (for example, 4 sec) and the rat presses the lever, food is delivered. If the signal is shorter or longer than this duration, no food is delivered following a response. If the rat does not make a response in some short period (for example, 5 sec), the lever is withdrawn. After a relatively long intertrial interval (for example, 30 sec) another trial is begun. Typically, a session lasts 90 min and about half the signals are of the reinforced duration. The response measure is the probability of a response as a function of signal duration. The function usually rises to a maximum near the reinforced signal duration and then falls in a fairly symmetrical fashion with a slight positive skew when the data are plotted on a linear time scale. Results are shown in FIGURE 1 for 26 rats divided into five groups on the basis of overall response probability.[2]

For the *peak procedure,* a trial begins with the onset of a signal. On some trials food is primed after a particular interval (such as 50 sec) and delivered after the next response; on other trials no food is primed and the signal lasts for a relatively long period of time (such as 130 sec). After a long and variable intertrial interval (for example, a random interval with a mean of 130 sec), another trial is begun. The dependent variable is the response rate, sometimes relative to the maximum response

[a]The behavioral research reported in this paper has been supported by Grants BNS 79-0492 and BNS 82-09834 from the National Science Foundation and the pharmacologic research has been supported by Grant MH 37049 from the National Institute of Mental Health.

rate, as a function of the time since the trial began. The function typically rises to a maximum near the time that food may be delivered, and then it falls in a fairly symmetrical fashion with a slight positive skew to an asymptotic rate slightly above 0 when the data are plotted on a linear time scale.[3] Results for 10 rats are shown in FIGURE 2.[4]

For the *bisection procedure,* a signal is presented for some duration (between 2 and 8 sec, for example). Then two levers are inserted into the box. If the signal was short (for example, 2 sec) and a left response is made, or if the signal was long (for example,

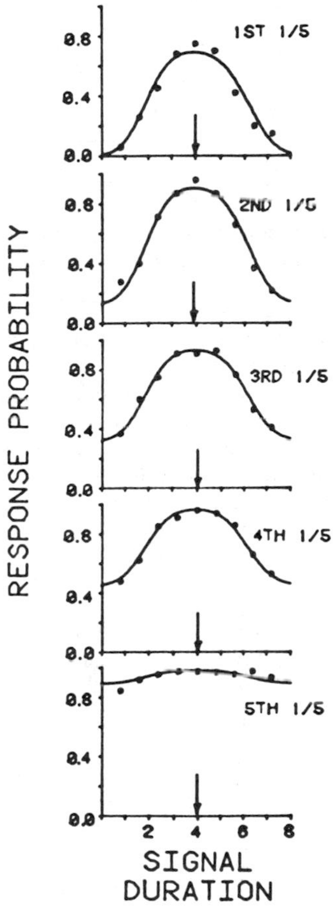

FIGURE 1. Temporal generalization procedure. Median probability of a response as a function of signal duration for five quintiles of individual subjects, ordered in terms of the overall probability of response. (From Church and Gibbon.[2] Reprinted by permission.)

8 sec) and a right response is made, food is delivered. No food is delivered after responses to signals intermediate in duration between the extreme signals. After an intertrial interval (for example, 30 sec), another trial is begun. The response measure is the probability of a right ("long") response as a function of signal duration. It typically rises in an ogival fashion that is fairly symmetrical on a logarithmic time axis. Our method[27] has been slightly modified from one used previously.[31] Results are shown in FIGURE 3 for groups of rats trained with extreme signals of 1 and 4 sec, 2 and 8 sec, and

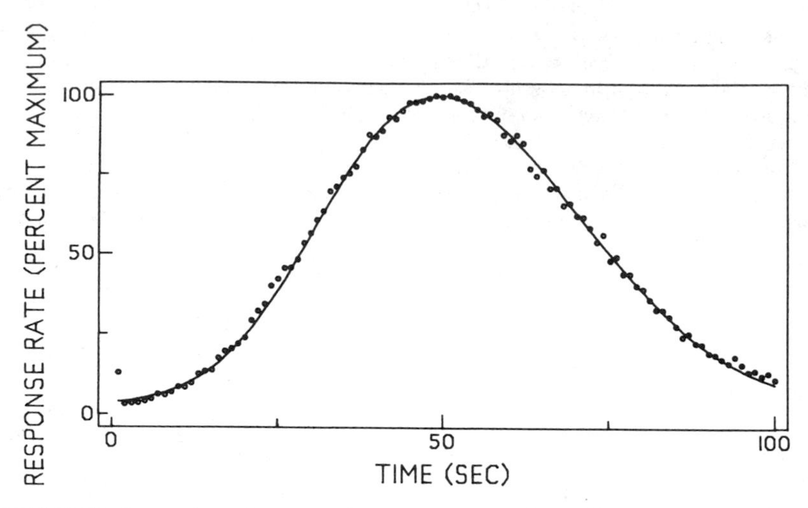

FIGURE 2. Peak procedure. Percent maximum response rate for median responses per minute as a function of time into the interval. (From Meck and Church.[4] Reprinted by permission.)

4 and 16 sec.[5] (A description and interpretation of the effects of drugs on results will be given later.)

These procedures differ in many ways, such as the number of reinforced signal durations (one or two), whether or not the animal has the opportunity to respond during the signal, the number of response alternatives (one or two), and the response measure (probability or rate), among other ways. Some characteristics of these three procedures are listed in TABLE 1. Many other combinations of these characteristics

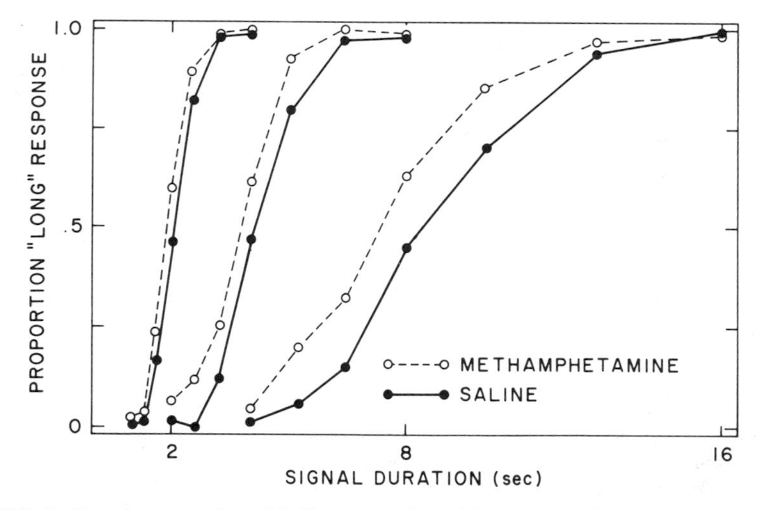

FIGURE 3. Bisection procedure. Median proportion of long response as a function of signal duration for three signal ranges (1 versus 4, 2 versus 8, and 4 versus 16 sec) for sessions with saline solution and methamphetamine. (From Maricq *et al.*[5] Reprinted by permission.)

define other useful timing procedures. For example, an alternative bisection procedure is one in which there are two reinforced signal durations and two response alternatives, but the animals have the opportunity to respond during the signals and response rate is monitored.[6] Other values of the listed characteristics also define reasonable procedures. For example, in the bisection procedure, latency to make a response is a response measure that is related in an orderly way to signal duration.[5]

In the temporal generalization and bisection procedures the animal times the signal duration and then makes a single response (yes–no or left–right). In the peak procedure the time of the response is the relevant characteristic. The properties that will be described are designed to apply to both signal timing and response timing procedures. One reason to believe that the same processes are used for signal and response timing is that they are mutually interfering. If a trial consists of a temporal generalization task in which the positive signal is t_1 sec, followed by a peak procedure in which food is sometimes delivered following a response t_2 sec after the lever is available, rats confuse the signal duration and the response duration.[33]

All of the properties described here have been studied with more than one procedure. A property that is identified in several procedures provides evidence that the property applies to the animal. Therefore, it is desirable to examine the performance of animals in several procedures with different characteristics. The development of a good reference experiment, however, is time-consuming, and it is

TABLE 1. Some Characteristics of Three Timing Processes

Procedure	No. of Reinforced Signal Durations	Response Opportunity during or after Signal	No. of Response Alternatives	Response Measure
Temporal generalization	1	After	1	Probability
Peak	1	During	1	Rate
Bisection	2	After	2	Probability

desirable to have extensive experience with any procedure to know what variables matter and to be able to recognize small deviations from the usual results. Thus, there are reasons to use few different procedures. As a compromise, we have restricted most of our experiments to the three procedures described above (temporal generalization, peak, and bisection), but we have studied most properties in more than one procedure. All of the properties described here have been reported in the research literature and many of them have been summarized.[7]

To describe properties of functional parts of an animal, it is necessary to develop a process model explanation. An explanation is a deductive exercise in which various facts are available and various rules can be used to deduce the empirical statements to be explained. In a process model if–then rules are strung together, at least in part, sequentially. That is to say, the input can activate some rule that activates some other rule before there is final activation of the output.

There are three types of processes: psychological, biological, and formal (FIG. 4.) A psychological process is one in which mental terms are used, such as subjective time, memory, and decision; a biological process model is one that refers to physiological events; a formal process model is one in which abstract, often mathematical, terms are used that do not necessarily refer either to mental or biological events. The three process models, however, are closely connected.

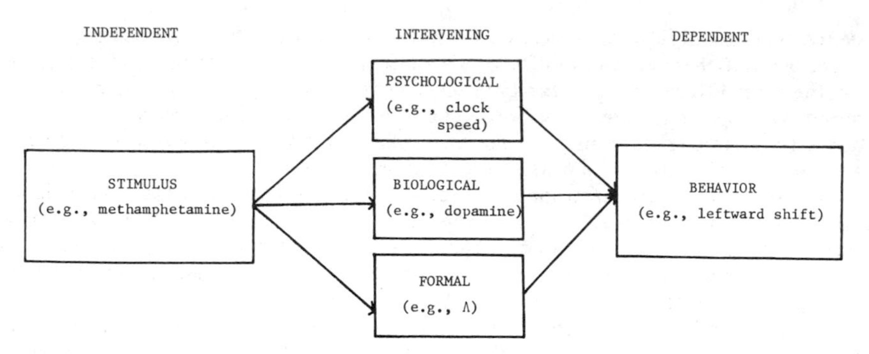

FIGURE 4. Three process models: psychological, biological, and formal.

Let us consider a specific result. In the temporal bisection procedure, under our standard conditions, we have found that the percentage of "long" response rises in an ogival manner as a function of signal duration, and that the point of indifference (50% "long" response) is near the geometric mean. When methamphetamine is administered, however, this psychophysical function shifts to the left in a proportional manner (FIG. 3.) Any drug effect on behavior can be interpreted in psychological, pharmacological, or mathematical terms. This particular drug effect on behavior can be interpreted in terms of a change in a psychological process (such as clock speed), a neurotransmitter (such as dopamine), or a parameter of a scalar timing model (such as Λ). If the effect of a particular drug can be interpreted either in psychological, pharmacologic, or mathematical terms, presumably there is an important relationship between psychological processes, neurotransmitters, and parameters of the formal model.

The psychological model that we previously described[8] and that we extended earlier in this volume[32] is composed of four major parts: clock, working memory, reference memory, and comparator. The clock is composed of a pacemaker, switch, and accumulator. FIGURE 5 shows one way in which these parts may be interrelated.

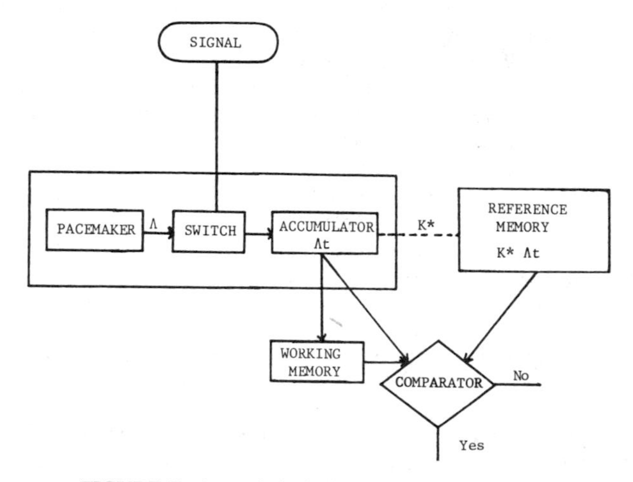

FIGURE 5. A psychological process model of timing.

The pacemaker emits pulses that are switched into an accumulator. The pulses related to the signal on the current trial are compared to a remembered number of pulses in reference memory that led to reinforcement. The comparator can combine the value from accumulator with a value in reference memory according to a response rule to make a decision. If it is close enough, a response is made; otherwise no response is made. If a response is made and reinforced, the value is stored in reference memory. A value in the accumulator (a perceptual store) may also be transferred to working memory (a short-term memory store) and the comparator can combine a value in working memory with a value in reference memory according to a response rule to make a decision. There may be variability at any stage, and there can be a transformation of the value between one stage and another. A multiplicative transformation, Λ, is shown between the pacemaker and switch, a value we will call "clock speed" and another multiplicative transformation, K*, is shown between working memory and reference memory, a value we will call the "memory constant."

The evaluation of a psychological model of this sort should involve something other than reasonableness and familiarity, which are probably closely related. It should be found to be necessary by the "double dissociation" test. That is to say, there should be some operations that change one of the processes (for example, Λ) and leave another one unaffected (for example, K*), and there should be some other operations that change the other process (K*) and leave the first one unaffected (Λ).

TABLE 2. Dissociation of Clock Speed (Λ) and the Memory Constant (K^*)

	Clock Speed	Memory Constant	Accumulator	Reference Memory	Maximum Response
Training	Λ_1	K_1^*	$\Lambda_1 t$	$K_1^* (\Lambda_1 t_{rf})$	$t = K_1^* t_{rf}$
Test	Λ_2	K_2^*	$\Lambda_2 t$	$K_1^* (\Lambda_1 t_{rf})$	$t = K_1^* (\Lambda_1/\Lambda_2) t_{rf}$

Meck[9] has provided a way to distinguish between clock speed (Λ) and the memory constant (K*). During training, a signal of t sec will produce a value of Λ_1 t in the accumulator, where Λ_1 is clock speed during training. The mean value in reference memory will be $K_1{}^* (\Lambda_1 t_{rf})$, where $K_1{}^*$ is the memory constant during training and t_{rf} is the signal duration that leads to reinforcement (TABLE 2, top row). If, on any given trial, the animal is maximally responsive at a time when the value in working memory equals the mean value in the accumulator, it is maximally responsive when:

$$t = K_1{}^* t_{rf} \tag{1}$$

During testing, a signal of t sec will produce a value of Λ_2 t in the accumulator where Λ_2 is the clock speed during testing. The mean value in reference memory, of course, will be the same as during training (TABLE 2, bottom row). If, on any given trial the animal is maximally responsive at a time when the value in the accumulator equals the mean value in reference memory, it is maximally responsive when:

$$t = K_1{}^* \Lambda_1/\Lambda_2 t_{rf}. \tag{2}$$

Suppose that there was an operation that affected only clock speed (Λ). During training it could not be detected in the mean (Eq. 1), but a change in clock speed could be detected during testing (Eq. 2). If an operation increased clock speed during training relative to testing, then $\Lambda_1 > \Lambda_2$, the animal would be maximally responsive during testing at a time *later* than the time of reinforcement.

Suppose, on the other hand, that there was an operation that affected only the memory constant (K^*). During training it could be detected in the mean (Eq. 1). If an operation increased the memory constant during training, the animal would be maximally responsive at a time later than the time of reinforcement. During testing, the effect of the memory constant during training (but not testing) would still be apparent.

Some operations differentially affect clock speed and others differentially affect the memory constant, as defined above[9] (FIG. 6). Methamphetamine apparently increases clock speed (Λ). There is no difference in the psychophysical function of rats trained with methamphetamine and saline solution, but there is a leftward shift (an increase in clock speed) when rats trained on saline solution are tested on methamphetamine. And there is a rightward shift of equivalent magnitude (a decrease in clock speed) when rats trained on methamphetamine are tested on saline solution.[9] Vasopressin, in contrast, apparently decreases the memory constant (K^*) (FIG. 7). The psychophysical function of rats trained with vasopressin is shifted to the left of the psychophysical function of rats trained with saline solution. The conditions of training, not testing, determined the performance during testing.

Thus, different drugs affect different parts of the process. This is important evidence that the parts are separate. Roberts has previously used the double dissociation method, with dependent variables rather than parameters, to provide evidence for the separateness of parts.[3]

PROPERTIES

We will now proceed through the psychological process model, commenting on some of the probable properties of the various parts. This section will provide a review of

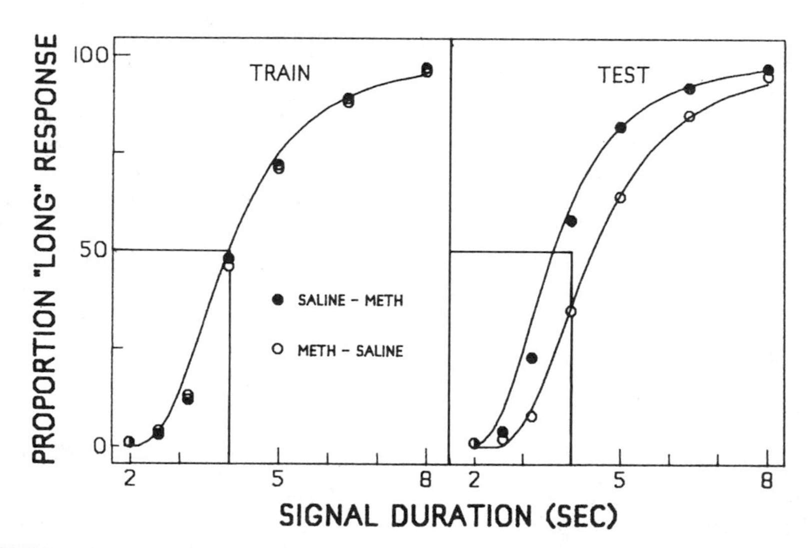

FIGURE 6. Selective adjustment of clock speed by methamphetamine. Median proportion long response as a function of signal duration during training and testing. (From Meck.[9] Reprinted by permission.)

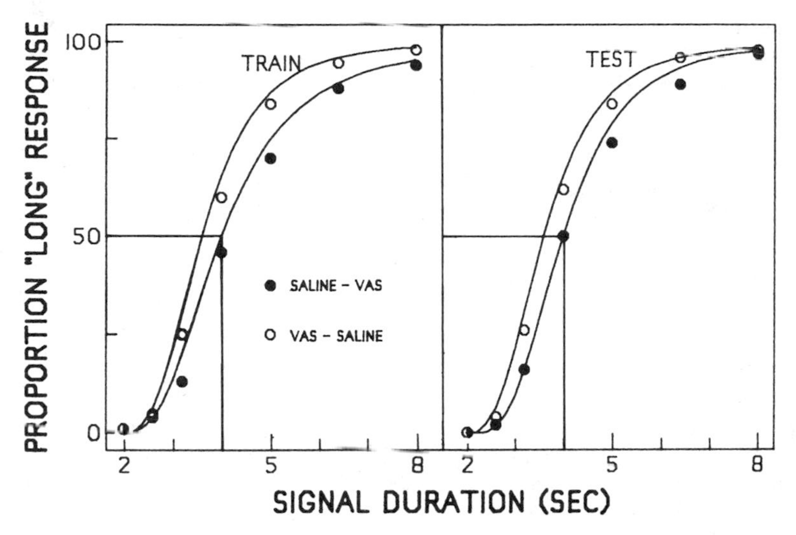

FIGURE 7. Selective adjustment of memory constant by vasopressin. Median proportion long response as a function of signal duration during training and testing. (Adapted from Meck.[9])

evidence that there are operations that affect single parts of the proposed psychological processing model, and it will describe some of the properties of these parts. A property of a part refers to a quality that is found in all exemplars of this part, but which is not a defining characteristic.

Pacemaker

The defining characteristic of the pacemaker is that it is an internal mechanism that generates pulses.

One property of the pacemaker is that its mean rate can be controlled by various drug, dietary, and environmental manipulations. For example, as we have seen, methamphetamine produces a leftward shift in the psychophysical function that relates probability of a "long" response to signal duration.[5,9,10] The leftward shift produced by methamphetamine has also been observed in the peak procedure.[5] In a psychological model, both of these can be interpreted as follows: Methamphetamine increases pacemaker speed, and this produces a leftward shift of the function. Other drugs, such as haloperidol, can produce a rightward shift in the psychophysical function that relates probability of a long response to signal duration.[9,10] In a psychological model, this can be interpreted as follows: Haloperidol decreases pacemaker speed and this produces a rightward shift of the function. Apparently, the pacemaker rate varies with the effective level of dopamine. In a biological model, the interpretation is as follows: Drugs (for example, methamphetamine) that increase the release or decrease the reuptake of dopamine increase the effective level of dopamine and produce a leftward shift in the psychophysical function; drugs (such as haloperidol) that block dopamine receptors decrease the effective level of dopamine and produce a rightward shift in the psychophysical function.

Similar effects can be produced by dietary influences. Prefeeding the standard diet to rats slows down the pacemaker, that is, leads to a rightward shift in the peak procedure.[3] The standard diet is high in carbohydrate, and prefeeding with a carbohydrate (sucrose) also decreases pacemaker speed; prefeeding with a protein (casein), however, increases pacemaker speed, that is, leads to a leftward shift.[11] Footshock stress produces shifts in the bisection function that can be interpreted psychologically as follows: There is an increase in pacemaker speed during stress and a decrease in pacemaker speed (below baseline) when stress is removed.[9] Thus, one property of the pacemaker is that its mean rate can be influenced by pharmacologic manipulations induced by drugs, diet, and other means.

To account for the ability of an animal to discriminate small differences in time, the pacemaker must be rapid; to account for the reliability of temporal discriminations, the pacemaker must be stable and relatively unaffected by environmental change. In quantitative applications of scalar timing theory,[32] we have assumed that the pacemaker generates a distribution of pulses that is Poisson, and that there is some drift between trials such that the pulse rate is normally distributed with some mean and standard deviation. Although we can estimate the ratio of the standard deviation to the mean (the coefficient of variation) and the amount of change in the mean clock speed, we do not currently have a way to measure the absolute clock rate. Treisman,[34] in this volume, proposes that the alpha rhythm may provide such a measure; we are exploring the use of lick rate of the rat. In any case, a measure of the pacemaker and of temporal judgment should (a) be affected by the same manipulations to an equivalent amount, and (b) moment-to-moment variation in a measure of the pacemaker should correlate with moment-to-moment variation in temporal judgment.

Switch

The defining characteristic of the switch is that it gates pulses from the pacemaker to the accumulator. In many cases, as described in this volume,[32] there is a short latent period between the onset of a stimulus and a switch closure (t_1) that exceeds the latent period between the termination of the stimulus and the reopening of the switch (t_2).[8,12,13] Meck[37] found that the difference ($T_0 = t_1 - t_2$) is affected by the nature of

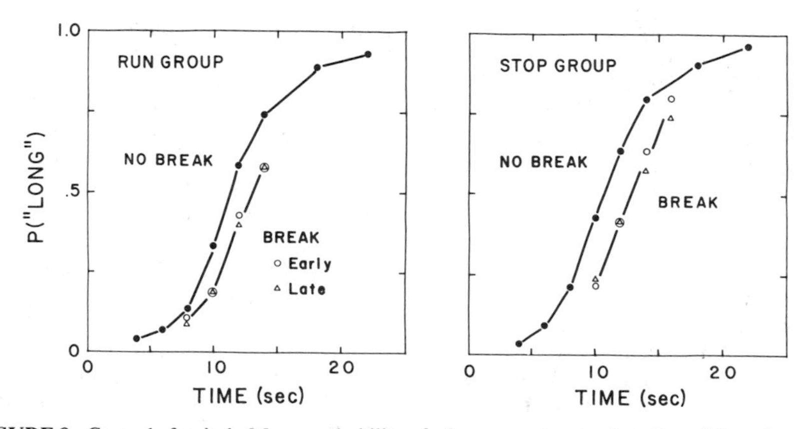

FIGURE 8. Control of switch. Mean probability of a long response as a function of time since the start of signal on trials with and without a 2-sec break. (From Roberts and Church.[15] Reprinted by permission.)

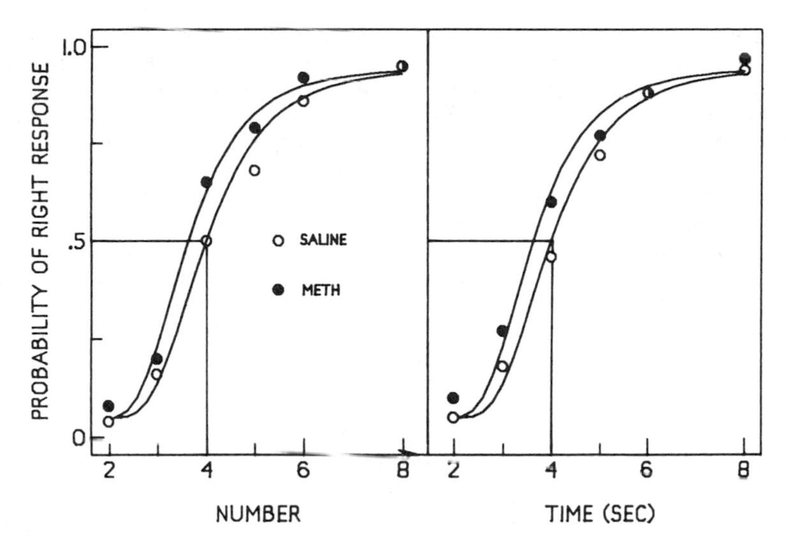

FIGURE 9. Event mode. Median proportion long response as a function of number and duration with methamphetamine and saline solution. (From Meck and Church.[17] Reprinted by permission.)

the stimulus (light versus sound) and the state of preparation for the signal (expected versus unexpected). Another example will be shown in FIGURE 10.

The main property of the switch is that it can be operated in various modes. Consider a signal that consists of a stimulus that is on for duration *a*, off for duration *b*, and on again for duration *c*. For what duration does the switch gate pulses? We have defined the following modes: In run mode, the duration is a + b + c, that is, the duration from signal onset until the end of the trial. In stop mode, the duration is a + c, that is, the duration that the stimulus is on. In our typical situations, rats initially adopt the stop mode. That is to say, when trained with continuous signals and shifted to signals with gaps, animals time the duration that the stimulus is on, and ignore the gaps. Thus the pacemaker puts pulses into the accumulator during the first stimulus segment, does not put pulses into the accumulator during the gap, but adds to the pulses in the accumulator during the second stimulus segment. This has been demonstrated in the bisection procedure[14,15] and in the peak procedure.[3,13] FIGURE 8 shows an example from the bisection procedure. Performance on trials with a signal with a 2-sec gap (peak) was similar to that on trials with a signal that was 2 sec shorter, that is, the rats appeared to be timing the total time the stimulus was on and ignoring the gap.

The switch can also operate in an event mode. In event mode the switch is closed for some short, fixed period of time following each stimulus onset. Thus, in event mode, the animal uses the internal clock as a counter.[16,17] Even when temporal cues are controlled, an animal can classify a stimulus sequence by the number of elements.[17,18,30,35] In one experiment[17] signals were presented in which a white-noise went on and off several times. On half the trials the signal lasted for 4 sec and consisted of 2 to 8 cycles; on the other half of the trials the signal consisted of 4 cycles and lasted 2 to 8 sec. A left lever response was reinforced after either a 2-cycle or 2-sec signal; a right lever response was reinforced after either an 8-cycle or 8-sec signal. Results are shown in FIGURE 9. The psychophysical functions for a 4:1 ratio of number and time are equivalent, and they are similarly affected by methamphetamine. This suggests that the same process is used for timing and counting. With one method we obtained an estimate of the length of time that the switch was closed at the onset of each event.[17] It was estimated to be about 200 msec. In event mode, the closure of the switch provides the signal for opening the switch. If the signal is expected, t_1 approaches 0, so the 200 msec becomes an estimate of t_2, the time required to open the switch.

To summarize, when a signal consists of several stimuli, the switch can close at the onset of a stimulus and open at (a) the beginning of each stimulus (event mode), (b) the end of each stimulus (stop mode), or (c) the end of the trial (run mode).

With differential reinforcement, rats can use the run, stop, or event mode. Perhaps the switch response obeys the same laws of reinforcement as overt motor responses[15] or the animals maintain a representation of all these times and use only the one that is reinforced. In either case, the switch mode can be transferred from one modality to another.[16,19,20,21,29]

Accumulator

The defining characteristic of an accumulator is that it holds the sum of the pulses. Thus, by definition, it is an "up" counter that increments in arithmetic units in an absolute manner. Evidence for such an accumulator comes from the "time left" procedure: When the time left in an elapsing interval is equal to a standard interval, animals are indifferent between the intervals.[22]

Further support for an accumulator is provided by the "shift" procedure.[3,15] In the shift procedure, a response is reinforced after one time in the presence of a signal in one modality, and the response is reinforced after a different time in the presence of a signal in another modality. For example, an animal could be reinforced for a response after 20 sec of sound and after 40 sec of light. Then, on some test trials, sound could be presented for some period of time (for example, 10 sec) and then the signal could be shifted to light. If the animal used an absolute, arithmetic, up-counter, the accumulator would contain the number of pulses appropriate for 10 sec and the remembered time of reinforcement would be 40 sec, so the peak response rate would occur after another 30 sec. Performance of rats in the shift procedure has been interpreted as evidence that this does occur.

Initially, we thought that the accumulator operated in the interval from seconds to minutes, but there is now some evidence that the same processes may occur at much shorter and much longer intervals. For example, the same processes may occur in the circadian range, as described by Terman *et al.*[36] in this volume, and at fractions of a second when repeated cycles are presented.[13,16]

Working Memory

The defining characteristic of working memory is that it stores information about the current trial in the absence of the signal.

One of the properties of working memory is that it can be reset quickly. For example, in the bisection procedure a rat can be trained to make a left response after a 2-sec signal and a right response after an 8-sec signal. Then, on some test trials, the animal may be given a preset signal of some duration followed after a brief interval, not with the usual opportunity to respond, but with another signal duration. The animal can learn to use the second signal without any interference from the first signal.[23]

We assume that working memory is used only if there is an interval between stimulus termination and the opportunity to respond; otherwise the sensory store (the accumulator) is used. With a lesion of the fimbria fornix, which cuts connections to the hippocampus, either information does not transfer between the accumulator and the working memory, information does not transfer between working memory and the comparator, or rats reset the working memory after a short gap in the signal. For example, rats were trained on a 20-sec peak procedure. Then signals were presented for 10 sec followed by a gap of 5 sec and then a continuation of the signal. Control rats stop the clock during the gap (that is, open the switch), so that the maximum response rate with a 5-sec gap would occur 5 sec later than usual (Fig. 10). Note that for the control rats the peak response rate without a gap is about 20 sec, the time reinforcement was

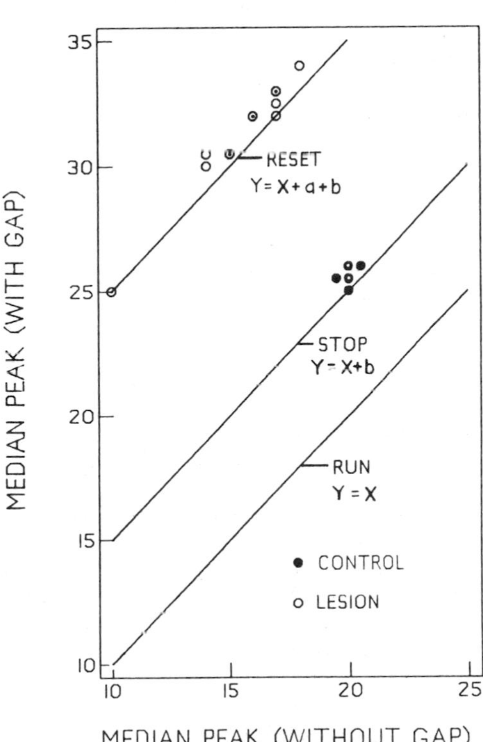

FIGURE 10. Reset. Peak times for control rats and rats with fimbria fornix lesions on the 20-sec peak procedure during unreinforced trials with and without 5-sec gaps. (From Meck *et al.*[13] Reprinted by permission.)

sometimes available, and that the peak response rate with a 5-sec gap is about 25 sec. Thus, control rats have a peak response rate about 5 sec later with a 5-sec gap than without it, so that all subjects are close to the "stop" line.

Rats with a fimbria fornix lesion, however, with a 5-sec gap, appear to be unaffected by the signal prior to the gap and thus their maximum response occurs 15 sec later than usual. The rats with fimbria fornix lesions are close to the "reset" line; they have a peak response rate about 15 sec later after a gap than with a gap.

There are two additional points to note about FIGURE 10. First, data points are typically about half a second above the stop or reset line. This presumably reflects the difference between the latent period required to close the switch at the onset of the stimulus and the latent period required to reopen the switch at the termination of the stimulus, a property that has already been described. Second, without a gap, the rats with fimbria fornix lesions had peaks that were reliably less than 20 sec, the time that reinforcement was sometimes available. The effect of the lesion, like the effect of some drugs, led to a permanent distortion in the remembered time of reinforcement. Reinforced durations were remembered as shorter than they actually were. This will be discussed further as a property of reference memory. A gap, however, added about 15 sec to each function. All subjects with fimbria fornix lesions acted as if they were timing only the second segment.

When a retention interval is placed between the time of exposure to a signal and the opportunity to respond, there is some decrement in performance that is presumably due to decay in working memory. The question is whether this decay is in the time dimension, that is, is the remembered time of reinforcement shorter if the delay is longer? We tested for this, but obtained no evidence for it: The point of indifference was approximately at the geometric mean regardless of the duration of the retention interval.[23] Therefore, we concluded that decay from working memory is on a dimension orthogonal to time. (But this view has been challenged.[24])

Reference Memory

The defining characteristic of reference memory is that it stores information about past trials and their consequences permanently.

Previously we have seen that various drugs and lesions can influence the remembered time of reinforcement. These include the following: Physostigmine produces a leftward shift and atropine produces a rightward shift. Physostigmine inhibits the acetylcholine-degrading enzyme, which increases the effective level of acetylcholine, and atropine blocks acetylcholine receptors, which reduces the effective level of acetylcholine.[9] Thus, it is plausible that the effective level of acetylcholine controls a memory constant. Vasopressin and oxytocin, two neuropeptides, also decrease the memory constant.[9]

Various lesions can also affect the remembered time of reinforcement. As we noted, a fimbria fornix lesion produces a leftward shift, that is, decreases the remembered time of reinforcement; large electrolytic lesions of the frontal cortex produce a rightward shift in the functions, that is, increase the remembered time of reinforcement.[25] Such changes can also be produced by dietary means. Choline administered in a saccharine-flavored solution prior to a session can lead to a maintained leftward shift in the response function.[11]

The change is in the transformation of the value in working memory to the value in reference memory (that is, storage), not in the transformation of a value in reference memory to the comparator (that is, retrieval). This is known since drug administration during training, not testing, was relevant. This also demonstrates that the original time value that was transformed by K^* in the storage process is not recovered by the inverse transformation ($1/K^*$) in retrieval. There are marked individual differences in the memory storage constant. Although the central tendency of a large group of rats is close to the actual time of reinforcement, some rats have a memory storage constant reliably less than 1.0 and some have a constant reliably greater than 1.0.[26]

Comparator

The defining characteristic of the comparator is that it determines a response on the basis of a decision rule which involves a comparison between a value in the accumulator or working memory with a value from reference memory. Thus, it must contain two time values and a response rule.

In many situations a measure of responding is a function of signal duration expressed as a proportion of the reinforced signal duration. This superposition result occurs when times are expressed on a subjective scale. To obtain the superposition result, the decision rule must be relative (a ratio) rather than absolute (a difference).[8,32] The scalar timing model includes a relative proximity rule with a normally distributed threshold that has some mean and standard deviation.

This completes the review of the properties of the parts of the psychological process used for timing intervals. Each of these parts represents a potential source of variance and at least one source of scalar variance is required. Because of the form of the observed response functions, there is reason to believe that at least one asymmetrical source of scalar variance (such as clock speed) and one symmetrical source of scalar variance (such as response threshold) is required.[8,32]

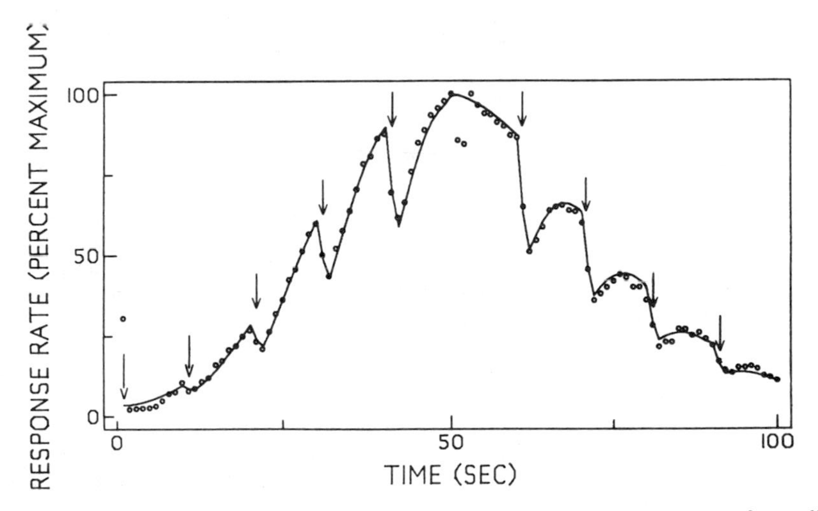

FIGURE 11. Simultaneous temporal processing. Percent of maximum response rate for median responses per minute as a function of time into the interval. Arrows indicate the presentation time of 1-sec segment signals. (From Meck and Church.[4] Reprinted by permission.)

SIMULTANEOUS TEMPORAL PROCESSING

Previously I had thought that a single timing process would be sufficient to explain data from animal timing experiments. I based my belief on two facts: First, in the shift experiment, when the signal was changed from sound to light, the value from reference memory was changed, but the clock value was not. Thus, it appeared that the same clock was used for different modalities. The second reason for my belief in a single timing process was because of the positive results from cross-modal transfer. A rat trained on a timing task with signals of one modality immediately performs accurately when the signal modality is changed. This has been shown for various procedures and modalities.[13,16,17,19,20,21,29]

Evidence of simultaneous temporal processing demonstrates that animals must have something more than the single timing process outlined so far. FIGURE 11 shows the response rate of rats on a 50-sec peak procedure, signaled by light, with a 1-sec noise signal presented every 10 sec (except at 50 sec).[4] The complex function is well fit by a scalar timing model in which the same model is used for the 50-sec and 10-sec timing process and the same parameter values are used for the 50-sec function whether or not the 10-sec signals were present. Thus, rats can time two signals simultaneously without interference. We have also found that rats can time and count simultaneously without interference.[17] At a minimum, rats must have several switch–accumulator parts to handle these tasks.

ATTENTION

In a timing task, animals are not always attentive to time. In several procedures we have found it useful to assume that on some proportion of the trials the animals were attending to time and using the scalar timing model, and on the remaining trials the animals responded with some constant bias unrelated to signal duration.[2,9,12] This idea has been by Heinemann *et al.*[28] Some information regarding selective attention has been provided by Meck.[37]

SUMMARY

Evidence has been cited for the following properties of the parts of the psychological process used for timing intervals:

1. The pacemaker has a mean rate that can be varied by drugs, diet, and stress.
2. The switch has a latency to operate and it can be operated in various modes, such as run, stop, and reset.
3. The accumulator times up, in absolute, arithmetic units.
4. Working memory can be reset on command or, after lesions have been created in the fimbria fornix, when there is a gap in a signal.
5. The transformation from the accumulator to reference memory is done with a multiplicative constant that is affected by drugs, lesions, and individual differences.
6. The comparator uses a ratio between the value in the accumulator (or working memory) and reference memory.

Finally, there must be multiple switch–accumulator modules to handle simultaneous temporal processing; and the psychological timing process may be used on some occasions and not on others.

ACKNOWLEDGEMENTS

All of my research on the properties of the internal clock has been done in close collaboration with John Gibbon, Warren H. Meck, or Seth Roberts. Many of the ideas expressed in this paper were originated or refined by one of these colleagues, although I have the responsibility for their expression in this chapter.

REFERENCES

1. CHURCH, R. M. 1983. The influence of computers on psychological research: A case study. Behav. Meth. Res. Instrum. **15:** 117–126.

2. CHURCH, R. M. & J. GIBBON 1982. Temporal generalization. J. Exp. Psychol. Anim. Behav. Processes **8:** 165–186.
3. ROBERTS, S. 1981. Isolation of an internal clock. J. Exp. Psychol. Anim. Behav. Processes **7:** 242–268.
4. MECK, W. H. & R. M. CHURCH. Simultaneous temporal processing. J. Exp. Psychol. Anim. Behav. Processes. **10:** 1–29.
5. MARICQ, A. V., S. ROBERTS & R. M. CHURCH. 1981. Methamphetamine and time estimation. J. Exp. Psychol. Anim. Behav. Processes. **7:** 18–30.
6. PLATT, R. J. & E. R. DAVIES. 1983. Bisection of temporal intervals by pigeons. J. Exp. Psychol. Anim. Behav. Processes. **9:** 160–170.
7. ROBERTS, S. Properties and function of an internal clock. *In* Animal Cognition and Behavior. R. L. Mellgren, Ed. North-Holland. Amsterdam. In press.
8. GIBBON, J. & R. M. CHURCH. Sources of variance in information processing theories of timing. *In* Animal Cognition. H. L. Roitblat, T. G. Bever & H. S. Terrace, Eds. Erlbaum. Hillsdale, NJ. In press.
9. MECK, W. H. 1983. Selective adjustment of the speed of internal clock and memory processes. J. Exp. Psychol. Anim. Behav. Processes **9:** 171–201.
10. MARICQ, A. V. & R. M. CHURCH. 1983. The differential effects of haloperidol and methamphetamine on time estimation in the rat. Psychopharmacology **79:** 10–15.
11. MECK, W. H. & R. M. CHURCH. Nutrients that modify internal clock and memory storage speeds. Paper to be presented at meeting of Psychonomic Society, San Diego, November, 1983.
12. CHURCH, R. M., D. J. GETTY & N. D. LERNER. 1976. Duration discrimination by rats. J. Exp. Psychol. Anim. Behav. Processes. **2:** 303–312.
13. MECK, W. H., R. M. CHURCH & D. S. OLTON. 1984. Hippocampus, time, and memory. Behav. Neurosci. **98:** 3–22.
14. CHURCH, R. M. 1978. The internal clock. *In* Cognitive Processes in Animal Behavior. S. H. Hulse, H. Fowler & W. K. Honig, Eds. Erlbaum. Hillsdale, NJ.
15. ROBERTS, S. & R. M. CHURCH, 1978. Control of an internal clock. J. Exp. Psychol. Anim. Behav. Processes. **4:** 318–337.
16. CHURCH, R. M. & W. H. MECK. The numerical attribute of stimuli. *In* Animal Cognition. H. L. Roitblat, T. G. Bever & H. S. Terrace, Eds. Erlbaum. Hillsdale, NJ. In press.
17. MECK, W. H. & R. M. CHURCH. 1983. A mode control model of counting and timing processes. J. Exp. Psychol. Anim. Behav. Processes. **9:** 320–334.
18. FERNANDES, D. M. & R. M. CHURCH. 1982. Discrimination of number and sequential events by rats. Anim. Learn. Behav. **10:** 171–176.
19. CHURCH, R. M. & W. H. MECK. Acquisition and cross-modal transfer of classification rules for temporal intervals. *In* The Quantitative Analyses of Behavior: Acquisition Processes. M. Commons, A. Wagner & R. Herrnstein, Eds. Vol. IV. Ballinger. Cambridge, MA. In press.
20. MECK, W. H. & R. M. CHURCH. 1982. Abstraction of temporal attributes. J. Exp. Psychol. Anim. Behav. Processes **8:** 226–243.
21. MECK, W. H. & R. M. CHURCH. 1982. Discrimination of intertrial intervals in cross-modal transfer of duration. Bull. Psychon. Soc. **19:** 234–236.
22. GIBBON, J. & R. M. CHURCH. 1981. Time left: Linear versus logarithmic subjective time. J. Exp. Psychol. Anim. Behav. Processes. **7:** 87–108.
23. CHURCH, R. M. 1980. Short-term memory for time intervals. Learn. Motiv. **11:** 208–219.
24. SPETCH, M. L. & D. M. WILKIE. 1983. Subjective shortening: A model of pigeons' memory for event duration. J. Exp. Psychol. Anim. Behav. Processes. **9:** 14–30.
25. MARICQ, V. 1978. Some effects of lesions on the prefrontal cortex on timing behavior in the rat. Unpublished honors thesis, Brown University.
26. MECK, W. H. & R. M. CHURCH. Individual differences in memory storage speed. Unpublished manuscript.
27. CHURCH, R. M. & M. Z. DELUTY. 1977. The bisection of temporal intervals. J. Exp. Psychol. Anim. Behav. Processes. **3:** 216–228.
28. HEINEMANN, E. G., E. AVIN, M. A. SULLIVAN & S. CHASE. 1969. Analysis of stimulus generalization with a psychophysical method. J. Exp. Psychol. **80:**215–224.

29. ROBERTS, S. 1982. Cross-modal use of an internal clock. J. Exp. Psychol. Anim. Behav. Processes. **8:** 2–22.
30. DAVIS, H. & J. MEMMOTT. 1982. Counting behavior in animals: A critical evaluation. Psycholog. Bull. **92:** 547–571.
31. STUBBS, D. A. 1976. Scaling of stimulus duration by pigeons. J. Exp. Anal. Behav. **26:** 15–25.
32. GIBBON, J., R. M. CHURCH & W. H. MECK. 1984. Scalar timing in memory. This volume.
33. MECK, W. H., F. KOMEILY-ZADEH & R. M. CHURCH. Interference of signal timing by response timing. Paper presented at meeting of the Psychonomic Society, Philadelphia, November, 1981.
34. TREISMAN, M. 1984. Temporal rhythms and cerebral rhythms. This volume.
35. CHURCH, R. M. & W. H. MECK. The numerical attribute of stimuli. Presented at the Harry Frank Guggenheim Conference on Animal Cognition, Columbia University, June 2–4, 1983.
36. TERMAN, M., J. GIBBON, S. FAIRHURST & A. WARING. 1984. Daily meal anticipation: Interaction of circadian and interval timing. This volume.
37. MECK, W. H. 1984. Internal clock, memory and attentional processes in animal time discrimination. This volume.

Retrospective Duration Judgments of a Hypnotic Time Interval

SCOTT W. BROWN

Department of Psychology
University of Maine
Orono, Maine 04469

Previous research has shown that hypnotic time intervals are retrospectively judged to be of much shorter duration than are nonhypnotic intervals. Various studies have attempted to uncover a relationship between hypnotizability and judgments of hypnotic time, but the results have been ambiguous and difficult to interpret. It has been suggested that absorbtion, rather than hypnotizability per se, may be the important factor. According to this view, the greater one's capacity to become immersed in the hypnotic experience, the greater the shortening of perceived time.

Some attempts have also been made to use hypnotic amnesia as a vehicle to assess the role of memory in retrospective time judgments. A memory-storage model of perceived time predicts that intervals associated with relatively large amounts of information in memory would be judged longer than intervals associated with relatively small amounts of information in memory. Unfortunately, the results of these hypnotic amnesia studies are unclear and may be subject to methodological artifacts. The present investigation was designed to explore these issues further.

Eight separate groups of subjects (total N = 216) completed the Tellegen Absorbtion Scale prior to administration of the Harvard Group Scale of Hypnotic Susceptibility, Form A (HGSHS:A). The final item of the HGSHS:A consists of a posthypnotic amnesia suggestion, which is cancelled after a 3-min recall period that occurs after subjects are awakened from hypnosis. Alternate groups of subjects were asked to retrospectively estimate the duration of the 38-min hypnosis session either just before or just after the amnesia suggestion was cancelled.

On the basis of their HGSHS:A scores, subjects were assigned to low (0–3), medium (4–8), or high (9–12) hypnotizability groups. A 2 × 3 (amnesia condition × hypnotizability) unequal *ns* ANOVA (unweighted means analysis) was performed on the time-judgment data. This analysis revealed a significant main effect for hypnotizability ($F = 4.57$, d.f. = 2; $p < .01$); these results are displayed in FIGURE 1. A Newman–Keuls multiple comparison test (adjusted for unequal *ns*) showed that subjects with both high and medium hypnotizability made shorter time judgments than did the subjects with a low degree of hypnotizability ($p < .01$ for both comparisons). Thus, although all subjects underestimated the interval by an average of 54% (a finding consistent with other research), the degree of underestimation was greater for the subjects with high and medium degrees of hypnotizability.

Neither the main effect for amnesia nor the interaction attained significance. A Pearson correlation between recall performance and time judgments yielded a coefficient of .03; thus, any predictions based on a memory-storage model of perceived duration were not confirmed.

A Pearson correlation was also computed between absorbtion scores and time judgments. The resulting coefficient of .03 lends little support to a hypothesized relationship between absorbtive ability and perceived duration.

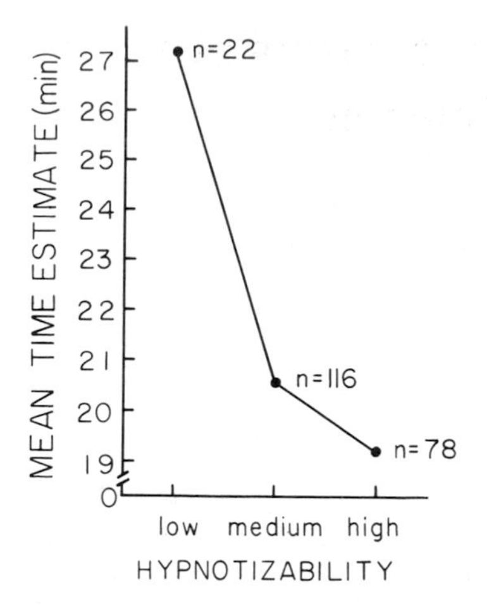

FIGURE 1. Time judgments of a 38-min hypnotic time interval as a function of hypnotizability.

Future research must attempt to isolate the critical variables responsible for the temporal underestimations exhibited by hypnotically responsive subjects. Perhaps the hypnotic interval creates a more passive/receptive attitude, or a lower arousal level, for the highly hypnotizable subjects and this in turn influences their perception of time. In any case, further research on hypnotic time may lead to a better understanding of the processes involved in time perception under normal conditions.

Scalar Timing and the Spatial Organization of Behavior Between Reward Presentations

F. R. CAMPAGNONI AND P. S. COHEN

Department of Psychology
Northeastern University
Boston, Massachusetts 02115

When a reinforcing event is scheduled periodically, behavior becomes organized around the reward deliveries in a distinctive manner. Late in the interreinforcer interval, as reward delivery becomes imminent, subjects engage predominately in activities directed toward acquisition of the reinforcer. In the period following reward presentation, when additional reward is unavailable, subjects typically turn or move away from the reinforcer dispenser and engage in activities directed toward other parts of the experimental environment.[1] Withdrawal from the area where reward is procured may be maintained, in part, by the change in visual stimulation accompanying such behavior.[2] This is consistent with evidence that subjects receiving periodic reward will initiate a change in stimulus conditions associated with reward when its presentation is unlikely.[3]

Using pigeons, we have demonstrated[4] that when an explicit change in stimulus conditions (chamber illumination and keylight color) is made contingent upon a subject's being in a portion of the chamber farthest removed from the reinforcer dispenser, then there is an increase in the probability of this behavior without a significant change in its temporal organization. In this same study, we also observed that the probability of location away from the reinforcer site appears to be referenced to proportional rather than absolute time within the interval between reward presentations. The following report summarizes this timing aspect of our findings.

Six White King pigeons maintained at 80% of their *ad lib* body weight were exposed to either a fixed time (response independent, two subjects) or a fixed interval (response–dependent, four subjects) schedule of grain deliveries. A standard operant chamber with a translucent response key mounted above the grain dispenser was fitted with photocells allowing the birds' location to be recorded in one of three areas; front (near the grain dispenser), rear (wall opposite grain dispenser), and middle (between front and rear photocells). Spatial location was sampled each second. Each session consisted of sixteen 7-sec grain deliveries. Subjects were exposed to an ascending and descending logarithmic series of interreinforcer intervals (30, 60, 120, 240, 480 seconds). Since the data were essentially recovered on the descending series, only the ascending series is presented. The data presented are based on the last five of 15 sessions at each value. For all sessions, interruption of the rear photobeam resulted in a change in chamber stimuli (six houselights rather than one were illuminated and the key changed from green to red) for the duration of time that the rear photobeam was interrupted.

FIGURE 1 depicts the proportion of total samples with location in the rear as a function of the interreinforcer interval (normalized functions)[5] for each subject at each interval value. One subject did not reliably leave the vicinity of the reinforcer dispenser and its data have been excluded from the analysis. Although tending to flatten

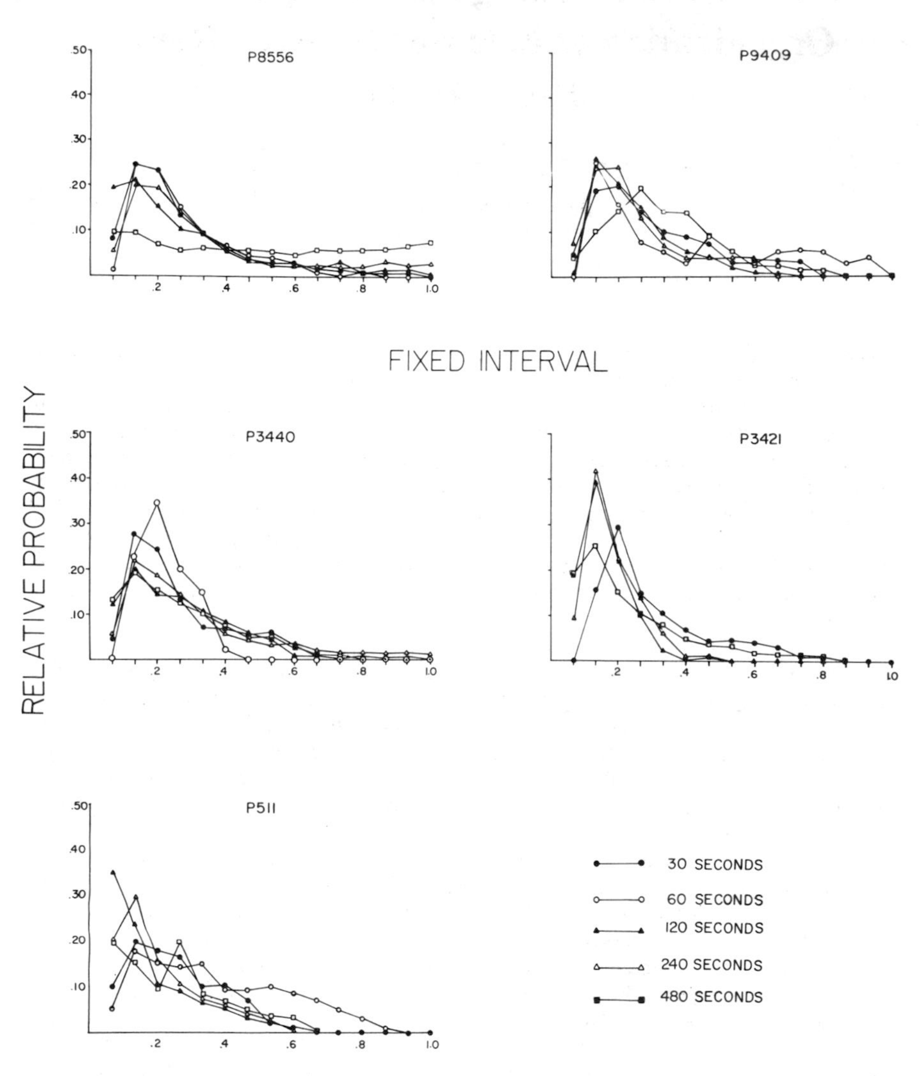

PROPORTION OF INTERVAL

FIGURE 1. Normalized functions depicting proportion of total samples with location in the rear of the chamber as a function of proportion of interreinforcer interval.

somewhat at the higher interval values, the functions, both within and between animals, superimpose on one another for a wide range of interval values. The subjects were most likely to be in the rear of the chamber after 15–25% of the interval had elapsed, regardless of the interreinforcer interval value. Thus, the probability of location in the rear was a function of the proportion of the interval that had elapsed between reward deliveries. This suggests that time allocation in the rear, like operant behavior and general activity,[5] involves a timing process similar to that proposed by scalar expectancy theory.[5]

REFERENCES

1. Staddon, J. E. R. & V. L. Simmelhag. 1971. The "superstition" experiment: A re-examination of its implications for the principles of adaptive behavior. Psychol. Rev. **78**: 3–43.
2. Falk, J. L. 1977. The origin and function of adjunctive behavior. Anim. Learn. Behav. **5**: 325-335.
3. Brown, T. G. & R. K. Flory, 1972. Schedule-induced escape from fixed-interval reinforcement. J. Exp. Anal. Behav. **17**: 395–403.
4. Cohen, P. S. & F. R. Campagnoni. Is time spent away from the reinforcer dispenser analogous to time out? Presented at a meeting of the American Psychological Association, Los Angeles, California, August 1981.
5. Gibbon, J. 1977. Scalar expectancy theory and Weber's law in animal timing. Psychol. Rev. **84**: 279–325.

Foraging in the Laboratory: Effects of Session Length

SUSAN M. DOW

Department of Zoology
University of Bristol
Bristol BS8 1UG, England

INTRODUCTION

Optimality models have become widespread in recent years in the search to predict the tradeoff between costs and benefits to the individual. When these models are applied to foraging behavior, the currencies used are frequently time and energy.

In a test of optimal foraging theory, Krebs, Kacelnik, and Taylor[1] considered two strategies for a forager facing "patches" of food of unknown density: (a) "immediate maximizing," and (b) "optimal sampling." The optimal sampler essentially divides its foraging session into a sampling and an exploitation phase. This requires a knowledge of time available for foraging since the optimal point to switch from sampling to exploitation is dependent on this.

The aim of the experiment reported was to distinguish between the immediate maximizing and the optimal sampling hypotheses by varying the length of foraging sessions for pigeons under concurrent random-ratio schedules of reinforcement. If the birds behaved like optimal samplers, varying session length should change the pattern of switches between schedules, involving the short-term sacrifice of exploitation. Although session length was determined by the number of pecks, the birds had demonstrated stable individual response rates in a similar experiment.

METHOD

A commercial pigeon operant test chamber was used with two keys available to the birds. The side key was lit either red or green associated with random-ratio schedules. The ranges of the averages of the schedules were 10–35 and 20–50 for the better and poorer patch, respectively; and the value for the poorer was at least twice that of the better. The color associated with the better patch was usually changed after each session. The white, center key acted as a changeover key[2] on FR 1.

Sessions were of two lengths, designated long and short, ending after 1024 and 256 pecks, respectively.

Twelve pigeons of racing stock, with previous simulated foraging experience (similar to "long" sessions), divided into two groups, had either 15 long sessions followed by 15 short and then a further 15 long, or 15 short, then 15 long, and a further 15 short.

RESULTS

Quantitative results are presented as medians (across birds) of means across the first 256 pecks of the last eight sessions in each block of sessions as a function of session

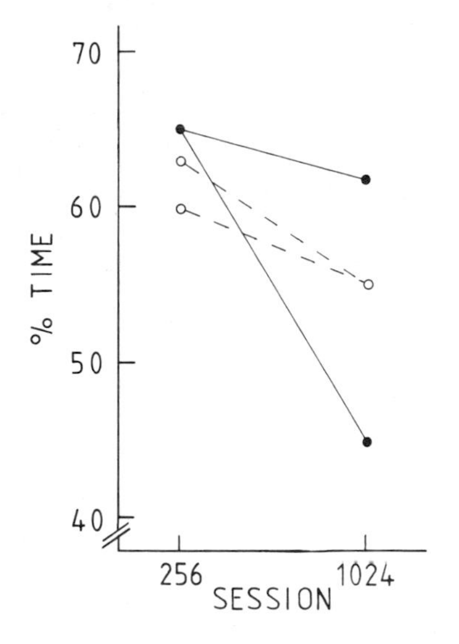

FIGURE 1. Percentage of time in the first 256 pecks spent in the better patch against session length. In this and in subsequent illustrations ●—● = birds following sequence of long–short–long; O---O = birds following sequence of short–long–short.

length (see FIGURES 1–4). The proportion of time spent in the better patch was higher in short sessions (FIG. 1), although this fell just short of significance (Wilcoxon test). Very similar results were obtained for proportion of pecks (FIG. 2). FIGURE 3 shows median total reinforcements obtained from the first 256 pecks; significantly fewer were obtained in this part of the long sessions than in the short (Wilcoxon test; $p < 0.05$, two-tailed).

The median number of changes in the first 256 pecks was 2.69 in long and 1.31 per session in short sessions. This tendency for fewer changeovers in short sessions was

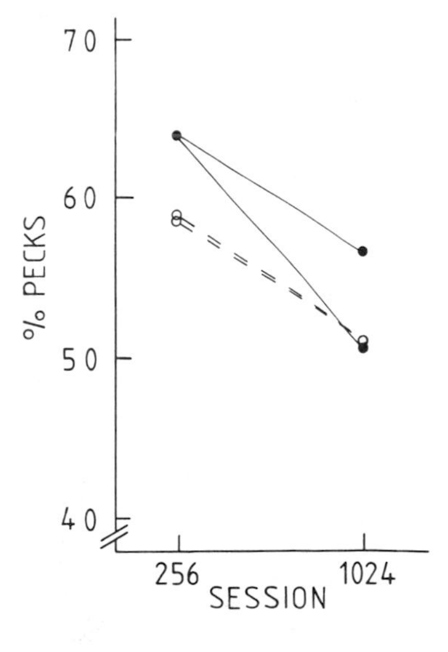

FIGURE 2. Percentage of first 256 pecks spent in the better patch against session length.

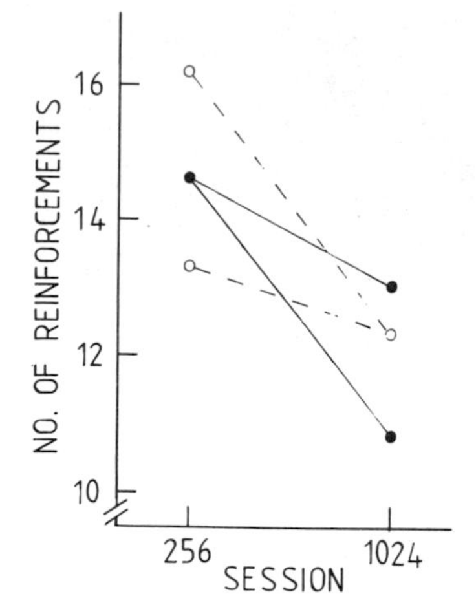

FIGURE 3. Number of reinforcements gained in 256 pecks against session length.

significant ($p < 0.05$, two-tailed signs test); see FIGURE 4 for details of each block of sessions.

DISCUSSION

The qualitative prediction of optimal sampling theory, that sampling would continue for longer in longer sessions was confirmed in that relative time allocation and response rates during the first 256 pecks favored the better patch in the short sessions. Although this trend fell short of significance, the birds did significantly "delay gratification" (FIG. 3). The immediate maximizing hypothesis can give no account of the poor reinforcement rate at the beginning of the long sessions, nor the decrease in rate of changes between schedules seen in the short sessions.

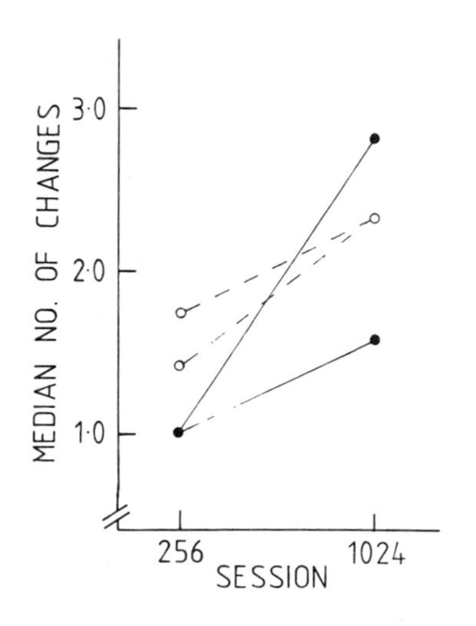

FIGURE 4. Number of changes made in first 256 pecks against session length.

REFERENCES

1. KREBS, J. R., A. KACELNIK & P. TAYLOR. 1978. Test of optimal sampling by foraging great tits. Nature (London) **275:** 27–31.
2. FINDLEY, J. D. 1958. Preference and switching under concurrent scheduling. J. Exp. Anal. Behav. **1:** 123–144.

Time-Order Discrimination of Sequences of Four Events[a]

STEPHEN R. GRICE

Department of Psychology
McMaster University
Hamilton, Ontario L8S 4K1, Canada

When a subject reports the order of a stimulus sequence that contains more than two events, different kinds of incorrect order reports become possible. With a sequence of four events a subject can produce a correct report or one of 23 different kinds of incorrect reports. This experiment collected these different kinds of reports and then compared the obtained proportions of these different kinds of reports with the proportions predicted by five different models of time-order discrimination.

The stimulus display was four small lights (*A*, *B*, *C*, *D*) arranged in a square. The fixation point was in the center of the square, thus:

A　　*B*

+

C　　*D*

The stimulus lights were viewed monocularly with the right eye at a distance of 67 cm. Horizontal and vertical distance between the lights was 7.0 cm. The subtended visual angle for 7.0 cm was 10.5 centiradians (6 degrees). The fixation point was a red light-emitting diode. The four stimulus lights were yellow light-emitting diodes with a luminance of about 4300 cd/m^2 (1350 millilamberts). The background was flat black with a luminance of about 10 cd/m^2 (3 millilamberts). In each trial, lights *A*, *B*, *C*, and *D* switched on simultaneously, remained on for about 3 sec, and then shut off in one of their possible orders. These shut-offs are labeled here *offsets*. The subject's task on each trial was to report the order of the offsets.

There were two kinds of trials: (1) trials in which only two of the four stimulus lights shut off; the inter-offset interval used in these trials was 40 msec; (2) trials in which all four of these lights shut off; the inter-offset interval was 40 msec on some of these trials and 80 msec on other trials. Three subjects participated in the experiment. For each of the inter-offset intervals used, each subject saw each of the possible two-offset combinations ten times and each of the possible four-offset combinations ten times.

Each order report was encoded as a string of digits. Within a string of digits, the *value* of a digit indicated the actual order of the offsets, and the *position* of a digit indicated the order reported by the subject. The encoding system is demonstrated in TABLE 1. The order reports most frequently obtained in this experiment are shown in TABLE 2.

[a] A complete description of this study is available as a circulating thesis through interlibrary loan from Scott Library, York University, Downsview, Ontario, M3J 1P3. Copies are also available from the Department of Psychology, McMaster University, Hamilton, Ontario, L8S 4K1, Canada.

TABLE 1. Encoded Order Report Examples

Actual Offset Order	Reported Offset Order	Encoded Report
AD	AD	12
AD	DA	21
BD	DB	21
ABCD	ABCD	1234
ABCD	ABDC	1243
CABD	CADB	1243

TABLE 2. Obtained Order Report Proportions

Interstimulus Interval (msec)	Order Report				
	12	1234	1243	1324	2134
40	.68	.24	.14	.09	.09
80	—	.49	.19	.06	.09

NOTE. Results are collapsed across three subjects. *N* for report 12 is 360. *N* for other reports is 720.

The five models of time-order discrimination were then tested against the obtained proportions. Two of the tested models were simple quantum models, two were attention-switching quantum models, and one was a perceptual latency model. None of these models fit the obtained data well.

The obtained proportions were also used to generate suggestions for two new models of time-order discrimination. One model is a revision of the perceptual latency model. The other model postulates brain cells—*subtracters*—which respond to the differences in activity between the brain cells that are directly activated by each of the stimulus events.

An Endogenous Metric for the Control of Perception of Brief Temporal Intervals

P. A. HANCOCK[a]

Motor Behavior Laboratory
Institute for Child Behavior and Development
University of Illinois at Urbana-Champaign
Champaign, Illinois 61820

The biochemical clock hypothesis, advanced by Hoagland[1] in 1933, suggested the existence of a unitary neurophysiological process which subsumed the human perception of brief temporal intervals. The resultant, necessary, and exclusive linear relationship required by this construct was supported by early observations, as shown in FIGURE 1. In subsequent investigations, although a general linear trend has been affirmed, the lack of consistency across individuals has refuted the notion of a simple governing metabolic pacemaker as originally envisaged.[2,3]

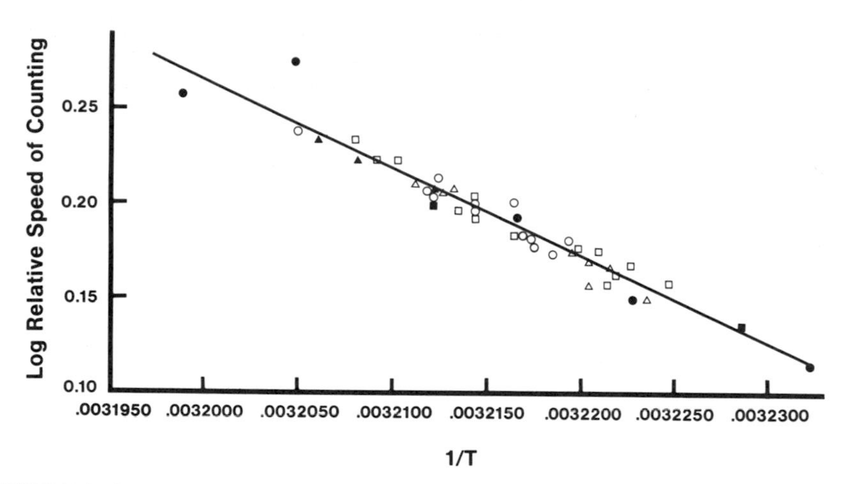

FIGURE 1. Composite of data from Hoagland[1] and François[5] showing the relationship between log relative speed of counting and the reciprocal of absolute temperature, an Arrhenius plot. Note that body temperature ranges from 36.3°C (97.3°F) at *right* to 39.5°C (103.1°F) (extreme *left* data point). The data are for six individuals. (From Hoagland.[1] Reproduced by permission.)

The present experiments were predicated upon this search for an endogenous and ubiquitous temporal mechanism. Specifically, the study examined the effects of selective head-temperature elevation and depression upon time judgment. In the first experiment, each of twelve subjects undertook one presentational order of four thermal conditions. In each condition, subjects produced operative estimates of 1-, 11- and 41-sec periods, with forty trials on each period. In the control condition, subjects produced estimates while seated in a sound- and light-proof room. In the placebo

[a]Present address: Department of Safety Sciences, Institute for Safety and Systems Management, University of Southern California, Los Angeles, California 90089.

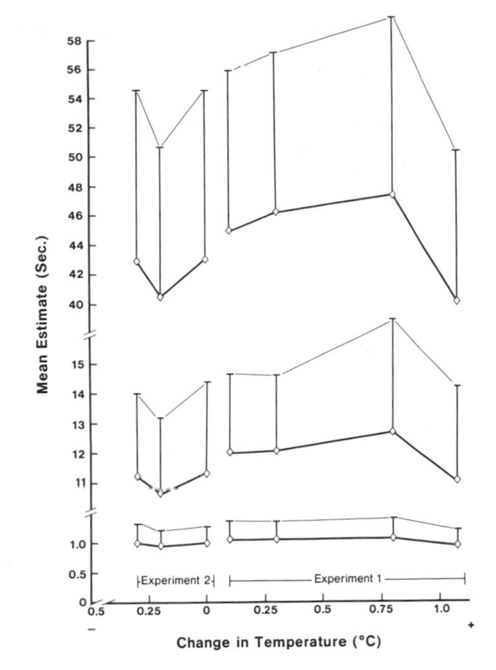

FIGURE 2. Overall means (*diamond symbols*) and standard deviations (*vertical bars*) by period against increase and decrease in deep meatal temperature. Results are for the mean of twelve subjects in each experiment.

condition, a heating helmet was worn but not activated, while in the two heat conditions prescribed temperature increases, measured in the deep auditory meatus, were stabilized above initial monitored baseline values prior to experimental commencement. The pattern of results for the mean estimation of all periods, under each thermal condition, was consistent across subjects, although absolute magnitude of the estimate varied with the individual. Analysis distinguished the mean at the highest elevation, 1.1°C, as significantly shorter than the three alternate unvarying means (FIG. 2). In the second experiment, twelve different male subjects performed the same task under control, placebo, and cold conditions. The use of a water-cooled unit provided insufficient temperature depression, leaving results somewhat equivocal for this experiment (FIG. 2).

Despite the latter result, overall findings suggest a nonlinear relationship between head temperature and duration estimation. This contradicts the direct isomorphism relating physiological change and temporal percept, as implied by the biochemical clock hypothesis. However, this is based upon nonlinearity rather than subject inconsistency. Consequently, the present study implies the existence of a thermally sensitive endogenous mechanism involved in the perception of brief temporal intervals. How such a mechanism is related to circadian variation in time judgments[4] and underlies or interacts with certain alternate cognitive functions is the subject of continuing investigation.

REFERENCES

1. HOAGLAND, H. 1933. The physiological control of judgments of duration: Evidence for a chemical clock. J. Gen. Psychol. **9:** 267–287.
2. BELL, C. R. 1966. Control of time estimation by a chemical clock. Nature **210:** 1189–1190.

3. Fox, R. H., P. A. Bradbury, I. F. G. Hampton & C. F. Legg. 1967. Time judgment and body temperature. J. Exp. Psychol. **75:** 88–96.
4. Pfaff, D. 1968. Effects of temperature and time of day on time judgments. J. Exp. Psychol. **76:** 419–422.
5. François, M. 1927. Contribution à l'étude du sens du temps. La température interne, comme facteur de variation de l'appréciation subjective des durées. Ann. Psychol. **28:** 186–204.

The Role of Temporal Factors in Multiple Schedules

ALISON D. HASSIN-HERMAN

Department of Psychology
Queens College
City University of New York
Flushing, New York 11367

The control of a pigeon's keypeck response by a stimulus–reinforcer (S–SR) relation was investigated in a four-component multiple schedule. Pigeons were exposed to a multiple schedule containing two fixed two-component sequences. Each sequence was differentially cued during the initial component (6 sec) of each sequence. Control of behavior in the initial component by an S–SR relation was demonstrated. That is, when an S-SR relation was introduced by the elimination of reinforcement (Ext) in the terminal component of one sequence, responding in the initial component of the other (RI) sequence increased. Brown *et al.*[1] proposed that responding during one component cue was dependent upon the delay to food signaled by that cue relative to the delays to food signaled from each of the other cues in the experimental setting. The component cue that signals a relatively high density of reinforcement concurrently signals a relatively short delay to food. Hence, high rates of responding would be expected in a component that was associated with a relatively short delay to food as well as a relatively high density of reinforcement.

In Experiment 1, keypecking emerged in the initial component that was associated with a relatively short delay to food in the terminal component (RI sequence). With an S–SR relation still in force, responding varied inversely with the duration of the initial

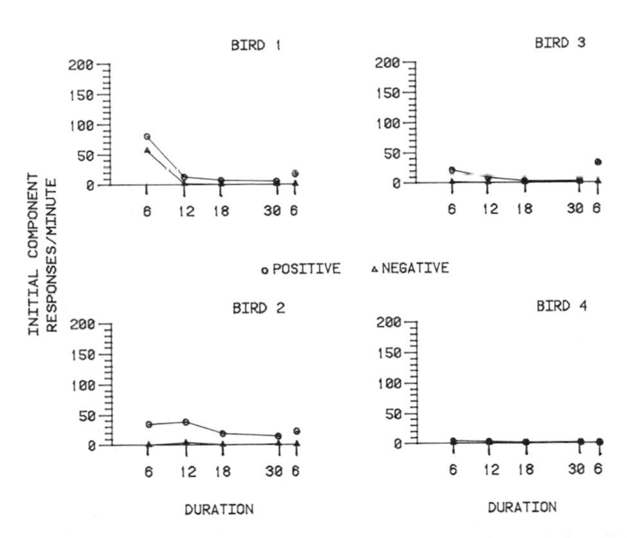

FIGURE 1. Mean response rates of the last ten sessions of phases VI and VII of Experiment 1. Each graph presents one bird's rate of responding during the initial component for each manipulation of the initial component duration.

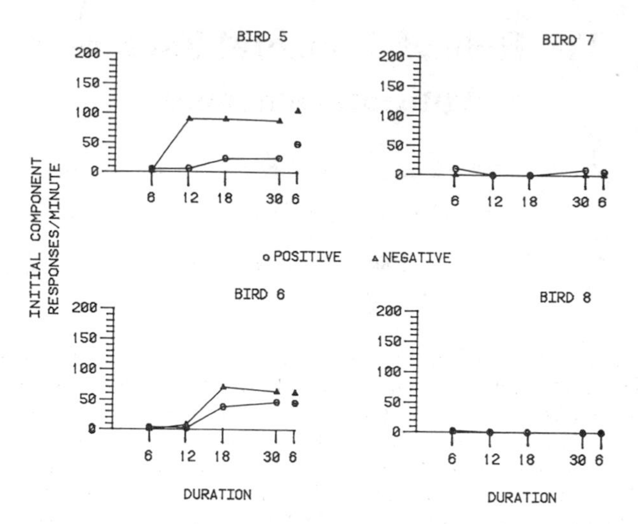

FIGURE 2. Mean response rates of the last ten sessions of phases VI and VII of Experiment 2. Each graph presents one bird's rate of responding during the initial component for each manipulation of the initial component duration.

components as well as their signaled delays to food. FIGURE 1 presents the mean response rates during the initial component duration.

In Experiment 2, responding during the initial component was influenced by the addition of a response-reinforcer (R–SR) relation in these components. Contrary to the findings of Experiment 1, responding that occured in the initial component of each sequence increased as the duration of the initial components increased (FIG. 2). Under

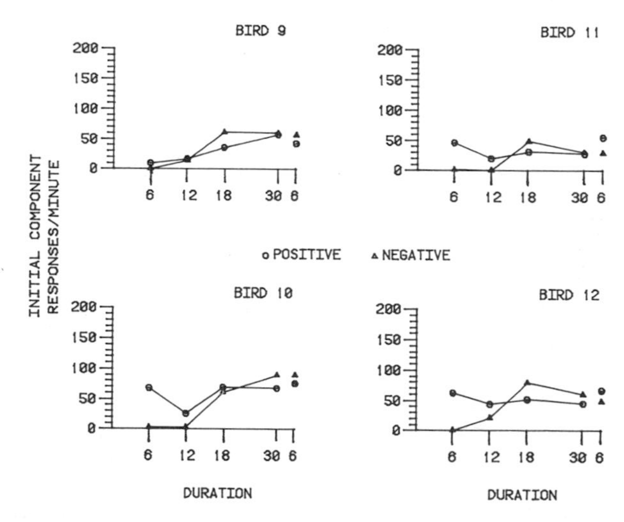

FIGURE 3. Mean response rates of the last ten sessions of phases VI and VII of Experiment 3. Each graph presents one bird's rate of responding during the initial component for each manipulation of the initial component duration.

the extended durations, responding during the initial component of the extinction sequence was higher than the responding during the initial component of the RI sequence.

Experiment 3 differed from Experiment 2 in the addition of discriminative stimuli to the terminal component of each sequence. As in Experiment 2, responding in the initial component of each sequence increased as the initial component duration increased (FIG. 3).

The data of Experiments 2 and 3 stand in opposition to those of Experiment 1, but these findings are similar to those of current research in multiple schedule paradigms (such as that of Williams[2]). These data also demonstrated a higher response rate in a component that preceded an extinction component than in a component that preceded a component containing a reinforcement schedule. The present investigation reveals responding in the initial component to be influenced by both component duration and the schedule of reinforcement. Differences in the results of the present study and those of Williams[2] and Brown *et al.*[1] appear to be the result of these two variables. These findings suggest the joint control of behavior by both S–SR and R–SR relations. At short durations, the S–SR relation was responsible for the appearance of keypecking in the initial component. At long durations, the S–SR relation waned in its control and the R–SR relation exerted its own control over responding.

REFERENCES

1. BROWN, B. L., N. S. HEMMES, D. A. COLEMAN, A. HASSIN & E. GOLDHAMMER. 1982. Specification of the stimulus–reinforcer relation in multiple schedules: Delay and probability of reinforcement. Anim. Learn. Behav. **10:** 365–376.
2. WILLIAMS, B. A. 1979. Contrast, component duration, and the following schedule of reinforcement. J. Exp. Psychol. Anim. Behav. Processes. **5:** 379–396.

Discrimination of Temporal Components of Acoustic Patterns by Birds

JOHN HUMPAL AND JEFFREY A. CYNX

Department of Psychology
The Johns Hopkins University
Baltimore, Maryland 21218

Two female European starlings (*Sturnus vulgaris*) learned a discrimination between an isochronous, or rhythmic, and a random pattern of tones. Two transfer patterns replaced the isochronous pattern with partly randomized patterns. The birds were able to maintain the discrimination well above chance levels, although with some deficit relative to initial training.

For initial discrimination training, two acoustic patterns, shown at the top and bottom of FIGURE 1, were constructed out of 2000-Hz sinewaves played at 70 ± 5 dB SPL. The starling was placed in a cage suspended in the center of an IAC acoustic chamber where she was confronted with a response panel of three response disks in a horizontal row, and a food magazine below the disks. To commence a trial, subjects were required to peck and extinguish the lit center disk. A peck to the center key resulted in the playing, with equal probability, of either the isochronous or random pattern for 4 sec. After the 4-sec listening period, the two side keys lit up. If an isochronous pattern was presented, a peck to the left key resulted in extinction of the key lights and pattern and presentation of food. A peck to the right key caused

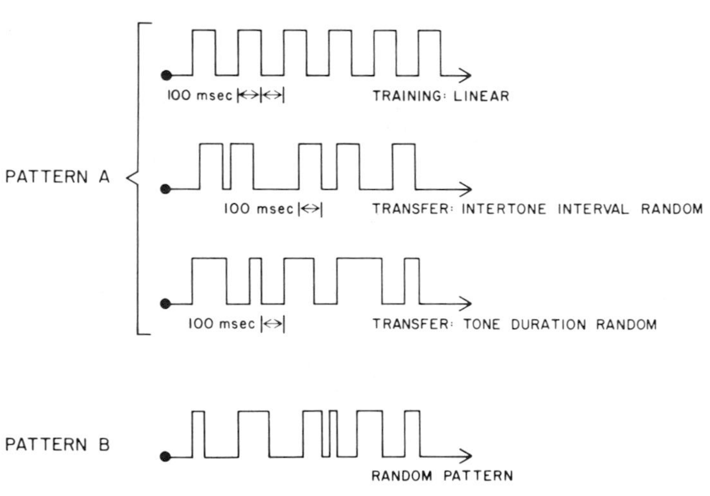

FIGURE 1. The isochronous and random patterns of initial training are shown in the top and bottom patterns of the figure, respectively. A pattern with a random intertone duration and constant (100 msec) tone duration is shown in the second trace. The third pattern is a sample of a pattern with random tone durations and 100-msec intertone intervals.

TABLE 1. Repeated Measures Analysis of Variance Showing Test of Subjects by Pattern Design

Source	Subjects	d.f.	M.S.	F
Birds	0.26	1	0.26	<1.0
Patterns	1272.77	2	636.39	18.11[a]
Interaction	228.52	2	114.26	3.25
Error	281.17	8	35.15	—
Totals	1782.72	13	—	—

[a] $p < .01$.

extinction of all lights and stimuli for a 10-sec time-out period. The reverse contingencies held if a random pattern was played. After reaching criterion on daily, ½-hr sessions, the isochronous pattern was altered in transfer tests by randomizing either tone or intertone durations. Birds were exposed to both types of transfer for two sessions each in counterbalanced order.

Performance for both starlings on the transfer patterns declined relative to baseline performance on the training pattern. For the randomized intertone interval, the birds performed at 67.0% correct; performance on the randomized tone duration was 65.25% correct. Baseline performance was 86.67% correct. As can be seen in TABLE 1 and FIGURE 2, this deficit was reflected by a significant difference for type of pattern ($F = 18.11$; d.f. $= 2,8$; $p < .01$). No other significant effects were found, although performance for all tests was well above chance (TABLE 2).

These results indicate that the starlings did not rely on the isochronicity of the *whole* pattern to discriminate it from a completely random pattern. Instead, the birds responded to the periodicity of only one component of the pattern. That component, either the tone or intertone interval, need only retain a constant duration throughout the pattern. The degradation of isochronicity by way of randomization of a formerly constant temporal interval in a well-learned pattern allows for the ranking of patterns

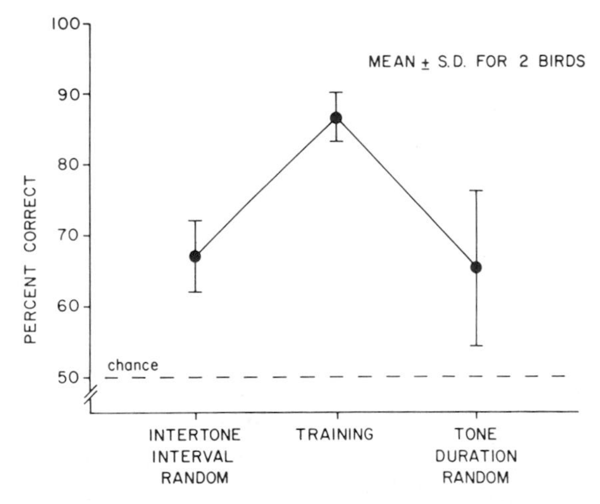

FIGURE 2. Performance of two birds on baseline discrimination and with randomizing of tone and intertone.

TABLE 2. Independent One-Way *T* Tests Comparing Test Performance and Chance

Test	d.f.	*t* Test	
Intertone interval randomized	3	6.94	$p < .005$
Tone duration randomized	3	2.80	$p < .05$
Training	3	25.64	$p < .001$

along an ordinal scale, the dimension of which ranges from random to isochronous temporal intervals. The fact that the starlings' performance also permits ranking in accordance with the regularity of the patterns suggests that the birds may employ an ordinal scale of isochronicity in the discrimination task we use. The results of these transfers argue for further experiments, perhaps employing a probabilistic schedule of the occurrence of intervals of constant duration.

The Processing of Temporal Information

J. L. JACKSON, J. A. MICHON, AND A. VERMEEREN

Institute of Experimental Psychology
University of Groningen
Haren, the Netherlands

Our research sets out to explore whether temporal information is dealt with as an inevitable automatic byproduct of information processing in general, or whether temporal information processing in itself requires a specific cognitive effort, such as use of adequate rehearsal and/or encoding strategies.

In examining this question three response measures found to vary in level of difficulty will be used:

1. *Order judgments:* Which of two items appeared earlier in a series?
2. *Lag judgments:* How many items were presented between two such items?
3. *Position judgments:* Give the absolute positions of items in a series.

An attempt is made to compare and evaluate these measures in relation to the underlying processes.

In the first experiment to be described, the question asked was: What effect does depth of processing have on the retention of temporal information? It is known that deeper processing improves recall; therefore, does it also increase the retention of temporal information?

Subjects were shown a series of 25 concrete words. In the "deep" condition, they were asked whether the word fitted a sentence; while in the "shallow" condition, the task was to check the word for a given letter. After presentation, an unexpected temporal-order task (either order, lag, or position judgment) was given. A control group performed a free recall task. The results show that although semantic processing produces moderate item recall, it does not produce adequate order retention. Performance in all temporal tasks is very low, barely reaching chance level in the easiest task, namely, order judgments. These results suggest that deliberate effort such as use of adequate encoding and/or rehearsal strategies is necessary for temporal coding to take place.

The second experiment takes a closer look at rehearsal strategies, which are known to increase with age. The question explored is whether the processing of temporal information is influenced by such developmental trends.

Subjects in this experiment were 5- and 11-year-old children who were presented with series of either 7 or 28 pictures. Half of each age group was given a serial rehearsal instruction, while the other half received no such instruction. The position task which followed presentation was expected. (To match the short 7-picture series, the long series were blocked into 7 units of 4 positions each.)

Although both age groups performed better on short series, a developmental effect in position judgments was found, suggesting that as rehearsal processing increases, so too does temporal information. Because of poor fulfillment of rehearsal instructions by 5-year-old children, the expected improvement over the no-instruction group was not found. With 11-year-olds, however, an effect of rehearsal instruction was indeed found in long series, which strongly suggests that the processing of temporal information requires deliberate effort.

Though *post hoc* order judgments derived from position judgments show a slight developmental trend, the most striking effect is the high level of performance in all groups (80–97.5% correct). An important question arises: Are intentional order judgments comparable to those derived from position judgments? Are the underlying processes identical?

An evaluation of temporal information measures such as the relationship between order, lag, and position judgments and the search for one underlying processing model is one of our main research goals.

The third experiment describes one attempt in this direction. It explores whether or not an extra operation is involved in making random-order (test pairs presented either in order *a–b* or *b–a*) as opposed to fixed-order (test pairs always presented in order *a–b*) lag judgments. A second question asks whether such lag judgments, known to be difficult, can be improved through training that emphasizes use of adequate strategies and also asks whether this training is equally effective for young and old subjects (average age 20 and 60, respectively).

Subjects were presented with series of 12 abstract words and they knew that they would be asked to make either order, fixed-order, or random-order lag judgments. Speed and accuracy of responses were measured. The results show that training in use of strategies suitable for judging lags, that is, the use of deliberate processing, does indeed improve performance. An effect of age was most clearly visible in the speed of responses, with older subjects taking an average 1.5 sec longer than the younger subjects. The more interesting finding, however, was that the pattern of results was similar. This pattern suggests that random-order lag judgments may indeed require an extra operation as opposed to fixed-order lag judgments, namely, ordering the pairs before further analyses.

The three experiments described therefore strongly suggest that temporal information processing requires a specific cognitive effort and also show that it may be possible to integrate various types of temporal judgments into one processing model.

Effect of Photoperiod on Timing of Incubation in Ring Doves

ROBERT M. KAHN

Department of Psychology
Columbia University
New York, New York 10027

The incubation pattern of the ring dove provides a model system for analyzing the properties of a socially mediated circadian behavior. The pattern of a given pair remains stable from day to day, with the male sitting for a block of time in the middle of the day, and the female sitting the rest of the time (FIG. 1). The timing of the sitting bouts is dependent upon a circadian system. The pattern persists, and free-runs, in constant dim light (0.5 lux), and with a brief transition period, the doves phase-shift their incubation pattern in response to a phase-shift of the light:dark regimen.[1] Other studies suggest that whereas the onset of incubation by each mate occurs at a particular time of day (controlled by the circadian mechanism), the end of the sitting bout by each mate is determined by a measure of duration (controlled by an interval timing mechanism).[2]

In the 14:10 light:dark (LD) cycle used in previous experiments, the incubating male and female share the daylight hours (FIG. 1). In the present series of experiments we asked how sitting duration, and the coupling of morning and afternoon nest exchanges to dawn and dusk, are affected by photoperiod. One possibility is that doves continue to split the daylight hours. Alternatively, the male and female may each sit for a fixed duration irrespective of daylength. In addition, the phase angle of the morning (female-to-male) and/or afternoon (male-to-female) nest exchange may be conserved with respect to dawn and/or dusk. Doves were paired in a 14:10 light:dark regimen. On day 1 of incubation, the birds were placed in one of five light:dark cycles: 6:18, 10:14, 14:10, 18:6, or 22:2 ($N = 6$ in each condition). The identity of the incubating bird was monitored for 25 days by a computer-controlled radio telemetry system, with a time-lapse video-recorder providing supplementary data as needed.

There was an overall effect of photoperiod on the duration of the male sitting bout and therefore, on the female sitting bout as well (measured by calculating the difference between male onset and female onset for each pair; FIG. 2); (One-way ANOVA, $F(4, 21) = 7.21$, $p < .01$). Individual paired comparisons using Student's t test indicate that only the 6:18 and 22:2 groups differed significantly from the other LD conditions. In LD 6:18, the males' daily sitting bout usually began soon after lights-on, and ended around the time of lights-off; (6:18 versus 10:14: $t(9) = 4.0$, $p < .01$). It is possible that the duration of male sitting in this photoperiod was constrained by an aversion on the part of male doves to incubating during the dark phase. In LD 22:2 (not shown in FIGURE 2), male sitting duration was 10.9 ($\pm$.6) hours, significantly longer than in any other group (22:2 versus 18:6: $t(9) = 3.45$, $p < .01$). However, the characteristic organization of incubation was disrupted in this condition, with male sitting divided into two or three long bouts instead of the usual single main bout. Neither the morning nor the afternoon nest transition maintained a constant phase angle with respect to dawn or dusk (95% confidence intervals of the slopes of the regression lines [phase-angle as a function of photoperiod] for onsets of incubation:

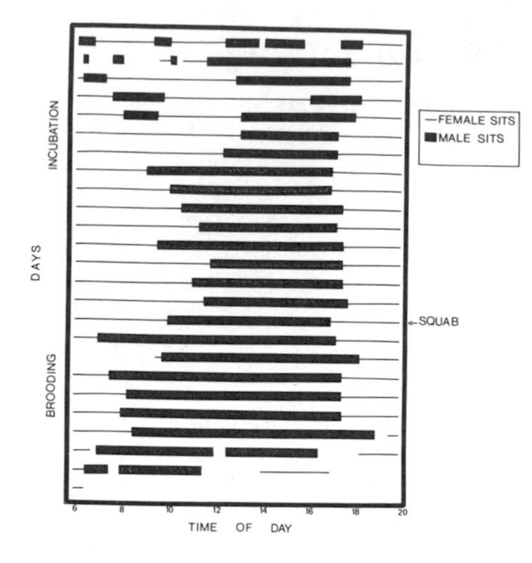

FIGURE 1. Daily timing of incubation on the 25 days after egg-laying by a representative pair of doves housed in LD 14:10. Only the daylight hours are shown. In all cases, the female sat through the night. (Modified from Ball and Silver.[3])

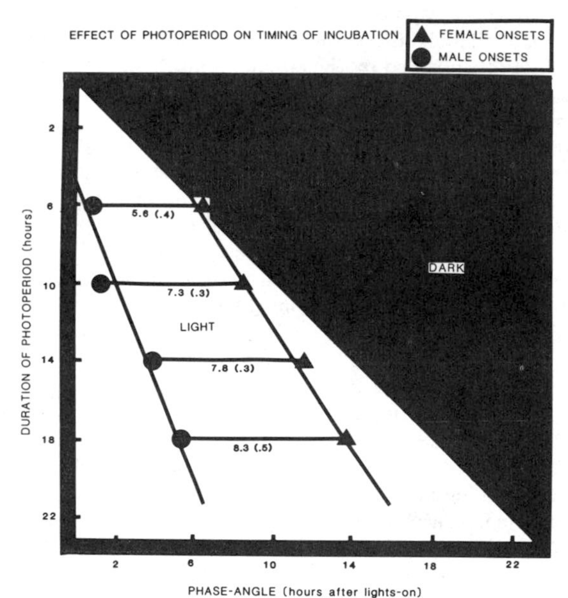

FIGURE 2. Mean onset of incubation in relation to light and dark portions of the LD cycle. Mean duration of the male sitting bout (standard error of the mean in brackets) is also shown for each condition. For each photoperiod listed on the vertical scale, the diagonal line represents lights-off, and the vertical sides of the square the time of lights-on. The "glitch" in the diagonal line at photoperiod 6 reflects the actual light:dark cycle in that experiment (6.3:17.7). Least-square regression lines for male and female onsets are shown. The organization of the figure is modified from Elliott.[4] Criteria for male onset: the start of the first sitting bout of 20 minutes or more following lights-on. Criteria for female onset: the start of the first sitting bout of 20 minutes or more following lights-on that was not followed by any significant male bouts (20 minutes or more) in that cycle (that is, until the next lights-on). Data are based on day 6 of incubation until nest abandoning started (around day 21). Data from pairs with highly irregular sitting patterns (for example, if the male regularly sat through the night) were not tabulated of (N = 2).

male: $.33 < b < .52$; female: $.55 < b < .76$, (d.f. = 20). The lines representing lights-on and lights-off have slopes of 0 and 1, respectively; FIG. 2).

While a constant phase relation was not found, the phase-angle of the female onset was somewhat better conserved with respect to dusk than to dawn throughout the range of photoperiods employed. In photoperiods under 12 hours, male onset seems to be tracking dawn, rather than the dusk, signal, but a more definite conclusion requires testing in additional photoperiods. Especially in light of the wide range of photoperiods employed, the most notable aspect of the results must be how little sitting duration changed in four of the five photoperiods. Socially modulated circadian and interval-timing mechanisms, operating in both mates, enable ring doves to evolve a stable sitting rhythm over a considerable range of photoperiods.

REFERENCES

1. SILVER, R. & S. NELSON. Social factors influence circadian cycles in parental behavior in doves. Physiol. Behav. In press.
2. GIBBON, J., M. MORRELL & R. SILVER. 1983. Two kinds of timing in the circadian incubation rhythm of the ring dove. Submitted for publication.
3. BALL, G. F. & R. SILVER. 1983. Timing of incubation bouts by ring doves (*Streptopelia risoria*). J. Comp. Psychol. **97:** 213–225.
4. ELLIOTT, J. A. 1976. Circadian rhythms and photoperiodic time measurements in mammals. Fed. Proc. **35:** 2339–2346.

Duration Perception and Auditory Masking

HOWARD J. KALLMAN AND MARYELLEN D. MORRIS

Department of Psychology
State University of New York at Albany
Albany, New York 12222

Massaro and his colleagues have developed a model of auditory information processing that is based, in part, on the results of backward recognition masking experiments.[1] The model assumes that the features of a presented sound are initially stored in a preperceptual auditory store. Perception of the sound involves reading out the features from this store and synthesizing a perceptual unit. If a second sound (a backward mask) occurs shortly after a target, it is assumed to replace the target in preperceptual storage, thereby terminating perceptual processing of the target. The perceptual clarity of the target sound thus depends on the amount of processing time available prior to mask presentation.

This model has been applied to the perception of auditory duration.[2] The model assumes that a subject's ability to discriminate auditory durations improves as the interstimulus interval (ISI) separating a target sound from a backward mask increases. In addition to the loss of discriminability with short ISIs, Massaro and Idson's data suggested that the perceived durations of both long and short target sounds are shortened when processing time is limited by a backward mask.

In the present study, the effects of forward as well as backward masks were assessed. If the effect of the backward mask is presumed to result from its' terminating the processing of the preceding target, there is little reason to expect comparable forward masking. Backward and forward mask conditions were presented in separate blocks to ten undergraduates. The subject's task was to judge whether a target tone (which had a duration of either 38 or 72 msec) was relatively long or short. After the response, the subject received feedback indicating the correct answer. The target and the 100-msec mask had frequencies of 800 Hz and were separated by a silent ISI of either 20, 40, 60, 80, 160, 250, or 500 msec, although on backward mask trials a no-mask condition was substituted for the 500-msec ISI. The target was presented at either 71 or 81 dB(A) and the mask was always presented at 81 dB(A). All experimental conditions occurred randomly within a block of trials, except that forward and backward mask conditions were blocked and counterbalanced.

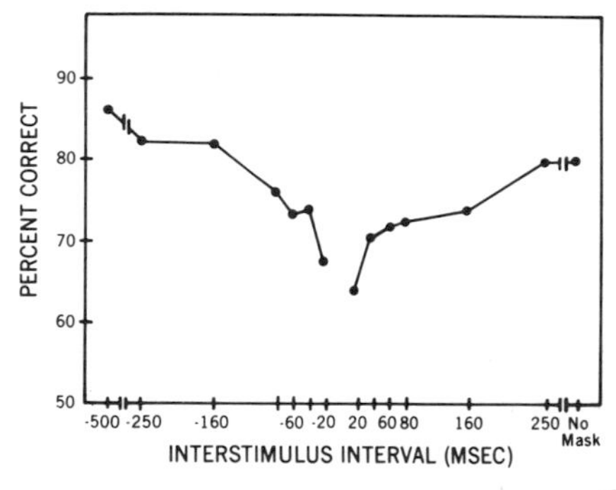

FIGURE 1. Percentages correct as a function of the silent interstimulus interval (ISI) between the test tone and mask. Positive ISI values refer to backward masking trials and negative ISI values refer to forward masking trials.

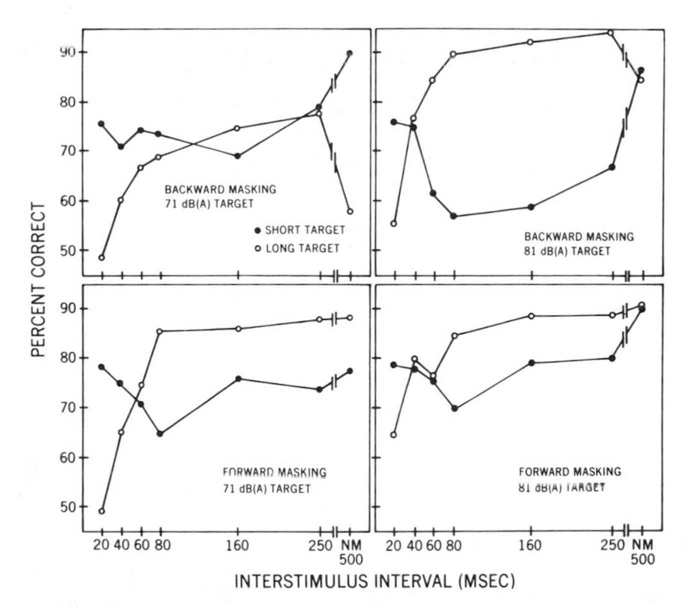

FIGURE 2. Percentages correct as a function of ISI for each combination of mask condition and target amplitude; the separate curves plot percentages correct for the short and long test tones.

On backward mask trials, Massaro and Idson's results were generally replicated. The overall percentage of correct duration judgments increased with increases in ISI (FIG. 1). FIGURE 2 shows that the effect of ISI differed substantially for short and long targets. These data replicate the results of Massaro and Idson,[2] who suggested that the perceived duration as well as clarity of a target sound increases as the ISI separating the backward mask from the target increases. Although the data from backward mask trials replicated the results reported by Massaro and Idson, the data from forward mask trials are not predicted by their theory. Indeed, the amount of forward masking found was about as great as the amount of backward masking. FIGURE 2 shows that the interaction between ISI and target duration was similar in form on backward and forward masking trials. There was also evidence that the 81-dB(A) target was perceived as longer than the 71-dB(A) target; this was evident on no-mask trials as well as on mask trials.

Any adequate theory of duration masking will have to account for the finding that forward masking is approximately equal to backward masking of auditory duration. This finding is problematic to an interruption theory of duration masking and suggests that alternative explanations of auditory duration masking—for example, explanations that incorporate integrative processes—need to be developed.

REFERENCES

1. KALLMAN, H. J. & D. W. MASSARO. 1983. Backward masking, the suffix effect, and preperceptual storage. J. Exp. Psych. Learn. Memory Cognition **9:** 312–327.
2. MASSARO, D. W. & W. L. IDSON. 1976. Temporal course of perceived auditory duration. Percept. Psychophys. **20:** 331–352.

Familiar Melodies Seem Shorter, Not Longer, When Played Backwards: Data and Theory

KATHLEEN KOWAL

Department of Psychology
University of North Carolina at Wilmington
Wilmington, North Carolina 28403

The apparent duration of events is often reported to decrease as their familiarity, predictability, or number increases.[1,2] This finding has been challenged,[3,4] however, and may be due to the specific methods and stimuli used in testing, it may result from a misconstrual of subjects' perceptions of the events, or it may reflect insufficient data points to establish firm psychometric functions.

In the studies reviewed here, the events were sequences of musical notes which varied in familiarity, organization, and predictability; a total of 300 subjects judged (by magnitude estimation or by verbal estimation) the duration, number, familiarity, organization, and predictability of those events in 12 separate experiments. The event sequences varied from 8 to 57 sec in duration and were either musical melodies, or the same notes in reverse sequence. Melodies and reverse melodies were judged both independently and in repeated measures designs.

Musical melodies were judged significantly more familiar, more organized, and more predictable than were reverse sequences. With respect to estimates of duration or number, results were unequivocal: The more familiar, organized, and predictable melodies were judged as being longer in duration and as having more notes than the less familiar, less organized, and less predictable reverse sequences. Both magnitude estimations and verbal estimations of duration reflected stimulus duration: Correlation coefficients relating perceived duration to duration ranged from .83 to .97. Means of individual exponents derived from magnitude estimations were .99 for melodies and .95 for reverse sequences, which are within the range reported in the literature,[2] and

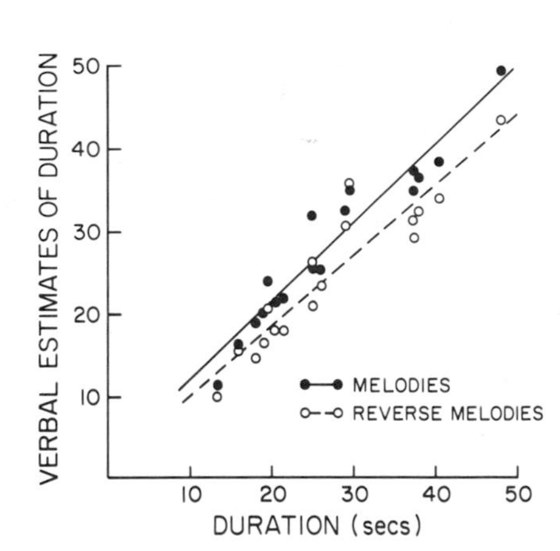

FIGURE 1. Mean verbal estimates of duration for 17 durations of melodies and 17 durations of reverse melodies obtained from 34 observers. Best-fitting straight lines are derived from the mean estimates.

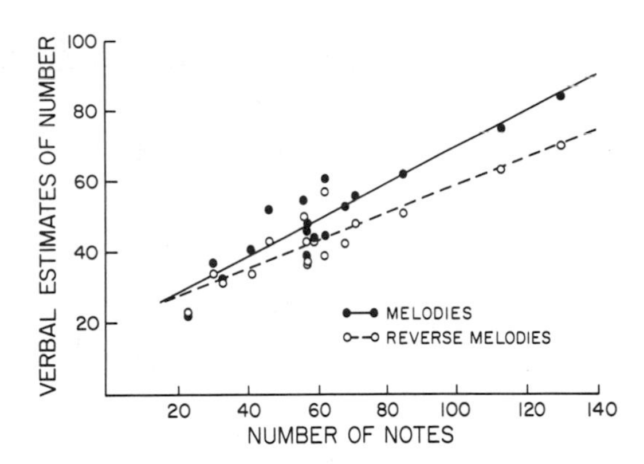

FIGURE 2. Mean verbal estimates of number of notes for 17 melodies and 17 reverse melodies obtained from 22 observers. Best-fitting straight lines are derived from the mean estimates.

showed no consistent effects of organization or familiarity (although the estimates themselves were significantly greater for melodies).

When measured with naturalistic stimulus events, therefore, over a wide range of durations, perceived duration bears an orderly and lawful relationship to manipulated and judged familiarity, to actual duration, and to manipulated and judged number of components (that is, notes). But it is not the relationship so widely cited in the literature: Apparent duration increases, rather than decreases, with increased familarity and organization.

The overall pattern of results is consistent with the idea that apparent duration is affected by apparent (as distinct from real) numerosity. With the present events, familiar melodies permit the perceiver to regenerate more notes, and durations therefore seem longer than with unfamiliar reverse sequences. In other cases, organizational structure might lower perceived numerosity (such as phonemes in speech) and consequently lower perceived duration. By this theory, therefore, effects of organization on events' duration would depend on how organization affects apparent numerosity, and its effects upon perceived number must be known before predictions about time perception can be made.

REFERENCES

1. FRAISSE, P. 1963. The Psychology of Time. Harper & Row. New York, NY.
2. SCHIFFMAN, H. R. & D. J. BOBKO. 1974. Effects of stimulus complexity on the perception of brief temporal intervals. J. Exp. Psychol. **103:** 156–159.
3. KOWAL, K. 1981. Growth of apparent duration: Effect of melodic and non-melodic tonal variation. Percept. Motor Skills **52:** 803–817.
4. SCHIFFMAN, H. R. & D. J. BOBKO. 1977. The role of number and familiarity of stimuli in the perception of brief temporal intervals. Am. J. Psychol. **90:** 85–93.

Tactile Temporal Acuities

EUGENE C. LECHELT

Department of Psychology
University of Alberta
Edmonton, Alberta, T6G 2E9 Canada

Basic research in cutaneous communication and sensory substitution systems that convey "visual information" via tactile stimulation has clearly shown that, in order for the substituted visual information to have tactile significance of credible ecological validity, it is critical that the tactile stimulation follow with great fidelity the actual temporal dynamics of stimulus change in the visual environment. Consequently, the delineation of tactile temporal resolutions, that is, acuities, is a most basic perceptual issue and is of central importance in the design of tactile sensory substitution systems which best incorporate the temporal discriminatory capabilities of the human skin.

Experiments were conducted to specify the effects of strategic variables on thresholds for temporal "discontinuity" and two-pulse "discreteness" in two-pulse tactile sequences. Specifically, two square-wave mechanical taps were delivered with variable interstimulus intervals (ISI values) to the pads of the distal phalanges of either the middle or ring fingers of both hands and 1 cm on either side of the midline of the center of the dorsal surface of each forearm. Two levels of stimulus intensity were employed: 7 dB SL and 14 dB SL.

Two thresholds were determined for four highly trained observers at every single locus, that is, the two pulses were delivered successively to a single finger or forearm location, as well as for dual loci conditions, that is, the two pulses were delivered successively to both fingers on each hand or both loci on either forearm. Bilateral finger and forearm conditions were also tested.

Subjects served in each of 24 stimulus location times stimulus intensity conditions according to different randomized orders. Within each condition, five ISI values were randomly repeated 50 times. In the first "discontinuity" experiment, subjects were instructed to report either "constant" (if the two-pulse sequence felt continuous) or "discontinuous" (if any irregularity in the sequence was perceived). For the second two-pulse "discreteness" experiment, an analogous procedure was used, but subjects responded either "two" when they perceived two discrete pulses or "one" when they did not perceive the pulses to be temporally discrete. The particular ISI values chosen for each condition were based on extensive pilot work and provided for a distribution of near 0% to near 100% reports of "discontinuity" (experiment I) or "two" pulses (experiment II).

By means of a constant stimulus method and a two-alternative forced-choice response paradigm, thresholds for temporal "discontinuity" and temporal "discreteness" were determined by converting the averaged proportion of "discontinuous" and "two" responses to Z scores. The method of least squares was then used to obtain the 75% thresholds, that is, the ISI at which subjects responded "discontinuous" or "two" 75% of the time, for each treatment condition. T values were also computed to test for the significance of threshold differences between selected treatment conditions as illustrated in FIGURE 1. The major findings were that reliably lower "discontinuous" and two-pulse "discreteness" thresholds were obtained under conditions of high stimulus intensity, at finger compared to forearm loci, and for unilateral dual loci compared to bilateral dual loci stimulation sequences. Furthermore, a high correlation

(ρ = .90) between the "discontinuity" and "discreteness" thresholds across the conditions of stimulus location times stimulus intensity suggests that the less-well-documented instance of temporal discontinuity in a simple tactile stimulus energy package is indeed a reliable and even more acute specification of temporal resolution than is two-pulse discrimination.

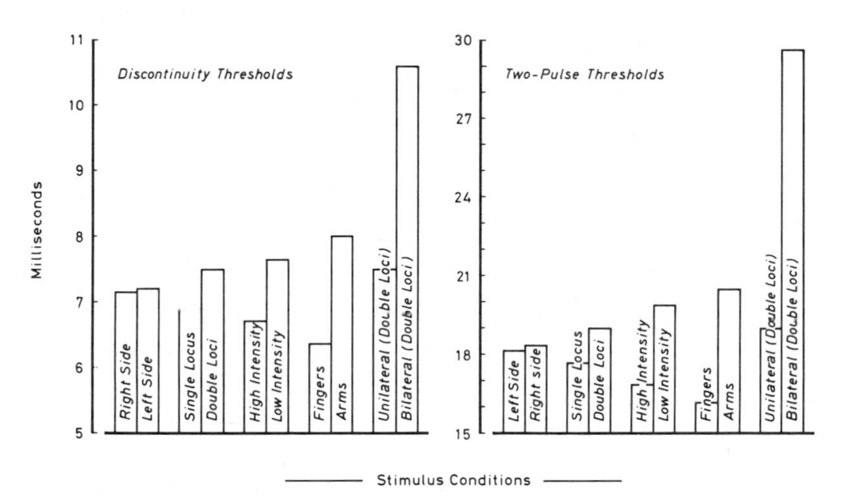

FIGURE 1. Discontinuity and two-pulse thresholds (in msec) shown separately for specific manipulations of stimulation parameters. These values (thresholds) represent averages across imbedded conditions of stimulation.

The data are relevant in examining the role of sensory peripheral versus more central mechanisms, the significance of peripheral nerve innervation density, the involvement of more cognitively based attentional factors, and laterality differences in perceptually defined acuities of stimulus timing differences. At an applied level, the data have implications for the design of tactile/vibratory sensory substitution systems for the sensorily handicapped.

Lick Rate and the Circadian Rhythm of Water Intake in the Rat: Effects of Deuterium Oxide[a]

DIOMEDES E. LOGOTHETIS,[b] ZIAD BOULOS,[b] AND MICHAEL TERMAN[c]

Department of Psychology
Northeastern University
Boston, Massachusetts 02115

Deuterium oxide (D_2O, heavy water) consistently and reproducibly lengthens the period of circadian oscillations in plant and animal species.[2,3,6,7,11] D_2O also lengthens the periods of several high-frequency (ultradian) rhythms, including the electric-organ discharge of a gymnotid fish, and the respiratory and cardiac cycles of crustaceans.[9]

Licking in many mammals is a highly stereotyped behavior that occurs at a relatively constant rate; in the rat, rates of 5–8 licks per second have been obtained with a variety of drinking solutions under several deprivation levels.[5] Although some investigators have concluded that lick rate is invariant, it is now clear that some variables can exert small but significant effects.[5] The present study was undertaken to compare the effects of D_2O on the period of the circadian rhythm of water intake and on concurrently measured lick rates in the rat.

Four adult male Long–Evans rats were maintained under constant dim red light (0.7 lux) in individual environmental chambers. Our methods for recording and analysis of circadian drinking rhythms have been described in detail.[1] In brief: licks at the drinking spout interrupted the infrared photobeam of a drinkometer, and were recorded on cumulative recorders and printing counters which were reset at hourly intervals. The hourly totals were displayed graphically, and were subjected to spectral analysis for quantification of free-running periods. Lick rate determinations were made by means of a PDP-12 computer, which recorded interlick intervals (ILIs) in 20-msec bins.

The rats were first given free access to drinking water, followed by three concentrations of D_2O in water (10%, 30%, 50% by volume), in different orders, and finally returned to water. Each of these five conditions was maintained for 30 days or longer. ILI samples were recorded for 3–5 consecutive days, towards the end of each condition. ILI distributions were obtained for each sample, and the mean ILI was calculated for all intervals between 100–220 msec.

D_2O resulted in a dose-dependent lengthening of the period of the circadian water-intake rhythm (FIGURE 1A and B). The changes in period took place gradually, over several days. The magnitude of the effect was similar to that reported for locomotor activity rhythms in several rodent species.[7,8,11] In three of the four animals,

[a]This research was supported by Biomedical Sciences Support Grant RR07143 and by Grant 27442 from the National Institutes of Mental Health.

[b]Present address: Department of Physiology and Biophysics, Harvard Medical School, Boston, Massachusetts 02115

[c]Present address: Department of Psychophysiology, New York State Psychiatric Institute, 722 West 168th Street, New York, New York 10032.

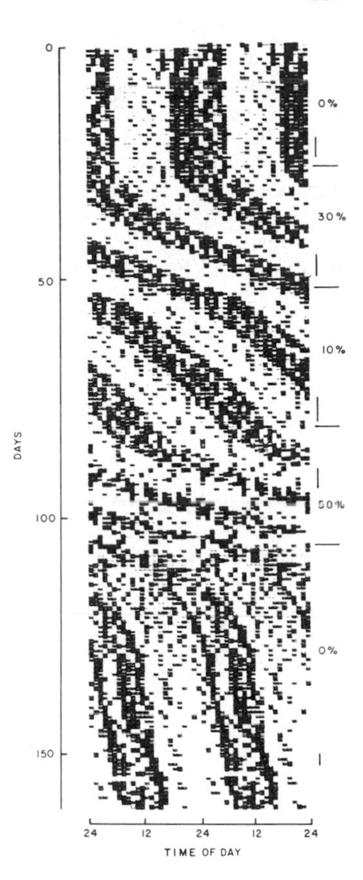

FIGURE 1A. Ranked quartile plot of drinking behavior for one rat. Each hour of the day is represented by a symbol, the size of which is proportional to the total number of licks for that hour. The data are double-plotted on the abscissa, and successive days of the experiment appear on the ordinate. D_2O concentrations are shown on the right of the plot. Vertical lines on the right indicate the times of the ILI samples.

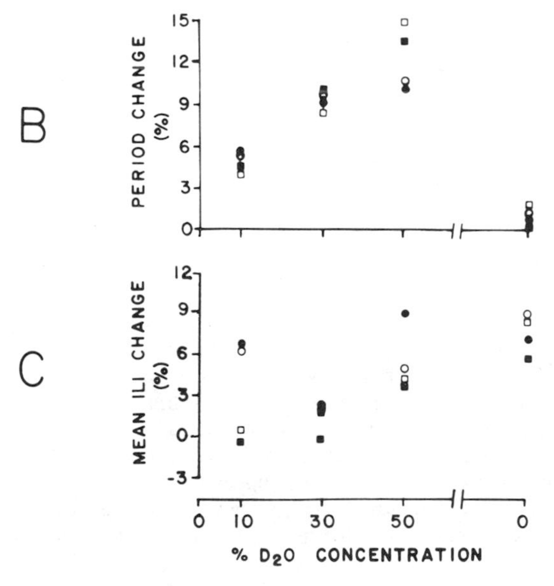

FIGURE 1B AND C. Percent change in circadian period and in mean ILI (relative to the initial water condition) as a function of D_2O concentration for each of the four rats. Standard errors of the ILI means ranged between 0.10 and 0.30 msec.

the circadian drinking pattern was disrupted, and daily intake levels were reduced towards the end of the 50% D_2O condition. Upon return to water, the free-running periods shortened again (range, 24.15–24.55 hr), although they remained slightly longer than during the initial water condition (range, 24.05–24.20 hr). This difference may either represent an aftereffect of D_2O consumption or be due to prolonged exposure to constant lighting conditions.

In general, the mean ILI was also longer under D_2O than during the prior water condition, but the changes were not uniform and showed no clear relation to D_2O concentration. Furthermore, ILIs remained elevated after the rats were returned to

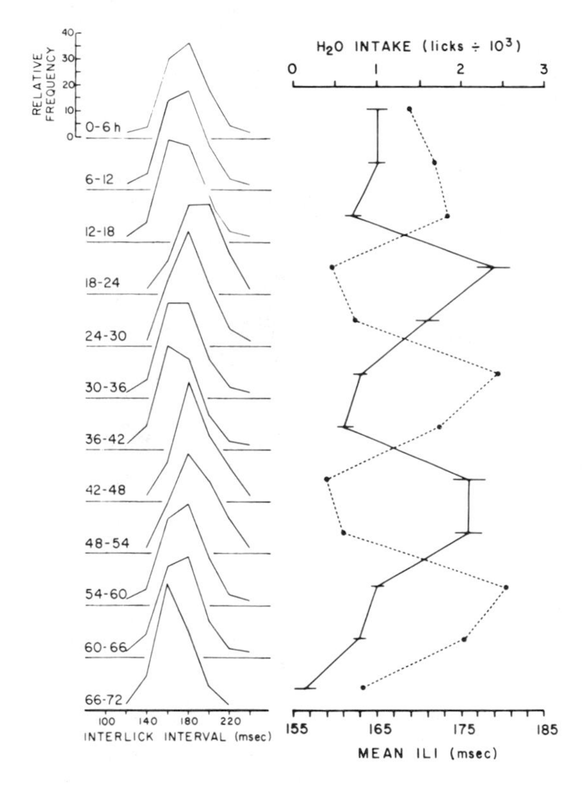

FIGURE 1D. ILI distributions (*left*) and mean ILI (±2 SEM) and total water intake (*right*) for successive 6-hr blocks from one rat under constant lighting conditions.

water (FIGURE 1C). This apparent lability of licking rates sampled at monthly intervals made it impossible to assess any effects of D_2O.

Within each sample, however, lick rates were found to be highly stable from one day to the next, and to display a circadian rhythm synchronized with that of water intake (FIGURE 1D). In all cases, lick rates were found to be faster (shorter ILIs) at the times of maximal water intake. Previous studies have demonstrated higher lick rates during the dark than during the light segment of a daily light–dark cycle in rats,[4] mice,[10] and opossums.[6] Our results indicate that these differences reflect a true circadian rhythm which free-runs under constant lighting conditions.

This circadian modulation of the frequency of the ultradian licking rhythm therefore contributes to the circadian rhythm of water intake, but the contribution is relatively minor. Circadian peak-to-trough differences in lick rate were approximately 10%, whereas water intake changed by about 400%. Thus, the circadian rhythm of water intake is mainly a function of the duration, rather than the rate, of licking.

SUMMARY

Deuterium oxide (D_2O) caused a dose-dependent, reversible increase in the period of free-running circadian rhythms of water intake in rats. Concurrently measured licking rates were generally slower during D_2O administration, but the effect was not dose-related and lick rates remained low after D_2O was discontinued. A circadian rhythm in lick rate synchronized with that of water intake was also observed.

ACKNOWLEDGMENTS

We are grateful to J.S. Terman and G. Ruben for their assistance in data collection and analysis.

REFERENCES

1. Boulos, Z., A. M. Rosenwasser & M. Terman. 1980. Feeding schedules and the circadian organization of behavior in the rat. Behav. Brain Res. **1:** 39–66.
2. Bruce, V. G. & C. S. Pittendrigh. 1960. An effect of heavy water on the phase and period of the circadian rhythm in *Euglena*. J. Cell. Comp. Physiol. **56:** 25–31.
3. Bünning, E. & J. Baltes. 1963. Zur Wirkung von schwerem Wasser auf die endogene Tagesrhythmik. Naturwissenschaften **50:** 622.
4. Cone, A. L. & D. M. Cone. 1973. Variability in the burst lick rate of albino rats as a function of sex, time of day, and exposure to the test situation. Bull. Psychon. Soc. **2:** 283–284.
5. Cone, D. M. 1974. Do mammals lick at a constant rate? A critical review of the literature. Psychol. Rec. **24:** 353–364.
6. Cone, D. M., A. L. Cone, A. J. Golden & S. L. Sanders. 1973. Differential lick rates in opossum: A challenge to the invariance hypothesis. Psychol. Rec. **23:** 343–347.
7. Daan, S. & C. S. Pittendrigh. 1976. A functional analysis of circadian pacemakers in nocturnal rodents. III. Heavy water and constant light: Homeostasis of frequency? J. Comp. Physiol. **106:** 267–290.
8. Dowse, H. B. & J. D. Palmer. 1972. The chronomutagenic effect of deuterium oxide on the period and entrainment of a biological rhythm. Biol. Bull. **143:** 513–524.
9. Enright, J. T. 1971. Heavy water slows biological timing processes. Z. Vergl. Physiol. **72:** 1–16.
10. Murakami, H. 1977. Rhythmometry on licking rate of the mouse. Physiol. Behav. **19:** 735–738.
11. Suter, R. B. & K. S. Rawson. 1968. Circadian activity rhythm of the deermouse, *Peromyscus:* Effect of deuterium oxide. Science **160:** 1011–1014.

Self-Control and Responding during Reinforcement Delay

TELMO E. PEÑA-CORREAL AND A. W. LOGUE

Department of Psychology
State University of New York at Stony Brook
Stony Brook, New York 11794

Self-control has been defined as the choice of a delayed large reinforcer over a less-delayed small reinforcer.[1,2] Mazur and Logue[3] increased self-control in pigeons by first giving the pigeons a choice between large or small equally delayed reinforcers. Then they slowly removed the delay to the small reinforcer. The pigeons continued to choose the large, delayed reinforcer more often than did a control group not exposed to this fading procedure. Grosch and Neuringer[4] showed that pecking a different key while pigeons were waiting could also improve self-control. In many real-life situations, individuals can make various responses during the large reinforcer delay, including changing their choice. However, many previous experiments that have attempted to train self-control have involved irreversible choices of the large reinforcer.[2,3,5] Therefore, it would be helpful in understanding self-control to design a procedure in which self-control is trained while subjects have the opportunity to change their choice during the large-reinforcer delay or to make other responses.

PROCEDURE AND RESULTS

Eight food-deprived pigeons received either 6- or 2-sec food reinforcers, each delayed .1 sec, for pecks on a left or a right key, respectively. Over a period of about 400 daily sessions, the delay to the 6-sec reinforcer was increased to 6-sec. During the large-reinforcer delays, pecks on the right key immediately delivered the small reinforcer, while pecks on the left key had no effect. In the last condition the contingencies for pecking the two keys were reversed in order to test for position bias.

FIGURE 1 summarizes the results of the self-control training conditions. The circles indicate the number of initial large-reinforcer choices and the plus signs the final large-reinforcer choices when they were different from the initial large-reinforcer choices. Two pigeons (16 and 58) maintained a substantial number of initial, large-reinforcer choices throughout all of the conditions; three pigeons (18, 26, 59) occasionally chose the large reinforcer during the last conditions, but showed a gradual decrease in the number of large-reinforcer choices; and three pigeons (17, 19, 55) exhibited an abrupt decrease in the number of large-reinforcer choices to near zero.

FIGURE 2 indicates the number of initial (open columns) and final (striped column) large-reinforcer choices during the reversal condition for each one of the subjects. The number of large-reinforcer choices is almost zero for four of the subjects, even though those subjects showed a relatively high number of initial large-reinforcer choices.

TABLE 1 shows that the percentage of initial, large-reinforcer choices in this experiment is not significantly different from fading groups of previous experiments, but differs from the percentage of final, large-reinforcer choices in this experiment and from the percentage of large-reinforcer choices of no-fading groups in previous studies.

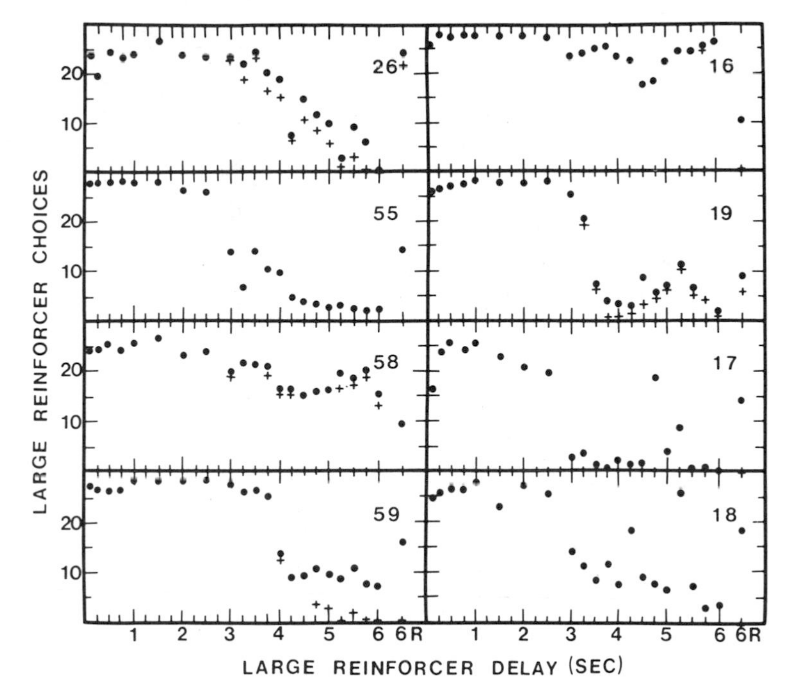

FIGURE 1. The mean number of large-reinforcer choices in the last 5 days of each condition for each experimental subject. 6R refers to the reversal condition. The *circles* represent the initial large-reinforcer choices and the *plus signs* represent the final large-reinforcer choices when they were different from the initial choices.

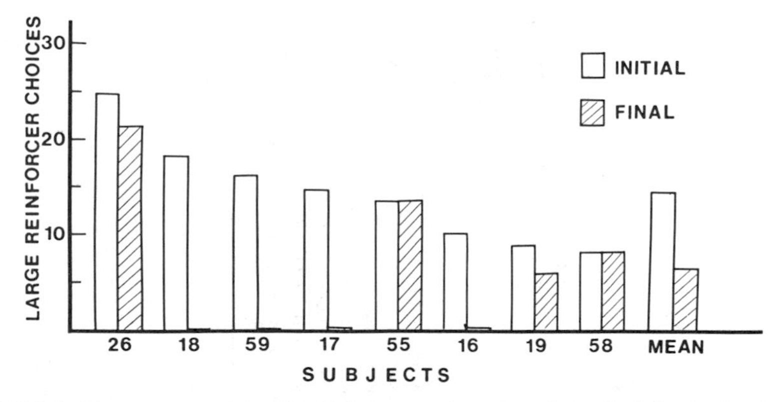

FIGURE 2. The mean number of initial (*open columns*) and the final (*striped columns*) large-reinforcer choices during the reversal condition for each one of the subjects.

TABLE 1. Comparison of Different Groups in the Number of Large-Reinforcer Choices (Significance of *t* Tests)

	Mazur and Logue[3] Fading	Logue *et al.*[5] Fading-A	Logue *et al.*[5] Fading-B	Pena and Logue Initial	Pena and Logue Final	Mazur and Logue[3] Controls	Logue *et al.*[5] (unpublished) Controls
Mazur and Logue[3] Fading	—	No	No	No	No	Yes	No
Logue *et al.*[5] Fading-A	No	—	No	No	No	Yes	No
Logue *et al.*[5] Fading-B	No	No	—	No	No	No	No
Pena and Logue Initial	No	No	No	—	Yes	Yes	Yes
Pena and Logue Final	No	No	No	No	—	No	No
Mazur and Logue[3] Controls	Yes	Yes	No	Yes	No	—	No
Logue *et al.* (unpublished) Controls	No	No	No	No	No	No	

Large-reinforcer choices and responses per second on the large-reinforcer key during the delay showed positive correlations.

Training self-control with responding during the delay is possible for at least some pigeons. These findings extend those of Mazur and Logue[3] and Logue *et al.*,[5] in which the small-reinforcer delay was gradually decreased rather than the large-reinforcer delay being increased, as was the case here. However, allowing a change of choice during the large-reinforcer delay in the present experiment probably decreased the effectiveness of the training by producing lower self-control (measured by the final large-reinforcer choices). Finally, the positive correlations between the number of large-reinforcer choices and the rate of large-reinforcer key pecks during the large-reinforcer delays supports Grosch and Neuringer's finding[4] that activity during the delay can facilitate self-control.

REFERENCES

1. AINSLIE, G. W. 1974. Impulse control in pigeons. J. Exp. Anal. Behav. **21:** 485–489.
2. RACHLIN, H. C. & L. GREEN. 1972. Commitment, choice and self-control. J. Exp. Anal. Behav. **17:** 15–22.
3. MAZUR, J. E. & A. W. LOGUE. 1978. Choice in a "self-control" paradigm: Effects of a fading procedure. J. Exp. Anal. Behav. **30:** 11–17.
4. GROSCH, J. & A. NEURINGER. 1981. Self-control in pigeons under the Mischel paradigm. J. Exp. Anal. Behav. **35:** 3–21.
5. LOGUE, A. W., M. RODRIGUEZ, T. E. PEÑA & B. MAURO. Quantification of individual differences in self-control. *In* Quantitative Analyses of Behavior, Vol. 5, M. L. Commons, J. A. Nevin & H. C. Rachlin, Eds. Ballinger. Cambridge, MA. In press.

Time Discrimination versus Time Regulation: A Study on Cats

F. MACAR, N. VITTON, AND J. REQUIN
Centre National de la Recherche Scientifique
Marseille, France

INTRODUCTION

This study analyzes the extent to which conditioning schedules in which temporal performances are elicited through different methods produce comparable performance levels. Five cats were successfully submitted to two schedules that exemplify two major categories of temporal conditioning techniques, and, more generally, two ways of dealing with time—discrimination of external duration and regulation of action. The major distinction between these procedures concerns the extent to which internal inhibition (in the Pavlovian sense) interacts with the performance. Temporal regulation tasks suppose that the response is postponed after a certain delay, and is precisely positioned on the temporal continuum. Discrimination tasks require a choice response that has no temporal link with the discrimination performance itself. Similar performances from the same animal in both procedures would constitute an argument in favor of homogeneity of time measurement processes, whereas dissimilar performances could provide indications about the sources of heterogeneity, particularly as concerns the role of internal inhibition.

METHOD

Discrimination Schedule (D): Comparison of Empty Durations

Each trial successively comprised: a standard or a comparison empty interval delimited by 50-msec auditory clicks (S1 and S2), a 2-sec delay, and an 8-sec auditory stimulus (S3) indicating the opportunity of a response (response on a left lever was correct after the standard interval, response on a right lever was correct after the comparison interval). The standard interval was always 4 sec long; the comparison interval was reduced from 10 to 5 sec by steps of 1 sec in successive blocks of five sessions. Thus, stimulus differences of 6 to 1 sec were provided.

Regulation Schedule (R): Differential Reinforcement of Response Latency

In each trial, a response on the only lever was correct if it followed a 50-msec auditory click (S1) by at least 4 sec. The upper limit for reinforcement availability was progressively reduced from 10 to 4.5 sec by steps of 1 (and finally 0.5) sec in successive blocks of five sessions. Thus, a "limited hold" (LH) of 6 to 0.5 sec was provided.

In both schedules, a trial was followed by 5- to 15-sec intertrial intervals, distributed randomly with 2-sec steps. Only responses during S3 were reinforced, but a response at any moment of the trial or the intertrial interval interrupted it and reset the program at the onset of the next intertrial interval.

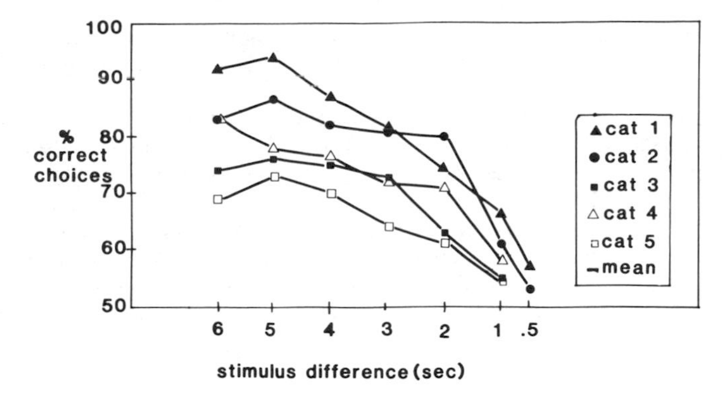

FIGURE 1. Psychometric functions obtained from each cat. *Abscissa:* successive differences between comparison and standard stimulus (in seconds). *Ordinate:* percentage of correct responses, emitted on either lever, referred to the total number of responses emitted during S3. Data are averaged by blocks of five sessions (the last five sessions with the comparison stimulus of 10 seconds are considered for the 6-sec difference).

RESULTS

The psychometric function obtained for each cat in schedule D is presented in FIGURE 1 (percentage of correct choices in function of stimulus difference). Consistent interindividual differences appeared throughout the experiment. The highest levels of performance were obtained from cats 1 and 2. The just-noticeable difference (JND), classically computed as the difference between the .75 quartile and the point of subjective equality (found in average at a stimulus difference of 0.5 to 1 sec), was about 2 sec in average. Weber fractions ranged between .22 (cat 2) and .80 (cat 5). The highest levels of performance in schedule D corresponded to the 5-sec rather than the 6-sec stimulus difference, because the latter supposed a comparison stimulus of 10 sec

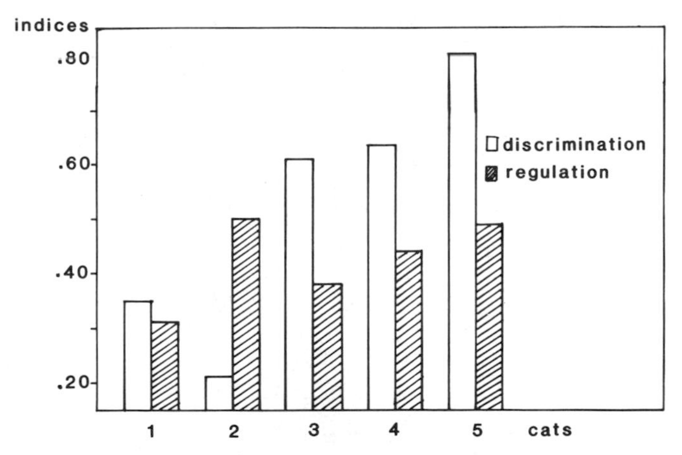

FIGURE 2. Comparison of Weber fraction (from the discrimination schedule) and of s/T index (from the regulation schedule) in each cat. *Ordinate:* value of the indices; *Abscissa:* cats.

during which the cats often presented collateral behavior that impaired the performance. This suggests difficulties in internal inhibition when long delays were imposed.

In schedule R, the percentage of correct responses in function of LH duration and the distribution of inter-response times indicate that cat 1, but not cat 2, remained the best subject. The ratios between the dispersion and the central tendency measures of response latency (s/T), roughly constant throughout the experiment, ranged between .31 and .49. The s/T index can be compared to the Weber fraction yielded by discrimination schedules.[1] This comparison (FIG. 2) reveals that, with the exception of cat 2, the levels of performance were ordered in the same way in both schedules, from cat 1 (the best level) to cat 5, and that higher levels vere obtained in the regulation schedule.

DISCUSSION

1. The relation observed between the levels of performance obtained from the same animal in both schedules—with the exception of cat 2—suggests that mostly homogeneous mechanisms are elicited in temporal discriminations and regulations.

2. The higher levels of performance obtained from most cats in the regulation schedule suggest that this species is relatively good at mastering the inhibitory components of the regulation procedures. Temporal regulations may be more familiar for cats than discrimination of external durations, in function of their behavioral repertoire (hunting and watching activities, for example).

3. Several indices indicate that internal inhibition intervened in the discrimination schedule as well as in the regulation one, although the inhibitory load was certainly heavier in the latter. This suggests that, as far as durations of a few seconds are concerned, time measurement mechanisms constantly involve internal inhibition processes, with only quantitative differences.

REFERENCE

1. CATANIA, A. C. 1970. Reinforcement schedules and psychophysical judgement: A study of some temporal properties of behavior. *In* The Theory of Reinforcement Schedules. W. N. Schoenfeld, Ed.: 1–42. Prentice-Hall. Englewood Cliffs, NJ.

The Effect of Variations in Reinforcement Probability on Preference in an Elapsed-Time Procedure

CHARLOTTE MANDELL

Department of Psychology
University of Lowell
Lowell, Massachusetts 01854

Given a choice between two responses, both producing reinforcement after a fixed temporal interval, most subjects would certainly prefer the response alternative with the higher probability of reinforcement. It is not clear, however, whether such preferences are best understood as changes in response bias or, rather, as changes in estimated time to reinforcement. This question was explored by modifying a procedure in which subjects chose between a long and short interval after varying portions of the long interval had elapsed.[1] In the present study the probability of reinforcement that terminated the short interval was varied across conditions.

Suppose that reinforcement was programmed after 60 sec for response A and after 10 sec for response B. If a choice between A and B were offered at the start of both intervals, subjects should prefer B. If B were offered after 50 sec had elapsed (such that 10 sec remained for A and B), then either response should be chosen equally often. If B

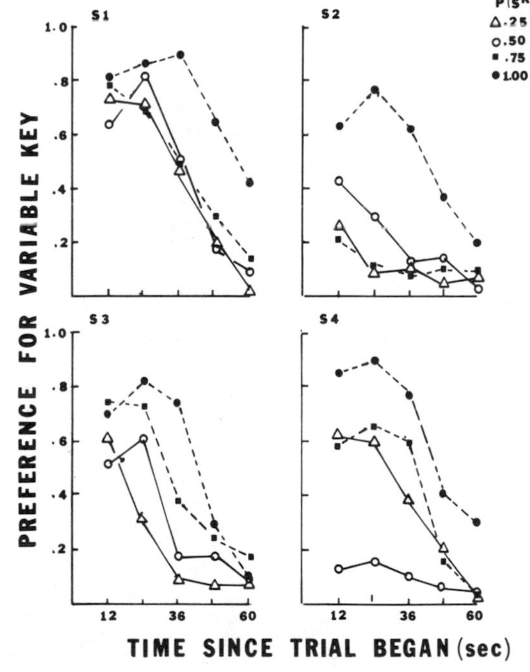

FIGURE 1. Preference for the variable side key as a function of the elapsed trial time.

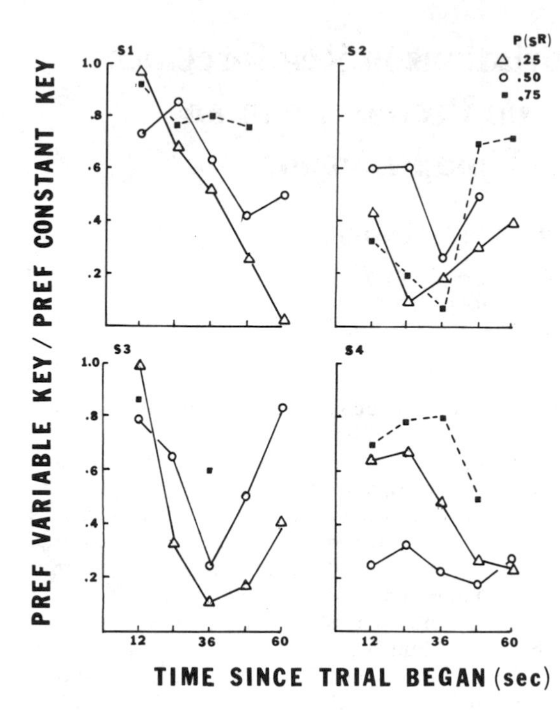

FIGURE 2. Preference for the variable side key relative to preference for the constant side key as a function of the elapsed trial time. (Only decrements in preference are indicated).

were offered after more than 50 sec had elapsed, then preference should shift to A. Consider what might now happen if the probability of reinforcement for B were reduced to .50. If the subject estimated the time to reinforcement as 20 rather than 10 sec, the preference shift from B to A should occur earlier (after 40 sec had elapsed). If, however, response bias was shifted, the probability of choosing B should be lowered as an equal proportion of the 100% reinforcement case throughout the 60-sec interval.

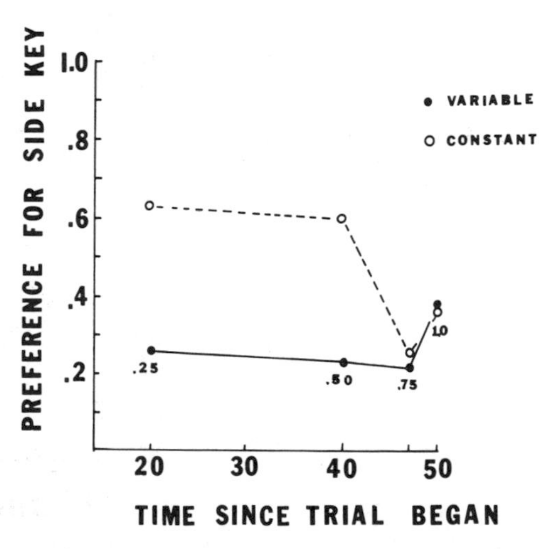

FIGURE 3. Side-key preference when time remaining in the trial equals average time to reinforcement on the variable side key.

To test these predictions, four pigeons were trained on a fixed-interval 60-sec schedule programmed on the center key of a three-key pigeon chamber. On 75% of the trials, subjects were presented with a concurrently available fixed-interval 10-sec schedule on the side keys of the chamber. The opportunity to choose, signaled by the illumination of one or both of the side keys, lasted 2 sec and could occur within each 60-sec trial. One of the side keys always produced reinforcement after 10 sec (constant key). The other produced reinforcement probabilistically (variable key). A single peck to the side key turned off the center key and caused the trial to terminate with the 10-sec interval.

The probability of choosing the shorter interval within five successive segments of the 60-sec trial is shown in FIGURE 1. Decreasing the reinforcement probability side-key preference throughout the interval. FIGURE 2, in which preference for the variable side key is shown relative to preference for the constant key, shows that preference was not reduced uniformly; the relative preference decreased through the 60-sec interval as preference for the variable key began to decline, and then increased as preference for the constant key followed. This suggests that the estimated time to reinforcement was increased by the operation of reducing reinforcement probability. Moreover, FIGURE 3, in which preference is shown when the time remaining in the interval equaled the average time to reinforcement on the center key, suggests that (allowing for response bias) the estimated time to reinforcement is adequately described by the arithmetic mean of the individual reinforcement intervals.

REFERENCE

1. GIBBON, J. & R. CHURCH. 1981. Time left: Linear vs. logarithmic subjective time. J. Exp. Psychol. Anim. Behav. Processes **7:** 87–108.

Timing of Initiation and Termination of Dual Manual Movements

MICHAEL PETERS

Department of Psychology
University of Guelph
Guelph, Ontario, Canada

A very simple dual motor task was used: subjects were asked to tap as quickly as possible with one hand while tapping slowly and regularly, at their own chosen pace, with the other hand. Subjects were asked to begin the performance with one hand, bringing the other hand in at a later point. Thus, four combinations (right fast/left slow; right slow/left fast; left fast/right slow; left slow/right fast) were tested. All taps activated switch closures that were recorded in real time via computer. The performance measure was the variability of intertap intervals during the time when both hands were active (TABLE 1). The 28 right-handed subjects showed two powerful effects. First, performance was better with the right fast/left slow combination than with the left fast/right slow combination, regardless of which hand began the task. This effect was a replication of an asymmetry observed earlier[1] with a different task. Second, subjects performed very much worse when the slow hand commenced the task, regardless of which hand commenced.

The first effect is interpreted as showing that the nonpreferred hand is capable of relatively independent performance while direct attention is focused on the preferred hand (when the preferred hand moves quickly). The reverse is not the case. The effect has relevance to the nature of dual movement during the playing of musical instruments; that is, the basic pattern of melody being played by the preferred (usually right) hand while the left provides support, or the predilection for rubato being carried out by the preferred (right) hand. No argument is made here that practice cannot overcome this natural asymmetry, but it is suggested that construction and mode of play on many musical instruments reflect the natural asymmetry described here.

The second effect is very powerful indeed. When the beginning pace is slow, subjects have a very hard time bringing in the fast hand. The converse is not true. It is suggested that the beginning hand sets the temporal resolution against which the flow of movement initiations is counted off. While a fast pace allows the counting off of

TABLE 1. Mean Standard Deviation of Intertap Intervals (msec) for 28 Right-handed Subjects Performing the Dual Task

Hand		Leading Hand[a]	Trailing Hand
Left	$\overline{X}$	82.8	174.0
	S.D.	48.7	149.9
Right[b]	$\overline{X}$	70.8	138.3
	S.D.	47.5	117.9

NOTE: When the hand begins the task, this is denoted by "lead" and when it begins after the other hand, this is denoted by "trail." All values refer to the fast performance mode for the portion of the trial where both hands are active. Lower means = better performance.

[a]Leading hand < trailing hand; $F = 34.9$; df 1/27; $p < .000001$.

[b]Right < left; $F = 6.17$; df 1/27; $p < .018$.

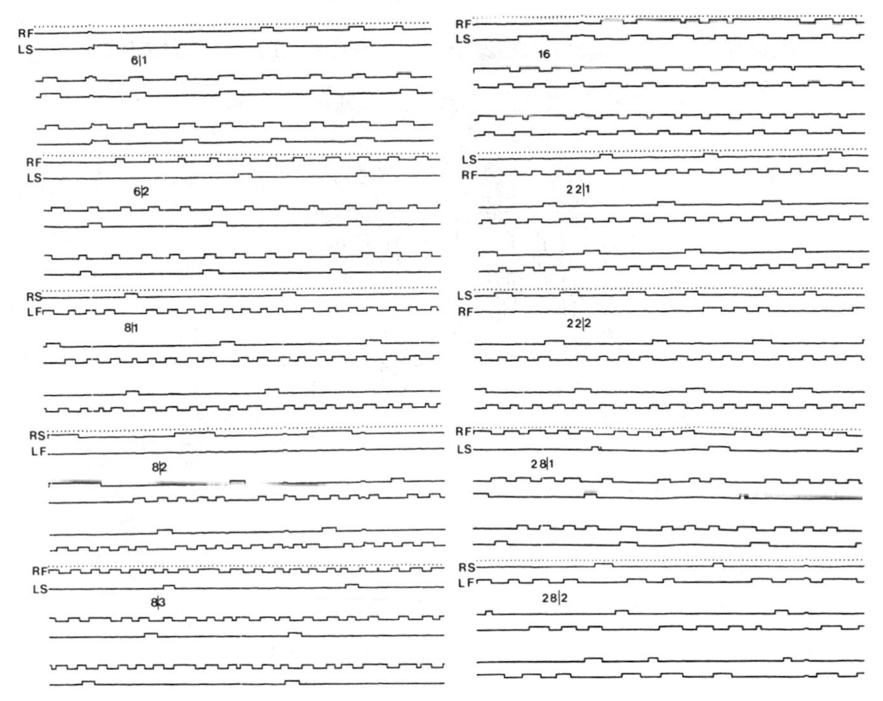

FIGURE 1. Computer-generated performance tracks for the two hands. Distance between dots represents about 30 msec. Onset of a response is an upward deflection, while offset is a downward deflection. The beginning hand is identified by the first deflections in a track. Each record represents continuous performance for the two hands (R = right, L = left, F = fast, S = slow). Effects seen are time-sharing (subject 8), concurrent movement (6), main effect of slow versus fast hand beginning (6/1 versus 6/2), performance breakdown (16) and RF/LS versus LF/RS performance asymmetry (28/1 versus 28/2).

slower movements, a slow pace does not permit the counting off of fast movement initiations. (Analogy: when counting time in seconds, hours can be counted off, but the reverse is not true). Preliminary analysis shows the subjects can perform the task as requested only by switching attention to the fast hand (FIGURE 1, 8/2) or by adhering to a rigid counting strategy with complete interdependence of the movements of the two hands (FIG. 1, 6/1).

The simple task used here allows identification of basic effects that likely go undetected in more complex tasks, but which are nevertheless important in their performance. TABLE 1 gives basic comparisons, while FIGURE 1 illustrates a variety of performance patterns. Effects seen in the figure are: time-sharing, concurrent movement, perturbations in the flow of movement of one hand caused by movements in the other, the main effects of slow versus fast hand beginning the task, and better performance with the RF/LS than with the LF/RS combination.

REFERENCE

1. PETERS, M. 1981. Attentional asymmetries during concurrent bimanual performance. Q. J. Exp. Psychol. **33A**: 95–103.

Explicit Counting and Time-Order Errors in Duration Discrimination

WILLIAM M. PETRUSIC

Department of Psychology
Carleton University
Ottawa, Ontario, K1S 5B6 Canada

The present experiments investigated the effects of explicit counting on duration discrimination, with a view to characterizing the nature of the counting process and determining the effects of counting on the accuracy of discrimination. As well, and of particular importance, was the possible effect of counting on the time-order error for duration. As repeated demonstrations have shown,[1,2] discriminability depends on presentation order. Typically, when two intervals of length 4000 msec and 4200 msec are presented, accuracy may be nearly 100% in the order 4000, 4200 but drop to 30% in the order 4200, 4000. Consequently, the experiments to be reported were designed to determine whether the time-order errors typically present in duration discrimination would also be present when explicit counting was required.

In the first experiment, the method of constant stimuli was used with a 4000-msec standard and seven variable stimuli having values equal to 4000 ± 200k msec, where $k = 0, 1, 2$ or 3. The standard preceded the variable stimulus on one-half of the trials under each of the instructions "choose the shorter" and "choose the longer." On the remaining trials, the variable stimulus preceded the standard. Each of the 28 types of trial (two instructions times seven stimulus pairs in each of two presentation orders) was presented exactly ten times during the experimental session. Five subjects were randomly assigned to the count condition and five to the no-count condition. Subjects in the count condition were explicitly instructed to count during stimulus presentation, to use this count as the basis for discriminating durations, and to report their counts at the completion of the trial. Subjects in the no-count condition were given the normal instructions not to count.

Overall mean accuracy in the no-count condition was 61.6%, and in the count condition accuracy improved to 85.2%. Moreover, this substantial increase in discriminative accuracy was found for each duration pair, as shown in FIGURE 1. However, instructions to count failed to remove the time-order error, as is evident from the plots of the psychometric functions presented in FIGURE 2. FIGURE 3 shows, entirely in accord with the sizeable negative time-order errors evident, that for each stimulus, the reported count was greater when that stimulus was the second presented than when it was the first presented.

To test for the possibility that some factor consequential to the decision process was involved in the alteration of the counts to the second-presented stimulus, a second experiment was conducted; subjects were asked to count during the two intervals of a stimulus pair, but comparison was neither requested nor mentioned. The results of this experiment, based on ten subjects and using precisely the same design as in the previous experiment, show that the reported counts were strongly influenced by presentation order (Fig. 4). Since discrimination was not required, it is evident that the process of comparison is not necessary for the occurrence of the time-order error.

In conclusion, since time-order errors were found in the absence of discrimination, models that conceptualize these errors in terms of some form of bias in the comparison

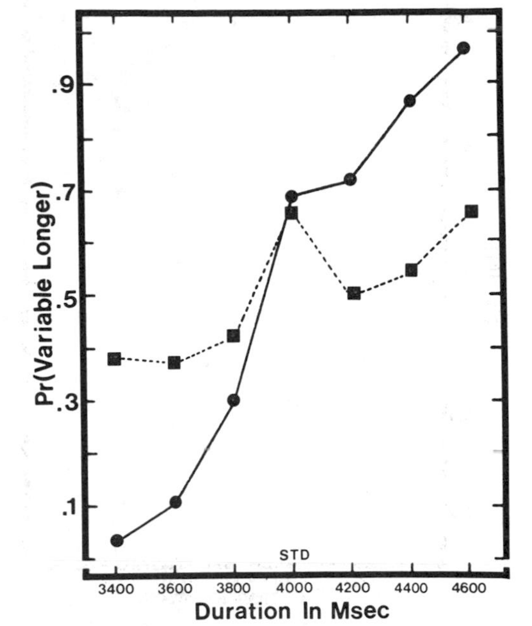

FIGURE 1. Psychometric functions for experiment 1. Proportion of times the variable stimulus is judged to be longer than the standard stimulus in the count condition (*circles*) and the more typical no-count condition (*squares*).

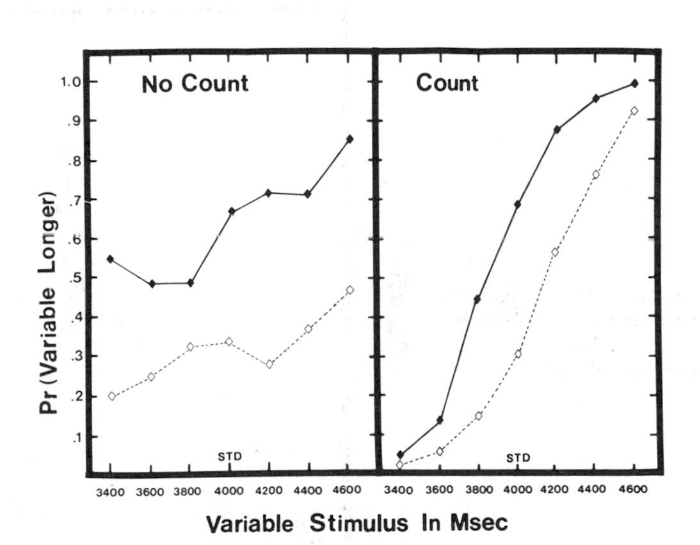

FIGURE 2. Psychometric functions for experiment 1 conditioned on presentation order. Proportion of times the variable stimulus was judged greater than the standard stimulus when the standard was presented first (*solid line*) and when it was presented second (*broken line*) for the no-count (*left*) and count (*right*) conditions.

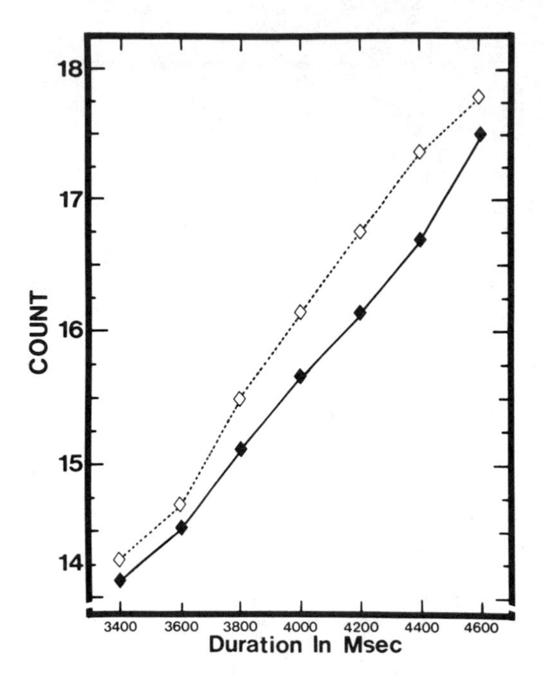

FIGURE 3. Mean count reported for each stimulus in experiment 1 by subjects in the counting condition plotted separately for each stimulus when it was the first presented (*solid line*) and when it was the second presented (*broken line*) of the pair.

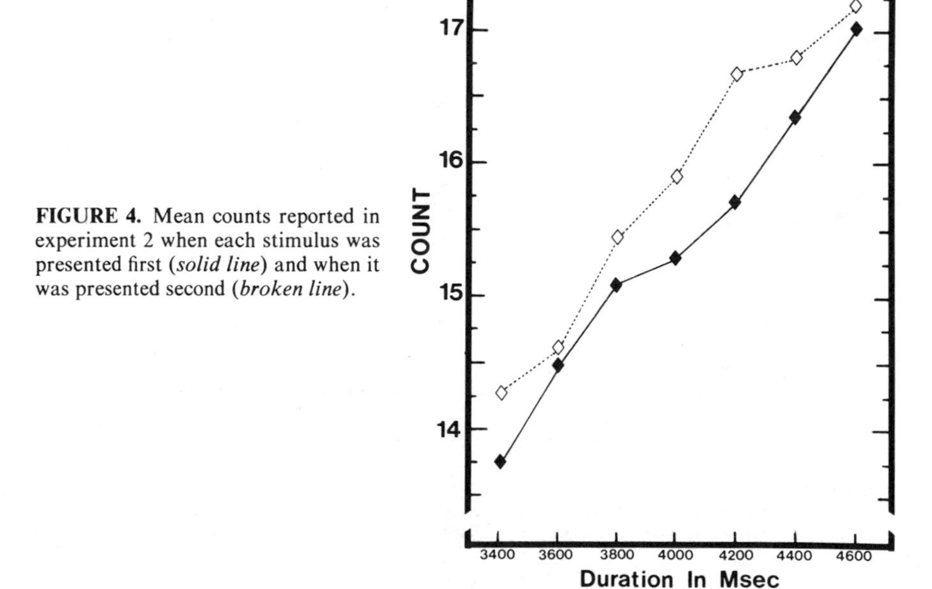

FIGURE 4. Mean counts reported in experiment 2 when each stimulus was presented first (*solid line*) and when it was presented second (*broken line*).

process must be rejected. As well, simple Poisson process models of the activity of the "internal clock" or "central timekeeper" are also rejected since these models require the independence assumption. The present findings strongly encourage the formulation of models of sufficient complexity to admit dynamic stimulus representations sensitive to local dependencies.

REFERENCES

1. ALLAN, L. G. 1979. The perception of time. Percept. Psychophys. **26:** 340–354.
2. JAMIESON, D. G. & W. M. PETRUSIC. 1975. Presentation order effects in duration discrimination. Percept. Psychophys. **17:** 197–202.

Rats Remember the Circadian Phase of Feeding

ALAN M. ROSENWASSER

Department of Psychology
University of Pennsylvania
Philadelphia, Pennsylvania 19104

Schedules that provide only limited daily access to food are capable of synchronizing a component of the rat's circadian activity pattern. When food is available for a limited time each day, robust wheel-running activity occurs during the hours preceding each daily feeding. Such "food-anticipatory activity rhythms" (FARs) are displayed independently of the light-entrainable circadian activity rhythm. Nevertheless, much evidence suggests that FARs do depend on an underlying circadian oscillator: FARs occur only when the period of the feeding schedule is close to 24 hours, and may persist for a few days after the termination of the feeding schedule. While even this temporary persistence of the FAR is suggestive of an underlying oscillator, the significance of the eventual damping is uncertain. This result could indicate that FARs depend on a damped oscillator, but it is clear that the *overt* expression of the FAR may also be lost for other reasons. It is possible that the expression of the underlying oscillator in the behavioral activity profile may be suppressed, while the oscillator itself continues to produce self-sustaining oscillations, in the absence of a feeding schedule.

I tested this hypothesis using a "memory" paradigm in which rats were subjected to periods of complete food deprivation many days after the damping of overt food-anticipatory activity. Groups of animals were allowed 2 hours of access to food each day under either light–dark cycles (LD), constant light (LL), or constant darkness (DD). The food-access period occurred in the middle of the light segment for the LD animals, when they are normally least active. After at least 1 month of scheduled feeding, the animals were transferred to a condition of free food-access. On the same day, the rats under LD were either continued on the same lighting cycle, subjected to a 10-hr phase delay of the LD cycle, or transferred to LL. The animals that had been under LL and DD were all continued under the same lighting conditions. At various intervals after the beginning of free feeding, all animals were subjected to 4- to 6-day periods of complete food deprivation. The longest interval tested was 50 days, and some animals were subjected to repeated deprivation periods separated by intervening periods of free feeding.

All animals showed robust FARs during the feeding schedule, regardless of lighting condition, while the photic activity rhythms showed either normal entrainment (in LD animals) or free-running rhythms (in LL and DD animals). The FARs damped almost immediately at the beginning of free feeding. During the deprivation periods, all animals that had been under LD during the feeding schedule showed robust deprivation-induced increases in activity which occurred at the same phase, relative to the photic activity rhythm, as did the previously scheduled daily feedings. This effect occurred in LD animals with entrained, phase-shifted, or free-running photic activity rhythms. Therefore, the animals appeared to learn and remember the relative circadian phase of the feeding schedule across the intervening free-feeding period. On the other hand, the animals that had been under LL or DD during the feeding schedule did not show convincing evidence that they could remember the phase of the feeding schedule, even though they showed clear FARs during the feeding schedule. Since the FAR and the photic rhythm had different periods during the feeding schedule, the phase relationship between the two rhythms was constantly changing.

These results support the hypothesis that the rat possesses a self-sustaining food-entrainable circadian oscillator. This oscillator seems to be flexibly coupled to the photic oscillator: both rhythms phase-shift and free-run in parallel during free-feeding conditions, but only if they had identical periods during the feeding schedule. Such flexible coupling between circadian oscillators may allow the animal to synchronize activity with rhythmic factors in the environment, both stable (LD cycle) and unstable (food availability), and may function as a "continuously consulted clock" in the adaptive temporal coordination of behavior.

Temporal Organization and Intermodality in Duration Discrimination of Short Empty Intervals

ROBERT ROUSSEAU, JOCELYN POIRIER, AND GÉRARD TREMBLAY

Department of Psychology
Université Laval
Sainte-Foy, Québec, Canada

Psychophysical models of duration discrimination commonly assume performance to only be dependant on the temporal extent of the intervals to be discriminated. In fact, discrimination of empty intervals longer than 100 msec is reported to be insensitive to variations in nontemporal stimulus dimensions of the markers that define the intervals. A notable exception comes from studies in which empty intervals are marked by a tone–light sequence (intermodal intervals). These intervals induce a large decrement in performance by comparison with levels observed with auditory or visual intramodal intervals. The present paper investigates the possibility that the decrement originates from the fact that the markers in intermodal intervals lack the property of intramodal markers to organize into patterns that could carry temporal information.

Such perceptual patterns result from a temporal organization that develops spontaneously and automatically when brief stimuli are presented in close temporal succession. Furthermore, pattern quality will improve with marker similarity and proximity. Consequently, intermodal intervals and/or longer intervals should generate patterns of lower quality than that of intramodal and/or shorter intervals.

FIGURE 1 describes an experimental paradigm developed to test the hypothesis that intramodal auditory intervals do form such perceptual patterns. In an intermodal duration discrimination task, an interfering signal (T_2) is presented t msec following T_1. In order to stress spontaneous aspect of the organization, subjects are instructed not to pay attention to the T_2 tone. Moreover the T_1–T_2 interval varies over trials and is independent of the T_1–L interval.

If auditory organization occurs automatically, intramodal temporal information will be forced on the intermodal information and will very likely influence the discrimination response. This influence will be assessed by the observed relationship between the magnitude of the T_1–T_2 interval and the proportion of "long" responses, $P(R_1)$. It will be further tested with two levels of duration defined by the midpoint, M.P. between d_0 and d_1, in which case it should decrease with an increase in M.P. due to the reduction in proximity.

EXPERIMENT 1

Experiment 1 is a direct evaluation of the automatic perceptual organization of auditory empty intervals. Four trained subjects ran through eight 280-trial sessions, alternating between control and experimental sessions.

The results, presented in FIGURE 2, show $P(R_1)$ to be strongly influenced by T_1–T_2 at M.P. = 250 msec, but not at M.P. = 1000 msec.

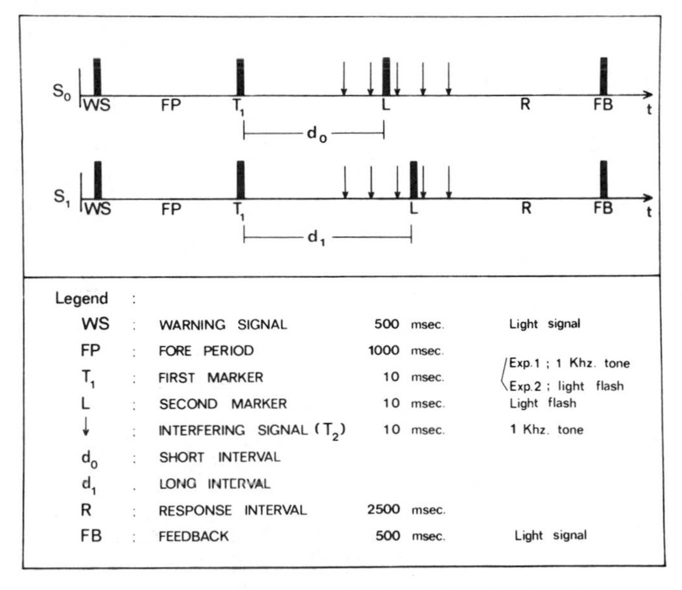

FIGURE 1. Experimental procedure for "short" (S_0) and "long" (S_1) trials in experiment 1 and experiment 2.

EXPERIMENT 2

In experiment 2, an intramodal visual discrimination situation, in which T_2 occurs as in experiment 1, controls for the possibility that T_2 by itself is responsible for the data reported in FIGURE 2. Eight trained subjects, four at each M.P. value, ran over eight sessions divided into one control block and two experimental blocks.

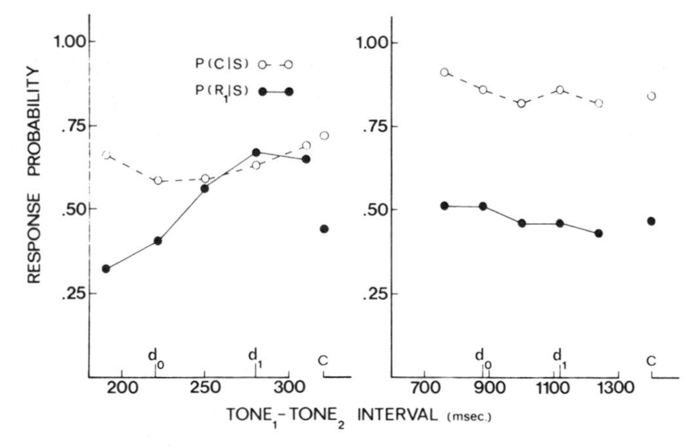

FIGURE 2. P(c) and P(R_1) averaged over four subjects for experiment 1 for M.P. = 1000 and 250 msec (864 trials per experimental point).

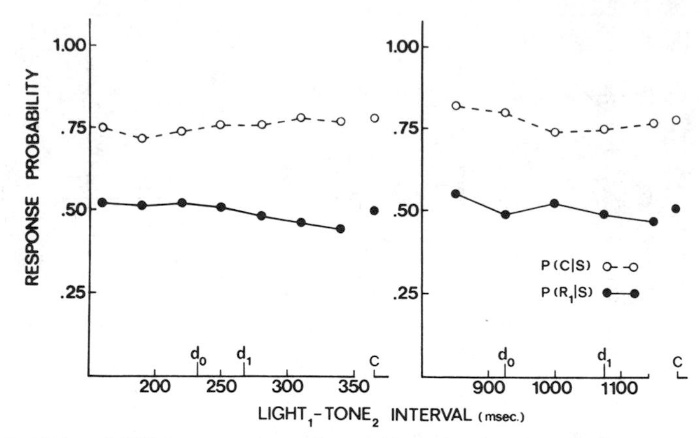

FIGURE 3. P(c) and P(R_1) averaged over four subjects for experiment 2 for M.P. = 1000 and 250 msec (768 trials per experimental point).

It is readily apparent from the data in FIGURE 3 that the presentation of T_2 around the second light flash produces no observable effect on the discrimination performance.

CONCLUSION

The present results give a strong support to the hypothesis that spontaneous organization of auditory patterns does occur and that it carries temporal information. Therefore, duration discrimination of empty intervals will depend not only on the interval itself, but also on the total pattern supporting the interval.

An Adaptive Counter Model for Time Estimation

ROBERT ROUSSEAU AND DANIEL PICARD

École de Psychologie
Université Laval
Québec, Canada G1K 7P4

EDGARD PITRE

St-Lawrence Campus
Champlain Regional College
Sainte-Foy, Québec, Canada

The main objection to Poisson counter models for time estimation originates from their prediction that $\mathrm{VAR}(T) = KT$, since empirical data generally show $\mathrm{SD}(T) = KT$ (at least up to $T = 2$ sec). Furthermore, as is the case for other psychophysical models, Poisson models do not account for the effect of cognitive variables on time estimation. FIGURE 1 is a description of a modified Poisson counter model compatible with Weber's law and able to take into account selective attention. The model has been tested with two experiments in which subjects produce temporal intervals (by finger-tapping) adjusted by feedback around a target duration, T.

EXPERIMENT 1

Experiment 1 is a test of the generality of Weber's law ($\mathrm{SD}(T) = KT$) over an extended range of T values (0.5–10 sec).

Five subjects produced 60 series of 20 intervals for each T. A session consisted of a training block with feedback followed by a series of blocks without feedback. During each session T remained constant.

FIGURE 2 shows $\mathrm{SD}/\overline{T}$ VS $\overline{T}$ (mean production) as being almost constant [$F(4, 16) = 1.44$, $p > .25$], which is consistent with Weber's law. A linear curve-fitting test applied to the VAR VS $\overline{T}$ function yields an $r^2 = .945$, whereas for the SD VS $\overline{T}$ function, $r^2 = .998$ with $\mathrm{SD} = .0637\,(T + 218)$. This should lead to the rejection of a standard Poisson counter model. However, if λ becomes a free parameter and $p = 1$, it can be shown that:

$$\frac{\mathrm{SD}}{T} = \frac{\sqrt{\mathrm{VAR}(W_N)}}{\mathrm{E}(W_N)} = \frac{\sqrt{N/\lambda^2}}{N/\lambda} = \frac{1}{\sqrt{N}} = \frac{1}{\sqrt{\lambda T}}, \text{ since } E[N] = \lambda T.$$

Thus, assuming Weber's law:

$$\frac{\mathrm{SD}}{T} = K = \frac{1}{\sqrt{\lambda T}} \text{ and then } \lambda = \frac{1}{K^2 T} \text{ and } \lambda T = \frac{1}{K^2}.$$

Estimates of λ, λT and the expected value of the interarrival times between pulses,

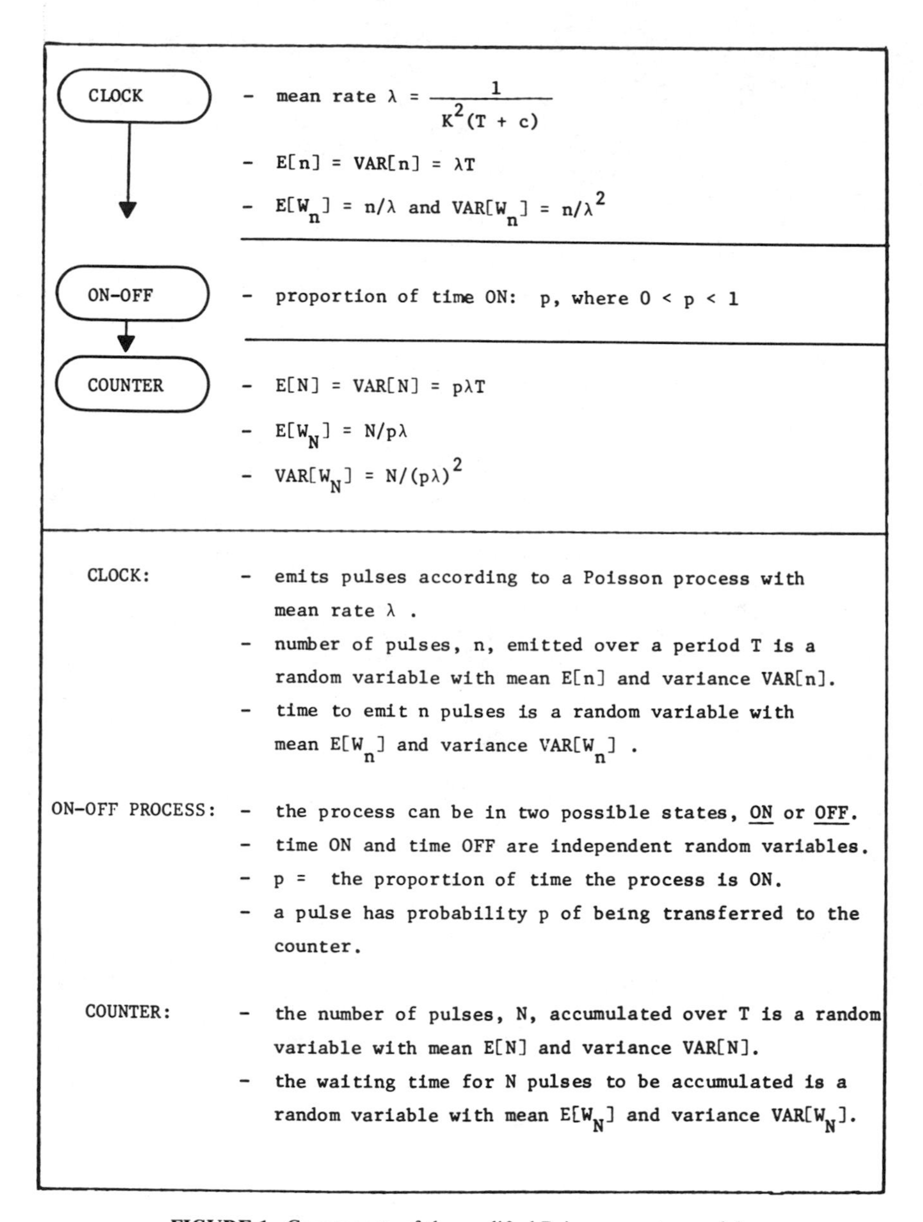

FIGURE 1. Components of the modified Poisson counter model.

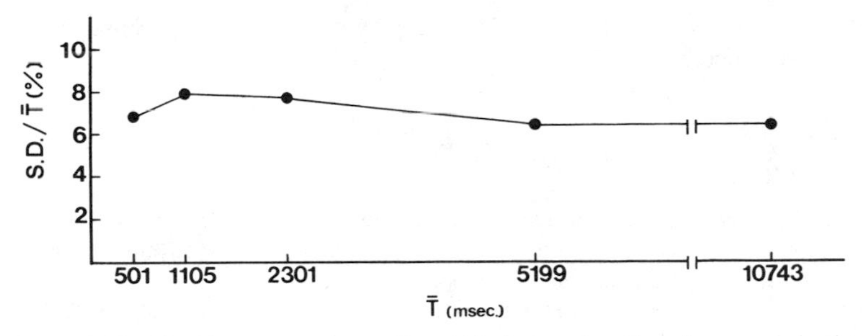

FIGURE 2. Weber fraction averaged over five subjects as a function of mean production (300 estimates per point).

TABLE 1. Estimates of λ, λT and E[IAT] from Simple or Generalized Weber's Law with $K = .0637$, $c = 218$, and $p = 1$ for Target Durations, T, Ranging from 0.5 to 10 sec

Target Duration (sec)	$SD = KT$			$SD = K(T + c)$		
	λ (N pulses/sec)	E[IAT] (msec)	λT (N)	λ (N pulses/sec)	E[IAT] (msec)	$\lambda(T + c)$ (N)
0.5	493	2.03	246	343	2.91	246
1.0	246	4.06	246	202	4.94	246
2.0	123	8.12	246	111	9.00	246
5.0	49	20.30	246	47	21.17	246
10.0	25	40.60	246	24	41.50	246

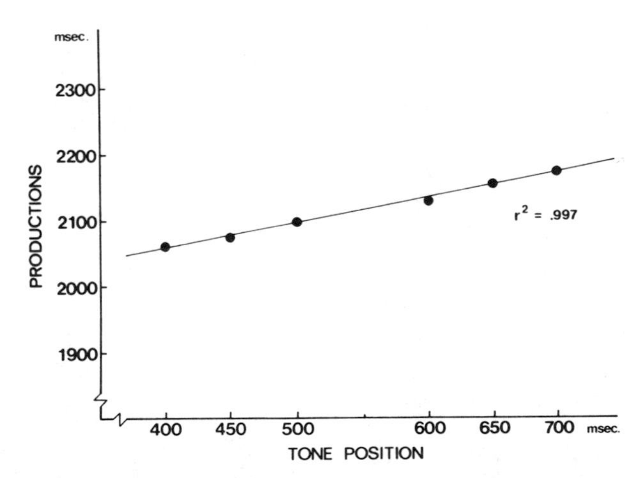

FIGURE 3. Mean production averaged over 5 subjects as a function of tone position, d_i (800 trials per point).

$E[\text{IAT}]$, are presented in TABLE 1. Values of λ range from 490 pulses/sec to 25 pulses/sec as T increases.

EXPERIMENT 2

The aim of experiment 2 is to show that a Poisson counter model can account for data obtained under conditions where a subject has to share attention between a temporal production and a concurrent nontemporal task.

In such a task, a production is defined as the time taken to accumulate a criterion number of pulses, $E[W_{Nc}]$. It is the summation of the time taken to *emit* n_c pulses ($E[W_{nc}]$) and the time to emit n' pulses ($E[W_{n'}]$) lost during time-sharing. Thus, $E[W_{Nc}] = E[W_{nc}] + E[W_{n'}] = n_c/\lambda + n'/\lambda$.

Since $E[n_c] = \lambda t_c$ and $E[n'] = (1 - p)\lambda t'$ then $E[W_{Nc}] = t_c + (1 - p)t'$, where t' is the period during which time-sharing is required.

Five trained subjects produced intervals around 2 sec while discriminating the frequency of a 10-msec tone occurring d_i msec after the onset of the interval. The discrimination response ended the interval. d_i varied randomly over trials. It is assumed that time-sharing between the clock and the auditory channel occurs over d_i; therefore, $t' = d_i$.

FIGURE 3 shows mean production as a linear function of d_i [$F(1, 4) = 53.96$, $p < .01$] with intercept at 1907 msec and slope $(1 - p)$ of .38.

CONCLUSION

A modified Poisson counter model with adjustable clock frequency and pulse accumulation under attentional control makes use of the basic mathematical properties of stochastic counters to account for basic psychophysical and cognitive aspects of time estimation.

The Decision Rule in Temporal Bisection

STEPHEN F. SIEGEL AND RUSSELL M. CHURCH

Walter S. Hunter Laboratory of Psychology
Brown University
Providence, Rhode Island 02912

At the present time, the model that best describes the behavior of animals in a temporal bisection task is the scalar expectancy theory.[2] This model, with a similarity rule and referents known exactly, has been found to account for more than 99% of the variance.[1] A consequence of the model (from Equations 17 and 43 in Gibbon[2])

$$P\,(T; S, L) = \phi\left[\sqrt{\frac{SL}{\sigma_T^2\, b\beta}} - \frac{1}{\gamma}\right]$$

where $p(T; S, L)$ is the probability of a short response given reinforcement of a short response after duration S and reinforcement of a long response after duration L. Phi (ϕ) is the cumulative normal distribution function, gamma (μ) = σ/μ and is a measure of sensitivity to time, b is bias, and β is the payoff differential. The model makes the following predictions: (1) The point of indifference (PI) should be at the geometric mean of S and L when $b\beta = 1$; (2) pairs of S and L with equal products should yield identical psychophysical functions (probability of a long response plotted against signal duration) because S and L enter into the computations by way of their product only; and (3) signals longer than L should be called long no less often than L and signals shorter than S should be called long no more often than S.

The experiment reported here tested these three predictions in the following manner: Twelve rats were randomly divided into two groups of six rats. In both groups the possible signal durations were 0.5, 1, 2, 2.8, 4, 5.7, 8, 16, and 32 sec. S refers to the signal duration for which a short response was reinforced; L refers to the signal duration for which a long response was reinforced. In the 2–8 group $S = 2$ and $L = 8$; in the 1–16 group $S = 1$ and $L = 16$. Since both pairs of S and L durations have equal products, according to the model they should yield identical functions with the PI at 4.00 sec.

The results were as follows: (1) The 2–8 group had a mean PI of 3.60, which was significantly less than 4.00 ($t[5] = -6.29, p < .01$), and the 1–16 group had a mean PI of 2.53, which was also significantly less than 4.00 ($t[5] = -17.25, p < .001$). (2) FIGURE 1 shows that the psychophysical functions for groups with equal products of S and L were very different from each other. For example, the mean PIs of each group were significantly different from each other ($t[10] = 10.05, p < .001$). (3) For the 2–8 group, the most extreme signals were called long significantly closer to 50% than the reinforced signals were called long ($t[5] = 10.81, p < .001$). This difference was not significant for the 1–16 group ($t[5] = 2.11, p > .05$. TABLES 1 and 2 show the mean percentage long response, PI, difference limen (DL), and Weber fraction (WF) for each subject in the 2–8 group and the 1–16 group, respectively.

The data reported here are consistent with those of Raslear,[3] who, using a L:S ratio of 100:1, found that the PI was not at the geometric mean, and Platt and Davis,[4] who found that for L:S ratios greater than 4:1 the PI was not at the geometric mean. An attempt will be made to fit these data with the proximity decision rule model that is

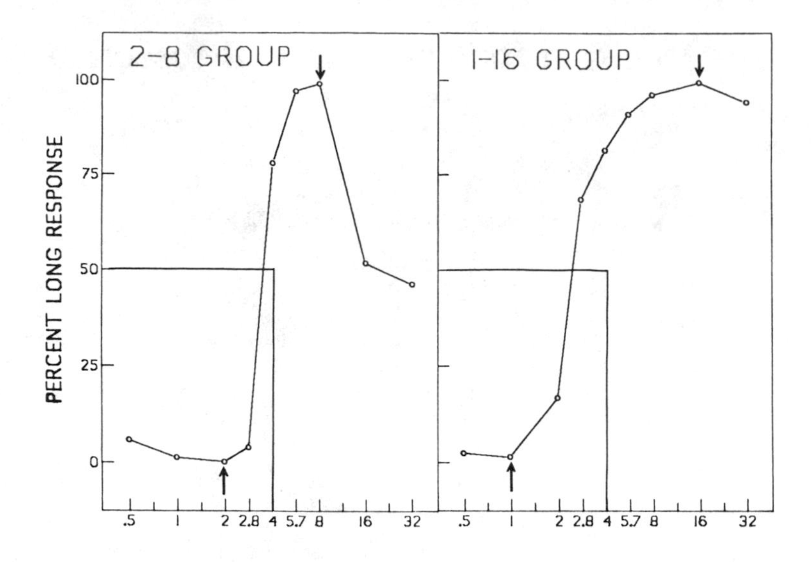

FIGURE 1. The *left panel* shows the median percentage long response as a function of signal duration for the 2–8 group; the *right panel* shows the median percentage long response as a function of signal duration for the 1–16 group. The arrows indicate the signal durations that could lead to reinforcement. The intersection of the lines drawn from 50% on the *y*-axis and 4 sec on the *x*-axis is the predicted PI. Note that: (1) the observed PIs are less than the geometric mean (4 sec); (2) the functions for the 2–8 and 1–16 groups are different; and (3) for signal durations less than *S* and greater than *L*, the percentage long response is closer to 50% than it is for *S* and *L*.

TABLE 1. Mean Percentage Long Response in 2–8 Group

Signal (sec)	Subject 1	2	3	4	5	6	Median
0.5	3.8	14.1	2.9	7.9	25.4	2.3	5.9
1.0	5.9	14.5	0	1.0	1.5	1.3	1.4
2.0	0	10.0	0	0.2	0.3	2.8	0.3
2.8	0	6.3	0	5.9	2.3	5.8	4.1
4.0	83.6	85.6	72.3	89.8	60.8	60.0	78.0
5.7	97.6	95.2	96.6	98.5	97.7	93.7	97.1
8.0	98.7	97.1	99.5	99.8	99.7	98.0	99.1
16.0	46.3	57.6	42.9	37.0	72.1	87.9	52.0
32.0	33.3	72.5	48.8	44.2	76.6	39.7	46.5
PI	3.51	3.46	3.63	3.43	3.78	3.78	3.57
DL	.36	.38	.42	.36	.51	.55	.40
WF	.10	.11	.11	.10	.14	.15	.11

NOTE: The PI is defined as the signal duration (sec) corresponding to 50% long response based on interpolation from a straight line drawn between the two adjacent points with the steepest slope. The DL is defined as ½ the difference of the points that correspond to 75% long response and 25% long response. The WF is defined as DL/PI.

TABLE 2. Mean Percentage Long Response in 1–16 Group

Signal (sec)	Subject 1	2	3	4	5	6	Median
0.5	1.1	1.1	14.9	4.1	0	4.1	2.6
1.0	2.7	0.9	1.4	0.7	1.6	5.2	1.5
2.0	10.8	1.8	14.8	18.8	32.3	37.5	16.8
2.8	60.9	46.0	68.2	68.9	79.4	69.2	68.6
4.0	76.9	86.4	97.0	71.6	91.2	69.3	81.7
5.7	89.2	93.1	98.0	93.1	88.8	80.3	91.2
8.0	92.0	96.6	100	95.9	100	84.6	96.3
16.0	99.3	100	100	98.3	99.7	91.1	99.5
32.0	69.1	76.8	98.7	98.6	90.8	97.8	94.3
PI	2.63	2.87	2.53	2.50	2.31	2.32	2.52
DL	.40	.45	.38	.45	.42	.63	.44
WF	.15	.16	.15	.18	.18	.27	.17

currently used to fit the temporal generalization procedure.[5] This would be a step in the development of a unified theory of timing.

SUMMARY

Under some conditions a similarity decision rule accounts for more than 99% of the variance of performance in a temporal bisection task.[1] When the range of signal durations is large (for example, 16 to 1), or unreinforced signal durations are presented that are more extreme than the reinforced values, the similarity decision rule does not fit the data. A proximity decision rule may be able to account for the data.

REFERENCES

1. MECK, W. H. 1983. Selective adjustment of the speed of internal clock and memory processes. J. Exp. Psychol. Anim. Behav. Processes **9:** 171–201.
2. GIBBON, J. 1981. On the form and location of the psychometric bisection function for time. J. Math. Psychol. **24:** 58–87.
3. RASLEAR, T. G. 1983. A test of the Pfanzagl bisection model in rats. J. Exp. Psychol. Anim. Behav. Processes **9:** 49–62.
4. PLATT, J. R. & E. R. DAVIS. 1983. Bisection of temporal intervals by pigeons. J. Exp. Psychol. Anim. Behav. Processes **9:** 160–170.
5. CHURCH, R. M. & J. GIBBON. 1982. Temporal generalization. J. Exp. Psychol. Anim. Behav. Processes **8:** 165–186.

Temporal Patterning and Selective Attention Effects on the Human Evoked Response

JUNE J. SKELLY,[a] ANTHONY RIZZUTO,[b]
AND GLENN WILSON[c]

[a]*Systems Research Laboratories, Inc.*
Dayton, Ohio 45440

[b]*Department of Psychology*
Bowling Green University
Bowling Green, Ohio 43402

[c]*Department of Psychology*
Wittenberg University
Springfield, Ohio 45501

This research examined the influence of temporally patterned visual sequences upon human evoked potentials. Of special interest was the *a priori* role of temporal context on selective attending. We posed the question, "Does processing differently timed visual sequences *prior* to a spatially fixed decision point affect the evoked response to the decision stimulus?"

To answer this question, we systematically varied the timing patterns controlling the sequential appearances of an element moving across a CRT along either of two oblique vectors. One timing pattern was a regular rhythm with equal temporal intervals, and the other was an irregular rhythm composed of three different temporal intervals (FIGURE 1). We manipulated attention by requiring subjects to just track the sequence (passive mode) or to selectively attend to "same" or "different" trials

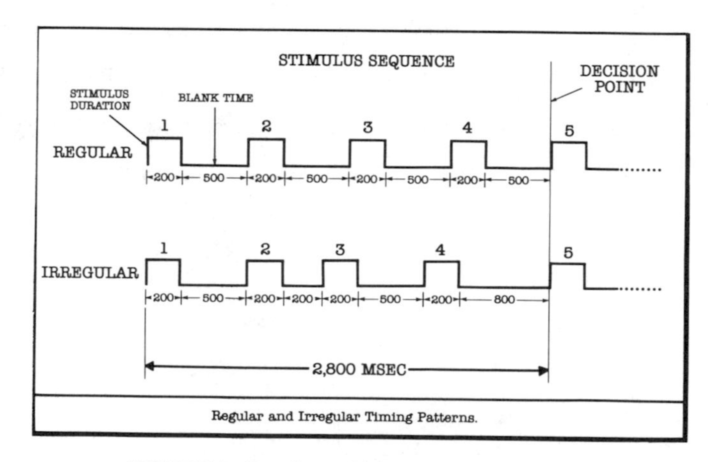

FIGURE 1. Regular and irregular timing patterns.

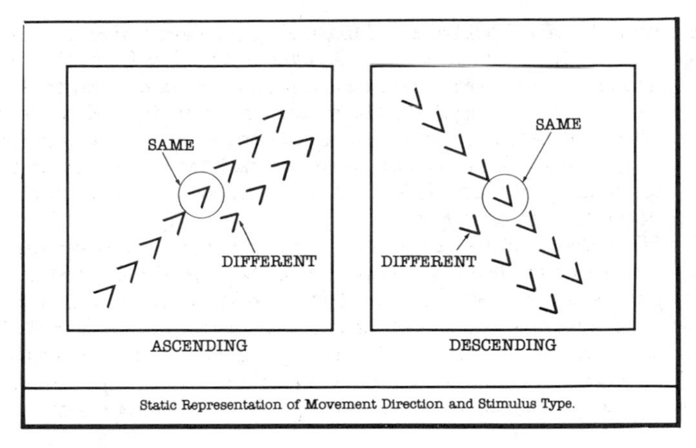

FIGURE 2. Static representation of movement direction and stimulus type.

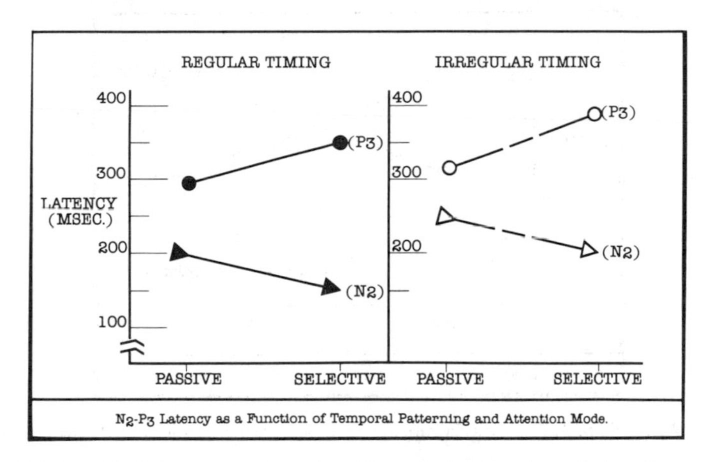

FIGURE 3. N_2–P_3 latency as a function of temporal patterning and attention mode.

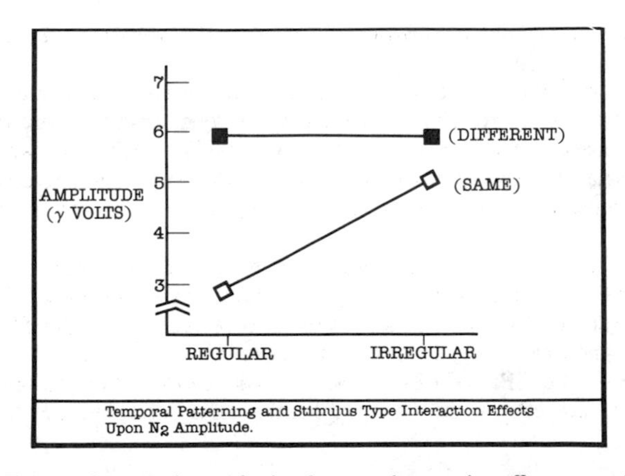

FIGURE 4. Temporal patterning and stimulus-type interaction effects upon N_2 amplitude.

(selective mode). A "same" trial referred to the element maintaining movement along the straight line, while a "different" trial denoted a deviation from the straight line (FIGURE 2). In the selective attention condition, subjects either counted the trials when the element passed through a circle on the vector (same) or dropped below the circle (different). Same and different trials were equiprobable, as was the occurrence of a particular timing pattern. Subjects were asked to track the entire sequence on every trial, and the evoked potentials (EPs) were recorded on each trial to the element appearance at the decision point (circle).

Results of the data analysis revealed that the timing pattern of stimulus appearances and the attention mode differentially affected both amplitude and latency of the N_2–P_3 complex of the evoked potential. EP latencies at the decision point were dramatically influenced by the preceding temporal pattern of each trial. Irregular sequencing produced longer latencies of both N_2 and P_3 as compared to the regular timing in both passive and selective attention conditions. Further analysis within each timing condition revealed that N_2 appeared earlier in the selective attention condition

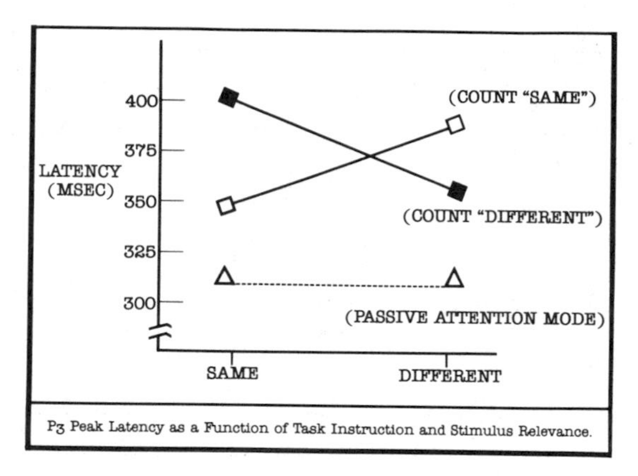

FIGURE 5. P_3 peak latency as a function of task instruction and stimulus relevance.

than in the passive attention condition. However, the opposite was true for the P_3 (FIGURE 3).

The timing structure interacted with stimulus classification to affect the amplitude of N_2 when subjects were required to selectively attend. Irregular and regular timing produced equivalent effects when the stimulus was "different." However, when the timing was regular and the stimulus was "same," N_2 amplitude decreased sharply as compared to irregular timing. Amplitude of the P_3 was not affected by this interaction (FIGURE 4).

P_3 latency, but not amplitude, was sensitive to stimulus relevance. When the stimulus matched the task instruction (count same count different), latencies were shorter than when the stimulus was irrelevant to instruction (FIGURE 5).

These data indicate that: (1) temporal patterning has a powerful effect on the N_2–P_3 complex of the EP; and (2) temporal context interacts with task demands to produce differential effects on the amplitudes and latencies of the N_2 and P_3 components.

Chronological Knowledge: The Cognitive Integration of Temporal Information in Pigeons

RON WEISMAN

Department of Psychology
Queen's University
Kingston, Ontario, Canada

Delayed sequence discriminations (DSD) present one sequence as positive (followed by reward) and other sequences as negative (not followed by reward). In general the DSD task involves a search for one sequence among many. In experiment 1 a two-event sample sequence preceded the test stimulus and in experiment 2 a three-event sample sequence preceded the test stimulus in a delayed discrimination. The events were red and yellow overhead lights. Half the birds had red as *A* (one event in the reinforced sequence) and yellow as *B* (a second event in the reinforced sequence). The remaining birds had yellow as *A* and red as *B*. *X* was the absence of either colored lamp for all birds. In experiment 1, the sequences were all two-event arrangements of *A*, *B*, and *X*; that is, *XX*, *AX*, *BX*, *XA*, *BA*, *AA*, *XB*, and *BB* were negative sequences and *AB* was the positive sequence. In experiment 2, the sequences were all three-event arrangements of *A* and *B*; that is, *AAA*, *BAA*, *ABA*, *BBA*, *BBB*, *ABB*, and *AAB* were the negative sequences and *BAB* was the positive sequence.

The discriminative performance of 26 pigeons in experiment 1 and of 18 pigeons in experiment 2 was used to establish reliable rankings of the sequences. The obtained order of the sequences was $XX = AX < BX < XA < AA < BA < XB < BB < AB(+)$ in experiment 1, and $AAA < BAA < ABA = BBA < BBB < ABB = AAB < BAB(+)$ in experiment 2. Adding models for the integration of decisions about successive events were evaluated after assigning algebraic values to the sequences in each experiment, for example, the value of *BA* in experiment 1 is the sum of the values of *B* as the first event and *A* as the second event: $V(BA) = V(B1) + V(A2)$, and the value of *BAA* in

TABLE 1. The Two-Event DSD Task

Rank	Sequence	Addition	Implies
1.5	XX	V (X1) + V (X2)	
1.5	AX	V (A1) + V (X2)	
3	BX	V (A1) + V (X2)	$V (X1) = V (A1) < V (B1)$
4	XA	V (X1) + V (A2)	
5	AA	V (A1) + V (A2)	
6	BA	V (B1) + V (A2)	$V (X1) < V (A1) < V (B1)$
7	XB	V (X1) + V (B2)	
8	BB	V (B1) + V (B2)	
9	AB	V (A1) + V (B2)	$V (X1) < V (B1) < V (A1)$

NOTES: 1. Differences between ordinally adjacent sequences are reliable, df. = 25, p = .05.

2. Ignore cases where the value of A1, V (A1), equals the value of B1, V (B1), because of possible low power.

3. Falsification of any adding model: the value of A1, V (A1), cannot be *both* greater and lesser than the value of B1, V (B1).

TABLE 2. The Three-Event DSD Task

Rank	Sequence	Adding Rule	Implies
1	AAA	V (A1) + V (A2) +V (A3)	
2	BAA	V (B1) + V (A2) +V (A3)	V (A1) < V (B1)
3.5	ABA	V (A1) + V (B2) +V (A3)	
3.5	BBA	V (B1) + V (B2) +V (A3)	V (A2) < V (B2)
5	BBB	V (B1) + V (B1) +V (B3)	
6.5	ABB	V (A1) + V (B2) +V (B3)	V (A1) > V (B1)
6.5	AAB	V (A1) + V (A2) +V (B3)	
8	BAB	V (B1) + V (A2) +V (B3)	V (A2) > V (B2)

NOTES: 1. Differences between ordinally adjacent sequences are reliable, d.f. = 17, p. = .05.
2. Ignore cases where the value of A, V (A) equals the value of B, V (B), because of possible low power.
3. Falsification of any adding model: the values of A1 and A2, V (A1) and V (A2), cannot be *both* greater and lesser than the values of B1 and B2, V (B1) and V (B2).

experiment 2 is the sum of the values of B as the first event, A as the second event, and A as the third event: $V(BAA) = V(B1) + V(A2) + V(A3)$, as shown in TABLES 1 and 2. Simple algebraic evaluation found the adding models logically inconsistent (see the tables). In order to generate the obtained ranking of the sequences, the value of $A1$ in experiment 1, and of $A1$ and $A2$ in experiment 2, had to vary from sequence to sequence, sometimes greater and sometimes lesser than the value of $B1$ in experiment 1, and of $B1$ and $B2$ in experiment 2. Many forms of the multiplying model were also falsified by the ordinal pattern of the obtained results, because under transformation many forms of the multiplying model must generate parallel functions without crossovers. However, the results of both experiments corroborate forms of the multiplying model that assign negative values to events ordered incorrectly and positive values to events ordered correctly.

A Computer-Based Portable Keyboard Monitor for Studying Timing Performance in Pianists

CRAIG MINOR, MARK TODOROVICH, AND JAMES BOYK

Division of Engineering and Applied Science, and Division of Humanities and Social Science California Institute of Technology Pasadena California 91125

GEORGE P. MOORE[a]

Departments of Biomedical and Electrical Engineering University of Southern California Los Angeles, California 90089

As a preliminary step in the study of timing in musical performance, we have developed a low-cost, portable computer-based data acquisition system which can be quickly and unobtrusively connected to pianos with standard keyboard dimensions.

The system is based on several components linked in a configuration that allows the identity and time of activation and release of each keyboard note to be coded and stored in a memory buffer and sequentially transferred to a host computer for decoding, processing, and analysis.

Sixty-four notes of the piano are monitored by two 32-note keyswitch boards (Marantz Piano Co., Inc., Morgantown, NC) which are inserted under the keyboard. Each key, when depressed, closes a spring switch; release opens the switch. At preselected time intervals (typically every millisecond or every 10 milliseconds), a 6502 microprocessor chip sends an interrogatory command to the switch register, which responds with a coded number sent to the microprocessor summarizing the status (up or down) of each of the 64 notes being monitored. This code is compared to the previous coded signal to determine whether the status of any note has changed. If a change of

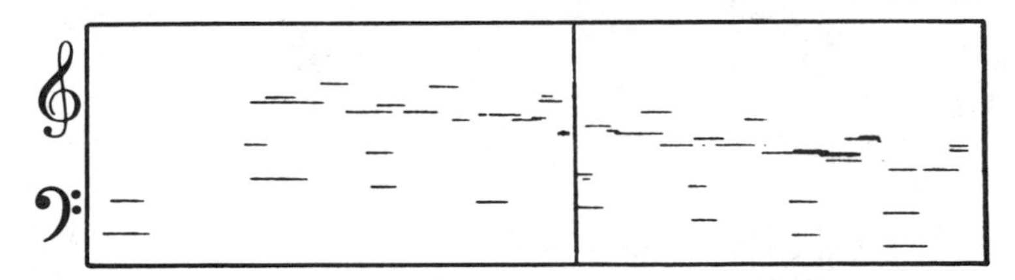

FIGURE 1. Computer display of two consecutive bars of Chopin's *Fantasie Impromptu* (Opus 66). Bar lines have been added. Each note is represented by a horizontal line whose distance above the baseline is proportional to its pitch; the length of each line is proportional to the note's duration, the beginning and end marking the moments of key activation and release.

[a]To whom correspondence should be sent.

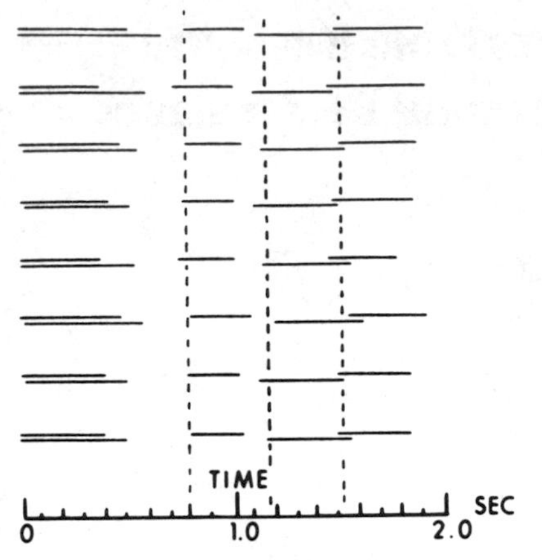

FIGURE 2. Computer display of eight consecutive bars of a passage in which the right hand plays three notes to the left hand's two. Each has been aligned on the first note struck in each measure. This display permits an easy visual assessment of the variability of performance. The mathematically "correct" points of temporal subdivision have been added (*dotted lines*). Despite the measure-to-measure variability of this performance, the average timing of each note in the right and left hands was almost exactly that demanded by mathematical correctness.

status has occurred, the identity of all notes with a status change is determined, and that information, together with the time of the change is stored sequentially in the memory buffer of the microprocessor.

As the performance continues, the contents of the microprocessor buffer are transmitted to a portable host computer (Attache, Otrona Corp., Boulder, CO) via a serial interface line, and stored in the memory or the storage disks of the host. Subsequently, the host computer reconstructs the exact sequence of note-production by the pianist, displaying the sequence as a series of marks on the computer screen which summarize, in a global way, the temporal and chromatic features of the music that has been played (FIGURE 1). An audio synthesis chip within the host computer sounds each note as it is displayed.

Additional programs calculate the statistics of time intervals between selected beats within or between measures, and the timing relations between subdivisions of the temporal pattern as actually performed by the right and left hands. These are then compared to the time divisions formally dictated by the printed score. An example from such a study is shown in FIGURE 2, where eight consecutive measures have been aligned vertically to exhibit the variability of performance of a 3-against-2 rhythm in the right and left hands.

Index of Contributors